Fourth Edition

Resource and Environmental Management in Canada

Addressing Conflict and Uncertainty

Edited by Bruce Mitchell

OXFORD

UNIVERSITY PRESS

OXFORD
UNIVERSITY PRESS

8 Sampson Mews, Suite 204, Don Mills, Ontario M3C 0H5
www.oupcanada.com

Oxford University Press is a department of the University of Oxford.
It furthers the University's objective of excellence in research, scholarship,
and education by publishing worldwide in

Oxford   New York

Auckland   Cape Town   Dar es Salaam   Hong Kong   Karachi
Kuala Lumpur   Madrid   Melbourne   Mexico City   Nairobi
New Delhi   Shanghai   Taipei   Toronto

With offices in

Argentina   Austria   Brazil   Chile   Czech Republic   France   Greece
Guatemala   Hungary   Italy   Japan   Poland   Portugal   Singapore
South Korea   Switzerland   Thailand   Turkey   Ukraine   Vietnam

Oxford is a trade mark of Oxford University Press in the UK
and in certain other countries

Published in Canada by Oxford University Press

Library and Archives Canada Cataloguing in Publication

Resource and environmental management in Canada : addressing conflict
and uncertainty / edited by Bruce Mitchell. – 4th ed.

Includes index.

ISBN 978-0-19-543128-5

1. Natural resources—Canada—Management. 2. Environmental management—Canada.
3. Sustainable development—Canada. I. Mitchell, Bruce, 1944–

HC113.5.R47 2010   333.70971   C2009-907229-7

Cover images: Forest © Nicholas Monu/iStockphoto; Home Construction © Kenneth C. Zirkel/iStockphoto;
Books © Nadezda Firsova/iStockphoto; Burning Fireplace © Jenny Swanson/iStockphoto;
Wood Chips at Sawmill © Phil Augustavo /iStockphoto

Oxford University Press is committed to our environment. This book is printed on
Forest Stewardship Council certified paper, harvested from a responsibly managed forest,
which contains 100% post-consumer waste.

FSC

Mixed Sources
Product group from well-managed
forests, controlled sources and
recycled wood or fibre

Cert no. SW-COC-000952
www.fsc.org
© 1996 Forest Stewardship Council

The production of *Resource and Environmental Management in Canada* on
Rolland Enviro 100 print paper instead of paper made from virgin fibres
reduces Oxford's ecological footprint by:

Trees: 44
Solid waste: 1,261 kg
Water: 119,310 L
Air emissions: 2,770 kg

Printed and bound in Canada.

2 3 4 – 13 12 11

Contents

List of Boxes, Figures, and Tables

Boxes

Figures

Tables

Notes on Contributors

Mark Anielski, MSc, is an economist and author of *The Economics of Happiness: Building Genuine Wealth* (Gabriola Island, BC: New Society Publishers, 2007). Mark is a well-being economist, entrepreneur, and professor of corporate social responsibility at the University of Alberta, Edmonton, and president of Anielski Management Inc., which specializes in helping communities measure and develop economies of well-being. He has pioneered natural capital accounting in Canada and has developed alternative measures of economic progress, including the US Genuine Progress Indicator (GPI), the Alberta GPI Sustainable Well-being measurement system, and the Edmonton GPI. For 14 years, he served as senior economic policy advisor to the Alberta government. Between 2003 and 2007, he was a senior economic counsel to the Government of China related to efforts to adopt 'green' GDP accounting and new measures of sustainable well-being. He is the past-president of the Canadian Society for Ecological Economics and a Senior Fellow with the Oakland-based economic think-tank Redefining Progress. He lives in Edmonton, Alberta. For more information about the Genuine Wealth model and applications to measuring ecological health, visit www.anielski.com or www.genuinewealth.net.

Annie L. Booth, PhD, is an associate professor in the Ecosystem Science and Management Program at the University of Northern British Columbia. Her PhD is from the University of Wisconsin in Land Resources. Her research interests include First Nations' resource management, community-based resource management, environmental ethics, community sustainability, and women in environmental studies. She volunteers with the North Caribou SPCA and is owned by a dog and a cat.

Mike Brklacich, PhD, is a professor and chair of the Department of Geography and Environmental Studies, Carleton University, Ottawa. His research over the past 25 years has focused on human dimensions of global environmental change (GEC), with specific interests in vulnerability of human systems to multiple stressors, food security, and the capacity of human systems to adapt to GEC. He has contributed to the GEC and Human Security project, the GEC and Food Systems projects, the Intergovernmental Panel on Climatic Change, and the Inter-American Institute for Global Change Research.

Christopher Bryant, PhD, is a geography professor at the Université de Montréal and director of the Sustainable Development and Territorial Dynamics Laboratory. Past-chair (2000–6) of the IGU Commission on the Sustainability of Rural Systems, he is currently editor of the *Canadian Journal of Regional Science*. His research interests include the dynamics and management of urban fringe communities, adaptation to climate change, and local development. He is involved in research action approaches with local and regional players in these domains.

Susan Cartwright holds a Masters degree in Environmental Studies (Geography) from the University of Waterloo and began her federal public service career in 1981. She worked in the Department of Industry, Trade and Commerce and then the Department of Foreign Affairs and International Trade for 22 years, including assignments in Kenya, Australia, Nigeria, India, Hungary, Slovenia, and Albania. In 2002, she moved to the Department of Fisheries and Oceans and then the Treasury Board Secretariat. In 2006, she was named associate deputy minister at Health Canada, and in 2007 she joined the Privy Council Office as the foreign and defence policy advisor to the prime minister. In 2008, she was named deputy national security advisor to the prime minister.

Ratana Chuenpagdee, PhD, is Canada research chair in natural resource sustainability and community development at Memorial University, St John's, Newfoundland and Labrador. With an interdisciplinary background in marine science, fisheries biology, resource economics, and coastal zone management, she has conducted research related to a range of topics, including environmental valuation, marine protected areas, visualization and decision-support tools, and governance in Thailand, Cambodia, Mexico, South Africa, Malawi, the US, and Canada.

Debbe Crandall has been involved with Save the Oak Ridges Moraine Coalition (STORM) since 1990 and its executive director since 1999. She has represented STORM on a number of provincial initiatives concerned with landscape-scale conservation planning and regional growth management issues. Debbe was appointed in 2001 to the Oak Ridges Moraine Advisory Panel, which developed the framework for Oak Ridges Moraine legislation, to the Central Ontario Smart Growth Panel in 2002, and to the Greenbelt Council in 2006 and is currently the chair of the Oak Ridges Moraine Foundation. She has a BSc with a major in geology and remains hopeful that she will soon complete her MES at the University of Waterloo. Debbe is a long-time resident of Caledon, Ontario.

Philip Dearden, PhD, is professor and chair of the Department of Geography at the University of Victoria. He is a member of the World Commission on Protected Areas and the co-editor of Oxford's *Parks and Protected Areas in Canada*. He has extensive overseas experience in PA management and is also the leader of the MPA Working Group for the Ocean Management Research Network and co-chair of Parks Canada's NMCA Science Advisory Group.

Rob de Loë, PhD, is the university research chair in water policy and governance at the University of Waterloo and director of the multi-university Water Policy and Governance Group. Previously, he held the Canada research chair in water management at the University of Guelph. During the past two decades, Rob has studied and written extensively about water security, water allocation, source water protection, and climate change adaptation. He draws on his research to provide policy advice to a wide range of government and non-government organizations in Canada at scales ranging from local watersheds to the nation.

Alan Diduck is an associate professor in the Environmental Studies Program at the University of Winnipeg. His research focuses on public involvement and social learning in resource and environmental management. He holds a PhD in geography from the University of Waterloo and MNRM and LLB degrees from the University of Manitoba. From 1990 to 1997, he was executive director of the Community Legal Education Association, a community-based organization promoting the development of civic and legal competence.

Meinhard Doelle, LLM, LLB, is an associate professor at Dalhousie Law School, where he specializes in environmental law. He is the associate director of the Marine and Environmental Law

Institute. From 1996 to 2001, he was the executive director of Clean Nova Scotia. Prior to this, he served as drafter of the NS Environment Act and worked for the Canadian Environmental Assessment Agency during the passage of the Canadian Environmental Assessment Act (CEAA). Professor Doelle served on the federal Minister's Regulatory Advisory Committee on CEAA from 2000 to 2008. He has acted for proponents and intervenors in numerous environmental assessments. In 2008, he was appointed as a panel member to the Lower Churchill joint panel review in Labrador. He has written on a variety of environmental law topics, including climate change, energy law, invasive species, environmental assessments, and public participation in environmental decision-making. His recent book on CEAA is entitled *The Federal Environmental Assessment Process: A Guide and Critique.*

Anthony H.J. (Tony) Dorcey, MSc, is a professor in the Institute for Resources, Environment and Sustainability and the School of Community and Regional Planning (SCARP) at the University of British Columbia. His teaching and research focus on the use of negotiation, facilitation, and mediation in sustainability governance. Over the past two decades, he has served as director of SCARP, a member of the BC Round Table on Environment and Economy, inaugural chair of the Fraser Basin Management Board, and a facilitator of global multi-stakeholder dialogues for organizations including the International Union for the Conservation of Nature and the World Bank (http://www.tonydorcey.ca).

Dianne Draper, PhD, is professor in the Department of Geography at the University of Calgary. She is recognized for her research in planning and policy in water resources management, including coastal zone and fisheries management, and sustainable tourism,

ecotourism, and tourism growth management. She is the founding author of *Our Environment: A Canadian Perspective.* Her current research focuses on governance and quality of life issues in communities working toward sustainability and on managing growth and its impacts on communities, water resources, and parks and protected areas.

O.P. Dwivedi, PhD, LLD. (Hon), Dr Env.S. (Hon), Fellow of the Royal Society of Canada, is University Professor Emeritus, Department of Political Science, University of Guelph, where he has taught since 1967. He has written, co-authored, and edited/co-edited 35 books, including *Public Service Accountability; Public Administration in World Perspective; Environmental Policies in the Third World; Sustainable Development and Canada; Managing Development in a Global Context;* and *Public Administration in Transition,* and more than 120 chapters/articles in professional journals and scholarly books. His international consultancy assignments include working for the World Bank, UNESCO, the World Health Organization, the UN Economic and Social Commission for Asia and the Pacific (UN ESCAP), the International Development Research Centre (IDRC), and the Canadian International Development Agency (CIDA). He is a past president of the Canadian Political Science Association and former president of the Canadian Asian Studies Association. In 2005, he was awarded the Order of Canada by the Government of Canada.

Patricia Fitzpatrick is an assistant professor of geography at the University of Winnipeg. She received her PhD in geography from the University of Waterloo. Her research considers the changing nature of resource management in Canada, focusing on environmental assessments of mineral and energy projects.

Graham J. Forbes, PhD, is a professor in the Faculty of Forestry and Environmental Management and the Biology Department at the University of New Brunswick and director of the New Brunswick Cooperative Fish and Wildlife Research Unit and the Sir James Dunn Wildlife Research Centre. Over the past 20 years, Graham and his graduate students have conducted wildlife research on a range of species, including ungulates, black bear, wolves, mustelids, turtles, forest and wetland birds, small mammals, and bats. He served on the New Brunswick Wildlife Trust, the Terrestrial Mammal Subcommittee of the Committee on the Status of Endangered Wildlife in Canada (COSEWIC) and was chair of the Biodiversity Committee of the Fundy Model Forest and director of the Greater Fundy Ecosystem Research Group.

Michael Fox, PhD, is professor and head of the Geography and Environment Department at Mount Allison University in Sackville, New Brunswick. Over the years, he has been a middle and high school teacher in geography and environmental studies as well as a secondary and post-secondary administrator. He currently teaches introductory human–environment relations, community planning, and environmental and resource management, and his research has been largely devoted to educational policy and environmental management issues.

Alison Gill, PhD, is a professor at Simon Fraser University and holds a joint appointment in the Department of Geography and the School of Resource and Environmental Management. Following her earlier research on socio-behavioural aspects of new resource town planning, she has, over the past 20 years, focused on tourism and resort community planning, especially in mountains and coastal areas.

Travis Gliedt, MAES, is a PhD candidate in the Department of Geography and Environmental Management at the University of Waterloo. His research interests include community energy management, energy policy, and environmental entrepreneurship.

Kevin Hanna, PhD, is an associate professor in the Department of Geography and Environmental Studies at Wilfrid Laurier University. His research centres on integrated approaches to natural resource management, forestry communities, and regional land-use planning. He is the editor of the books *Environmental Impact Assessment, Practice and Participation*; *Integrated Resource and Environmental Management: Concepts and Practice* (with D.S. Slocombe) (both from Oxford University Press); and *Parks and Protected Areas: Design and Policy* (with Douglas Clark and D.S. Slocombe) (Routledge).

Roger Hayter, PhD, is an economic geographer in the Department of Geography at Simon Fraser University in Burnaby, with research interests in institutional approaches, industrial location dynamics, regional development, and the restructuring of the Canadian, especially British Columbia's, forest industries. On the latter theme, he has authored several books, including *Flexible Crossroads: The Restructuring of British Columbia's Forest Industries* (University of British Columbia Press 2001).

Reid Kreutzwiser, PhD, is professor of geography at the University of Guelph. His research interests broadly concern the role of institutional factors in resource and environmental management. Specifically, current research addresses the effectiveness of Canadian water allocation arrangements, governance for drinking water source protection, and private water well stewardship in Ontario. Much of his research involves collaboration with government agencies and non-government organizations.

Mary Louise McAllister is associate professor of environment and resource studies at the University of Waterloo. She received her BA (Hons.), MA, and PhD in political studies from Queen's University at Kingston. She has written monographs and articles about mineral and environmental policy, mining communities, and local government.

Gordon McBean, PhD, is a professor of geography and political sciences at the University of Western Ontario and director, policy studies, at the Institute for Catastrophic Loss Reduction. He was formerly professor of atmospheric and oceanographic sciences at the University of British Columbia and an assistant deputy minister with Environment Canada. He is a Member of the Order of Canada and a Fellow of the Royal Society of Canada and shared the 2007 Nobel Peace Prize as a significant contributor to the Intergovernmental Panel on Climate Change. He received a MSc in meteorology from McGill University and a PhD in physics and oceanography from the University of British Columbia.

Dan McCarthy, PhD, is an assistant professor in environment and resource studies and social innovation generation at the University of Waterloo. He specializes in systems thinking, social innovation, and environmental assessment and planning and works collaboratively with stakeholders on the Oak Ridges Moraine, Ontario, and with First Nations communities on the west coast of James Bay.

Virginia Maclaren, PhD, is an associate professor in the Department of Geography and Program in Planning at the University of Toronto. She teaches graduate and undergraduate courses in waste management and is a member of the Board of Waste Diversion Ontario, a non-Crown corporation established to develop, implement, and operate waste diversion programs in Ontario. Her research has focused on waste management issues in Southeast Asia and on landfill opposition in Canada.

Sonya Meek, MA, is the national past president of the Canadian Water Resources Association and manager of the Watershed Planning Group at the Toronto and Region Conservation Authority.

Bruce Mitchell, PhD, is professor of geography and environmental management and associate provost, Academic and Student Affairs, at the University of Waterloo. His research focuses on policy and governance aspects of water management, with particular attention to the concept of 'integrated water resource management'. He has conducted research in Australia, China, Indonesia, and Nigeria as well as in Canada. He is a Fellow of the Royal Society of Canada and a Fellow of the International Water Resources Association.

Linda Mortsch, MES, is a researcher with the Adaptation and Impacts Research Division of Environment Canada and is also an adjunct in the Faculty of Environment at the University of Waterloo. She has more than 20 years of research experience in the climate change impact assessment field. Her research interests include climate change vulnerability, impact, and adaptation assessments in wetlands, water resources and the urban environment, and climate change scenario development. For the 2007 Intergovernmental Panel on Climate Change *Fourth Assessment Report*, she was the co-ordinating lead author for the North America chapter.

Bram Noble, PhD, is an associate professor in the Department of Geography and Planning and in the School of Environment and Sustainability at the University of Saskatchewan. His research interests are in resource and environmental management, particularly impact assessment and the implications of resource development. He is author of *Introduction to Environmental Impact Assessment: A Guide to Principles and Practices*, published by Oxford University Press in 2006.

Paul Parker, PhD, is a professor in the Department of Geography and Environmental Management at the University of Waterloo. His research interests include residential energy efficiency, renewable energy, and community energy planning. He has participated in many energy studies, ranging from Canadian cities to remote First Nations communities and from Australian coalfields to the Japanese solar industry.

David J. Rapport, PhD, FLS, is principal of EcoHealth Consulting and one of the developers of the field of ecosystem health. He is past founding president of the International Society for Ecosystem Health and co-founder of the Program in Ecosystem Health in the Faculty of Medicine at the University of Western Ontario. Dr Rapport held the Eco-Research Chair at the University of Guelph. He co-authored the first *State of Environment Report* for Canada. He now lives on Salt Spring Island, BC.

Maureen Reed, PhD, is professor in the School of Environment and Sustainability and the Department of Geography and Planning at the University of Saskatchewan. She studies community involvement in ecosystem management and focuses attention on rural communities in and near national parks, forests, and biosphere reserves. Part of her research program has been devoted to

understanding the differential effects in the creation and implementation of environmental policies and management programs for women and men who live in resource-dependent communities.

John Sinclair, PhD, is a professor at the Natural Resources Institute, University of Manitoba. His main research interest focuses on community involvement and learning in the process of resource and environmental decision-making. He is very active in the area of environmental assessment (EA) law and policy. Through a current research grant, he is considering how constructs such as critical EA education and transformative learning can contribute to a conceptual framework for participation in community EA. He is also working on the application of EA to hydro development in the Indian Himalayas. John has been a member of the Canadian Environment Network's EA caucus for many years and is also currently a member of the Canadian Environmental Assessment Agency's Regulatory Advisory Committee.

Norm Skelton, BA, is a graduate student attending the University of Northern British Columbia. He received a Bachelor of Arts degree in May 2002 with majors in anthropology and political science. His interests and study areas include justice, governance, and cultural dynamics. Norm returned to school in his mid-40s after working in a variety of fields. His Master of Arts thesis is based in cross-cultural concepts of justice and a comparative study of justice in communities and academia.

Scott Slocombe, PhD, is professor of geography and environmental studies at Wilfrid Laurier University. His primary research interest is planning and management of large, complex regions. This includes systems and ecosystems approaches in resource and environmental management, protected areas, environmental assessment, and policy. Much of his research has taken place in western and northern Canada, Australia, and the Great Lakes Basin.

Chui-Ling Tam, PhD, is assistant professor of development studies at the University of Calgary. Her primary concern is the communicative politics embedded in environmental management and resource development. Her current research is focused on the intersection of political ecology, communication geographies, environmental management, and participatory development in Southeast Asia. She is also interested in innovative pedagogy, primarily collaborative, multidisciplinary experiential field schools and knowledge-sharing among private and public spheres.

Iain Wallace, PhD, is a professor in the Department of Geography and Environmental Studies, Carleton University, Ottawa. His interest in the society–natural environment interface lies behind his ongoing research into economic globalization in the anthropocene era, human dimensions of global environmental change, global agribusiness systems, and theology, epistemology, and the nature of geography.

Graham Whitelaw, PhD, is an associate professor in the School of Environmental Studies and School of Urban and Regional Planning at Queen's University, Kingston, Ontario. His research interests focus on environmental assessment, land-use planning, and First Nations.

Preface

The Nature and Structure of This Book

As explained in the introduction to the previous editions, a book addressing resource and environmental management in Canada might be focused and organized in various ways. The intent here has been to accomplish a number of objectives. The first is to demonstrate the reality of the resource and environmental management process, to stress the mix of technical and non-technical considerations that influence the decision process, to illustrate the variety of strategies that are used, and to emphasize the trade-offs that must be reached to satisfy diverse and legitimate societal interests.

Second, to provide a focus for the book, 'conflict' and 'uncertainty' were chosen as key themes. Each is central in resource and environmental management. Different understandings, values, interests, and contexts lead to varying preferences regarding goals and objectives, alternative interpretations of information and evidence, and differing ideas regarding appropriate strategies and actions. Indeed, resource and environmental management often seems to be centred on the resolution of conflict generated by diverse regional, sectoral, substantive, or ideological perspectives. Uncertainty also is a basic feature of resource and environmental management. Whether dealing with biophysical or social phenomena, we often do not have adequate knowledge or understanding. Yet pressure builds, and action must be taken, even when analysts and managers are not at all sure about present or future conditions. The dual focus here on conflict and uncertainty is a reminder that perfect solutions to complex problems are rare.

More common is a mix of imperfect responses from which we must choose.

Third, the book is oriented toward introductory university students, although we expect that it will also be useful in senior university courses. A key concern was that the writing should be relatively free of jargon and that concepts should be presented in a straightforward manner. When it was difficult to avoid technical terms, authors were asked to ensure that these terms were clearly explained for the reader.

Fourth, it was clear at the outset that no matter how the book was organized or what the chapters covered, others might take a different view as to the appropriate organization and content. In that regard, several points are worth mentioning here. The organization of the book into three main parts—'Emerging and Current Concerns', 'Enduring Concerns', and 'Responses'—may well be imperfect, since inevitably the chapters in the first two sections include consideration of responses. Furthermore, what is emerging and what is enduring becomes a judgment call and dependent on a time frame. Nevertheless, some resource and environmental management issues related to fisheries, agriculture, forestry, and mining have been ongoing since before Canada became a nation. Others, such as ecosystem health, globalization, climate change, First Nations rights, gender, climate change, and water or energy security, have emerged more recently. The intent is to highlight concerns that are *relatively* recent in the first section and to address those that have attracted attention for a longer period of time in the second section. Again, there can be some overlap between the two sections. For example, the fisheries, agriculture, forestry, and mining issues discussed in the second section are obviously affected by globalization and other topics addressed in the first section.

In the fourth edition, the chapters that appeared in the third edition have been revised, some very significantly. In addition, new chapters and/or new authors have been added (Chapters 1, 2, 5, 6, 9, 10, 13, and 17). A continuing feature, however, is the set of guest statements prepared by academics and practitioners, reflecting on their experience in translating concepts into action. The third edition contained seven such statements, whereas this edition includes 11, all but one written by new authors. As a result, the reader will find the fourth edition both updated and expanded in coverage.

As noted in the introduction to the previous editions, in any book made up of original essays there is always a risk of omissions or overlaps. In a field as broad as resource and environmental management, it is inevitable that readers will find that some topics, problems, or approaches have not been covered. Regarding any omissions, it is hoped that readers will be stimulated to pursue further reading and study on topics not covered here because of space limitations. Regarding overlap, all of the authors were aware of the material to be covered in other chapters. In a number of instances, they exchanged their outlines and drafts to ensure complementarity. At the same time, given the prominence of some

issues and problems, it is not surprising that some are treated in more than one chapter. This can be seen as a positive feature, underlining the pervasiveness of some issues and concepts and highlighting that they can and should be considered from different perspectives.

I would like to thank all of the authors who contributed chapters, either by revising their contributions to the third edition or by writing new chapters for this fourth edition. I also would like to thank Frances Hannigan, who formatted the text and helped with figures.

As in all my book projects, Joan Mitchell provided valuable assistance and advice as well as offering patience and understanding, for which I am very grateful.

Nancy Hofmann of Statistics Canada was very helpful regarding insight about environmental indicators.

At Oxford University Press, constructive assistance was provided by Jacqueline Mason, sponsoring editor, Dina Theleritis, developmental editor, and Jodi Lewchuk, supervising editor. Excellent copyediting was completed by Dorothy Turnbull.

Bruce Mitchell
University of Waterloo
November 2009

Introduction

Policy and Practice—Issues, Challenges, and Opportunities

Bruce Mitchell

Learning Objectives

- To understand the roles and responsibilities of various levels of government related to resource and environmental management.
- To appreciate the emerging and evolving roles of First Nations and civil society.
- To recognize the implications of the evolution toward globalization and market-oriented policies and practices.
- To grasp the consequences of the global economic difficulties that emerged in the fourth quarter of 2008.
- To discern various dimensions of 'conflict' and 'uncertainty'.
- To become familiar with the mix of 'best practices' that should characterize resource and environmental management.

CONTEXT

In the first section of this introduction, attention is directed to the significance of federal, provincial, and municipal roles, First Nations roles, civil society, regional interests and perspectives, globalization, the consequences of the economic downturn that began in late 2008, and competing agendas.

Federal, Provincial, and Municipal Roles

Canada is a federated state, with power and authority shared between federal and provincial governments. Municipal governments receive their power and authority from provincial legislatures.

The Canada Act, 1982, differentiates between proprietary rights and legislative authority relative to natural resources and the environment. Through the Canada Act, the provinces have ownership of, or proprietary rights to, all Crown lands and natural resources not specifically given over to private ownership, except for the Canadian North (north of 60 degrees latitude), where the federal government has proprietary rights to land and resources until territories become equivalent to provinces and to resources found on or under sea beds off the coasts of Canada (although some provinces are challenging this right).

In contrast to proprietary rights, legislative authority is divided between the federal and provincial governments, often becoming a significant source of conflict. The federal government has statutory jurisdiction over trade and commerce, giving it substantial control over both interprovincial and export trade of resources (oil, natural gas, water). Provinces such as Alberta, Saskatchewan, and British Columbia, which have oil and natural gas, often object to the federal government becoming involved in setting prices and determining buyers, arguing that such matters are within provincial authority because of their responsibility for property and civil rights and also because these matters are of local interest. The federal government has also used its legislative authority for navigation, shipping, and fisheries to create pollution regulations—even though water within provinces is under the authority of the provinces. As Mitchell and Sewell (1981, 7) concluded almost 30 years ago, 'the ambiguity and inconsistencies in jurisdiction over resources has been a major and continuing issue in Canada.' That conclusion is still valid today. Such ambiguity and inconsistencies create major challenges to establishing *national approaches* (combined federal, provincial, and territorial) to deal with resource and environmental issues. The different proprietary and legislative roles are examined in more detail in Chapter 2 by O.P. Dwivedi.

As already noted, municipalities receive their authority from provincial legislatures. In the early to mid-1990s, many provincial governments began to download traditional provincial responsibilities to municipalities. The publicized rationale was that such downloading was consistent with the principle of *subsidiarity*, which stipulates that decisions should be taken at the level closest to where services are used or received and therefore at the level where consequences are most noticeable. While such a principle is logical, another interpretation has been that the primary motive for downloading was the desire of provincial governments to shift the cost of many responsibilities to lower levels of government as part of a strategy to reduce provincial debts and deficits. Whatever the motivation, the outcome has been that municipalities have become much more significant players in natural resource and environmental management, since in many instances, provinces have withdrawn from related management activities. One manifestation of this has been the preparation of sustainable development strategies for implementation at the local level, along with various initiatives to create 'green' communities.

Aboriginal Peoples' Roles

Aboriginal peoples have argued that they have had rights to land and natural resources 'from time immemorial' and that even when treaties were signed with European nations, the Aboriginal peoples retained their traditional rights. However, the courts, and federal and provincial governments, have interpreted Aboriginal rights very narrowly under treaties and Canadian legislation, leading to ongoing conflict between Aboriginal peoples and Canadian governments.

However, starting in the early 1980s, significant changes began to occur relative to recognition of the collective rights of Aboriginals. When the Canadian Constitution was repatriated in 1982 and the British North America Act became the Canada Act, existing Aboriginal and treaty rights were recognized. The implication was that Aboriginal peoples have special collective rights under law because they are Aboriginal peoples (Chapter 4, by Booth and Skelton).

Such Aboriginal rights are gradually being recognized under law. A ruling in 1992 on a challenge by an Aboriginal person, Ronald Sparrow, regarding fishing rights in British Columbia became one benchmark confirming the special rights of Aboriginal people. Sparrow had fished out of season and used gear that violated conservation regulations as a deliberate challenge. Sparrow's right to fish was upheld in the courts, and as Wolfe-Keddie (1995, 59) remarked, that decision 'not only confirms Aboriginal and treaty rights to resource use but also gives weight to Aboriginal demands to participate as major decision makers in resource management.'

An important implication is that while federal, provincial, and territorial governments share jurisdiction and responsibility over natural resources, in the future they will often have to negotiate with Aboriginal leaders in reaching decisions related to resource and environmental management. More insight into the role and significance of Aboriginal people can be found in Chapter 4.

Civil Society

Many citizens and interest groups want to be more directly involved in planning and decision-making related to natural resources and the environment as scepticism about the commitment and capability of governments has grown or as different groups seek to lobby governments so that decisions will support their priorities (see Chapter 18). Woodrow (1980, 39–41) noted that three basic civil society groups can be identified:

1. *Promoters*, who advocate policies that support economic growth, streamline and reduce regulations, and create an encouraging climate for investment and private-sector entrepreneurship activities. This category is represented by groups such as the Canadian Manufacturers Association, the Canadian Chamber of Commerce, and the Conference Board, as well as more focused organizations such as the Canadian Association of Petroleum Producers, the Canadian Pulp and Paper Association, and the Fraser Institute.

2. *Conservationists and environmentalists*, who support restraints and regulations to control or avoid wasteful exploitation of natural resources and the environment. Representative groups are the Canadian Arctic Resources Committee, the Canadian Environmental Law Association, the Canadian Wildlife Federation, the David Suzuki Foundation, Greenpeace, Pollution Probe, and the Sierra Club of Canada.

3. *Technologists*, who are mainly interested in natural resources and the environment and whose interest is to ensure 'good' management of them. They are represented by groups such as the Canadian Water Resources Association.

With such a mix of interests and perspectives, the potential for conflict is clear, given that so many interest groups often have mutually exclusive ideas about how resource and environmental management should be conducted.

Regional Interests and Perspectives

Compounding the fragmented responsibilities among federal, provincial, and territorial governments are differing regional perspectives, which lead to various perceptions regarding what the issues are or what action should be taken. For Atlantic Canada, the closure of the northern cod fishery in 1992 was devastating for the livelihood of many small coastal communities, especially in Newfoundland (see Chapter 8). With a close tie to home communities, many Newfoundlanders believed priority should be given to developing alternative economic opportunities so that people could continue to live in their communities. In other parts of the country, however, some believed that the small coastal communities were no longer viable and that spending public funds to extend their life would be futile.

Other examples of differing regional needs and interests can be identified. Ontario as a net energy importer has different priorities regarding appropriate prices for crude oil or natural gas than those of Alberta, British Columbia, and Saskatchewan as net energy exporters. Softwood lumber is a major issue for British Columbia, and agriculture is a high priority for Alberta, Saskatchewan, and Manitoba, while Ontario is more likely be concerned about decisions affecting the automotive and other manufacturing industries. Such differing interests create significant challenges for a federal government as it negotiates with the government of the United States about limits on greenhouse gas emissions, tariffs or taxes on softwood lumber, financial support programs for farmers, and arrangements for an automotive industry tightly interconnected with operations in the United States.

Debate over the Kyoto Protocol to reduce greenhouse gas emissions provided an excellent example of these divergent regional interests. Under the protocol, Canada made a commitment to reduce its greenhouse gas emissions to 6 per cent below the 1990 level by the period 2008 to 2012. Canada ratified the protocol after much debate, and part of the difficulty in reaching a national consensus came from widely divergent views in various regions:

- Alberta: Provincial politicians argued that the provisions in Kyoto would cause a huge negative impact on the economy of Alberta, given its reliance on oil and natural gas industries.
- Quebec: The provincial government was a strong supporter of Kyoto, and the National Assembly in Quebec City unanimously passed a resolution to support it. Provincial leaders acknowledged that because Quebec relies heavily on hydroelectricity, viewed as a clean source of power, a decision to support Kyoto was more straightforward for Quebec than for Alberta.
- Manitoba: The premier supported Kyoto, given the potentially devastating implications of climate change for agriculture in the province and the importance of hydropower in Manitoba. At a First Ministers' meeting, the premier was quoted as saying that climate change, significantly driven by the reliance of economies on oil and gas, would create desert-like conditions in Manitoba and that he did not want Manitoba to move from raising cattle to raising camels.
- Nunavut: The premier's view was that climate change had already been destroying the environment of his territory and that such destruction could never be compensated for through equalization payments from the 'have' provinces that benefited from their reliance on fossil-fuel energy. Therefore, he supported Kyoto.

In terms of sectoral interests, polls suggested that the general public was strongly in favour of Canada ratifying the Kyoto Protocol, while many in the business community opposed it. With such conflicting perspectives, it was a challenge to develop a national position regarding the protocol (see Chapter 6).

Globalization

Factors related to globalization and transnational approaches to trade, economic development, and environmental management have been increasing in importance, as noted in Chapters 1 and 2. Examples are numerous, ranging from the Canada–United States Free Trade Agreement to the Kyoto Protocol. A consequence is that frequently, Canadian decision-makers do not control policies and decisions with major implications for the way in which natural resources are used and managed in this country.

One example is the steady movement toward more arrangements for 'free trade', the purpose of which is to stimulate economic growth, productivity, and better living standards by eliminating restrictive regulations between and among nations. Frequently, these arrangements are silent on environmental regulations or controls. Decisions made in countries with lower wages or fewer environmental regulations than Canada can sometimes result in multinational firms choosing to make new investments there rather than in Canada.

Economic Downturn, 2008–

The economic downturn that began in the final quarter of 2008, triggered by the housing market correction and the sub-prime mortgage crisis in the United States, contributed to a 40 per cent decline in the Toronto stock market (TSX) over the last three months of 2008 relative to the beginning of the same year. This dramatic and significant drop was led in Canada by declines in natural resource stocks as demand for oil and metals fell sharply. Reflecting this fall in demand, while the price of oil exceeded $100 per barrel in early 2008 and climbed to a peak of more than $147 a barrel during July of the same year, by the beginning of 2009 the price had dropped to $46 a barrel.

The stocks of financial institutions also declined significantly in value because banks and other financial institutions experienced significant losses as a result of the housing market collapse in the US. And house prices in Canada were not exempt from the drop, decreasing an average of 11 per cent across the country relative to peak prices in May 2008. The drop was highest in Vancouver, the most expensive real estate market in the country. Then in early 2009, the federal governments in the United States and Canada offered unprecedented financial support to the automotive industry, based on a belief that the industry was too large and critical to be allowed to fail.

Canadians began to consider the prospect of a recession or perhaps even a depression. Many working people wondered whether they would continue to hold a job or have to join the growing number of unemployed. Many people at or approaching retirement began to worry that their investments or pensions would not be sufficient. In such an economic climate, it becomes increasingly difficult to maintain a focus on balancing economic growth and protection of the environment.

Competing Agendas

The discussion in the above sections highlights that many issues compete for the attention of policy- and decision-makers. Indeed, not only does competition exist among different resource and environmental issues, but they also have to compete with other societal issues, such as health care, education, security in the wake of 9/11, and economic well-being. As a result, Downs's (1972) 'issue-attention cycle' is useful to keep in mind, since it reminds us that every issue literally has to struggle to remain on the radar screens of key policy- and decision-makers (Figure I.1).

In Downs's view, every issue has the potential to pass through five stages: (1) *pre-problem*, when there is general lack of awareness of a problem except on the part of a few experts; (2) *alarmed discovery and euphoric enthusiasm*, during which the public becomes aware of an issue and pressures decision-makers to take action; (3) *realization of the cost of significant progress*, during which both decision-makers and the public become aware of the costs that could be involved, as occurred with the Kyoto Protocol in the early 2000s in Canada; (4) *gradual decline of public interest* as a result of appreciation of the costs of action and also the emergence of competing issues reaching stage 2 of the issue-attention cycle; and (5) *post-problem*,

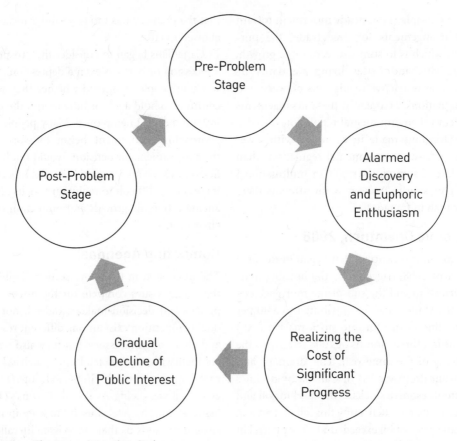

Figure I.1: The Issue-Attention Cycle

during which the issue remains of interest to those with expertise related to it, who strive to shift the issue back to stage 2.

The issue-attention cycle reminds us that those concerned about a particular issue have to work diligently to maintain its position on policy agendas, because supporters of other issues are always striving to replace it or at least surpass it in order of priority. Furthermore, advocates must keep in mind that rarely are there sufficient funds and human resources available for decision-makers to allocate to all the issues that deserve them. Thus, priorities must be established, trade-offs made, and choices taken. All of this often occurs against a backdrop of much

uncertainty, which is also an important contextual component and, along with conflict, is reviewed in the next section.

CONFLICT AND UNCERTAINTY
Conflict

> a fight or struggle, especially a prolonged
> one; a disagreement, dispute or quarrel;
> . . . the clashing of opposed principles,
> statements, etc.
>
> *[Barnhart 1975, v. 1, 443]*

Dyer (2008, ix–xiii), in his book on 'climate wars', highlights that much of resource and environmental management involves conflict. For example, he explains that he decided to write his book because of his growing appreciation that one of the first and most significant consequences of climate change on humans would be an acute and serious crisis related to food supplies. And, he observed, since 'eating regularly is a non-negotiable activity', one likely outcome would be that 'countries that cannot feed their people are unlikely to be "reasonable" about it.' Continuing, he also observed that with every degree increase of temperature due to climate change, there would be increasing mass movements of populations, more failed and failing states, and increased probability of internal and international wars. Regarding the prospect of such climate wars, he noted that 'if they become big and frequent enough, [they] will sabotage the global cooperation that is the only way to stop the temperature from continuing to climb.' Dyer's concern about possible conflict is confirmed by Wiener (2009, 32), who examined how China can become engaged in mitigation initiatives to reduce greenhouse gas emissions. His views are presented in the accompanying box.

Conflict may emerge in many forms. Dorcey (1986, 39–40) has usefully identified four different types (Table I.1). He also provided an example to illustrate how these different types can occur either on their own or in combination. His example considered negotiations between a regulatory agency and a forestry company about the implications of storing logs in an estuary.

At the *cognitive* level, biologists with the regulatory agency might have collected data indicating that storing logs in the estuary would adversely affect juvenile salmon. In contrast, using different data collection and analytical methods, biologists for the company concluded that storage of logs posed no significant threat to the fish. Both parties might share a genuine concern about the well-being of the juvenile salmon, but in this situation they had different understandings of the nature of the possible threat to salmon from storing logs in the estuary.

Regarding *values*, it is possible that both parties shared a concern about the well-being of the fish and also that each of their data supported a common conclusion that a threat did indeed exist for the juvenile fish. However, distinctly different viewpoints could still exist regarding the appropriate trade-offs between the protection of the fish and the creation of jobs for loggers and mill workers at the forestry company. In this manner, different values could create conflict, even though at a cognitive level there was no dispute.

Box I.1 Climate Change and Conflict in China

Research in climate history has begun to suggest that rapid climate changes in China in past millennia have been associated with wars and the ends of ruling dynasties. Along similar lines, traditional Chinese beliefs (expounded by the ancient philosopher Dong Zhongshu) linked extreme environmental events to impending political upheaval. Although modernization may have reduced the intensity of these beliefs in China, gradual political liberalization may have allowed these ideas to become more prevalent, or at least more openly acknowledged.

Table I.1 Four Types of Conflict

Cognitive	Results from different understandings of a situation. Such differences may lead to different technical judgments.
Value	Results from different judgments about the ends to be achieved, even when there is agreement regarding the consequences of alternative ends.
Interests	Occurs as a result of disagreement about the distribution of benefits and costs. In other words, who should be the beneficiaries, and who should pay the costs?
Behavioural	Emerges related to the personalities and circumstances of the interested parties. This reflects a history of previous and ongoing relationships among those involved in the conflict, which may colour their attitudes toward each other and toward possible solutions.

Conflict could occur even if the two parties agreed that the storing of logs could threaten the fish and had reached a mutually acceptable decision about the most appropriate mix between environmental quality and economic growth. An *interest*-based conflict would emerge if disagreement arose over who should pay for moving the log booms to preserve the fishery habitat. In this situation, conflict would not be based on cognitive or value issues but rather on different interests regarding responsibility for costs.

Transcending the above types of conflicts, *behavioural* aspects could confound matters even after common positions were established regarding the problem and the information, the desirable ends, and the sharing of benefits and costs. Previous decisions regarding logging practices and fishing, lack of communication, and mistrust and emotions could combine to make it difficult for the two sides to trust each other and therefore reach an agreement.

The four kinds of conflict do not have to occur in isolation. In any disagreement, more than one source of conflict could be involved. As a result, it is important that resource and environmental managers be aware of the different types of conflict and able to recognize them in management situations. Being aware of the causes of conflict is a first step toward developing a solution.

Uncertainty

not known with certainty; not finally established; in doubt; dubious; likely to change, not reliable; not constant, varying; not settled or fixed, indeterminate
[Barnhart 1975, v. 2, 2254].

With regard to climate change, Dyer (2008, xii) highlights that uncertainty is a fundamental characteristic. For example, he remarks that 'this thing [climate change] is coming at us a whole lot faster than the publicly acknowledged wisdom has it.' Furthermore, he suggests that it is 'unrealistic to believe that we are really going to meet those deadlines' (e.g., in reducing greenhouse gas emissions).

Uncertainty is the second theme in this book. This concept was chosen because in most situations, there is imperfect knowledge or understanding in resource management. Nevertheless, decisions must be taken. As a result, resource or environmental managers often make decisions without knowing the full implications or consequences of their choices. Recognizing the importance of uncertainty is thus essential. At the same time, however, it is also necessary to recognize that the concept of *uncertainty* in itself does not encompass all the situations that might be encountered. In that regard, Wynne (1992, 114) identified four kinds of 'uncertainty' (Table I.2).

Table I.2 Four Kinds of Uncertainty

Risk	Know the odds or probability.
Uncertainty	Do not know the odds or probability. May know the key variables and their parameters.
Ignorance	Do not know what we should know. Do not even know what questions we should be asking.
Indeterminacy	Causal chains or networks are open. Understanding not possible.

Wynne suggested that we should talk about *risk* when the behaviour of a system is basically known and the probability of various outcomes can be defined and quantified. An example would be estimating the risk of a specified magnitude of flood based on a long hydrological record for a river system. With a lengthy time series of information, the risk could be calculated. There would always be a possibility of error, but there could be considerable confidence in the estimate.

In contrast, *uncertainty* occurs if the behaviour of the system is not known and therefore it is not possible to estimate the probability of a given outcome. In the example of estimating the magnitude of a flood, the resource managers could well know the key variables contributing to a flood event, but if inadequate data were available, then it would be difficult if not impossible to estimate the probability with any confidence.

Often, however, resource managers must deal with the more difficult issue of *ignorance*—in other words, failure to recognize an existing problem. In such situations, we are unaware of a possible problem and therefore fail to even consider it. Two examples are the lack of awareness of acid rain in Ontario and Quebec in the 1960s and failure to acknowledge climate change at a global scale during the 1970s.

Indeterminacy, a fourth kind of uncertainty, arises as a result of lack of understanding of cause-and-effect relationships. Much work in risk assessment assumes that uncertainty arises because of incomplete definition or understanding of cause-and-effect relationships in ecosystems. The belief is that more scientific work should be able to reduce or eliminate the uncertainties. However, some people, such as Waldrop (1992), have argued that because of the complexity of ecosystems, it may be unrealistic to anticipate that humankind will ever be able to understand some systems. Others, such as Gleick (1987), have argued that many systems appear to be better characterized by chaos than by order and stability. In such situations, Wynne argues that managers may have to accept the indeterminacy of knowledge and understanding.

Wynne's discussion of the various kinds of uncertainty has been echoed by other writers such as Dovers and Handmer (1992) and Gibson (1992). For the purposes of this book, it is important to appreciate that humans do not understand many aspects of ecosystems, and in some instances the prospects for gaining understanding are poor, as noted in Chapter 3. Nevertheless, management decisions must still be made in the face of both uncertainty and conflict. As a result, resource and environmental managers often find themselves involved in situations in which the issues and the solutions do not appear in black and white but rather in shades of grey, many of which are notably fuzzy. This is reality, and resource and environmental managers must be able to operate under such conditions.

BEST PRACTICES

This section looks at 'best practices', or aspects expected to be present when a resource or environmental issue is being managed: (1) vision, (2) legitimacy, (3) systems perspective, (4) adaptation and resilience, (5) partnerships, (6) impact and risk assessment, and (7) communication.

1. Vision

If a situation is to be improved or a problem resolved, an obvious prerequisite is a clearly identified desirable future condition. A vision identifies the future desirable condition. It is similar to planning a journey. If you do not know what your future destination is or the purpose of your journey, it is will be difficult for you to know which route to select.

Several important points must be kept in mind when developing a vision. First, it is vital to differentiate among the most probable, the most desirable, and the most feasible future conditions. In resource and environmental management, most attention is often allocated to trying to determine the most probable future, which leads to emphasis on *forecasting*. In forecasting, analysts seek to understand the characteristics of the present condition, make assumptions that could characterize the future (e.g., the cost of petroleum, the impact of climate change on food production), and then aim to identify the most likely conditions at some time in the future.

However, the most probable future will not automatically be the most desirable future. Furthermore, human beings have the capacity to intervene and change conditions that will shape the future. As a result, many believe that at least as much or more time needs to be allocated to exploring possible desirable futures, and this is normally done through *backcasting*.

In backcasting, an ideal future condition is identified—say, 20 years from now—and then the focus shifts to determining what decisions must be made between now and that future time to ensure that the desirable future is realized. Crafting a scenario is one technique often used to build or create alternative future outcomes and to identify the paths that will need to be created for any one of them to be achieved.

Finally, it is important to consider what is feasible or practical. Some desirable future conditions may be beyond our grasp. There is nothing wrong in envisioning 'stretch' or 'aspirational' desired futures, but that kind of vision also calls for a 'reality check' to ensure that societal resources will not be expended in pursuit of a goal that is not achievable, given current knowledge and technology.

Moreover, given divergent and competing values, interests, and needs in a society, various stakeholders will likely have different and indeed sometimes mutually exclusive visions for desirable future conditions. As a result, developing a vision that will be generally endorsed and supported can be a formidable task, which reinforces the importance of the ability to manage and resolve conflicts that can emerge.

Sustainable development, or sustainability, often constitute the basic building blocks for a vision related to resource and environmental management. The World Commission on Environment and Development (1987) popularized the concept of sustainable development in its report, *Our Common Future*. The best known statement from *Our Common Future* is that 'sustainable development is development that meets the needs of the present without compromising the ability of future generations to meet their own needs' (1987, 43). However, other statements from the report also merit our attention, including:

- A fundamental need is to integrate economic and ecological considerations in decision-making because they are integrated in the workings of the real world.
- The objective of sustainable development and the integrated nature of global environment/development challenges pose problems for institutions, national and international, which were established on the basis of narrow preoccupations and compartmentalized concerns. Challenges are both interdependent and integrated, requiring comprehensive approaches and popular participation. The real world of interlocked economic and ecological systems will not change; the policies and institutions concerned must.
- The law alone cannot enforce the common interest. Community knowledge and support are also required, which entails greater public participation in the decisions that affect the environment. This is best secured by decentralizing the management of resources upon which local communities depend and giving these communities an effective say over the use of those resources.
- No single blueprint of sustainability will be found, since economic and social systems and ecological conditions differ widely. Each nation, region, and community will have to create a custom-designed approach.
- The search for common interest would be less difficult if all development and environment problems had solutions that would leave everyone better off. This is seldom the case; there are usually winners and losers.

In June 1992, the United Nations Conference on Environment and Development (the Earth Summit) took place in Rio de Janeiro, involving representatives from 179 countries. The purpose was to establish international agreements respecting the interests of all and protecting the integrity of global environmental and development systems, building on the concept of sustainable development as articulated by the World Commission on Environment and Development.

Ten years later, in 2002, the Johannesburg Summit was held from 26 August to 4 September in South Africa. Officially known as the World Summit on Sustainable Development, it was designed to document progress since the Rio Summit of 1992, particularly to focus on developing tangible steps and identifying quantifiable targets for effective implementation of Agenda 21, the strategy for sustainable development in the twenty-first century prepared at Rio. The reports from the Rio and Johannesburg events are worth reviewing to see what was viewed as a desirable future for humankind.

2. Legitimacy

Instruments should be in place to provide legitimacy for a vision and to facilitate the implementation of goals and objectives related to the vision. Such instruments include political commitment, statutory foundation, administrative arrangements, financial support, and stakeholder support. The more instruments in place, the more likely that a vision will have credibility or legitimacy.

Commitment from senior elected officials provides credibility, since such people are in a position to determine priorities and to decide which activities will receive funding. Without political support, anyone striving to realize the aspirations of a vision will have a challenging road to travel, because they will face a never-ending struggle to gain the attention of key decision-makers and thereby keep the vision as a priority for action. This situation reinforces the significance of the issue-attention cycle discussed earlier in this chapter.

A statutory or legislative base also helps. If an agency has a legal responsibility for a task or function, it is more difficult to ignore or set aside that task or function. In contrast, if an implementing agency has responsibility for more tasks than it has resources to support and thus has to make choices, it is quite probable that non-statutory responsibilities will receive lower priority.

Once political commitment and statutory authority are in place, explicit and specific administrative directives and structures provide further support for the vision and guide people with responsibility for it. Without such directives and structures, it is too easy for individuals to interpret political statements or legislation in a way that supports their own needs or interests. This can sometimes result in a failure to follow the spirit or intent of a vision.

Without financial resources, it is normally difficult to move ahead with needed initiatives. Initiatives can be supported at varying levels (e.g., transportation can be facilitated by a Porsche, a Pria, or a bicycle), so choices exist related to what level of quality or excellence will be achieved. If a reasonable level of funding is not available, legitimacy or credibility will be low. If people sense an unwillingness to provide adequate funding, they will tend to conclude that the vision is not of high priority.

At the beginning of this section, it was pointed out that political support is important, and it usually is. However, politicians represent just one stakeholder in most resource and environmental management situations. To achieve a high level of credibility, support should be widespread across the continuum of stakeholders, including the private sector, environmental groups, and the general public. The broader the base of support for a vision or initiative, the greater its credibility will be.

3. Systems Perspective

Resource and environmental systems consist of many components or subsystems, with numerous interconnections and linkages. For example, addressing the issue of food security requires attention to matters as diverse as the impacts of climate change, the nature and extent of poverty, the nature of arrangements for land tenure and ownership, choices about fuel types, the impacts of credit and other financial arrangements, and the capacity of extension programs to diffuse technology and other innovations to farmers. A systems perspective directs us to take a holistic view, even though that raises the risk that so many variables and relationships may be identified that it will be difficult to develop implementable strategies.

Without a systems perspective, however, the danger is that managers become locked into 'sectors' or 'silos' and only consider matters viewed as relevant to their very circumscribed interests. A frequent outcome is that the actions of different resource agencies work at cross-purposes, or undercut one another, or lead to undesirable inefficiencies. For example, a road may be resurfaced by one agency, and then within months the newly surfaced road is dug up because another agency decides to replace aging water, sewerage, or power lines. In this example, different agencies with different functions exist to ensure appropriate specialized expertise to deal with different public functions: transportation, water and sewerage, and electricity services. If each agency proceeds with its work without regard for the activities of the other, however, the kind of inefficiency outlined above is a possible outcome. Alternatively, a land-based agency might pursue activities that cause problems or constraints for a water-based agency and vice versa. If each agency took a broader perspective—that is, a systems view—one would expect that they could more readily harmonize their respective activities to support rather than conflict with each another.

A major challenge for a systems approach is to identify the appropriate spatial unit for management. Whether at the federal, provincial, territorial, or municipal level, spatial units for management are usually based on political or administrative units that generally bear no relationship to ecological systems. For example, rivers are often used as boundaries between municipalities or provinces, whereas a logical ecosystem spatial unit for water management is the river basin or catchment. Furthermore, communities are often located adjacent to rivers, resulting in upstream communities having different approaches to stewardship of the river than downstream communities. One solution is the creation of basin-wide authorities, such as the conservation authorities in Ontario, the Meewasin Valley Authority in Saskatchewan, and the Fraser Basin Council in British Columbia.

4. Adaptation and Resilience

The concept of *adaptive environmental management* was developed some time ago, led by Holling (1978; 1986) and his colleagues. The starting point was their belief that policies and approaches related to resource management should be able to cope with the uncertain and the unexpected—and even the unknown. As a result, they argued that resource managers should try to resolve each problem in the spirit of a rigorous experiment. Systematic monitoring would catch errors or mistakes, providing new information that would enable managers to modify subsequent initiatives. In this manner, 'failures' generate new information and insight, leading to new knowledge. For this to happen, however, individual managers have to be willing to acknowledge their errors or mistakes rather than adopting the usual stance of not openly acknowledging that a management action has not had the desired outcome. Such an approach would require a major culture shift in many organizations, because it is rare for managers to admit that their strategies or practices have been ineffective or have failed. For example, how often have you heard a senior politician state publicly that one of the policies or programs for which she or he is responsible had failed?

Holling and others believe that with an adaptive approach, greater resilience will be created for natural, economic, and social systems. That is, if we learn how to adjust and adapt, natural and societal systems will be better able to 'bounce back' or 'recalibrate' relative to changing circumstances or sudden shocks. The alternative is to be brittle and therefore more vulnerable or fragile when evolving circumstances make behaviour that worked well in the past no longer relevant.

An analogy sometimes offered is the predilection of military planners to develop and perfect strategies and tactics with reference to previous battles and skirmishes, which may not work well if the opponent's behaviour is different in the future. The failure of the Maginot Line in eastern France at the start of World War II is a good example in this regard. France developed a trench system between itself and Germany based on the tactics of the First World War. However, in 1939, the panzer divisions of the German army had developed different tactics and drove their tanks around rather than through the Maginot Line fortifications, making them ineffective and perhaps irrelevant.

5. Partnerships

Given growing complexity, considerable uncertainty, and rapidly changing conditions, many agree that it is unrealistic to expect that any person, group, or organization can have all the knowledge and understanding required for effective resource and environmental management. As a result, it is increasingly expected that partnerships among the public and private sectors, civil society, and Aboriginal peoples will be needed to develop strategies and initiatives that will meet expectations regarding economic development, environmental integrity, and social acceptability.

Many benefits can be realized from partners collaborating to resolve problems or take advantage of opportunities. For example, Himmelman (1996, 29) has indicated that with collaboration, it is possible to improve decisions through exchanging information, modifying activities in light of others' needs, sharing resources, and enhancing the capacity of others to achieve mutual benefit and to realize common goals or purposes. Other advantages have been identified by Selin and Chavez (1995), including joint decision-making that reflects sharing of power and stakeholders accepting collective responsibility for their actions and the outcomes. In this context, 'stakeholders' are generally understood to include (1) any public agency with prescribed management responsibilities, (2) all interests significantly affected by a decision, and (3) all parties who might intervene in the decision-making process to facilitate, block, or delay it.

Creating and using partnerships is not problem-free, however. For example, public agencies often have legal responsibilities and functions, and it is difficult for them to turn them over to other stakeholders if they are to be held legally accountable for them. Also, achieving a common vision or consensus among many stakeholders is often problematic, creating the potential for unreasonable lengths of time spent in searching for common ground among stakeholders with diverse, and often conflicting,

values, expectations, and needs. And there is danger that some stakeholders will become disenchanted with engaging in a partnership if those with legal responsibilities are consistently the final arbitrators or decision-makers, especially when decisions do not incorporate the views of a particular group. The reality is, however, that 'hard choices' are often needed, and it is not possible to accommodate all interests and needs, resulting in 'winners' and 'losers'. In brief, life is not always fair.

For partnerships to be effective and flourish, there needs to be trust, understanding, and mutual respect among participants. When trust and mutual respect are not present, partnerships can quickly be derailed when inevitable problems or bumps are encountered, especially when jointly developed and implemented solutions turn out to have unexpected and/or unintended negative consequences. Significant time is often required for the necessary trust and mutual understanding to grow, and the difficulty will be that in the interim, problems may become worse. And it is always possible that some stakeholders have no interest in collaborating with others and instead pursue solutions that meet only their needs without regard to the needs of others. When some groups have no interest in seriously attempting to understand and meet the needs of others, the likelihood of a successful partnership becomes very low.

One type of partnership is known as 'co-management', a situation in which there is actual reallocation of power and authority to citizen groups and away from elected officials or technical experts. In Canada, co-management arrangements have most often been created for management of forest and wildlife resources and, more recently, fisheries (see Chapters 4, 8, 10, 11, and 19). Co-management arrangements are likely to be used more extensively in the future.

6. Impact and Risk Assessment

Impact assessment has existed in Canada for about 35 years. The basic intent is to ensure that due consideration is given to the possible environmental impacts of proposed policies, plans, and projects. Initially, attention focused only on projects, but over time the focus has broadened to include attention to the impacts of policies and plans. When policies and plans are assessed for their environmental impacts, the process is often referred to as 'strategic impact assessment' or as 'sustainability impact assessment'. From the perspective of 'best practices', resource and environmental decision-makers should consider the economic, environmental, and social implications of any possible initiative or decision.

It also has been recognized that while many small or modest projects may be innocuous on a case-by-case basis, there is a real likelihood that the cumulative impact of many small initiatives can be significant. Such recognition has led to the development of what is referred to as 'cumulative impact assessment', in which the purpose is to ensure that many small initiatives with relatively minor impacts do not result in a cumulative set of initiatives with significant impacts. This challenge is considered in much more detail in Chapter 17.

Risk assessment can be part of, or a foundation for, impact assessment, since it focuses on determining the probability or likelihood of an environmentally or socially negative event of some specified magnitude and the costs of mitigating the consequences. Risks are estimates, leaving room for error, given that the future cannot be known perfectly. It was because of such uncertainty that the *precautionary principle* was endorsed at the Earth Summit of 1992. The precautionary principle stipulates that when there is a risk of serious or irreversible damage to the environment, lack of full scientific certainty regarding the extent or possibility of risk shall not be used as a reason for postponing cost-effective measures to prevent environmental degradation. In other words, according to this principle, decision-makers and managers should always err on the side of caution when deciding what actions to take or initiatives to approve.

7. Communication

If a vision is to be developed, legitimacy created, a systems perspective used, adaptation and resilience achieved, partnerships established, and risk assessed, communication of understanding based on science and experiential knowledge is essential. Indeed, some argue that there can never be too much time allocated to communication. Nonetheless, if all effort is focused on communication at the expense of attempting to clarify the nature of problems or opportunities and developing strategies to address them, communication will likely come across as empty or hollow to many people.

To help develop communication among stakeholders, Carpenter (1995) recommended giving attention to four complementary questions, regardless of the audience:

1. What do we know, with what accuracy, and how confident are we about our data?
2. What don't we know, and why are we uncertain?
3. What could we know, with more time, money, and talent?
4. What should we know in order to act in the face of uncertainty?

If, in addition to asking the above four questions, one were to focus on the possibility of various types of conflict and the underlying reasons for it, as reflected in the four types of conflict depicted in Table I.1, planners, managers, and decision-makers would increase the likelihood that everyone will understand the nature of problems to be resolved or opportunities to be pursued.

CONCLUSION

In this chapter, the first objective was to highlight the main contextual aspects of resource and environmental management in Canada. A key aspect is the many jurisdictions involved, including federal, provincial, territorial, and municipal governments as well as Aboriginal peoples. This mix creates potential for vertical (federal to provincial to municipal) as well as horizontal (within one level of government, among different line agencies) fragmentation. The challenge is exacerbated when Aboriginal peoples and civil society groups are factored in. Thus, there is a real need to search for and achieve co-ordination and harmonization of policies, programs, and projects—often easier said than realized.

The difficulty of achieving co-ordination and harmonization becomes more daunting because of differing regional interests and needs across the country and the pressures (and opportunities) created by globalization. A new, prominent variable emerging since the third edition of this book is the economic downturn that began in the last quarter of 2008. The very serious economic conditions have created difficulties worldwide and have made it a real challenge to determine how to keep resource and environmental issues on the issue-attention cycle.

Conflict clearly takes many forms, and in this chapter four different types have been identified: cognitive, value, interest, and behavioural. In parallel are four kinds of uncertainty: risk, uncertainty, ignorance, and indeterminacy. The different types of conflict and uncertainty can interact in various permutations and combinations and require systematic attention on the part of resource and environmental analysts and managers.

In determining strategies to resolve conflict and uncertainty, it is helpful to know about the nature of best practice attributes. Here, attention has been given to vision, legitimacy, systems perspective, adaptation and resilience, partnerships, impact and risk assessment, and communication. When the best of each of these aspects can be brought to bear on issues, problems, and opportunities, the likelihood of successful resolution should be enhanced. In the chapters and guest statements in this book, you will be introduced to various interpretations and applications of these ideas.

REVIEW QUESTIONS

1. What are the main contextual aspects of Canadian and global resource and environmental issues?
2. Explain the significance of the issue-attention cycle.
3. Explain the four different types of conflict and of uncertainty, and outline their significance in the context of a resource or environmental issue in your community.
4. Outline the main features of the seven best practices outlined in this chapter, and give at least one example in which each has been applied effectively.

REFERENCES AND FURTHER READINGS

Barnhart, C.L. 1975. *The World Book Dictionary*, v. 1 and 2. Chicago: Field Enterprises Educational Corporation.

Beazley, K., and R. Boardman, eds. 2001. *Politics of the Wild: Canada and Endangered Species*. Toronto: Oxford University Press.

Bernstein, S., and C. Gore. 2001. 'Policy implications of the Kyoto Protocol for Canada'. *Isuma* 2 (4): 26–36.

Carpenter, R.A. 1995. 'Communicating environmental science uncertainties'. *Environmental Professional* 17: 127–36.

Dale, A. 2001. *At the Edge: Sustainable Development in the 21st Century*. Vancouver: University of British Columbia Press.

Dearden, P., and B. Mitchell. 2009. *Environmental Change and Challenge: A Canadian Perspective*. Toronto: Oxford University Press.

de Loë, R., R. Kreutzwiser, and L. Moraru. 2001. 'Adaptation options for the near term: Climate change and the Canadian water sector'. *Global Environmental Change* 11 (3): 231–45.

DeVille, A., and R. Harding. 1997. *Applying the Precautionary Principle*. Sydney: Federation Press.

Dorcey, A.H.J. 1986. *Bargaining in the Governance of Pacific Coastal Resources: Research and Reform*. Vancouver: Westwater Research Centre, University of British Columbia.

Dovers, S.R., and J.W. Handmer. 1992. 'Uncertainty, sustainability and change'. *Global Environmental Change* 2 (1): 262–76.

Downs, A. 1972. 'Up and down with ecology—The "issue-attention" cycle'. *Public Interest* 28 (summer): 38–50.

Draper, D., and M.G. Reed. 2009. *Our Environment: A Canadian Perspective*. Scarborough, ON: ITP Nelson.

Dwivedi, O.P., et al., eds. 2001. *Sustainable Development and Canada: National and International Perspectives*. Peterborough, ON: Broadview Press.

Dyer, G. 2008. *Climate Wars*. Toronto: Random House.

Garner, R. 1999. 'Biodiversity since Rio'. *Environmental Politics* 8 (2): 148–52.

Gibson, R.B. 1992. 'Respecting ignorance and uncertainty'. In E. Lykke, ed., *Achieving Environmental Goals: The Concept and Practice of Environmental Performance Review*, 158–78. London: Belhaven Press.

Gleick, J. 1987. *Chaos: Making a New Science*. Toronto: Penguin.

Gullett, W. 1997. 'Environmental protection and the "precautionary principle": A response to scientific uncertainty in environmental management'. *Environmental and Planning Law Journal* 14 (1): 52–69.

Harding, R., and E. Fisher, eds. 1999. *Perspectives on the Precautionary Principle*. Sydney: Federation Press.

Himmelman, A.T. 1996. 'On the theory and practice of transformational collaboration: From social service to social justice'. In C. Huxham, ed., *Creating Collaboration Advantage*, 19–43. Thousand Oaks, CA: Sage.

Hodgins, B.W., and J. Benidickson. 1989. *The Temagami Experience: Recreation, Resources, and Aboriginal Rights in the Northern Ontario Wilderness*. Toronto: University of Toronto Press.

Jaccard, M., J. Nyboer, and B. Sadownik. 2002. *The Cost of Climate Policy*. Vancouver: University of British Columbia Press.

Krajnc, A. 2000. 'Wither Ontario's environment? Neo-conservatism and the decline of the environmental ministry'. *Canadian Public Policy* 26 (1): 111–27.

Krause, P., et al. 2001. 'Achievements of the Grand River Conservation Authority, Ontario, Canada'. *Water Science and Technology* 43 (1): 45–55.

Laing, R.D. 2002. *Report of the Commission of Inquiry into Matters relating to the Safety of the Public Drinking Water in the City of North Battleford, Saskatchewan*. Regina: Queen's Printer for Saskatchewan.

McKenzie, J.I. 2002. *Environmental Politics in Canada: Managing the Commons into the Twenty-First Century*. Toronto: Oxford University Press.

Mercredi, O., and M.E. Turpel. 1993. *In the Rapids: Navigating the Future of First Nations*. Toronto: Viking.

Mitchell, B., and W.R.D. Sewell. 1981. 'The emerging scene'. In B. Mitchell and W.R.D. Sewell, eds, *Canadian Resource Policies: Problems and Prospects*, 1–23. Toronto: Methuen.

O'Connor, D.R. 2002a. *Part One—Report of the Walkerton Inquiry: The Events of May 2000 and Related Issues*. Toronto: Queen's Printer for Ontario.

————. 2002b. *Part Two—Report of the Walkerton Inquiry: A Strategy for Safe Drinking Water.* Toronto: Queen's Printer for Ontario.

O'Riordan, T., and A. Jordan. 1995. 'The precautionary principle in contemporary environmental politics'. *Environmental Values* 4: 191–212.

Parson, E.A. 2000. 'Environmental trends and environmental governance in Canada'. *Canadian Public Policy* Supplement (August): S123–43.

Rollings-Magnusson, S., and R.C. Magnusson. 2000. 'The Kyoto Protocol: Implications of a flawed but important environmental policy'. *Canadian Public Policy* 26 (3): 347–59.

Selin, S., and D. Chavez. 1995. 'Developing a collaborative model for environmental planning and management'. *Environmental Management* 19: 189–95.

Sproule-Jones, M. 2002. *Restoration of the Great Lakes: Promises, Practices, Performances.* Vancouver: University of British Columbia Press.

Torrie, R., R. Parfett, and P. Steenhof. 2002. *Kyoto and Beyond: The Low Emission Path to Innovation and Efficiency.* Vancouver: David Suzuki Foundation; Ottawa: Climate Action Network.

van Kooten, G.C., and G. Hauer. 2001. 'Global climate change: Canadian policy and the role of terrestrial ecosystems'. *Canadian Public Policy* 27 (3): 267–78.

VanNijnatten, D.L., and R. Boardman, eds. 2002. *Canadian Environmental Policy: Context and Cases.* 2nd edn. Toronto: Oxford University Press.

Waldrop, M.M. 1992. *Complexity: The Emerging Science at the Edge of Order and Chaos.* New York: Simon and Schuster.

Wiener, J.B. 2008. 'Climate change policy and policy change in China'. *UCLA Law Review* 55: 1805–26.

————. 2009. 'Engaging China on climate policy'. *Resources* 171: 29–33.

Winfield, M. 1994. 'The ultimate horizontal issue: The environmental policy process experiences of Alberta and Ontario, 1971–1993'. *Canadian Journal of Political Science* 27 (1): 129–52.

Wolfe-Keddie, J. 1995. 'First Nations' sovereignty and land claims: Implications for resource management'. In B. Mitchell, ed., *Resource and Environmental Management in Canada: Addressing Conflict and Uncertainty*, 55–79. Toronto: Oxford University Press.

Woodrow, R.B. 1980. 'Resources and environmental policy-making at the national level: The search for focus'. In O.P. Dwivedi, ed., *Resources and the Environment: Policy Perspectives for Canada*, 23–48. Toronto: McClelland and Stewart.

Woolard, R.F., and A.S. Ostry, eds. 2000. *Fatal Consumption: Rethinking Sustainable Development.* Vancouver: University of British Columbia Press.

World Commission on Environment and Development. 1987. *Our Common Future.* Toronto: Oxford University Press.

World Summit on Sustainable Development. 2002. 'Sustainable development—The government's approach'. http://www.sustainable-development. gov.uk/wssd/wssd1.html.

Wynne, B. 1992. 'Uncertainty and environmental learning: Reconceiving science and policy in the preventive paradigm'. *Global Environmental Change* 2 (2): 111–27.

PART 1

Emerging and Current Concerns

The overriding themes for this book are *conflict* and *uncertainty*. In Part 1, authors address both themes in various contexts. The topics in this section were included because they are relatively 'new' on the policy agenda or because their intensity has become much higher.

The first two chapters, new in this edition, address aspects of globalization. Chui-Ling Tam explores how resource development and environmental management in Canada are affected by globalization and how Canada actively participates to enhance resource and environmental management on a global scale. In her guest statement, Susan Cartwright highlights four related challenges and opportunities for policy-makers. O.P. Dwivedi considers the role and significance of *governance* in the next chapter. He concludes that the new world of shared power politics requires new global governance structures, because existing international institutions, policies, and programs do not adequately reflect today's complex realities.

In the third chapter, David Rapport argues for an integrative approach to environment and resource management. He then considers how an *ecosystem health model* is well suited to meet this need. Mark Anielski's guest statement outlines the Genuine Wealth Model designed to measure human, social, natural, and financial wealth.

Annie Booth and Norm Skelton in Chapter 4 consider the roles of Aboriginal peoples with regard to environment and resources. They examine the concerns of Aboriginal peoples regarding access and rights to natural resources as well as the differences between Aboriginal rights and title.

In the early part of the twenty-first century, climate change, energy, and water issues have gained prominence. Each of these issues is addressed in the remaining chapters of Part 1. In Chapter 5, another new chapter, Gordon McBean focuses on climate change and how societies can mitigate and adapt to it. He views climate change as both a long-term and a global issue, thereby raising issues about intergenerational and international equity. In her guest statement, Linda Mortsch argues that debate is required regarding perceptions, social values and expectations, priorities and trade-offs that society will choose, risks and changes it will accept, and how mitigation and adaptation strategies will be implemented.

Energy is the focus in Chapter 6, another new chapter. Paul Parker and Travis Gliedt describe and analyze the uneven global distribution of energy resources, the use of global markets to balance the demand and supply of energy commodities, and the influence of economic and political factors on prices. They then consider the desire for security of energy supplies and creation of systems that focus on local energy resources.

In the seventh chapter, Reid Kreutzwiser and Rob de Loë examine water security concerns in Canada and how they are manifested in water allocation and water quality management. Sonya Meek's guest statement examines integrated watershed planning in urbanizing environments. She argues that a multi-stakeholder approach is necessary to gain community support for actions.

1

Canada in the World: Globalization, Development, and Environment

Chui-Ling Tam

Learning Objectives

- To understand the interrelatedness of globalization and development.
- To recognize old and emergent strands in environmental globalization.
- To appreciate Canada's resource development in historical context.
- To assess possibilities for Canadian engagement in overseas resource management and development challenges.
- To be informed about the level of Canadian commitment to environmentally responsible development.

GLOBALIZATION, RESOURCE DEVELOPMENT, AND CHANGE

These are interesting times. When the new millennium began, the global human population had just breached six billion, the Kyoto Protocol was three years old, and countries around the world were still in the process of ratifying it in an effort to combat global warming (Worldwatch Institute 2001). Fast forward one decade, and life on Earth seems grimmer than ever: a world population approaching seven billion (UNFPA 2008), a notable lack of co-ordination and implementation on key Kyoto requirements to reduce greenhouse gas emissions, and the dubious distinction of the past decade experiencing eight of the 10 warmest years since global temperatures were recorded (Worldwatch Institute 2009).

By 2050, the human population on the Earth is projected to top nine billion (UNESA 2009), which will mean more demand for food, water, building materials for shelter, energy to fuel homes, food production, industrial production, and private and public transportation and the attendant ills of pollution and environmental degradation. Fears about the viability of life on Earth, the future of the world's children, resource sufficiency and distribution, access to education, and poverty alleviation can all be connected to unease over the state of the environment and, in particular, climate change. States the Human Development Report (UNDP 2007, 1): 'Climate change is the defining human development issue of our generation' (see also Chapter 5).

While the above tone is alarmist, it speaks to a need to engage critically with conflict and

GUEST STATEMENT

Challenges for Today's Public Policy-Makers

Susan M.W. Cartwright

It is tempting to say that each generation of public policy-makers faces greater challenges than their predecessors did. Many challenges are decades, even centuries, old but I see four interrelated challenges that are greater and that have implications and offer opportunities for policy-makers: speed, scale, complexity, and impact.

Policy-makers must deal with *speed*: of information flows and the transfer of ideas, of the evolution of events and the movement of people and things across huge distances and national borders, and of the actions expected of governments. Speed, combined with the increasingly complex inter-relationships characteristic of globalization, means that developments can now spread farther and more quickly. In some instances, population and economic growth lead to accelerating speed—consider the rise in the demand for water. It is often difficult for governments to keep up—public policy processes can be unwieldy and cumbersome. At times, under political or media pressure, the focus becomes the need for speed itself at the expense of a considered policy response. Examples of speed, which previous generations could only have imagined, include real-time consultations, news from conflict zones available in minutes, Internet campaigns mobilized within hours, and the spread of the global financial crisis over a matter of weeks in late 2008.

Scale (both temporal and spatial) also presents new challenges. In some cases, scale may not really have changed; we now know more, and more is visible. But in other cases, scale has increased along with the impact of human activity, population

growth, and the aspirations of citizens. Examples include climate change, the exploitation of natural resources, and information (an unprecedented amount available from an unparalleled number of sources and accessible to more people than ever).

Such increased scale calls for new responses. Supra-national organizations (such as the European Union, the United Nations, and the World Bank) are one response to large-scale, interconnected issues and a readiness for collective action. For other issues, sub-national entities have been the answer. Society and governments increasingly look to non-government actors and foster partnerships with the voluntary and commercial sectors (which deliver more than just increased resources). An example is the public–private alliance to battle major diseases in the developing world. There are obvious challenges to such collective action. A successful global response to climate change, for instance, demands action from all major players, requiring that the international community bridge the divide in means and aspirations that exists among nations and commits to collective, as well as national, effort.

As a result of population growth, mobility, corporate linkages, easy communications, trade, and other phenomena of globalization, as well as our improved understanding of the interrelationships among issues and the consequences of our actions, many challenges and issues are now more *complex*. This requires better recognition of systems (the 'whole'). It calls for allowing systems to guide planning, managing, and responding. It also requires effective communication of complexity to citizens to garner support for policy decisions and a concerted effort by politicians, the media, and academics to understand and reflect complexity. The global financial crisis, which began in the latter part of 2008, offers an excellent example of a complex issue that began locally, rapidly became global, featured intricate public and private linkages, and demanded concerted international and national action.

The *impact* of public policy is not always what was intended and desired. The impact of actors, which governments may neither control nor influence, can require government intervention and simultaneously limit a government's options. Commodity prices, set internationally, affect Canadian industries. The international price of oil affects the budgets of individual consumers as well as a government's tools to influence consumption. It also has an impact on overall economic growth and the pricing of products and services. Currency values and stock market activity in distant capitals ripple through Canadian industrial sectors. Our understanding of impact, while incomplete, grows as we learn from actions and decisions in Canada and in other countries. For example, after decades of providing development assistance around the world, understanding its impact helps to ensure that well-intentioned projects deliver measurable and enduring results for their intended beneficiaries.

A cursory look at the world shows that in addition to these four challenges, we continue to live with uncertainty and conflict. *Uncertainty* is likely to remain our travelling companion. For instance, India and China's evolution is changing the world. Canada is changing as its population becomes more diverse and urban and as traditional industries are replaced by new ones. Similarly, *conflict*, as old as history itself, seems likely to intensify as populations increase, demand grows, and resources become increasingly scarce. And it is not just conflict over resources; other kinds of conflict frequently limit our options and our capacity to act.

The public sector has already changed to meet these challenges. It will need to continue doing so in order to foster agility and responsiveness, without sacrificing the quality of public policy. This carries implications for public sector organization and governance, consultation, and increasingly complex regulatory systems, for example. Governments are grappling with managing a rapidly increasing amount of information, including keeping some of it secure—and their success so far has been mixed. If we are to continue to respond to the needs and aspirations of Canadians and those who depend on our international activities, then we must also strive to create an environment in the public sector that fosters innovation, tolerates reasonable risk, and allows learning from experience.

uncertainty in the context of environmental stress in a rapidly changing and increasingly connected world. McLuhan's (1967) 'global village' has arrived, enabled by technological advances that allow goods, money, and people to conquer time and space. In a world of increasing integration and interaction, conflict can be seen not only as a destructive phenomenon to overcome but also as an opportunity for learning and transformation.

Tsing (2005) argues eloquently for an ethnography of global connection, demonstrating how we can explore the global implications of and linkages to specific local circumstances, such as the destruction of Indonesia's rainforests, by using the traditional research methods of anthropology—observation, participation, immersion in the field, thick description of the minutiae of daily life. Tsing (2005, 5) frames cross-cultural and long-distance interactions within the metaphor of 'friction' as a way to view the 'new era of global motion'. Thus, friction is a means of perceiving the points of engagement among a messy constellation of environmental saviours, users, and abusers—nature lovers, conservation agencies, extractive industries, investment capital, village elders, indigenous medicinal experts, and researchers, to name a few. It is a useful reminder that the challenges of environmental sustainability and development necessity are intertwined, that the global nature of environmental stress is often acted out and *lived* locally, and that such encounters can yield unexpected outcomes and insights about mobility, culture, and the choices we make in a world where power relations are constantly shifting (see also Chapter 2).

The state of the world in these, our interesting, times requires a critical engagement with environment, development, and globalization. Increasingly, scholars and activists promote critical globalization studies (Appelbaum and Robinson 2005) to address the intertwined challenges of neo-liberalism, power and conflict, global and local governance (discussed in Chapter 2), identity politics, and environmental change. *Critical globalization studies* and its predecessor, critical development studies, embrace reflexivity and evaluate the significance of particular social arrangements; this awareness of how personal assumptions, biases, and affinities contribute to the construction of meanings reminds us that environmental challenges and responses are *shaped* by actors who do not—indeed, perhaps cannot—see the multiple facets of a problem objectively.

Neo-liberalism promotes the unrestricted exchange of goods and services to create wealth within a self-regulating global market of demand and supply; its emphasis on economic growth through production and consumption has contributed to the degradation of natural resources

on a global scale. *Identity politics* matter, because by identifying the processes by which particular groups of individuals are marginalized, we can hopefully better understand and address environmental justice at both local and global scales. In the context of the twenty-first century, *global–local structures and processes* are of paramount concern, and scholars need to recognize their embeddedness in global society and relevance to everyday struggle (Appelbaum and Robinson 2005).

Global society is constantly faced with the speed at which information travels in a deeply connected world, increasingly complex interrelationships resulting from mobility, encounters, and the relations thus formed, the greater scale of environmental challenge and response, and uncertainty over the impact of public policy (see the guest statement by Susan Cartwright).

What is Canada's role and place in this world of rapid change? This chapter focuses on the ways in which resource development and environmental management in Canada are affected by globalization and on how (and whether) Canada serves as an active participant to enhance resource and environmental management on a global scale. Canada is an active contributor to collaborative and participatory strategies, informed by its history of managing environmental change and conflict. Arguably, Canada's relevance to global environmental development is its locally tested responses to environmental friction.

Implications for Resource and Environmental Management

As intertwined phenomena, globalization and sustainable development require us to engage critically with global–local socio-economic processes over time (Borghesi and Vercelli 2008; see also Chapter 2). In other words, the geographical location and historical period of particular environmental and resource governance matters. Our understanding of the place of environment vis-à-vis development and globalization must be informed by global and domestic histories and the vying for prominence—and arguably dominance—of particular ideologies, political priorities, environmental knowledges, and resource management strategies.

What is often neglected in the discussion of globalization, especially in its corporate function, is its relation to ecological change. Ecological change, whether natural or human-induced, almost inevitably has an impact on the livelihoods of the disadvantaged: the poor, indigenous peoples, and women suffer multiple jeopardies of oppression (Bhavnani, Foran, and Talcott 2005), often in relation to particular struggles over livelihood and access to environmental resources (Rocheleau, Thomas-Slayter, and Wangari 1996; Resurreccion and Elmhirst 2008; see also Chapters 4 and 19). But historically, globalization has resulted in economic growth and poverty alleviation for the nations that took part in it; the challenge lies in finding or creating the right environmental and social conditions that allow globalization to benefit society broadly and minimize the inequalities endured by those less skilled, empowered, or willing to embrace it (Borghesi and Vercelli 2008). This is a vision of globalization as good change, to co-opt Chambers's (1997) definition of development.

Recognition of the linkages among globalization, development, and ecological change invites us to engage with broader questions about an environmental justice that is—as Bhavnani, Foran, and Talcott (2005) would argue—'socialist, green, antiracist, and feminist'. Industry, arguably the dominant economic force in the world, also has a stake in good change. In recent decades, the proliferation of voluntary codes of conduct, the adoption of standards under the International Organization for Standardization (ISO), and the widening of discussion about corporate social responsibility (CSR) all point to a shifting awareness that environmental management is a corporate concern as well as a corporate responsibility.

Globalization is mainly identified with the 1980s when the resurgence of neo-liberalism was accompanied by an easing of state intervention in economic activity and a push toward freer markets, both in capitalist regimes and in post-communist states. The onset of widespread economic and political globalization became popularly identified with the Washington Consensus, a concept and critique introduced by John Williamson in 1989 to explain the dominance of US policies of global capital accumulation that were adopted and enforced through unilateral and multilateral consent and coercion (Williamson 1993; Harvey 2005).

Globalization thus was a realization of the global village first conceptualized by Marshall McLuhan in his seminal work *The Medium Is the Massage*[1] (1967) and has come to be known as a phenomenon of increasing global integration and homogeneity through a process of time–space compression. This compression is enabled by ever more efficient and accessible communications and transportation technologies.

The global mobility of goods, capital, people, and information in the pursuit of capital accumulation via environmental resource extraction and conversion has a long history rooted in the age of imperialism and renewed during the early 'developmentalism' of the post–Second World War era (when development was decided and implemented by international agencies and nation-states) and the late-twentieth-century neo-liberalism resurgence of heightened global financial mobility. The now moribund cod fisheries off the Grand Banks of Newfoundland, the questionable outcomes of the Green Revolution in agriculture in India, and indigenous displacement from spaces of petroleum and forestry extraction in Borneo, Canada, and the Amazon River Basin are the end result of development-as-globalization gone amok. Such outcomes show the flawed face of globalization that counters its image as a benign force of liberalism in the world (Milanovic 2003).

Harvey (2000, 60) sees globalization as an historical 'process of production of uneven temporal and geographical development' in pursuit of capital accumulation. Globalization thus has broad implications for and impacts on environmental management and resource development. Nor are those connections 'new'. The globalized environment discourse has been part of the popular imagination since the Bretton Woods institutions—namely, the International Bank for Reconstruction and Development (latterly known as the World Bank) and the International Monetary Fund—were charged in the 1940s with postwar reconstruction (Stevis 2005). The global environment concept has been reincarnated in subsequent global initiatives, such as the 1968 UNESCO Biosphere Conference and the various Earth Summits, the Brundtland Commission's *Our Common Future*, the 1997 Kyoto Protocol, and the marketing of climate change as the biggest development issue of our times. The proliferation of protected areas worldwide is evidence of acceptable environmental globalization (Zimmerer 2006) as opposed to voracious economic globalization.

As an historical project, development—and its alter ego, globalization—has largely focused on quantitative growth, based on the logic of ever-expanding capital. As new paradigms emerged to refocus development upon societal change, the emphasis has been on non-material improvements such as human knowledge, civility, and artistic forms of expression. Such development considerations have notably defined the significance of environment as a limit to growth. However, the recognition

of the interconnectedness of environment and development—most famously encapsulated by the Brundtland Commission's adoption of 'sustainable development' (WCED 1987)—has led to a flowering of critical engagement with the intersection of development and environment, as witnessed by the increased attention to these dual phenomena among educational institutions. Given development's checkered history, Becker and Jahn (1998) suggest that the term is outdated and inadequate to deal with the challenges of global and regional ecological crisis. Instead, they propose that the defining concept going forward should be 'socio-ecological transformation'.

Globalization, Uncertainty, and Conflict

Globalization as a neo-liberal project and ideology encourages homogenization but at the same time acts as a destabilizing force because it foments resistance among those who are marginalized by its manifestations. Examples are the surge in protesters at meetings of the World Trade Organization (WTO) and the notable alliances among Northern and Southern environmentalists, indigenous peoples dispossessed by timber extraction, and female garment-trade workers. Globalization encourages core–periphery relations through uneven development, creating spaces of relative wealth and privilege in rich countries and rich urban centres, juxtaposed against poor countries and rural communities. At the individual level, threat to livelihood is at issue. At a societal level, economic and social resilience is at issue. At a state level, distribution of impact and benefits, questions over ownership and sovereignty, are at issue.

McGrew (2000) has identified four dimensions of globalization: *extensity* (the spread of transboundary activity), *intensity* (the deepening integration of far-flung locales within the world system), *velocity* (the faster interactions of people, capital, and information across time and space enabled by advances in communication and transportation technologies), and *impact* (in that wider, deeper, and faster integration into the world system results in the near impossibility of anyone being unaffected by events and decisions occurring in distant parts of the world). Globalization is driven by economic, technological, political, and cultural shifts (McGrew 2000, citing Walker 1988), as shown in Table 1.1.

Globalization as an economic, political, and cultural force breeds uncertainty and conflict as resource strategies, practices, and benefits change. In its most basic form, globalization increases homogeneity around the world; it threatens the environment through the pursuit of ever-expanding capitalism but at the same time presents opportunities for positive change (Eberts 2004). Examples abound, not only in the conflicts over migratory resources such as the Atlantic cod

fishery but also global efforts to unite in the face of global threat, such as the Brundtland report, *Our Common Future* (WCED 1987), the various Earth Summits held in recent decades, and continued efforts—if rather limited in measurable success—to curtail emissions of carbon dioxide through international instruments such as the Kyoto Protocol, as well as internal measures in countries that refused to ratify the Kyoto Protocol, such as the United States and Australia.

Globalization and development are connected and rooted in a colonial project that ostensibly rewards poor countries that integrate into a world economic system premised on market accountability as good governance and the allocation and legitimization of property rights (McMichael 2005; Woodhouse and Chimhowu 2005) to the detriment of older indigenous systems such as common property resources. Globalization has reached its most expansive form in the past three decades, but it shares shifting

Table 1.1 Dimensions and Drivers of Globalization

Dimensions of Globalization	Drivers of Globalization
1. **Transboundary activity (extensity):** a stretching of social, political, and economic activities across political frontiers, regions, and continents.	1. **Capitalist expansion (economic shifts):** expressed increasingly in the Information Age in the need of business, large and small, to compete in regional and global markets.
2. **Interconnectedness (intensity):** the growing magnitude of flows of trade, investment, finance, migration, culture, etc.	2. **Industrialism, information (technological shifts):** the move away from manufacturing-based to service-based economies (i.e., a shift to a post-industrial economy) and the informatics greatly facilitates globalization in every domain from the economic to the criminal.
3. **Faster interactions (velocity):** speeding up of global interactions and processes as more sophisticated transportation and communication systems increase diffusion of ideas, goods, information, capital, and people.	3. **Less state intervention (political shifts):** a dramatic shift away from state intervention to the market; deregulation, privatization, and economic liberalization make economies and societies more open to the world.
4. **Impact over distance (impact):** the effects of distant events can be highly significant elsewhere; local developments can have enormous global consequences. Boundaries between domestic matters and global affairs become blurred.	4. **Civic awareness (cultural shifts):** elites and civil society become aware that the fate of nations and communities is increasingly bound up with the dynamics of the global economy and global environment.

Source: Adapted from McGrew 2000.

ground with development as a project of domination through capital accumulation and the spread of neo-liberal ideology (Harvey 2005).

Popular discourse associates development with a post-colonial and post–Second World War drive toward stability, modernization, and improvement of the human condition; the failures of this period gave rise to leftist criticism of the ideology and foci of development and the developmental state. However, it is useful to remember that paradigmatic shifts in the theory and practice of development have occurred in particular moments of world history, profound geopolitical change, and personal struggle (Kothari 2005). It is that struggle that has given prominence to counter-narratives about gender, environment, and indigeneity. And while globalization has often been pitched as antithetical to environmental care and sustainable development, a number of scholars are attempting to transcend such polarized positions to provide more nuanced understandings of globalizing processes of human–environment interaction that produce spaces for change, whether good or ill (see Tsing 2005; Zimmerer 2006; Borghesi and Vercelli 2008).

As Bhavnani, Foran, and Talcott (2005, 324) remind us:

[W]hat could make globalization 'unnatural' and 'not-inevitable' is development. Despite the failures of development, people on the ground everywhere engage with its contradictions, using development to create possibilities for social transformation and the redistribution of wealth and resources.

THE CANADIAN EXPERIENCE

A History of Resource Extraction

Arguably, globalization has figured in resource development in Canada since the first merchant fishers came to Newfoundland to fish for cod in the sixteenth and seventeenth centuries. They followed in the wake of the explorer John Cabot, who remarked in 1497 that the sea was so thick with cod that 'they virtually blocked his ship' (Greenpeace n.d.). The cod fishery is matched in pedigree only by the fur trade and the birth in 1670 of the country's oldest company, the Hudson's Bay Company. These two resource industries launched a pattern of local resource extraction in a conquered periphery territory to service demand in the political, economic, and cultural core of the Old World in western Europe. These patterns of servicing the core are still etched into the Canadian economic landscape, and this history as a resource periphery makes Canada particularly susceptible to global economic, political, technological, and cultural shifts.

Politico-economic interventions, such as US import restrictions and stumping duties on Canadian timber, have circumvented provisions under the North American Free Trade Agreement, thus *levelling the playing field* in the US to ensure that US forest producers remained competitive against Canadian resource imports. International capital flows and capricious world commodities prices are dictated as much by supply as market sentiment. One result was the collapse of global petroleum products prices in the early part of 2009 in what some pundits call the first great global depression since the *real* Great Depression of the 1930s. Alberta, which in the past 20 years steadily emerged as the most economically productive province as a result of a take-off in oil sands explorations and development, has now fallen into its largest annual deficit in history at $2.2 billion and is in the unusual position of seeking an extra $700 million in transfer payments from the federal government after years of complaining of being banker to the poorer provinces (*Edmonton Journal* 2009, B5).

The preceding examples point to the depth of Canada's integration in the global economy and the vulnerability of a resource economy that remains largely undiversified, notwithstanding active domestic manufacturing, service, and cultural industries. Another important feature of any economic activity is its susceptibility to changing consumer demand. For instance, fishing by both artisanal fishers off the Grand Banks of Newfoundland and the rapacious extraction by commercial fishers under domestic and foreign flags eventually led the Department of Fisheries and Oceans (DFO) to declare a moratorium on all commercial Atlantic cod fishing in 1992. The direct impact was the loss of about 40,000 jobs; the indirect impacts were the slow decline of fishing communities, social disintegration as families coped with changed household circumstances, and intraprovincial outmigration in search of work (see also Chapter 8). As the cod fishery flounders, the increasing global appetite for marine food and wild-caught game has environmentalists concerned about the quality, quantity, and viability of fish populations around the world (Earle 1995).

Globalization, it must be remembered, is ultimately about increasing homogeneity through increased interaction. As such, the mobility of goods, people, and capital—as well as the knowledge, cultural practices, and ideologies that travel with them—is a critical feature of globalization. The speed and scale of information travel can hugely influence environmental practices. For instance, an organized global social movement and information campaign during the 1980s in protest against the harvesting of the fluffy white pelts of baby harp seals in eastern Canada led to a change in global consumer preferences for seal fur and meat, contributing to the collapse of the seal industry and its official ban by the federal government in December 1987. The Royal Commission on Seals and the Sealing Industry of Canada, established in 1984,

found that clubbing, when properly performed, is at least as humane as killing methods in commercial slaughterhouses, challenging one of the key claims of hunt opponents, who argued that clubbing was not properly performed and led to unnecessary suffering (Royal Commission on Seals and the Sealing Industry of Canada 1986). It did not matter—the global perception of seal hunting as cruel sport had become entrenched.

Lessons for Resource Management

The slow collapse of Canada's Atlantic cod fishery in 1992 was accompanied by continued attempts during the 1980s by DFO policy-makers, scientists, and the fishing industry to assess the actual state of the fishery resource. This period roughly coincided with the beginnings of the latest cyclical economic boom in the Alberta oil industry that seems to have now run its course. Boom-and-bust cycles are a common feature of resource-reliant economies. The permanent collapse of Newfoundland's cod industry and the cyclical thinning of Alberta's oil-based capital flows both serve as cautionary tales of the dangers of resource dependence.

More than half a century ago, Innis (1946; 1950) posited that staples of knowledge created spaces of power; thus, the political economy of communication amid technological advances would directly contribute to environmental resource extraction and degradation. He predicted in the 1930s that faster, broader access to US markets, in terms of communication and transport, would encourage rampant development of the Atlantic Canada cod fishery (Innis 1940). His prediction proved disastrously true. The Staples Theory that has come to be identified with Innis is still relevant to understanding the Canadian resource economy and arguably the effect of globalization on resource-dependent nations generally (Norcliffe 2001).

The Staples Theory asserts that national economic and social development is built on the generation of national income through the export of raw materials or semi-processed materials. Resource economies that produce such low-value goods cannot command favourable terms of trade because the high-value-added manufactured products, and the accompanying individual employment income and collective sales revenue, become lost to the resource producer. Therefore, dependency on resource extraction makes an economy intrinsically vulnerable to shocks because the economy is not adequately diversified.

There are inherent dangers to producing goods to meet global resource demand, as Harold Innis warned. Resource degradation is one possible end-result of global engagement. The question now is: has Canada learned? Its record is patchy. Controversial projects such as the Three Gorges Dam in China, for which the Canadian International Development Agency (CIDA) provided financial support in 1988, only to withdraw four years later in a hail of criticism from environmentalists, point to a need for constant evaluation and vigilance in the pursuit of sustainable practice. The creation of the James Bay hydroelectric projects in northern Quebec on traditional Cree lands left an indelible mark on Canada's environmental record after a litany of protest, citizen engagement, litigation, and ultimately development. James Bay also produced watershed moments in Canadian environmental history, first by establishing the first 'modern' treaty recognizing indigenous land rights and second by conceding, two decades later, that further hydroelectric expansion of the area was unacceptable.

Despite its patchy record, arguably Canada has established environmental standards of care, behaviour, and management, specifically in the areas of environmental impact assessment, integrated resource management, collaborative and community-based resource management, and innovative wildlife management schemes. The Berger Commission was a watershed in environmental assessment in Canada, since it was the first time that Aboriginal communities had any significant input into the development and management of their ancestral territories (Bocking 2007), and has been heralded as 'Canada's Native charter of rights' because it established the validity of Aboriginal settlement and required that a pipeline could be built only after the impacts on northern inhabitants, livelihoods, and environment could be determined and found acceptable. The Berger Report is significant in that it demonstrates that post-colonial states can open avenues to redress past and present environmental injustice. Subsequently, the Mackenzie Valley Five-Year Action Plan permanently protected 40 million hectares of culturally and ecologically significant land before construction begins on a $7-billion natural gas pipeline in the Mackenzie River Valley in the Northwest Territories (Pelley 2005). With the Berger Report following just two years after the James Bay and Northern Quebec Agreement of 1975, it would be fair to say that political recognition of Aboriginal priorities, beliefs, and resource management systems had entered into the Canadian consciousness.

Clashes over old-growth forests in Clayoquot Sound, British Columbia, were significant in the global reach of environmental and Aboriginal civil society networks and the emergence of the first coherent attempts at collaborative forestry management and the incorporation of indigenous knowledge (see also Chapter 12). Years of organized protest prompted the BC government to create, in 1993, the Clayoquot Sound Science Panel (CSSP), which recognized three broad classes of values: ecological services such as air quality, specific objects such as trees, and spiritual and cultural values. The CSSP issued a series of groundbreaking reports in 1995 that outlined a vision for collaborative forestry management that respected the interests of companies, workers, residents, environmentalists, and the five First Nations in the Clayoquot Sound area (Iisaak n.d.). Significantly, large commercial foresters are no longer active in Clayoquot Sound; the main forestry company is the First Nations company (Iisaak n.d.; see also Chapter 12). Clayoquot Sound, like James Bay, demonstrates the increasing sophistication of First Nations' organization, and it also contributes to models of forestry management and knowledge as part of an evolving environmental globalization.

The Banff-Bow Valley Task Force was created in March 1994 and 27 months later made 500 recommendations as part of a 50-year vision emphasizing ecological integrity. Key recommendations included curtailing human settlement and population growth in the town of Banff and the hamlet of Lake Louise. Wildlife corridors were established, with miles of fence lining either side of the TransCanada Highway to funnel wildlife over 'bridges' straddling the highway. Banff, the birthplace of Canada's national parks system and a UNESCO World Heritage Site, enjoys an iconic place in Canadian and world history and has reinforced its stature as an example of responsible human–nature coexistence.

The Diavik Diamond Mine is an interesting example of environmental stewardship with resource extraction (see also Chapters 13 and 16). Given the proliferation of 'conflict diamonds', which are commonly extracted by forced or ill-paid labour for the purpose of funding civil war, such as in Sierra Leone, Diavik Diamonds has successfully tapped into a global niche for rare gems that are harvested responsibly. Rio Tinto Diamonds and its subsidiary, Diavik Diamond Mines Inc., are signatories to the Canadian government's Voluntary Code of Conduct for Authenticating Canadian Diamond Claims. Diavik Diamonds taps into a global environmental conscience that prizes responsibility and respect for indigenous traditions, as is evident from this self-description on its corporate website homepage:

> Canada's premier diamond producer, creating a legacy of responsible safety, environmental and employee development practice and enduring community benefit. We believe in using resources wisely, continuing a tradition which has been practiced for centuries by the people of the North [Rio Tinto 2009].

Along Canada's border with the United States, a highly cooperative arrangement for transboundary water management in the Great Lakes region of Ontario and Quebec has been in place for 100 years (Heinmiller 2007; see also Chapter 7). One of its enduring and active legacies is the International Joint Commission (IJC), a bilateral body that manages conflict and investigates complaints concerning use of the shared Great Lakes and other water resources.

At a global scale, Canada, as represented by former Liberal Prime Minister Jean Chrétien, found the political will to sign the Kyoto Protocol on 29 April 1998, followed by formal ratification some four years later on 17 December 2002. Little progress has been made on national guidelines for the reduction of greenhouse gas emissions under the Conservative government, which came to power on 23 January 2006, and it is difficult to evaluate at this stage the global political will to make good on the promises of Kyoto.

A Leader in Protected Areas Management?

Looking at its track record, it can be argued that Canada has emerged as a global leader in protected areas management (see also Chapter 12).

Table 1.2 Significant Moments in Canada's Resource Economy and Environmental Evolution

Initiative or Event	Outcome	Significance in Environmental Globalization
United States–Canadian Transboundary Water Management (Boundary Waters Treaty 1909)	A regulatory process for water diversions from boundary waters. Created International Joint Commission, an international organization that investigates and reports on questions referred to it by US and Canada.	Early example of transboundary cooperation and management of a common resource. Shows challenges of environment and development in Canada are non-local in scope.
James Bay Hydroelectric Project I and James Bay II, Quebec, 1971 and 1994	James Bay I: James Bay and Northern Quebec Agreement of 1975. James Bay II: indefinite suspension of Great Whale project to expand existing hydroelectricity operation.	Indigenous rights to decision-making power over project monitoring and expansion in traditional lands. Effective partnership among indigenous and environmental actors.
Berger Commission (Mackenzie Valley Pipeline Inquiry 1977)	Mackenzie Valley Five-Year Action Plan to permanently protect 40 million hectares of culturally and ecologically significant land before construction begins on $7-billion natural gas pipeline in the Mackenzie River Valley	Recognition for indigenous peoples. A Native 'charter of rights'. Indigenous land protection takes precedence over economic prerogative.
Clayoquot Sound Scientific Panel (CSSP) 1993	CSSP recommends forestry management be premised on ecological integrity, partnership, human and spiritual value of old-growth forests. Followed in 2003 by Clayoquot Sound Land and Resource Management Plan.	Emergence of co-management structure, indigenous management of forestry resources, and benefits of wielding global environmental networks and global media to broadcast an environmental cause.
Banff Bow Valley Study 1996	Wildlife corridor established. Restrictions on population growth and urbanization.	Example of multi-party consultative process and multiple-use park management. Iconic global status of Banff and Canadian Rockies lends weight to call to preserve wilderness and wild species.
Kyoto Protocol 1997	International climate change treaty that is signed by Canada in 1998, then ratified by Canada in 2002.	The Kyoto Protocol, in and of itself, is a remarkable example of multinational negotiation and cooperation. Global recognition of climate change threat, but efforts at international action are slow.
Diavik Diamond Mine 1999	Second Canadian diamond company. Designed to meet environmental standards in Canada. Provides employment in economically depressed area to an economically depressed group of Indigenous peoples.	Indigenous peoples involved in planning and consultation, will be part of monitoring processes to ensure ecologically sound practice. Demonstrates an alternative to conflict diamonds harvested for the purpose of financing and maintaining civil war.

Its national parks system dates back to the establishment in 1885 of what is now Banff National Park in Alberta; Banff remains the crown jewel of Canada's parks system and is a proven site of successful integrated use and human–nature cohabitation. Tourism underpins the Banff economy; sustainable mountain tourism and communities is both its purpose and an engine of innovation for resource and environmental management (Draper 2000). Against the backdrop of a global shift toward protected areas management as a form of environmental globalization (Zimmerer 2006), Canada has seen its national and provincial parks system grow rapidly since the turn of the millennium. The past decade has been a period of accelerated change for protected areas. As of the year 2000, an estimated 6.84 per cent of Canadian ecosystems were protected in Canada, with more areas currently under consideration for future designated protected status; this is reflective of a global trend toward increasing protected lands (Dearden and Dempsey 2004).

Protected areas have emerged over the past three decades as arguably the most environmentally friendly face of globalization. This environmental globalization reflects the tension between older, more established and acceptable environmental globalization and other emergent globalizing forces (Zimmerer 2006). The globalism of environment, as discussed earlier in this chapter, is not new. Globalization discourses of conservation and environmental protection have two distinct features: that environmental conservation is a specific component of globalization and that globalization has forced conservation to engage with other types of resource use, as witnessed by the popularity of integrated management and multiple-use zoning (Zimmerer 2006).

The spread of environmental globalization through the activities of global conservation organizations such as the World Wildlife Fund and the incorporation of protected areas into national economic strategies, such as in Indonesia and Thailand, requires that we revisit the significance of other globalizations that are political, cultural, and ecological; the interpretation of globalization as an economic force is increasingly limited. Zimmerer (2006) offers a useful typology of the effects of this newer globalization:

1. Conservation and globalization have led to new spatial arrangements and conservation territories that involve not just physical spaces of activity but also social spaces of network interaction. These new conservation spaces require creative management strategies to navigate the diverse peoples and priorities that are present. For example, Indonesia's Wakatobi National Park (WNP), established in 1996 with an integrated multiple-use zoning system, began a rezonation exercise after 10 years to address local dissatisfaction with the implementation of the park. Consultations were held among WNP residents, the new regional government development planning agency, the Nature Conservancy, and WWF, among others (Figure 1.1).

2. New spatial scales of importance to environmental management have emerged, which can inform global organizations and allow them to work more effectively across scales. For example, the trend toward recognizing and involving particular social constituencies and communities such as indigenous peoples in contested spaces, such as Wakatobi, which requires co-ordination that transcends traditional urban–rural arrangements, and the unprecedented degree of importance of transboundary disputes, management, and linkages, such as the Yellowstone to Yukon conservation initiative—known commonly

Source: travelib asia/Alamy

Figure 1.1 A local fisherman arrives with his catch at the market in Ambuea village, Wakatobi National Park, Indonesia. Fish stock degradation and waste management are chronic challenges.

as Y2Y—extending from the Yukon in Canada to Yellowstone National Park in the US (see Box 1.1); the Mekong River basin that straddles Thailand, Laos, Cambodia, and Vietnam; and the 100,000-square-kilometre Great Limpopo Transfrontier Park that was signed into existence by a treaty among South Africa, Zimbabwe, and Mozambique in 2002 but has yet to formally open.

3. There has been decentralization of environmental governance because of the incorporation of the local scale through collaborative, participatory, and community-based approaches. A prime example of this trend is seen in forest management in Clayoquot Sound, BC.

Value of Indigeneity

First Nations in Canada have wielded increasing influence over environmental policy in the past two decades and have made alliances with environmentalists over specific environmental issues (Poelzer 2002). Examples of such partnerships include Clayoquot Sound. Aboriginal peoples' environmentalism is often modelled on various ways of valuing nature as a spiritual, political, and economic resource. Thus, many of the First Nations' struggles over the environment are not only about protection and conservation but also about title to lands and resources as well as rights to govern those lands and resources (Poelzer 2002).

Box 1.1 Connectivity Is Not Just about Wildlife and Landscape—It's About People Too

Enabling human communities to thrive while maintaining the natural abundance of the Yellowstone to Yukon region is the vision—and challenge—of Y2Y. Aboriginal communities have lived here and depended upon the abundance of the land from time immemorial. Today, this landscape remains important to all who live in and around the region. Humans and their activities may threaten connectivity in the region, but humans are also part of the solution for restoring or maintaining landscape connections. Y2Y works to promote best practices for use of the landscape so wildlife and humans can successfully coexist in this region (Y2Y 2009).

The emergence of First Nations as a political community can be traced to the more permissive environment for dissent that emerged in Western countries after the Second World War. At the same time, the postwar era saw supranational initiatives such as the Universal Declaration of Human Rights and the International Labour Organization's Convention on Indigenous Peoples. Indigenous leaders in Canada and elsewhere were able to take advantage of this new political mood to forge global alliances and to publicly embarrass the Canadian government for its treatment of its First People. The new dissent-permitting society, international pressure, better Aboriginal political organization, and legal and constitutional changes have moved Aboriginal people from the periphery of environmental policy in Canada to the core of governance (Poelzer 2002).

It is significant that the $13.3-billion proposed expansion of the James Bay hydroelectric project was postponed indefinitely in November 1994. The suspension of the Great Whale project (James Bay II) was, from a First Nations' perspective, hugely significant after its failure to stop the James Bay I project announced in 1971. Although unable to stop the first project, the James Bay Cree were able to negotiate the James Bay and Northern Quebec Agreement of 1975 with Quebec and the federal government, creating the first 'modern' treaty that 'provided for land ownership, advisory environmental and resource co-management, local and regional self-government, and monetary compensation for the Cree' (Poelzer 2002, 94). During the intervening two decades between James Bay I and James Bay II, the Cree had learned to organize politically and to strategically deploy international pressure and alliances among the Canadian mainstream. What they achieved, since 1975, was to change the scale of debate by tapping into cultural shifts in the global perception of indigenous peoples and to make their story heard beyond the confines of Quebec. By staging a local hydroelectric project as a national and international issue, by reframing the meaning of James Bay from energy security to racial and environmental injustice, the Cree achieved a much different outcome during the second battle over James Bay.

Meanwhile, advances in indigenous representation and participation are reflected in the Berger Report, the CSSP, and latterly, the gradual exodus of big-name forestry producers such as MacMillan Bloedel and their replacement with smaller-scale Aboriginal foresters such as Iisaak.

Diavik Diamond Mines and Iisaak in Clayo-quot Sound are specific examples of indigenous management premised on the incorporation of indigenous knowledge as *valuable* knowledge. This idea of indigeneity as valid and valuable is an historically new perception and mirrors multi-sited instances of the political and cultural shift of indigenous peoples and livelihoods from the periphery of economic development (see also Chapter 4). In this regard, the revaluing of First Nations into Canadian resource and environmental management reinforces and is strengthened by a wider trend of environmental globalization as social justice.

An Enduring Resource Economy

Canada has endured many hard lessons about resource dependency. Typically, as economies develop, they shift into more advanced states of industrialization. The automotive industry in southwestern Ontario and the US midwest, notwithstanding the slowdown and plant closings during the current global economic downturn, is evidence of a thriving manufacturing sector. Parts of the Canadian economy have entered a post-industrial state of service and knowledge industries led by Research in Motion, music and cinema in Toronto, Montreal, and Vancouver, and call centres in previously depressed regions of eastern Canada that are benefiting from the technological shifts of an increasingly connected world.

However, it is hard to deny that Canada's huge landmass, the tar under Alberta's boreal forests, potash in Saskatchewan, diamonds in the Northwest Territories, fisheries on the east and west coasts speak to an enormous natural resource wealth that is highly sought after in local and global markets. It is little surprise that Canada remains an enduring resource economy, notwithstanding diversification. At present, some 2000 Canadian communities are considered resource-reliant—i.e., with 50 per cent of local economic activity committed to resource extraction and production (Natural Resources Canada 2006). Alberta has endured several oil price–related cyclical slowdowns since oil was discovered in Leduc in 1947 and is now deeply entrenched in yet another after years of double-digit economic growth. In British Columbia, with the exception of the densely populated and urban Lower Mainland, fisheries (in particular, salmon aquaculture), tourism, and timber dominate the economy (see also Chapter 10). In Newfoundland, the collapse of the once seemingly inexhaustible cod fish stock was alleviated somewhat by the introduction of shellfish production; now the big economic story in the province is the Hibernia offshore oil project, which began production in 1997. The Hibernia offshore field is the fifth largest oilfield ever discovered in Canada (Hibernia n.d.), with estimated recoverable oil reserves of 1.2 billion barrels and the potential to rise to 1.9 billion barrels (Reuters 2007).

Canada's entrenched dependence on resource development is linked into a global demand for its products: timber, fish, minerals and gems, oil and gas. In 2009, Innis's cogent observation of Canada's economy of staples holds as true as it did more than half a century ago. Indeed, in an age of increasing economic, technological, political, cultural, and environmental integration, Canada is arguably even more thoroughly embedded in global chains of production and consumption than it was as a young colony. It is also more thoroughly embedded in global cultural shifts and the 'new' environmental consciousness through the speed of mobility of people, money, and ideas.

Mobility is enabled by modern communication avenues such as the world wide web, the proliferation of real-time media, and globally dispersed networks of socio-cultural and environmental resistance. The speed and intensity of material and non-material mobility is a deeply important feature of the new environmental globalization of political organization, networks of resistance and collaboration, the valuing of people and nature, and the management and development strategies that ensue. Global mobility contributes to the instabilities identified in Staples Theory but contrarily also provides opportunities for policy change and knowledge-sharing to improve human engagement with the natural environment.

CANADA AS GLOBAL ACTOR

The new environmental globalization is a promising avenue for Canadian institutions. In particular, federal agencies are leading the way in promoting research at the intersection of development and environment. There has been a recent flourishing of interdisciplinary activity in Canadian universities, partly in answer to the challenge of meeting post-secondary students' awareness of the essential connection between development and environment. An increasing number of Canadian scholars are openly addressing environment and development as *twinned* concerns; this seems particularly disingenuous, given that resource exploitation has so long been a feature of Canadian economic development. The question is not so much that it is time to consider environment, development, and globalization as a single unit of analysis but why it has taken so long to do so. After all, was not environmental and resource management in Canada and elsewhere, historically, fundamentally oriented toward economic growth? Were not imperialism, colonialism, developmentalism, and now globalization fundamentally based in an ideology of liberalism favouring ever-expanding capital accumulation?

Approaches to environment and development and their connection to this 'new' age of globalization in which we now live are evolving. And Canadian institutions are evolving with them.

Agencies

Agencies such as the International Development and Research Centre (IDRC) place increasing focus on environment, sustainability, participatory development, and resource management. Significantly, one of the IDRC's three main areas of program support is Environment and Natural Resources Management; its other two areas complement each other—Information and Communication Technologies for Development (ICT4D) and Social and Economic Equity (SEE). One of the greatest challenges of doing research in environment and development, as in much research with an ostensibly prescriptive agenda, is achieving relevance. The IDRC, for instance, has increasingly come to assess and embrace the potential of the projects it funds to influence actual policy and practice.

Carden (2004) notes that the IDRC, as a key granting institution supporting Canadian research and international knowledge transfer, continues to extend its evaluation of research in hope of guiding future studies on policy influence. Increasingly, researchers are expected not only to produce sound research but also to draft sound recommendations that can be implemented to achieve substantive learning and development outcomes. In other words, research proposals are evaluated largely on the *usefulness* of their work. There are two key features of knowledge utilization: the cultural differences between researchers and policy-makers and the 'enlightenment' function of research in policymaking (Carden 2004, 138). Ultimately, the relevance of the IDRC as a research and policy-influencing agency depends on its taking into account how systems of policy, research, support, and implementation evolve.

The Canadian International Development Agency (CIDA) is another giant in the country's efforts to make positive change beyond its borders. CIDA typically funds collaborative research projects with a prescriptive mandate: to improve capacity-building through the study of particular processes and activities that can alleviate the suffering of those less well-off. Poverty reduction, economic and technological efficiency, social change through education and gender equality initiatives, and health care all fall within the purview of CIDA. It is notable, therefore, how often CIDA-funded projects include an environmental focus—e.g., the now-ended Collaborative Environmental Project in Indonesia (CEPI). Under its environmental projects banner, CIDA most recently has offered internships with a mixture of organizations including Oxfam-Quebec, Canadian universities (Dalhousie, Acadia), Canadian colleges (College of the Rockies, Algonquin), and developing country institutions (Vietnam Provincial Environmental Governance, Africa Community Technical Service Society, Post-Tsunami Restoration), among others (CIDA 2009). It also funds a range of initiatives officially termed 'energy, production and distribution', 'agriculture', 'fishing', 'food aid and security'—all these have relevance to environment.

Interdisciplinary and Multidisciplinary Programs

The increasing focus on the intersection of environment, development, and globalization is marked by a proliferation of resource and environmental management schools and environment and development schools. Most of these are interdisciplinary programs in which particular research fields straddle a number of disciplines; other partnerships and initiatives between academic units might be more multidisciplinary in scope in that researchers are embedded in a particular discipline and work with others located in other disciplines. Such initiatives are now in place at 40 Canadian postsecondary institutions across nine provinces, spurred by administrators' responsiveness to students' requests for 'progressive environmental policies and programs for their campuses' (AJ environmental directory 2007, 15).

Across Canadian universities, sustainability has been embraced both as pedagogy and as institutional practice. Sustainability initiatives include retrofitting energy-efficient lighting, conserving water, bicycle sharing, battery recycling, and seminars, workshops, and teachings on reducing vehicular transportation and CO_2 emissions to combat climate change. The University of Calgary launched North America's first free campus bicycling program in 2005. At the University of Waterloo, the Faculty of Environment in partnership with St Paul's College launched a new international development program in 2008 with an explicit focus on environmental sustainability. And at the University of Toronto, the Centre for Environment and the Jane Goodall Institute of Canada jointly hosted the university's first annual environment and development seminar in April 2009.

The University of Calgary offered its first multidisciplinary conservation and development group study program during May 2009. The Conservation and Development in Amazonia-Peru program, offering three courses in biology, environmental governance, and participatory development, is believed to be the first of its kind in Canada to explicitly engage with the global nature of environmental development challenges and to do so in a context that allows for simultaneous learning at the physical–social science interface. The inaugural offering saw 18 undergraduate students from diverse disciplines come together to engage in biological monitoring in Pacaya-Samiria National Reserve under the direction of local field biologists, site visits to and informal interviews with indigenous communities inside and outside the reserve, and instructor-designed self-guided tours of the city of Iquitos (see Figure 1.2), which functions as the gateway to the Peruvian Amazon and is a centre of economic activity for trade in Amazon timber and wild species caught for food and decorative purposes. The three instructors (of whom one is the author of this chapter) designed co-ordinated lectures and field activities that integrated the three courses and organized students in multidisciplinary 'expert' teams to encourage critical thinking. Each multidisciplinary team

member began the program as an expert in a particular field (indigenous, rainforest land use, socio-economic, political, and ecosystem), contributing a piece of a group effort; by the end of the program, the student teams were far more accomplished in presenting the integration of different perspectives and expertise within environment and development challenges. Significantly, students who began the month-long program with a faith in conservation spaces came away with a heightened appreciation of protected areas as *lived* spaces where local peoples face daily challenges to economic, food, and health security. Such challenges were highlighted by unusually high levels of annual seasonal flooding in the Amazon River Basin

Source: © Paolo Aguilar/epa/Corbis

Figure 1.2 Houses in the Balen district of Iquitos, capital of the Peruvian province of Loreto. In addition to plagues, bad weather, and difficult socioeconomic circumstances that constantly affect the region, unusually high seasonal flooding in 2009 caused local food shortages, water contamination, and a flu epidemic.

that contributed to local food shortages and flu epidemics in the villages visited. The program received funding from the Alberta Education and Technology (AET) Grant for International Learning to defray students' participation costs.

Against this context of pedagogical innovation, it should come as no surprise that Canadian university students are keenly interested in the intersection of development, environment, and globalization. 'Environment' was among the top 10 topics of interest in 2005 and in 2002–3 for undergraduate students in development studies programs in Canada, whereas 'globalization' dropped off the top 10, to be replaced by a more nuanced interest in specific types of global development challenges such as 'relations between industrialized and developed countries', 'human rights, conflict, and cooperation', and 'regional development, urbanization, migration' (Child and Manion 2004; Grey, McLellan, and Pena 2005). Concerns about culture, participation, and education also figure prominently in the top 10 topics of the new millennium.

Engagement by Canadian Scholars in Developing Countries

Relevant environment and development research is conducted against the particular context of the time and place. It is useful here to remember the evolving history of globalization as development and the need for engagement with environmental concerns of local significance and global reach.

Tsing (2005, 1) suggests that regardless of our place in the world, there are universal aspirations that can only be enacted through understanding and navigating 'the sticky materiality of practical encounters'. Her call for a deeper engagement with local experience is located within the broader context of global environmentalism, particular historical encounters, capitalism as global ideology, and chains of commodity and cultural production. Tsing's conception addresses the particular partnerships among the environmental movement and Indonesian student nature lovers and village leaders, but she paints a masterful picture of the subtleties of place and identity within international relations that serves as a reminder of the interconnections among environmentalists, investors, policy-makers, funding agencies, and scholars in the Global North and South. The intersections among these various forces are points of friction, and friction generates not only conflict but collaboration and hope.

It is useful, then, to consider the influence of Canadian scholars, environmentalists, and development specialists beyond their own borders. How do we deal with the *messiness* of resource competition, environmental uncertainty, development needs, economic prerogatives, and identity politics? How do we manage and govern amid these conflicting interests, and what lessons do we have to offer to resolve those conflicts?

I suggest that among our greatest strengths as Canadian scholars of environmental management and resource development is an ever-deepening engagement with complexity, integration, collaboration, and participation. CEPI, a consortium led by the University of Waterloo and York University, sent graduate students to the islands of Bali and Sulawesi for 12 years from 1990 to 2002, generated much collaborative research into environmental management, gender training, and longitudinal research, and provided an intellectual home for Indonesian scholars who were able to complete a Canadian graduate education that could be put to use in their home country. Among the outputs from the CEPI Bali and Sulawesi projects were the edited collections by Martopo and Mitchell (1995), and Wismer, Babcock, and Nurkin (2005) and numerous graduate theses and related journal publications.

Knowledge has limited use if it is not shared. Therefore, engagement in new communicative strategies is essential. The IDRC's strong logistical support for published research is commendable, but we must remain vigilant about environmental education and communication to disseminate the message. At Canadian post-secondary institutions, students have exposure to a range of internships, both paid and unpaid, that allow them to experience development studies 'in the field', and increasingly students are able to combine an environmental focus into such work. An initiative popular with Canadian development studies students is Engineers Without Borders (EWB). Students take internships with EWB and are relocated to a developing country, typically for four months, to help develop and implement projects related to water, waste disposal, and energy, but they also get involved in social issues. EWB's model of placing Canadian students among local families is also an excellent means of encouraging Canadian students to observe, participate in, and *live* the minutiae of daily life in their host countries. The effect is, arguably, to engage Canadians in an 'ethnography of global connection' (Tsing 2005) by allowing them to observe and *talk through* the day's events and insights with their foreign host families.

Knowledge-sharing, collaborative research projects, and student placements in foreign countries are all features of an environmental globalization that—I submit—is intrinsically based on the communication of environment and development ideas enabled by a globalized world of increased integration. We can travel to other nations to learn about other people's lives because *we can*; we have the technology, we have the money, and we have the intellectual and logistical infrastructure through our institutions of research, practice, and education.

Is Canada an Environmental Leader?

Canada has much to offer as an environment and development actor. It has a long history of integration in the global economy, largely through its wealth of natural resources and their sale and latterly through its scholarly and pedagogical activities.

Is Canada an environmental leader? The jury is still out. Our development trajectory has been rather predictable. Historically, economies take off with the exploitation of a comparative advantage—i.e., some combination of land, capital, and labour that allows for the efficient production of a good for maximum profit. Canada's comparative advantage historically was its fish, forests, and furs. Our staples economy has been integrated into global demand markets for 500 years, and as seen now in the sharp decline of the Alberta oil economy and Ontario's manufacturing economy—as well as property price decline across the globe—we remain dependent on the healthy performance of resource and commodities markets.

In terms of environmental and resource management, I would propose that among our bleakest modern failures is the collapse of the Atlantic cod fishery. But we have also made advances with our Aboriginal peoples, engaged with and recognized the value of indigenous knowledge and resource management systems, and produced some watershed moments regarding indigenous land claims and a Native charter of rights. We are leaders in integrated, multiple-use parks management and have demonstrated that environmental stewardship and human economic activity can coexist, quite literally, in 'spaces of hope' (Harvey 2000) such as Banff National Park.

As a nation, we have had significant successes but clearly some failures. I would suggest that our successes relate to the scaling-up of locally significant environmental development initiatives that can inform sustainable practice in other parts of the world. Canada benefits from the longest peaceful border with another nation—the US—and the International Joint Commission formed in the wake of the 1909 Transboundary Waters Treaty remains an example of bilateral management and conflict resolution. But our record in terms of scaling down global initiatives such as the Kyoto Protocol and implementing them locally still leaves much to be desired. This is not to downplay the complex political negotiation among Canadian provinces and the international community or the challenges of addressing the position held by the US, Australia, and Alberta that federally enforced carbon emissions reduction targets are less effective than locally specific corporate incentives and initiatives. The fact remains that on 'the defining human development issue of our generation' (UNDP 2007, 1), our performance vis-à-vis climate change debate and action has been poor.

On the positive side, Canada's role in the new environmental globalization involves technology and knowledge transfer, with Canada as communicator and educator. There are signs that Canadian efforts are bearing fruit in terms of foreign adoption of Canadian resource management strategies, such as integrated management in Wakatobi, environmental consultation, and waste management strategies modelled on the wet–dry recycling technology pioneered in Guelph, Ontario, in the 1990s.

CONCLUSION

Globalization can be seen as not only threat but also opportunity, given the right conditions. In historical terms, economic globalization has generally increased wealth for the nations that participate in it, but it creates greater inequality between them and the states less able to take part, whether by intent or structural constraints (Borghesi and Vercelli 2008).

This chapter does not attempt to assert a position on good or bad globalization vis-à-vis resource development—clearly, there are benefits and limitations. Rather, it is argued here that Canada's history of resource development, and its evolution and progression in resource and environmental management, reflects an ongoing engagement with conflict and uncertainty. The lessons learned are pluralistic perspectives on resource and environmental management, economic and social development, participation or marginalization through environmental consultation and political recognition, and accommodation of different perspectives and priorities through integrated, and multiple-use management and development of natural resources.

The Canadian experience promotes information-sharing on resource management strategy and thus contributes to environmental communication and education on a global scale—i.e., resource management training, resource management practice, recognition of multiple perspectives and priorities, and inventive and responsive strategies for cooperation and reconciliation.

Looking forward, the challenges and opportunities at the intersection of globalization, development, and environment are Canada's continued dependence on its natural resources as a source of economic wealth. The economy of staples is unstable. We have yet, as a nation, to decouple from the global chains of production and consumption that caught our east coast fisheries, cut down our forests, and stripped the furs from our wildlife. We have not adequately answered the call for reducing carbon emissions to contribute to a global effort to slow down climate change. But we are connected. Through our resource production, management and development strategies, and research and education, we are integrated into a globalized world of increasing breadth beyond borders, of interconnectedness, of speed of information transfer, and scale of impact. The local has global relevance—and vice versa.

Ultimately, globalization is not exclusively antithetical to environment, nor does it only generate bad development. It is an opportunity for exchange, for growth that is more than economic, for environment that is not only valued as resource, and for development that is ecologically sound and socially responsible.

FROM THE FIELD

Field research is challenging. Time is always too short, organizing meetings is a fine craft, and the demands of the physical environment can be overwhelming. These challenges are magnified when one is working with local and/or indigenous residents in a foreign country with attendant language and cultural barriers. In such circumstances, field research becomes a real test of character and commitment.

During my numerous research visits to Indonesia, my latest foray into the Peruvian Amazon as a field school instructor, and my burgeoning comparative North-South research program, it has become clear that *how* researchers conduct themselves as scholars and as guests and the appropriateness of their behaviour in *particular spaces* matters. Place and interaction—the place-based politics of scholarly engagement—are critical to gathering data successfully, translating socio-ecological environments among new scholars, and building cross-boundary relationships that foster respect, trust, collaboration, and relevance. The importance of communicating well, of understanding, navigating, and manipulating the complexities of communication, is central for high-quality prescriptive research in environment and development.

The words and deeds of individuals or institutions affect the dissemination, development, and implementation of resource and environmental management strategies at multiple scales. This is true of my field research and field teaching experiences: it is all too easy to arrive in a foreign country with the notion that the environment needs protecting above all else. That view shapes the research topics that we consider valid, our assumptions about what 'the natives' know and should care about, and the research questions to pursue.

To borrow a Buddhist concept, the key is *mindfulness*, being ever aware of oneself and the world around us. In the field, I try to critically engage with the global–local connections of environmental decision-making and prevailing environmental fads, trends, and paradigms. My aim is to challenge environmental norms in their global reach and local impact. Resource and environmental management is embedded in a specific historical and geopolitical context. We need to be mindful of the global and transboundary influences on particular environmental strategies. For instance, in Indonesia, the global shift toward conservation and the eminence of the resource scarcity discourse have led to a growth in marine protected areas created with assumed benefit to the local communities, which must curtail their resource extraction (e.g., fishing) and adapt to often limited options as tourism providers.

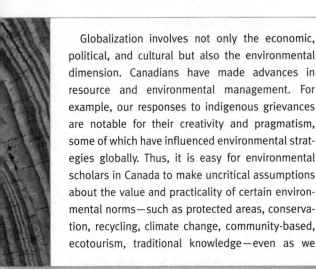

Globalization involves not only the economic, political, and cultural but also the environmental dimension. Canadians have made advances in resource and environmental management. For example, our responses to indigenous grievances are notable for their creativity and pragmatism, some of which have influenced environmental strategies globally. Thus, it is easy for environmental scholars in Canada to make uncritical assumptions about the value and practicality of certain environmental norms—such as protected areas, conservation, recycling, climate change, community-based, ecotourism, traditional knowledge—even as we challenge the efficacy of particular strategies and responses. When doing so, researchers risk limiting the scope of debate and analysis.

Whether conducting field research or teaching students in the field, I advocate three rules for environmental scholarship: be aware of one's *positionality* as a communicator—i.e., one's world view and relation to others within their socio-political context; be mindful of the *local–global intersections* of environmental policy and strategy; and be ever willing to *listen* to narratives potentially dissonant with prevailing environmental norms.

—CHUI-LING TAM

Note

1. McLuhan originally intended the title of his book to be *The Medium Is the Message*, but a typing error during the printing of the book resulted in *The Medium Is the Massage*. McLuhan chose to keep the error in the title because of its implied ambiguity and because the altered title reflected his message.

REVIEW QUESTIONS

1. What are the meanings of globalization?
2. What were the significant moments, historically, in Canada's resource economy?
3. What are Canada's key contributions to environmental management and development in global terms?
4. How does the Canadian research and educational system enable study of globalization, development, and environment?
5. What are the opportunities for post-secondary student engagement in environment and development?

REFERENCES

'AJ environmental education directory 2007: Green campus life and learning'. *Alternatives Journal* 35 (5): 15–40.

Appelbaum, R., and W. Robinson, eds. 2005. *Critical Globalization Studies*. New York: Routledge.

Becker, E., and T. Jahn. 1998. 'Growth or development?' In R. Keil, D. Bell, P. Penz, and L. Fawcett, eds, *Political Ecology: Global and Local*, 68–83. London and New York: Routledge.

Bhavnani, K., J. Foran, and M. Talcott. 2005. 'The red, the green, the black, and the purple: Reclaiming development, resisting globalization'. In R.

Appelbaum and W. Robinson, eds, *Critical Globalization Studies*, 323–32. New York: Routledge.

Bocking, S. 2007. 'Thomas Berger's unfinished revolution'. *Alternatives Journal* 33 (2/3): 50–1.

Borghesi, S., and A. Vercelli. 2008. *Global Sustainability: Social and Environmental Conditions*. Basingstoke, UK: Palgrave MacMillan.

Carden, F. 2004. 'Issues in assessing the policy influence of research'. *International Social Science Journal* 179: 135–51.

Chambers, R. 1997. *Whose Reality Counts? Putting the First Last*. London: Intermediate Technology Publications.

Child, K., and C. Manion. 2004. 'A survey of upper-year students in international development studies'. *Canadian Journal of Development Studies* 25 (1): 167–86.

CIDA (Canadian International Development Agency). 2009. 'Project browser'. http://www.acdi-cida.gc.ca/cidaweb/cpo.nsf/vWebProjBySectorEn?OpenView&Start=1&Count=1000&Collapse=13#13.

Dearden, P., and J. Dempsey. 2004. 'Protected areas in Canada: Decade of change'. *The Canadian Geographer* 48 (2): 225–39.

Draper, D. 2000. 'Sustainable mountain communities: Balancing tourism development and environmental protection in Banff and Banff National Park, Canada'. *Ambio* 29 (7): 408–15.

Earle, S. 1995. *Sea Change: A Message of the Oceans*. New York: Fawcett Columbine.

Eberts, D. 2004. 'Globalization and neo-conservatism: Implications for resource and environmental management'. In B. Mitchell, ed., *Resource and Environmental Management in Canada: Addressing Conflict and Uncertainty*, 3rd ed, 54–79. Toronto: Oxford University Press.

Edmonton Journal. 2009. 'Province chases $700M from Ottawa; Premier vows to lobby federal government for millions more in transfer payments'. 9 April: B5.

Greenpeace. n.d. 'Canada Atlantic fisheries collapse'. http://archive.greenpeace.org/comms/cbio/cancod.html.

Grey, S., K. McLellan, and E. Pena. 2005. 'Voices of the present, visions of the future: An exploration of undergraduate international development studies in Canada'. *Undercurrent* 2 (2): 57–79.

Harvey, D. 2000. Spaces of Hope. Berkeley: University of California Press.

———. 2005. 'From globalization to the new imperialism'. In R. Appelbaum and W. Robinson, eds, *Critical Globalization Studies*, 91–100. New York: Routledge.

Heinmiller, B.T. 2007. 'Do intergovernmental institutions matter? The case of water diversion regulation in the Great Lakes basin'. *Governance: An International Journal of Policy, Administration, and Institutions* 20 (4): 655–74.

Hibernia. n.d. 'The Hibernia Reservoir'. http://www.hibernia.ca/html/about_hibernia/index.html.

Innis, H.A. 1940. *The Cod Fisheries: The History of an International Economy*. Toronto: Ryerson Press.

———. 1946. *Political Economy in the Modern State*. Toronto: Ryerson Press.

———. 1950. *Empire and Communications*. Oxford: Clarendon Press.

Iisaak Forest Resources Ltd. n.d. 'Clayoquot Sound background'. http://www.iisaak.com.

Kothari, U., ed. 2005. *A Radical History of Development Studies: Individuals, Institutions and Ideologies*. London: Zed Books.

Martopo, S., and B. Mitchell, eds. 1995. *Bali: Balancing Environment, Economy and Culture*. Publication Series no. 44. Waterloo, ON: University of Waterloo, Department of Geography.

McGrew, A. 2000. 'Sustainable globalization? The global politics of development and exclusion in the new world order'. In T. Allen and A. Thomas, eds, *Poverty and Development: Into the 21st Century*, 345–64. Oxford: Oxford University Press.

McLuhan, M. 1967. *The Medium Is the Massage: An Inventory of Effects*. New York: Bantam Books.

McMichael, P. 2005. 'Globalization and development studies'. In R. Appelbaum and W. Robinson, eds, *Critical Globalization Studies*, 111–20. New York: Routledge.

Milanovic, B. 2003. 'The two faces of globalization: Against globalization as we know it'. *World Development* 31 (4): 667–83.

Natural Resources Canada. 2006. 'All resource-reliant communities, 2001'. *The Atlas of Canada*. http://atlas.nrcan.gc.ca/site/english/maps/economic/rdc2001/rdcall.

Norcliffe, G. 2001. 'Canada in a global economy'. *The Canadian Geographer* 45 (1): 14–30.

Pelley, J. 2005. 'Conservation first in Canadian Arctic'. *Environmental Science and Technology* 15 March: 127A.

Poelzer, G. 2002. 'Aboriginal peoples and environmental policy in Canada'. In D.L. VanNijnatten and R. Boardman, eds, *Canadian Environmental Policy: Context and Cases*, 87–106. Toronto: Oxford University Press.

Resurreccion, B., and R. Elmhirst. 2008. *Gender and Natural Resource Management: Livelihoods, Mobility and Interventions*. London: Earthscan.

Reuters. 2007. 'Canada Hibernia oil output declines as field ages'. 22 October. http://www.reuters.com/article/companyNewsAndPR/idUSN2245014620071022.

Rio Tinto. 2009. 'The Diavik Diamond Mine'. http://www.diavik.ca.

Rocheleau, D., B. Thomas-Slayter, and E. Wangari. 1996. *Feminist Political Ecology: Global Issues and Local Experiences*. London and New York: Routledge.

Royal Commission on Seals and the Sealing Industry of Canada. 1986. *Seals and Sealing in Canada: Report of the Royal Commission*. Ottawa: Minister of Supply and Services Canada.

Stevis, D. 2005. 'The globalizations of the environment'. *Globalizations* 2 (3): 323–33.

Tsing, A.L. 2005. *Friction: An Ethnography of Global Connection*. Princeton, NJ: Princeton University Press.

UNDP (United Nations Development Programme). 2007. *Human Development Report 2007/2008*. New York: UNDP.

UNESA (United Nations Department of Economic and Social Affairs). 2009. 'World population prospects: The 2008 revision'. New York: UNESA. http://esa.un.org/unpd/wpp2008/index.htm.

UNFPA (United Nations Population Fund). 2008. *State of World Population 2008*. New York: UNFPA.

Walker, R.B.J. 1988. *One World, Many Worlds: Struggling for a Just World Peace*. Boulder, CO: Lynne Rienner.

Williamson, J. 1993. 'Democracy and the "Washington Consensus"'. *World Development* 21 (8): 1329–36.

Wismer, S., T. Babcock, and R. Nurkin, eds. 2005. *From Sky to Sea: Environment and Development in Sulawesi*. Geography Publication Series no. 61. Waterloo, ON: University of Waterloo, Department of Geography.

Woodhouse, P., and A. Chimhowu. 2005. 'Development studies, nature and natural resources: Changing narratives and discursive practices'. In U. Kothari, ed., *A Radical History of Development Studies: Individuals, Institutions and Ideologies*, 180–99. London: Zed Books.

WCED (World Commission on Environment and Development). 1987. *Our Common Future*. Oxford and New York: Oxford University Press.

Worldwatch Institute. 2001. *State of the World 2001*. Washington: Worldwatch Institute.

———. 2009. *State of the World 2009*. Washington: Worldwatch Institute.

Y2Y (Yellowstone to Yukon Conservation Initiative). 2009. 'We're all about connectivity'. http://www.y2y.net.

Zimmerer, K.S., ed. 2006. *Globalization and New Geographies of Conservation*. Chicago: University of Chicago Press.

2

Governance for Environment and Resources: The World Remains Oceans Apart

O.P. Dwivedi

Learning Objectives

- To be aware of the meaning, context, and nature of globalization and good governance.
- To understand how growing resource scarcity would affect human security, which could result in worldwide conflicts and even wars among nations.
- To appreciate the need for a dialogue between the rich North and the poor South.
- To grasp the nature of the imminent challenges facing our planet Earth as we embark on the twenty-first century.

GLOBAL GOVERNANCE: THE CONTEXT

Globalization during the twenty-first century is different from the old-fashioned seventeenth-century colonialism and nineteenth-century British imperialism. Essentially, this new type of globalization is stronger and more insidious because of political power supported by technology, economic wealth, and a solid demographic base. Moreover, it is not only about trade and economics, important as they are, but even more about culture.

Irrespective of the definition used for 'globalization', the context, the structure, the processes, and the effects of governance and administration are decisively influenced by it (for further details

on the matter of definition and interpretation, please see the next section as well as Chapter 1). Furthermore, the circumstances of governance are increasingly defined by parameters beyond the confines of the nation-state. So are goals, resources (human, material, and 'semiotics'—national symbols), communications, and performance. The same is true regarding the impact of policy decisions, non-decisions, actions, and inactions in the context of governance.

The context of governance also encompasses interwoven domestic and extraterritorial dimensions (beyond the traditional sphere of public sector management). In an era of growing interdependence but also of mutual vulnerability, issues of domestic and international micro and macro human as well as environmental security are interconnected. At the centre of this global–local interface, there is also an emerging global consciousness that clean government is imperative for ushering in the era of good governance. However, the rise of *amoralism* in governance has further muddied the waters of the worldwide impact of globalization, a matter to be discussed later in detail.

Globalization did give us, for some years, a uni-polar world, with the United States as the only superpower able to create its own reality and also to be unilateralist. Nevertheless, US supremacy is coming to an end with the resurgence of China and India; the power equation is slowly but dramatically changing as it is becoming more diffused among a group of nations such as the European Union, China, India, Japan, Russia (as the major natural-resource exploitative power), Brazil, and South Africa, to name a few.

The new global power structure is multilevel and multi-polar, as well as a complex and differential one. This also means that the present world is more difficult to manage compared to the situation the West envisaged and became accustomed to during the last century. This new world of shared power politics requires new structures of global governance, because neither the existing international institutions (those created as the post-1945 international order) nor their policies and programs for common well-being reflect today's complex realities. Three of these realities must be understood: (1) the emergence of global governance; (2) mounting resource scarcity and the need to arrest the continuing deterioration of our planet's ecology, which poses a major threat to human security and the viability of our world; and (3) environmental insecurity, the greatest challenge. In this chapter, I examine these three aspects, which also relate to the two main themes of this book—conflict and uncertainty—ending with some concluding observations.

GOVERNANCE

Governance emerged during the 1980s as a new paradigm denoting something much more than 'government' and replacing the traditional meaning of the term 'government'. The emphasis of this new paradigm is on reforming the management structures and processes of most Western nations. At the time, such reforms were considered part of a revolutionary change in the management of governmental affairs that involved a 'paradigm shift' from the Weberian model of bureaucracy (a dominant model of the twentieth century characterized by government administration) to the New Public Management (NPM), or the 'new managerialism' (Saint-Martin 1998).

The rise of the NPM was closely related to the election of right-of-centre politicians such as Margaret Thatcher of Britain, Ronald Reagan of the US, Jacques Chirac of France, Brian Mulroney of Canada, and Malcolm Fraser of Australia. These leaders wanted to restrain or cut back public service spending and employment and to roll back the boundaries of the welfare state. They also thought that over the years, the exercise of political power to control and allocate state resources had created imbalances, causing poverty and bad governance. For them and the academics who supported their ideology, the term 'government' was seen as too restrictive. Bureaucracy was also seen as tardy, inefficient, and unresponsive as well as suffering from systematic rigidities, needless complexity, and over-centralization. Thus, new emphasis was given to the term 'governance', which essentially meant a minimal state with emphasis on less government through privatization of government operations, ensuring de-bureaucratization, treating citizens as clients, and using private-sector techniques for achieving results (Dwivedi and Gow 1999). Soon, the ideas imported from business management dominated the governmental reform policy agenda of OECD countries.

Governance is essentially the exercise of economic, political, and administrative power and authority to manage public affairs at all levels of government, and in that exercise of power, it freely permits citizens and various groups to articulate their interests, to exercise their legal rights, to meet their obligations, and to settle their differences in a manner that ensures sustenance and improvement in the quality of their lives. A broader interpretation for the term governance was subsequently put forward: 'governance includes a full range of activities involving all stakeholders in a country such as: all governmental institutions (legislative, executive and administrative, judicial, para-statals [public enterprises or government corporations]), political parties, interest groups, non-governmental organizations (NGOs) (including civil society), private sector, and the public at large' (Frederickson 1997, 86). Hyden and Court (2002, 19) provide a sharper definition of the concept of governance: 'Governance refers to the formation and stewardship of the formal and informal rules that regulate the public realm, the arena in which state as well as economic and societal actors interact to make decisions.' In essence, the term 'governance' implies a high complexity of activities, pluralistic in nature, inclusive in decision-making, set in a multi-institutional organizational context, empowering the weaker sections of a society, and geared to achieving the generally accepted common good. On the other hand, governance requires the appropriate and ethical use of power and authority, because poor governance leads to unsustainable development and misery all round.

At the same time, the era of governance also ushered in the need for legitimacy, transparency, accountability, and morality (Dwivedi 2002).

For example, Sen (1999, 10) suggests the following five attributes as crucial for nurturing good governance: political freedoms, economic amenities, social opportunities, transparency guarantees, and protective security. To these one may also include the sharing of power between the public and private sectors, with civil society, and with political and religious as well as social organizations. It is a kind of governance in which societal needs and issues are not left entirely with the state machinery; rather, there is explicit collaboration to jointly manage such issues.

Nevertheless, when it comes to establishing a good governance regime, certain elements as suggested by Hyden (1992, 7) are highly desirable: degree of *trust*, *reciprocity* of relationship between government and civil society, degree of *accountability*, and *legitimacy* derived from the nature of authority wielded.

1. *Trust* is essential among governing leaders and elites about the nature and purposes of the state, including rules and practices of socio-political behaviour. Without trust in the political system, individuals and other interest groups will have no reason to engage in active political life. Public trust also helps to create an environment in which stakeholders are able to interact across the public, private, and community sectors to form alliances and seek change in the governing process.
2. *Reciprocity* is necessary within a civil society because it permits associations, political parties, and other interest groups to promote their interests through competition, negotiations, and conflict resolution.
3. *Accountability* forces those governing to be held accountable and to act transparently through institutionalized processes (such as fair elections, public oversight, and referenda).
4. The *legitimacy* of political leaders depends on how they use the *power* and *authority* to

create policy and to implement programs by creating conditions in the polity in which the first three criteria are sustained. Public confidence and trust in the process of governance is maintained only when it demonstrates a higher moral tone drawing on spirituality of action and, most importantly, when it tries to sustain the public good.

Good governance is easier to define than to operationalize. Of course, the era of governance has intertwined politics and state administration so closely that the insidious aspects of politics have often overwhelmed the administrative culture of bureaucracy by requiring officials to justify ends by any means. The sponsorship scandal during the Jean Chrétien regime illustrates this phenomenon.

Good Governance as a Foundation for Operationalizing the Integration of Human and Ecosystem Well-being

People normally agree that good governance is essential for the well-being of all. Hyden (1992), UNDP (1997), Dwivedi (2002), and Dwivedi and Mishra (2007) have all argued that good governance is an instrument by which people and societies can realize their needs and aspirations. How can that be possible? Although there is no general agreement among nations and cultures as to what constitutes good governance because of differences in cultural norms, style of doing things, expectations of people from their governments, and the quality of environment surrounding them, they all usually agree that without good governance, effective, efficient, and equitable delivery of public services is not possible.

The term 'good' is value-laden and involves a comparison between two things or systems against some standard of measure; a system of governance is considered good if it exhibits certain fundamental characteristics. Perhaps the United Nations Development Programme (UNDP) offers the most comprehensive definition and an idealistic model. Good governance is, among other things, participatory, transparent, and accountable. It is also effective and equitable. And it promotes the rule of law. Good governance also ensures that political, social, and economic priorities are based on broad consensus in society and that the voices of the poorest and the vulnerable are heard in decision-making regarding the allocation of development resources (UNDP 1997).

From the above, the following characteristics of good governance can be suggested:

1. *strategic vision* by the leaders toward broad-range long-term perspectives on sustainable human development;
2. *legitimacy* of the governing process, which allows those who govern to derive authority and power from legitimate constitutional instruments of governance;
3. *rule of law* that is enforced impartially;
4. *effective and efficient responsibility and accountability* of institutions and the statecraft that meet the basic needs of all by using state-controlled resources to their optimum accountability;
5. *transparency* for access to governing process (including institutions and information sources);
6. *equity* assured to all individuals so that they have the opportunity to improve their well-being;
7. *responsiveness* of institutions to the needs of all stakeholders;
8. *consensus* among different and differing interests in the society;
9. *public participation* in decision-making; and

10. *stewardship of governance* in which governing leaders and elites dedicate their lives to service to the public and in which amoralism does not reign supreme.

Good governance and sustainable human development, especially for developing nations, also requires conscientious attempts to eliminate poverty, sustain livelihoods, fulfill basic needs, and design an administrative system that is clean and open (Dwivedi 2008, 67).

RESOURCE SCARCITY AND HUMAN SECURITY

Environmental insecurity is caused by resource scarcity. Homer-Dixon (1994) identifies three ways that humans cause a scarcity of renewable resources. The first is a reduction in the quality and quantity of renewable resources at higher rates, undercutting natural renewal (supply-induced scarcity). The second is increased population growth or per capita consumption (demand-induced scarcity), and the third is unequal resource access (structural scarcity). Resource scarcity can be experienced as a result of climatic changes, decreased agricultural production, decreased economic productivity, population displacement, disrupted institutions, and social tensions. Given the relationship between conflict and resource scarcity, it is clear that environmental security is an important feature of current social, economic, and political trends.

Nef (2008) argues that the environment is the most important national security issue of the early twenty-first century, pointing to the political and strategic impact of surging populations, the spread of disease, deforestation and soil erosion, water depletion, air pollution, and, possibly, rising sea levels that could lead to mass migrations and increasing group conflicts. Water, he argues, will be in dangerously short supply in areas such as the Middle East, central Asia, and the southwestern United States. Water scarcity could cause wars between nations not only over access to fresh water but also over the damming of different parts of rivers (as evident in the continuing conflicts between India and Pakistan, and Bangladesh and India). On the other hand, because of the melting of Antarctica and northern glaciers, people living in low-lying areas may be forced to move when sea levels rise.

Let us consider the example of water as one of the most scarce resources. Although most of our Earth's surface is covered by water, only 3 per cent can be used as fresh water. And from this small proportion, the needs of households, industry, irrigation, and the environment must be met. As the population increases (it increased by three billion people during the past 30 years and is expected to grow by another three billion during the next 50 years), all these people will need water for drinking and household chores, greater food production, and industrial development. Water scarcity, coupled with poverty, will not only increase the incidence of malnutrition, general diseases, and starvation but could lead to major conflicts and wars, particularly when rivers and lakes cross national boundaries. That is why both the scarcity and abundance of natural resources (for example, to many, countries like Brazil, Russia, and Canada appear to be blessed by an abundance of natural resources such as water, forests, and minerals) could be possible causes of violent conflict on our planet as resource-scarce nations look for ways to deal with such calamities as tsunamis, earthquakes, avalanches, floods, and droughts. All such perils cause loss of life, disease, and disruptions in social and economic structures and also divert precious government resources from normal nation-building activities. Conflict-oriented disruptions include problems such as food crops destroyed as a war tactic and landmines hidden in the fields and forests on which people depend for their livelihoods.

Understanding environmental security within the context of human security moves the idea of environmental security from a state-centred perspective to the environmental security of individual humans. Forging a new understanding of environmental security through a human security lens brings people into international discussions and highlights issues and concerns about the safety and environmental security of people, not just nation-states. The role of the state is still central, but the emphasis shifts from protecting the state from environmental insecurities to protecting citizens. The role of the state then becomes geared to ensuring that people have the capacity and opportunities to access resources and to manage or control these resources in a sustainable way. Most importantly, it means that civil society becomes involved in the decision-making process about how resources are distributed, protected, regenerated, and controlled.

Environmental Security as a Human Security Concern

Nef (2005), writing about human security concerns, argues that a policy shift occurred regarding environmental security because of several factors: the end of the Cold War, which opened new conceptual vistas and paradigms; highly publicized disasters, such as the *Exxon Valdez* oil spill off the coast of Alaska; and the realization that resource scarcity would damage even advanced industrial economies. These events coincided with a rise in concern about the global nature of ecological issues, moving beyond transborder pollution and resource-sharing questions and into the even more complex realm of ozone-layer depletion and global warming.

The importance of non-state actors, both profit-oriented and issue-related, also became increasingly obvious, another radical departure from traditional security based on the territorial integrity of a state. For example, multinational corporations fear the expropriation of their property and 'unfair' taxation policies, which is one reason that they pushed, mostly through the World Trade Organization (WTO), for a Multilateral Agreement on Investment (MAI). NGOs, meanwhile, might define security as protection from arbitrary state arrest or torture, freedom from coerced gender roles, or elimination of landmines.

The idea of directly linking the environment to security concerns has also been articulated by Gleick (1991), who identified primary environmental threats to security. Natural resource acquisitions can be strategic goals in themselves, often pursued as part of military strategy. Resources can be utilized as military tools ('eco-sabotage' is gaining currency as a concept, but it also has the sinister implication of using food and water as weapons). Finally, various disruptions to environmental services, such as water supply, are obvious threats to the well-being of citizens. According to this perspective, it is necessary to view environmental threats within their proper context—as challenges to the national interest—but they can also be seen as broader threats to the interests of humanity itself.

ENVIRONMENTAL RELATIONS: OBLIGATORY DIALOGUE BETWEEN NORTH AND SOUTH

The end of the Cold War has given humanity an opportunity for a fundamental rethinking about the nature of relations among nations. That rethinking ought to be directed toward the planetary-proportion threats of continuing poverty, ecological disasters, and global warming. In our world, poverty prevents one-fourth of humanity from achieving even its basic needs (adequate food, safe and sufficient water, primary health care, education, shelter). As long as economic disparity remains, stress on the environment will continue; for the poor, the struggle for survival overrides concern for the environment. At the same time, although for different reasons, both North and South continue to exhaust Earth's resources. That is why environmental concern ought to be acknowledged as a fundamental point of agreement in their relationship.

A global environmentally sustainable development policy requires the consideration of all and concern for all. Until now, we have had a sharp discrepancy between rich and poor, between developing and affluent, between North and South, and between First and Third Worlds; this gap is not closing, despite rhetoric to the contrary. It is based on the differences in economic progress and material well-being between people living in two separate worlds, isolated from each other not only because of geographical boundaries, trade barriers, and economic prosperity but also because of the political dominance of one over the other. However, various aspects of the environmental crisis, such as ozone depletion, the greenhouse effect, disappearing biodiversity, and need to protect world commons, do not respect such barriers and borders. We know that ecological disasters have ways of affecting even those who may not live in close proximity to them. Thus, old distinctions that marginalized the poor and the undeveloped, while keeping the wealthy secured and privileged, are outmoded; both groups have to share the same planet and the world's dwindling resources. Although realization of this fact is perhaps much more apparent in the poorer sections of our global society than in the richer parts, a similar sensitivity is arising among the people of the North to the idea that no one group can live in isolation from the other.

For the South, poverty alleviation is the most crucial task, while for the North, ozone depletion, deforestation, and population growth in the South seem more important global issues. This difference in perspective was first highlighted by Indira Gandhi, prime minister of India, at the 1972 UN Conference on Human Environment in Stockholm: 'On the one hand the rich look askance at our continuing poverty—on the other, they warn us against their own methods. We do not wish to impoverish the environment any further and yet we cannot for a moment forget the grim poverty of large numbers of people. Are not poverty and needs the greatest polluters?' (Gandhi 1972, 2).

The World Bank has repeatedly reported that millions of people in developing nations are still struggling with poverty, hunger, squalor, and poor health care. To such people, environmental problems are actually afflictions of overdevelopment in the industrialized North, although they are now convinced that environmental protection itself is also a threat. This perspective gives them a different view about population growth and the perpetuation of poverty. The following discussion explores these two aspects, population growth and poverty, and their impact on the environment.

Population Growth and Environmental Degradation

Demography—the size of the population and its rate of growth—is a principal factor in environmental degradation, since the faster the population increases, the greater the depletion of natural resources, thus causing environmental degradation. But a reduction in population increase, if that is the answer, is difficult to achieve, even if some authoritarian measures are adopted, as happened in India during the regime of Indira Gandhi between 1975 and 1977.

Demographers have estimated that even in the unlikely event that we manage to immediately reduce world fertility to a level of simple replacement (of the order of two children per couple) and maintain this level in the future, world population, because of the present age distribution and expected increase in life expectancy, would continue to grow for approximately a century before levelling off at a number almost twice the present figure (Cabre 1993, 2).

However, predictions such as this have not always proven accurate. In India, for example, it was argued in the 1970s and 1980s that its population would grow beyond the nation's carrying capacity (Commission 1992, 33). Yet by the 1990s, India had enough food surplus to export. Through better land use, soil management, increased levels of agricultural inputs, and improvements in seed quality, nations such as India could have the capacity to feed many more millions of people. Thus, population increase is not always the main culprit in environmental damage.

Instead, the acquisitive materialism and wasteful ways in the North might be a more significant cause of environmental problems. Cabre (1993, 2) gives one glaring example from the American past: during a single decade, 1870 to 1880, the skeletons of 750,000 bison were loaded at one particular railway station in the US West, bound for fertilizer factories on the east coast. But beyond population growth and unsustainable lifestyles as key factors in environmental degradation, we should look for another culprit: poverty.

Poverty and Pollution

For hundreds of millions of people, life remains a constant struggle for daily survival. For them, nature and its resources are the only means by which they can survive, and only when poverty is eradicated and living conditions become tolerable can they think about environmental protection. As Shridath Ramphal (1992, 17),

former secretary general of the Commonwealth of Nations and a member of the Brundtland Commission noted: 'If poverty is not tackled, it will be extremely difficult to achieve agreement on solutions to major environmental problems. Mass poverty, in itself unacceptable and unnecessary, both adds to and is made worse by environmental stress. . . . That is why the global policy dialogue must integrate environment and development.'

At the World Summit for Sustainable Development in 2002 in Johannesburg, the issue of poverty and the environment was on agenda. However, the summit did not become a turning point, because the global gathering of leaders could not respond seriously to: (1) the injustice of continuing poverty and poor development in most of the world; (2) prevailing disparity between the rich (who have relatively easy access to resources) and what is left for the poor; (3) the need to commit specific funds to help improve the lives of the marginalized people on the planet; and (4) a continuing assault on the ecological well-being of the planet.

It was clear that rich nations were not about to make commitments that would undermine their position of privilege and respond with urgency to global ecological threats. And it was also clear that multinational corporations, which benefit most from the current economic model of governance, would not let their patrons—rich countries—barter away their global power and influence. The Johannesburg World Summit was intended to create 'a new ethic of global sustainability, a new common plan of action, and to agree on developing a greater sense of mutual care and responsibility' (Dwivedi and Khator 2006, 131). However, no global agenda for poverty alleviation emerged to lead to a global strategy for sustainable well-being. Consequently, pressure on natural resources continues not in the South alone but in both North and South because of people's consumptive behaviour. It is no surprise, then, that a poor nation would probably emphasize poverty alleviation and providing basic needs with appropriate financial assistance above all and only after that turn attention to protection of the environment.

THE WORLD IS STILL OCEANS APART!

What are the specific challenges that confront our planet as we face the twenty-first century? This section includes observations on a wide range of issues affecting all of humanity, around the globe.

1. *Limiting ecological footprints:* Sustainability requires more human action to limit ecological footprints rather than relying on technical solutions that might simply reduce some specific harm to the environment. While technology has been and should prove to be helpful, the main culprit in environmental degradation is humanity's unquenchable appetite for material goods and demand for related services.

   As global citizens, we have a shared responsibility for the common good, because living on the same planet, we share a common destiny and therefore we must pay attention to the ethics of ecological sustainability. Ever-increasing demands for material goods and the wasteful ways of consumption in the West are being copied in the rest of the world: what would be the result if nations such as China, India, and the countries of South America and Africa were to modernize and their middle classes were to pursue similar consumptive behaviour patterns? Unless the rising consumption can be curbed—and first within the wealthy nations—the sustainability of our planet will remain in jeopardy. The desire to accumulate more must give way to desiring less. This will require a change in lifestyle, because the grim reality in the twenty-first century so far is that the globalization of American consumerism is leading us toward an unsustainable future and ecological catastrophe.

2. *Planetary survival through ecological diversity:* In March 2006, the United Nations released a report, Global Biodiversity Outlook 2, that painted a grim picture of life on Earth, with a current rate of extinction 1000 times greater than in the past (UNEP 2006). That spiralling rate of biodiversity loss is mainly due to the global demand for biological resources, which exceeds the planet's capacity to renew them by 20 per cent, the introduction of invasive alien species (including bio-seeds), nutrient loading, and climate change. There is an urgent need to take unprecedented efforts worldwide. Diversity is directly linked to adaptability and survivability of species. For example, people migrating from one geographic system were able to adapt to another totally different kind of environment, and this has been happening since the beginning of human history. The new environment in which those migrant people sought stability and prosperity was born out of diversity. It is the diversity that impels us to adapt and seek benefit from new surroundings. On the other hand, uniformity creates dependency, inflexibility, and inability to adapt to new and challenging situations. Human ingenuity is based on such challenges: in the absence of such challenges, creativity, genius, and immunity fade away. The strongest societies are the most diverse, and the same is true of ecosystems.

3. *A futuristic model of sustainable development:* Sustainable development cannot be perpetual growth if one considers that world resources

are limited and can be exhausted. Yet it cannot be 'zero growth' either. So we have to ask: what kind of purposeful growth (even negative growth) should we plan for the twenty-first century, and can such growth be accommodated within the Earth's existing resources and the space required for managing waste created in its wake?

Writing a 30-year update of their book *Limits to Growth* (published in 1972), which forced the world to consider the wasteful ways of consumerism and materialism and highlighted a frightening scenario of impending ecological catastrophe, the three original authors revised their assessment. They suggested in 2004 a model of a sustainable society: '(a) A sustainable society would not lock the poor permanently in their poverty; (b) a sustainable state would not be a society of despondency and stagnancy, unemployment and bankruptcy that current economic systems experience when their growth is interrupted; (c) a sustainable world would not and could not be a rigid one, with population or production or anything else held pathologically constant; (d) a sustainable world would need rules, laws, standards, boundaries, social agreements, and social constraints. . . . [Those] roles for sustainability, like every workable social rule, would be put into place not to destroy freedoms, but to create freedoms or to protect them; and (e) there is no reason for a sustainable society to be uniform. As in nature, diversity in a human society would be both a cause of and a result of sustainability. . . . Cultural variety, autonomy, freedom, and self-determination could be greater, not less, in such a world.' (Meadows, Randers, and Meadows 2004, 255).

Meadows, Randers, and Meadows also argued that in order for such a model to function, people would have to control their

'unquenchable' appetite for material things by finding non-material ways to satisfy them. This will be a real challenge for the generations living in the twenty-first century: 'not only to bring their ecological footprint below the earth's limits, but to do so while restructuring their inner and outer worlds' (Meadows, Randers, and Meadows 2004, 263).

4. *The impact of globalization on sustainability of our global village:* Globally, about 1.1 billion people lack access to safe drinking water, while 2.6 billion people lack access to safe sanitation. One consequence is that there are 'about 4 billion cases of diarrhoea per year, which cause 1.8 million deaths, mostly among children under five' (Trace 2005, 12). Sanitation, as an example, is too often a forgotten problem globally. Of course, globalization offers great opportunities, but at the same time, we should be aware of its negative, disruptive, and marginalizing effects.

Globalization appears to have divided the world between the connected minority, which has a monopoly on almost everything, and the majority on the fringes, who are isolated and have practically nothing. Globalization seems to have created an international order in which a group of privileged people controls virtually all political and economic power, while the rest remain poor and marginalized.

Would it not be fair and just for the global actors of international development aid to show more sensitivity, vision, and the right kind of leadership? Perhaps globalization ought to become a process whereby all the citizens of this global village feel they are on equal terms in sharing the prosperity, the natural resources, as well as the liabilities confronting everyone. Without such a collaborative and cooperative partnership between North and South, will the twenty-first century be any different from what humanity experienced during previous centuries?

5. *Is globalization the answer?* Until relatively recently, praise for globalization was almost universal. For example, Thomas Friedman (1999, 8–9) argued in his book *The Lexus and the Olive* that countries that tried to avoid globalization became poorer. He pointed out that whereas in 1975, only 8 per cent of countries had open markets, amounting to $23 billion in direct foreign investment, the proportion had grown to 28 per cent by 1997, with $644 billion in foreign investment; thus, he concluded, nations that hesitate to adopt or that avoid globalization are doomed to poverty. Similarly, the World Bank asserted that since 1980, globalizers enjoyed growth rates three times higher than those of non-globalizers (World Bank 2000). The bank further reported that with globalization, countries were compelled to improve public services, provide better social security nets, and even enhance environmental protection. Ignored by most was the World Bank's observation that the average income in the 20 richest countries was 37 times higher than the average income in the poorest 20—a gap that had nearly doubled over the previous 20 years (World Bank 2000, 3).

Despite the developed world's stated commitment to globalization and free markets, the trade barriers imposed by the rich nations remain an impediment to poverty alleviation. In 2003, *The New York Times* reported that the industrialized nations provided $320 billion in subsidies to their farmers compared to $50 billion in international aid (*The New York Times* 2003). For example, the United States government gave a $3-billion subsidy to its 25,000 cotton farmers in 2004, thereby distorting world cotton prices (Werlin 2004, 1034). If farm subsidies were reduced, about 114 million people in poor countries could be pulled out of poverty because their products could be appropriately priced and exported to the countries of the North.

Moreover, despite globalization and a growth in national income, poverty remains substantive. For example, in 2000, Mexico, after joining NAFTA (the North American Free Trade Agreement), had a $20 billion trade surplus with the US. Yet, 42 per cent of the Mexican population was reported to be living below the poverty line, an increase of nearly 75 per cent since 1989 (Werlin 2004, 1035). Clearly, globalization has not contributed to a reduction in poverty.

6. *Protect both human-created and nature-gifted biosphere diversity:* Globalization, as noted above, is creating a trend toward the homogenization of many activities, such as management of business (including interfering with the management of state machinery), trade and commerce, education and research, communications, and technology transfer. This trend brings, on the one hand, uniformity of context, perspective, and style of doing things. On the other hand, it occurs at the expense of the world's diversity, not only of thought and action but also of species and grain stocks.

This trend toward a monoculture of thoughts and actions may also weaken nations less commercially powerful. As more and more genetically modified organisms (GMOs), biotechnology, and biogenetics are pushed by the multinational corporations of the West on developing nations, indigenous variety and diversity is going to lose out. Sustainable development means protecting the world's creative and indigenous variety and diversity by developing socially just and ecologically sustainable means for the conservation and use of biodiversity through imaginative environmental governance (UNESCO 2002).

7. *Sustaining human development—by balancing Western lifestyles and the needs of the South:* Progress in human development has been extraordinary during the past 50 years because on average, people in developing nations are healthier, wealthier, better fed, and more literate. And this has all happened during my lifetime. Life expectancy has risen, great advances have been made in primary education, and food sufficiency has been achieved in several countries.

Nevertheless, wide disparities remain. For example, the amount spent by Europeans on mineral water in one year is enough to provide primary education in developing countries for the next 10 years. One billion people worldwide still cannot read and write, two-thirds of whom are women. In terms of wealth, the income gap between the top 20 per cent of nations and the bottom 20 per cent rose from a proportion of 30:1 in 1960 to 78:1 in 1994 (Dwivedi and Khator 2006). Thus, human development has not kept an even pace.

Sen (1999) has defined human development as the fundamental freedom that enhances people's choices and thus raises their level of well-being. From this perspective, substantive freedom includes the capacity to avoid deprivations such as starvation, under-nourishment, or premature mortality. It also includes acquiring sufficient basic education and skills to be gainfully employed, as well as the freedom to participate in political, economic, and social systems. It means building up capacity for people to make their own decisions.

At the same time, it must be noted that human development and economic growth are mutually reinforcing: for development to be sustainable, both aspects should accelerate in tandem. This requires that (1) citizens receive basic services (such as education, primary health care, an adequate supply of food, clean water, and sanitation); (2) people participate in the implementation as well as in the design of developmental programs created in their name if the resources are to benefit the most needy (especially women and other marginalized persons); (3) there is recognition of the need for greater cooperation between all sectors of human development, including spiritual well-being; and (4) nations become proactive in the global economy, because globalization disproportionately favours those who have expertise, power, and the capacity to compete in the global market (Kagia 2002).

8. *Has sustainable development as a concept lost its shine?* Sustainable development has come under criticism from many quarters. The main criticism is its alleged vagueness—that the concept means everything to everyone because it connects everything and becomes what any group of people wishes it to be. From multinational corporations, chemical companies, nuclear power generating plants, pesticide manufacturers, and loggers to mining and aggregate extraction operators, everyone uses the term to suit their needs and promote their products.

To critics, therefore, the concept 'has become a cover for inaction and a black hole for resources' (Victor 2006, 92). Moreover, there is a continuing and growing schism between the industrialized nations and developing countries with respect to the duty of all states to protect the environment versus the claim of poorer nations that they are sovereign when it comes to exploiting their own resources as they see fit.

But there are the three pillars of sustainability about which all agree: poverty alleviation, social justice (including the well-being of all), and environmental protection. Thus, despite the claim that the concept has become meaningless, fallen prey to special interest groups, and promoted false universal goals (Victor 2006, 103), it remains the only concept that has brought the entire international community together and still carries worldwide relevance and appeal.

9. *Needed: an Earth Charter to protect the environment:* By the time we entered the third millennium, it was increasingly clear that many of our values were totally inadequate for long-term survival and the development of a sustainable future in which all species (not only human beings) would be taken care of. This was evident from the emergence of a wide spectrum of challenges to the traditional materialistic view.

For centuries, guided by Western culture, people placed their faith in the power of science and technology to bring about material progress. Only recently have we come to understand that material prosperity should not be an end in itself. Slowly, a realization is also emerging that spirituality and control of one's desires can create more lasting happiness than acquisitive materialism can.

However, such a realization has yet to enter the domain of governmental policy or the corporate world, where spiritual perspectives are generally ignored. The economic criteria that place no value on the commons (the air, water, oceans—even outer space) and that use concepts such as cost-benefit analysis, law of supply and demand, rate of return, and land as commodity have been based on the delusion that humanity operates independently of the cultural and spiritual domain.

Until now, we have taken a great deal from our Mother Earth. We have given little thought to putting limits on our plundering and ravaging instincts. Without a change in

our current value system, there is little hope of correcting the present environmental problems. Slowly, however, a heightened consciousness is emerging for the formation of a new international environmental rights regime, including instilling respect for Mother Earth and care for all species. This was highlighted by the World Commission on Environment and Development, which stated in its report, *Our Common Future,* 'The Earth is one but the world is not. We all depend on one biosphere for sustaining our lives. Yet each community, each country, strives for survival and prosperity with little regard for its impact on others' (WCED 1987, 27). With this admonition as inspiration, initiatives were undertaken to draft an Earth Charter as an 'ark of hope' toward a peaceful and sustainable world.

An Earth Charter is an appropriate instrument for helping to empower that consciousness. Clearly, without a universally accepted regime for environmental protection, world resources will continue to be depleted, and the quality of life for the majority of people on Earth will remain poor. For this reason, we urgently need a holistic vision of basic ethical principles supported by broadly accepted tenets and practical guidelines to govern the behaviour of people and states in their relations with each other and the Earth. This is not a new view; there have been persistent calls for such a charter from various environmental NGOs, international organizations, and reports such as *Our Common Future* by the World Commission on Environment and Development and *Caring for the Earth* by the International Union for the Conservation of Nature and Natural Resources (IUCN). Moreover, 'the Earth Charter is a reminder of our moral duty to leave a healthy legacy for future generations by not only protecting the environment from the harmful ways of our activities, but also by attempting to restore the status quo of two generations ago' (Dwivedi and Khator 2007, 1030).

The Earth Charter, in brief, consists of 16 principles:

i. respect Earth and life in all its diversity;

ii. care for the community of life with understanding, compassion, and love;

iii. build democratic societies that are just, participatory, sustainable, and peaceful;

iv. secure Earth's bounty for present and future generations;

v. protect and restore the integrity of Earth's ecological systems, with special concerns for biological diversity and the natural processes that sustain life;

vi. prevent harm as the best method for environmental protection and, when knowledge is limited, apply a precautionary approach;

vii. adopt patterns of production, consumption, and reproduction that safeguard Earth's regenerative capacities, human rights, and community well-being;

viii. advance the study of ecological sustainability and promote exchange and wide application of the knowledge required;

ix. eradicate poverty as an ethical, social, and environmental imperative;

x. ensure that economic activities and institutions at all levels promote human development in an equitable and sustainable manner;

xi. affirm gender equality and equity as prerequisites to sustainable development and ensure universal access to education, health care, and economic opportunity;

xii. uphold the right of all, without discrimination, to a natural and social environment supportive of human dignity, bodily health, and spiritual well-being, with special attention to the rights of indigenous people and minorities;

xiii. strengthen democratic institutions at all levels, and provide transparency and accountability in governance, inclusive participation in decision-making, and access to justice;

xiv. integrate into formal education and life-long learning the knowledge, values, and skills needed for a sustainable way of life;

xv. treat all living beings with respect and consideration; and

xvi. promote a culture of tolerance, non-violence, and peace (Earth Charter 2009).

10. *Politics is extremely important:* It is not an exaggeration to say that solutions to all the major problems lie ultimately in the realm of politics. Despite the fact that some of these problems may be largely scientific in nature, others 'economic', and still others 'philosophic' or 'moral', in the end all require political decisions to deal with them and to specify the type and timing of solutions. The reason it took a long time for the environment to receive the public attention and resources it deserves is that it had to compete with other issues of public concern, which confirms the value of the 'issue-attention cycle' described in the introduction to this book.

11. *Mutual vulnerability in the global village:* It is now generally acknowledged that because we live in a global compact, we are mutually interdependent. While one sector might appear more 'secure' than others, whatever happens to another sector, no matter how seemingly insignificant, often has the potential for catastrophic effects on the entire system, including those deemed less vulnerable.

Consider 11 September 2001 as a case in point. Events in distant lands brought about the terrible incident in the US, and from there it affected the entire world. We are just beginning to understand this vulnerability with regard to terrorism; however, it applies in the environmental, economic, social, and cultural spheres as well. The absence of good governance and leaders in many nations has bred insecurity, rebellion, deprivation, corruption, and poverty. Although a kind of global integration in the economic and commercial sector has emerged, it has turned out to be unbalanced, with uneven participation on the part of poorer countries and their people. Before unbalanced growth leads to further volatile situations and hostilities, greater income disparities, higher incidence of poverty, and civil strife, good leaders will be needed to set the system right.

These leaders will have to be trained, especially in democratic governance. And it is here that the West could contribute something concrete. The survivability and sustainability of our global village means interdependence, and interdependence requires international cooperation. Thus, it is in the best interest of everyone that effective democratic institutions are nurtured by strengthening the democratic fabric worldwide. For this to happen, a reservoir of good leaders is needed.

12. *Breaking boundaries: building bridges of collaboration between the North and the South:* In 2004, the United States allocated $36 billion for homeland security and about $360 billion for the military; contrast this with the $11.3 billion it expends on international aid each year (Brusasco-Mackenzie 2004, 12). Billions more have been spent by the 'coalition of the willing' in the Iraq war. On the other hand, only $54 billion to $64 billion

would be needed annually to cut world poverty in half by 2015 (Devrajan, Miller, and Swenson 2002).

Clearly, there is a need to change the militaristic definition of security to the broader concept of the 'well-being of all'. The world population currently stands at about 6.5 billion, and it could rise to about 8.9 billion by 2050 (Worldwatch Institute 2004). If 831 million people across the world remained hungry and malnourished in 2003, what will happen later on when the expected further loss of croplands as a result of degradation, soil erosion, and climatic changes occurs during the first half of the twenty-first century? As discussed earlier, the world now confronts a problem of water scarcity, and with increased competition between urban and agricultural water needs, there is a serious risk that sufficient water might not be available to serve the two sectors. From the Murray-Darling River basin, to Ganges water, to the Nile River Basin, to the Red River Delta, water insecurity is causing national nightmares. For the people affected and for their well-being, instead of nations spending billions on military ventures, a new model of development is needed, one that could ensure human security and planetary well-being. It is also clear that while the North has been able to achieve a higher level of environmental protection than the South, the nations of the North have also missed the opportunity to promote sustainability and equity since the Rio Summit of 1992. As a result, there is an urgent need to build bridges of collaboration between North and South, specifically by working toward achieving the UN Millennium Development Goals (MDGs).

13. *The effects of a financial crisis of planetary proportions:* It is hard to believe that the US

economic meltdown of late 2008, which humbled Wall Street and Canary Wharf, could create a crisis in virtually all of the countries on this planet. For developing countries such as Bangladesh, Nigeria, Mexico, and Central American nations, there came in the wake of the crisis in the West a plunge in exchange rates and an astronomical rise in commodity prices that led to a local credit squeeze, loss of jobs, and stalled entrepreneurial growth.

The downturn was the first major financial crisis to have a direct and immediate impact on the very poor of the world, with loss of industrial orders, elimination of jobs, and decline in the availability of funding for developmental works. The meltdown not only killed jobs, savings, property values, and business in the wealthy West, it also affected the people of developing nations even more dramatically. Remittances of earnings by foreign or 'guest' workers, the largest single source of economic growth in many countries, shrank to a minimum, severely hurting the economies of poor regions. For example, in Bangladesh, remittances amounted to 11 per cent of that country's GDP in 2007 (Saunders 2008).

Exchange rate changes, decline in direct foreign investment flows, slow demand for garment and electronic products, rise in the cost of living, and overall increase in unemployment—all of these factors are side-effects of globalization. While the beneficial effects of globalization created economic growth for many poor nations, when the economic downturn came, global interconnection also led to deep webs of unemployment and financial dependency. The crisis that started in the US moved around the entire world, affecting issues such as sustaining natural resources and protecting the environment.

As far as Canada was concerned, there was more dismal or distressing news for the country toward the end of 2008 than there had been during most years in the immediate past. From sustained budget surpluses, the country faced a deficit, and the green awakening championed by the Leader of the Opposition, Stéphane Dion, stalled when his Liberal party lost to the Conservatives in the federal election held in the midst of the market meltdown. As always happens when the economy declines, environment is an early casualty. Thus, by the end of 2008, few senior government officials were talking about global warming and green issues.

14. *Unprecedented heat to trigger a global food crisis:* Global warming's effect on agriculture is likely to be its greatest threat to humanity, greater than the submerging of coastal areas as a result of the melting of ice sheets in the Arctic and Antarctica (see also Chapter 9). *The Globe and Mail* reported a study conducted by Battisti and Naylor revealing that by the middle of the twenty-first century, extreme heat will wither crops that are heat-sensitive, such as wheat, rice, and corn, and will have dire consequences for other food crops, particularly in the tropics and subtropics where people are already malnourished (Mittelstaedt 2009). Although warmer temperatures may extend the growing season in Canada and other northern countries such as Russia, incremental crop production would not balance the deficit in the rest of the world. In addition to an impact on cereals, other natural resources such as forests and grazing areas will also be affected negatively. One way to prevent a global food crisis will be to breed heat-tolerant varieties, but most efforts should be directed at gaseous-emission control.

15. *Good governance is the key to sustainable development:* Although 'politics determines all', operationalization of politics in the form of good governance is the key to improving quality of life, securing human and environmental security, and attaining sustainable growth, thereby reducing poverty. The objective of good governance is to create conditions that will ensure that a society has access to nourishing food, clean water and sanitation, education, health care, security, and justice so that people may lead dignified and meaningful lives.

One of the keys to attaining good governance is the development of a trusted, trustworthy, and caring professional public service guided by honest political leaders. Nothing destroys the credibility of public institutions more than corruption in government, because it destroys confidence, undermines people's faith, diverts public resources to questionable ventures, mocks the application of laws and the judicial process, and casts deep shadows over the legitimacy of government policies and programs (Caiden and Dwivedi 2001, 245). Clearly, the prudent way forward for sustainable development and the well-being of all must be based on promoting good governance—as well as controlling corruption, which is the most insidious impediment to achieving good governance (Dwivedi and Mishra 2007).

16. *Think globally but act nationally:* While financial crises are unpredictable in timing and scope, this is not true of environmental crises (unless we are talking about natural disasters). International treaties, conventions, and agreements may appear strong on paper and at meetings where nations and governments are exhorted to undertake definite actions. However, the answers to environmental crises do not rest with international institutions but squarely with nations and their regulatory agencies. International agencies do not have any direct input into the policy of a nation; national governments, even after signing treaties and conventions, have often ignored them or diverted funds elsewhere in the name of national interest or exigencies. The Harper government in Canada, for example, shied away from implementing the Kyoto Protocol during the 2007–8 period. Even a global environmental regulator, unless it operates at the level of broad policy principles, will not be effective. Thus, the best regulation, implementation, and supervision and monitoring of any environmental issue (as recommended at the international level) will have to be at the national level. Preventing a crisis should be the mantra, which requires co-ordination among countries but not necessarily the intervention of any supranational agency. Finally, the prospects of a major policy shift and persuasion to bring about changes are greatest at the regional (such as under NAFTA), not the global, level.

17. *Emerging new world order and the environment:* It appears that a new world order may be in the offing. The Group of 20 (called G-20) Summit on 2 April 2009 decided to launch a new era of global governance by replacing the Washington Consensus and institutions created during the Bretton Woods summit of 1944 (the World Bank and the International Monetary Fund). The G-20 declared that these policies and institutions were dominated by the logic and structures of the West, overwhelmingly by the United States, and that such logic had defined the way the world had been financially managed (Reguly and Laghi 2009). Gradually, it seems that a new consciousness of

cooperation, collaboration, and awareness of mutual vulnerability is emerging, aimed at achieving a sustainable future for lasting peace and stability.

18. *Global governance for environmental protection:* If, as mentioned above, a new world order might emerge, what could it be called? Kofi Annan, the former secretary-general of the United Nations, explained it in the following manner: 'While the postwar multilateral system made it possible for the new globalization to emerge and flourish, globalization, in turn, has progressively rendered its designs antiquated. Simply put, our postwar institutions were built for an international world, but we now live in a global world' (Annan 2000). Living in a global world requires a global governing system that includes an interaction among such actors as nation-states, international governmental organizations (IGOs), multinational corporations (MNCs), and international non-governmental organizations (INGOs). That interaction would be based on a common regulatory regime that would enable these actors to operate in an orderly fashion and in accordance with the established norms of the international system (Halabi 2004, 21). The regulatory domain would include managing such issues as climate change, resource scarcity, and human well-being.

But could a global governing system succeed in saving the world from further environmental crisis and resource scarcity? Yes, it is possible, as demonstrated by various highly industrialized and urbanized countries as they successfully implement necessary pollution control policies; other countries could follow suit. However, without some form of global regulatory regime, most countries would not give the environment

the priority it deserves. That is why a mechanism would have to be established that could determine a universal environmental policy agenda, mediate among nations, and then broker deals among national governments, MNCs, IGOs, and INGOs at both the regional and global level with a view to building its own global regulatory policy. For this to happen, there is an urgent need to mobilize efforts toward establishing global governance by restructuring the existing international system.

CONCLUSION

In an ideal world, environmental and human security problems would already have been solved; the global population would have fallen to two billion; energy and natural resources would be evenly shared; technology would be harnessed to reduce pollution and resource depletion to the carrying capacity of our planet; and no one nation or group of nations would be dominant and privileged. But such an ideal is still a distant dream.

Instead, we should understand that the environmental crisis has given us a profound opportunity to restore and strengthen international cooperation for worldwide sustainable resources development. It is an opportunity for world leaders (from both the North and the South), as well as local people and their organizations, to develop planetary-level strategies of survival and sustainability. For such strategies to be realized, control on consumption levels in the North is required, as well as redressing inequitable and unsustainable growth in the South. While priorities concerning sustainable development and environmental security differ between North and South, they must reflect that there is only one global village that we all inhabit and therefore, no one community, nation, or region of this 'village' should enjoy first-class quality of life while the rest continues to suffer from poverty and underdevelopment.

Unless we reach a proper balance, this world can be neither sustainable nor developed. That is why people living in all parts of the world, if united in the enterprise of protecting our common heritage and if they can combine their concerns, talents, energies, and resources, can build a common future on Earth in which they share their fortune (whether meagre or in abundance) to equalize opportunities among all their neighbours and shared among all regions of our planet.

Globalization is criticized on many accounts, but in terms of environmental protection, it is a positive development, because in order to develop a common strategy for a sustainable future, much will depend on the way that people living 'oceans apart' can still perceive a common future for the world and act collectively to foster the convergence of one shared fundamental

FROM THE FIELD

The following three perspectives, based on my life experiences, have shaped my attitudes toward the environment, resulting in research and advice to provincial, federal, and international agencies:

1. In 1969, while visiting my hometown in India, I was astonished to see that the big pond behind our home was full of algae, leaving no place to frolic in the water and swim because factory waste was being dumped untreated into it. I further learned that despite public complaints, nothing was happening because of collusion between the factory owner and local government officials. Such official apathy, weak compliance to regulations, and the absence of eco-justice impelled me to think about environmental protection and green governance. On returning to Canada, I found that the federal government was discussing the Canada Water Act; this prompted me to start researching Canadian environmental policy and programs, resulting in my first book, *Protecting the Environment: Issues and Choices*, published in 1974.

2. In 1970, the International Joint Commission (IJC) published a report indicating that the lower Great Lakes, and especially Lake Erie, were heavily polluted. Lake Erie was said to be on the verge of 'dying'. I undertook a field trip around Lake Erie, noticing places near Sandusky, Cleveland, and Buffalo that were heavily polluted and had been declared not suitable even for swimming. The IJC report forced the Canadian and US governments to work together, and thus on 12 April 1972, the US president and Canadian prime minister signed the Great Lakes Water Quality Agreement. That agreement required the provincial governments of Ontario and Quebec, supported by federal funding, to develop and implement a comprehensive program for water quality control. Eventually, the water quality of our Great Lakes (particularly of Lake Erie) improved. Thus, my first Lake Erie field trip, and subsequent work on IJC and the federal–provincial environmental programs, strengthened the sustainability of my research on environmental policy, management, and compliance.

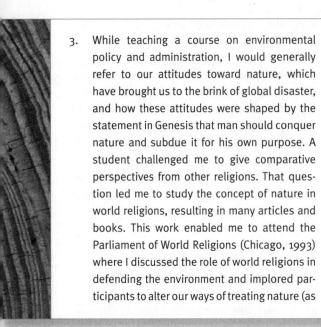

3. While teaching a course on environmental policy and administration, I would generally refer to our attitudes toward nature, which have brought us to the brink of global disaster, and how these attitudes were shaped by the statement in Genesis that man should conquer nature and subdue it for his own purpose. A student challenged me to give comparative perspectives from other religions. That question led me to study the concept of nature in world religions, resulting in many articles and books. This work enabled me to attend the Parliament of World Religions (Chicago, 1993) where I discussed the role of world religions in defending the environment and implored participants to alter our ways of treating nature (as it is no longer a 'free' commodity, to be squandered away) and to care for it. Subsequently, I took part in drafting an Earth Charter (from the Asian perspective). Obviously, we need to support and strengthen the Earth Charter with the goal of having it accepted as a UN document.

My basic philosophy: our planet is in trouble, and it will take an amazingly large number of people to strengthen the sustainable way of living. For this to happen, everyone (including policy-planners, environmental activists, and even spiritual leaders) must rally together to fulfill our sacred duty: serving humanity and caring for Creation.

O.P. DWIVEDI

REVIEW QUESTIONS

1. What do the two terms 'governance' and 'good governance' mean for you? Select a number of attributes of good governance, and try to see how they could be operationalized to enhance the quality of governance in Canada.

2. Globalization now affects all of humanity. In what ways can globalization become a process by which all the citizens of this global village feel that they are equal in sharing prosperity as well as the liabilities of managing resources and confronting environmental challenges?

3. Environmental security is intimately linked to human security, particularly when depletion of natural resources (causing deforestation, land degradation, water shortage, damage to oceans through the dumping of human and industrial waste and other contaminants, and climate changes) is linked to food insecurity, bringing about a deterioration of health standards and social cohesion, which often results in political turmoil that in turn can create not just national but also global political crises. Can you substantiate the link between environmental security and human security? What are the key elements of mutual vulnerability facing the world, and what can be done to resolve them?

4. The growing squalor of the many, which makes the prosperity of the few possible, has divided the world into the 'rich' and the 'poor', continuing a sharp discrepancy between the two groups concerning economic progress and environmental sustainability. For the poor in the South, alleviating poverty, eradicating hunger, and improving health, rather than environmental protection, are the most crucial tasks. How can this difference in perspective be overcome in order to reach global agreement on resource conservation and environmental protection?

5. Despite the fact that our world remains 'oceans apart', this chapter has suggested the need for a futuristic model of sustainable development that would balance needs and wants as well as realization that without cooperation and collaboration among world stakeholders (through such means as the Earth Charter) and without a planetary-level strategy for survival and sustainability, our future remains problematic. What are the main pros and cons in this statement?

REFERENCES

Annan, Kofi. 2000. 'We the people: The role of the United Nations in the 21st Century'. Speech introducing the Millennium Development Goals. New York: United Nations.

Brusasco-Mackenzie, M. 2004. 'Environmental security: A view from Europe'. Environmental Change and Security Project, 10: 12–18. Washington: Woodrow Wilson International Center for Scholars.

Cabre, A. 1993. 'Population growth and environmental degradation'. Barcelona: Centre UNESCO de Catalunya.

Caiden, G.E., and O.P. Dwivedi. 2001. 'Official ethics and corruption'. In G.E. Caiden, O.P. Dwivedi, and J.G. Jabbra, eds, *Where Corruption Lives*, 245–55. Bloomfield, CT: Kumarian Press.

Commission on Developing Countries and Global Change. 1992. *For Earth's Sake*. Ottawa: International Development Research Centre.

Devarajan, S., M.J. Miller, and E.V. Swanson. 2002. 'Goals for development: History, prospects and costs'. Washington: World Bank. http://econ.worldbank.org/files/13269_wps2819.pdf.

Dwivedi, O.P. 2002. 'On common good and good governance: An alternative approach'. In D. Olowu and S. Sako, eds, *Better Governance and Public Policy*, 35–51. Bloomfield, CT: Kumarian Press.

———. 2008. 'The well-being of nations: Reflections on integrating the human and ecosystem well-being'. In D.K. Vajpeyi and R. Khator, eds, *Globalization, Governance and Technology: Changes and Alternatives*, 60–75. New Delhi: Deep and Deep Publications.

Dwivedi, O.P., and J.I. Gow. 1999. *From Bureaucracy to Public Management: The Administrative Culture of the Federal Government of Canada*. Peterborough, ON: Broadview Press.

Dwivedi, O.P., and R. Khator. 2006. 'Sustaining the development: The road from Stockholm to Johannesburg'. In G.M. Mudacumura, D. Mebratu, and M. Shamsul Haque, eds, *Sustainable Development Policy and Administration*, 113–33. Boca Raton, FL: Taylor and Francis.

———. 2007. 'The Earth Charter: Towards a new global environmental ethic'. in A. Farazmand and J. Pinkowski, eds, *Handbook of Globalization, Governance, and Public Administration*, 1019–34. Boca Raton, FL: Taylor and Francis.

Dwivedi, O.P., and D.S. Mishra. 2007. 'Good governance: A model for India'. In A. Farazmand and J. Pinkowski, eds, *Handbook of Globalization, Governance, and Public Administration*, 701–41. Boca Raton, FL: Taylor and Francis.

Earth Charter, The. 2009. http://www.earthcharterinaction.org/contents/pages/Read-the-Charter.html.

Frederickson, H.G. 1997. *The Spirit of Public Administration*. San Francisco: Jossey-Bass.

Friedman, T.L. 1999. *The Lotus and the Olive Tree*. New York: Anchor Books.

Gandhi, Indira. 1972. 'Address at the UN Conference on Human Environment, Stockholm, Sweden', delivered on 14 June, published by the Department of Science and Technology as Agenda Notes for NCEPC Meeting, 28–9 July 1972. New Delhi: Government of India, Department of Science and Technology.

Gleick, P. 1991. 'Environment and security: The clear connections'. *Bulletin of the Atomic Scientists* 47 (3): 16–21.

Halabi, Y. 2004. 'The expansion of global governance into the Third World: Altruism, realism, or constructivism?' *International Studies Review* 6: 21–48.

Homer-Dixon, T.F. 1994. 'Environmental scarcities and violent conflict: Evidence from cases'. *International Security* 19 (1): 5–40.

Hyden, G. 1992. 'The study of governance'. In G. Hyden and M. Bratton, eds, *Governance and Politics in Africa*, 1–26. Boulder, CO: Lynne Rienner.

Hyden, G., and J. Court. 2002. 'Comparing governance across countries and over time: Conceptual challenges'. In D. Olowu and S. Sako, eds, *Better Governance and Public Policy: Capacity Building and Democratic Renewal in Africa*, 13–34. Bloomfield, CT: Kumarian Press.

Kagia, R. 2002. 'Prospects for accelerating human development in the twenty-first century'. In H. van Ginkel, B. Barrett, J. Court, and J. Velasquez, eds, *Human Development and the Environment: Challenges for the United Nations in the New Millennium*, 63–75. Tokyo: United Nations University Press.

Meadows, D., J. Randers, and D. Meadows. 2004. *Limits to Growth: The 30-Year Update*. White River Junction, VT: Chelsea Green Publishing.

Mittelstaedt, M. 2009. 'Unprecedented heat will trigger global food crisis'. *The Globe and Mail* 9 January: A4.

Nef, J. 2005. 'Third systems, human security and sustainable development'. Paper for presentation at a Symposium on Sustainable Development and Globalization, organized by the Globalization Research Center, University of South Florida, Tampa, 7 April.

————. 2008. 'Political economy of globalization, exclusion, and human insecurity in the Americas: Historical and structural perspectives'. In D.K. Vajpeyi and R. Khator, eds, *Globalization, Governance and Technology: Changes and Alternatives*, 141–69. New Delhi: Deep and Deep Publications.

New York Times. 2003. 'Harvesting poverty: The rigged trade game'. Editorial. 20 July: 10.

Ramphal, S. 1992. *Our Country, the Planet*. Washington: Island Press.

Reguly, E., and B. Laghi. 2009. 'G-20 ushers in a "new world order"'. *The Globe and Mail* 3 April: A1.

Saint-Martin, D. 1998. 'The new managerialism and the policy influence of consultants in government: An historical–institutionalist analysis of Britain, Canada and France'. *Governance: An International Journal of Policy and Administration* 11 (3): 319–56.

Saunders, D. 2008. 'Trickledown meltdown'. *The Globe and Mail* 27 December: B4.

Sen, A. 1999. *Development as Freedom: Human Capability and Global Need*. New York: Alfred Knopf.

Trace, Simon. 2005. 'Sanitation: No silver bullets, but reasons for hope'. *Environment Matters at the World Bank* (annual review): 12–13.

UNDP (United Nations Development Programme). 1997. 'Discussion Paper 3 on corruption and good governance'. Prepared by the Management Development and Governance Division, Bureau of Public Policy and Programme Support. New York: UNDP.

UNEP (United Nations Environment Programme). 2006. 'Global biodiversity outlook 2'. Convention on Biological Diversity, Montreal. http://www.cbd.int/doc/gbo2/cbd-gbo2-en.pdf.

UNESCO. 2002. 'Enhancing global sustainability'. Report of the Preparatory Committee for the World Summit on Sustainable Development, New York, 25 March. http://undesdoc.unesco.org/images/0012/001253/125351e.pdf.

Victor, D.G. 2006. 'Recovering sustainable development'. *Foreign Affairs* 85 (1): 91–103.

WCED (World Commission on Environment and Development). 1987. *Our Common Future*. New York: Oxford University Press.

Werlin, H.Z.H. 2004. 'The benefits of globalization: Why more for South Korea than Mexico'. *International Journal of Public Administration* 27 (13/14): 1031–59.

World Bank. 2000. *World Development Report*. Washington: World Bank.

Worldwatch Institute. 2004. *State of the World*. Washington: Worldwatch Institute.

3

How Healthy Are Our Ecosystems?

David J. Rapport

Learning Objectives

- To define ecosystem health.
- To identify key indicators of the health of terrestrial and aquatic ecosystems and stresses from human activities that compromise ecosystem health.
- To understand barriers and opportunities for managing for healthy ecosystems.

THE LEGACY OF ANTHROPOGENIC STRESS

It is something of an enigma how a country such as Canada, abundantly blessed in natural resources and a leader in environmental sciences and in the development of statistical systems for tracking the impact of human activities on the environment, can nonetheless find many of its ecosystems in precarious condition. More than two decades have passed since Canada produced its first *State of the Environment Report* (Bird and Rapport 1986)—a report hailed by scientists and politicians alike as the most comprehensive and authoritative report ever produced on the state of our environment. The report provided ample documentation that human activities had compromised the health of our ecosystems, as evidenced by clear signs of pathology in our forest, prairie, sub-arctic, agricultural, freshwater, and marine systems. Based on this report, the minister of environment gave Canada 'failing

marks' in nearly every aspect of environmental stewardship.

The 1986 *State of the Environment Report for Canada* (SOE) and accompanying statistical compendium, *Human Activities and the Environment* (Statistics Canada 1986), catalyzed multiple initiatives designed to provide ongoing monitoring on the state of the environment in relation to human activity. The 1986 SOE report was followed by two subsequent comprehensive assessments (Environment Canada 1991; 1996) and by periodic reports by Statistics Canada on human activity and the environment.[1]

The SOE reports, the statistical compendia produced by Statistics Canada, and many other assessments of our environment carried out by federal and provincial agencies collectively present a disquieting picture. Taken as a whole, the environment and resource assessments suggest that Canada, at best, is 'treading water' and at worst has lost significant ground as a steward of its ecosystems and natural resources. To be sure, the picture is a mixed one. For some environmental concerns, there have been improvements—particularly in reducing the release of persistent organic pollutants and stratospheric-ozone-depleting substances (CFCs). Considerable efforts have also been made to reduce the over-harvesting that has compromised our marine fisheries. The fact remains, however, that we have not done enough. By and large, our ecosystems continue to show signs of degradation: for example, western old growth coastal rainforests have been severely depleted by logging and development (BC MFR 2006); marine fisheries on both the Atlantic and Pacific coasts remain in critical condition (Myers and Worm 2003; see also Chapter 8); inland waters, including our Great Lakes, are considered to be at if not beyond the tipping point (Bails et al 2005; see also Chapter 7). Many of our environmental and resource woes cannot be entirely attributed to surprise, or to lack of awareness, or to the lack of capacity to document sources of anthropogenic stress on ecosystems that cause breakdown. Rather, they are largely attributable to insufficient societal forces (administrative actions, actions of civil society, and 'political will') to 'connect the dots' and act on clear evidence that anthropogenic stresses on our ecosystems have resulted in ecosystem degradation, thus compromising economic, social, and ecological well-being and sustainability.

Canada is by no means alone in this failure. The continued degradation of global and regional environments (MEA 2005; UNEP 2007) is the result of actions or inactions by most of the world's nations—generally speaking, with economically more privileged nations adding the greater share of stress. What is disappointing about Canada, however, is that despite this country's world leadership in the development of statistical systems to track the impact of human activities on the environment and its significant leadership in the development of the field of ecosystem health, it is no further ahead than other nations in terms of environmental and resource stewardship.

There are signs of renewed interest in Canada in monitoring 'environmental quality' and in relating environmental quality to societal well-being. If these efforts are to be meaningful, a science-based ecosystem approach will be essential. In this chapter, I suggest that monitoring and assessing the health of Canadian ecosystems has proved practical and feasible and is indispensable in providing guidance in order to achieve long-term sustainability. Thus, I argue that while 'ecosystem health' has become a guiding force in a number of environmental monitoring programs in Canada, it needs to be given more than 'lip service'; it needs to become the mainstay of our environmental and natural resource assessment policy.

In the following sections, I first examine the need for an integrative approach to environment and resource management and show how an ecosystem health model well suits this need. Then I examine what constitutes ecosystem health—showing that 'health' is an objective property of ecosystems on a rigorous scientific basis. Next I describe the main features of Statistics Canada's Pressure-State-Response model, a model developed in the mid-1970s that has been adopted by most international organizations and countries today as a basis for reporting on human activities and the environment. This model is ideally suited to assessing the health of Canadian ecosystems. I then turn to describing examples of prominent Canadian monitoring programs that have been designed to assess the health of our terrestrial and aquatic ecosystems. These programs amply demonstrate the potential for assessing ecosystem health on a regional basis. I conclude by looking at the role of conflict and uncertainty regarding efforts to improve the health of our ecosystems.

A HOLISTIC APPROACH TO ENVIRONMENT AND RESOURCE MANAGEMENT

The history of our failure to arrest ecosystem degradation speaks loudly to the need for a new approach, one that rejects our preoccupation with immediate short-term returns from economic activities to the detriment of our environment and instead looks to the broader interests of long-term sustainability. What is needed for effective environmental assessment and management is a central focus, one that motivates mapping, monitoring, modelling, and management (Rapport and Moll 2000).

If we are to achieve the sustainability goals of the World Commission on Environment and Development (WCED 1987), then we must better document the ways in which human activities cumulatively and synergistically have led to the degradation of our ecosystems and take preventive and remedial actions to redress the situation. The motivation for this undertaking rests in the growing awareness that degradation of our ecosystems has significant costs for all species with which we share the planet and in particular to our well-being in its many dimensions (physical, psychological, socio-economic, and ecological).

Over the past half century, while we have greatly improved our technical capability to monitor the environment, particularly with developments in remote sensing and computation, the 'big picture' often remains elusive. We are generally data-rich but often synthesis-poor. We generally lack an overarching framework to put together the many facets of the jigsaw puzzle: the particulars of how the many stresses from human activities (anthropogenic stresses) interact and how the cumulative impact of those stresses disables ecosystem functions and compromises the essential biotic and abiotic structure of healthy ecosystems (Hildén and Rapport 1993). Unless we have this full understanding, we find ourselves in the position of presiding over our own demise while following disparate environmental trends on high-definition computer screens.

Canada has led the development of statistical frameworks for relating human activities to the environment (Rapport and Friend 1979; Friend and Rapport 1991). It has also been a leader in understanding the behaviour of ecosystems under stress and the mechanisms whereby human actions can progressively disable ecosystems (Rapport, Thorpe, and Regier 1979, 1985; Schindler 1987; Regier and Hartman 1973; Myers and Worm 2003). What is needed now is to capitalize on these significant advances and adopt a synthetic approach to monitoring and assessing the interactions between humans and the environment, an approach that provides clear guidance for public policy on how to sustain our ecosystems and natural resources while satisfying the basic economic, social, and cultural needs of Canadians (Rapport and Regier 1980; Rapport and Singh 2006).

I argue that the field of ecosystem health, which in its broadest sense is about achieving human needs in a manner consistent with maintaining the vitality of ecosystems, should become the main focus of environmental and resource policy. Canada hosted the NATO Advanced Research Workshop on Monitoring and Assessing the Health of Large-Scale Ecosystems (Rapport, Gaudet, and Calow 1995) and is well positioned to take the critical next steps.

WHAT CONSTITUTES ECOSYSTEM HEALTH?

Although early critiques of ecosystem health suggested that the concept was merely a metaphor, it has been shown that 'health' is an objectively measurable property at all levels of biological organization (Costanza, Norton, and Haskell 1992; Mageau, Costanza, and Ulanowicz 1995; Rapport, Regier, and Hutchinson 1985; Rapport et al. 1998a, 1998b; Rapport 1989a, 1989b; Rapport, Costanza, and McMichael 1998; Rapport et al. 2003). The health of ecosystems is not tied to individual or societal preferences or tastes. Rather, it can be objectively assessed based on changes in ecosystem structure and function (Wichert and Rapport 1998; Rapport and Whitford 1999). While the mechanisms governing the organization of ecosystems (Levin 1998) are of a very different character from those governing the organization at the level of populations or individuals (Rapport, Regier, and Hutchinson 1985), ecosystems, like populations and individuals, can show dysfunction—that is, clear evidence of failing health. While it is sometimes argued that the 'health' of an ecosystem can only be evaluated with reference to human goals (for example, from a forester's perspective, a healthy forest is one that sustains harvests of merchantable timber; from a conservationist's perspective, a healthy forest is one that sustains biodiversity), we suggest that both these views are limited and in ecosystem health terms are part of a larger picture. Healthy forests sustain both economic productivity and biodiversity and have many other functions as well. Just as human health should not be judged on limited information such as blood pressure and lung function, so too ecosystem health is assessed on the basis of a broad spectrum of measures.

Ecosystems pose special challenges in terms of both description and defining their state of health (Rapport 1995). To begin with, ecosystem boundaries are to some extent arbitrary, and they are generally highly dynamic systems, presenting an ever-changing kaleidoscope. Further, they comprise myriad species (many of which are invisible, such as in terrestrial systems, the important below-ground bacteria and fungi and the small invertebrates that ply the soils). Yet despite these vast complexities, there remain macro-level patterns by which we are able to distinguish ecosystems from one another (boreal forests, for example, appear very different from grasslands) and at a finer level are able to distinguish between healthy and unhealthy ecosystems.

Healthy ecosystems have three main attributes: *organization*—the capacity to maintain their biotic structure (i.e., their characteristic biological diversity, their interactions between species and with the abiotic environment); *vitality* (or vigour)—the capacity to maintain their level of biological productivity; and *resilience*—the capacity of ecosystems to rebound from perturbations such as those caused by fire, flood, drought, and so forth (Rapport 1989a; Mageau, Costanza, and Ulanowicz 1995; Rapport et al. 1998a; Rapport, Costanza, and McMichael 1998).

Since humans are part of ecosystems—so much so that we refer to many ecosystems as 'human-dominated' (Vitousek et al. 1997)—a robust definition of ecosystem health cannot be restricted to only the ecological dimensions; it must also recognize the human component. Nielsen (1999) succinctly accomplished this, defining ecosystem health in a broad socio-ecological context as the capacity for maintaining biological and social organization on the one hand and the ability to achieve reasonable and sustainable human goals on the other. Achieving ecosystem health is also recognized as integral to achieving public health (Arya et al. 2009).

Neess (1974) early on provided a list of properties of healthy ecosystems, most of which are indeed aspects of organization, vitality, and resilience. His list included:

1. energetic, in that natural systemic processes are strong and not severely constrained (this relates to the property of 'vitality');

2. self-organizing, in an emerging, evolving way (this relates directly to the property of 'organization');
3. self-defending against the introduction of foreign elements (this relates to the property of 'resilience');
4. able to recover from occasional stress or natural calamities (again relating to the property of 'resilience');
5. aesthetically attractive (relating from the human perspective to the property of 'organization'); and
6. productive of goods and services valued by humans (again relating to the property of 'vitality' but from a socio-economic perspective).

Given the dynamic character of ecosystems on time scales ranging from seasonal to many decades, assessing ecosystem health depends on identifying key features that distinguish normal ecosystem fluctuations and transformations from those that are pathological or 'out of bounds'—not always a simple task (Rapport and Regier 1995). Here we turn to a statistical framework that has found wide application in identifying the key kinds of information necessary to evaluate the health of ecosystems and landscapes at scales ranging from local to national and international.

THE STATISTICS CANADA FRAMEWORK FOR REPORTING ON THE STATE OF ENVIRONMENT

In the mid-1970s, Statistics Canada initiated its program in environmental statistics, an entirely new undertaking for an agency that traditionally focused on economic and demographic data. To lead this initiative, the agency required someone with a background in both economics and ecology in order to develop an integrative approach that incorporated these two critical dimensions. In the mid-1970s, there was a dearth of such models and frameworks, and data pertaining to the environment were generally restricted to statistics on harvest (fish landings, timber harvest, and so forth), threats (e.g., fire, insect, pest, and parasite damage) and the quality of air and water. It was obvious that this was a very fragmented approach and that a collaborative effort between agency scientists and leading academics was needed in order to forge a science-based holistic approach to human activity and the environment.[2]

The Statistics Canada model, known initially as the Stress-Response Environmental Statistical System (SRESS), revolutionized the field of environmental statistics and state of environment reporting. It transformed the field from a haphazard collection of environmental information on air, water, and land to a fully integrated system with a well-developed taxonomy of anthropogenic stresses as well as an identification of key signs of ecosystem breakdown (Rapport and Friend 1979). It incorporated human activity as part of and not separate from ecosystems. By identifying the pressures (stresses), the drivers generating the pressures, key indicators of ecosystem response to stress (i.e., indicators of ecosystem breakdown), and societal responses to environmental change, SRESS was capable of providing a diagnosis of the state of health of Canada's major ecosystems. By 1977 (pre-publication), SRESS had been adopted by the OECD (Organisation for Economic Co-operation and Development), and shortly thereafter it was taken up by the UN Statistical Bureau. It then became widely known as the 'Pressure-State-Response' (PSR) model. The European Environment Agency began to make use of PSR from the early 1990s under the acronym of DPSIR (Drivers, Pressures, State, Impact, Response) (http://en.wikipedia.org/wiki/dpsir). But while the acronyms changed (from SRESS to PSR to DPSIR), the basic Statistics Canada framework remains as it was first developed (Rapport and Singh 2006). Today, the SRESS (the PSR model) serves as the basis for state of environment reporting by many governments and international organizations worldwide (World Bank 2006).

The initial application of the SRESS was to examine historical changes in the state of the Laurentian Lower Great Lakes for which there existed long-term data on important components, such as water quality and fisheries. Drawing from these sources and historical reviews (e.g., Regier and Hartman 1973), SRESS yielded an integrated statistical picture of the degradation of the Lower Great Lakes over the past century (Rapport 1983a; Rapport and Regier 1983).

SRESS also served as the foundation for state of environment (SOE) reporting in Canada (Bird and Rapport 1986), as well as underpinning Statistics Canada's publication series *Human Activities and the Environment* (Statistics Canada 1986). These twin undertakings required constructing a new geography of Canada, one not based on jurisdictional boundaries but rather on eco-cultural zones (that is, on criteria of the natural environment and human activity). At the largest scale, Canada was partitioned into 15 terrestrial ecozones and three (now expanded to five) marine ecozones.

The first Canadian SOE report (Bird and Rapport 1986) portrayed, on both ecozone and watershed bases, the major sources of anthropogenic stress (i.e., the release of waste residuals, physical restructuring [land-use change], introduction of exotic [non-native and often invasive] species, over-harvesting, and extreme natural events)[3] as well as key signs of ecosystem degradation (i.e., loss of biotic diversity, increased dominance by non-native species, reduced primary and secondary productivity, regression to earlier stages of ecosystem development,

increased disease prevalence, loss of soil fertility, and reduced capacity to rebound after perturbations such as fire, flood, or drought). One of the main achievements of the SOE report was that it allowed for making the crucial connection between stresses on ecosystems from human activities and changes in the structure and function of these systems.[4] The report also touched on other important topics, such as the health impacts on Canadians from adverse environmental change, government responses to environmental concerns, and the attitudes of Canadians regarding environmental issues. The 1986 report was followed by two additional comprehensive assessments (Environment Canada 1991; 1996) and inspired new monitoring and synthesis initiatives by various agencies, such as the work of the Marine Environmental Quality (MEQ) Working Group of Environment Canada (Wells 2004).

MONITORING AND ASSESSING THE HEALTH OF OUR ECOSYSTEMS

Statistics Canada's SRESS/PSR model served as an important stepping stone to the full development of the ecosystem health perspective. By identifying a taxonomy of anthropogenic stresses on the environment and some of the major signs of ecosystem breakdown, the elements were in place for the full development of the concept of ecosystem health (Rapport, Thorpe, and Regier 1979; Rapport 1989a; Costanza, Norton, and Haskell 1992) and the seeding of a new generation of holistic approaches to mapping, monitoring, modelling, and managing for ecosystem health (Rapport 2007a, 2007b; Rapport et al. 1998a; Wells 2003, 2004). Today, there are dozens of federal and provincial programs that explicitly focus on the health of Canadian ecosystems. Nearly all are built on the foundation established by the SRESS/PSR model. The scale of these monitoring and assessment efforts varies considerably: from a local focus (that of a stream, a bay, a harbour, an estuary, or a small provincial park) to regional, national, and international levels. Indeed, programs for monitoring the health of ecosystems have become mainstream in North America, Europe, and Australia, and similar programs have been initiated in Asia, Africa, and the Middle East.

Below, I evaluate a few examples of monitoring the health of Canadian forest, freshwater, and marine ecosystems. I describe some of the indicators, the strengths and weaknesses of the monitoring efforts, and the degree to which effective actions have been taken to improve the health of ecosystems where they have been shown to be faltering.

Forest Ecosystems

The notion of what constitutes healthy forests has evolved considerably since the mid-1970s. At that time, a forest was deemed in 'good health' if it was relatively free of infestation by forest insects and diseases, a view that has not altogether disappeared today. Old-growth forests were generally considered 'unhealthy' because they contain a high percentage of so-called 'overmature' trees—that is, old growth that naturally shows signs of decay. Even today, reports on 'forest health' focus almost exclusively on pest burdens. The series of reports on forest health conditions in Ontario (OMNR 2006a) focuses on provincially significant forest pests and invasive species, with only passing mention of other considerations that go into an ecosystem health assessment.

The emphasis on pests and other threats to various tree species reflects a resource focus rather than an ecosystem focus within the forestry sector. In this context, a healthy stand is one in which there are vigorous and productive trees, in which there is rapid biomass accumulation and nutrient cycling, in which insect and diseases are at bay, and in which there is normal ecological succession and stand development and 'high vitality' (Kimmins 1996). In contrast, an ecosystem focus would examine the health of the forest in terms of its multi-layered complex canopy structure with large gaps and uneven-aged trees; large, dead, deformed, broken, and partly decayed trees; and logs that provide much needed habitat for forest animals, both rare and specialist species (Kimmins 1996).

In recent years, the Canadian Forest Service (CFS), which falls under Natural Resources Canada, has moved toward adopting an ecosystem approach, characterizing a healthy forest as 'one that maintains and sustains desirable ecosystem functions and processes. . . . A forest ecosystem may be considered healthy when inherent ecological processes are operational within a natural range of variability' (Allen 2001, 30; CFS 1999). This definition still contains some ambiguity, however, as to what are '*desirable* ecosystem functions and processes'. If they are limited only to 'functions and processes that are (economically) useful to humans', it remains a resource-based approach; however, to the extent that it includes functions and processes that sustain biodiversity, soil formation, hydrological regulation, nutrient cycling, temperature modification, water quality, and pathogen regulation, including human pathogens, then it becomes an ecosystem perspective (Wilcox and Ellis 2006; Patz et al. 2004). As the CFS approach has evolved (see below), it appears to be more and more aligned to an ecosystem approach.

Assessing Forest Ecosystem Health

Although provincial reports under the title 'forest health' are largely concerned with forest pests, provincial reports on the 'state of the forests' cover a broader territory, more in line with an ecosystem health model. A major stimulus to moving in this direction came from the Montreal Process[5] (http://www.rinya.maff.go.jp/mpci/criteria_e.html)[6].

The process identified the following criteria for sustainability of forest ecosystems:

1. conservation of biological diversity;
2. maintenance of productive capacity of forest ecosystems;
3. maintenance of forest ecosystem health and vitality;
4. conservation and maintenance of soil and water resources;
5. maintenance of forest contribution to global carbon cycles;
6. maintenance and enhancement of long-term multiple socio-economic benefits to meet the needs of societies; and
7. legal, policy, and institutional framework.

Of the seven, criterion 3 relates directly to ecosystem health, while others relate indirectly (i.e., criteria 1 and 4 relate to 'organization', criteria 2 and 5 relate to 'vitality', and 6 and 7 bring into focus the social dimensions). Thus, the Montreal Process takes an all-encompassing view of forest sustainability. Its main drawback is the unwieldy number of indicators employed, creating a policy-making nightmare. As a study of transformation of the Carolinian deciduous forest in southwestern Ontario demonstrates, forest ecosystems can be adequately assessed with only a handful of well-chosen indicators covering both ecological and social concerns[7] (Box 3.1).

However, the Montreal Process helped to stimulate federal and provincial initiatives in Canada for assessing the state of our forests. The Canadian Council of Forest Ministers identified six major criteria (CCFM 2003) comprising what has also turned out to be an unwieldy number

of specific indicators: biological diversity (eight indicators), ecosystem condition and productivity (five indicators), soil and water (three indicators), role in global ecological cycles (four indicators), economic and social benefits (13 indicators), and society's responsibility (12 indicators).

In the 1990s, the CFS initiated several programs to comprehensively assess forest ecosystem health. The Forest Health Network had three objectives: first, to assess the effects of stresses such as the invasion of exotic species and climate change on the health of Canada's forests; second, to establish criteria and indicators for managing forests sustainably; and third, to provide national inventory data on forests and their health using remote sensing techniques. A Monitoring and Analysis Program was charged with reporting on forest stressors and responses in order to provide a foundation for national forest policy in Canada. A National Forest Health Database was designed for indicator data on biodiversity, sapling and tree regeneration, soil health, and the history of insects and disease. CFS focuses on important aspects of forest health assessment, including symptoms of forest deterioration, determinants of forest condition, and consequences of forest decline for the well-being of Canadians (CFS 1999, 5).[8]

Currently, several provincial governments produce comprehensive reports on the state of their forests. Ontario's *State of the Forest Report 2006* (OMNR 2006b) (http://www.mnr.gov.on.ca/MNR_E005126.pdf) is in close alignment with the Montreal Process and the CCFM recommendations on criteria and indicators. OMNR criteria for forest ecosystem sustainability include: conserving biological diversity, maintaining and enhancing forest ecosystem condition and productivity, protecting and conserving forest soil and water resources, and monitoring forest contributions to global ecological cycles (Miller and Nelson 2003).

Box 3.1 Carolinian Forests of Southwestern Ontario

Taking an ecosystem health (EH) approach to forest management begins with the identification of a suite of indicators for monitoring forest condition. A study of Pinery Provincial Park was conducted to determine: (1) which indicators of forest condition are sensitive to stresses acting on the forest system and (2) how lay society defines forests and forest health and values forests (Patel and Rapport 2000). The study examined the history of the major stressors (deer browsing, prescribed burn, visitor use, and trails) and looked at the sensitivity of six indicators of forest condition: species richness, plant density, species cover, proportion of native species, height of individual plants, and foliar damage by insect herbivory.

Principal findings were that stem density, species cover, and proportion of native species were the most sensitive indicators for deer browsing, with only the latter escaping negative impacts. Generally, all indicators were affected by prescribed burns, with the proportion of native species declining in burned areas and stem density and cover increasing. Foliar herbivory did not vary with trail use, but high trail use did tend to influence the proportion of native species, species cover, and species density. Finally, native species, species richness, and stem density were all sensitive to distance from trail edge, with native species increasing with distance, species richness declining, and cover and seedling height remaining effectively unchanged.

Overall, the most sensitive indicators to the suite of stresses at the Pinery were stem density, cover, and proportion of native species. From focus group discussions with varying interest groups and informal surveys of park tourists, it was observed that lay perceptions of the main measures of forest health did overlap substantially with the chosen indicators. These findings are encouraging, since 'the condition of forest ecosystems, ecosystem services obtained from forests, and management and policy decisions related to forests, are all underpinned by the values that society places on forests' (Patel et al. 1999, 239). When these values and, more importantly, society's perceptions of forest health coincide with scientific understanding, sound management strategies and policy can be synthesized for both people and forests.

Alberta and British Columbia also produce comprehensive reports on their forests. The Alberta 2006 report entitled 'Forest health in Alberta' (Government of Alberta 2006) has on its cover its forest health vision: 'A healthy forest environment that provides sustainable fiber resources and a diverse forest ecosystem that supports biodiversity and critical wildlife habitats'. Yet this document deals with only a very narrow interpretation of 'forest health', focusing on the status of forest insect pests, disease, and alien invasive plants—with virtually no information relating to 'a diverse forest ecosystem that supports biodiversity and critical wildlife habitats'.

The state of the forest report for British Columbia, 'The state of British Columbia's forests' (BC MFR 2006) is outstanding in providing synthetic and detailed data on biodiversity, forest cover, age classes, and so forth. Data are geographically coded and portrayed through a series of interactive on-line maps by biogeoclimatic zones. For each base map, the user can overlay various data sets such as biodiversity, old growth, protected areas, agricultural areas, and private land. Further, the report is broad-based, encompassing environmental, economic, social, and governance aspects.

Its overall objective is to provide information that will be useful to decision-makers and the public in enhancing the resilience of the province's forest ecosystems. The report does not shy away from reaching some disquieting conclusions. It concludes, for example, that some indicators of forest health are in decline—namely, species diversity and ecosystem dynamics—while others, such as ecosystem diversity and genetic diversity, present a 'mixed picture'. The report is wanting in regard to developing linkages between human-induced stress and forest condition. For example, how do forest practices affect the state of the forests and neighbouring ecosystems? The report does mention some details in passing, stating that only 3 per cent of the coastal Douglas fir forests in the age class of 250 years or older still remain 'owing to development of agriculture and logging'.

Managing for Forest Ecosystem Health

Although many agencies in Canada have mandates for sustaining forest ecosystems, there is no common vision as to what this means or how to achieve it. The Canada Forest Accord (CFA) (2003–8) (http://nfsc.forest.ca/accords/accord3. html) was an attempt to establish 'common ground' among more than 60 organizations, including federal, provincial, and academic institutions, industry, and NGOs, by articulating a common vision, namely: 'The long-term health of Canada's forest will be maintained and enhanced, for the benefit of all living things, and for the social, cultural, environmental and economic well-being of all Canadians now and in the future.'

The CFA attempts to be 'all things to all people'. It speaks of 'stewardship of the entire forest', 'managing Canada's forests through an ecosystem-based approach', and 'comprehensive national forest reporting systems', while at the same time referring to 'developing markets and increasing the value of all forest products and services', 'adopting policies and practices that support forest-based community sustainability', and so forth. Of course, taken individually, each of these objectives may be seen as worthy. The key question is: how to achieve all these objectives while maintaining the underlying health and resilience of Canada's forest ecosystems?[9]

Techniques for carrying out broad-scale monitoring and assessment of the health of Canadian forest ecosystems have advanced considerably with applications of remote sensing and advanced computation. Yet these tools have seldom been applied to document the pressures (stresses) on forest ecosystems, especially from forestry practices themselves—for example, from over-harvesting, fire suppression, habitat fragmentation, introduction of invasive species, and herbicide and pesticide use. Also missing are the links between the state of the forests and its bearing on human well-being. State of the forest reports will remain weak and incomplete if they do not attempt to associate change in forest health with human pressures on the forest through over-harvesting and other means. State of environment reports should also draw attention to some of the underlying problems, particularly our economic system that depends on ever-increasing use and exploitation of natural resources. Much more attention needs to be given to actions designed to safeguard the health of our forests and their effectiveness.

The Laurentian Great Lakes Ecosystem

When it comes to threats to the health of our aquatic ecosystems, one generally thinks of pollution. Indeed, contaminants, particularly persistent organic pollutants (POPs, such as DDT, PCBs), have been a chronic problem since the 1960s (Carson 1962; Colburn 2001). Excess nutrient loading has also become a major problem with widespread overuse of chemical fertilizers. Nutrient loading often triggers algal blooms, which can be toxic to humans and other organisms (Arya et al. 2009). Excess nutrient loading also triggers anoxic conditions, resulting in 'dead zones' that are practically devoid of life.

But such considerations are far from the only source of problems in our aquatic ecosystems. Other major influences that compromise the health of aquatic ecosystems include overharvesting (of fish and other aquatic resources), physical restructuring (including land-use changes such as wetland drainage), purposeful or accidental introduction of non-native species (Regier and Hartman 1973; Rapport and Regier 1980; Rapport 1983a; Harris et al. 1988), controls of water levels and flow regimes, and temperature changes resulting from climate warming.

Assessing Great Lakes Ecosystem Health

Numerous agencies are involved in various aspects of mapping, monitoring, modelling, and management of the Laurentian Great Lakes. Prominent among them are the International Joint Commission (IJC) on Boundary Waters (initially set up to regulate navigation on waters shared between the US and Canada, but since 1972 also focused on environmental issues), the Great Lakes Fishery Commission (whose records of the Great Lakes fishery go back to the late nineteenth century), and the Great Lakes Basin Commission, emphasizing the integration of many interests from a basin-wide perspective. Further, because of the widely acknowledged need for definitive indicators by which to measure progress toward restoration of the Laurentian Great Lakes, the US Environmental

Protection Agency (US EPA) and Environment Canada established the State of the Lakes Ecosystem Conference (SOLEC), which was charged with providing the basis for 'coordinated and consistent reporting on the state of health of the Great Lakes Basin ecosystem'.[10]

SOLEC's mission statement importantly refers explicitly to the *health* of the Great Lakes Basin ecosystem, while the Great Lakes Water Quality Agreement (GLWQA) of 1978 refers to *integrity*.[11] This prompts the question: are these concepts interchangeable? Strictly speaking, they are not, although in common use the distinctions have become blurred. 'Health', as we have already discussed in this chapter, refers to the capacity of an ecosystem to maintain its structure and functions and to be resilient to perturbations. 'Integrity' refers to the state of an ecosystem that would prevail if there were no human influences (Rapport et al. 1998b). Thus, for the Laurentian Great Lakes, where anthropogenic stress has been pervasive since European settlement (Regier and Hartman 1973), restoration and maintenance of ecosystem *health* is an option; restoration of ecological *integrity* is not—if for no other reason than that key native fauna are now extinct as a consequence of anthropogenic stress.[12]

As with other large-scale ecosystems, monitoring and assessing the health of the Great Lakes poses significant scientific, technological, and resource challenges. The SOLEC process has identified some 80 indicators and has issued a number of reports on the status of and trends for the Great Lakes Basin as a whole, as well as individual lakes, and in the process has identified data gaps.[13]

SOLEC's framework for 'indicator categories' and examples of core indicators are given in Box 3.2. The large number of SOLEC indicators is a barrier to synthetic and definitive conclusions. For example, for 2008 SOLEC reported that some conditions or areas were deemed 'good' while

Box 3.2 An Overview of the SOLEC Indicators

The State of the Lakes Ecosystem Conference (SOLEC) has facilitated the development of indicators utilizing the framework of the State-Pressure-Response model (Bertram and Stadler-Salt, 2000). The SOLEC model evaluates indicators using four criteria: (1) is the indicator required? (2) will the suite of indicators be adequate? (3) will monitoring the indicator be feasible? (4) can the indicator be understood by both the scientific and the lay communities?

SOLEC indicators, falling under seven broad categories, aim to assess the health of open and near-shore waters, coastal wetlands, the near-shore terrestrial zone, land use, human health, society, and the unbounded ecosystem itself. Indicators of pressure comprise nutrient concentrations and loadings, sediment flowing into coastal wetlands, water level fluctuations, land conversion, and levels of *E. coli* and faecal coliform bacteria in recreational waters. The indicators of the state of the ecosystem include near-shore habitat, abundance of exotic plants, urban density, geographic trends in disease incidence, and economic prosperity. Indicators of human response (action) include gain in restored coastal wetlands, brownfield redevelopment, and community-based stewardship activities. A more recent initiative is the development of indicators of societal responsibility, which focus on 'shared governance and responsible management'. These indicators are particularly aimed at measuring the socio-cultural response of local communities and include an evaluation of the non-biophysical parameters of ecosystem health.

For a comprehensive description of SOLEC criteria and more information about how the Pressure-State-Response model of appraising ecosystem health is being applied in the Great Lakes, refer to http://www.on.ec.gc.ca/solec/pdf/mainpaper-v4.pdf.

others were deemed 'poor'; some indicators showed improvement, others showed degradation (Environment Canada and US Environmental Protection Agency 2009). Pruning the list of more than 80 indicators down to a more manageable number appears essential if SOLEC is to accomplish its goal—which is to assess the overall health of the lakes.

Managing for Great Lakes Ecosystem Health: A Prescription

One of the most synthetic ecosystem health assessments of the Great Lakes is contained in the white paper prepared by a group of experts on the Great Lakes and endorsed by more than 200 others (Bails et al. 2005).[14]

The white paper does what many of the reports on the health of large-scale ecosystems mentioned above have failed to do: namely, to interrelate critical indicators of stress on ecosystems with indicators of ecosystem transformation. The paper reaches firm conclusions on the state of the Great Lakes ecosystem: that it exhibits many signs of pathology as a result of cumulative anthropogenic stresses, including 'toxic contaminants, invasive species (such as the sea lamprey), nutrient loading, shoreline and upland land-use changes, and hydraulic modification', as well as from over-harvesting by commercial fisheries. As evidence of ecosystem breakdown, the report cites the persistence of an anoxic/hypoxic zone in the central basin of

Lake Erie, eutrophication in Saginaw and Green Bay (Lake Michigan), rapid disappearance of the once abundant amphipods in the genus *Diporeia* in all areas of the lakes except for Lake Superior, recent declines in growth, condition, and numbers of lake whitefish in Lake Michigan and portions of Lake Huron, and elimination of the macrophyte (rooted plant) community and simplification of the benthic food web in areas of Lake Erie and Lake Ontario.

Other signs of ecosystem distress mentioned in this document include extirpation or major declines in important native species (e.g., lake trout and deepwater ciscoes); prevalence of invasive species (e.g., sea lamprey, alewives, and rainbow smelt); widespread reproductive failures of native species, including lake trout, sturgeon, lake herring; fouling of coastlines resulting in beach closings and loss of fish and waterfowl habitat; toxic contamination of fish; loss of coastal wetlands (including 90 per cent of pre-settlement wetlands along the Lake Huron and Lake Erie corridor); recent introductions of aquatic invasive species (e.g., zebra and quagga mussels); decreased populations of benthic organisms, causing a decline in health of lake whitefish; and loss of water quality (e.g., increase of algal blooms, type E botulism in fish and waterfowl, contamination of drinking water). A major factor contributing to the decline in health of the Great Lakes is the loss of self-regulation mechanisms, particularly the capacity of near-shore communities to mitigate stress (Bails et al. 2005).

The prescription for Great Lakes restoration includes the following:

- restoring and enhancing the self-regulating mechanisms of the Great Lakes by focusing on the health of key geographic areas;
- remediation of existing and prevention of major new perturbations from human activities;

- protecting existing healthy elements by adopting sustainable land- and water-use practices in the basin that maintain the long-term health of the Great Lakes ecosystem and associated benefits; and
- monitoring for ecosystem health and signs of effectiveness of restoration and protection efforts.

The authors find that: 'There is compelling evidence that many parts of the Great Lakes are at or beyond the tipping point' and warn that in the absence of immediate and sustained actions, the lakes could suffer 'irreversible and catastrophic damage'. The white paper offers strong support for the application of an ecosystem health approach both as a means for diagnosing the condition of the Great Lakes and to better understand the self-regulatory mechanisms that have become disabled as a result of cumulative impacts of anthropogenic stress. Superimposed on the primary interactive stresses are 'broader, large-scale changes in global and regional climate', resulting in elevated water temperatures and increases in the potential for hypoxic areas ('dead zones'). Global warming has also contributed to more rapid spring river and stream runoff, increasing stream bank erosion and damaging potential fish and wildlife habitat.

In their diagnosis of what has caused the virtual collapse of the Great Lakes Basin ecosystem, the authors point to well-established findings that multiple stresses on ecosystems generally have compounding effects (Regier and Hartman 1973; Rapport and Regier 1995). In particular, the Great Lakes have suffered from the cumulative effects of overfishing (beginning in the late 1800s and continuing to the mid-twentieth century), introduction of invasive species (by the mid-twentieth century), nutrient loading (particularly in Green Bay, Saginaw Bay, and the western and central basins of Lake

Erie), release of toxic chemicals (especially from the concentrations of heavy industry along the Niagara, Detroit, and Fox rivers), and physical restructuring (causing major disturbance in the basin, fragmentation of habitat, wetland loss, shoreline restructuring, and so forth). Damage to the near-shore has resulted in reduction of the capability of the Great Lakes to buffer the effects of stress within the basin and thus a loss of ecosystem resilience.

The Role of Governance

The transformation of the Laurentian Great Lakes from their essentially pristine state at the time of European settlement (see, for example, Regier and Hartman 1973) to their current highly degraded condition calls for new forms of governance. Currently, under a multiplicity of authorities from two federal governments (Canada and the US), eight states, and two provinces (Ontario and Quebec) (see also Chapters 7 and 15), there is bound to be conflict, confusion, and difficulty in co-ordinating actions. Each authority tends to spawn its own bewildering number of programs to tackle specific issues, largely in isolation from the work of others. According to a 2003 report of the US General Accounting Office (US GAO 2003), 151 federal programs and 51 state programs were supporting environmental restoration within the Great Lakes Basin. The GAO report cites a 'lack of any overarching approach' by which to co-ordinate program activities in support of Great Lakes restoration, as well as a lack of co-ordinated monitoring to determine basin-wide progress toward meeting restoration goals, as major impediments to effective action (US GAO 2003; Bails et al. 2005).

One effort to harmonize governance between two of the Great Lakes authorities is the Canada–Ontario Agreement (COA) respecting the Great Lakes Basin ecosystem. The COA is guided by the vision of a 'healthy, prosperous and sustainable

Great Lakes Basin for present and future generations'. Through its four annexes, the agreement establishes the priorities and goals for the environmental protection and rehabilitation of the Great Lakes that are subscribed to by both the federal and provincial governments. Their joint recommendations are due in late 2009. This constitutes a small but significant effort toward more collaborative and co-ordinated management of the Great Lakes Basin.

In summary, the Laurentian Great Lakes offer a sobering example that even the largest of our freshwater ecosystems can find itself at or beyond the tipping point. The Great Lakes have succumbed to the cumulative impacts of human activities over two centuries. Although billions of dollars have been expended in efforts to reduce some of the abuses through remedial action programs—e.g., reduction of contaminants and improvements in shoreline habitat (http://www.great-lakes.net/envt/pollution/aoc.html) (see also Chapter 7)—much more must be done if the health of the system is to be restored. The GAO (2003) study emphasizes the need for co-ordinated monitoring and action within an overarching framework. The white paper (Bails et al. 2005) suggests that the overarching framework already exists (i.e., ecosystem health), but the critical actions to implement the framework remain lacking.

Other Freshwater Ecosystems

In addition to the extensive monitoring of the Laurentian Great Lakes, a number of programs have been designed to assess the health of other Canadian freshwater ecosystems. Below are several examples.

The Northwest Territories and Alberta's Northern River Basins Study

The Northern River Basins Study (NRBS 1996) (http://www3.gov.ab.ca/env/water/nrbs/toc.html) was designed to bridge the information gap with respect to the relationship between human activities and the health of the Peace, Athabasca, and Slave river basins within the Northwest Territories and the province of Alberta. This region covers 580,000 square kilometres, with a diverse landscape comprising forests, marshes, and grasslands. The river basins are affected by numerous anthropogenic stresses from agriculture to pulp and paper mills to open-pit coal mines and, since the 1990s, the massive extraction of oil from the Athabasca oil sands.

Initiated in 1991 and completed in 1996, the NRBS investigated the presence of contaminants in the environment, drinking water quality, the region's hydrology, nutrient cycling, food chains, uses of the rivers, ecosystem health, and traditional knowledge. Ecosystem health was one of the key foci. The NRBS found numerous areas of environmental concern, resulting in about two dozen specific recommendations for mitigating adverse impacts on the basins from human activity. The general restoration strategy included: (1) identifying ecosystem goals supported by scientific literature and community goals; (2) developing specific management objectives into an ecosystem strategy; (3) selecting appropriate ecosystem health indicators, including those to assess socio-economic, biological, cultural, and chemical parameters of health; (4) monitoring and assessing the state of the chosen indicators; and (5) taking appropriate action. The final report of the NRBS can be found at http://www3.gov.ab.ca/env/water/nrbs/toc.html.

The Fraser Basin Council

The Fraser Basin Council (FBC) promotes actions to restore health to the major river system and its estuary (http://www.fraserbasin.bc.ca/about_us/history.html). Both the federal and provincial governments have long been aware that the Fraser River had become highly degraded, particularly as a result of unregulated release of waste residuals from sources including sawmills, pulp and paper production, mining, municipal sewage, agricultural waste, and urban runoff. Thus, the river was identified as part of Canada's Green Plan, which established a Fraser River Action Plan (FRAP). Under FRAP, targets were set for restoration of the health of this large watershed. In 1992, the Fraser Basin Management Program built on the work of FRAP and developed a strategic plan, the Charter for Sustainability. Since 1997, the FBC has become the major advocate for the sustainability of the basin, placing emphasis on actions to restore the health of the basin and its ecosystems.

The common feature of the programs highlighted above is that they are designed to monitor and assess the health of freshwater ecosystems. However, there remain significant barriers to achieving this goal. In part, the barriers are due to the lack of a synthetic approach to ecosystem health monitoring; in part, the issue is lack of effective governance, particularly when overlapping and conflicting authorities are vested in numerous agencies; and in part, the obstacles are due to ineffective administrative practices, lack of co-ordinated efforts within civil society, and outright lack of political will.

Marine Ecosystems

Several Canadian initiatives on both the Atlantic and the Pacific coasts focus on monitoring, assessing, and improving the health of marine ecosystems.

The Atlantic Coastal Action Program: Addressing the Health of Atlantic Coastal Environments

Facing seriously degraded coastal environments and failing socio-economic conditions in Canadian Atlantic coastal communities, Environment Canada founded the Atlantic Coastal

Action Program (ACAP) in 1991 (http://www.ns.ec.gc.ca/community/acap/index_e.html). The objective is to encourage local communities to restore vigour and health to the region's ecosystems (including its human communities) through the implementation of Comprehensive Environmental Management Plans (CEMPs). Of primary concern are water quality relating to concentrations of domestic sewage and toxics and how these residuals affect the biophysical environment and recreation; natural habitat and how it is threatened by anthropogenic activity; atmospheric emissions relating to air quality and climate change; and the economic, social, and cultural characteristics of the region, which have been damaged by ecosystem transformation.

The Bay of Fundy/Gulf of Maine Ecosystem

The Bay of Fundy/Gulf of Maine forms part of the continental shelf of eastern Canada and New England. The bay and the gulf form one large marine ecosystem comprising 180,000 square kilometres, of which 16,000 square kilometres form the bay portion. This large marine ecosystem is relatively shallow—no more than 200 metres in depth for the most part. It has the world's largest tidal range, exceeding 16 metres from low to high tide in Minas Basin at the head of the bay. At peak flood, the tidal flow into the Gulf of Maine is more than 25 million cubic metres, equivalent to 2,000 times the average discharge of the St Lawrence River.[15]

There have been a variety of programs, many involving cooperative efforts between Canada and the US (since they share these waters), to assess the health of the bay/gulf ecosystem. Wells (2004) provides an excellent account of how the Statistics Canada framework and SOE report (Bird and Rapport 1986), combined with the development of the ecosystem health concept, motivated Environment Canada and other agencies to establish programs for assessing and evaluating the health of this ecosystem.

Among the many human activities that place the bay/gulf at risk are physical changes to shorelines (particularly barriers such as causeways and dykes), increased coastal development, increase in aquaculture activity, and increased small boat traffic in bays and inlets. Among the most notable changes in the ecosystem are changes in sediment patterns in estuaries, contaminants in sediments and tissues, reduction in fish stocks, and reduction in seabird populations (phalaropes) and marine mammals (right whales) (Wells 2004). More specific changes indicating deterioration in health include reduced abundance of herring and wild Atlantic salmon *Salmo salar*), changes in distribution of seabirds such as red-necked phalarope (*Phalaropus lobatus*), high levels of contaminants in biota (e.g., copper in crustaceans, PCBs in birds and mammals), and a reduced area of salt-marsh in the upper bay (Wells 2004).

On the basis of these and other findings, the consensus expert opinion is that there are pervasive signs of ecosystem pathology. For example, the 2002 Gulf of Maine Forum Protecting our Coastal and Offshore Waters[16] examined 64 potential concerns, of which 28 were judged to be severe, 18 were judged moderate to severe, 15 moderate, and only three were rated as 'no problem'. Among the many severe problems was the state of aquatic habitat, particularly benthic habitat, seagrass, wetlands, breeding areas, and spawning areas. Moderate to severe problems included shifts in populations, loss of biodiversity, changes in species dominance, invasive species, and overfishing. Collectively, these assessments served as a 'call to action' and have triggered both preventive and restoration programs, such as relocating sea lanes away from the feeding areas of the northern right whale, remediation of unused salt-marsh, improved sewage treatment for wastes discharged to the sea, and improved aquaculture practices.

The Puget Sound/Georgia Basin Ecosystem

The Puget Sound/Georgia Basin marine ecosystem encompasses the waters of Georgia Strait, Puget Sound, and their link to the Pacific through the 153-kilometre-long Strait of Juan de Fuca. The basin, essentially an inland sea, comprises a mosaic of complex benthic habitats, extensive tidal flats, and eelgrass beds. Many rivers (including the Fraser) and streams empty into the basin and host five species of salmon—although many populations, which a decade or so ago were in great abundance, have plummeted. The basin is home to a number of large marine mammals, including killer whales (*Orcinus orca*), northern sea lions (*Eumetopias jubata*), California sea lions (*Zalophus californianus*), harbour seals (*Phoca vitulina*), and northern elephant seals (*Mirounga angustirostris*) (Tsao et al. 2005). The human population in the basin exceeded seven million in 2009 and is expected to reach nine million by 2020 (NOAA 2007).

Not surprisingly, in view of the population pressures on this region, a recent assessment concludes:

In the past two centuries, humans have become another driver of large-scale ecosystem change and have disrupted or altered many processes that sustain the species and services of the Puget Sound Georgia Basin ecosystem. Humans have eliminated or impaired habitat through the modification of rivers, shorelines and marshes, have harvested some species to critically low levels, intentionally or accidentally introduced non-native and invasive species, and deposited toxic chemicals and concentrated nutrients into the marine waters. In recent decades efforts to ameliorate these effects have been initiated, and examples of localized successes can be found in many parts of the Puget Sound Georgia Basin' [NOAA 2007, 1].

While there are indeed 'local successes', they are overwhelmed by the bigger picture. A US EPA review (http://www.epa.gov/region10/psgb/indicators) shows that of nine groups of indicators of the health of this regional system, five show 'conditions worsening' and the remaining four show 'insufficient progress'. Among those in decline are marine water quality, marine species at risk, river–stream–lake quality, and urbanization and forest change.

Despite this gloomy assessment, the Georgia Basin Action Plan (GBAP) has as its focus *Sustaining a Healthy Ecosystem and Healthy Communities 2003–2008*.[17] This document includes a framework for action between the province of British Columbia and various Canadian federal departments—all with a shared vision of 'healthy, productive and sustainable ecosystems and communities in the Georgia Basin'. The shared vision is a place '[w]here people . . . can breathe clean air, drink clean water . . . enjoy unparalleled vistas of mountain, ocean and shore . . . where the integrity of natural ecosystems is protected and the future of forests, wildlife, fish and marine mammals is secure . . . where our common goal of sustainable communities—embracing individual and social well-being, economic opportunities and environmental quality—can be realized by today's generation and those yet to come.' The question is: can such a vision be realized, and how?

Under the heading 'Protecting this special place is our shared responsibility', each partner pledges to carry out its part of the bargain: Environment Canada, to preserve and enhance the quality of the natural environment; Fisheries and Oceans Canada, to manage Canada's oceans, the conservation and sustainable use of fisheries resources and habitats that support them; Parks Canada, to protect and present nationally significant examples of Canada's natural and cultural heritage; the BC Ministry of Water, Land and Air Protection, to ensure clean and safe water, land, and air and to maintain and restore the natural diversity of ecosystems; and the BC Ministry of Sustainable Resource Management, to provide strong provincial leadership to achieve sustainable development of the province's land and resources while protecting environmental values.

However, simply pledging co-ordination and cooperation among the partners does not guarantee that sufficient action will be taken to achieve the shared vision. Indeed, most assessments point to continuing degradation of the basin. The need for fast action with respect to Puget Sound (and the same applies to the Georgia Basin) was clearly spelled out by the director of the Puget Sound Action Team (Gaydos and Karlsen 2003, 4):

> We must first seek to *stop the bleeding* . . . end the ongoing losses and degradation that we are clearly aware of. Then we need to *heal the wounds*. This translates into clean-up and restoration. Then the patient needs to learn to *live a healthy lifestyle*. In the case of the Puget Sound region, this is about developing and adopting new practice . . .

This holds true not only for Canadian marine ecosystems but for many marine (and other) ecosystems around the world (Rapport 1989b; Hildén and Rapport 1993; Birkett and Rapport 1996).

CONFLICT, UNCERTAINTY AND FUTURE CHALLENGES

In the previous sections, we reviewed the concept of ecosystem health and the various programs for assessing and monitoring the health of our terrestrial, freshwater, and marine ecosystems. Here we address the thorny issues of conflict and uncertainty from an ecosystem health perspective.

Conflict over resource use and management is often portrayed as one of 'environment' versus 'economy'. Those favouring economic development are seen as justifying environmental degradation as a necessary 'trade-off' to achieve economic goals. Those favouring the environment are often seen as standing in the way of economic development. From an ecosystem health perspective, however, the question that Aldo Leopold posed in the early 1940s remains paramount: 'how to humanly occupy the Earth without rendering it dysfunctional?' Trade-offs between economy and environment generally produce a slippery slope to ongoing ecological degradation and ultimately to economic and social bankruptcy. If trade-offs of environment for economy continue, our ecosystems by degrees are likely to suffer a fate similar to what has already befallen the Laurentian Great Lakes. That such trade-offs are often accepted by the body politic suggests that while maintaining healthy ecosystems is increasingly part of the mandate of our federal and provincial agencies, it is not yet part of everyday practice.

Uncertainty is inherent in the dynamics of large-scale ecosystems; this source of uncertainty is compounded by uncertainty as to how human activities might influence the ecosystem over time. Uncertainty, however, should not be used as an excuse for inaction—which allows well-documented sources of anthropogenic stress to continue to place ecosystem and human health at risk. Here the need for vigorous application of the precautionary principle arises.

Conflict and uncertainty are central to the problematic of environmental management. Take, for example, the unprecedented outbreak of the mountain pine beetle (MPB) (*Dendroctonus ponderosae*) in the forests of central British Columbia. By 2009, the MPB had deforested more than seven million hectares of land over a region spanning the entire southern half of British Columbia (except for the western mountain range), particularly in the Vanderhoof/Prince George area.[18] The MPB epidemic has also crossed BC's border with Alberta to the east and its border with the US to the south. Despite a decade of research, there remains considerable uncertainty and conflicting views over the primary causes. Foresters tend to claim that the beetle, which is indigenous to the forests, undergoes natural cycles and that in recent times these cycles have been disrupted by global warming, which has led to drier summers (rendering trees more vulnerable to MPB attack) and warmer winters (reducing the over-winter kill of MPB larvae). Environmentalists point to forestry practices that through fire suppression and silviculture, have increased the older-aged stands of lodgepole pine[19]—the preferred food of MPB larvae.[20] Both views have merit. An increase in preferred food supply, an increase in the vulnerability of trees to infestation, and an increase in over-winter survival of MPB larvae have likely combined to create the 'perfect storm' for sparking the catastrophic MPB outbreak in central BC.

Conflict and uncertainty also abound regarding how to ameliorate the situation: how to limit the growing forest fire risk resulting from the large number of dry and dead standing trees; how to manage the economic consequences that have triggered a classic 'boom-and-bust' cycle;[21] how local communities might survive in an environment that has become heavily degraded—with loss of tree cover, rising water tables, increased soil erosion, threats to drinking water quality (many residents and even town sites rely on well water), loss of aesthetic value, and so forth. Many of the proposed 'solutions' remain weighted heavily toward short-term benefits (fire protection, salvage logging), with little attention to what is required to restore the long-term health of the regional ecosystem.

This situation further demonstrates that the main preoccupation in Canada (and in much of the rest of the world) remains rooted in short-term economic gains rather than in the long-term health of ecosystems. In the words of Ola Ullsten, former prime minister of Sweden and co-chair of the World Commission on Forests and Sustainability, 'It is our failure to take this holistic view that has led to a situation where all the Earth's ecosystems are being mismanaged. . . . It is also a system which in its reporting of progress does not distinguish between a sustainable and non-sustainable use of natural resources because only measurable market values such as the cost of timber show up in the [accounting] books' (Ullsten 1998, 131).

From an enlightened self-interest perspective, there should be no objection to policies that foster healthy ecosystems (Rapport and Singh 2006; Rapport 2007a, 2007b). From a scientific perspective, the way forward is through taking a transdisciplinary approach, one that transcends disciplinary boundaries and integrates across many fields of knowledge (Somerville and Rapport 2000). And while there are no blueprints for transdisciplinarity, the starting point is to consciously take into account various dimensions, including ecological, economic, public health, cultural, and governance. In a study of the transformation of the Carolinian forest in southwestern Ontario, where an effort was made to fuse the social and biophysical domains, it was shown that societal values and the biophysical requirements for healthy ecosystems were highly congruent. The aspects of ecosystems most valued by stakeholders were also those most indicative of ecosystem health (Patel et al. 1999). From a social-political perspective, the way forward must be to rally public support and processes for taking better care of the health of our ecosystems. A powerful vehicle for bringing increased attention to the benefits of healthy ecosystems is to articulate the many linkages between human health and ecosystem health (Arya et al. 2009; Rapport et al. 2009). Further, there is a need to foster an interactive social learning process whereby the linkages between human actions, ecosystem condition, and impacts on human well-being are explored and the effectiveness of societal interventions, as well as key indicators of ecosystem health, are continuously being evaluated. (See also Chapter 18).

Mark Anielski's guest statement in this chapter suggests the need for merging the considerations that go into the assessments of the health of our ecosystems into an accounting system that integrates the ecological and economic dimensions of our society. He cogently argues for focusing on Genuine Wealth, which includes as a critical component the health of our ecosystems, rather than restricting, as we do at present,

our accounts to material wealth. Anielski recognizes that monetary values (so often used as measures of 'natural capital') are inadequate in many cases, since many of the properties of healthy ecosystems have no monetary equivalent and therefore attributed values are largely spurious. I believe we must accept that while the economy of humankind is part of the economy of nature, the latter is not measurable in monetary terms except in highly restricted cases. In terms of a system of linked economic–ecological accounts, it is probably more feasible and defensible to represent the total system in biophysical and energy terms, without the need for monetary equivalents.

A commitment to maintain, restore, and enhance the health of our ecosystems is a stated objective of many of our provincial and federal natural resource and environmental agencies. Yet, the use of the term 'ecosystem health' in some cases may be little more than 'greenwashing', because activities that are causing grave damage are allowed to continue. The well-being of Canadians, in terms of public health, economic sustainability, and cultural and spiritual values, is intimately tied to reversing this trend and restoring health and vitality to our ecosystems. The way forward is not only to embrace the concept of ecosystem health, as in the stated goals of many agency programs, but also to take urgent preventive and remedial action where evidence strongly suggests that the health of our ecosystems remains compromised.

GUEST STATEMENT

Genuine Wealth and Measuring Ecological Health

Mark Anielski

A business could not function without a balance sheet showing the condition of its assets and liabilities. Yet we manage our nation, our provinces, and our communities without a balance sheet—that is, without any account of the physical or monetary condition of our natural, human, and social capital.

The implication of operating our nation without an account of these 'real capital assets' is that we could be degrading our living capital in the pursuit of economic growth, measured by Gross Domestic Product (GDP). For example, we could destroy a watershed or wetland in the short-term pursuit of another barrel of oil, without any account of the loss of this natural capital asset either on the books of industry or the GDP of the nation. Only when these natural assets are degraded or destroyed might we appreciate their true ecological and economic value.

To rectify this shortcoming of national accounting and conventional economics, I have proposed and developed the Genuine Wealth model for measuring and managing the assets and liabilities of our human (i.e., people), social (i.e., relationships), natural (i.e., the environment), produced (i.e., buildings), and financial (i.e., money) capital that matter most to our lives (Anielski 2007). The model measures what matters most to people's values—with the word 'value' properly defined in Latin as meaning 'to be strong or worthy'. The word 'wealth' comes from Old English, meaning 'the conditions of well-being'. The model of Genuine Wealth attempts to measure the conditions of well-being that contribute and align most closely to people's values and determinants of happiness.

The Genuine Wealth model is a practical tool for decision-making that sees all five forms of capital as interrelated or integrated and as together constituting the well-being of a society. Ideally, all five should be maintained in a state that ensures a sustainable flow of benefits (services) for a flourishing society and a healthy ecosystem. The model is based on ecological and economic principles that view the economy as a wholly owned subsidiary of the ecosystem; without natural capital, there would be no economy or material well-being.

The Genuine Wealth model begins with an inventory and assessment of the physical and qualitative conditions of the wealth of nature: land, air, and water, including proxies of the integrity of ecological goods and services. In addition, the model determines a monetary value associated with each of the physical attributes of natural capital, if appropriate. For example, it is possible to determine the economic value of a watershed or wetland in terms of how much it would cost to replace ecological services with a water treatment facility. It is also important to account for the environmental depreciation costs (e.g., pollution costs), which are normally counted as additions to GDP. This approach to adjusting the GDP for natural capital depreciation costs is part of the Genuine Progress Indicator (GPI) accounting protocols advocated by several organizations, including the Pembina Institute (Anielski et al. 2001) and GPI Atlantic in Halifax.

The practicality of such natural capital accounting was demonstrated in research by my colleague Sara Wilson, an ecological economist, regarding the boreal ecosystem of Canada and the Mackenzie watershed in northern Canada (Anielski and Wilson 2005; 2007). Both studies demonstrated that it is possible both to measure the state of well-being and integrity of ecosystems and to account for their economic or monetary value in the context of conventional GDP measures. It is also possible to translate the loss of health or integrity of ecosystems due to human economic activity to natural depreciation costs that ultimately impose short-term or long-term economic costs on society.

Ecosystem valuation and natural capital accounting practices can help decision-makers take the true value of nature more fully into account. However, at the same time, it is important to understand the limitations of ecosystem valuation. Such valuation does not fully represent the values of all ecosystem services and therefore should not be the only tool used in decision-making. Ecological, cultural, and social values associated with the sustainability of ecosystems need to be considered in policy decisions. Measures of ecosystem health based on proxies (e.g., indicator species) of ecological integrity are also needed, informed by ecologists, natural scientists, local representatives, and economists.

Unfortunately, no national or provincial commitment to natural capital accounting exists for managing for sustainability the genuine natural wealth of our nation. This will require an act of political will, making the establishment of an integrated balance sheet for the nation both prudent and possible.

FROM THE FIELD

While advances in the science of environmental management are a necessary condition of maintaining the viability of our natural resources and ecosystems, they are not sufficient. They must be complemented by a system of governance that takes on the responsibility to implement practices for the public good, even if they go against special interests. 'Politics' has been characterized as the 'choice between the unpalatable and the disastrous'. When it comes to the environment, the choice has too often favoured the 'disastrous'.

In what has become a global struggle to achieve better balance between humans and their ecosystems, Canada should be able to provide much needed leadership in that it pioneered the statistical system for assessing the relationship between human activities and the environment within an integrated ecosystem approach. Yet our environmental record remains dismal. We remain weak in our commitment to reducing greenhouse gas emissions. We have failed to take decisive action to reduce the degradation of the Great Lakes. We have presided over the failure to sustain cod stocks on our east coast and wild salmon stocks on our west coast. Our forests are virtually in shambles—with the almost total loss of the old-growth coastal and interior forests of British Columbia, for example. Our marine ecosystems have become severely degraded owing to forestry practices, pollution, urban development, and other sources of stress.

Most of our government actions on the environment have de facto been little more than 'smoke and mirrors'. While calling for public and high-level science input, our governments routinely ignore the advice when it counters prevailing practice. Clear-cut logging is dressed up as contributing to a 'living legacy' forest. Carbon credit offsets encourage continued fossil-fuel consumption today with only a thin promise of repayment in the indefinite future, and so forth. In a nutshell, our federal and provincial governments continue to favour unfettered resource exploitation at the cost of the health of our ecosystems, our public health, and our social well-being. And while a few highly publicized cases of heavily polluted areas are beginning to be addressed—such as the Sydney Tar Ponds, after many decades of stalling—other major pollution disasters are allowed to aggressively expand, as in the case of the massive oil sands development in Alberta.

While some corporations have opted for policies in keeping with 'corporate social responsibility', such behaviour is the exception rather than the rule. Most efforts amount to little more than thinly disguised 'green-washing'. At the same time, it would be too easy to lay blame entirely on governments yielding to 'corporate greed'. Collusion between government and industry is not the whole story. We as consumers, and through our lifestyle choices, bear heavy responsibility for damage to our ecosystems. Our excesses in consumption add directly to the pressures that our local ecosystems (and the global biosphere) cannot withstand. We must become far more sceptical of marketing ploys to consume more and of entertainments that contribute greatly to environmental damage. The health of our ecosystems continues to be in failing condition. To turn this around, governments, industry, and consumers must radically change their behaviours to those that will support the health and vitality of our ecosystems. The days of trade-offs between environment and economy must come to an end. Life is non-negotiable.

—David Rapport

Acknowledgements

The author thanks Dr Luisa Maffi and Dr Bruce Mitchell for critical review and editorial assistance. He is also most grateful to Dr Mikael Hildén, Dr Henry Regier, Dr Rick Moll, and Nils Odén-Rapport for their very helpful suggestions.

Notes

1. Statistics Canada initially published a statistical compendium under the title *Human Activity and the Environment* to coincide with the first three SOE reports (1986, 1991, 1996). Currently, it publishes an annual statistical compendium on human activity and the environment, the latest edition being 2007–8 (http://www.statcan.gc.ca/pub/16-201-x/2007000/5212593-eng.htm). The series of reports are based on the Pressure-State-Response model developed by Statistics Canada (Rapport and Friend 1979).

2. The Statistics Canada approach was developed through the close collaboration between the author and Anthony Friend, an economist/statistician with Statistics Canada, and collaborations with colleagues at the University of Toronto, particularly Henry Regier, a population and fisheries ecologist, Tom Hutchinson, a terrestrial ecologist, and Ted Munn, an atmospheric scientist.

3. At first it was thought that all extreme natural events, such as unusual floods, prolonged droughts, volcanic activity, and tsunamis, were entirely 'forces of nature' outside human control. Today we recognize that at least some of them, such as weather-related extremes, are in part triggered by global warming, which in turn is related

to the release of greenhouse gases through fossil-fuel consumption. Extreme events such as floods and tsunamis are made significantly worse by environmental degradation such as deforestation and removal of coastal habitat.

4. Before publication, the first SOE report for Canada was reviewed and scrutinized by a distinguished committee comprising senior scientists of Canada.

5. Member countries, which account for 90 per cent of the world's temperate and boreal forests, are: Argentina, Australia, Canada, Chile, China, Japan, Korea, Mexico, New Zealand, the Russian Federation, the US, and Uruguay.

6. The process was launched following the 1992 United Nations Conference on Environment and Development (UNCED). Under the process, a Working Group on Criteria and Indicators (WGCI) for the Conservation and Sustainable Management of Temperate and Boreal Forests was established in 1994. The WGCI identified seven criteria for sustainability, as well as 67 specific indicators by which to evaluate the state of forests (http://www.rinya.maff.go.jp/mpci/meetings/an-6.pdf).

7. This project was funded by the Richard Ivey Foundation. David Rapport of the University of Guelph and John Eyles of McMaster University were co-principal investigators.

8. It assists the provinces through research programs focusing on: assessing the health of forests with respect to the effects of climate change, invasive species, etc.; establishing criteria and indicators for sustainable forest management; and using remote sensing techniques to provide inventory data on the state of forests (http://cfs.nrcan.gc.ca/forestresearch/subjects/forestconditions).

9. Kimmins (1996) suggests that forestry practice has contributed to: depleting nutrient reserves and soil fertility so as to impair tree nutrition, vigour, and resistance to insects and disease; depleting organic matter reserves so as to impair soil moisture and soil chemistry; increasing the incidence and risk of tree pathogens and insect enemies to the level at which ecosystem processes are altered beyond the normal range characteristic of that stage of succession; eliminating early serial stages or preventing later stages of succession necessary for the maintenance of soil fertility and physical and biological conditions; eliminating,

or reducing below critical levels, organisms that play a key and indispensable role in controlling mammalian, insect, microbial, and other organisms that would impair the health and vigour of the ecosystem; and eliminating or reducing the frequency of natural disturbance (through means such as fire suppression), which in many forest ecosystems is essential to maintaining forest health.

10. http://www.iisd.org/measure/compendium/DisplayInitiative.aspx?id=1458.

11. The goal of the GLWQA is 'to restore and maintain the chemical, physical and biological integrity of the waters of the Great Lakes Basin Ecosystem' (GLWQA 1978, Article II). See also Chapter 7.

12. The distinction between 'integrity' and 'health' has become blurred in recent years. Parks Canada now defines 'integrity' in nearly the same terms as one defines 'health': namely, as 'a term used to describe ecosystems that are self-sustaining and self-regulating. For example, they have complete food webs, a full complement of native species that can maintain their populations, and naturally functioning ecological processes (energy flow, nutrient and water cycles, etc.)' (http://www.pc.gc.ca/apprendre-learn/prof/itm1-con/on/eco/eco1_e.asp). This definition would make the meaning of 'ecological integrity' virtually indistinguishable from that of 'ecosystem health'. However, according to the Panel on Ecological Integrity of Canada's National Parks, '[a]n ecosystem has integrity when it is deemed characteristic for its natural region, including the composition and abundance of native species and biological communities, rates of change and supporting processes. In plain language, ecosystems have integrity when they have their native components (plants, animals and other organisms) and processes (such as growth and reproduction) intact' (http://www.pc.gc.ca/docs/pc/rpts/ie-ei/report-rapport_1_e.asp). By this more accurate definition, most ecosystems today lack 'integrity'.

13. In addition to an overview of the Great Lakes Basin, SOLEC (2009 draft report) makes a lake-by-lake assessment. For Lake Erie, for example, the status of both open lake and the near-shore was judged to be 'poor'; in Lake Ontario, the open lake was judged to be 'good', but near-shore was 'poor'. The report goes on to examine the various

pressures contributing to eutrophication of the Great Lakes and makes suggestions for management of this problem. Surprisingly, we learn that although phosphorus-loading remains a key issue for the health of the Great Lakes, data needed to support loadings calculations have not been collected since 1991 in all lakes except Erie, where loadings information is available up to 2002.

14. http://www.healthylakes.org/site_upload/upload/prescriptionforgreatlakes.pdf.

15. Figures taken from *The Canadian Encyclopedia* (Bay of Fundy and Gulf of Maine), available at http://www.thecanadianencyclopedia.com/index.cfm?PgNm=TCE&Params=A1SEC821005.

16. http://www.gulfofmainesummit.org/pdf_presentations/SHAW_PRESENTATION_SUMMIT.pdf.

17. http://www.pyr.ec.gc.ca/georgiabasin/reports/action_plan_2003/actionplan_e.pdf.

18. http://www.pc.gc.ca/apprendre-learn/prof/sub/mpb-ddp/page3_e.asp.

19. Owing to fire suppression, BC now has roughly three times more mature lodgepole pine stands than it had 90 years ago as the efficiency of equipment and techniques for preventing wildfires has greatly increased (http://www.for.gov.bc.ca/hfp/mountain_pine_beetle/facts.htm#responding).

20. When female MPBs attack a tree to lay eggs in galleries just under the bark, they release pheromones that attract other females to do the same. During this process, a fungus (which stains the wood blue) is injected by the MPB, which impedes the natural protective response of the tree. Over-wintering larvae, on emergence, feed on tree sap, which deprives the tree of fluids and nutrients and results in tree death within a matter of weeks.

21. This 'boom-and-bust' cycle was triggered by anthropogenic stresses. In other forest ecosystems, there are natural boom-and-bust cycles, and conflict arises when humans attempt to demand stability in systems inherently dynamic. Solutions in such cases must be socio-ecological—that is, neither purely ecological nor purely socio-economic.

REVIEW QUESTIONS

1. How would you describe a 'healthy ecosystem'?
2. What are some of the key signs of healthy ecosystems?
3. What are the primary causes of ecosystem pathology?
4. What are the main features of statistical frameworks for reporting on the state of regional ecosystems?
5. How can we best insure the health of our ecosystems?

REFERENCES

Allen, E. 2001. 'Forest health assessment in Canada'. *Ecosystem Health* 7 (1): 28–34.

Anielski, M. 2007. *The Economics of Happiness: Building Genuine Wealth*. Gabriola Island, BC: New Society Publishers.

Anielski, M., et al. 2001. *Alberta Sustainability Trends 2000: Genuine Progress Indicators Report 1961 to 1999*. Calgary: Pembina Institute for Appropriate Development.

Anielski, M., and S. Wilson. 2005. *Counting Canada's Natural Capital: Assessing the Real Value of Canada's Ecosystem Services*. Prepared for the Pembina Institute and the Canadian Boreal Initiative. (Updated (2009) versions of this report are available at http://www.borealcanada.ca/research-e.php).

———. 2007. *The Real Wealth of the Mackenzie Region: Assessing the Natural Capital Values of a Northern Boreal Ecosystem*. Prepared for the Canadian Boreal Initiative. (Updated (2009) versions of this report are available at http://www.boreal-canada.ca/research-e.php).

Arya, N., et al. 2009. 'Time for an ecosystem approach to public health? Lessons from two infectious disease outbreaks in Canada'. *Global Public Health* 4 (1): 31–49.

Bails, J., et al. 2005. 'Prescription for Great lakes ecosystem protection and restoration: Avoiding the tipping point of irreversible changes'. December 2005, endorsements as of May 2006. http://www.miseagrant.umich.edu/downloads/habitat/PrescriptionforGreatLakes.pdf.

Bertram, P., and N. Stadler-Salt. 2000. 'Selection of indicators for Great Lakes Basin ecosystem'. http://www.on.ec.gc.ca/solec/pdf/mainpaper-v4.pdf.

BC MFR (BC Ministry of Forests and Range). 2006. 'The state of British Columbia's forests'. http://www.for.gov.bc.ca/hfp/sof/2006.

Bird, P.M., and D.J. Rapport. 1986. *State of the Environment Report for Canada*. Ottawa: Canadian Government Publishing Centre.

Birkett, S., and D.J. Rapport. 1996. 'Comparing the health of two large marine ecosystems: The Gulf of Mexico and the Baltic Sea'. *Ecosystem Health* 2 (2): 127–44.

Carson, R. 1962. *Silent Spring*. New York: Houghton Mifflin.

CCFM (Canadian Council of Forest Ministers). 2003. 'Defining Sustainable Forest Management in Canada: Criteria and Indicators'. http://srd.alberta.ca/forests/pdf/ForestHealth%20AnnualReportfinal2006.pdf.

CFS (Canadian Forest Service), Science Branch. 1999. *Forest Health: Context for the Canadian Forest Service's Science Program*. Science Program Context Paper. Ottawa: Natural Resources Canada.

Colburn, T. 2001. 'It's time to say "no" to toxic hitchhikers'. *Ecosystem Health* 7 (4): 192–4.

Costanza, R., G. Norton, and B. Haskell, eds. 1992. *Ecosystem Health: New Goals for Environmental Management*. Washington: Island Press.

Costanza, R., et al. 2000. 'Managing our environmental portfolio'. *Bioscience* 50 (2): 149–55.

Environment Canada. 1991. *The State of Canada's Environment*. Ottawa: Government of Canada.

———. 1996. *The State of Canada's Environment*. Ottawa: Government of Canada.

Environment Canada and US Environmental Protection Agency. 2009. 'State of the Great Lakes 2009: Highlights'. http://binational.net/solec/sogl2009/sogl_2009_h_en.pdf.

Friend, A.M., and D.J. Rapport. 1991. 'Evolution of macro-information systems for sustainable development'. *Ecological Economics* 3: 59–76.

Gaydos, J.K., and E. Karlsen. 2003. 'Securing a sustainable future for the Georgia Basin/Puget Sound region'. Georgia Basin/Puget Sound Research Conference. http://www.vetmed.ucdavis.edu/whc/seadoc/pdfs/GaydosKarlsen2003.pdf.

GLWQA (Great Lakes Water Quality Agreement). 1978. *Revised Great Lakes Water Quality Agreement of 1978 as Amended by Protocol Signed November 18, 1987*. Washington and Ottawa: International Joint Commission.

Government of Alberta. 2006. 'Forest health in Alberta'. http://srd.alberta.ca/forests/pdf/ForestHealth%20AnnualReportfinal2006.pdf.

Harris, H.J., et al. 1988. 'Importance of the nearshore area for sustainable redevelopment in the Great Lakes with observations on the Baltic Sea'. *Ambio* 17: 112–20.

Hildén, M., and D.J. Rapport. 1993. 'Four centuries of cumulative cultural impact on a Finnish river and its estuary: An ecosystem health approach'. *Journal of Aquatic Ecosystem Health* 2: 261–75.

Kimmins, J.P. 1996. 'The health and integrity of forest ecosystems: Are they threatened by forestry?' *Ecosystem Health* 2 (1): 5–18.

Levin, S.A. 1998. 'Ecosystems and the biosphere as complex adaptive systems'. *Ecosystems* 1 (5) 431–6.

Mageau, M.T., R. Costanza, and R.E. Ulanowicz. 1995. 'The development and initial testing of a quantitative assessment of ecosystem health'. *Ecosystem Health* 1: 201–13.

MEA (Millennium Ecosystem Assessment). 2005. *Ecosystems and Human Well-being: Current State and Trends*. Washington: Island Press.

Miller, R.J., and C. Nelson. 2003. 'Development of a criteria and indicators framework in Ontario'. Paper presented to the Twelfth World Forestry Congress. http://www.fao.org/DOCREP/ARTICLE/WFC/XII/0390-B1.HTM.

Myers R.A, and B. Worm. 2003. 'Rapid worldwide depletion of predatory fish communities'. *Nature* 423: 280–3.

Neess, J. 1974. 'Protection and preservation of lakes'. Paper presented at the Conference on Lake Protection and Management, University of Wisconsin, Madison.

Nielsen, N.O. 1999. 'The meaning of health'. *Ecosystem Health* 5 (2): 65–6.

NOAA (National Oceanographic and Atmospheric Administration). 2007. 'Sound science: Synthesizing ecological and socio-economic information about the Puget Sound ecosystem'. http://www. nwfsc.noaa.gov/research/shared/sound_science/documents/sound_science_finalweb.pdf.

NRBS (Northern River Basins Study) Board. 1996. 'Northern River Basins Study final report'. http://www3.gov.ab.ca/env/water/nrbs/toc.html.

OMNR (Ontario Ministry of Natural Resources). 2006a. *Forest Health Conditions in Ontario 2006.* T.A. Scarr, J. Hopkin, and J. Pollard, eds. Toronto: Queen's Printer for Ontario.

———. 2006b. 'State of the forest report' (executive summary). http://www.mnr.gov.on.ca/202348.pdf.

Patel, A., et al. 1999. 'Forests and societal values: Comparing scientific and public perception of forest health'. *The Environmentalist* 19: 239–49.

Patel, A., and D.J. Rapport. 2000. 'Assessing the impacts of deer browsing, prescribed burns, visitor use, and trails on an oak-pine forest: Pinery Provincial Park, Ontario, Canada'. *Natural Areas Journal* 20 (3): 250–60.

Patz, J.A, et al. 2004. 'Unhealthy landscapes: Policy recommendations on land use change and infectious disease emergence'. *Environmental Health Perspectives* 112 (10): 1092–8.

Rapport, D.J. 1983a. 'Ecosystem medicine'. In J.C. Calhoun, ed., *Perspectives on Adaptation, Environment and Population,* 180–9. New York: Praeger.

———. 1983b. 'The Stress-Response Environmental Statistical System and its applicability to the Laurentian Lower Great Lakes'. *Statistical Journal of the United Nations* ECE 1: 377–405.

———. 1989a. 'What constitutes ecosystem health?' *Perspectives in Biology and Medicine* 33: 120–32.

———. 1989b. 'Symptoms of pathology in the Gulf of Bothnia (Baltic Sea): Ecosystem response to stress from human activity'. *Biological Journal of the Linnean Society* 37: 33–49.

———. 1995. 'Ecosystem health: More than a metaphor?' *Environmental Values* 4: 287–309.

———. 1998. 'Evaluating landscape health: Integrating societal goals and bio-physical process'. *Journal of Environmental Management* 53 (1): 1–15.

———. 2007a . 'Healthy ecosystems: An evolving paradigm'. In J. Pretty et al., eds, *The Sage Handbook of Environment and Society,* 431–41. London: Sage.

———. 2007b. 'Sustainability science: An EcoHealth approach'. *Sustainability Science* 2: 177–84.

Rapport, D.J., et al., eds. 1998a. *Ecosystem Health.* Oxford: Blackwell Science.

———, eds. 2003. *Managing for Healthy Ecosystems.* Boca Raton, FL: CRC Press.

———. 2009. 'The impact of anthropogenic stress at global and regional scales on biodiversity and human health'. In S. Osvaldo, L. Meyerson, and C. Parmesan, eds, *Biodiversity Change and Human Health,* 41–60. Washington: Island Press.

Rapport, D.J., R. Costanza, and A.J. McMichael. 1998. 'Assessing ecosystem health'. *Trends in Ecology and Evolution* 13 (10): 397–401.

Rapport, D.J., and A.M. Friend. 1979. *Towards a Comprehensive Framework for Environmental Statistics: A Stress-Response Approach.* Ottawa: Statistics Canada.

Rapport, D.J., C. Gaudet, and P. Calow, eds. 1995. *Evaluating and Monitoring the Health of Large-Scale Ecosystems.* Heidelberg: Springer-Verlag.

Rapport, D.J., and R. Moll. 2000. 'Applications of ecological theory and modelling to assess ecosystem health'. In S.E. Jorgensen and F. Muller, eds, *Handbook of Ecosystem Theories,* 487–96. Boca Raton, FL: CRC Press.

Rapport, D.J., and H.A. Regier. 1980. 'An ecological approach to environmental information'. *Ambio* 9: 22–7.

———. 1983. 'Ecological information services for the Great Lakes'. In J.F. Bell and T. Atterburg, eds, *Renewable Resource Inventories in Monitoring Changes and Trends,* 493–7. Corvallis: College of Forestry, Oregon State University.

———. 1995. 'Disturbance and stress effects on ecological systems'. In B.C. Patten and S.E. Jorgensen, eds, *Complex Ecology: The Part-Whole Relation in Ecosystems,* 397–414. Englewood Cliffs, NJ: Prentice Hall.

Rapport, D.J., H.A. Regier, and T.C. Hutchinson. 1985. 'Ecosystem behavior under stress'. *American Naturalist* 125 (5): 617-40.

Rapport, D.J., and A. Singh. 2006. 'An EcoHealth based framework for state of environment reporting'. *Ecological Indicators* 6: 409–28.

Rapport, D.J., C. Thorpe, and H.A. Regier. 1979. 'Ecosystem medicine'. *Bulletin of the Ecological Society of America* 60: 180–2.

Rapport, D.J., and W.G. Whitford. 1999. 'How ecosystems respond to stress: Common properties of arid and aquatic ecosystems'. *Bioscience* 49 (3): 193–203.

Regier, H.A., and W.L. Hartman. 1973. 'Lake Erie's fish community: 150 years of cultural stresses'. *Science* 180: 1248–55.

Schindler, D.W. 1987. 'Detecting ecosystem responses to anthropogenic stress'. *Canadian Journal of Fisheries and Aquatic Sciences* 44 (supplement 1): 6–25.

Somerville, M.A., and D.J. Rapport, eds. 2000. *Transdisciplinarity: ReCreating Integrated Knowledge.* Oxford: EOLSS Publishers.

Statistics Canada. 1986 . *Human Activities and the Environment.* Ottawa: Government of Canada.

Tsao, C.-H., L.E. Morgan, and S. Maxwell. 'The Puget Sound/Georgia Basin region selected as a priority conservation area in the Baja California to Bering Sea initiative'. Proceedings of the 2005 Puget Sound Georgia Basin Research Conference. http://www.mcbi.org/publications/pub_pdfs/Tsao_et_al_2005.pdf.

Ullsten, O. 1998. 'The world forest crisis: The five imperatives for a sustainable future'. *Ecosystem Health* 4 (2): 130–3.

UNEP (United Nations Environment Programme). 2007. *Global Environment Outlook.* Valletta, Malta: Progress Press.

US GAO (US General Accounting Office.) 2003. *Great Lakes: An Overall Strategy and Indicators for Measuring Progress Are Needed to Better Achieve Restoration Goals.* Washington: GAO-03-515.

Vitousek, P.M., et al. 1997. 'Human domination of earth's ecosystems'. *Science* 277 (5325): 494–9.

Wells, P. 2003. 'Assessing health of the Bay of Fundy'. *Marine Pollution Bulletin* 46 (9): 1059–77.

———. 2004. 'Assessing the health of the Bay of Fundy—Concepts and framework'. In P.G. Wells et al., eds, *Health of the Bay of Fundy: Assessing Key Issues. Proceedings of the 5th Bay of Fundy Science Workshop and Coastal Forum 'Taking the Pulse of the Bay'*, Wolfville, NS, 13–16 May, 1059–77. Occasional Report no. 21. Dartmouth, NS: Environment Canada–Atlantic Region.

Wichert, G.A., and D.J. Rapport. 1998. 'Fish community structure as a measure of degradation and rehabilitation of riparian systems in an agricultural drainage basin'. *Environmental Management* 22 (3): 425–43.

Wilcox, B.A., and B. Ellis. 2006. 'Forests and emerging infectious diseases of humans'. *Unasylva* 57/2 (224): 11–18.

World Bank. 2006. *Measuring Coral Reef Ecosystem Health: Integrating Societal Dimensions.* Report #36623. Washington: World Bank.

WCED (World Commission on Environment and Development). 1987. *Our Common Future.* Oxford: Oxford University Press.

4

First Nations' Access and Rights to Resources

Annie L. Booth and Norman W. Skelton

Learning Objectives

- To understand First Nation concerns with regard to access and rights to natural resources.
- To understand the definitions of and differences between Aboriginal rights and title.
- To identify the significant laws and legal cases affecting First Nation rights and access to resources.

Annie was once trying to explain the First Nations concept of land to a workshop full of non-Aboriginal resource managers. After a mental struggle, she looked at the First Nation co-facilitator, who without hesitation said quietly, 'It's home.' That statement is only one of the sources of conflict and uncertainty over First Nations' access and rights to natural resources. Land and resources are far more than merely something to exploit, as is often the case in non-Aboriginal society.

Complicating the situation is the fact that First Nations are a not-quite-conquered people surrounded by a growing, resource-hungry non-Aboriginal population whose actions are tempered only by an odd streak of altruism as articulated, poorly perhaps, in a series of federal laws and court decisions over First Nation rights and resource access. In other words, the subject of First Nations' access and rights to resources is a complex and messy one, not readily compressed into one book chapter. Further, as is the case with most Canadian resource issues, First Nations issues are continuing to evolve, guaranteeing that what is written here will be somewhat out-of-date by the time the book is published. If we could say, for once and certain, that *this* is the case regarding First Nations and rights to resources, resource management decisions would be clearer. As things stand, we can expect such decisions to remain complex well into the future, appropriate for a book on conflict and uncertainty.

As of this writing, the situation is complicated further as First Nations, angered by encroachments into their traditional territories, launch not just legal challenges but actual confrontational protests. In Ontario, beginning in 2005, the Six Nations of Grand River have held sometimes violent demonstrations and set up blockades near the community of Caledonia over a land claim. The land was destined for residential development, but the Six Nations maintain that their title was never extinguished (http://auto_sol.tao.ca/node/view/2012). British Columbia has witnessed protests over fishing rights (with non-Aboriginals dumping fish on the steps of the Legislature to protest Aboriginal commercial fishing rights), logging, and mining in traditional territories and may see protests erupt over the proposed 'Site C' dam in northeastern BC. As First Nations feel increasingly that their constitutional and treaty-guaranteed rights to live a traditional lifestyle are threatened, and as the non-Aboriginal economy continues to push for expanded resource development to meet Canada's (and the world's) insatiable demand for raw resources, conflicts, sometimes violent ones, will become more common.[1] Court cases, land claim negotiations, and legal change may be too little, too late in these circumstances.

Students are cautioned that this chapter is neither exhaustive nor written by legal experts. Nonetheless, we have summarized what we think are key points relevant to First Nations' rights and access to natural resources. Further, everyone seems to have a different opinion about how to interpret the treaties, law, and court cases presented here; when discussing First Nation issues, the question to ask is 'according to whom?' This chapter represents our understanding only.

A note is required on terminology. In Canada, the terms 'Indian' and 'Aboriginal' are recognized political and legal terms for First Nations. 'Indian' as a term generally refers to Aboriginal peoples, not Inuit or Métis, but people recognized as members of an Indian Band. 'Aboriginal' is used in a broad fashion to refer to all three peoples recognized by the Canadian government: Indian, Inuit, and Métis. The term 'First Nations' has no legal basis but is used politically and popularly. It is an imprecise term. Depending on what part of Canada we are in, 'First Nations' can refer only to Indians or can include the Inuit or Métis. We use these three terms interchangeably. However, we try to use the legal terms in the context of a legal discussion. The term 'Native' is no longer used politically or legally.

WHO ARE THE FIRST NATIONS (OR LIES, DAMN LIES, AND STATISTICS)?

This is a complex question but a key concept in understanding why First Nations' rights to resources are important and complex issues. Where their ancestors originated and when they became residents of North America are questions still being debated among archaeologists and the First Nations themselves. Archaeologists tend to suggest an origin in central Asia and a migration into North America between 10,000 and 20,000 years ago (Dickason 1992). Some First Nations argue a figure closer to 40,000 to 50,000 years ago. However, traditional First Nations state that according to their religious doctrine, they originated in North America and were placed there to live within specific relationships with the natural world (ibid.). All arguments suggest a long-term occupancy, which morally, as well as legally, might be expected to confer substantial claim to land and resources.

Bolstering this claim is well-documented evidence that First Nations used and/or occupied most of the landscape of North America, although population levels and land use varied widely depending on the climate and ecological productivity of a region. More temperate and productive areas tended to have larger populations and more settled lifestyles (Dickason 1992).

It is important to recognize that First Nations were, and are, very different Nations, with different languages, customs, cultures, histories, and religions. Lumping all Canadian First Nations into a single group is as inaccurate, and misleading, as lumping the Scots and the Poles together as 'Europeans'. It can be done, but it tells you little of much use. To understand land and resource issues from a First Nations' perspective, you need to acknowledge that each cultural perspective is unique. From a Canadian legal perspective, however, Aboriginal peoples are a single entity and are treated as such. Legal challenges initiated by individual Nations regarding their rights to lands and resources are considered to be legal precedent for all First Nations.

It is also important to recognize that Canadian Aboriginals were never formally conquered by the invading Europeans (as they were in the United States). Rather, their loss of control over land came through subtler means: legal concessions through treaties (with key exceptions), ties of dependency generated by participation in the fur trade, the loss of both people and culture to the ravages of introduced diseases and alcoholism, the betrayal of France (which ceded Canada to Britain), and finally, an inability to keep up with the rates of immigration. By the time Canada became a sovereign country, the original inhabitants were a vastly reduced, subjugated minority, popularly expected to disappear by the beginning of the twentieth century. Even the right to determine who they were now belonged to the government of Canada.

The Canadian government and its responsible ministry, Indian and Northern Affairs Canada (INAC), legally recognizes three groups of Aboriginal peoples: status Indians, Inuit, and Métis (a person of historically mixed Aboriginal and European blood) (INAC 2008). Both legally recognized Inuit and status Indians receive special benefits, such as free health care and certain rights to hunt and gather for subsistence purposes on Crown lands. However, the government also determines who holds status as an Indian or Inuit. Originally, this was done through a census of individuals of recognized Bands (although many individuals and some Bands were missed).

Their descendants were also recognized by the government but could become disenfranchised (or lose their status) through a number of actions, such as an Aboriginal woman marrying a non-Aboriginal, buying property, attending a university, or serving in the army. These individuals, as well as those who were missed in the original survey, were not legally recognized (although they might be recognized by the Nation they belonged to) and were denied benefits, including the right to subsistence hunting and gathering. Amendments to the Indian Act in 1985 changed this situation, and many individuals have applied to be legally reinstated. Only legally recognized Aboriginal peoples hold claim, in the governments' eyes, to rights to natural resources. Others of First Nation descent hold only the rights of other Canadians.

Modern Aboriginal peoples represent a small percentage of the Canadian population. INAC reports, based on the 2001 census, the following population numbers: 1,064,300 Aboriginal people, including 633,600 registered (status) Indians, 46,200 Inuit, and 274,200 Métis (INAC 2007). While this seems like a small number, it is important to realize that the population has more than doubled in 20 years and that INAC projects Aboriginal population numbers to grow sharply: it projects 920,100 Indians, 74,800 Inuit, 376,500 Métis, and a total of 1,566,900 Aboriginal people by 2026 (INAC 2007). It is also important to recognize that in 2001, 48 per cent of the Aboriginal population was under 25 years of age. Not only is the population growing rapidly, but such a young population means rapidly growing demand for jobs, housing, resources, and land in the future. This fact must be taken into account when Aboriginal communities plan for their future.

Ontario has the largest concentration of status Indians, 22.8 per cent in 2001. British Columbia and Manitoba are next, with 16.4 and 15.9 per cent, respectively. However, British Columbia has more recognized Bands, or separate political entities, to work with—198 Bands to Ontario's 126. In total, 612 Bands are recognized by the Canadian government. They are scattered across 2666 reserves, or lands held by the Canadian government in trust for the Bands (excluding Nunavut). Although the reserve land totals more than 3 million hectares, for a number of reasons this amount of land is inadequate to meet the needs of a young, growing First Nation population. Thus, many Nations have undertaken legal action to claim additional lands and to gain self-government. Self-government and land negotiations are taking place in all provinces and territories of Canada. A complete (and current) list can be viewed at INAC's website (www.ainc-inac.gc.ca).

The situation of the Métis requires a special note. Since the 1982 Constitution, Métis have been recognized as one of the three Aboriginal peoples of Canada. By definition, a Métis person is of historically mixed Aboriginal and European blood, generally a situation considered as having arisen largely in the Prairies. Until 1982, however, the Métis had no special legal standing and were not included in most legal considerations. Since their recognition in 1982 as an Aboriginal group, it has been argued that they too possess Aboriginal rights.

In recent years, Métis have launched court cases to prove that they have the same rights to hunt, fish, and gather as other Aboriginal groups do, and in some cases they have been included within modern land claims. In 2001, the Ontario Court of Appeals ruled that Métis held the right to hunt and fish outside of normal hunting seasons and without licences, as do status Indians (*Powley v. Ontario*). Interim harvesting agreements are in place in several provinces and have been upheld in a couple of legal challenges in Alberta (*R. v. Kelley*[2]) and Ontario (*R. v. Laurin, Lemieux and Lemieux*[3]); both judgments

were handed down in 2007. Another court case, however, focused on the issue of identity. In *R. v. Hopper*,[4] the judge ruled that someone seeking to exercise traditional rights must have a proven ancestral heritage to a rights-bearing community (as outlined in *R. v. Powley* [2003]); if there is no historic rights-bearing community, there cannot be a contemporary rights-bearing community. Further, identification as a member of a rights-bearing community must be accepted by a modern Aboriginal or Métis community. Finally, a 2006 court case in Labrador established that the government of Newfoundland and Labrador did indeed have an obligation to consult the Labrador Métis Nation regarding proposed industrial developments, a significant victory for the Métis.[5] However, as Métis rights remain in an undefined and murky state at present, we have made the decision to generally not include the Métis in this discussion, although you are strongly advised not to ignore the Métis in the future.

As stated in the introduction, and as indicated by these basic statistics, the situation of First Nations with regard to land and natural resources is complex—reflecting this book's themes of conflict and uncertainty. First Nation governments must cope with a rapidly growing and youthful population requiring greater access to land and resources that can generate employment and income to communities. However, land claims, questions of Aboriginal rights and title, and demands for self-government complicate the situation.

ABORIGINAL RIGHTS TO RESOURCES

Access to resources and *rights* to resources are two different but related issues. Access and rights to resources were traditionally controlled by the individual Nations that occupied the territory. When Europeans arrived, they often found access and rights to resources controlled through family (or clan or house) affiliations. Academics have long debated whether these Aboriginal institutions were traditional or a reaction to the European fur trade. However, for the time, they were functional institutions for controlling access to resources, documented in Nations ranging from the Cree (Speck and Eisley 1942) to the Carrier (Dickason 1992; McMillan 1995). These institutions were reasonably effective in ensuring that a family had regular access to necessary resources while also ensuring that the resources were not subject to uncontrolled exploitation. Historical resource rights, however, are a story for another time. Here, we are interested in modern rights to resources.

When talking about First Nations' rights to natural resources, several terms are thrown around: Aboriginal rights, Aboriginal title, and Aboriginal land claims. These are different, if poorly defined, ideas. Until recent court cases, Aboriginal rights and Aboriginal title often were taken to mean the same thing. Nadasdy (2002, 248) points out that 'Aboriginal title', while taken to form the legal basis for First Nations' land claims, never appears in Section 35 of the Canadian Constitution. The Constitution, rather, recognizes existing 'Aboriginal and treaty rights'. McNeil (1997) argues that recent court decisions have treated Aboriginal title as a subset of rights—in other words, the ability of First Nations to hold title to land stems from a larger collection of rights. Kulchyski (1994, 10), on the other hand, argues that Aboriginal title 'is the basis of all other Aboriginal rights; that all the other political and property rights flow from the doctrine of prior occupancy and the title to land'. Both title and rights are understood to have originated from the uncontested fact that the Aboriginal Nations were present prior to the arrival of the Europeans.

The *Calder* legal decision (1973) confirmed that Aboriginal title had never been extinguished by European occupation. *Delgamuukw* (1997), while legally determining nothing, is held out as the case that made the rationale for Aboriginal title, since the Supreme Court judge in the case chose to define title as the right to the land itself—in other words, as a property right rather than simply rights to usage. Further, if title is demonstrated, according to *Delgamuukw*, uses of that land could not be inconsistent with Aboriginal use of the land. In other words, if First Nations hunted as part of Aboriginal relations with the land, a use could not render the land unsuitable for hunting. The federal government must deliberately, through the law, choose to extinguish title, or the First Nations must explicitly agree to extinguish title (for example, through a land claim agreement) according to the 1982 Constitution. However, the concepts remain subject to ongoing legal definition. Further, each Nation faces an uphill battle in demonstrating that they possess Aboriginal title to a piece of land. Recent court cases have refused to discuss title, however. In court cases, rights and title have become separate legal concepts and can exist independently of each other.

What is clear, at least regarding First Nations' rights to natural resources, is the following: First, since the Constitution of 1982 (Section 35), the federal government has recognized that First Nations have rights of access to and use of certain resources. These rights derive from traditional practices and traditions, or they derive from rights specifically guaranteed in historical or modern treaties. While First Nations and their lands (reserves) are a federal responsibility, and while their rights are upheld by federal law, until otherwise demonstrated, these rights can be subject to provincial legislation. This is a crucial issue, since many natural resources, including wildlife, freshwater fish, forests, and minerals are under provincial jurisdiction; others, such as ocean or anadromous fish, are a federal concern. Thus, the exercise of Aboriginal rights crosses the federal–provincial jurisdictional line. Section 88 of the Indian Act states that all provincial laws apply to First Nations unless those laws interfere with treaty rights, conflict with federal laws, or otherwise intrude into areas covered by the Indian Act. Demonstrating that treaty rights *do* supersede provincial regulations has been the content of several court cases.

Other court cases have determined that any infringement upon established rights must be avoided unless there are 'compelling and substantial' reasons (*Sparrow*, cited in Chandran 2002, 4). If those reasons can be demonstrated, the government can infringe on Aboriginal rights, including limiting access to or use of natural resources. However, recent court cases have limited the government's right to infringe. In *Tsilhqot'in Nation v. British Columbia* (2008), the BC Supreme Court found that forest harvesting and silviculture activities did infringe upon the Tsilhqot'in Nation's hunting and trapping rights by depleting species, both in their diversity and numbers, through direct mortality and the destruction of species habitat. Since the provincial government lacked a solid database on species numbers and types, their efforts at consultation could not justify the infringement. In *Mikisew Cree First Nation v. Canada (Minister of Canadian Heritage)* (2005) the court determined that under Treaty 8, a historical treaty, the First Nation's right to hunt was subject to express geographic limitations related to land required or taken up for settlement, mining, lumbering, and trading or, in this case, a road, an implied geographic limitation. However, the key finding was that the 'Crown must however act in good faith and not take up so much land that no meaningful right to hunt remains.'

In general, Aboriginal rights to resources have been restricted to use for subsistence or traditional purposes or as otherwise necessary for cultural survival. However, at least a few cases have considered the issue of commercial rights (*Gladstone, Van der Peet* and *Kapp*), with mixed results. *Gladstone* and *Van der Peet* found no rights to a commercial application of fishing. *Kapp* upheld a limited right to a commercial fishery in British Columbia.

Second, as mentioned earlier, First Nations might have some claim to hold title on some lands, with rights to manage the resources thereon. Hence the process of land claims. These rights and claims to title have been defined and determined through a complex set of legislation, treaties between First Nations and the government of the day, and a number of lengthy court cases. Title, as defined in *Delgamuukw*,

> entails a right to the land itself, whereby a First Nation possesses the right to exclusively use the land however it sees fit subject to the limitation that the land not be used in a way that destroys the people's relationship to the land. An important aspect of Aboriginal title defined in this way is that it does not restrict Aboriginal rights to land and resources to traditional uses [Chandran 2002, 7].

Title is generally held to stem (although courts have disagreed on this) from the Royal Proclamation of 1763, which stated that 'any Lands whatever, which, having been ceded to or purchased by Us [George III of Britain] . . . are reserved to said Indians' (cited in McMillan 1995, 320). Thus, as has been argued by First Nations in court, unless a treaty explicitly extinguished title, Aboriginal title still exists on lands not covered by such treaties. First Nations have sought to have their title affirmed through both the courts (*Calder* and *Delgamuukw* were both about title) as well as, after 1974, through a modern treaty process. Gaining title to traditional lands would be, and is, a significant achievement for a First Nation, since it theoretically grants them the right to develop, manage, and, if they choose, benefit commercially from the natural resources on the land. Title would come closest to returning self-sufficiency and cultural integrity to First Nations. Naturally, gaining title is difficult and controversial.

Access to natural resources, then, can come through two sources: Aboriginal rights and Aboriginal title, sometimes simultaneously. Both can be defined through legislation, treaties, and legal challenges. A third issue has made its way into the courts, however, in recent years, and that is the duty of the Crown (the federal or provincial governments) to consult with First Nations (and the Métis, since *Labrador Metis Nation*) prior to activities occurring within their territory (lands where they hold rights by treaty or by tradition). *Haida Nation v. BC* (2004) and *Taku River Tlingit First Nation v. BC and Redfern Resources* (2004) outlined and confirmed the duty of the Crown to consult with First Nations whose traditional rights are affected by activities conducted either by government or a third party. However, the court did not outline clearly what that duty was or what accommodation of traditional rights was required. In other words, more cases will be required to further describe the nature of consultation and accommodation. *Haida* did one further thing: it established that the duty to consult rested with the government and not with the third party, unless otherwise provided for (the British Columbia government had attempted to delegate consultation responsibilities to the industry).

We will look briefly at all three with respect to access to natural resources.

First Nations and the Law

The subject of First Nations and the law is at least a 14-volume set and cannot be addressed completely here. However, some key federal legislation will be mentioned.

The Royal Proclamation (1763)

The Royal Proclamation of 1763 was the basis for much subsequent legislation regarding First Nations. As mentioned earlier, it first identified what has since been called Aboriginal title, which Aboriginals hold through use and occupancy. The proclamation also set the stage for the reserve system. Finally, the proclamation set up a process for extinguishing title by permitting the sale of Aboriginal land only to the Crown and established the government's fiduciary relationship with their wards, the First Nations (McPherson and Rabb 1993). Fiduciary responsibility is a concept meaning that the federal government has a responsibility to act in the best interests of First Nations in any dealings they undertake on their behalf.

The Constitution Act (1867)

Among other things, this legislation made the federal government responsible for Indians and reserves.

The Indian Act (1876)

This first consolidated Act laid out who was recognized as an Indian by the government as well as the conditions by which Indians could become enfranchised, or Canadian citizens—in other words, the acts by which a First Nation person could lose government recognition as an Indian.

The Act was revised in 1985: the enfranchisement process was deleted, and the process of registration was begun. To become registered, a person must demonstrate that at least one of his or her biological parents was registered or a member of an Indian Band recognized by the government in 1985. This resulted in a sharp increase in the Aboriginal population as those who had lost their status were re-admitted. These 'new' First Nations and their children must be provided for through access to resources and economic development.

Section 88 of the Act also affirms that First Nations are subject to all provincial regulations that do not interfere with treaty rights, conflict with federal laws, or overlap with Indian Act matters.

The Constitution Act (1982)

Section 35 recognizes and constitutionally guarantees Aboriginal and treaty rights of Aboriginal peoples. The Act not only recognizes Indians as Aboriginals but includes Inuit and Métis within the category of Aboriginal, thereby recognizing their Aboriginal and treaty rights. Rights gained by way of land claim agreements are explicitly recognized, and therefore protected, as treaty rights. Chandran (2002, 6) comments that it is only since the Constitution affirmed treaty rights that federal, provincial, and territorial prosecutions for violations of various wildlife legislation have declined. The Constitution also recognizes the federal Crown's fiduciary responsibility to Aboriginal peoples.

Resources Rights and Treaties

Treaties are legally binding agreements between recognized governments. For First Nations, treaties were, and are, generally negotiated by the representatives of recognized Bands and occasionally larger councils made up of different Bands and representatives of sovereign nations, including Britain and Canada. While in modern times, in British Columbia for example, the provincial government has sometimes initiated and directed treaty negotiations, in the end the federal government has to agree to any treaty. Treaties are held to be binding on provinces in some cases, such as the Natural Resource Transfer Agreements. The Natural Resource Transfer Agreements with the three Prairie provinces impose constitutional obligations to uphold treaty rights in exchange for specific rights to the land and its resources (Bartlett 1991).

The decision to negotiate 'modern' treaties in the form of land claims after 1974 is an interesting one, since it was the beginning of an implicit recognition, or re-recognition, of First Nations as sovereign nations. Treaties can only be negotiated between independent governments.

Treaties between First Nations and governments come in two strata: 'historical' (or 'numbered') treaties and 'modern' treaties. Historical treaties include those signed between First Nations, Britain, and later the Canadian government prior to 1924, after which no treaties were negotiated. Modern treaties are the modern land claim negotiations, which began in 1974. We will look first at the significance of the historical treaties.

The first treaties were signed in the Atlantic provinces with the British government, seeking 'peace and friendship'. However, increasing European settlement turned treaty-making into land grabs. Small treaties were later signed that ceded lands in the Ontario region between 1780 and 1850. In only a few cases did First Nations hold on to hunting and fishing rights, and many lost both home and a living. The first large-scale treaties were negotiated in 1850 in the upper Great Lakes region. In exchange for a reserve, annual cash payments, and a lump sum, First Nations surrendered their lands. Significantly, they retained the right to hunt and fish over all unoccupied Crown lands.

Also beginning in 1850, the Douglas Treaties were negotiated for parts of southern Vancouver Island. Again, in return for surrendering their land, which became 'the entire property of the White people for ever', the Indians were assured that they would be at 'liberty to hunt over the unoccupied lands, and to carry on fisheries as formerly' (McMillan 1995, 318). The British North America Act of 1867 made the federal government responsible for First Nations and their reserves as well as treaty negotiations. However, all subsequent treaties (15 in all) followed similar patterns: lands were ceded in exchange for dedicated reserve lands and the right to hunt and fish over unoccupied Crown lands. The last 'historical' treaty was negotiated in 1923, extinguishing all Aboriginal claims in southern Ontario.

From the government's perspective, the treaties effectively extinguished Aboriginal *title* to the lands. The property became that of the federal government. It is unclear, and still debated, as to what the treaties meant from a First Nations' perspective. As Nadasdy (2002) and others (including Booth and Jacobs [1990] and Booth [1998], [2007]) argue, the concept of property, in a European sense, was and perhaps still is literally foreign to First Nations. Many First Nations argue that having never owned the land, they could never sell it. Unfortunately, this too is another story but one worth investigating. In this story, the treaties should have ended matters. They did not (otherwise, this would be an easier chapter to write): they became instead one piece of a complex situation.

First, historical treaties were not negotiated throughout Canada: missed were Quebec, most of British Columbia, most of the Atlantic Provinces, part of the Yukon, and part of the Northwest Territories. These areas became the subject of the modern quest for treaties: land claims. Second, many reserve lands promised in historical treaties never materialized (in addition to a number of other promises; these too are the subject of modern land claims, as well as of some key legal court cases such as *Delgamuukw*). Finally, the rights to hunt and fish over unoccupied Crown lands promised in all treaties were honoured more in the breach than in the fulfilment for many First Nations. And this failure has also precipitated a number of court cases.

Modern treaties, also known as land claims, have been under negotiation in Canada since 1974 when Ottawa opened the Office of Native Claims, prompted by a court case brought by the Nisga'a Tribal Council (British Columbia). Two types of claims are considered: *comprehensive*, which are based on Aboriginal title, and *specific*, based on specific unfulfilled obligations. Most land claims are comprehensive (McMillan 1995).

Land claims were also prompted by the dire straits in which many First Nations found themselves. In areas not covered by treaties, most First Nations had been forcibly settled on small amounts of land, chosen by the provincial or federal governments and managed by the federal government. They had no control over, or technical ownership of, reserve lands. Some Aboriginal communities never received reserve status and thus have no protection at all (as in the well-publicized case of the Lubicon Cree in Alberta [Brown 1996]). In non-treaty areas, many received lands on a per capita basis, amounting to very small allotments in comparison to lands in treaty areas— in some cases, just the land their houses sat upon. Poverty was, and is, endemic and epidemic, leading to considerable social ills. Many First Nations hoped that pursuing land claims would give them social security, some control over their own lives, and a land base from which revenue and employment could result through natural resource exploitation. Those with unfulfilled treaties pursued specific claims under the same hope.

Modern treaties have been concluded with many Aboriginal groups that had originally been overlooked. In 1975, a land claim was settled with the Cree and Inuit of northern Quebec (the James Bay Agreement) and extended in 1978 to include the Naskapi Innu. In exchange for surrendering Aboriginal title to the lands, these First Nations received a cash settlement ($225 million over 20 years), ownership of certain lands, and exclusive hunting, trapping, and fishing rights over a larger area. However, the settlement was achieved specifically so that the Quebec government could proceed with the hydroelectric development known as the Great Whale. It is worth noting that the first portion of the development began in 1971 without any consideration of land claim settlement or compensation. Threatened with extensive flooding of their traditional territory and loss of their hunting and fishing lands, the Cree obtained an injunction to halt the project. The 1975 James Bay Agreement cleared the way for the project to continue. Many felt, however, that the amount of compensation did not make up for the loss of traditional lands and key subsistence areas (McCutcheon 1991; Jenson and Papillon 2000).

Later, the Quebec government proposed additional dam construction that would flood a greater area, including some lands covered by the 1975 agreement. Protests by the Cree and Innu resulted in more injunctions, an international publicity campaign, a crisis during the Quebec referendum on sovereignty in 1995, lawsuits, and ongoing bitterness. This issue, the lawsuits, and the land claim may have finally been resolved in 2002 when the Cree signed an agreement with Quebec for $2.25 billion (US) in exchange for permitting ongoing hydroelectric development. The James Bay Cree will receive the payments over 50 years (the first came in 2008) and have gained greater say over logging

in their remaining territory, as well as jobs with the power authority (Associated Press 2002). It may, however, represent only a partial resolution, because the agreement was not universally endorsed by the Cree. Overall, 70 per cent of those who voted approved the deal, but approval in some of the nine communities involved ran as low as 50 per cent. Nor does money completely compensate for a loss of living, history, or homeland (Krauss 2002). The decision will significantly affect the Cree and Innu's ability to access natural resources such as fish and game, demonstrating that rights do not always translate into access. In 2007, the Grand Council of the Cree (Quebec) held a referendum on an 'Agreement concerning a New Relationship (Paix des Braves) between le Gouvernement du Quebec and the Crees of Quebec', which was approved. The agreement seems intended to advance provisions on social, economic, and environmental concerns briefly addressed in the settlement and to place the Grand Council on a stronger footing when dealing with the provincial government. It will be interesting to see whether the province of Quebec signs the agreement (see the website of the Grand Council of Crees, http://www.gcc.ca, for further information).

Another significant land claim settlement was negotiated in the Northwest Territories with the establishment of the territory of Nunavut in 1993. This development represents a very clear case of Inuit having their Aboriginal title recognized. Nunavut has its own legislative assembly, holds unrestricted harvesting rights within the territory, includes water and mineral rights within 10 per cent of the land base in the land title, and requires substantial consultation with wildlife and environmental co-management boards, which have significant Inuit representation (Chandran 2002). These are all important firsts. The Nunavut settlement is also significant in terms of the size of the territory, encompassing approximately 350,000 square kilometres of land, the largest settlement in Canada.

The agreement explicitly recognizes the need to incorporate Inuit culture and traditions within economic development goals—hence the emphasis on unrestricted harvesting rights. Further, the Inuit may lease surface rights for resource development and share in royalties from oil and gas developments on Crown land in the territory. This provides both unprecedented economic opportunity and unprecedented challenges, including reconciling traditional cultural norms with something as potentially destructive as oil, gas, and mineral development (O'Faircheallaigh 1998; Arnott 2000). Further, as is the case for almost all Aboriginal peoples, the residents of Nunavut expect to be on the land long after the oil and gas run out. Other, long-term sources of economic self-sufficiency must be sought, but what else is available? As well, leases in place at the time that the settlement was reached must be honoured, and the revenues and resources they cover will continue to depart, since the leases give little back to residents. Finally, as the 2001 squabble between Nunavut and the Northwest Territories over the potential location of the Alaska oil pipeline demonstrated, the residents of Nunavut must compete with the rest of Canada for economic development opportunities, and there is as yet no answer to the question of whether they will be able to do so. While their long-desired title grants them rights, it also gives them responsibility for the consequences of the decisions they make.

The establishment of Nunavut was a significant settlement of an outstanding land claim, but it is unlikely to be repeated. Nunavut could be contemplated because it encompassed a very sparsely populated area, with Inuit as a majority of the population. No major metropolitan areas exist within the territory. There are no significant resources within its boundaries with the exception of oil, gas, and minerals, and the Crown only lost access to 10 per cent of the potential resources and a share of royalties. Nor was the former Northwest Territories a major player in federal politics. Thus, we are not likely to see a similar settlement elsewhere in the country.

The Yukon Territory also chose an innovative, if problematic, strategy to try and settle claims. The 1993 Umbrella Final Agreement with the Council for Yukon Indians tried to provide a comprehensive and detailed framework upon which individual agreements with Nations would subsequently be negotiated. While an agreement exists in principle, Nadasdy (2002) argues that those principles create conflict within the Nations because they attempt to impose external ideas of property rights on Aboriginals. Some individuals accept the situation; others are torn by the loss of their cultural understanding of the land and its inhabitants. On the other hand, it does represent a compromise attempt to encompass the particular needs and values of the individual Nations.

The agreement sets out principles, including the crucial right to hunt throughout the territory regardless of who obtains the land title (Council for Yukon Indians 1993, 158–60). What the treaty does not give is the right to sell edible parts—the meat—commercially except to other Yukon Aboriginals. Nor, aside from legally registered trap-lines, are the First Nations allowed to sell non-edible parts such as skins and furs. Thus, unlike the Nunavut settlement, the Yukon agreements will probably limit First Nations in their ability to exploit natural resources for anything other than government-defined subsistence, a limitation on title that will restrict economic development.

Several other modern land claims have been settled, although we will not discuss them in detail. They include the Western

Arctic (Inuvialuit) Settlement Region (1984), the Gwich'in Comprehensive Agreement (1992), and the Sahtu Dene and Métis Comprehensive Agreement (1993). In general, title was explicitly extinguished, but subsistence hunting and fishing was guaranteed. In several areas, co-management boards have been established to attempt to ensure that Aboriginal concerns are heard and incorporated in decisions affecting resource development and their impacts on wildlife (see Kendrick (2000) for a discussion of the generally successful Beverly-Qamanirjuaq Caribou Management Board).

Our last look at land claims will be the situation in British Columbia, which as of this writing is controversial. With the exception of a portion of northeastern BC and southern Vancouver Island, treaties were never signed with First Nations. Thus, perhaps title was never extinguished. The first to query this was the Nisga'a Nation in *Calder*; the 1973 decision confirmed that title had never been extinguished and prompted the establishment of the Office of Land Claims. After well over 100 years of constant and often illegal effort, the Nisga'a succeeded in reaching the first modern treaty settlement in British Columbia. The agreement-in-principle was reached in 1996 and finally ratified by the Government of Canada, the Government of British Columbia, and the Nisga'a Nation in 1998 (although not unanimously by any of the parties).

Unlike other modern treaties, the Nisga'a agreement did not force extinguishment of Aboriginal title (although there is a provision for it under specific circumstances). As one person described it, 'We did not stoop to beg for self government. We negotiated for recognition that we were a self-governing and sovereign nation at the time of contact with the white man' (Robinson 2002, 191). This hard-won recognition of title grants the Nisga'a 1930 square kilometres of

their traditional territory, including sub-surface mineral rights. Fifty-six recognized reserves outside the agreement boundary were converted to fee simple lands owned by the Nisga'a. Fee simple stands for a simple ownership such as any homeowner holds. These lands will be owned communally by the Nisga'a people with title vested in the Nisga'a government. Both will be subject to provincial statutes.

The Nisga'a agreement has several provisions affecting their access to resources. Existing tenures on Nisga'a lands are protected, meaning that timber, including some very valuable wood, continues to leave Nisga'a hands (McKee 1996, 65–7; Nisga'a Agreement-in-Principle 1996). Woodlot licences and agricultural leases on Crown land in the territory remain the province's. However, on the remaining land, the Nisga'a will have a free hand to negotiate subsequent timber leases, subject to meeting or exceeding the industry standards required by the BC government. Similarly, they may negotiate mineral leases.

Salmon on the Pacific coast represent a more contentious issue, however, and the federal government, through the Department of Fisheries and Oceans, retains responsibility for management and conservation. The Nisga'a agreement instead grants a specific catch allotment of pink and sockeye runs (15 and 13 per cent, respectively), or about 18 per cent of the entire Nass River allowable catch. In addition, the agreement makes provision for an allocation of sockeye and pink for commercial purposes outside of treaty agreements and gives the right to sell catch commercially, subject to monitoring. An annual allotment is also given for halibut, oolichan, and shellfish, but only for personal consumption. Steelhead allotments are granted annually when the stock is strong enough for a harvest (steelhead are a very vulnerable species). Finally, almost $12 million was committed to assist Nisga'a in purchasing vessels and licences to participate (or continue to participate) in commercial ocean fishing.

Wildlife and migratory birds may be harvested, subject to conservation needs as determined by a co-management committee, but only for personal use. No commercial sale is permitted. However, the public is permitted to access Nisga'a lands for legal hunting, fishing, and recreation, subject to safety and environmental considerations, which will make for interesting management challenges.

The Nisga'a case presents a complex mix of opportunity and challenge but does offer some economic security with respect to access and rights to natural resources. However, although ratified, the settlement has been the subject of legal challenges, including one by the Liberal party of BC (quietly dropped after their election to government in 2000) and by members of the Nisga'a Nation. Litigation was launched by clan chiefs whose traditional territories fell outside of the negotiated boundaries and who subsequently ended up landless. While this might seem a small matter to outsiders, losing territories held by families for generations was a severe personal and cultural shock to many.

Further, there appear to be cultural issues affecting the internal division of resource allocations. We were told of families not wishing to allow community harvesting to take place on their traditional territories or, conversely, seeing the timber rights as remaining with their family (this source wished to remain unidentified). This poor integration of cultural norms into Western-style property settlements (which is what land claims essentially are) may continue to present management challenges for the Nisga'a Nation, not just with regard to timber but to mineral rights, pine mushrooms, and medicinal plants as well. Since the treaty's implementation, little has been heard in Nisga'a Nation about any difficulties with the treaty.

While a significant decision, the Nisga'a agreement was never part of the mass treaty talks entered into by the province of BC in 1994. Under growing pressure from legal challenges, including one by the Gitxsan and Wet'suwet'en hereditary chiefs that culminated in *Delgamuukw*, from increasing Aboriginal protests over resource harvesting, and from the federal government, British Columbia established the BC Treaty Office and undertook to negotiate land claims with First Nations willing to come to the treaty table. The treaties were to be jointly negotiated among a Nation or group of Nations, the province, and the federal government. Nations that requested it were given loans (to be paid back at the end of negotiations) to prepare for negotiations. The negotiations were non-binding in the sense that a Nation could choose to walk away from a negotiation and launch a lawsuit over Aboriginal title instead, as a few have done. One Nation went back and forth a few times before remaining with treaty negotiations.

Theoretically up for grabs is much of the province; the popular figure is 110 per cent, given overlapping Aboriginal territorial claims. This is not quite accurate, because the northeastern corner is covered by Treaty 8. Further, in 2001 the McLeod Lake Band petitioned to join Treaty 8, with certain updates of its provisions. This gives them a set amount of land over which they have some jurisdiction, as well as monies. These lands are no longer under claim. However, the 110 per cent figure, in addition to other factors, has made the treaty process a highly controversial one within the province.

One area of increasing controversy is that in spite of millions of dollars and a great deal of time spent, as of 2008 only two treaties had actually reached a conclusion, those of the Tsawwassen and Maa-nulth Nations. Both the public and Aboriginals accuse the provincial government of negotiating in bad faith. In turn, the government argues that demands on the negotiation tables far exceed what they can accept in terms of money and lands. Complicating outcomes is the fact that not all Nations or tribal councils are actually in the treaty process: if all the ones in the process were to be settled tomorrow, a significant number of Nations would still have no treaty or agreement and would retain a viable claim on provincial lands. In the meantime, several Nations have accused resource extraction industries, particularly forest companies, of liquidating their assets—rapidly moving to lease and cut timber on lands that may be lost to treaty settlements, leaving the Nations with denuded landscapes. At this writing, however, much of the province of BC is in land claim limbo.

To assist with resource management during the land claims process, interim agreements were negotiated with many First Nations. For example, forest licensees are expected to 'meaningfully consult' with a First Nation if it appears that a forest plan will infringe on Aboriginal rights as outlined by the Constitution and ongoing legal cases (British Columbia Ministry of Forestry 1998). Infringements had to be either modified to accommodate the right or justified to the government. The rights protected, however, extended only to subsistence, traditional practices, or cultural practices that had been practiced before European contact. So, for example, modern trapping for commercial purposes was generally not considered a protected right (it being held until recently that commercial trapping did not occur until after the European arrival), and a company was not liable for infringement if a traditional trap-line was logged. Nor was much comment offered on the availability of the resource: if logging practices changed habitat so that species were no longer available to be hunted, it created a murky legal area subject to debates between wildlife managers and hungry hunters.

The interim agreements did not address the First Nations' ability to participate in resource extraction commercially, since such rights were poorly defined by the courts. As discussed earlier, one court case (*Tsilhqot'in Nation v. British Columbia*) did address poor wildlife management practices because they affected the Nation's access to wildlife for hunting purposes.

In term of rights and access to resources, the treaty negotiations in British Columbia appear to leave most First Nations in rather poor shape. Our statement in the 2004 version of this book still holds true: it appears to be a race between negotiations for the land and resources and the ability of industry to extract the resources.

Treaties, then, have a significant say in First Nations' rights to resources, if not access. However, historical treaty rights to resources have been more often ignored than not. Cases brought before the courts have been instrumental in regaining and redefining Aboriginal rights to resources, including, recently, the right to profit commercially, as well as confirming the unextinguished existence of Aboriginal title. We now turn to the courts.

First Nations and the Courts: Rights as Determined by Legal Argument (and Argument, and Argument . . .)

In addition to, and an integral part of, the proclamations, treaties, and other legislation of the British and, later, Canadian governments and their system of laws, several Supreme Court of Canada (SCC) cases are key in understanding the issue of First Nations access to resources. The following will be discussed here: *St. Catharine's Milling and Lumber Company* (1888), *Calder* (1973), *Bob and White* (1965), *Guerin* (1985), *Siuoi* (1990), *Sparrow* (1990), and *Delgamuukw* (1997). Several other cases will also be discussed to outline the evolution of First Nations' rights to resource access. Aspects of the Royal Proclamation of 1763, the various Indian Acts, the Natural Resource Transfer Agreements, and the Canadian Constitution are included in the rights evolution.

In the late nineteenth century, a resource dispute occurred between the Government of Canada and the Government of Ontario over access to resources. Ontario claimed a provincial right under the British North America Act 1867. The federal government held a responsibility under the terms of their treaty (3) with the Ojibway Nation. An overseas legislative body known as the Judicial Committee of the Privy Council (JCPC) ruled that provincial rights take precedence over treaty rights. The decision implied that treaty rights are terminated by the creation of a province-level government.

However, with the replacement of the overseas JCPC, the Supreme Court of Canada ruled, in *R. v. Bob and White*, that First Nations hold operative and pre-existing rights. The decision opened the possibility for one First Nation to proceed with the *Calder* case of 1973. The *Calder* case differed from that of the Ojibway Nation in *St. Catharine's Milling and Lumber*. The Nisga'a and Canada had never signed a treaty. Mr Calder, and many others, sought a treaty for almost 100 years before finally signing the Nisga'a Treaty Agreement of 2001. Mr Calder lost the decision when the tie-breaking seventh vote, utilizing a legal technicality, came down against the claim. However, the case and the surrounding publicity forced the Canadian government to begin negotiations with First Nations in earnest.

The next major step was taken when a First Nation, the federal government, and private enterprise came together in *R. v. Guerin* (1985). The federal government had leased land claimed by the Nation to the Shaugnessy Golf Club. The decision noted that land rights pre-existed not only through the Royal Proclamation and

the Indian Act but through 'any other executive order or legislative provision' (Coates 2000, 86). Further, all Aboriginal groups had unspecified legal rights to reserve lands and traditional territories.[6]

Sparrow (1990) expanded the definition of rights when the courts stated that rights could 'evolve over time'. Any government infringement of rights must have a compelling objective, and any interpretation of rights must be broad and give 'substantial recognition' to Aboriginal rights that had not been specifically extinguished by a treaty (Isaac 1999, 308; Coates 2000, 88–9). In the same year, the Huron Wendat of central Canada were convicted for camping in a park and cutting trees in the park while practising religious rites and traditional customs (Isaac 1999, 56–7; Coates 2000, 87–8). The decision found that at the time of the treaty (1760), the Huron Wendat were an 'independent nation'. The rights under the treaty are not restricted by the Royal Proclamation and 'like instruments' but are, in fact, in addition to these agreements (Isaac 1999, 156). More important, the court found that rights are not extinguished by a lack of sustained use.[7]

In one view, 40,000 years of occupancy by First Nations in their traditional territory and the rights to those territories lies in the Royal Proclamation of 1763. The Aboriginal right of occupancy, according to the proclamation, can only be alienated to the Crown. The Royal Proclamation of 1763 is considered a critical legal document in the development and evolution of court decisions on Aboriginal access to resources:

> the foundation for historical and contemporary treaty rights and a pivotal first illustration of Britain's commitment to signing land surrender treaties prior to occupation. Although the British government considered Aboriginal territory to be 'fully British', the Proclamation

sought to limit settlement on the lands specifically 'reserved' for First Nations people as hunting territories. Through the Act, it is explicit that the colonial authorities recognized the importance of negotiating land treaties with First Nations and, in advance of settlement, protecting the area of Aboriginal use [Coates 2000, 80–1].

Many cases involving the Aboriginal right to 'subsistence' access to resources lie in the Natural Resources Transfer Agreements (NRTA). Of particular note in the cases reviewed here is paragraph 12 of the agreements in Alberta and British Columbia and paragraph 13 in the Manitoba agreement. The NRTA were designed to ensure that First Nations would have access to hunting, trapping, fishing, and game for subsistence and support. The agreements support the rights to access 'at all seasons of the year on the lands specified' (Isaac 1999, 313), and contrary provincial laws are inapplicable. However, the agreements restrict access to resources used for food purposes. By this interpretation, 'commercial' rights were extinguished. Other relevant cases include *R. v. Horseman* (1990), *R. v. Badger* (1996), *R. v. McIntyre* (1992), *R. v. Alexson* (1990), and *R. v. Horse* (1988).[8] The Constitution Act 1982, Section 35(1), affirms that '[t]he existing Aboriginal and treaty rights of the Aboriginal peoples of Canada are hereby recognized and affirmed.' These legislative actions by the Canadian government opened the way for the 1997 Supreme Court of Canada decision found in *R. v. Delgamuukw*.

The Gitskan and Wet'suwet'en sought a decision affirming jurisdictional and Aboriginal rights and ownership of their traditional territory in northwest British Columbia. The British Columbia Supreme Court recognized unextinguished, non-exclusive Aboriginal rights. In the court's judgment, the evidence did not 'provide a substantive basis for right of ownership and jurisdiction over the lands' (Isaac 1999, 65). The Nations, noting that they had never signed a treaty, claimed title to their land. In 1990, BC Court Judge Alan McEachern ruled against the Nations in a pejorative and dismissive decision that concluded all rights had been extinguished by provincial and federal governmental action. McEachern also decided that the Nations could maintain traditional fishing, hunting, and gathering but had no other rights of significance.

The SCC overturned the McEachern decision, ordering a new trial and noting that oral testimony must be given considerable attention in First Nations cases (McEachern had mostly ignored oral testimony). Importantly, the Supreme Court solidified the Crown's responsibility and obligation to consult with First Nations in negotiations on land claims and concluded that the Crown must work in good faith with First Nations. The SCC decision also found that the right to land is protected under Section 35 of the Constitution Act, 1982, and that rights to harvest traditional resources extend to contemporary purposes, although these rights are not unregulated. The judgment also held that the use of lands by First Nations must not undermine the 'special bond' that the First Nations have with their land. First Nations must prove exclusivity in use and occupancy of the land, and the burden of such proof lies with the First Nation (Coates 2000, 90–1). Although the *Delgamuukw* decision greatly broadened the legal opportunities for First Nations to find some form of justice, Canadian courts and First Nations have had a long history of dispute regarding rights to access to resources.

The 1888 case involving *St. Catharine's Milling*,[9] the first case to address the effects and meaning of the Royal Proclamation of 1763 (Isaac 1999, 23), originated from a dispute between the federal government[10] and the government of

Ontario to control Crown lands and resources.[11] Under the terms of Treaty 3 signed between the federal government and the Ojibway Nation, Ottawa claimed federal regulatory right. Ontario, however, claimed provincial ownership of the resources. The Judicial Committee of the Privy Council judged that the province had rights over any Aboriginal claim, in effect stating that Aboriginal rights and claims were terminated with the creation of a provincial government. Forty years later, in 1928, Mi'kmaq Chief Syliboy[12] was convicted of contravening the Lands and Forests Act for possessing pelts that he claimed were rightfully his under the Treaty of 1752 right to 'hunt and trap at all times'. The negative decision reinforced the supremacy of provincial regulations over treaty rights (Coates 2000, 82–3).

It must be remembered, when dealing with legislation and First Nations today, that the Indian Act during the first half of the twentieth century prohibited 'raising funds to advance claims':

Every person who, without the consent of the Superintendent General expressed in writing, receives, obtains, solicits or requests from any Indian any payment of contribution or promise of any payment . . . shall be guilty of an offence and liable upon summary conviction for each such offence to a penalty not exceeding two hundred dollars and not less than fifty dollars or to imprisonment for any term not exceeding two months.[13]

The restriction was repealed by subsection 123(2), c.29, S.C.C. in 1951. Once free of this restrictive legislation, First Nations peoples entered the court system in earnest. In 1965, Clifford White and David Bob were both charged with hunting out of season and hunting without a permit. Both were members of Vancouver Island's Nanaimo Indian Band. Their legal rights were subject to the Douglas treaties

of the mid-nineteenth century. Here the SCC agreed with the court of appeal that the men were under treaty as defined by the Indian Act and this decision 'clarified aboriginal harvesting rights under the Douglas treaties' (Coates 2000, 84). Consequently, the White and Bob case gave new importance and relevance to nineteenth-century treaties: notably, they are effective.[14]

The SCC recognition of operative and pre-existing rights for First Nations in *Bob and White* was followed by the SCC *Calder* decision in 1973. The Nisga'a, after losing in the British Columbia courts, appealed to the SCC. There was no treaty between Canada and the Nisga'a Nation, although the Nisga'a had been pursuing such a relation for a century. Frank Calder, as early as 1960, had petitioned the federal government to recognize Nisga'a land ownership of their traditional territories. The SCC decision split. One group of three agreed that Aboriginal rights had existed but were extinguished by Britain and the colony of British Columbia prior to Confederation. The second group of three agreed that in the absence of a treaty, Aboriginal rights to land and resources had not been surrendered. The deciding seventh vote, dependent on a legal technicality, resulted in the loss of the Nisga'a claim. However, the decision resulted in then Prime Minister Pierre Trudeau and Indian Affairs Minister Jean Chrétien undertaking land claims negotiations with all First Nations without treaty agreements. The *Calder* decision 'radically transformed the legal and moral foundation of Aboriginal rights in Canada, politicizing and publicizing claims and entitlements that had long been ignored by the political and legal system in the country' (Coates 2000, 86).

Although the 1985 *Guerin* case primarily involved the leasing of Aboriginal land by the federal government to a private enterprise, the decision by Chief Justice Brian Dickson noted that First Nations had land rights that pre-existed not only in the Royal Proclamation and the Indian Act but in 'any other executive order or legislative provision' (Coates 2000, 86). Further, the decision stated that these rights were not dependent on acts of legislation. In fact, all Aboriginal groups had unspecified legal rights to reserve lands and traditional territories. The SCC decision in the *Simon* case confirmed the protection of treaty rights under Section 35 of the Constitution Act, 1982. James Simon claimed that the Mi'kmaq Treaty of 1752 'provided him with immunity from provincial hunting regulations'. The SCC overturned the lower court decisions, holding that the terms of the treaty of 1752 remain in force and effect (Coates 2000, 87).[15] This decision confirmed *sui generis* (prior) rights and required a 'liberal, generous, interpretive approach in favour of the Indian peoples concerned' (Isaac 1999, 156).

The use of a drift net that exceeded acceptable length limits resulted in a SCC ruling in favour of Ronald Sparrow in 1990. Mr Sparrow considered a length restriction an infringement on his Aboriginal right to fish for food. The ruling, in favour of Sparrow, fell under Section 35 of the Constitution Act of 1982. The ruling noted that rights could 'evolve over time' and that rights must be interpreted in a broad, not narrow, manner. Further, governmental regulation must have a 'compelling and substantial objective'. The Sparrow decision also confirmed the strength of a recently enacted Section 35 and its ability to give 'substantial recognition to aboriginal rights that had not been specifically extinguished by treaty' (Isaac 1999, 308; Coates 2000, 88–89).

The *Sparrow* decision was applied in the cases of *R. v. Little* and *R. v. Jack*.[16] The *Little* case involved fishing for food purposes and treaty rights under the Douglas Treaty of 1854. *R. v. Jack* involved fishing and the use of fish for ceremonial purposes. The court of appeal overturned the decisions as an unjustified infringement on

Mr Jack's right to fish. The decision also noted a failure on the part of the Department of Fisheries and Oceans (DFO) to properly consult with First Nations or give priority to Aboriginal rights (Isaac 1999, 309).

A series of cases involving right of access were heard from the 1990s onward. Fishing rights have been, and still are, one of the most frequent and important areas of legislation. Recent cases have involved decisions that found that a prohibition of fishing during 'closed' periods infringes the Aboriginal right to fish for food, including *R. v. Ogushing* (1998).[17] The justification of infringements of fishing rights for conservation purposes was central in *R. v. Couillonneur* (1997).[18] Mr Couillonneur used a fish net with a mesh smaller than allowed by regulation. The court found that this restriction has a minimal infringement on treaty rights, since there was no evidence of inflicting hardship. In *R. v. Cote* (1996),[19] the application of an access fee was not considered an infringement on food fishing. It was found that food fishing is integral to Mr Cote's distinctive culture, which showed continuity from pre-contact to the present.

The aspects of food fishing as integral to a distinctive culture and showing continuity from pre-contact to present appeared again in *R. v. Adams* (1996).[20] Regulations that subject fishing rights to ministerial discretion (with no criteria for the exercise of such discretion) imposed undue hardship on Mr Adams's preferred mode of exercising a right. The Adams case also found that an Aboriginal right is not dependent on Aboriginal title. If an Aboriginal group can prove an activity is integral to its culture, it is not necessary to prove title to the land. Adams's right to fish was therefore not dependent on his having title to the areas in which he was fishing (Isaac 1999, 309). Other cases (*R. v. Shipman*,[21] *R. v. Denault*,[22] *R. v. Martin*,[23] and *R. v. Bellehumeur*[24]) found that traditional rights were tied to the land, rather than being an inherent right regardless of location, and so must be practiced within an Aboriginal's territory.

A number of decisions have involved not only the right to fish for food but also the right to commercial fishing. *R. v. Hunt* (1995)[25] found that the Douglas Treaties cannot be construed as encompassing commercial/deep-ocean salmon fishing. *R. v. Gladue* (1996)[26] noted that although Treaty 6 included the right to fish commercially, the treaty had been extinguished by the Natural Resources Transfer Agreement. And again in *R. v. Sasakamoose* (1997),[27] fishing for commercial trade, barter, or sale was found not to be a defining characteristic of culture, and therefore commercial fishing was not established as an Aboriginal right. The court further claimed that had such a right existed, it had been extinguished by the Natural Resource Transfer Agreement. *R. v. Kapp*[28] supported an Aboriginal commercial fishery in British Columbia as constitutional; however, *Laxkw'alaams Indian Band v. Canada (Attorney General)*[29] found that clear evidence of a traditional (i.e., historical) practice of fish sales must be demonstrated in any claim to commercial rights.

Similarly, the cases of *Van der Peet*, *N.T.C. Smokehouse Ltd.*, and *Gladstone*[30], decided in 1996, clarified the government's ability to limit First Nations' rights and provided tests that a claim must pass before court recognition of the right is granted—i.e., the 'practice, tradition or custom' must be integral and significant to the culture and be in place prior to European arrival and must be consistent, not 'marginal or occasional'. *Van der Peet* involved fish caught under a food fish licence and then sold to a non-Aboriginal. The sale was prohibited by fishery regulations. The decision, in favour of the First Nation, recognized a 'dynamic right' that opened the possibility that the trade of resources might 'constitute an aboriginal right' (Coates 2000, 90).

The *N.T.C. Smokehouse Ltd.* case was based on the sale and resale of fish caught under a food licence and sold to a processor. The processed fish was then resold by the processor. Again, fishing regulations that prohibit the sale or purchase of food fish were employed. The Shesaht and Opetchesaht Band claimed that trade in fish was part of their distinctive culture. The court, however, found that the facts did not support the First Nation's claim and therefore the claim to a right to fish commercially could not be supported.

In the *Gladstone* decision, the Heilsuk were more successful in their claim that the harvesting and commercial trade of herring spawn were integral to their distinct culture. The court found that the commercial trade of herring spawn was an integral part of the Heilsuk culture prior to contact. The court determined that the regulations and legislation prior to 1982 did not demonstrate an intention to extinguish their right and that an attempt to limit the take of herring spawn is an interference. Although the court found that the evidence was insufficient and ordered a new trial, the case confirmed that the government must demonstrate that the process of allocating resources and the actual allocation must reflect the prior interest of Aboriginal rights-holders.

Two 1998 cases involved Mi'kmaq timber harvesting rights on lands leased to a timber corporation.[31] Drummer's Treaty states that the Mi'kmaq 'can harvest any and all trees they wish on Crown lands', a *sui generis* right. The Forests Act[32] was no longer applicable. Further, the court argued that the *sui generis* right to harvest and sell timber need not be grounded in title.

Mr Paul claimed that he could harvest trees without the standard harvesting licence, 'based on the powers conveyed to him and his community under the treaties signed between the British and the Mi'kmaq and Maliseet Nations'

(Coates 2000, 96). Provincial Court Judge Frederick Arsenault found that 'the treaties clearly authorized commercial harvesting and that treaties took precedence over provincial legislation' (Coates 2000, 96). Arsenault further claimed that Aboriginal rights were not extinguished by the 'mere existence' of timber licences granted by the Crown. His decision was appealed, and in 1997 Justice John Turnbull upheld the decision, finding no evidence of surrender of rights or privileges. His decision was overturned by the New Brunswick Court of Appeal, and a further appeal to the SCC was denied. Mr Paul lost his case, but these decisions provided the spark for First Nations in the Maritimes to press for rights to commercial access (Coates 2000, 97–105).

Donald Marshall, Jr, was arrested in 1993 for fishing in a closed season, without a licence, and for selling his catch illegally. Marshall claimed the right to commercial fishing under treaties signed with Britain in 1760 and 1761. The case was heard in 1997[33] when the treaty right to commercial fishing was claimed as part of the Truckhouse system, a system of commercial trade using licensed Mi'kmaq traders. The Truckhouse system ultimately failed, but Marshall claimed that the commercial right remained. The Nova Scotia Court of Appeal (NSCA) disagreed, claiming that the Truckhouse system was an 'expedient', not 'an open ended guarantee'. In 1999, the SCC ruled that the First Nations' right under the treaties was not unlimited but that a 'moderate income' could be derived from commercial sale. Marshall was acquitted of all charges (Coates 2000, 3–20).[34]

However, what is interesting in this case is that shortly after the original decision was delivered by the SCC, shooting broke out between Aboriginal and non-Aboriginal lobster fishers (Burnt Church, Nova Scotia). Shortly afterwards, the SCC reversed its finding, declaring that they had made a mistake. It was the first time that this had happened in Canada, and one lawyer of our acquaintance has pointed out that it means that the Supreme Court can no longer be considered impartial: it can be held hostage by violent protest (Shawana 2002).

A QUESTION OF ACCESS: DO RIGHTS REALLY EQUAL GETTING YOUR SHARE?

Treaties and court cases have established that Aboriginal peoples have rights to resources, including, in some cases such as fisheries, a right to commercial exploitation. However, having the right is not the same as access. Further, the morass of treaties and court cases has only poorly considered some resources, including forests and minerals, except where laid out in certain modern treaties, such as the Nunavut and Nisga'a land claim settlements. This also results in a significant impediment to access. Rights and title do not always equal access.

Access to natural resources is an issue for Aboriginal peoples in several different ways. We will only consider a few. The first question is, does the resource fall within Aboriginal rights? Forests are a key example of this question of access. Aboriginal and treaty rights, to date, tend to be limited to resources that were used for subsistence purposes. While First Nations have rights to harvest timber for subsistence purposes, they do not have rights—yet—to participate in commercial forestry, except those of any other commercial operator (see Booth and Skelton in press). This is unfortunate, because across Canada, many reserves are surrounded by harvestable timber and many have community members who have worked in the timber industry. As the National Aboriginal Forestry Association (NAFA 1993) has pointed out, forestry is a logical choice for economic development on the reserve.

However, much of the country with a viable forest cover is licensed to international corporations. This has made it very difficult for First Nation–owned logging companies to break into the forestry business (NAFA 1995; 1996; 2000). Two notable exceptions in British Columbia include the Tl'azt'en Nation (Booth and Skelton in press) and the Nuu-Chul-Nuuth. Both, however, came about as the result of protest that forced the government into permitting tenures to pass to the Nations. Other potential tenures are available, including participation in model forests, woodlot licences, and, in British Columbia, community forests (see also Chapters 10 and 19). As of 2008, the BC government was offering special tenure rights to First Nations, but they remain small in volume. However, despite years of work toward gaining timber licences, the concentration of timber rights in the hands of large multinationals and the fact that timber is not reflected in Aboriginal rights will continue to limit access to this most valuable resource.

Another limit to access is availability. Since 1990, *Sparrow* has guaranteed access to fisheries by First Nations for subsistence purposes. *Gladstone* provided for the ability of Aboriginals to sell fish under specific circumstances. *Marshall*, however, makes clear, as did *Sparrow*, that conservation of the resource is of primary importance. Given the state of fisheries on both the east and west coasts of Canada at the beginning of the twenty-first century, the demands of conservation may well limit access to the resource regardless of rights (see Chapter 8). Indeed, the threat of extinction that faces many stocks of Pacific salmon suggests that if we continue to exploit the salmon resource as we currently do, First Nations' rights may become a moot point (Newell 1996).

CONCLUSION

The issue of Aboriginal peoples' rights to natural resources is a clear example of the themes of this book: conflict and uncertainty. It is a complex mix of history, law, court challenges, and judges' decisions, most of which remain in process and unsettled. Canada's First Nations continue to push to regain what they see as their rightful title to a land they have lived with for thousands of years, as well as to gain their rights of access to the resources attached to that land. Sometimes that pushing takes the form of violent conflict, as at Burnt Church over the *Marshall* decision or at Caledonia, ON, over access in general. Complicating matters are the differences between Aboriginal cultures and the Western culture by which they are surrounded: a culture that continues to consume resources at a increasing rate even while the Nations struggle for access, leaving us with uncertainty regarding the future of relations between Aboriginal peoples and other Canadians. The problem is complex and difficult and one that resource managers—perhaps yourself—will need to consider, confront, and work to resolve. While we must continue to deal with ongoing uncertainty and conflict, understanding and interest are the first steps in solving these problems. So, too, is our willingness to learn.

FROM THE FIELD

As a transplanted American, I find I have somewhat different views from those of my Canadian colleagues on a lot of issues, including environmental ones. While things are changing, it has always seemed to me that although Canada is truly rich in resources, in wild places, and in wildlife—much more so than the lower 48 American states—Canadians are far too complacent about protecting and preserving them. What the Americans lack in resources they make up for in policy to try and preserve them, despite certain presidents. Part of that is because some Americans are very good at criticizing their government and being vocal about changing things (sometimes not in ways I agree with, but . . .). I want to see more vocal Canadians demanding that their governments take better action on protecting the environment and not trusting their government to do so on their behalf. As I say, that is changing, and I see it in the students I teach, which is most hope-inspiring.

The other groups I see becoming increasingly vocal are the indigenous peoples in Canada and elsewhere, and that is also a good thing. Again, I think that Canadians are often complacent when they view their record on dealing with indigenous groups, but I think that complacency is misplaced. Linking with indigenous groups to better manage environmental decisions is vitally important. Indigenous Nations are often the first to experience the damage caused by or the consequences of poor environmental management or trade-offs between development and protection, and if there is no space on the land for their children and their cultural practices, there certainly will not be for the rest of us soon either.

Some of the most important learning experiences I have had, and the most profound, have been from getting out into the field and listening to people. Researchers are often not good at this. We tend to tell people what we think as experts and to put their experiences into our own theoretical explanations. But theory doesn't always tell you about people's lives. Learning to hear someone is the most important research skill you can develop. I once sat with another researcher on the shore of Lake Superior and listened to an Ojibwe spiritual leader tell about things that were important to him, to his culture, and to his spirituality. We sat for several hours, listening. When he finished and we went to leave, only an hour had passed. He apologized for literally moving us out of time, but he said it was important to do so, so that we could hear what he had to say and researchers were always in a hurry! I have no theoretical explanation for this, but I have tried to be less in a hurry since then when I listen. Not everyone has his power, but they do equally need to be heard.

When we learn to hear, whether it is people, animals, the land, or sub-atomic particles, that's when our research comes alive and has meaning. So go listen. And then find a voice.

—Annie L. Booth

Notes

1. For example, the authors are working with one First Nation in northeastern British Columbia that in 2008 was faced with about 30 existing and proposed industrial developments such as mines, pipelines, and wind farms within their traditional lands in addition to forestry and hydroelectric developments.
2. *R. v. Kelley* (2007) Alta. Q.B. C.N.L.R., 2, p. 332.
3. 'Metis National Council president applauds Ontario Metis court victory on harvesting rights'. http://www.turtleisland.org/discussion/viewtopic.php?p=8344.
4. *R. v. Hopper* (2008) N.B.C.A. C.N.L.R., 3, p. 337.
5. *The Labrador Metis Nation v. Labrador and Newfoundland.* http://www.turtleisland.org/discussion/viewtopic.php?p=7245.
6. The SCC decision in the Simon case confirmed the protection of treaty rights under Section 35 of the Constitution Act, 1982. James Simon claimed that the Mi'kmaq Treaty of 1752 'provided him with immunity from provincial hunting regulations'. The SCC overturned the lower court decisions, holding that the terms of the treaty of 1752 remain in force and effect.
7. *R. v. Siuoi* (1990) 3 C.N.L.R. 127 (SCC).
8. *R. v. Horseman* (1990) 3 C.N.L.R. 95; *R. v. Badger* (1996) 1 S.C.R. 771; *R. v. McIntyre* (1992) 3 C.N.L.R. 113 (Sask. C.A.); *R. v. Alexson* (1990) 4 C.N.L.R. 28 (Alta. Q.B.); *R. v. Horse* (1988) 2 C.N.L.R. 112.
9. *St. Catharine's Milling and Lumber Company v. R.,* 2 C.N.L.C. 541 (JCPC). The Earl of Selborne, Lord Watson, Lord Hobhouse, Sir Barnes Peacock, Sir Montague E. Smith, and Sir Richard Couch, 12 December 1888.
10. British North America Act 1867, Section 91.
11. British North America Act 1867, Section 92.
12. *Nova Scotia County Court, R. v. Syliboy,* 10 September 1928.
13. S. 141m c. 98, R.S.C. 1927 141. 1927, c. 32, s. 6.
14. *R. v. White and Bob,* 6 C.N.L.C. 684 (1965).
15. See also *Blueberry River Indian Band v. Canada* (DIAND) (1996) 2 C.N.L.R. 25 (SCC) where land surrendered 'in trust' was sold for less than its real value by the Crown. The sale breached the Crown's fiduciary duty.
16. *R. v. Little* (1996) 2 C.N.L.R. 136 (B.C.C.A.) and *R.V. Jack* (1996) 2 C.N.L.R. 113 (B.C.C.A.).
17. *R. v. Ogushing* (1998) 4 C.N.L.R. 236 (C.Q. [Crim. Div.]).
18. *R. v. Couillonneur* (1997) 1 C.N.L.R. 130 (Sask. Prov. Ct.).
19. *R. v. Cote* (1996) 4 C.N.L.R. 26 (SCC).
20. *R. v. Adams* (1996) 4 C.N.L.R. 1 (SCC).
21. *R. v. Shipman* (2004) Ont. Ct. J. C.N.L.R. 2, p. 288.
22. *R. v. Denault* (2008) B.C. Prov. Ct. C.N.L.R., 1, p. 87.
23. *R. v. Martin* (2008) Alta. Q.B. C.N.L.R., 2, p. 335.
24. *R. v. Bellehumeur* (2008) Sask. Prov. Ct. C.N.L.R., 2, p. 335.
25. *R. v. Hunt* (1995) 3 C.N.L.R. 135 (B.C. Prov. Ct.).
26. *R. v. Gladue* (1996) 1 C.N.L.R. 153 (Alta. C.A.).
27. *R. v. Sasakamoose* (1997) 3 C.N.L.R. 270 (Sask. Prov. Ct.).
28. *R. v. Kapp* (2008) S.C.C. C.N.L.R., 3, p. 346.
29. *Laxkw'alaams Indian Band v. Canada (Attorney General)* (2008) B.C.S.C. C.N.L.R., 3, p. 158.
30. *R. v. Van der Peet* (1996) 4 C.N.L.R. 177 (SCC), *N.T.C. Smokehouse Ltd.* (1996) 4 C.N.L.R. 130 (SCC), and *R. v. Gladstone* (1996) 4 C.N.L.R. 65 (SCC).
31. *R. v. [Peter] Paul* (1998) 1 C.N.L.R. 209 (N.B.Q.B) and *R. v. Peter Paul* (1998) 3 C.N.L.R. 221 (N.B.C.A.).
32. Crown Lands and Forests Act, S.N.B. 1980, c.C-38.1.
33. *R. v. Marshall* (1997) 3 C.N.L.R. 209 (N.S.C.A.).
34. See also www.mcgill.ca/mqup/marshall.

REVIEW QUESTIONS

1. Distinguish among Aboriginal rights, title, and access.
2. What principles concerning Aboriginal access to resources have been established by the law courts?
3. Why should we be concerned with Aboriginal access to natural resources and a land base?

REFERENCES

Arnott, S. 2000. *Aboriginal Communities and Non-renewable Resource Development Issue Identification Paper*. Ottawa: National Round Table on the Environment and the Economy.

Associated Press. 2002. 'James Bay Cree approve deal with Quebec on hydropower development'. 5 February. Environmental News Network. enn.com/news/wire-stories/2002/02/02052002/ap_46302.asp.

Autonomy and Solidarity: Six Nations Caledonia Resource Page'. http://auto_sol.tao.ca/node/view/2012.

Bartlett, R.H. 1991. *Resource Development and Aboriginal Land Rights*. Ottawa: Canadian Institute of Resources Law.

Booth, A. 1998. 'Putting "forestry" and "community" into First Nations' resource management'. *Forestry Chronicle* 74 (3): 347–52.

———. 2007. 'Environment and nature in Native American thought'. In H. Selin, ed., *Encyclopaedia of the History of Non-Western Science: Natural Sciences, Technology and Medicine*, 798–810. Heidelberg: Springer Verlag.

Booth, A., and H. Jacobs. 1990. 'Ties that bind: Native American beliefs as a foundation for environmental consciousness'. *Environmental Ethics* 12´ (1): 27–43.

Booth, A., and N. Skelton. In press. '"There's a conflict right there": Integrating indigenous community values into commercial forestry in the Tl'azt'en Nation'. *Society and Natural Resources*.

British Columbia Court of Appeals. http://www.courts.gov.bc.ca.

British Columbia Ministry of Forests. 1998. *Consultation Guidelines*. Victoria: Ministry of Forests.

Brown, R. 1996. 'The exploitation of the oil and gas frontier: Its impact on Lubicon Lake Cree Women'. In C. Miller and P. Church, eds, *Women of the First Nations*, 151–65. Winnipeg: University of Manitoba.

Chandran, C. 2002. 'First Nations and natural resources—The Canadian context'. http://www.firstpeoples.org/land_rights/canada/summary_of_land_rights/fnnr.htm.

Coates, K. 2000. *The Marshall Decision and Native Rights*. Montreal: McGill-Queen's University Press.

Council for Yukon Indians. 1993. *Umbrella Final Agreement between the Government of Canada, the Council for Yukon Indians and the Government of the Yukon*. Ottawa: Minister of Indian Affairs and Northern Development.

Dickason, O.P. 1992. *Canada's First Nations*. Toronto: Oxford University Press.

Grand Council of Crees (Quebec). Website. http://www.gcc.ca.

INAC (Indian and Northern Affairs Canada). 2007. 'Aboriginal demography—Population, household and family projections 2001–2026'. http://www.ainc-inac.gc.ca/ai/rs/pubs/re/abd/abd-eng.asp.

———. 2008. Website. http://www.ainc-inac.gc.ca.

Isaac, Thomas. 1999. *Aboriginal Law: Cases, Materials and Commentary*. Saskatoon: Purich Publishing.

Jensen, J., and M. Papillon. 2000. 'Challenging the citizenship regime: The James Bay Cree and transnational action'. *Politics and Society* 28 (2): 245–64.

Kendrick, A. 2000. 'Community perceptions of the Beverly-Qamanirjuaq Caribou Management Board'. *The Canadian Journal of Native Studies* 20 (1): 1–33.

Krauss, C. 2002. 'Will the flood wash away the Crees' birthright?' *New York Times* 27 February. www.nytimes.com/2002/02/27/international/americas/27QUEB.html?.

Kulchyski, P. 1994. 'Thesis on Aboriginal rights'. In Peter Kulchyski, ed., *Unjust Relations: Aboriginal Rights in Canadian Courts*, 1–20. Toronto: Oxford University Press.

McCutcheon, S. 1991. *Electric Rivers: The Story of the James Bay Project*. New York: Black Rose Books.

McKee, C. 1996. *Treaty Talks in British Columbia*. Vancouver: University of British Columbia Press.

McMillan, A.D. 1995. *Native Peoples and Cultures of Canada*. Vancouver: Douglas and McIntyre.

McNeil, K. 1997. 'Aboriginal title and Aboriginal rights: What's the connection?' *Alberta Law Review* 36 (1): 117–48.

McPherson, D.H., and J.D. Rabb. 1993. *Indian from the Inside: A Study in Ethno-Metaphysics*. Occasional Paper no. 14. Thunder Bay: Centre for Northern Studies, Lakehead University.

Nadasdy, Paul. 2002. '"Property" and Aboriginal land claims in the Canadian subarctic: Some theoretical considerations'. *American Anthropologist* 104 (1): 247–61.

NAFA (National Aboriginal Forestry Association). 1993. *Forest Lands and Resources for Aboriginal People*. Ottawa: NAFA.

———. 1996. *Aboriginal Forest-Based Ecological Knowledge in Canada*. Ottawa: NAFA.

———. 1995. *Aboriginal Participation in Forest Management: Not Just Another Stakeholder*. Ottawa: NAFA.

———. 2000. *Implementing National Forest Strategy Commitments*. Ottawa: NAFA.

Newell, D. 1996. "'The highest right that a man hath": Maritime property rights regimes and BC First Nations'. *Native Studies Review* 11 (1): 49–64.

Nisga'a Agreement-in-Principle in Brief. 1996. Ottawa: Government of Canada.

O'Faircheallaigh, C. 1998. 'Resource development and inequality in indigenous societies'. *World Development* 26 (3): 381–94.

Robinson, R. 2002. 'Nisga'a patience: Negotiating our way into Canada'. In J. Bird, L. Land, and M. Macadam, eds, *Nation to Nation: Aboriginal Sovereignty and the Future of Canada*, 186–94. Toronto: Public Justice Resource Centre.

Shawana, P. 2002. Personal communication. Prince George, BC.

Speck, F., and L. Eisley. 1942. 'Montagnais-Naskapi Bands and family hunting districts of the central and southern Labrador peninsula'. *Proceedings of the American Philosophical Society* 85: 215–42.

ALSO RECOMMENDED

Battiste, M., and J.Y. Henderson. 2000. *Protecting Indigenous Knowledge and Heritage*. Saskatoon: Purich Press.

Berkes, F. 1999. *Sacred Ecology*. Philadelphia: Taylor and Francis.

Grim, J.A., ed. 2001. *Indigenous Traditions and Ecology*. Cambridge: Harvard University Press.

Krech, S. 1999. *The Ecological Indian: Myth and History*. New York: W.W. Norton.

Nelson, R.K. 1983. *Make Prayers to the Raven*. Chicago: University of Chicago Press.

5

Climate Change, Adaptation, and Mitigation

Gordon McBean

Learning Objectives

- To understand the role of human activities in a complex global system.
- To place the issue of climate change in a broader global societal context.
- To address the international framework within which Canada and other countries should be taking action.
- To identify the action strategies.
- To place in context the role of uncertainty and conflict between vested interests.
- To recognize the unfortunate role that Canada is playing in this global issue.

INTRODUCTION

Climate, which defines our natural ecosystems and plays a major role in our national psyche, is usually defined as the statistics of weather. Extreme weather and climate events stress Canadian ecosystems and communities, and they are likely to increase in frequency and intensity as a result of climate change (Lemmen, Warren, and Lacroix 2008). Flooding has been the cause of some of Canada's worst disasters. In 1996, after several days of heavy rain, a dam broke, causing severe flooding in the Saguenay–Lac-Saint-Jean region in southern Quebec. More than 1000 homes were destroyed; the evacuation of 16,000 people was required, and damages in excess of $800 million resulted (Brooks 2008). Similarly,

major flooding along the Red River in Manitoba in 1997 prompted the evacuation of 25,000 residents and caused damage of approximately $800 million (Morris-Oswald and Sinclair 2005). Major flooding of the Red River in April 2009 was a reminder, even after mitigation actions taken after the 1997 flood, that occupants of the flood plain remain vulnerable.

In January 1998, eastern Ontario, southern Quebec, and parts of the Maritime provinces experienced six days of heavy freezing rain, resulting in millions without power, at least 28 deaths, and more than 900 injuries (Kerry et al. 1999). Estimated total costs associated with the disaster exceeded $5 billion. Hurricane Juan in 2003 resulted in eight deaths and about $200 million in damage in Atlantic Canada (Lemmen,

Warren, and Lacroix 2008). In December 2006, British Columbia's Lower Mainland and Vancouver Island were struck by a series of storms more powerful than any experienced in the province's recorded history (Environment Canada 2007). Extreme weather refers to infrequent but significant departures from a location's normal weather conditions. These anomalous events—hot and cold temperatures, severe thunderstorms, ice storms, blizzards, windstorms, tornadoes, and hail—exceed the range of weather intensity that a location normally experiences (Berry, McBean, and Séguin 2008). They are specific manifestations of climate that are potentially harmful for people and property.

The droughts on the Prairie provinces created major economic and social costs for several years in this century. Wildfires pose a significant hazard, as evidenced in British Columbia in the summer of 2003 (British Columbia 2004). There has been an unprecedented outbreak of the mountain pine beetle in British Columbia, which encompassed more than 9.2 million hectares of forest in 2006 and is now spreading eastward into Alberta (see also Chapter 10). Although fire suppression and other factors have contributed to this outbreak, the recent predominance of hot summers that favour beetle reproduction and mild winters that allow their offspring to survive have been critical factors, and a changing climate is playing a role (Lemmen, Warren, and Lacroix 2008).

This chapter is about climate change and how societies can mitigate and adapt to its changes. Climate change needs to be seen in the context of other societal and environmental issues and national security (Hulme 2009). Its long-term and global nature leads to concerns about intergenerational and international equity. The result is conflict over how to value and protect the 'rights' of future generations in comparison to the aspirations of present generations and the 'rights' of developing countries

to have higher standards of living while reducing the impacts on the climate system. Conflict, in the face of uncertainty, will be a recurring theme of this chapter.

THE SCIENTIFIC BASIS FOR CONCERN

The global climate system is a complex interaction of the atmosphere, oceans, land surfaces, sea ice, glaciers, and ecosystems (Le Treut et al. 2007; Weaver 2008; Hulme 2009). Understanding of the climate system is based on centuries of scientific study going back to Archimedes and Newton. Fourier (1824) wrote a paper hypothesizing that the atmosphere blocks outgoing radiation from the Earth and re-radiates a portion of it back, thereby warming the planet. The sun, with a temperature of 5700°C, emits short-wave radiative energy, much as visible light. The Earth, with an average temperature of about 15°C, emits longer infrared-wave energy, invisible but still transferring energy. Intercepted incoming solar radiation is partially reflected off clouds, snow and ice, and desert surfaces so that 31 per cent of the energy is reflected back to space. The remainder is absorbed by the climate system, mainly oceans and forest canopies, and heats it. Changes to these characteristics of the Earth will change the climate by reducing or increasing the amount of solar radiation absorbed either at the surface or in the atmosphere.

Whereas the short-wave solar radiative energy is only slightly absorbed by the atmosphere, the longer-wavelength Earth radiative energy is significantly absorbed by the water vapour, carbon dioxide, methane, and nitrous oxide gases in the atmosphere. The greenhouse effect resulting from these greenhouse gases (named in analogy with plant greenhouses) keeps the lower atmosphere and Earth's surface warmer than they would otherwise be. The result is that the

Earth as seen from space appears to be $-18°C$, compared to the actual surface temperature of $15°C$. This warming of $33°C$ is due to the natural greenhouse effect, which has resulted in an appropriate temperature for the evolution of life as we know it.

In terms of its contribution to the Earth's greenhouse warming, water vapour is the most important gas, contributing about 65 per cent of the natural greenhouse effect. Carbon dioxide is next at about 25 per cent, and the other gases contribute the remaining 10 per cent or so, with methane and nitrous oxide being the most important. However, the amount of atmospheric water vapour depends on its temperature, and the atmosphere is already near saturation. If water is added to the atmosphere, it will precipitate out in about 10 days, what we call the time scale for water vapour in the atmosphere.

Plants take up CO_2 from the atmosphere and grow and give back CO_2 when they die or decompose. Over long periods of time, these processes are in approximate balance; over even longer periods, the carbon is sequestered beneath the surface of the Earth and becomes the fossil fuel that we now extract from those burial vaults and, in a sense, re-inject into the atmosphere. The atmosphere and oceans also exchange CO_2, with a small net transfer into the oceans. Small floating plants and animals in the upper oceans absorb CO_2, and when they die, their carcasses sink to the depths of the ocean and gradually become part of the ocean floor. This gradual transfer of carbon from the atmosphere into the upper ocean and then to burial at depth is slow to the extent that the time required for 50 per cent of the additional CO_2 to be removed is between 50 and 200 years.

Methane is removed from the atmosphere by chemical reactions, with an equivalent removal time of 8 to 12 years. The time scales, in round numbers, are 10 days for water vapour, 10 years for methane, and 100 years for CO_2. Analyses of the transport of trace components within the atmosphere show that it takes from two to four years for a gas to mix around the globe. Since water vapour's lifetime is 10 days, it does not travel far, while methane and certainly CO_2 are in the atmosphere long enough to be globally well-mixed. From the policy point of view, this makes CO_2 and methane global issues, since the location of the source becomes very quickly irrelevant.

Much of the basis for our understanding of the climate system comes from systematic measurement programs. In 1957, the International Geophysical Year, the first systematic measurements of atmospheric ozone and carbon dioxide were undertaken. Revelle and Suess (1957, 18–27), arguing that the oceans could not absorb the human emissions of CO_2 as fast as they were being produced, stated that '[h]uman beings are now carrying out a large-scale geophysical experiment of a kind that could not have happened in the past nor be reproduced in the future.' The global atmospheric CO_2 concentration monitoring program continues today.

Palaeoclimate studies (Jansen et al. 2007) of ice cores from Antarctica, using oxygen isotope ratios to deduce the surrounding oceanic temperatures and trapped air bubbles for past atmospheric CO_2 concentrations, have documented the temperatures and CO_2 values over about the past million years, showing regular variations between cold ice age conditions and warm interglacial eras with periodicity of about 100,000 years. Estimated global mean temperatures vary from near present values to about $5°C$ colder when major ice sheets covered most of what is now Canada. These variations in global temperatures are primarily due to the variations in the Earth's orbit around the sun and in the tilt of the North Pole. Changes in summertime solar radiation that when reduced fail to melt the ice formed during the previous winter resulted in the gradual accumulation of ice sheets. The CO_2 concentrations varied from near 300 parts per million (ppm) during the warm periods to about 180 ppm during the peak of the ice ages. These results put upper and lower limits on the natural variations in global temperature and CO_2 concentrations over the past million years. The magnitude of the cooling during the ice ages is larger than can be explained only by reductions in solar input due to the orbital variations. When the oceans cooled, they absorbed more CO_2, reducing the atmospheric concentrations and the greenhouse effect, leading to further cooling. Understanding and being able to model these past climatic variations gives the climate science community increased confidence in the ability to project changes in the future.

Atmospheric concentrations of CO_2 from about 10,000 years ago until about 1800 were close to 280 ppm. They then started to rise, reaching about 315 ppm 50 years ago, and in 2008 were about 385 ppm. The increase in atmospheric CO_2 over the past 200 years is due to the injection of fossil fuel carbon, not due to changes in the emissions from living biomass or other natural factors.

The first sophisticated atmospheric modelling studies aimed at investigating the climatic consequences of increasing atmospheric CO_2 by Manabe and Weatherald (1967) in the 1960s led them to conclude that doubling the CO_2 content in the atmosphere would have the effect of raising the global atmospheric temperature by about 2°C. Over the following 40 years, the global scientific community has built upon the foundations laid by these scientific leaders, and major advances in scientific knowledge and understanding have been the result. How much is known and still unknown? How much may be unknowable? The following sections discuss how these questions have been answered.

BECOMING AN INTERNATIONAL ISSUE

In 1985, scientists brought together by the World Meteorological Organization (WMO), the UN Environment Programme (UNEP), and the International Council of Scientific Unions (ICSU, now renamed the International Council for Science) agreed on a concluding statement: 'Many important economic and social decisions are being made today on long-term projects, all based on the assumption that past climatic data, without modification, are a reliable guide to the future. This is no longer a good assumption' (Bruce 2001, 3). This conclusion set in place a sequence of activities that moved climate change forward on the public agenda. Other studies such as Broecker (1987, 123), which emphasized the possible 'unpleasant surprises in the greenhouse', added to the concern.

In 1988 in Toronto, Prime Minister Brian Mulroney and Norwegian Prime Minister Gro Bruntland, opened the historic conference Our Changing Atmosphere: Implications for Global Security. It marked the first time that heads of government had addressed the issue of climate change. The conference summary statement opened with: 'Humanity is conducting an unintended, uncontrolled, globally pervasive experiment whose ultimate consequences could be second only to a global nuclear war.' The Second World Climate Conference in late 1990 called for an international convention to address the threat of climate change. International negotiations then resulted in the United Nations Framework Convention on Climate Change (UNFCCC—the Climate Convention), signed at the 1992 Earth Summit by many heads of state and government, including Prime Minister Mulroney and President George H.W. Bush. The Climate Convention formally entered into force in 1994 and has now been ratified by almost all countries.

Through its process of ratification in 1994, Canada took on a binding commitment.

UN Framework Convention on Climate Change

The United Nations Framework Convention on Climate Change (UNFCCC 1992) has as its objective (Article 2):

> . . . stabilization of greenhouse gas concentrations in the atmosphere at a level that would prevent *dangerous* anthropogenic *interference* with the climate system. Such a level should be achieved within a time frame sufficient to allow ecosystems to adapt naturally to climate change, to ensure that food production is not threatened, and to enable economic development to proceed in a sustainable manner.

The words 'dangerous' and 'interference' have been emphasized to stress the connection with the sense of security, which is also emphasized by the reference to the protection of food production and enabling economic development to proceed.

Although the focus of the Climate Convention is on emission reductions, it also contains sections on cooperation in preparing for adaptation, on research and systematic observations, and on education, training, and public awareness. The Climate Convention addresses the two fundamental policy responses to the risk associated with climate change—emissions reductions (usually called mitigation) and adaptation.

Mitigation involves efforts to stabilize or reduce greenhouse gas emissions to slow or stall changes in climate and is the central focus of most national policies concerning climate change, largely driven by international agreements such as the Kyoto Protocol, which set binding emissions-reduction targets for signatory states. The other policy response, *adaptation*,

recognizes that despite even the most ambitious efforts to reduce greenhouse gas emissions by Canada and other states, a significant degree of climate change is inevitable (Burton 2008). Furthermore, it acknowledges that the climate already exerts significant pressure on physical, social, and economic systems, which has not been sufficiently addressed. Strategic approaches to mitigation can have adaptive benefits and vice versa. The Climate Convention, under Article 3 on principles, endorses the *precautionary principle*, stating that lack of full scientific certainty should not be used as a reason for postponing action. It was also agreed that sustainable development was a principle of the convention and an issue for climate change.

The Intergovernmental Panel on Climate Change

As the issue of climate change became more politically sensitive, several countries were concerned about the need for more formal assessment processes. As a result, in November 1988, the Intergovernmental Panel on Climate Change (IPCC) (Bruce 2001) was created jointly by the World Meteorological Organization and the United Nations Environment Programme. The panel does not do research but assesses and synthesizes the policy-relevant results of peer-reviewed published research. Its assessments are policy-relevant but not policy-prescriptive and are reviewed by peers and governments.

The *First IPCC Assessment Report* was presented to the Second World Climate Conference in 1990. The *Second Assessment Report* was completed in 1995 and was important in leading to the decision by governments to adopt the Kyoto Protocol of 1997. The *Third Assessment Report* was completed in 2001 and the fourth in 2007. This last report is the foundation for most of the information on climate change, its impacts, and possible response strategies presented in this chapter.

The IPCC is currently structured as a bureau and three working groups: I on the physical science basis; II on impacts, adaptation, and vulnerability; and III on mitigation (emissions reductions). Its procedures to ensure unbiased consideration of the full range of scientific results—and to appear to be doing so—were put in place at the beginning and have also evolved. The texts of the chapters it produces are the responsibility of the lead author team and are extensively peer- and government-reviewed. The summaries for policy-makers for each working group report and an overall IPCC synthesis report are approved line-by-line at plenary meetings of governments. All reports are available on the IPCC website (www.ipcc.ch). The Climate Convention has recognized the role of the IPCC in providing the scientific foundation for its work.

The IPCC's Role in Communicating Certainty and Uncertainty

The scientific community needs to better communicate certainty and uncertainty regarding climate change (McBean 2009b; see also Linda Mortsch's guest statement). One of the most important scientific issues, in terms of motivating international action on climate change, is the detection of climate change and attribution of the change to anthropogenic or human actions. The IPCC *First Assessment Report* (Houghton, Jenkins, and Ephraums 1990, xiii) stated: 'this warming is broadly consistent with the predictions of climate models but it is also of the same magnitude as natural variability'. Five years later, the *Second Assessment Report* concluded that 'The balance of evidence suggests that there is a discernible human influence on global climate' (Houghton et al. 1996, 4). This statement reflects the tenuous nature of the conclusion but indicates that the human-induced signal was starting to show.

The IPCC statement led to a large controversy, with challenges from some critics who 'claimed that the IPCC had inappropriately altered a key chapter for political reasons' and 'had corrupted the peer review process' (Edwards and Schneider 2001, 219). It provided an opportunity for

GUEST STATEMENT

Communicating about Human-Caused Climate Change— The Evolving Challenge

Linda Mortsch

As a researcher involved in climate change impact and adaptation assessments for more than two decades, I have participated in the challenging, sometimes frustrating, but necessary task of communicating about climate change. For me, the communication process is fundamental to raising awareness, developing understanding, and motivating action. As the issue has matured, the nature of the dialogue has also evolved regarding who is engaged, what are the salient issues and associated questions, and what are the information needs.

Addressing climate change requires a discourse broader than establishing scientific facts and technical solutions. Debate is required regarding perceptions, social values and expectations, priorities and trade-offs society will choose, risks and changes it will accept, and how mitigation and adaptation strategies will be implemented. Communication helps to foster social learning in 'publics' (e.g., government representatives, industry, non-governmental organizations, professionals, the general public) so that they have the understanding, information, and requisite skills to participate in debate on climate change, appreciate the consequences, and develop and implement response strategies.

Awareness or people's ability and/or willingness to accept an environmental issue, such as climate change, may be hindered by 'cognitive uncertainty'. People struggle to assimilate new ideas, vocabulary, and mental images. 'Effective' communication, and the dialogue it engenders, presents and reconciles different, often contradictory, information on climate change, its impacts and severity, and what should be done (Andrey and Mortsch 2000). Ideally, it assists in developing a common perception or shared framing of the problem from which cooperation and collective action can be developed and conflict reduced (Nilsson 1993). The four assessment reports from the Intergovernmental Panel on Climate Change (IPCC) since 1990 have been an important, credible means of raising and sustaining awareness of climate change through assessments of the science (IPCC 1995; 2001; 2007a). At the same time, there has been 'celebritization' of climate change—a rise in high-profile individuals (e.g., Michael Crichton, Arnold Schwarzenegger, Leonardo DiCaprio) who have inserted their voices into the dialogue on climate change through the mass media (Boykoff and Goodman 2008). The most notable is Al Gore, whose film *An Inconvenient Truth*, some argue, contributed to greater public awareness and shifts in the politics of climate change (Boykoff 2007a).

The complexity and many uncertainties in the climate change issue make it difficult to understand well and to develop a sense of competence regarding what is happening, the severity of the impacts, and the range of potential responses, their costs and urgency for action. Problem-solving is inhibited by 'environmental uncertainty'. Most want to act on the basis of definitive scientific evidence so that unambiguous, defensible, and 'correct' decisions can be made (Conn and Feimer 1985). Information on climate change is presented as trends, impacts, risks, vulnerabilities, scenarios, and ranges; however, the science cannot provide precision and certainty. Adding to the confusion are various 'publics' with competing agendas that capitalize on uncertainty in the science to promote or discredit the issue (Boykoff 2007b; McCright and Dunlap 2000; 2003).

Communication that builds understanding provides factual information because it is easier to accept new facts than to accept new opinions. It summarizes what we know based on the weight of scientific evidence, what is the nature of the uncertainties, what are the available options and risks, the costs and benefits, and how they are distributed within society. For example, the IPCC developed terminology to clarify language with respect to confidence in being correct and the likelihood of an occurrence or outcome[1] and used it when presenting key findings. The IPCC assessment reports surveyed and assessed research globally and delineated where there were contentious issues and uncertainties as well as where the preponderance of evidence could lead to strong statements such as 'warming of the climate system is unequivocal' (IPCC 2007a, 2).

Action on climate change is likely to require widespread social change, with the most important related to behavioural and organizational change (O'Neill and Nicholson-Cole 2009). However, the willingness to undertake these changes suffers from a 'social dilemma' (Milinski et al. 2006; 2008) in that individuals are reluctant to forfeit short-term gains for long-term collective interests if there is uncertainty about the commitment of others to act. This issue is most significant with mitigation actions in which many people with competing interests are involved over numerous jurisdictions: it is difficult to build trust, identify common values and goals, and create consensus. In contrast, the benefits of adaptation actions most often directly accrue to the locale where they are implemented. Communication can foster progression from a reluctance to act, to acknowledging the need to act, to a willingness to act or an informed consent to

certain climate change mitigation and adaptation policy measures. Many professional and government organizations are now preparing documents to guide assessment of impacts and the development of adaptation and mitigation actions (e.g., Bizikova, Neale, and Burton 2008). Examples of innovative programs and local, national, and international mitigation and adaptation activities help others to learn and undertake action.

In this context, 'effective' communication becomes an important part of the science–policy link in adaptive planning and management.

---

those whose coal- and oil-producing interests were potentially negatively affected to highlight claims of uncertainty and inappropriate processes. Uncertainty and conflict were again partners. The *Third Assessment Report* (IPCC 2001, 10) statement 'There is new and stronger evidence that most of the warming observed over the last 50 years is attributable to human activities' was affirmative and quantitative. However, the statement did not say that the evidence was firm or decisive, only new and stronger. The *Fourth Assessment Report* (IPCC 2007a, 5, 10) concluded that '[w]arming of the climate system is unequivocal' and that 'most of the observed increase in global average temperatures since the mid-20th century is very likely due to the observed increase in anthropogenic greenhouse gas concentrations.'

In summary, there has been a progression in the scientific statements corresponding to this question—from 'could be' and 'alternatively', to 'balance of evidence suggests', to 'new and stronger evidence', to 'unequivocal' and 'very likely due', demonstrating increasing confidence in the conclusion that human activities are the cause of the climate change over the past few decades.

In preparing the 2007 assessment, the IPCC put special emphasis on the treatment of

uncertainties (Solomon et al. 2007, 22). Consistent guidelines were used throughout the report. It noted that uncertainties can be due to 'value uncertainties' arising from the incomplete determination of particular values or results (when data are inaccurate or not fully representative of the phenomenon of interest) and 'structural uncertainties' arising from an incomplete understanding of the processes that control particular values or results (when the conceptual framework or model used for analysis does not include all the relevant processes or relationships). Statistical techniques and results expressed probabilistically can address the value uncertainties, while the structural uncertainties are generally described by giving the authors' collective judgment of their confidence in the correctness of a result.

The uncertainty guidance provided for the *Fourth Assessment Report* draws, for the first time, a careful distinction between levels of confidence in scientific understanding and the likelihood of specific results (Solomon et al. 2007). For confidence, the terminology was to be based on the degree of confidence in being correct. Hence, very high confidence meant at least a 9 out of 10 chance of being correct, based on the expert judgment of the scientific group, and low confidence was about a 2 out of 10 chance of being correct. The standard terms used to define the likelihood of an outcome or result when this could be estimated probabilistically ranged from virtually certain, >99 per cent; very likely, >90 per cent; likely, >66 per cent, to exceptionally unlikely, <1 per cent. The usual approach was to give values in terms of assessed best estimates and their uncertainty ranges at 90 per cent confidence intervals (i.e., there is an estimated 5 per cent likelihood of the value being below the lower end of the range or above the upper end of the range).

The 2007 IPCC Assessment and Related Information

The 2007 IPCC *Fourth Assessment Report* presented a thorough analysis of climate and climate change based on an assessment by the world's leading climate scientists. Working Group I reported that the planet has been warming at a linear rate of 0.07°C per decade over the past 100 years (Figure 5.1) (0.7°C warming in the past 100 years) and that rate has increased to 0.18°C per decade over the past 25 years (0.45°C warming in the past 25 years). These are global averages, including the oceanic areas, which are warming more slowly. Northern hemisphere land areas have been warming by about 0.3°C per decade over the past 25 years. From 1900 to 2005, there was a general increase in precipitation over land north of 30°N, but there have been downward trends in precipitation in the tropics since the 1970s (Trenberth et al. 2007). Temperatures and precipitation increased in most parts of Canada

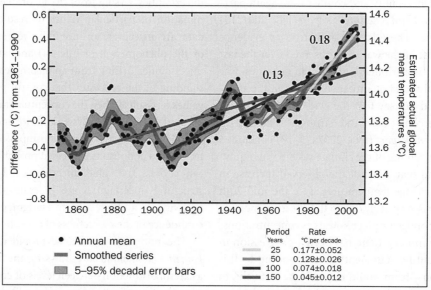

Solomon et al. 2007, 37.

Figure 5.1 Annual global mean temperatures (black dots) with linear fits to the data. The left-hand axis shows temperature anomalies relative to the 1961 to 1990 average, and the right-hand axis shows estimated actual temperatures, both in °C. Linear trends are shown for the past 25, 50, 100, and 150 years. The smoothed series curve shows decadal variations, with the decadal 90% error range shown as a paler band about that line. The total temperature increase from the period 1850 to 1899 to the period 2001 to 2005 is 0.76°C ± 0.19°C.

during the twentieth century (Vincent, Wijn-gaarden, and Hopkinson 2007). It is also likely that there have been increases in the number of heavy precipitation events (the most intense 5 per cent) within many land regions, even in those where there has been a reduction in total precipitation (Zhang, Hogg, and Mekis 2001). There have been changes in atmospheric circulation such that the mid-latitude westerly winds have generally increased in both hemispheres. Based on its analysis of global observations of the climate system, the IPCC concluded that 'warming of the climate system is unequivocal, as is now evident from observations of increases in global average air and ocean temperatures, widespread melting of snow and ice, and rising global average sea level' (IPCC 2007a, 5).

Detection and Attribution of Climate Change

The detection of climate change and its attribution to possible factors has been carefully analyzed (Hegerl et al. 2007) using well-tested coupled atmosphere–ocean–climate system models. Comparisons of simulations done with multi-model ensembles using external forcings (drivers of change in the climate system) that were natural (solar variations, volcanoes) (lower part of Figure 5.2) are compared with anthropogenic (human-caused factors—changes in greenhouse gases, aerosols, land-use changes) plus natural forcing (upper part of Figure 5.2). The models simulate very well the observed variability and changes of global mean temperature from 1900 until the 1950s. This provides additional evidence of the capacity of these models to realistically simulate the climate. However, climate models are only able to reproduce observed global mean temperature changes over the latter part of the twentieth century when they include anthropogenic forcings, and they fail to do so when they exclude

anthropogenic forcings. This is evidence for the influence of humans on global climate. Further evidence is seen in spatial patterns of temperature and other climate indicators.

The IPCC (2007a, 10) conclusion is: 'Most of the observed increase in global average temperatures since the mid-20th century is very likely due to the observed increase in anthropogenic greenhouse gas concentrations.' Other indicators of climate change, such as climate extremes and variability, were also shown to be influenced by anthropogenic influences. In addition, 'a global assessment of data since 1970 has shown it is likely that anthropogenic warming has had a discernible influence on many physical and biological systems.' (IPCC 2007b, 9) In essence, climate change is happening, it is affecting natural systems, and human activities are the primary cause.

Analyses of past climatic changes also help to improve our understanding of the climate system. For example, the sensitivity of the climate system, defined as the global mean change in temperature corresponding to doubling the atmospheric concentrations of greenhouse gases, is concluded to be likely between 2°C and 4.5°C with a most likely value of approximately 3°C.

Complexities and Uncertainties in Climate Science

As the climate warms, the atmosphere can retain more water vapour, thus enhancing the greenhouse effect. This is the water-vapour positive feedback effect. The warming also results in the melting of some snow and reducing sea and lake ice coverage. Snow and ice reflect incoming solar radiation and cool the climate; reducing snow and ice coverage results in more solar radiation being absorbed at the surface and a warmer climate—snow–ice feedback.

Clouds, in their varying shapes, sizes, and lifetimes, provide another set of complicated

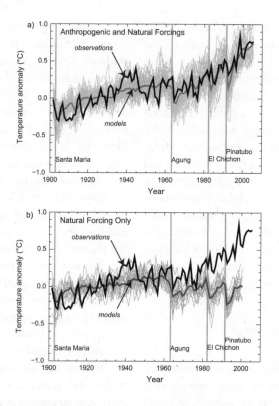

Source: Solomon et al. 2007, 62.

Figure 5.2 Detection and attribution of climate change. Comparison between global mean surface temperature anomalies (°C) from observations (black) and Atmosphere–Ocean General Circulation Model (AOGCM) simulations forced with (a) both anthropogenic and natural forcings and (b) natural forcings only. All data are shown as global mean temperature anomalies relative to the period 1901 to 1950, as observed (black; Hadley Centre, Climatic Research Unit gridded surface temperature data set [Had[CRUT3]; Brohan et al. 2006) and, in (a), as obtained from 58 simulations produced by 14 models with both anthropogenic and natural forcings. The multi-model ensemble mean is shown as a medium-grey curve, and individual simulations are shown as light-grey curves. Vertical grey lines indicate the timing of major volcanic events. Those simulations that ended before 2005 were extended to 2005 by using the first few years of the IPCC *Special Report on Emission Scenarios (SRES)* A1B scenario simulations that continued from the respective twentieth-century simulations, where available. The simulated global mean temperature anomalies in (b) are from 19 simulations produced by five models with natural forcings only. The multi-model ensemble mean is shown as a medium-grey curve, and individual simulations are shown as light-grey curves.

feedbacks. If enough additional water vapour is added to the atmosphere, then more clouds form. More high clouds generally have the effect of warming the climate, since they are transparent to solar radiation but opaque to long-wave radiation. If more low clouds are formed, the result is the opposite—cooling—because they reflect solar radiation. The role of clouds and their modelling in climate models is one of the principal remaining uncertainties in projecting climate change due to increased greenhouse gas concentrations.

Another area of concern is the role that natural systems—the oceans and the terrestrial ecosystems—play in the carbon cycle. The general scientific conclusion is that as the climate warms, the natural greenhouse gas cycles will be altered such that they amplify the rate of climate change (Cox et al. 2000). There is additional uncertainty in scaling from global to regional and local results as well as in taking the implications of extreme events into account (Christenson et al., 2007).

These uncertainties have led to conflict. Many groups have a vested interest in the fossil-based energy sector, and any action to reduce the use of fossil fuels will affect them negatively. Some scientists feel that these uncertainties are such that the relationship between fossil fuels and a changing climate has not been demonstrated. As a result, various alliances have formed that aggressively question the science and reject it as a basis for action (e.g., Essex and McKitrick 2002; Cato Institute 2009). Solomon (2008) has written about some of these scientists, but also see the book review by McBean (2008). Hulme (2009) discusses this further.

Projecting the Future

Prediction of the future states of natural, environmental, social, and economic systems requires predictions across a wide range of natural and social sciences. Prediction is the process of looking ahead on the basis of incomplete knowledge of the present and with incomplete understanding of how the system works (McBean 2007). The prediction accuracy depends on the cumulative uncertainties in each component of the prediction system; this has been referred to as the 'cascade of uncertainty' (IPCC 2004, 25). When people consider a prediction, both of a natural occurrence and of its response to human influences, accurate,

they are more likely to take action. For example, if smog is forecast for today, individuals will tend to respond in limited ways, such as reducing their exposure to the smog. If smog is forecast over the next few days, individuals, industry, and governments will respond by reducing emissions and hence reducing the smog level. These predictions of future states can or should lead to actions that change the outcome. Thus, even in the face of predictions considered accurate, people can make choices, and the choices they make can render the prediction wrong.

In order to predict climate change, it is necessary to predict what choices humans will make in terms of the emission of greenhouse gases, land-use change, and deforestation and then to predict the way the climate will evolve in the future as a result of these choices. Again, human choices may make predictions wrong; our future climate is not predetermined but the result of collective choices. Since prediction usually implies one specific outcome in the future, the term 'projection' is commonly used to accommodate the fact that there is a range of possible futures, each one dependent on collective human choices. These choices can result in conflict stemming from differing values and visions of the future in different parts of the world.

There is now 'considerable confidence that climate models provide credible quantitative estimates of future climate change, particularly at continental scales and above' (Randall et al. 2007, 591). These models are based on accepted physical principles and have demonstrated ability to reproduce observed features of current climate and of past climate changes. They provide opportunities and challenges for climate prediction, recognizing existing uncertainties that are larger for some variables, such as precipitation, than for others, such as temperature. Uncertainties in the models are certainly a basis for conflict over response strategies (Hulme 2009).

The importance of human activities can be assessed through a series of 'what if, then' scenarios. Using assumptions about population growth, industrial transformation, degrees of internationalization, and other factors, national and global emissions of greenhouse gases were projected for many different socio-economic scenarios (Naki´cenovi´c and Swart 2000). An illustrative way of looking at this is to consider the following equation for the change of emissions in the future that is related to other changes:

$$\Delta CO_2 \text{ emissions} = \Delta(pop)^*(\Delta(GDP/pop)^* \Delta(energy/GDP)^*\Delta(CO_2/energy)$$

where: pop is population and GDP is gross domestic production (wealth by many definitions). Hence, changes in future emissions will depend on growth in population, per capita wealth, energy intensity (how much energy is needed to produce the wealth), and carbon intensity (what fraction of the energy is derived from fossil or carbon-based energy sources). From a climate change policy perspective, the main focus is on how to reduce energy intensity and the carbon intensity of the energy being used.

The left side of Figure 5.3 shows a range of emission scenarios up to 2100. For some scenarios, such as A2, emissions continue to rise from 40 $GtCO_{2eq}$/yr to close to 140 $GtCO_{2eq}$/yr by the end of the century.[2] Others, such as B1, assume actions to reduce energy and carbon intensity leading to peak emissions of about 60 $GtCO_{2eq}$/yr at mid-century and declining to about 20 $GtCO_{2eq}$/yr at 2100. The IPCC does not recommend any specific scenario, consistent with its approach of being policy-relevant but not policy-prescriptive. Each emission scenario was then used with several different climate models to project ahead the atmospheric concentration of greenhouse gases, requiring coupled climate–carbon cycle models, and the changes in climate.

Changes in the Physical Climate System

The results for mean global temperature change are shown on the right-hand side of Figure 5.3. How will the climate warm over the next few decades? A lower limit is the artificial scenario for concentrations remaining unchanged at year 2000 values, resulting in warming at a rate of 0.1°C per decade. This is due to the slow response of the oceans adjusting to the accumulation of greenhouse gases that has already occurred. For more realistic emission scenarios, the projected warming over the next few decades will be about 0.2°C per decade, or a little larger than 0.18°C per decade observed over the past 25 years. By about 2040, the projected temperatures start to diverge, with the lowest emission scenario (in this group) leading to 1.8°C warming by 2100 in relation to the period 1980–99. The highest emission scenario results in 3.4°C warming by 2100. The uncertainties in the climate models are indicated by the vertical bars on the right. For the shown range of scenarios and including the uncertainties, the warming by 2100 is between 1°C and 6°C, or about 1.5–7°C relative to the pre-industrial global temperature. For any of the scenarios, warming continues for centuries to follow. This range is what might be called the 'human choice' or 'our' impact.

The spread of the envelope is due to uncertainty in both the emission scenario (a social sciences prediction) and in the climate models (a natural sciences prediction). It is appropriate to note that the uncertainty range does not include zero—i.e., the climate will warm. And the upper range is about the same as the global temperature difference between the present climate and that of an ice age. The warming coming out of the last ice age was about 5°C over 10,000 years, whereas the projected warming over the next 100 years is the same magnitude, a rate of change about 100 times faster, with large implications for natural ecosystems and

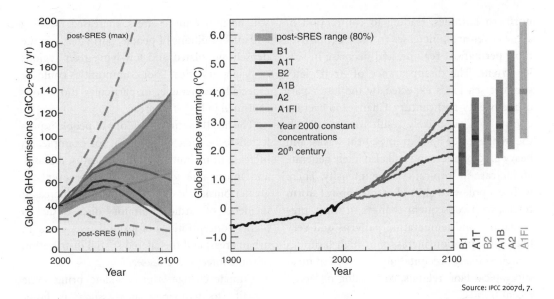

Source: IPCC 2007d, 7.

Figure 5.3 Emission scenarios and corresponding climate change. Left panel: Global GHG emissions (in CO$_{2\text{-eq}}$) in the absence of climate policies: six illustrative SRES marker scenarios (lines) and the 80th percentile range of recent scenarios published since SRES (post-SRES) (grey-shaded area). Dashed lines show the full range of post-SRES scenarios. The emissions cover CO$_2$, CH$_4$, N$_2$O, and F-gases. Right panel: solid lines are multi-model global averages of surface warming for scenarios A2, A1B, and B1, shown as continuations of the twentieth-century simulations. These projections also take into account emissions of short-lived GHGs and aerosols. The 'Year 2000 constant' line is not a scenario but is for Atmospheric-Ocean General Circulation Model (AOGCM) simulations where atmospheric concentrations are held constant at year 2000 values. The bars at the right of the figure indicate the best estimate (solid line within each bar) and the *likely* range assessed for the six SRES marker scenarios at 2090–9. All temperatures are relative to the period 1980–99.

resources. And although the uncertainty in the climate models is important, uncertainties about how global human societies will respond to this challenge are larger.

Together with these temperature changes will come other changes in the climate system. Sea-level rise is a critical parameter for the billions of people who live on or near the coasts. The IPCC projected a range of a 0.2- to 0.6-metre rise by 2100. Because of larger uncertainties in projecting the melting rates of Greenland and Antarctic ice sheets, they were excluded from the projections. Post–IPCC 2007 results (Rahmstorf

et al. 2007) indicate that the sea level is rising along the upper limit of the 2001 IPCC projections, leading to estimates of closer to a metre rise by 2100. If the Greenland ice sheet were to melt, sea level would rise globally about seven metres, inundating many coastal zones. There are major uncertainties as to if, when, and how this may happen.

Although global changes are important, regional changes are of most interest for resource management. The projected regional-scale climatic changes (IPCC 2007a) have the warming greatest over land and at most high northern latitudes, leading to contraction of snow-cover area, increases in thaw depth over most permafrost regions, and decrease in sea-ice extent. The disappearance of Arctic late-summer sea ice is expected by the latter part of the twenty-first century. Changes in average climate will very likely result in increases in the frequency of hot extremes, heat waves, and heavy precipitation and will also likely generate an increase in tropical cyclone intensity. There will be a poleward shift of extra-tropical storm tracks, with consequent changes in wind, precipitation, and temperature patterns and very likely precipitation increases in high latitudes as well as likely precipitation decreases in most subtropical land regions, continuing observed recent trends.

Climate Change Impacts— Around the Globe

The IPCC (2007b) has summarized the climate change impacts on both a sectoral and a regional basis. The impacts on water systems will be critically important for all sectors and regions. For many ecosystems, their ability to survive through a range of stresses, such as flooding, drought, wildfire, insects, land-use change, pollution, fragmentation, and overexploitation, will be diminished as climate change reduces their resilience. Approximately 20 to 30 per cent of plant and animal species assessed are likely to be at increased risk of extinction if increases in global average temperature exceed 1.5 to 2.5°C (see also Chapters 11 and 12). At lower latitudes, especially in seasonally dry and tropical regions, crop productivity is projected to decrease for even small local temperature increases (1 to 2°C), while it may increase slightly at mid to high latitudes, depending on the crop, and then decrease beyond that in some regions (see also Chapter 9). Because of sea-level rise and increasing storms, coastal regions and small-island states

will be under more stress. Increased flooding will affect millions of people, particularly in the densely populated and low-lying river regions of Asia and Africa. Poor communities could be especially vulnerable, in particular those concentrated in high-risk areas.

The health status of millions of people is projected to be affected through, for example, an increase in malnutrition; more deaths, diseases, and injury due to extreme weather events; a greater burden of diarrhoeal diseases; increased incidence of cardio-respiratory diseases due to higher concentrations of ground-level ozone in urban areas; and the altered spatial distribution of some infectious diseases.

Climate change is projected to bring some benefits to temperate areas, such as fewer deaths from cold exposure, and some mixed effects, such as changes in range and transmission potential of malaria in Africa. Overall, it is expected that benefits will be outweighed by the negative health effects of rising temperatures, especially in developing countries, and critically important will be factors that directly shape the health of populations, such as education, health care, public health initiatives, and infrastructure and economic development.

Climate Change as an Issue of National Security

In awarding the 2007 Nobel Peace Prize to the Intergovernmental Panel on Climate Change and Albert Gore, the Nobel Committee of the Norwegian Parliament stated that it 'is seeking to contribute to a sharper focus on the processes and decisions that appear to be necessary to protect the world's future climate, and thereby to reduce the threat to the security of mankind. Action is necessary now, before climate change moves beyond man's control' (Nobel Peace Prize 2007). In so doing, it placed climate change in the context of global peace and security.

Security as a concept has evolved and now includes the ability to defuse political and socio-economic crises through the use of development and environmental policy measures (see also Chapters 1 and 2). In this context, climate security can be defined as: 'that achieved through the implementation of measures that ensure the defence and maintenance of the social, political, and economic stability of a country and of the human population, including freedom from fear and want—both state and human security—from the affects of climate change and global-to-local response to it' (McBean 2009a, 10).

A related facet of climate change is the connection with human health about which the World Health Organization director-general (Chan 2008, 1) stated, 'climate change endangers health in fundamental ways' while dedicating the 2008 World Health Day to the impact of climate change on human health. Climate change can be viewed as a threat multiplier to the extent that superimposing climate change on the wide range of global trends, such as globalization and an aging society, and other security issues, such as terrorism and pandemics, can result in a higher overall impact.

The Council of the European Commission in 2008 adopted a report on the security implications of climate change (European Council 2008, 8) noting that 'the impact of climate change on international security is not a problem of the future but already of today and one which will stay with us.' The report (European Council 2008, 1) states that '[t]he risks posed by climate change are real and its impacts are already taking place. The UN estimates that all but one of its emergency appeals for humanitarian aid in 2007 were climate related. In 2007 the UN Security Council held its first debate on climate change and its implications for international security.'

The United Kingdom's national security strategy (United Kingdom 2008, 18) identified the security challenges as terrorism, weapons of mass destruction, transnational organized crime, global instability and conflict, failed and fragile states, and civil emergencies and then went on to say that climate change is 'potentially the greatest challenge to global stability and security and therefore to national security.' In April 2008, the German Advisory Council on Global Change (2008, 1) presented its report *Climate Change as a Security Risk*. The core message is that 'without resolute counteraction, climate change will overstretch many societies' adaptive capacities within the coming decades', which could result in destabilization and violence, jeopardizing national and international security to a new degree. They used the term *climate-induced conflict constellations* and identified four specific types: degradation of freshwater resources, decline in food production, increase in storm and flood disasters, and environmentally-induced migration.

Climate change is now seen as about conflicts in a much broader sense than earlier. The sense of conflict between fossil-based energy producers and users and the environment has been extended to conflict between states and peoples for human and national security based on emissions from one affecting all.

What Is Dangerous Interference with the Climate System?

A key question in addressing the UNFCCC (1992) target of avoiding dangerous anthropogenic interference with the climate system is 'what is dangerous?' The European Union and some states have adopted the target of 2°C warmer than pre-industrial global temperatures, based on the information provided through the IPCC and other assessments. The IPCC assessments provide information as to which global emission scenarios would meet this target within limits of uncertainty. However, since the publication of the 2007 IPCC *Fourth Assessment Report*, some studies have questioned whether that target is adequate.

Parry et al. (2008) analyzed the impacts on various sectors (Figure 5.4) from different levels of emission reductions. Although a 50 per cent reduction by 2050, based on meeting the target of 2°C relative to pre-industrial temperatures (or 1.4°C compared to 1980–99 values), seemed to avoid dangerous impacts, they noted two additional points. First, with the uncertainties involved in such projections, which are skewed toward larger changes, unacceptable impacts are possible. Second, because the climate system is still not in equilibrium with the emission reductions, one must really look at the impacts at 2100, with their associated uncertainties. Based on these calculations, a 50 per cent emissions reduction by 2050 was not deemed to be adequate to avoid dangerous impacts. Hansen et al. (2008, 1) concluded that 'CO$_2$ will need to be reduced from its current 385 ppm to at most 350 ppm.'

The specification of what is 'dangerous' depends on the personal, regional, and time perspective for the analysis—i.e., dangerous to whom, from what, and on what time scale. It will also depend on what is 'acceptable' risk in that there will always be uncertainty in the projections and the relationship between the danger and human vulnerability.

The Climate Convention and Its Processes

Canada has participated in all the Climate Convention's Conferences of the Parties, including the third in Kyoto and the thirteenth in Bali, which endorsed the Bali Action Plan. The 2009 Conference of the Parties in Copenhagen is expected to be critical in defining the post–Kyoto Protocol regime.

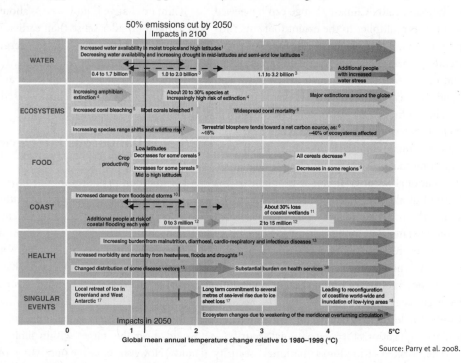

Source: Parry et al. 2008.

Figure 5.4 Selected global impacts from warming associated with various reductions in global greenhouse gas emission.

Conventions on Desertification and Biodiversity

Although most of the international focus is on the UN Framework Convention on Climate Change, other important and relevant international conventions, namely the UN Convention to Combat Desertification (UNCCD) and the UN Convention on Biological Diversity (UNCBD) were both debated at the Earth Summit in 1992. The UNCCD was ratified by Canada in 1995. The UNCCD objective is to combat desertification and mitigate drought. Canada ratified the Convention on Biological Diversity in 1992. The UNCBD objectives are the conservation of biological diversity and the sustainable use of its components (see Chapter 12). A changing climate will have clear implications for both desertification and biological diversity, and Canada, as a party to the three related conventions, has obligations in that regard.

International Agreements and Declarations

Whereas the conventions on climate change, desertification, and biological diversity are binding on governments in that they have undertaken a formal ratification process, international agreements and declarations do not hold the same legal weight. However, they are statements of governments' intentions and concerns, so they have political if not legal significance. The World Summit on Sustainable Development's Summit Plan of Implementation (UNWSSD 2002) linked climate change and international development in its strategy to meet the Millennium Development Goals (see Chapter 2). The signatories agreed on a series of actions, one of which included protecting and managing the natural resource base of economic and social development. In the report that followed the summit, strong connections were drawn between international development and climate change, which also demand action.

Adapting to a Changing Climate

Adaptation strategies for a changing climate are necessary (Burton 2008) and will need to be implemented as an ongoing process. The Canadian National Assessment (Lemmen, Warren, and Lacroix 2008, 3) defines adaptation as 'making adjustments in our decisions, activities and thinking because of observed or expected changes in climate, in order to moderate harm or take advantage of new opportunities'. Every nation and each community within nations have some inherent capacity to cope with variable weather and climate, but significant departures from normal conditions threaten to exceed this 'coping range'. The risks posed cannot be fully resolved through the autonomous actions of individuals and organizations and thus require state intervention through public policy. Climate adaptation policy is a course of action chosen by public authorities to mandate or facilitate adjustments to practices, processes, or structures, aimed at reducing current and future impacts of a changing climate.

Although climate hazards pose a potential threat, their associated impacts are largely determined by a community's vulnerability, which is a function of its exposure to climate hazards, its sensitivity to the stresses they impose, and its capacity to adapt to these stresses. Thus, the central goal of adaptation policy must be to reduce vulnerability (Burton et al. 2002). The vulnerability of communities to extreme weather events is not a fixed condition and can be reduced through actions that minimize exposure, reduce the sensitivity of people and systems, and strengthen the community's adaptive capacity.

Adaptive capacity refers to the ability to adjust practices, processes, or structures to moderate or offset potential negative impacts associated with climate change (Burton 2008). A community with a greater capacity to adapt is less vulnerable to negative impacts associated with climate change. In view of the broad implications of climate change, it is very important that the ministries of industry, economic development, and trade be involved in the development of the adaptation strategy. This is consistent with the usual intent of 'mainstreaming' climate change adaptation into all public and private sector decision-making.

Climate Change Impacts and Adaptation in Canada

There is a substantial body of information on climate change impacts on Canada from the 2007 IPCC assessment chapters on North America (Field et al. 2007) and the polar regions (Anisimov et al. 2007), the Arctic Climate Impact Assessment (ACIA 2005), and the National Assessment (Lemmen, Warren, and Lacroix 2008), which notes that many current climate risks will be exacerbated and new risks and opportunities,

with significant implications for communities, infrastructure, and ecosystems, will appear.

Figure 5.5 from Christensen et al. (2007, 889) shows the annual, winter, and summer changes in temperature and precipitation over North America. By 2080–99, for a medium (A1B) scenario, a 2.8°C global mean warming is projected, and Canada will generally warm more than that, with the Arctic regions warming about twice as much. The winter warming in Canada is projected to be 7°C or more in the High Arctic, 3 to 4°C in Ontario and southern Quebec, and about 2.5°C along the British Columbia coast. In the summer, the most warming is expected to occur over southern interior BC and the southwestern Prairies, reaching 4°C. A summer warming of between 2.5 and 3.5°C is expected over most of the rest of Canada, with the exception of the Arctic coast (1°C), where the absence of summertime sea ice greatly moderates it. This warming will be a continuation of the observed warming (Lemmen, Warren, and Lacroix 2008).

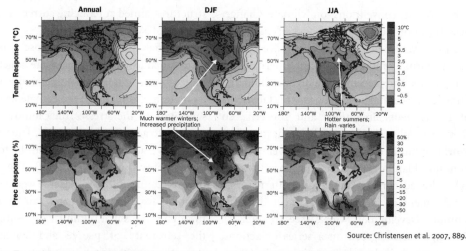

Source: Christensen et al. 2007, 889.

Figure 5.5 Temperature and precipitation changes over North America based on A1B scenario and an ensemble of model simulations. Top row: annual mean, DJF and JJA temperature change between 1980 to 1999 and 2080 to 2099, averaged over 21 models. Bottom row: same as top but for fractional change in precipitation.

All of Canada will see more annual total precipitation, from a 5 to 10 per cent increase in the south to a 30 to 50 percent increase in the north, and more precipitation will fall as rain than snow relative to present conditions. In the winter, the overall situation is similar, except that southern Canada increases are expected to be slightly higher (10 to 15 per cent). However, in the summer the southern BC interior and southwestern Prairies will see a decrease in precipitation of about 10 per cent, while most of the rest of southern Canada will see either no change or a small increase. Further north, the increases will be in the range of 10 to 15 per cent.

The vulnerability of Canada to extreme events has already been discussed. More frequent and intense extreme weather events, such as extreme rainfall, can be expected in Canada in the twenty-first century (Zwiers and Kharin 1998; Christensen et al. 2007). Warmer surface air temperatures could increase the frequency and areal extent of wildfires and lengthen the wildfire season (Flannigan et al. 2005). Tornadoes are a concern for communities in many parts of Canada, particularly the Prairie provinces and southern Ontario. Scientific opinion is mixed as to whether climate change will increase the frequency of tornadoes, but the destructive power of this hazard nevertheless demands strategies to reduce the risk to people and property (McBean 2005).

Climate change is also expected to increase the frequency of hot days in most parts of Canada. For example, under a medium emission scenario, by 2050 the number of days with a maximum temperature above 30°C increases in Victoria from 1 to about 5, in Calgary from 5 to almost 20, in Winnipeg from 14 to nearly 50, in Toronto from 12 to 35, and in Fredericton from 8 to about 25 (Hengeveld, Whitewood, and Fergusson 2005). Summer heat poses a significant risk to public health and safety, as was starkly demonstrated by the 2003 heat wave in western Europe, which was associated with more than 35,000 deaths (Poumadère et al. 2005). The risk is greater for those who live in densely populated urban neighbourhoods, in part because of the urban heat island effect whereby heat is absorbed by asphalt surfaces and various infrastructure materials, increasing the outdoor air temperature by 0.5 to 5°C. Extreme heat events can also have indirect negative health impacts. For instance, hot summer days are usually also smoggy days. The Canadian Medical Association (2008) projects that the annual number of deaths associated with acute effects of air pollution will increase by 2031 to almost 90,000 people because of the combined effects of projected health, demographic, and climate trends in Canada, as well as changes related to social conditions and infrastructure.

It is not just the extreme events that affect natural and human ecosystems. Melting permafrost in the Arctic is affecting infrastructure and transportation systems. Poor communities are especially vulnerable, since they tend to have more limited adaptive capacities and are more dependent on climate-sensitive resources such as local water and food supplies.

Integrating climate change into existing planning processes, often using risk management methods, is an effective approach toward addressing climate change adaptation in Canada. It is also important to recognize that climate change impacts elsewhere in the world and adaptation measures taken to address them will affect Canadian consumers, the competitiveness of some Canadian industries, and Canadian activities related to international development, aid, and peacekeeping. There are barriers to adaptation action that need to be addressed, including limited awareness and availability of information and decision-support tools. These deficiencies include an incomplete knowledge of health risks, uneven access to protective measures, limited awareness of best adaptation practices to protect health, and constraints on the ability of decision-makers to strengthen existing health protection programs or implement new ones (Health Canada 2008). Although there is considerable action at both federal and provincial levels, there is not yet a national or full provincial strategy in place on climate change adaptation.

Greenhouse Gas Emission Reductions

The analysis of the IPCC (2007c, 9) on mitigation of climate change through emission reductions showed that there was 'substantial economic potential for the mitigation of global GHG emissions over the coming decades that could offset the projected growth of global emissions or reduce emissions below current levels.' Looking to 2030, the economic models projected economic costs between a 3 per cent decrease and a small increase in global GDP, with significant regional differences in costs.

A comprehensive assessment of the economics of climate change was completed by Stern (2007, vi), who concluded, 'There is still time to avoid the worst impacts of climate change, if we take strong action now.' Based on economic models, Stern estimated that the overall costs and risks of climate change would be more than the costs of taking action, and he recommended action now. The conflicts between present and future generations, between different regions and states of development, and between energy-producing and most-impacted states will continue to play out in international and national negotiations.

The Global Carbon Project reported on a comparison of projected global fossil-fuel emissions and actual emissions up to 2007 (Global Carbon Project 2008) (Figure 5.6). Global carbon emissions have increased from 6.2 GtC per year (billions of tonnes of carbon or carbon equivalents) in 1990 to 8.5 GtC in 2007, a 38 per cent increase from the Kyoto reference year of 1990. The growth rate of emissions was 3.5 per cent per year for the period 2000 to 2007, an almost fourfold increase from 0.9 per cent per year between 1990 and 1999. Since 2004, global

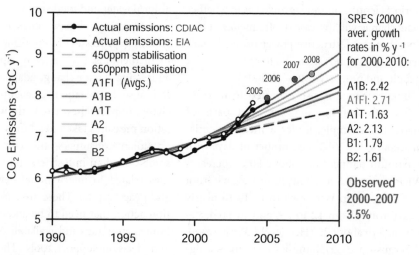

Global Carbon Project 2008.

Figure 5.6 Fossil fuel emissions: actual versus IPCC scenarios.

emissions have been about equal to or greater than the more pessimistic of the IPCC emission scenarios. Emission reduction strategies clearly have thus far been ineffective on a global scale. The IPCC report on climate change mitigation notes that longer-term efforts will be required to stabilize greenhouse gas emissions before they ultimately decline; the lower the stabilization level, the more quickly this peak and decline will occur (IPCC 2007c). Emission increases over the next two to three decades will have a major impact on opportunities to achieve stabilization.

Canada's Record on Emission Reductions—The Interplay of Science and Politics

Canada has signed and ratified both the Climate Convention and its Kyoto Protocol. Under the Kyoto Protocol, Canada agreed to reduce its greenhouse gas emissions, with respect to the 1990 reference year, by 6 per cent averaged over the six-year commitment period of 2008–12. The decision to do so and the processes have been thoroughly described by Simpson, Jaccard, and Rivers (2007) in a book appropriately titled *Hot Air*. By 2009, Canada had a sorry record in responding to the internationally agreed-upon need to reduce greenhouse gas emissions.

As can be seen from Figure 5.7, Canada's emissions had risen, by 2007, to 747 MT CO_{2eq} per year, 26.2 per cent above the 1990 reference level and 33.8 per cent above the Kyoto Protocol commitment. The energy sector was the source of 81 per cent of Canada's emissions, and Alberta had replaced Ontario as the largest source of emissions. When it ratified the Kyoto Protocol with a target of −6 per cent, Canada did accept a difficult target. For example, other Arctic Council countries' targets are: European Union countries (−8 per cent overall with an

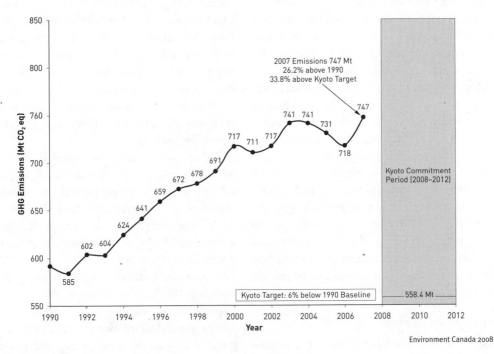

Environment Canada 2008.

Figure 5.7 Canada's greenhouse gas emissions 1990–2007.

internal decision to allocate targets of: Sweden +4 per cent; Finland 0 per cent; Denmark −21 per cent); Iceland +10 per cent; Norway +1 per cent; Russia 0 per cent. The United States' target was −7 per cent, but it did not ratify.

In terms of greenhouse gas emissions per person, Canada ranks among the highest. Based on 2004 data, Canada and the United States both produced about 20 tonnes per person. Germany, Japan, and the United Kingdom produced about one-half this amount, while the figure for China was about 2.6 tonnes per person and for India about 1 tonne per person. Canada has a sorry record in this regard, particularly when one considers that its emissions have increased about twice as much as those of the United States since 1990. Canada has been described on this issue as a 'rogue state' (Dyer 2008).

The reports of the Commissioner of the Environment and Sustainable Development (2006; 2008) were very critical of the federal government's policies and approaches toward reducing greenhouse gas emissions. The December 2008 report noted, with respect to the Turning the Corner (a plan to regulate greenhouse gas emissions and air pollutants) that the 'federal government cannot demonstrate that the results it has reported for the policy tools we (the Commissioner) examined have actually been achieved or that processes are in place to verify the results reported by the private sector.' The report also noted 'flawed' analyses with respect to emission reductions. The credibility of emission reduction plans depends in part on the validity of the economic forecasts that go into them, particularly for intensity-based targets, and these forecasts have been criticized. Overall, these reports do not inspire confidence that targets will actually be achieved.

Interplay on Science

There has been interplay between the climate change sceptics exploiting uncertainty and science advocates urging action. The sceptics contest that climate change is an issue, saying either that it is not happening or that if it is, it is not a result of human activities. Some say it would be good for us anyway (how can we complain about warmer winters?). Gelbspan (1998; 2004) has documented much of the early and international debate. Hulme (2009), with his provocative book title *Why We Disagree about Climate Change*, puts the debate in the context of values, fears, beliefs, the way science is done, and global governance.

Some of the interplay in Canada is highlighted by the following. Shortly after the election of Prime Minister Stephen Harper in 2006, *The Financial Post* (6 April 2006) ran a major story about 60 scientists (about 20 from Canada) who were sceptical of the climate science, calling on the prime minister to undertake a review of the scientific foundation of the federal government's climate change plans. A quick response (19 April 2006) came from 90 Canadian climate scientists urging the government to develop a national strategy on climate change, and two years later 130 Canadian climate scientists (Canadian Climate Science Leaders 2008) expressed concern that the climate was changing even faster than projected only a few years ago.

Other groups have also stated their opinions. In a policy sense, perhaps most important in this interplay have been the G-8 leaders meetings of 2005 and 2008. The G-8 leaders' 2005 communiqué stated, 'Climate change is a serious and long-term challenge that has the potential to affect every part of the globe.' The 2008 G-8 declaration (G-8 2008) stated, 'Recognizing the linkage between the potential impacts of climate change and development, mitigation and adaptation strategies should be pursued as part of development and poverty eradication efforts.'

A *Financial Times* (2009) article entitled 'Global warning—Fighting climate change has never been more important' commented:

Do not be misled by the recent cold winter in Europe and North America—or by this week's conference of vocal sceptics in New York [one national newspaper in Canada did highlight the sceptics' meeting]. Pay attention instead to the larger gathering in Copenhagen, where mainstream scientists issued a series of dire warnings that global warming is proceeding far faster than the scenarios published by the Intergovernmental Panel on Climate Change two years ago.

The International Scientific Congress on Climate Change (2009) held in Copenhagen with more than 2,500 delegates concluded that 'Temperature rises above 2°C[3] will be very difficult for contemporary societies to cope with, and will increase the level of climate disruption through the rest of the century.'

Canadian Politics

Canadian climate change policy dynamics have been dominated by a conflict mainly between the oil-producing provinces, primarily Alberta, and the others. British Columbia and Quebec have already implemented carbon taxes, and British Columbia, Manitoba, Ontario, and Quebec have joined the Western Climate Initiative led by western US states. A key aspect of Canada's federal position is the Alberta oil sands. Canada is the largest foreign supplier of energy to the US, and the output of the oil sands has been an increasing fraction of that energy. The US is well aware of accusations that both the Canadian oil sands and US coal represent a 'dirty' source of energy and also the call for 'clean' energy (see also Chapter 6). At the present time, it is not clear how this situation will evolve.

One political difficulty is the conflict between the desire of all politicians to be elected and, if elected, to be re-elected. The unfortunate reality is that a successful approach to elections is to provide the most tangible and visible benefits over a short term. The political process is very much influenced by lobbyists and vocal interest groups. The oil and gas lobby, based on the strength of its contributions to employment and to tax revenue, holds considerable strength within the political process. An analogy can be drawn with the reluctance of most governments to fund disaster risk reduction, preferring instead to pay, usually more substantially, when disasters strike (Henstra and McBean 2005). Climate change is a long-term intergenerational and international issue—one of equity and fairness. Actions taken to reduce greenhouse gas emissions will generate benefits—including economic benefits—in the future, but not before the next election.

Taxes, Economics, and Carbon

Another issue rests at the core of the political and economic systems in Canada and in its major trading partner to the south. The political process caters to voters' desire for lower taxes. At the same time, the market-based approach dominates the economy: in other words, the market should determine prices; the market-based approach is the most efficient; and governments should not intervene. Suppose your neighbour decided that rather than paying (through taxes) to have trash removed from his or her property, he or she would just dump it over the fence into your yard. You would then have three options: accept the situation without compensation; accept it but demand that your neighbour pay a fee; or ask your neighbour to desist and if necessary launch a legal process to halt the practice. If you choose to impose a fee, you will be creating a market, essentially putting a price on the trash.

If you choose the legal route, you will be creating a legal framework to govern the issue. This is feasible when the waste is obvious, the source and receptor (your yard) are clear, and you can work within legal and economic frameworks.

In the case of climate change, the sources may be clear, but they are not always recognized, and the immediate receptor is the atmosphere, which just carries it away. The result is climate change, which creates costs for others both locally and around the globe. The person or organization that used the fossil fuel and presumably gained some economic or social benefit from doing so is not held responsible for creating an economic and social cost for someone else through climate change. In economics, this is called an externality: a cost suffered by someone other than the responsible party. Stern (2007) and Sinn (2008) have called climate change the greatest and widest ranging market failure ever seen. When there is a market failure, is it not appropriate for governments to intervene?

Taxing people to ensure that they pay for the external costs they impose on others would seem fair and appropriate. But this kind of thinking is lost in market-based socio-economic approaches. Economics is about the efficient allocation of resources, which markets do wonderfully well—except when externalities or market failures are involved. Thus, making those responsible pay for externalities will improve market efficiency—and this can be done by putting a price on carbon. There are various approaches to reducing fossil-fuel emissions, but most do involve putting a price on carbon, either through a carbon tax or a cap-and-trade system, which allocates caps on emissions to various sectors or industries and then allows them to be traded on an open market. However, such a course of action is not acceptable to those who fundamentally oppose government intervention.

Reasons for Action

In addition to the scientific case for and international agreements on taking action on climate change, such action as promoting energy efficiency and imposing appropriate pricing on carbon can generate significant economic benefits. The pollutants produced from the combustion of fossil fuels also contribute in a major way to urban smog and other air and water pollution concerns. Adopting smart approaches toward reducing greenhouse gas emissions will therefore help to reduce pollution, generating health and economic benefits. The most important reason for Canada to take action on climate change is that it will be among the most affected countries. However, we cannot act alone. Because Canadian emissions are only about 2 per cent of global emissions, it is important that the major developing countries—China, India, Brazil, Indonesia, and others—be brought into a future emission reductions regime. But as long as Canada's record remains dismal in terms of per capita consumption of fossil fuels, it has very little leverage or moral standing to persuade these countries to do so.

In contrast to the situation in the early years of this century, the major driver for a Canadian plan for greenhouse gas emission reductions now comes from south of the border with the new administration of President Barack Obama. During his first visit to Canada (19 February 2009), President Obama and Prime Minister Harper held a joint press conference to talk about three issues: the global economic recession, cross-border cooperation on environmental protection and energy security, and priorities for international peace and security (McBean 2009c). Although they presented these three issues as separate, they are actually linked, both as key issues and as policy responses. President Obama put the issue in the right perspective by noting that climate change is 'an issue that, ultimately, the prime minister's children and my children are going to have to live with for many years', an issue with national and international consequences.

To deal with climate change, the Canadian government has been putting most of its faith in carbon capture and storage, a strategy also endorsed by the United States, but there are significant economic and technical issues yet to be resolved (*Economist* 2009). Capturing carbon will have to be part of the solution over the next few decades, but there is a strong need to develop parallel strategies and to look to the more distant future. In early 2009, US and Canadian emission reduction targets for 2020 were similar (0 per cent for the US and −3 per cent for Canada), adjusted to the 1990 reference year. As long as President George W. Bush was in power, Canada could perhaps legitimately argue that it was really premature to talk about harmonization with the United States—but that dynamic has changed with the new US administration, and it is not clear how the situation will evolve.

CONCLUSION: THE WAY AHEAD—INTERNATIONALLY AND NATIONALLY

Internationally, there are encouraging signs. The Bali Action Plan includes quantified emission reduction objectives agreed to by all developed country parties and appropriate mitigation actions agreed to developing country parties. The Bali Action Plan also included calls for enhanced action on adaptation and cooperation on an urgent basis. In addition, it was agreed that technology development and transfer of information for both mitigation and adaptation should be coupled with the provision of financial resources and investment.

In the United States, where significant climate change actions have already been taken at the state level, the federal government under President Obama has made combating climate change a priority. The appointment of a very strong team on climate change, comprising such members as a person responsible for energy and climate change, a scientific adviser, the secretary of energy, and the administrator of the National Oceanic and Atmospheric Administration, bodes very well for the future. Significant as well was the president's statement that '[s]cience and the scientific process must inform and guide decisions of my administration on . . . mitigation of the threat of climate change, and protection of national security.' This focus on science is a major step forward.

The Conference of the Parties (CoP15) in Copenhagen in late 2009 will be a critical event in the history of addressing climate change. The Danish minister for climate and energy, who will play a major role in moving the agenda of the CoP forward, has asked the question: 'Is it morally defendable to secure our own wellbeing at the potential expense of future generations—given what we know now about the costs of inaction?' (Hedegaard 2008, 2) Recognizing the long lives of greenhouse gases and the public infrastructure such as power, water, and transportation systems that is currently being constructed, she concluded, 'The clock is ticking, ladies and gentlemen! . . . I think it is fair to say that it is time to act.' Delegates to Climate Convention conferences need to listen to and rely on science and technological expertise when they debate and decide what is needed and what is possible; they need to listen to people and open their eyes to the impacts already occurring around the world.

Climate change should be positioned rightly as an issue of international security and of international and intergenerational equity. Ethics and action are essential for the sake of children and grandchildren wherever they may live.

FROM THE FIELD

At the end of August 2009, I boarded a plane to Geneva to attend and speak to the World Climate Conference 3 (WCC3). During the conference, there was time to reflect on the Second World Climate Conference (WCC2) held in the same congress centre in 1990—at which I had also been a scientific speaker—and compare WCC3 with it. The Second World Climate Conference was the first climate change conference to attract the participation of five heads of government and many ministers as well as to draw global media attention. The conference received the first scientific assessment from the Intergovernmental Panel on Climate Change, established in the same building two years earlier. The conference also called for the creation of a climate convention, which came into being two years later.

Nineteen years later, I reflect on the reality that, despite the Climate Convention, global emissions of greenhouse gas have been rising more rapidly, at a rate two to three times faster, in the years since 2000 than in the 1990s and that Canada has been one of the poorest performers in the world in this regard. With the climate now warming 'unequivocally', the focus of WCC3 is on climate information services to provide the basis for countries to adapt to and reduce their risk from climate change.

In 1990, the Canadian delegation was headed by the minister of environment and included members of Parliament from the other parties. In 2009, the Canadian delegation was headed by an assistant deputy minister and did not include any politicians.

For me personally, these 19 years involved much change. In 1990, I was an academic, a professor of atmospheric and oceanic sciences at the University of British Columbia, and participated in WCC2 as chair of the international scientific committee for the World Climate Research Programme. Between then and now, as an assistant deputy minister with Environment Canada, I had participated in the Climate Convention negotiations, including Kyoto. I returned to academia in 2000 as a professor of geography and political science. My focus is now on climate change adaptation and natural disaster risk reduction. How can vulnerable populations, mainly in developing countries along low-lying coasts or flood plains or in drought-prone areas, adapt to or even survive the impacts of a changing climate? Climate change acts as a threat multiplier, perhaps tipping already unstable states and societies. The concerns of the heads of governments from Ethiopia, Bangladesh, and many other countries were emphasized in their speeches and reflected on their faces as they spoke at WCC3.

I often tell my students about listening to the prime minister of Tuvalu at WCC2 refer to the entire population of his low-lying island state as an 'endangered species' because of sea-level rise. A few years ago, I read in the press that he was asking other countries to take in his people as the seas continued their rise, having lost hope that the major emitting states would take action in time. Now, in 2009, the UN secretary-general calls upon the leaders of governments to become global leaders and take action at the Climate Convention's Copenhagen meeting only months away. By the time you read this book, the result will be evident— I can only hope that my pessimism will prove unfounded.

—GORDON MCBEAN

Notes

1. http://www.ipcc.ch/activity/uncertaintyguidance note.pdf.
2. Note that 40 $GtCO_{2eq}$/yr = $11GtC_{eq}$ per year, where the $_{eq}$ subscript is used to designate the equivalent impact radiatively of CO_2, methane, and other greenhouse gases.
3. Referenced to pre-industrial temperatures, which means only 1.4°C more warming.

REVIEW QUESTIONS

1. Are humans causing climate change, and if so, how can this be demonstrated?
2. What has Canada's role been in this global issue?
3. What are possible response strategies, and what is the global framework for action?
4. How has uncertainty been used to foster conflict in climate change response strategies?

REFERENCES

ACIA. 2005. *Arctic Climate Impacts Assessment*. Cambridge: Cambridge University Press.

Andrey, J., and L. Mortsch. 2000. 'Communicating about climate change: Challenges and opportunities'. In D. Scott, et al., eds, *Climate Change Communication: Proceedings of the International Conference*, 22–4 June, Kitchener-Waterloo, ON.

Anisimov, O.A., et al. 2007. 'Polar regions (Arctic and Antarctic)'. In M.L. Parry et al., eds, *Climate Change 2007: Impacts, Adaptation and Vulnerability. Contribution of Working Group II to the Fourth Assessment Report IPCC*, 653–85. Cambridge: Cambridge University Press.

Berry, P., G. McBean, and J. Séguin. 2008. 'Vulnerabilities to natural hazards and extreme weather'. In J. Séguin, ed., *Health in a Changing Climate: A Canadian Assessment of Vulnerabilities and Adaptive Capacity*, 47–111. Ottawa: Health Canada.

Bizikova, L., T. Neale, and I. Burton. 2008. *Canadian Communities' Guidebook for Adaptation to Climate Change. Including an Approach to Generate Mitigation Co-benefits in the Context of Sustainable Development*. Vancouver: Environment Canada and University of British Columbia.

Boykoff, M.T. 2007a. 'Flogging a dead norm? Newspaper coverage of anthropogenic climate change in the United States and United Kingdom from 2003 to 2006'. *Area* 39 (4): 470–81.

———. 2007b. 'From convergence to contention: United States mass media representations of anthropogenic climate change science'. *Transactions of the Institute of British Geographers* 32 (4): 477–89.

Boykoff, M.T., and M.K. Goodman. 2008. 'Conspicuous redemption? Reflections on the promises and perils of the "celebritization" of climate change'. *Geoforum* doi:10.1016/j.geoforum.2008.04.006.

British Columbia. 2004. '*Firestorm 2003': Provincial Review*. Victoria: Government of British Columbia.

Broecker, W.S. 1987. 'Unpleasant surprises in the greenhouse'. *Nature* 328: 123–6.

Brohan, P., et al. 2006. 'Uncertainty estimates in regional and global observed temperature changes: A new dataset from 1850'. *Journal of Geophysical Research* 111 D12106, doi:10.1029/2005JD006548.

Brooks, G. 2008. 'Geomorphic effects and impacts from July 1996 severe flooding in the Saguenay area, Quebec'. 30 January 2008. http://gsc.nrcan.gc.ca/floods/saguenay1996/index_e.php.

Bruce, J.P. 2001. 'Intergovernmental Panel on Climate Change and the role of science in policy'. *Isuma* winter: 11–15.

Burton, I. 2008. 'Moving forward on adaptation'. In D.S. Lemmen, F.J. Warren, J. Lacroix, and E. Bush, eds, *From Impacts to Adaptation: Canada in a Changing Climate 2007*, 425–40. Ottawa: Government of Canada.

Burton, I., et al. 2002. 'From impacts assessment to adaptation priorities: The shaping of adaptation policy'. *Climate Policy* 2: 145–59.

Canadian Climate Science Leaders. 2008. 'An open letter on climate change science to all Canadian elected government leaders'. http://www.cmos.ca/whatsnew.html.

Canadian Medical Association. 2008. 'No breathing room: National illness costs of air pollution'. http://www.cma.ca.

Cato Institute. 2009. 'Climate change reality'. http://www.cato.org/special/climatechange.

Chan, M. 2008. 'Statement by WHO Director-General Dr Margaret Chan, 7 April 2008'. http://www.who.int.

Christensen, J.H., et al. 2007. 'Regional climate projections'. In S. Solomon, et al., eds, *Climate Change 2007: The Physical Science Basis. Contribution of Working Group I to the Fourth Assessment Report IPCC*, 849–940. Cambridge and New York: Cambridge University Press.

Commissioner of the Environment and Sustainable Development. 2006. 'September report of the Commissioner of the Environment and Sustainable Development'. http://www.oag-bvg.gc.ca/internet/English/parl_cesd_200609_e_936.html.

———. 2008. 'December report of the Commissioner of the Environment and Sustainable Development'. http://www.oag-bvg.gc.ca/internet/English/parl_cesd_200812_e_31872.html.

Conn, W.D., and N.R. Feinme. 1985. 'Communicating with the public on environmental risk: Integrating research and policy'. *Environmental Professional* 7: 39–47.

Cox, P.M., et al. 2000. 'Acceleration of global warming due to carbon-cycle feedbacks in a coupled climate model'. *Nature* 408 (6809): 184–7.

Dyer, Gwynne. 2008. *The Climate Wars*. Toronto: Random House Canada.

Economist. 2009. 'Politicians are pinning their hopes for delivery from global warming on a technology that is not quite airtight'. 5 March. http://www.economist.com/search/search (carbon capture and storage).

Edwards, P.N., and S.H. Schneider. 2001. 'Self-governance and peer review in science-for-policy: The case of the IPCC *Second Assessment Report*'. In C.A. Miller and P.N. Edwards, eds, *Changing the Atmosphere, Expert Knowledge and Environmental Governance*, 219–46. Cambridge, MA: MIT Press.

Environment Canada. 2007. 'The top ten Canadian weather stories for 2006'. http://www.msc-smc.ec.gc.ca/media/top10/2006/topten2006_e.html.

———. 2008. 'Canada's 2007 greenhouse gas inventory: A summary of trends. http:www.ec.gc.ca.

Essex, C., and R. McKitrick. 2002. *Taken by Storm: The Troubled Science, Policy and Politics of Global Warming*. Toronto: Key Porter Books.

European Council. 2008. 'Climate change and international security'. Paper from the High Representative and the European Commission to the European Council. Paper S113/08. 14 March. http:www.consilium.europa.eu/ueDocs/cms_Data/pressData/en/reports/99387.pdf.

Field, C.B., et al. 2007. 'North America'. In M.L. Parry et al., eds, *Climate Change 2007: Impacts, Adaptation and Vulnerability. Contribution of Working Group II to the Fourth Assessment Report IPCC*, 617–52. Cambridge: Cambridge University Press.

Financial Times. 2009. 'Global warning'. http://www.ft.com/cms/s/43a35e16-100a-11de-a8ae-0000779fd2ac,dwp_uuid=063fb9c2-3000-11da-ba9f-00000e2511c8.

Flannigan, M.D., et al. 2005. 'Future area burned in Canada'. *Climatic Change* 72 (1–2), 1–16.

Fourier, J.B.J. 1824. 'Remarques générales sur la température du globe terrestre et des espaces planétaires'. *Annales de chimie et de physique* 27: 136–67.

G-8. 2008. 'Hokkaido Toyako Summit Leaders Declaration. 8 July. http://www.g8summit.go.jp.

Gelbspan, R. 1998. *The Heat Is on: The Climate Crisis, the Cover-up, the Prescription*. Reading, MA: Perseus Publishing.

———. 2004. *Boiling Point: How Politicians, Big Oil and Coal, Journalists and Activists Are Fueling the Climate Crisis—And What We Can Do to Avert Disaster*. Reading, MA: Perseus Publishing.

German Advisory Council on Global Change. 2008. *Climate Change as a Security Risk*. London: Earthscan.

Global Carbon Project. 2008. 'Carbon budget and trends 2007'. http://www.globalcarbonproject.org/carbonbudget/07/index.htm.

Hansen, J., et al. 2008. 'Target atmospheric CO_2: Where should humanity aim?' http://arxiv.org/abs/0804.1126 and http://arxiv.org/abs/0804.1135.

Health Canada. 2008. *Health in a Changing Climate: A Canadian Assessment of Vulnerabilities and Adaptive Capacity*. ed. J. Séguin. Ottawa: Health Canada.

Hedegaard, C. 2008. 'Climate science and the need for action'. Address at the 33rd International Geological Congress, Oslo, 8 August. http://www.kemin.dk/en-us/theminister/speeches.

Hegerl, G.C., et al. 2007. 'Understanding and attributing climate change'. In S. Solomon et al., eds, *Climate Change 2007: The Physical Science Basis. Contribution of Working Group I to the Fourth Assessment Report IPCC*, 663–745. Cambridge and New York: Cambridge University Press.

Hengeveld, H., B. Whitewood, and A. Fergusson. 2005. *An Introduction to Climate Change: A Canadian Perspective*. Downsview, ON: Environment Canada.

Henstra, D., and G.A. McBean. 2005. 'Canadian disaster management policy: Moving toward a paradigm shift?' *Canadian Public Policy* 31 (3): 303–18.

Houghton, J.T., et al., eds. 1995. *Climate Change 1995: The Science of Climate Change. Contribution of Working Group I to the IPCC Second Assessment Report*. Cambridge: Cambridge University Press.

Houghton, J.T., G.J. Jenkins, and J.J. Ephraums, eds. 1990. *Climate Change, IPCC Scientific Assessment*. Cambridge: Cambridge University Press.

Hulme, M. 2009. *Why We Disagree about Climate Change*. Cambridge: Cambridge University Press.

International Scientific Congress on Climate Change. 2009. 'Key messages'. http://climatecongress.ku.dk.

IPCC (Intergovernmental Panel on Climate Change). 1995. 'IPCC Second Assessment Climate Change 1995'. http://www.ipcc.ch/pdf/climate-changes-1995/ipcc-2nd-assessment/2nd-assessment-en.pdf.

———. 2001. *Climate Change 2001: Synthesis Report. Summary for Policymakers*. Cambridge: Cambridge University Press.

———. 2004. 'IPCC Workshop on Describing Scientific Uncertainties in Climate Change to Support Analysis of Risk and of Options'. Workshop Report. http://ipcc-wg1.ucar.edu/meeting/URW/product/URW_Report_v2.pdf.

———. 2007a. 'Summary for policymakers'. In S. Solomon et al., eds, *Climate Change 2007: The Physical Science Basis. Contribution of Working Group I to the Fourth Assessment Report of the Intergovernmental Panel on Climate Change*, 1–18. Cambridge: Cambridge University Press.

———. 2007b. 'Summary for policymakers'. In M.L. Parry and O.F. Canziani, eds, *Climate Change 2007: Impacts, Adaptation and Vulnerability. Contribution of Working Group II to the Fourth Assessment Report of the Intergovernmental Panel on Climate Change*, 1–18. Cambridge: Cambridge University Press.

———. 2007c. 'Summary for Policymakers'. In B. Metz et al., eds, *Climate Change 2007: Mitigation of Climate Change. Contribution of Working Group III to the Fourth Assessment Report of the Intergovernmental Panel on Climate Change*, 1–23. Cambridge: Cambridge University Press.

———. 2007d. 'Synthesis report'. In R.K. Pachauri and A. Reisinger, eds, *Climate Change 2007. Contribution of Working Groups I, II and III to the Fourth Assessment Report of the Intergovernmental Panel on Climate Change*. Geneva: IPCC.

Jansen, E., et al. 2007. 'Palaeoclimate'. In S. Solomon et al., eds, *Climate Change 2007: The Physical Science Basis. Contribution of Working Group I to the Fourth Assessment Report IPCC*, 433–97. Cambridge and New York: Cambridge University Press.

Kerry, Mara, et al. 1999. 'Glazed over: Canada copes with the ice storm of 1998'. *Environment* 41 (1): 6–11, 28–32.

Lemmen, D.S., F.J. Warren, and J. Lacroix. 2008. 'Synthesis'. In D.S. Lemmen, F.J. Warren, J. Lacroix, and E. Bush, eds, *From Impacts to Adaptation: Canada in a Changing Climate 2007*, 1–20. Ottawa: Natural Resources Canada (www.nrcan.gc.ca).

Le Treut, H., et al. 2007. 'Historical overview of climate change'. In S. Solomon et al., eds, *Climate Change 2007: The Physical Science Basis. Contribution of Working Group I to the Fourth Assessment Report IPCC*, 93–127. Cambridge and New York: Cambridge University Press.

McBean, G.A. 2005. 'Risk mitigation strategies for tornadoes in the context of climate change and development'. *Mitigation and Adaptation Strategies for Global Change* 10 (3): 357–66.

———. 2007. 'Role of prediction in sustainable development and disaster management'. In H.G. Brauch, et al., eds, *Globalisation and Environmental Challenges: Reconceptualising Security in the 21st Century*, 929–38. Berlin: Hexagon Series on Human and Environmental Security and Peace, v. 3.

———. 2008. 'The danger of misinformation'. *Alternatives* 34 (4): 37.

———. 2009a. *Addressing Climate Change in the Context of Security Policy: Implications for Canada.* Ottawa: Conference Board of Canada.

———. 2009b. 'Communicating to policy makers climate science with its inherent uncertainties'. In V. Grover, ed., *Global Warming and Climate Change*, 577–94. Enfield, NH: Science Publishers.

———. 2009c. 'The environment and energy security: Obama and Harper have different takes'. *Policy Options* (April): 53–5.

McCright, A.M., and R.E. Dunlap. 2000. 'Challenging global warming as a social problem: An analysis of the conservative movement's counter-claims'. *Social Problems* 47 (4): 499–522.

———. 2003. 'Defeating Kyoto: The conservative movement's impact on US climate change policy'. *Social Problems* 50 (3): 348–73.

Manabe, S., and R.T. Weatherald. 1967. 'Thermal equilibrium of the atmosphere with a given distribution of relative humidity'. *Journal of Atmospheric Sciences* 24: 241–59.

Milinski, M., et al. 2006. 'Stabilizing the Earth's climate is not a losing game: Supporting evidence from public goods experiments'. *Proceedings of the National Academy of Sciences* 103 (11): 3994–8.

———. 2008. 'The collective-risk social dilemma and the prevention of simulated dangerous climate change'. *Proceedings of the National Academy of Sciences* 105 (7): 2291–94.

Morris-Oswald, Tonia, and A. John Sinclair. 2005. 'Values and floodplain management: Case studies from the Red River Basin, Canada'. *Environmental Hazards* 6 (1): 9–22.

Nakićenović, N., and R. Swart, eds. 2000. *Special Report on Emissions Scenarios.* A Special Report of Working Group III of the Intergovernmental Panel on Climate Change. Cambridge and New York: Cambridge University Press.

Nilsson, L.J. 1993. 'Situational and cognitive obstacles to political consensus on climatic change'. *Climatic Change* 25: 93–6.

Nobel Peace Prize. 2007. http://nobelprize.org/nobel_prizes/peace/laureates/2007/index.html.

O'Neill, S., and S. Nicholson-Cole. 2009. '"Fear won't do it": Promoting positive engagement with climate change through visual and iconic representations'. *Science Communication* 30 (3): 355–79.

Parry, M., et al. 2008. 'Climate policy: Squaring up to reality'. *Nature Reports: Climate Change* 2: 168–70.

Poumadère, M., et al. 2005. 'The 2003 heat wave in France: Dangerous climate change here and now'. *Risk Analysis* 25 (6): 1483–94.

Rahmstorf, S., et al. 2007. 'Recent climate observations compared to projections'. *Science* doi:10.1126/science.1136843.

Randall, D.A., et al. 2007. 'Climate models and their evaluation'. In S. Solomon et al., eds, *Climate Change 2007: The Physical Science Basis. Contribution of Working Group I to the Fourth Assessment Report IPCC*, 589–662. Cambridge and New York: Cambridge University Press.

Revelle, R., and H. Suess. 1957. 'Carbon dioxide exchange between the atmosphere and the ocean and the question of an increase of atmospheric CO_2 during the past decades'. *Tellus* 9: 18–27.

Simpson, J., M. Jaccard, and N. Rivers. 2007. *Hot Air: Meeting Canada's Climate Change Challenge.* Toronto: Douglas Gibson Books.

Sinn, H.W. 2008. 'Public policies against global warming: A supply side approach'. *International Tax Public Finance* 15: 360–94.

Solomon, L. 2008. *The Deniers: The World-Renowned Scientists Who Stood Up against Global Warming Hysteria, Political Persecution, and Fraud.* Minneapolis: Richard Vigilante Books.

Solomon, S., et al. 2007. 'Technical summary'. In S. Solomon et al., eds, *Climate Change 2007: The Physical Science Basis. Contribution of Working Group I to the Fourth Assessment Report IPCC.* Cambridge and New York: Cambridge University Press.

Stern, N.H. 2007. *The Economics of Climate Change: The Stern Review.* Cambridge: Cambridge University Press.

Trenberth, K.E., et al. 2007. 'Observations: Surface and atmospheric climate change'. In S. Solomon et al., eds, *Climate Change 2007: The Physical Science Basis. Contribution of Working Group I to the Fourth Assessment Report IPCC*, 235–336. Cambridge and New York: Cambridge University Press.

United Kingdom. 2008. *The National Security Strategy of the United Kingdom, Security in an Interdependent World.* Presented to Parliament by the prime minister, by command of Her Majesty. March. Document Cm 7291.

UNFCCC. (United Nations Framework Convention on Climate Change). 1992. http://www.unfccc.int.

UNWSSD (World Summit on Sustainable Development). 2002. http://www.un.org/esa/sustdev/documents/WSSD_POI_PD/English/WSSD_PlanImpl.pdf.

Vincent, L.A., W.A. Wijngaarden, and R. Hopkinson. 2007. 'Surface temperature and humidity trends in Canada for 1953–2005'. *Climate* 20: 5100–13.

Weaver, A.J. 2008. *Keeping Our Cool: Canada in a Warming World*. Toronto: Viking Canada.

Zhang, X., W.D. Hogg, and E. Mekis. 2001. 'Spatial and temporal characteristics of heavy precipitation events over Canada'. *Journal of Climate* 14: 1923–36.

Zwiers, F.W., and V.V. Kharin. 1998. 'Changes in the extremes of the climate simulated by CCCGCM2 under CO_2 doubling'. *Journal of Climate* 11: 2200–22.

6

Integrated Energy Resource Management

Paul Parker and Travis Gliedt

Learning Objectives

- To understand various types of conflict and uncertainty associated with energy resource management and its outcomes.
- To understand the uneven distribution of energy resources, the use of global markets to balance the demand and supply of energy commodities, and the resulting influence of economic and political factors on prices.
- To understand the social desire for security of energy supplies and the creation of systems that focus on local energy resources.
- To understand the role of policies and decision-makers at various levels and in various types of organizations.
- To recognize the need for an integrated approach that recognizes the competing goals of different stakeholders.

INTRODUCTION

In a world facing accelerated climate change, limited oil supplies, aging nuclear reactors, and innovative new technologies, energy decisions are critically important. Current energy systems need to be transformed into more sustainable forms through decisions and actions at all levels. Management choices range from which light bulb to purchase at the household level to which international option for a carbon tax or cap-and-trade system should be developed as part of the United Nations Framework Convention on Climate Change (UNFCCC). Management priorities change over time and are influenced by energy price fluctuations, energy security objectives, local energy resource availability, and the environmental need to reduce the carbon intensity of energy use.

Managing energy resources *sustainably* depends on decisions made at multiple locations and scales and must take into account demand and supply considerations. Uncertainty is high when trying to predict specific values such as the price of oil in a future year. Nevertheless, the factors that influence energy markets are well known and can be studied to understand the processes that create the outcomes of greatest interest to decision-makers. For example, energy demand is a function of population, economic, and technological changes over time and is moderated by policies. Energy supply is dictated by the natural endowments of energy resources such as fossil fuels and uranium, the prices of energy commodities, which in turn define available *reserves* (what is economically recoverable), and the technological capability to transform solar insolation and its derivatives (wind, wave, bioenergy) into useful energy available for work.

The key goal for energy resource managers is to construct a *sustainable energy future* based on major improvements in energy efficiency and conservation, a significantly larger reliance on renewable energy sources, and reduced environmental impacts of fossil and nuclear options (Boyle, Everett, and Ramage 2003). To help you understand how the preceding goal and three objectives could be achieved, in the following sections we review the environmental, social, and economic implications of international and Canadian energy supply, demand, and management. This discussion builds a case for the use of *integrated energy resource management* to slow climate change, address limited fossil fuel supplies, and enhance energy supply security.

ENERGY AND CLIMATE CHANGE: CONFLICTING OBJECTIVES AND UNCERTAIN FUTURES

Energy resource managers face conflicting objectives and many uncertainties. Energy is essential for basic services, such as cooking food, warming or cooling buildings, and providing transportation or producing goods and services for the economy. Low-cost energy is thus an important enabler for economic development. However, the desire for low-cost energy has resulted in systems for which external costs (e.g., social, health, and environmental costs) were excluded from the decision-making process. Large-scale energy systems with oil, coal, and natural gas were established as the most important global sources of energy. The combustion of billions of tonnes of these fossil fuels each year results in the release of billions of tonnes of greenhouse gases (28 billion tonnes in 2006 [IEA 2008a]). The direct link between energy choices and climate change impacts has been well established (IPCC 2007).

The traditional approach of viewing low-cost energy or environmental protection as conflicting objectives has created a global system with undesirable consequences. Policy-makers often address each issue independently, choosing to prioritize one at the expense of the other without recognizing their integrated nature. In this chapter, we propose a more comprehensive approach to recognizing the uncertainties faced and to identifying shared objectives that enable decision-makers to create future energy systems that achieve multiple objectives. The uncertainties that need to be recognized include price volatility, declining global oil supplies, and climate change. This section reviews each issue and highlights the linkage between climate change and energy futures.

Oil Price Uncertainty

The year 2008 highlighted the uncertainty in oil prices and volatility in the oil market (Figure 6.1). Oil is easily shipped from one country to another through pipelines or in large ocean vessels and has become the world's most valuable export commodity within an integrated global market. The global oil market and trade grew rapidly in the twentieth century. Many sophisticated players participate in the market, and analysts track multiple trends, yet 2008 demonstrated that almost all forecasts of oil prices, even just six months in advance, were wrong. Prices climbed to record levels in the first half of the year and

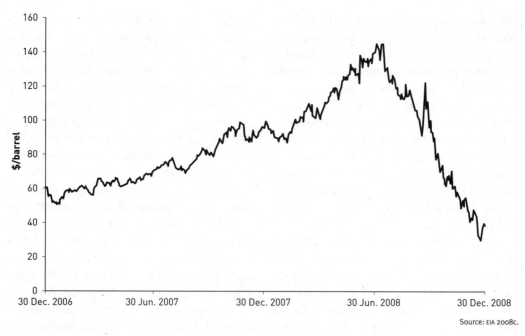

Source: EIA 2008c.

Figure 6.1 Daily Oil Spot Oil Prices for West Texas Intermediate: 2007–8.
Note: Price at Cushing, Oklahoma, storage and delivery point.

then tumbled from $147 per barrel to $40 per barrel. What caused these dramatic changes? In principle, market prices are determined at the equilibrium point where supply equals demand. If supply is fixed either because a point of peak production has been reached or infrastructure has reached its maximum capacity, then growing demand (for example, by rapid economic growth in China and India) will cause prices to rise. In contrast, if supply is stable when demand falls because of recession in a major market such as the United States, the price will fall. This simple economic explanation overlooks many important factors but highlights the extreme sensitivity of global energy markets to changes in marginal demand or supply and the impact of sales on commodity markets where speculative sales can far exceed the volume of fuel actually produced.

The dramatic changes in oil prices in 2008 should be seen in a longer term perspective. This was not the first time that wide swings in oil prices occurred. Figure 6.2 highlights key political events that are as important as economic factors in determining market perceptions and pressures on oil prices. The initiation of wars that restrict supplies from major Middle East exporting countries (Iran/Iraq War, Gulf War, Iraq War) or decisions by OPEC (Organization of Petroleum Exporting Countries) to reduce exports can trigger price rises. Conversely, when OPEC increased its exports in response to the 1997 Asian economic crisis, prices fell. The high degree of integration in international markets is demonstrated by the rapid transfer of the price signal throughout the world. Despite Canada being an oil exporter with ample domestic supplies, consumers immediately faced higher prices for gasoline when oil prices rose in mid-2008 and then saw the prices decline a few months later. Clearly, external factors can have

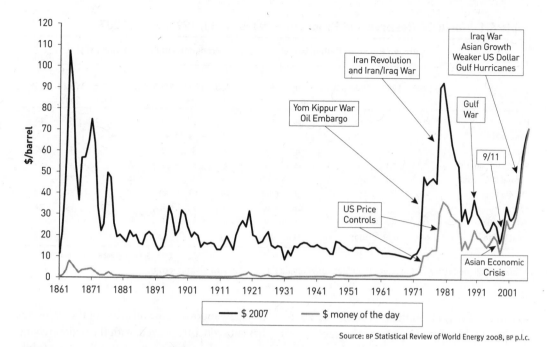

Source: BP Statistical Review of World Energy 2008, BP p.l.c.

Figure 6.2 Crude Oil Prices in US$ 2007 versus $ of the Day: 1861–2007.
Notes: 1861–1944 US average, 1945–83 Arabian Light posted at Ras Tanura, 1984–2007 Brent dated.
World events adapted from Williams 2008.

immediate impacts on local energy decisions. Prices provide the signal to decision-makers, but what are the energy resource trends that influence prices? A primary factor influencing the market is the supply of the commodity—oil in this case—so oil supplies are considered next.

Uncertain Reserves

It takes hundreds of millions of years to create oil, which is made when pressure and heat modify deposits of aquatic animals and plants buried under layers of mud (Boyle, Everett, and Ramage 2003). Deposits of various sizes are trapped in sedimentary rock basins. Extracting petroleum becomes more challenging over time as the largest and most accessible deposits are depleted and remaining deposits decrease in quality and accessibility. These challenges increase production costs and prices for refined products like gasoline

and diesel fuel. Oil is an essential ingredient in thousands of products, ranging from plastics to fertilizer, and is considered the 'lifeblood' of the global economy. Therefore, it is important to know how much oil is left in the world, where it is located, and how long it will last.

To estimate the availability of an energy resource, managers often use the reserve to production (R/P) ratio. The R/P ratio is defined as 'the number of years for which the current level of production of any energy and mineral can be sustained by its reserves' (Feygin and Satkin 2004, 57). The natural endowment of energy resources is determined by nature, but the energy *reserves* are the quantity of energy *resources* that can be extracted under current technological and economic conditions (Boyle, Everett, and Ramage 2003). Table 6.1 compares the proven reserves and production rates for oil from 1980 to 2007.

Table 6.1 World Oil Reserve and Production Rates: 1980, 1990, 2000, 2007

	Proven Reserves (billion barrels)						Production (million barrels daily)						R/P Ratio
	1980	1990	2000	2007	% Chg 00–07	% of World 2007	1980	1990	2000	2007	% Chg 00–07	% of World 2007	2007
US	37	34	30	29	−3	2	10	9	8	7	−11	9	12
Canada*	6.4	5.8	4.7	4.5	−4	0.4	1.5	1.3	1.6	1.6	0	2	8
Mexico	47	51	20	12	−40	1	2	3	3	3	1	4	10
S. & Cent. America	27	72	98	111	14	9	4	5	7	7	−3	8	46
Europe & Eurasia	98	80	109	144	32	12	15	16	15	18	19	22	22
Middle East	362	660	693	755	9	62	19	18	24	25	7	32	82
Africa	53	59	93	117	26	10	6	7	8	10	32	13	31
Asia Pacific	34	37	43	41	−5	3	5	7	8	8	0	10	14
Total World	665	998	1091	1215	11	100	63	65	74	80	8	100	42

Source: BP 2008; Statistics Canada 2008.

* Canada's oil production and reserves from Statistics Canada (Tables 130-0001 and 153-0013).

According to the BP 'Statistical review' (BP 2008), the world has approximately 42 years of oil reserves left at current production levels. However, the distribution of oil is not uniform, nor is production. When production is high relative to reserves, the share of remaining global reserves in that country decreases, as in the case of the United States (9 per cent of global production and 2 per cent of reserves) and Mexico (4 per cent of production and 1 per cent of reserves).

The global R/P ratio has remained almost unchanged at around 40 years for the past three decades as new reserves were discovered to offset rising production levels. This stable global pattern masks the experience of particular countries or regions where R/P ratios have declined substantially. For example, the oil reserves of the United States, Mexico, and the Asia Pacific region have declined and their R/P ratios fallen to 10 to 14 years. The conventional response to the decline in production in one region is to import the energy resource from another region. Examples include declining oil production in the United States and increasing imports. In northern Europe, oil production from the North

Sea grew following the oil crises of the 1970s but has subsequently peaked, with the region relying increasingly on imports. Globally, the Middle East holds more than 60 per cent of identified reserves and thus plays a critical role in future global oil supplies. This uneven geographic distribution of supply ensures that political as well as economic factors will influence price fluctuations in the future.

The implication of a declining R/P ratio is that supplies are in decline and adjustments are needed. By definition, non-renewable resources have a fixed endowment that limits the amount of total extraction possible. A famous yet controversial profile of expected extraction rates is Hubbert's Peak. Hubbert was an oil industry geologist who calculated that additional discoveries were declining despite continued drilling and exploration activity in the continental United States. He derived a bell-shaped production curve using historical discovery and extraction data up to the 1950s, indicating that the peak in US oil production would be reached in 1970 (Hubbert 1956). Despite disbelief among his peers at the time, history verified that his calculations were accurate in predicting peak continental US oil production to within a year. The resulting decline in American production led to greater reliance on imports from Canada, Mexico, Venezuela, and the Middle East. As Mexican production and exports decline in the early twenty-first century, further shifts in import patterns are required.

Uncertainty about the accuracy of reserve estimates has led some analysts to speculate that global peak oil production is imminent. Campbell and Laherrère (1998) predicted in 1998 that the global peak was only a decade away, while International Energy Agency reports gave much later estimates. Regardless of the precise date, the reality is that future energy systems will have to shift away from reliance on conventional oil.

At the national level, Canada's oil sands hold reserves far larger than its conventional oil reserves (see Table 6.2). Production levels from oil sands doubled between 2000 and 2007, so the R/P ratio was cut in half. However, at current production levels, oil from oil sand reserves could last for more than 300 years. Despite their large scale nationally, the Canadian oil sands equalled only 12 per cent of global conventional oil reserves in 2007 (BP 2008), so if they were extracted at the 2007 global production rate, they would only last

five years. Of course, such a massive increase in production is not feasible, and the R/P ratio does not predict actual exhaustion dates because production levels do not remain fixed.

Canadian conventional oil reserves would last eight years at current reserve and production levels despite the gradual shift in production from declining conventional Albertan oil fields to the frontier and offshore deposits where reserves have increased. Scaling up production of Alberta's oil sands is forecast to increase Canada's total oil production to 3.9 to 4.3 Mb/d in 2015 and 4.5 to 5 Mb/d in 2020, depending on the extent to which new pipelines are developed in North America (CAPP 2008). Oil sands production is expected to surpass conventional oil production as early as 2009. Developing Alberta's oil sands will undoubtedly improve energy supply security for Canada and North America. However, the environmental implications, including air and water pollution and increased greenhouse gas emissions, are significant and must be considered along with the economic benefits of the oil sands development. See Box 6.1 for a discussion of the environmental implications of Alberta's oil sands.

Canadian coal production increased in the 1980s when exports to Asian markets expanded but stabilized in the 1990s and early 2000s at

Table 6.2 Canadian Fossil Fuel Reserves and Production Rates: 1980, 1990, 2000, 2007

	Proven Reserves					Production					R/P Ratio	
	1980	1990	2000	2007	Units	1980	1990	2000	2007	Units	1980	2007
Oil[a]	6.4	5.8	4.7	4.5	bbl	1.5	1.3	1.6	1.6	mbl/d	12	8
Oil sands	n/a	n/a	163[e]	152	bbl	n/a	0.3[b]	0.6[b]	1.2[b]	mbl/d	n/a	347
Natural gas[d]	2.5	2.7	1.7	1.6	tm3	75	109	182	184	nm3/year	33	9
Coal[c]	4.2	6.6	4.7	4.4	billion tonnes	37	68	69	69	Mt/year	114	63

Source: Statistics Canada 2008; BP 2008.

[a] Oil production and reserves from Statistics Canada (Tables 130-0001 and 153-0013).

[b] Oil sands production from Statistics Canada.

[c] 'Recoverable' coal reserves from Statistics Canada (Tables 153-0017 and 153-0018) and coal production from Statistics Canada (Table 303-0016).

[d] Natural gas figures from BP.

[e] Oil sands reserves from BP.

Box 6.1 Alberta Oil Sands: Energy Security or Climate Change Nightmare?

Soderbergh, Robelius, and Aleklett (2007) assessed the ability of a potential immediate and massive scale-up of Alberta's oil sands production to delay the negative consequences of oil supply shortages. Even under optimistic assumptions, a massive scale-up would only produce 5 Mb/d by 2030 and have no effect on the declining bell of global peak oil. To make matters worse, 135 to 165 Mt CO_2e would be produced annually by 2018, depending on the degree to which bitumen replaces natural gas in production. This represents a 200 to 250 per cent increase over 2006 levels and is double business-as-usual forecasts, which range from 115 to 140 Mt CO_2e (Soderbergh, Robelius, and Aleklett 2007). A massive expansion will make it nearly impossible for Canada to meet its Kyoto target (Pembina Institute 2005; Sierra Club of Canada 2008). Oil sands also use three to six barrels of water to produce one barrel of oil and destroy land through deforestation and residue ponds created by the production process (Telfer and James 2008). Extracting oil sands may even require the building of nuclear plants in Alberta (Soderbergh, Robelius, and Aleklett 2007), which pose health and environmental challenges associated with storage and disposal. A comprehensive assessment of the options and impacts is required to enable well-informed decisions.

nearly 70 million tonnes per year. Recoverable reserves could support this level of output for more than 60 years. The abundant coal reserves stand in contrast to natural gas reserves, which would only last nine years if current production levels are maintained and no new reserves are found. Canada's production of natural gas is expected to peak in 2011 and then decline (Canada's Energy Outlook 2006). Canada is the second-largest natural gas exporter in the world after Russia (IEA 2008a). However, exports to the US are expected to decline over the next decade because of declining reserves (EIA 2007). An important question, then, is how fast we should deplete valuable energy reserves. Canada's natural gas R/P ratio is similar to the value for oil in the United States and indicates that either new reserves (e.g., from the Arctic) need to be discovered and brought on-line or the supply will have to come from overseas. New liquefied natural gas (LNG) terminals are proposed in eastern Canada to facilitate this shift to international sources.

There are two ways to increase or maintain the R/P ratio. The first is to increase the size of known reserves through discovery or new technological advances that allow more oil to be extracted economically. The second is to use less oil because of energy efficiency improvements, conservation, and substitution of other energy sources, thereby slowing consumption and the associated production rate. The latter option is also critical to mitigating climate change, an issue of international importance when we consider the conflicting objectives (energy-based growth versus emission reductions) that influence energy decisions.

Climate Change

The *Fourth Assessment Report* of the Intergovernmental Panel on Climate Change (IPCC 2007) outlines the consequences of not stabilizing greenhouse gas (GHG) concentrations in the atmosphere. Some changes may be 'abrupt or irreversible, depending upon the rate and magnitude of the climate change.' Two decades earlier, the 1988 World Conference on the Changing Atmosphere in Toronto declared that an 'unprecedented experiment' in human-induced climate change was underway with the potential to destroy the habitability of the planet (Toronto Conference 1988). These warnings highlight the importance of stabilizing GHG concentrations below a level above which there would be uncertain yet dangerous social, environmental, political, and economic consequences. Many dimensions of climate change are explored in Chapter 5, while the focus here is on aggregate GHG emission trends and energy choices.

Annual global GHG emissions grew by 70 per cent between 1970 and 2004, while CO_2 emissions grew by 80 per cent (IPCC 2007). Global CO_2 emissions associated with energy production exceeded 28,000 Mt in 2006 (IEA 2008a) and are expected to top 34,000 Mt in 2015 and 42,000 Mt in 2030 (EIA 2008a). GHG emissions from developing countries, especially China and India, are expected to continue rising (IEA 2008a) because of the use of coal for electricity generation and high economic growth rates (EIA 2008b).

To mitigate climate change and its associated impacts, annual GHG emissions must be reduced significantly to a point at which GHG concentrations in the atmosphere stabilize at a sustainable level. According to the *Stern Review* (2006), the required stabilization level is between 450 and 550 ppm CO_2e (e = the equivalent global warming potential of CO_2 for a collection of GHGs: methane, NOx, ozone, CFCs, CO_2). Estimates of the magnitude of GHG emission reductions necessary to achieve stabilization vary: an absolute reduction of 50 per cent (Pacala and Socolow 2004) or 80 per cent (*Stern Review* 2006) below current levels by the middle of this century or a 60 per cent reduction below 1990 levels (Barker

et al. 2001). These reductions are far larger than Canada's Kyoto commitment of 6 per cent below the 1990 level, or 30 per cent below 2005 levels (Simpson, Jaccard, and Rivers 2007), as GHG emissions in Canada have continued to rise.

To combat climate change, policy-makers around the world have considered an array of tools and options. The Kyoto Protocol included a proposed cap-and-trade system to connect energy management decisions with the environmental imperative of slowing climate change. Studies have shown, however, that even if the Kyoto targets were met, they would not achieve the 'ultimate' UNFCCC objective of stabilizing GHG concentrations in the atmosphere (McKibbin and Wilcoxen 2002; Jacoby and Ellerman 2004). While international talks are underway to craft a post-Kyoto framework, it is unlikely that an agreement will be reached with strict emission reduction requirements in the short-term. Long-term reduction targets are increasingly ambitious, but changing the direction of global emission trends requires immediate changes in domestic energy decisions.

Canada's energy consumption generated the seventh most CO_2 emissions in the world in 2006 (631 Mt), 50 per cent from oil, 30 per cent from natural gas, and 20 per cent from coal (EIA 2007). Canada creates four times more CO_2 emissions per capita (17 t CO_2) than the world average (4 t CO_2) and 35 per cent more than the OECD average (11 t CO_2), making it one of the worst per capita contributors to climate change after Australia, the United States, and several Middle Eastern oil-producing countries (IEA 2008a). According to Canada's Energy Outlook (2006), total GHG emissions are expected to grow to 828 Mt in 2010 and 897 Mt in 2020, exceeding the Kyoto target in 2010 by 47 per cent. Canada's total GHG emissions had already risen 25 per cent above 1990 levels by 2006, with the western provinces of Alberta and Saskatchewan leading the growth in emissions (Figure 6.3). Business-as-usual scenarios show that Canada's GHG emissions may exceed 1990 levels by 65 per cent in 2020 without policy interventions (NRTEE 2008). The transportation and fossil-fuel industries had the largest increase in energy consumption and associated GHG emissions over the 1990–2006 period (Figure 6.4). This pattern of rising total emissions was reversed in 2005 and 2006 with substantial reductions achieved in the electricity industry—e.g., coal-fired power plants were used less in Ontario as a result of provincial policy decisions.

Canadian federal governments have introduced climate change policies, but they have had limited impact (Doern 2005; Rivers and Jaccard 2005). The small reductions achieved in 2005 and 2006 (see Figure 6.4) are signs of the impact that energy resource managers and government policy can have. However, scenario studies show

that the federal government policies introduced in 2007 are not expected to meet their stated targets (Simpson, Jaccard, and Rivers 2007; NRTEE 2008), let alone the Kyoto target. Overall, Canadian climate policies have lacked cohesion between energy and climate policy objectives and sustainable development principles (Rowlands 2008; Liming, Haque, and Barg 2008). Policy integration must therefore be a priority. Before considering Canadian energy choices in more detail, we will look at a simple method to connect energy choices and emissions.

Connecting Energy and Emissions

Now that we have recognized uncertainty in the energy-related factors driving GHG emissions and the expected impacts, this section introduces one commonly used method to study the relationships among four major driving factors. The KAYA identity is an equation that defines

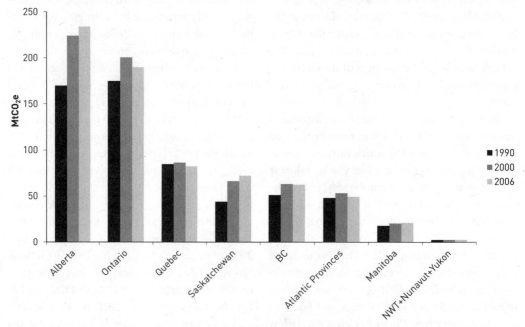

Figure 6.3 Canada's total GHG emissions by region: 1990, 2000, 2006.

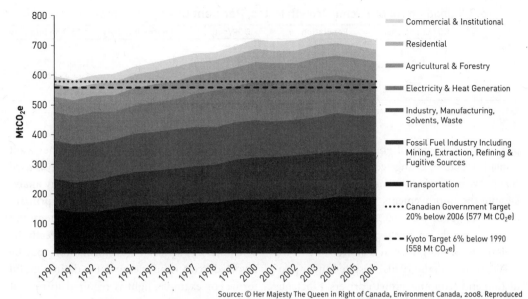

Source: © Her Majesty The Queen in Right of Canada, Environment Canada, 2008. Reproduced with permission of the Minister of Public Works and Government Services Canada.

Figure 6.4 Canada's total GHG emissions by sector: 1990–2006.
Note: Residential, commercial, institutional, and industrial sectors 'exclude' electricity emissions, which are accounted for in the electricity and heat generation category; Canadian government target of 20 per cent below 2006 by 2012 from Regulatory Framework for Air Emissions 2007.

the variable of interest (CO_2 emissions) and has been used in many studies including those by the IPCC in the development of its scenarios for comparing policy alternatives (IPCC 2001). The KAYA identity is a reformulation of Ehrlich's IPAT identity—I = P x A x T, or I = PAT—that was widely used in environmental impact debates in the 1970s. It states that environmental impact (I) is the product of three terms: (1) population (P), (2) affluence (A), and (3) technology (T) (Chertow 2001). In the revised equation, energy is given a central role in two of the driving factors under consideration—the energy intensity of the economy and the carbon intensity of energy:

CO_2 Emissions = Population × (GDP/Population) × (Energy/GDP) × (CO_2/Energy)
Or
$\Delta C/C \approx \Delta P/P + \Delta(G/P)/(G/P) + \Delta(E/G)/(E/G) + \Delta(C/E)/(C/E)$

Where:
C = CO_2 Emissions   P = Population
G = GDP   E = Energy

This simple representation enables energy resource managers to consider which factors can be changed to achieve a desired outcome. For example, if a country has a growing population and a growing economy per capita, then the two energy factors (the energy efficiency of the economy and the carbon intensity of energy sources) must be improved at a higher rate than the first two drivers. These relationships are easily illustrated (see Table 6.3). In the 1990s, Canada experienced annual population growth of 1 per cent and per capita economic growth of nearly 2 per cent per year. If other factors remained unchanged, emissions would rise 3 per cent annually. These growth pressures were partially offset by reductions in energy intensity,

Table 6.3 Canadian KAYA Factor Growth Rates, Per Cent Change per Year

	CO$_2$ Emissions	Population	GDP/cap	Energy Intensity	Carbon Intensity
1970s	2.9	1.4	2.8	−1.1	−0.2
1980s	0.4	1.2	1.6	−2.0	−0.5
1990s	1.8	1.0	1.9	−1.2	0.0
2000s*	2.1	1.0	1.5	−1.2	0.7

*2000s = 2000–6.

primarily a function of energy efficiency, with a small structural change component (−1.2 per cent per year). Carbon intensity remained steady. Despite the improvements in energy efficiency, the overall result was increased GHG emissions averaging nearly 2 per cent per year.

Emissions had grown at a slower rate in the 1980s (relative to the 1990s) despite similar population and economic growth drivers. Efficiency gains had been higher, and carbon intensity was reduced as a result of policies and decisions influenced by the high oil prices following the two oil crises during the 1970s. The 1988 Conference on a Changing Atmosphere had called for reductions in GHG emissions of 2 per cent per year in the 1990s through a 1 per cent annual gain in efficiency and a 1 per cent annual reduction through switching to less carbon-intensive fuels. However, given the size of the population and economic growth drivers facing Canada, the combined improvements in energy efficiency and carbon intensity must be 4 to 5 per cent per year to achieve an overall 2 per cent reduction target. Trends in the intensity of energy use and fuel switching based on the choices among available sources are thus of central importance to decision-makers seeking to reduce GHG emissions.

CANADIAN ENERGY CHOICES

Canada is fortunate in having a wide range of energy resources. However, energy resources are rarely chosen because of direct demand. Instead, people choose the energy services that they want to enjoy, and then a particular technology is used to meet that service. This leads to energy analysis being divided into two parts: the primary energy extracted and put into the system and the final-consumption energy used directly by consumers. A series of transformations may be involved to alter the energy to the desired useful form. For example, light is required to read at night, and electricity is a convenient way to provide energy to lighting devices. The electricity in turn may have come from any of several types of electric generating stations: coal, oil, gas, nuclear, hydro, wind, or solar. These transformations and energy flows are presented in Figure 6.5 to illustrate the many paths operating within the Canadian energy system. The left-hand side of the diagram shows the inputs of the major fossil fuels as well as electricity to constitute domestic energy production. The lower-left corner indicates that imports are then added to complete the domestic supply of energy. Domestic production forms the largest part of domestic supply in Canada, but the pattern varies across the country, with imports being more important in the eastern half of the country.

The right-hand side of Figure 6.5 indicates where the supply goes. Exports are substantial, but the larger share of domestic energy supplies is consumed to meet the demand for services by residential, agricultural, commercial, and industrial consumers, as well as transportation. Some energy is consumed directly by energy producers to run oil refineries and power stations

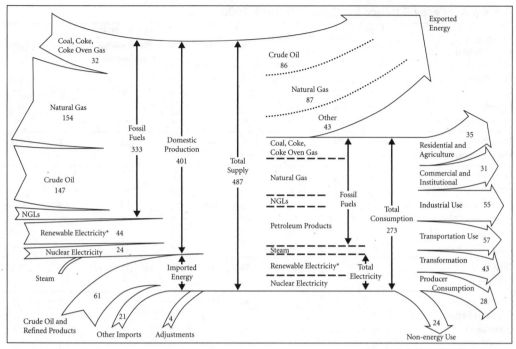

Sources: Environment Canada 2009; IEA 2009.

Figure 6.5 Energy flow diagram for Canada (Mtoe): 2005.
* Includes hydro, solar, geothermal, combustible materials, and waste.

(28 million tonnes of oil [Mtoe]), while other energy is lost through the inefficiency of transformations (43 Mtoe). The end result is that the amount used by final consumers as shown on the right-hand side is always less than the primary energy coming in as shown on the left-hand side of the diagram.

The pattern of energy flows depicted in Figure 6.5 is reported in a standard energy balance sheet for each country or jurisdiction. Table 6.4 gives the Canadian values for the energy balance and the respective energy flows in 2005. The production, trade, transformation, and consumption values are given for each fuel. Total energy production in Canada is equivalent to more than 400 Mtoe. Half of the total is exported, primarily from western Canada, while 82 Mtoe is imported, primarily into eastern Canada. The

resulting balance is the total primary energy supply for Canada. Major uses of each fuel are easily identified. Most coal goes to electricity plants, and most oil goes to petroleum refineries. Petroleum products account for nearly half of the total final consumption of energy in Canada, while the transportation sector consumes half of these products. Natural gas is consumed by the industrial, residential, and commercial sectors, but more than half of Canada's gas production is exported to the United States. Nuclear and hydro energy are used to generate electricity. The final major energy source is biofuels and waste used by industry, such as lumber mills. Other sources of energy (such as wind, solar, geothermal, and tidal) are so small (<1 Mtoe) that they do not register in Table 6.4 for 2005. Advocates for renewable energy propose that these sources

Table 6.4 Energy Balance for Canada (Mtoe): 2005

Supply and Consumption	Coal and Peat	Crude Oil	Petroleum Products	Gas	Nuclear	Hydro	Combustible Renewables and Waste	Electricity	Total*
Production	32	147	0	154	24	31	13	0	401
Imports	12	47	14	8	0	0	0	2	82
Exports	−17	−86	−22	−87	0	0	0	−4	−216
Stock changes	1	−2	0	6	0	0	0	0	5
Total Primary Energy Supply	28	106	−8	81	24	31	13	−2	272
Transfers	0	−3	6	0	0	0	0	0	3
Statistical differences	0.3	−2	2	−5	0	0	0	0	−4
Electricity plants	−24	0	−4	−5	−24	−31	−2	53	−37
Petroleum refineries	0	−102	105	−0.6	0	0	0	0	3
Own use	−0.1	0	−9	−15	0	0	0	−4	−28
Distribution losses	0	0	0	0	0	0	0	−4	−4
Total Final Consumption	3	0	92	51	0	0	11	44	201
Industry Sector	3	0	6	18	0	0	8	18	55
Transport Sector	0	0	52	4	0	0	0.2	0.4	57
Other Sectors	0	0	13	25	0	0	2	25	66
Residential	0	0	2	14	0	0	2	13	31
Commercial and Public Services	0	0	9	11	0	0	0	12	31
Agriculture/ Forestry/Fishing	0	0	2	0.4	0	0	0	0.9	4
Non-Energy Use	0.3	0	20	3	0	0	0	0	24
of which petrochemical feedstocks	0	0	12	3	0	0	0	0	15

Source: Electricity Information 2009 © OECD/IEA, Table 6.4, modified by the author.

Note: Factors less than 1.0 Mtoe have been removed; geothermal and solar production = 0.1 Mtoe and CHP = 0.8 Mtoe.

*Totals may not add up due to rounding.

become more important in Canada, as they have in other countries.

Renewable Energy and Economic Capacity

Renewable energy can help to reduce the carbon intensity of a nation's energy systems. However, one also needs the financial resources to invest in new technology. Figure 6.6 compares countries based on the size of GDP, CO_2 emissions, and installed renewable energy capacity. The world looks very different when the size of countries is based on the size of CO_2 emissions: the US, China, and Russia stand out. Comparing the CO_2 emissions map to the GDP map is even more enlightening. While the US is most evident on both maps, China, India, and Russia shrink on the GDP map in contrast to the increased size of Japan and the major European countries. The maps based on renewable energy reveal a different pattern. Germany, Spain, and the US have the largest installed wind capacity (MW). A

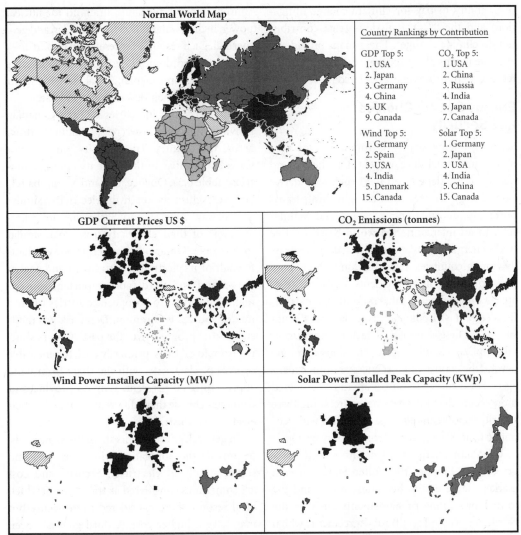

Figure 6.6 World maps by GDP, CO$_2$ emissions, and solar and wind capacity contribution.

similar pattern exists for installed solar capacity (KWp), where Germany, Japan, and the US are most evident.

When we compare the size of Canada on all four maps, we notice that Canada's wind and solar capacity are dwarfed relative to GDP and CO$_2$ emissions. This can be contrasted with Germany's size on the solar and wind maps, which is larger than their relative GDP or CO$_2$ emissions. While solar and wind capacity alone do not represent a country's total climate change mitigation effort, they symbolize the value a country places on premium sustainable energy technologies relative to the size of their economy (e.g., ability to pay for climate change mitigation) and their CO_2 emissions (e.g., contribution to climate change). These maps provide an effective means of comparing countries in visual terms on important energy-related measures.

Energy Supply Choices and Security

Securing energy supply is important for economic and social development. Our health and quality of life depend on the benefits we derive from energy services such as light, heat, transportation, information exchange, and ventilation. In an interconnected world based on free trade, energy resources are expected to flow easily among countries. However, long supply lines are also vulnerable to accidents, terrorism (e.g., attacks on pipelines or tankers), and regional conflicts (e.g., the Middle East, Nigeria) that disrupt supplies. These geopolitical factors can disrupt oil supplies, leading to price instability, and are beyond the control of most energy resource managers. Local energy resources are thus considered more secure than international sources, and these security factors can reinforce or conflict with economic factors when considering energy options.

Canadian energy markets are tightly integrated with those of the United States through markets, large established trade flows, and the formal provisions of agreements such as the North American Free Trade Agreement, which ensures that companies in either country are treated equally. However, the energy profiles of the two countries differ substantially, since the United States is a major importer of oil and gas while Canada is a major exporter.

Canada has a full range of energy resources (see Figure 6.5), but provincial resources vary widely. The challenge is to extract, develop, transform, store, and transport them from where they are found to where they are required. Canada's energy resources include natural gas, coal, conventional oil, oil sands, uranium for nuclear power, hydro (water power), wind, solar, and geothermal. Each resource represents an opportunity for export and domestic use as well as differing degrees of environmental, social, and economic implications of their development and use.

The electricity industries of each Canadian province illustrate how energy resources differ across the country. The diversity found across Canada is similar to the diversity among countries (Table 6.5). Quebec, BC, and Manitoba rely almost exclusively on hydroelectricity, similar to Norway. Alberta and Saskatchewan generate most of their electricity from coal, similar to Australia. Ontario uses nuclear power plants for half of its electricity, with hydro and coal as the second and third most important sources, making it similar to European countries where a mix of sources is common. Denmark has made substantial progress over the past 20 years shifting from fossil fuels, primarily coal, to renewable sources of electricity, with the share from wind approaching 20 per cent. Why do provinces or countries choose these different options for their electricity industries?

In general, governments share the same goals: to provide the population with a low-cost and secure supply of electricity. Security and cost are sometimes considered as trade-offs in which local resources are considered more secure but may have a higher cost. A third goal has been added in many jurisdictions: to minimize the environmental damage associated with electricity production and distribution. Different stakeholders may have different preferences among these goals, and the challenge becomes one of selecting the best solution.

The different solutions adopted by Canadian provinces reflect differences in their energy resource endowments, access to energy markets, and energy security priorities. BC has abundant rain and mountainous topography,

Table 6.5 Electricity Production by Fuel Type, Per Cent

	Nuclear	Hydro	Coal	Oil	Gas	Solar/ Wind*	Wood, Waste, Biomass	Geo thermal	Total
OECD	22	13	37	4	20	1	2	0	100
Non-OECD	5	21	45	8	20	0	0	0	100
Canada	16	58	17	2	5	0	1	0	100
Alberta	0	4	79	3	12	2	0	0	100
BC	0	93	0	0	6	0	1	0	100
Ontario	50	25	18	1	5	2	0	0	100
Quebec	3	95	0	2	0	1	0	0	100
Australia	0	6	79	1	12	1	1	0	100
Denmark	0	0	54	4	21	13	9	0	100
Finland	28	14	29	1	15	1	13	0	100
France	78	11	5	1	4	0	1	0	100
Germany	26	4	47	1	12	5	3	0	100
Iceland	0	73	0	0	0	0	0	27	100
Japan	28	9	27	11	23	0	2	0	100
Norway	0	98	0	0	0	1	0	0	100
Spain	20	10	23	8	30	9	1	0	100
USA	19	2	38	1	35	1	3	0	100
UK	19	7	49	2	20	1	2	0	100
China (excl. HK)	2	15	80	2	0	0	0	0	100

Sources: OECD and country data for 2006: derived from *Electricity Information 2009* © OECD/IEA, Tables 1.2 and 1.3, modified by the authors. Provincial data for 2004: BC Energy Plan 2007.

Note: Totals may not add up due to rounding.

*Includes tide, wave, ocean.

which creates opportunities for hydroelectricity. Alberta has abundant black coal resources. Saskatchewan has brown coal resources, which have less energy per tonne than black coal but can be combined with a power station adjacent to the mine to minimize transportation costs. Manitoba has hydroelectric installations in the northern portion of the province. Ontario has more limited conventional sources. Niagara Falls offered significant hydroelectric capacity, but demand grew in the twentieth century, so coal-fired power plants were added, with the coal imported from the United States. Instead of continuing to import larger amounts of fossil fuels, investments were then made in Canadian nuclear technology (CANDU reactors). Ontario focused on its financial and human resources

as a banking centre with advanced technology workers to become the centre of Canada's nuclear industry (18 of the 20 CANDU reactors were built at Pickering, Bruce, and Darlington, while New Brunswick and Quebec each had one reactor). Uranium was mined in Ontario and Saskatchewan, but the cost of the fuel was small compared to that of the technology.

Quebec and Labrador developed large hydroelectric facilities beyond the size of their demand, so they arranged long-distance transmission lines to export the excess electricity to New York City and the New England states. New Brunswick and Nova Scotia have modest coal reserves that were used in local power stations, but they also used imported oil. The higher cost of the small coal deposits led New Brunswick to diversify with a

nuclear plant and Nova Scotia to shift to importing some of the coal and all of the oil it used.
Prince Edward Island remained dependent on
oil-fired power plants, but the success of its wind
installations led to a change in policy to promote
renewable sources in the 2000s. Improvements
in commercial-scale wind turbines resulted in
the wind resources of each province being considered for development in the early twenty-first
century. Once again, some areas offered better potential than others. The resulting pattern
is that the primary source of electricity used by
consumers varies greatly across Canada.

Electricity grids started as local networks to
connect the source of electricity to the consumers. Electricity is a versatile form of energy with
many uses; however, it is not easily stored, so the
points of generation and consumption need to
be connected for instantaneous transfers. The
resulting grids have grown from a series of independent local grids to the current configuration
of North America being divided into four massive grids with significant interconnections that
facilitate the trade and sale of electricity among
jurisdictions.

The conventional model of large-scale electricity generation in central plants with large
networks of high-capacity transmission lines to
consumers is being challenged by the increasing
diversity of potential sources of electricity from
smaller and dispersed sources. This distributed
generation model promises reductions in line
losses because more of the electricity can be
generated on-site by technologies such as photovoltaics that convert sunlight to electricity. Local
energy sources may include solar electricity,
solar thermal, wind, geothermal, ground source
heat pumps, landfill gas, or biofuels. The shift to
reliance on many sources increases the need for
better electronic control and co-ordination systems. The continental grid is expected to retain
an important role in distributing excess production or meeting incremental demand, but more
energy is expected to be sourced locally.

At the national level, Canada's electricity supply is dominated by hydro, coal, and nuclear (see
Table 6.5). Canada is the seventh largest consumer of nuclear electricity (93 TWh) (Table
6.6) and has the eighth largest installed capacity
in the world (13 GW) (IEA 2008a). Canada ranks
third behind China and Brazil in global hydroelectricity consumption, with 368 TWh per year
(Table 6.6). However, Canada remains a minor
player in wind and solar electricity generation

Table 6.6 Nuclear and Hydroelectricity Consumption by World Rank (TWh)

	Nuclear (TWh)							Hydroelectricity (TWh)					
	1967	1977	1987	1997	2007	2007 Rank		1967	1977	1987	1997	2007	2007 Rank
US	8	264	479	662	849	1	China	19	46	100	196	483	1
France	3	18	266	396	440	2	Brazil	27	93	186	279	371	2
Japan	1	28	189	322	279	3	Canada	133	199	316	351	368	3
Russia	n/a	n/a	125	108	160	4	US	227	226	255	360	251	4
South Korea	–	0	39	77	143	5	Russia	n/a	n/a	163	158	179	5
Germany	2	41	142	170	140	6	Norway	53	72	104	111	135	6
Canada	–	27	77	83	93	7	India	22	45	49	70	122	7
Total World	42	535	1736	2392	2749		Total World	1021	1490	2063	2602	3134	

Source: BP 2008.

Note: Based on gross generation and not accounting for cross-border electricity supply.

capacity (i.e., <1 per cent), despite the potential to meet 20 per cent of total electricity demand with wind power (CANWEA 2008). The situation is similar to using solar thermal for water heating. Box 6.2 compares Canada to Germany and Japan, countries with similar solar potential that have followed different policy paths. The outcome: Germany and Japan are world leaders, while Canada is a solar laggard.

To this point, we have reviewed the challenges facing energy resource managers: climate change, finite fossil-fuel reserves, and energy supply security. Increasing renewable energy supply capacity is one way to improve the sustainability of energy systems while increasing domestic supply security. A second way is through energy demand management policies and practices (e.g., energy efficiency, conservation, fuel switching).

Equity must go hand-in-hand with sustainability for an energy system to be socially secure. Energy resource management can improve economic development, provide access to the benefits of energy, and generate the capacity to mitigate and adapt to climate change. In theory then, the job of the energy resource manager appears straightforward: choosing energy sources that meet the demand for energy services that provide economic, health, and social benefits to citizens, while ensuring that the choice of energy generates minimal GHG emissions and negative environmental impacts. But who are energy resource managers, and how do they make these choices?

Energy resource managers can be individuals or groups working alone or in partnerships. They possess the decision-making authority to influence energy supply and demand options within a society. Given the multi-scale challenges of climate change and energy-supply security, energy resource management must occur at multiple scales and in multiple sectors. The next section discusses the multi-scale governance of energy in Canada.

Box 6.2 Renewable Energy Experience: Germany and Japan versus Canada

Two countries with solar potential similar to Canada's have followed different solar policy paths.

Germany used feed-in-tariffs, stable long-term grants, long-term and low-interest loans, and partial debt write-offs to encourage solar installations. As a result, solar thermal installations increased from 26,000 (32.6 TJ/year) in 2000 to more than 100,000 (59.7 TJ/year) in 2006 (IEA 2007).

Japan introduced a subsidy system that paid 50 per cent of the cost for residential photovoltaic systems and then gradually reduced the subsidy to zero over a decade. However, the price to the consumer was stable because as the solar industry grew, the cost declined (Parker 2008).

The results in Germany and Japan included new employment in renewable energy industries to meet rising domestic and international demand.

Canada, on the other hand, introduced modest subsidies ($500 per system) as part of the ecoENERGY program to encourage demand for solar thermal systems. Some provinces have matched the federal contribution, but participation rates remain modest.

Ontario has introduced feed-in-tariff policies similar to those in Germany to see whether similar results can be achieved.

MULTI-SCALE GOVERNANCE

Governments of all scales play an important role in the development of sustainable energy systems through energy and climate change policy creation and implementation. Policies are important because 'nearly 90% of all anthropogenic GHG emissions in Canada result from the production and consumption of fossil fuels' (Liming, Haque, and Barg 2008, 92), mostly generated by private industry and citizen actions. Policies created at one scale (e.g., global) can influence the development of policies to achieve targets at other scales. For example, the Kyoto Protocol to the United Nations Framework Convention on Climate Change can influence the development of national policies such as a carbon tax. A national carbon tax or cap-and-trade system may persuade provinces to create their own policies, such as a conservation and demand management mandate. This may oblige municipal governments to support programs that encourage citizens and businesses to improve energy efficiency. Policies created at each scale are ultimately designed to encourage changes in individual citizen behaviour and corporate decisions.

Federal Policy: Canada

Canadian federal energy policy includes energy resource development, energy efficiency, science and technology, and climate change mitigation (NRCAN 2007). The Canadian government has jurisdiction over national trade and therefore a significant ability to influence energy and climate change decisions. However, federal energy policies tend to support non-renewable energy resource development, including a massive investment in Alberta's oil sands (Winfield 2008; Sierra Club of Canada 2008). While focusing on liberalizing the North American energy market and promoting Canada as an 'energy superpower' may have positive short-term economic benefits, it sours the political environment for climate change action and delays the transition to more sustainable energy choices (Brownsey 2007; Doern 2007).

Canada requires a climate change policy that is efficient, as measured by the amount of CO_2 reduced per public dollar, and effective in that the absolute reduction must be large enough to meet a given target (Stadler et al. 2007). A review of past and current Canadian federal policies reveals that neither criterion is being met (Simpson, Jaccard, and Rivers 2007). Oil sands developments in Alberta will contribute a significant portion of the expected growth in Canadian GHG emissions through 2020 (Persaud and Kumar 2001), making it nearly impossible for Canada to achieve absolute GHG emission reductions (Simpson, Jaccard, and Rivers 2007). As a result, Canadian energy policy may undermine long-term sustainable development objectives.

Cap-and-trade policies in Canada are for large final emitters only and are based on an intensity target to provide an incentive for economic growth while preventing industry from receiving credit for poor economic performance (Regulatory Framework for Air Emissions 2007). The use of intensity targets, however, may limit the environmental effectiveness of the cap-and-trade system (Simpson, Jaccard, and Rivers 2007). Using an absolute rather than an intensity target could improve the effectiveness of Canada's climate policy.

Canadian energy resource managers face tough choices. On one hand, economic benefits created by Alberta's oil sands are felt across Canada. On the other hand, economic benefits associated with mitigating climate change will be realized at some point in the future and may diminish if immediate action is not taken. The National Round Table on Environment and Economy (NRTEE 2008) argues that Canada must institute a policy that puts an economy-wide price on carbon immediately if the federal government's target of a 65 per cent below 2006 reduction by 2050 is to be achieved. The report claims that the longer Canada waits, the more difficult it will be to meet the target and the higher the compliance costs will be. There are three options to putting a price on carbon: (1) an emissions tax, (2) a cap-and-trade system, or (3) a hybrid system containing both options. Until one of these systems is implemented at the federal level, provincial and municipal governments must take the lead for climate change mitigation and sustainable energy development.

Provincial Policy: Ontario

Provincial governments are responsible for electricity supply, land use, and environmental health and safety. In contrast to the conflicts between energy development and climate change policies at the federal level, some provinces have integrated energy and climate change policies within a single plan. Manitoba and Quebec are good examples, along with the Ontario policies reviewed here. The electricity supply challenge and associated targets outlined by the Integrated Power System Plan (IPSP 2007) are co-ordinated with climate change policies and targets within Ontario's Action Plan on Climate Change (GoGreen 2007a; 2007b).

In June 2006, the Ontario government issued the Supply Mix Directive, outlining the government's 20-year electricity system management plan (Duncan 2006). This directive led to the creation of the IPSP (2007). The first objective is conservation and demand management (CDM). This includes energy efficiency, behavioural conservation, decentralized generation, and fuel switching. The IPSP includes $10 billion for conservation, or $500 million annually (OCB 2007). The second objective is to increase renewable electricity capacity from 8000 MW to 10,402

MW by 2010 and 15,700 MW by 2025. Although CDM and renewables are expected to contribute more than 20,000 MW to peak capacity in 2025, a significant supply gap will still exist. The third objective, therefore, involves maintaining Ontario's nuclear generation capacity. Ontario plans to refurbish Pickering B and expand Darlington nuclear facilities (OPG 2008). A final objective is to replace coal capacity with combined cycle gas turbines (CCGT) for intermediate needs and simple cycle gas turbines (SCGT) for peaking demand. Duff and Green (2008) conclude that Ontario's renewable electricity supply policies will reduce GHG emissions. However, non-profit environmental organizations have criticized the IPSP's inclusion of nuclear and have offered more ambitious alternative energy scenarios, as described in Box 6.3.

Ontario has used various policies to increase renewable electricity supply capacity, including renewable portfolio standards (RPS), bidding systems, feed-in-tariffs (FIT), and grant/subsidy/rebate programs (Rowlands 2007). Ontario's standard offer contract policy provides an incentive of 42 cents per kWh for solar photovoltaics (PV) and 11 cents per kWh for other renewable projects, with an additional peak performance incentive of 3.5 cents per kWh for hydro and biomass (OCB 2007; OPA 2008). The program targets renewable electricity generation technologies, ranging from small-scale residential solar PV projects (1 to 5 kW) to 10 MW wind farms.

Of the 915 MW capacity represented by the standard offer contracts executed in 2007, 63 per cent were for wind, 28 per cent for solar PV, 4 per cent for small hydro, and 6 per cent for biomass (OPA 2007). A similar program being developed will provide an incentive to small natural gas and waste-to-energy projects, and the Ontario Power Authority expects this to create 140 MW of capacity by 2010 (OCB 2007). The Green

Box 6.3 Alternative Scenarios to the IPSP

Non-profit organizations have been critical of the IPSP and have offered alternative scenarios that exclude nuclear and require an earlier coal phase-out date. Peters, Cobb, and Winfield (2007) offer two scenarios (soft green, deep green) that phase out coal and meet Ontario's expected supply gap without investing in additional nuclear capacity. Soft green expands renewable and CDM assumptions of IPSP by including an additional contribution from combined heat and power and a larger contribution from new hydro capacity. Deep green includes a 50 per cent larger wind capacity expansion than soft green and a greater focus on decentralized generation and solar than IPSP. The total capital costs (2007–27) of soft green ($109 billion) are similar to those of IPSP ($102–13 billion) but less than those of deep green ($128 billion) (Godin 2007). Both scenarios are expected to achieve larger emission reductions below 2004 levels in 2027 (IPSP: 21 to 26 Mt, 60 to 74 per cent reduction; soft green: 27 Mt, 77 per cent reduction; deep green: 28 Mt, 80 per cent reduction) (Godin 2007).

Ontario's Clean Air Alliance recommends that the Ontario government remove the 10 MW cap on the size of renewable energy projects that qualify under the standard offer program to create an incentive for a larger renewable energy investment from the private sector (Gibbons 2008). They also recommend a standard offer program for energy efficiency initiatives.

The trends of change in Ontario's electricity sources were demonstrated in 2008 when hydro provided a record 38.3 TWh (24 per cent), coal declined to 23.2 TWh (14.5 per cent), and wind supplied 1.4 TWh (0.9 per cent) (IESO 2009). Nuclear remained the most important source, but a reduction in total demand also indicated that the promotion of conservation could be achieving its goal.

Energy Act was introduced in 2009 to accelerate the introduction of additional renewable energy projects and to promote the creation of jobs in new 'green' sectors of the economy.

Ontario's Action Plan on Climate Change (GoGreen 2007a) is designed to combine the IPSP objectives and policies with broader energy management programs (e.g., Greenbelt, Places to Grow Act, home energy audits, building code changes to require new homes to meet R-2000 standards by 2011, Municipal Eco Challenge Fund, a North American automobile fuel economy standard, MoveOntario 2020 rapid transit plan, green jobs fund) to achieve a GHG emission reduction of 15 per cent below the 1990 level by 2020 and 80 per cent below 1990 by 2050. The

latter target would achieve a reduction of the order necessary to contribute to a stabilization of atmospheric CO_2 concentrations.

Ontario integrated conservation, efficiency, and renewable energy policies and targets from the IPSP into its climate change plan. The overall climate change targets are subdivided to clearly outline the proportion that each policy and sector is expected to achieve. Nearly half of the emission reductions required to meet the targets must come from the electricity sector (GoGreen 2007b). This represents an 80 per cent reduction in GHG emissions from fossil-fuel generation by 2014 (from 35 to 7 Mt). Achieving these ambitious provincial goals will only be possible with the help of municipal governments.

Municipal Policy

Local actions can act as a hedge against global and national policy uncertainty. Municipal governments can directly change energy choices by improving the sustainability of their corporate government operations and influencing actions within the broader community. Corporate policies and actions focus on infrastructure and building design (e.g., Leadership in Energy and Environmental Design [LEED]), green procurement, building and lighting retrofits, water conservation, municipal waste to energy, and green electricity purchases. Community policies include support for energy audits for homes and businesses and integrated land-use, energy, and transportation planning. Community energy plans or local action plans provide an important framework for addressing local energy conservation, renewable energy supply, and carbon reduction goals (Jaccard, Failing, and Berry 1997; ICLEI 2006; St Denis and Parker 2009).

Karlsson (2007, 122) argues that 'for issues with local–global linkages in both their effects (e.g., climate change) and driving forces (e.g., fossil fuel energy use), institutions and management at the local and global level need to be mutually supportive.' Municipal governments must therefore play a key role in climate change and energy management, as outlined by the United Nations' Local Agenda 21 program. The mission of Local Agenda 21 is to 'build a worldwide movement of local governments and associations dedicated to achieving sustainable development through participatory, multi-stakeholder sustainable development planning and implementation' (ICLEI 2008). Founded in 1990 to help municipal governments fulfill the Local Agenda 21 objectives, the International Council for Local Environmental Initiatives (ICLEI 2008) 'provides technical consulting, training, and information services to build capacity, share knowledge, and support local government in the implementation

of sustainable development' (ICLEI 2008). ICLEI's emphasis has shifted from '*agenda to action* to ensure an accelerated implementation of sustainable development' (ICLEI 2008).

One of ICLEI's action initiatives, called the Cities for Climate Protection (CCP) program, helps more than 800 municipal governments 'adopt policies and implement quantifiable measures to reduce local GHG emissions, improve air quality, and enhance urban liveability and sustainability' (ICLEI 2008). CCP uses a five-step framework that involves setting a baseline emissions inventory and forecast, adopting an emission reduction target, developing policies and measures to meet the desired GHG emission reduction, implementing these policies, and monitoring results through audits (ICLEI 2008). ICLEI (2007) estimates that participating cities are saving 60 million tons of CO_2e and $2.1 billion annually.

The local integration of land-use, energy, and transportation planning is important to building sustainable cities that are resilient to oil supply shocks and climate change policy responses. Karlenzig (2008) found that the US cities most prepared to handle oil price volatility, climate change regulations, and oil supply shocks had high levels of public transit ridership, well-organized and dense city centres, a high level of mixed land uses, and medium to high city population density. Cities with high telecommuting rates also performed better in the survey. The least prepared cities are characterized by higher sprawl, lower density, and less organized public transit systems. Many cities that ranked highly were also first-movers in climate change mitigation planning. For example, San Francisco, Seattle, and Portland rank near the top of Karlenzig's list and are also among the best North American cities in terms of GHG emission reductions relative to population and economic growth (Pander 2007). Portland limited absolute GHG emissions to 1 per cent above

1990 levels in 2005 despite high economic and population growth rates while reducing per capita emissions to 12.5 per cent below 1990 levels (ICLEI 2006).

In summary, municipal energy managers are recognizing the benefits of integration. London, UK, created an integrated energy management plan based on three objectives: (1) reduce the contribution to climate change, (2) eliminate fuel poverty through efficiency and conservation programs, and (3) use sustainable energy and building retrofits to drive economic development (Hammer 2008). Integrated land-use, energy, and transportation management principles and public–private partnerships are relied upon to facilitate the plan. Another example is PlusNetwork's strategic planning method (Seymoar 2008). This approach integrates sustainability principles from the built and natural environments, political governance, social systems, and economic development and was used to generate the imagineCalgary 100-year integrated urban sustainability plan (Calgary 2007). Some Toronto neighbourhoods are already demonstrating the benefits of integrated planning in the form of reduced energy use and GHG emissions (Norman, MacLean, and Kennedy 2006). Vancouver also incorporated integration principles into its climate change plans and ranks favourably against leading North American cities in terms of GHG emission reductions per capita (Box 6.4 outlines Vancouver's climate change achievements).

While governments at all levels are critical to integrated energy resource management, partnerships with First Nations, non-profit organizations, and businesses are also important.

First Nations Policies

First Nations traditionally relied upon biofuels as their primary energy source, but fossil fuels have gained dominance for transportation and

Box 6.4 Vancouver Case Study: Local Policy Integration

A panel of experts assembled in 2003 by the City of Vancouver created corporate (20 per cent below 1990 by 2010) and community (6 per cent below 1990 by 2012) targets for GHG emission reduction and action plans to achieve them. In 2007, council adopted new targets designed to reduce community emissions to 33 per cent below 2007 levels by 2020 and 80 per cent below 2007 levels by 2050. Although Vancouver had experienced significant population, employment, and car ownership growth since 1990, corporate emissions were reduced to 5 per cent below and community emissions to 5 per cent above 1990 levels in 2006. Vancouver's per capita emissions are 4.9 tonnes, 15 per cent below 1990 levels and substantially less than those of Toronto or Calgary. Most of the corporate reduction was achieved by building retrofits and landfill gas recovery. The community reduction was due largely to integrated land-use and transportation planning and residential and commercial energy conservation. Other initiatives include a $14-million residential district space and water heating system expected to reduce GHG emissions by 50 per cent, green procurement, a low-carbon taxi initiative, an idling bylaw, a re-urbanization program, and a number of sustainable transportation initiatives (Pander 2007).

other services. A new vision statement for First Nations' energy futures was proposed by the First Nations Energy Alliance (2007): 'Promote and foster the self reliance, self sufficiency and sustainability of First Nations and territories through the development of energy.'

In the case of remote First Nations communities, the cost of energy is high both in terms of purchase price and environmental impacts. Diesel is commonly used in electricity generators, with the fuel transported along ice roads in the winter or flown in by plane. These diesel systems are carbon-intensive and therefore contribute to climate change, which reduces the ice-road season and requires more fuel to be transported by plane. Many First Nations wish to change their energy systems to rely more on local sustainable sources based on respect for their lands and environment. A study of community energy plans across Canada, including remote northern communities, found that smaller communities were more likely to incorporate renewable energy sources than large cities (St Denis and Parker 2009). Wind has been identified as a promising energy source in many First Nations, with large grid-connected projects proposed in BC, Quebec, Ontario, and Saskatchewan, while research is underway to develop small wind solutions for remote communities (OCE 2008).

Non-profit Organizations

International non-profit municipal partnerships like CCP help to share best practices, promote municipal energy and climate change solutions, and frame the need for action based on ancillary benefits (Lindseth 2004; Bulkeley and Betsill 2005). Local partnerships among green non-profit organizations, universities, local governments, and utilities can provide effective means of engaging community members to take action, conserve energy, and reduce emissions (Parker, Rowlands, and Scott 2003). Local non-profit

environmental service organizations use participatory principles to build multi-sector partnerships that help to deliver energy management services and reduce GHG emissions. Organizations including Green Communities Canada (GCC 2008) use social capital networks to share ideas, techniques, and services and to facilitate community energy management. Environmental service organizations in the social economy are becoming more entrepreneurial, creating new programs to meet demands from local communities for more energy management services (Gliedt and Parker 2007).

Businesses

Many businesses are delivering energy management services, marketing green products, and using internal environmental management programs to improve the sustainability of their firms. Environmental entrepreneurship is driven by the desire of employees within organizations to reduce GHG emissions and operating costs. Some firms are creating new products in emerging sectors of the economy with potential to revolutionize energy and climate change management, while taking advantage of first-mover benefits to potentially make a profit in the long run. The following example illustrates the importance of business and entrepreneurship to integrated energy resource management.

A new company aims to make the world a 'Better Place' by reducing dependence on oil and tackling climate change (Better Place 2008). Partnerships with Denmark, Israel, Hawaii, and Ontario are being fostered and used to deliver an innovative sustainable transportation model based on electric cars. As explained on the company website, 'consumers subscribe to transportation as a service, much like they do today with mobile phones. Auto companies make the electric cars that plug in to the electric recharge network of charging stations and battery swap stations.

Energy companies provide the network's power through renewable energy projects. And Better Place provides the batteries to make owning an electric car affordable and convenient.' This integrated approach overcomes three challenges that have plagued the development of electric vehicles for decades: (1) charging stations provide the necessary supporting infrastructure, (2) battery 'change out' stations reduce the time to 'charge' a car to seconds rather than hours, and (3) partnerships among automakers, federal and state governments, and the supplier of the battery infrastructure increase the likelihood of success by spreading risk and increasing resilience to policy and price uncertainty.

In summary, policies and actions are emerging that influence changes in energy supply choices. Given that these actions and the challenges they address exist in and between different sectors, scales, and regions, energy resource managers must use the principles of *integration* and *collaboration* to overcome the uncertainty and conflict described earlier in this chapter.

INTEGRATED ENERGY RESOURCE MANAGEMENT

The goal of energy resource managers is to develop sustainable energy systems. To be sustainable, an energy system must:

- provide the services of energy to meet current and future generations' needs in an accessible, effective, equitable, and efficient manner;
- enable the stabilization of atmospheric concentrations of GHG;
- protect or restore the Earth's air, land, and water resources throughout its lifecycle;
- be safe today and result in no burdens of risk for future generations; and
- empower communities to live satisfying and healthy lives (Pembina Institute 2008).

Integrated energy resource management can help to achieve these goals by incorporating multiple-stakeholder perspectives and resources to spread risk and response capacity (Ruth and Coelho 2007).

Integration of energy resource management must occur along three problem and policy dimensions: (1) energy resource supply and climate change policy (Rowlands 2008), (2) climate change mitigation, adaptation, and sustainable development (Wilbanks and Sathaye 2007), and (3) climate change causes, vulnerabilities, and resource capacities to respond (Zahran et al. 2008). Sustainable economic development provides the capacity for climate change mitigation and adaptation through the development of renewable energy technologies, institutions, and human and social capital resources (Bizikova, Robinson, and Cohen 2007; Burch and Robinson 2007). Solving these integrated challenges requires the incorporation of bottom-up decision-making and the empowerment of communities (Wilson and McDaniels 2007).

Ontario's IPSP represents a first step toward an integrated approach to energy resource management. Climate change and energy supply capacity development goals and policies are incorporated within a single plan guided by an integrated set of sustainability principles (Rowlands 2008). Unlike the three traditional pillars of sustainability, which encourage trade-offs and create winners and losers (Gibson 2006), the principles used in the IPSP represent an integrated sustainability assessment:

- socio-ecological system integrity
- livelihood sufficiency and opportunity
- intragenerational equity
- intergenerational equity
- resource maintenance and efficiency
- democratic governance
- precaution and adaptation
- immediate and long-term integration

The three main challenges identified in this chapter are addressed simultaneously by the IPSP, and the development of sustainable communities provides the capacity to address change, uncertainty, complexity, and conflict across governance sectors and scales.

Energy and climate policies in Canada are fragmented and therefore less effective at creating sustainable energy systems than they could be. Canada should implement an integrated approach if it wishes to ensure energy supply security, adapt to declining fossil-fuel reserves, and contribute to the global effort to mitigate climate change. To start, Canada could develop a national strategy framework for energy sustainability (Liming, Haque, and Barg 2008). This national framework would help to integrate climate and energy policy into sustainability policy (Robinson et al. 2006; Burch and Robinson 2007) to co-ordinate the creation of sustainable energy systems. National policies should encourage and support local actions. More communities can be encouraged to prepare community energy plans with input from a cross-section of stakeholders in the public, private, and not-for-profit sectors. Targets for GHG emission reduction and renewable energy capacity development should be better co-ordinated among and between government scales. Finally, programs to achieve targets should be integrated so that federal, provincial, territorial, and local governments can take actions that do not overlap or undermine actions taken by each other.

CONCLUSION

This chapter has introduced *integrated energy resource management* as one approach to addressing climate change, fossil-fuel energy resource depletion, and energy security. The dynamics of energy supply and demand were discussed, including oil price and supply uncertainty, conflicting climate change and fossil-fuel energy resource development objectives, and the contribution of population growth, technological innovation, and economic development to future energy and emission pathways. Canadian energy resource supply and demand figures were contrasted with international examples, and interventions to improve the sustainability of energy systems at multiple scales and in multiple sectors were identified.

Energy resource managers can improve the sustainability of energy systems by integrating interventions and institutions between scales and sectors. Integration should be 'loosely' coupled and 'flexible' to ensure the exchange of ideas and information necessary for creativity and innovation but also to avoid unnecessary complexity that could inhibit the ability to adapt to uncertain and turbulent shocks such as peak oil and dangerous climatic events (Homer-Dixon 2007). Integration helps to enhance local capacity to generate 'green shoots' of renewable energy, energy efficiency, and green building and transportation systems development. Green shoots can be nurtured by government through policy interventions. However, the current global energy system is facing an uncertain future because of the converging climate and energy challenges outlined in this chapter. Therefore, strategic partnerships and social capital networks that connect local communities and their green shoots to each other may provide the resilience and regeneration capacity necessary to guide a successful transformation from an unsustainable energy system based largely on fossil fuels to a more resilient, equitable, and sustainable energy future.

FROM THE FIELD

Energy research is rapidly changing, so our field experience has covered a broad range of projects over the past 30 years. As an undergraduate in the 1970s, Paul studied the socio-economic impacts of the Ark in PEI where we thought this solar- and wind-powered house was the start of a new wave of housing, with renewables replacing conventional fuels. In the 1980s, conventional fuels dominated, and Paul's research turned to the impacts of coal mining on communities in Australia, the contrasting structures found in Japanese and European energy markets, and the implications for communities like Tumbler Ridge in BC.

Returning to Canada in the 1990s, Paul looked at energy as part of ecological footprint calculations. He joined a group of colleagues at the University of Waterloo (Dan Scott and Ian Rowlands) and in the community (Don Eaton) to start the Residential Energy Efficiency Project (REEP), with funding from the federal Climate Change Action Fund. We hired outstanding students (including Travis) and promoted residential evaluations that identified steps for households to take to reduce their energy consumption. Our argument was that the human dimensions of behavioural change and decision processes were as important as technical solutions to facilitate emission reductions. During 10 years, more than 8000 evaluations were conducted in our community, thousands of tons of annual emissions were avoided, and a new green community organization was created, Waterloo Region Green Solutions. The latest project is the REEP House, a century-old home being renovated to advanced twenty-first-century standards.

Conservation of electricity became a priority for the Ontario provincial government, and Paul joined colleagues on several projects for utilities and the Ontario Centres of Excellence to study the impact of interventions such as smart meters on various user groups. These studies led to a larger project, Energy Hub Management System, for households, farms, institutions, or businesses, led by Ian Rowlands.

Interest in renewable energy has grown, with students examining the profile of early adopters who purchased solar panels for their houses and also the barriers that prevented people from making the purchase decision. As an early solar adopter, Paul participated in the Alternative Energy Tour for the Healthy Communities Committee of Woolwich and shared his family's solar experience with hundreds of visitors.

Wind is a rapidly growing source of renewable energy in Canada, and remote communities have special needs and opportunities for smaller turbines that do not need heavy equipment for installation or maintenance. The conventional practice of flying in diesel fuel for electricity generators is both expensive and carbon-intensive. A new partnership among a remote First Nations community, industrial partners, and university researchers is seeking to turn the generators off at times of low demand and replace the diesel with renewable sources of energy. Remote Canada is blessed with wind, sun, and hydro, but the path to an energy transition from conventional to renewable sources remains challenging.

Travis and a new generation of students are working to build local capacity so that communities can change their energy trajectory by preparing community energy plans that incorporate demand-side management along with alternative sources and new technologies to meet community energy needs with fewer carbon emissions. Voluntary green electricity purchases illustrate one of the many strategies used by businesses, social economy organizations, government agencies, and individual households.

—PAUL PARKER and TRAVIS GLIEDT

REVIEW QUESTIONS

1. What government scales are best equipped to create, deliver, and enforce energy and climate change policies?
2. What sector(s) are best able to provide energy services, achieve GHG emission reductions, and develop sustainable energy supply capacity?
3. In which geographic regions should emission reductions occur, given that CO_2 reductions make the same contribution to *climate change mitigation* regardless of where they are achieved, while economic and social benefits of sustainable energy developments are often concentrated at the location of investment?
4. Should exporting or importing countries and regions pay for GHG emissions associated with energy transfers?
5. Should energy resource development or climate change mitigation be given priority in policy development?

REFERENCES

Barker, T., et al. 2001. 'GHG mitigation options to achieve Kyoto targets'. *International Environmental Agreements: Politics, Law and Economics* 1: 243–65.

BC Energy Plan. 2007. 'The BC Energy Plan: A vision for clean energy leadership'. BC Ministry of Energy, Mines and Petroleum Resources. http://www.energyplan.gov.bc.ca.

Better Place. 2008. 'Electric changes everything'. http://www.betterplace.com.

Bizikova, L., J. Robinson, and S. Cohen. 2007. 'Linking climate change and sustainable development at the local level'. *Climate Policy* 7: 271–7.

Boyle, G., B. Everett, and J. Ramage. 2003. *Energy Systems and Sustainability; Power for a Sustainable Future*. Glasgow: Oxford University Press.

BP. 2008. 'Statistical review of world energy. Full workbook of historical data'. http://www.bp.com/productlanding.do?categoryId=6929&contentId=7044622.

Brownsey, K. 2007. 'Energy shift: Canadian energy policy under the Harper Conservatives'. In B. Doern, ed., *The Harper Conservatives—Climate of Change. How Ottawa Spends*, 143–60. Montreal and Kingston: McGill-Queen's University Press.

Bulkeley, H., and M.M. Betsill. 2005. 'Rethinking sustainable cities: Multilevel management and the "urban" politics of climate change'. *Environmental Politics* 14 (1): 42–63.

Burch, S., and J. Robinson. 2007. 'A framework for explaining the links between capacity and action in response to global climate change'. *Climate Policy* 7: 304–16.

Calgary. 2007. 'ImagineCalgary plan for long range urban sustainability'. City of Calgary. http://www.imaginecalgary.ca/imagineCALGARY_long_range_plan.pdf.

Campbell, C., and J. Laherrère. 1998. 'The end of cheap oil'. *Scientific American* March: 78–83.

Canada's Energy Outlook. 2006. 'The reference case 2006'. Analysis and Modelling Division, Natural Resources Canada. http://www.nrcan.gc.ca/com/resoress/publications/peo/peo-eng.phpv.

CAPP (Canadian Association of Petroleum Producers). 2008. 'Crude oil forecast, markets and pipeline expansions'. http://www.capp.ca/GetDoc.aspx?DocID=123361.

CANWEA (Canadian Wind Energy Association). 2008. 'Canadian Wind Energy Association'. http://www.canwea.ca.

Carbon Dioxide Information Analysis Center. 2004. 'CO_2 emissions in metric tons of carbon'. US Department of Energy, Carbon Dioxide Information Analysis Center. http://cdiac.ornl.gov.

Chertow, M.R. 2001. 'The IPAT equation and its variants: Changing views of technology and environmental impact'. *Journal of Industrial Ecology* 4 (4): 13–29.

Doern, B. 2005. 'Canadian energy policy and the struggle for sustainable development: Political-economic context'. In B. Doern, ed., *Canadian Energy Policy and the Struggle for Sustainable Development*, 3–50. Toronto: University of Toronto Press.

———. 2007. 'The Harper Conservatives in power: Emissions impossible'. In B. Doern, ed., *The Harper Conservatives—Climate of Change. How Ottawa Spends*, 3–22. Montreal and Kingston: McGill-Queen's University Press.

Duff, D.G., and A.J. Green. 2008. 'A comparative evaluation of different policies to promote the generation of electricity from renewable sources'. In S. Bernstein, J. Brunnee, D.G. Duff, and A.J. Green, eds, *A Globally Integrated Climate Policy for Canada*, 222–46. Toronto: University of Toronto Press.

Duncan, D. 2006. 'IPSP Directive'. Ontario Ministry of Energy. http://www.energy.gov.on.ca/index.cfm?fuseaction=about.speeches &speech=13062006.

EIA (Energy Information Administration). 2007. 'Canada energy data, statistics and analysis—Oil, gas, electricity, coal'. Country analysis briefs. http://www.eia.doe.gov/emeu/cabs/Canada/pdf.pdf.

———. 2008a. 'Energy-related carbon dioxide emissions'. Chapter 7: International energy outlook. http://www.eia.doe.gov/oiaf/ieo/pdf/emissions.pdf.

———. 2008b. 'Coal'. Chapter 5: International energy outlook. http://www.eia.doe.gov/oiaf/ieo/pdf/coal.pdf.

———. 2008c. 'World crude oil prices'. *Official Energy Statistics for the US Government*. http://tonto.eia.doe.gov/dnav/pet/hist/rwtcd.htm.

Environment Canada. 2008. 'National inventory report: GHG sources and sinks in Canada, 1990–2006'. *The Canadian Government's Submission to the United Nations Framework Convention on Climate Change*. http://www.ec.gc.ca/pdb/ghg/inventory_report/2006_report/tdm-toc_eng.cfm.

———. 2009. 'National inventory report: GHG sources and sinks in Canada, 1990–2006'. *Comparison of Sectoral and Reference Approaches, Annex 4*. www.ec.gc.ca/pdb/GHG/inventory_report/2006_report/a4_eng.cfm.

Feygin, M., and R. Satkin. 2004. 'The reserves-to-production ratio and its proper interpretation'. *Natural Resources Research* 13 (1): 57–60.

First Nations Energy Alliance. 2007. 'First Nation opportunities: Powering our future'. Discussion Paper. http://www.firstnationsenergyalliance.org/FNEA_Forum_Final_Discussion_Paper.pdf.

GCC (Green Communities Canada). 2008. http://www.gca.ca/indexcms/index.php.

Gibbons, J. 2008. *Ontario's Green Future: How We Can Build a 100% Renewable Electricity Grid by 2027*. Toronto: Ontario Clean Air Alliance.

Gibson, R.B. 2006. 'Beyond the pillars: Sustainability assessment as a framework for effective integration of social, economic and ecological considerations in significant decision-making'. *Journal of Environmental Assessment Policy and Management* 8 (3): 259–80.

Gliedt, T., and P. Parker. 2007. 'Green community entrepreneurship: Creative destruction in the social economy'. *International Journal of Social Economics* 34 (8): 538–53.

Godin, M. 2007. 'Renewable is doable. A smarter energy plan for Ontario. Analysis and scenario modeling of the Ontario power system'. *Report 2. Portfire Associates and Emerging Energy Options WADE Canada*. Pembina Institute. http://www.renewableisdoable.com/pdf/analysis.pdf.

GoGreen. 2007a. 'Ontario's action plan on climate change: Go Green Ontario'. Government of Ontario. http://www.gogreenontario.ca/docs/actionplanonclimatechange.pdf.

———. 2007b. 'Ontario's GHG emissions targets: A technical brief'. Government of Ontario. http://www.gogreenontario.ca/docs/061807-Technical Brief.pdf.

Hammer, S.A. 2008. 'Renewable energy policymaking in New York and London: Lessons for other "world cities"?' In P. Droege, ed., *Urban Energy Transition: From Fossil Fuels to Renewable Power*, 143–72. New York: Elsevier.

Homer-Dixon, T. 2007. *The Upside of Down: Catastrophe, Creativity, and the Renewal of Civilization*. Toronto: Vintage Canada.

Hubbert, M.K. 1956. 'Nuclear energy and the fossil fuels'. Presented before the spring meeting of the Southern District, American Petroleum Institute, San Antonio, TX, 7–9 March.

ICLEI (International Council for Local Environmental Initiatives). 2006. 'Combating climate change: A comprehensive look at local climate protection programs'. *Cities for Climate Protection*. http://www.iclei.org/documents/USA/services/CaseStudiesDec2006_2_.pdf.

———. 2007. 'Progress report'. *Cities for Climate Protection*. http://www.iclei.org/documents/USA/documents/CCP/ICLEI-CCP_International_Report-2006.pdf.

———. 2008. 'How CCP works'. *Cities for Climate Protection*. http://www.iclei.org/index.php?id=810.

IEA (International Energy Agency). 2007. 'Renewables for heating and cooling: Untapped potential'. Paris: IEA. http://www.iea.org/textbase/nppdf/free/2007/Renewable_Heating_Cooling.pdf.

———. 2008a. *Key World Energy Statistics*. Paris: IEA.

———. 2008b. 'Electricity Information'. *IEA Statistics*. Paris: IEA.

———. 2009. *Energy Balance for Canada*. Paris: IEA.

IESO (Independent Electricity System Operator). 2009. 'Generation by fuel type'. http://www.ieso.ca.

IMF (International Monetary Fund). 2007. 'GDP current prices in US dollars'. IMF. http://www.imf.org/external/data.htm.

IPCC (Intergovernmental Panel on Climate Change). 2001. 'Special report on emissions scenarios'. IPCC. http://www.grida.no/publications/other/ipcc_sr.

———. 2007. 'Fourth assessment report, synthesis report. Summary for policy makers'. IPCC. http://www.ipcc.ch.

IPSP (Integrated Power System Plan). 2007. 'Integrated Power System Plan'. Exhibit B. Tab 1. Schedule 1. Ontario Power Authority. http://www.powerauthority.on.ca/IPSP/Page.asp?PageID=924&SiteNodeID=320.

Jaccard, M., L. Failing, and T. Berry. 1997. 'From equipment to infrastructure: Community energy management and GHG emission reduction'. *Energy Policy* 25 (13): 1065–74.

Jacoby, H.D., and A.D. Ellerman. 2004. 'The safety valve and climate policy'. *Energy Policy* 32: 481–91.

Karlenzig, W. 2008. 'Major US city post-oil preparedness ranking: Which cities and metro areas are best or least prepared for price volatility, supply shocks and climate change regulations?' *Common Current*. http://www.commoncurrent.com/publications.shtml.

Karlsson, S.I. 2007. 'Allocating responsibilities in multi-level management for sustainable development'. *International Journal of Social Economics* 34 (1–2): 103–26.

Liming, H., E. Haque, and S. Barg. 2008. 'Public policy discourse, planning and measures toward sustainable energy strategies in Canada'. *Renewable and Sustainable Energy Reviews* 12: 91–115.

Lindseth, G. 2004. 'The Cities for Climate Protection Campaign (CCPC) and the framing of local climate policy'. *Local Environment* 9 (4): 325–36.

McKibbin, W.J., and P.J. Wilcoxen. 2002. 'The role of economics in climate change policy'. *The Journal of Economic Perspectives* 16 (2): 107–29.

Mappingworlds. 2008. 'ShowWorld'. http://show.mappingworlds.com.

Norman, J., H.L. MacLean, and C.A. Kennedy. 2006. 'Comparing high and low residential density: Life-cycle analysis of energy use and GHG emissions'. *Journal of Urban Planning and Development* 132 (1): 10–21.

NRCAN (Natural Resources Canada). 2007. 'Energy'. Natural Resources Canada. http://www.nrcan-rncan.gc.ca/com/eneene/index-eng.php.

NRTEE (National Round Table on the Environment and the Economy). 2008. 'Getting to 2050: Canada's transition to a low-emission future: Advice for long-term reductions of GHG and air pollutants'. NRTEE. http://www.nrtee-trnee.ca/eng/publications/getting-to-2050/intro-page-getting-to-2050-eng.html.

OCB (Ontario Conservation Bureau). 2007. 'Taking action'. *Annual report 2007*. Ontario's Chief Energy Conservation Officer. Ontario Power Authority. http://www.conservationbureau.on.ca/Storage/18/2340_AR2007_26Oct07 (Final).pdf.

OCE (Ontario's Centres of Excellence). 'Centres of Excellence for Energy'. http://www.oce-ontario.org/Pages/COEEnergy.aspx?COE=EN.

OPA (Ontario Power Authority). 2007. 'A progress report on renewable energy standard offer program'. Ontario Power Authority. http://www.powerauthority.on.ca/SOP/Storage/60/5606_RESOP_Dec._2007_report.pdf.

———. 2008. 'How the renewable energy standard offer program works'. *Information for Renewable Energy Generators. Ontario's Standard Offer Program*. Ontario Power Authority. http://www.powerauthority.on.ca/SOP/Page.asp?PageID=122&ContentID=3937&SiteNodeID=170&BL_ExpandID=158.

OPG (Ontario Power Generation). 2008. 'Darlington new build and Pickering B study'. Ontario Power Generation. http://www.opg.com/power/nuclear/darlington/d_overview.asp.

Pacala, S., and R. Socolow. 2004. 'Stabilization wedges: Solving the climate problem for the next 50 years with current technologies'. *Science* 305: 968–72.

Pander, S. 2007. 'Climate protection progress report and action plans'. City of Vancouver. http://vancouver.ca/sustainability/climate_protection.htm.

Parker, P. 2008. 'Residential solar photovoltaic market stimulation: Japanese and Australian lessons for Canada'. *Renewable and Sustainable Energy Reviews* 12 (7): 1944–58.

Parker, P., I.H. Rowlands, and D. Scott. 2003. 'Innovations to reduce residential energy use and carbon emissions: An integrated approach'. *The Canadian Geographer* 47 (2): 169–84.

Pembina Institute. 2005. 'Oil sands fever: The environmental implications of Canada's oil sands rush'. Pembina Institute. http://www.oilsandswatch.org/album/osf-slide-show/projector.php.

———. 2008. 'Our approach: What is a sustainable energy system?' Pembina Institute. http://www.pembina.org/about/approach.

Persaud, A.J., and U. Kumar. 2001. 'An eclectic approach in energy forecasting: A case of Natural Resources Canada's (NRCan's) oil and gas outlook'. *Energy Policy* 29: 303–13.

Peters, R., P. Cobb, and M. Winfield. 2007. 'Renewable is doable: Analysis of resource potential and scenario assumptions'. Pembina Institute. http://www.renewableisdoable.com/pdf/resources.pdf.

Regulatory Framework for Air Emissions. 2007. 'Regulatory Framework for Air Emissions'. Environment Canada.ttp://www.ec.gc.ca/doc/media/m_124/toc_eng.htm.

Rivers, N., and M. Jaccard. 2005. 'Canada's efforts towards GHG emission reduction: A case study on the limits of voluntary action and subsidies'. *International Journal of Global Energy Issues* 23 (4): 307–23.

Robinson, J., et al. 2006. 'Climate change and sustainable development: Realizing the opportunity'. *A Journal of the Human Environment* 35 (1): 1–8.

Rowlands, I.H. 2007. 'The development of renewable electricity policy in the province of Ontario: The influence of ideas and timing'. *Review of Policy Research* 24 (3): 185–207.

———. 2008. 'Integrating climate policy and energy policy'. In S. Bernstein, J. Brunnee, D.G. Duff, and A.J. Green. *A Globally Integrated Climate Policy for Canada*, 293–314. Toronto: University of Toronto Press.

Ruth, M., and D. Coelho. 2007. 'Understanding and managing the complexity of urban systems under climate change'. *Climate Policy* 7 (4): 317–36.

Seymoar, N.K. 2008. 'The sustainable cities: Plus planning cycle'. International Center for Sustainable Cities. http://www.icsc.ca/content/dm documents/plus_planning_cycle.pdf.

Sierra Club of Canada. 2008. 'Mackenzie Valley pipeline and Alberta tar sands'. Sierra Club of Canada. http://www.sierraclub.ca/national/programs/atmosphere-energy/energy-onslaught/campaign.shtml?x=307.

Simpson, J., M. Jaccard, and N. Rivers. 2007. *Hot Air: Meeting Canada's Climate Change Challenge*. Toronto: McClelland and Stewart.

Soderbergh, B., F. Robelius, and K. Aleklett. 2007. 'A crash programme scenario for the Canadian oil sands industry'. *Energy Policy* 35: 1931–47.

Stadler, M., et al. 2007. 'Policy strategies and paths to promote sustainable energy systems—The dynamic *Invert* simulation tool'. *Energy Policy* 35: 597–608.

Statistics Canada. 2008. 'Oil production and reserve database'. International oil production and proved reserves table 130-0001; established crude oil reserves table 153-0013. CANSIM socioeconomic database.

St Denis, G., and P. Parker. 2009. 'Community energy planning in Canada: The role of renewable energy'. *Renewable and Sustainable Energy Reviews* RSER 629: 15.

Stern Review. 2006. '*Stern Review* on the economics of climate change'. Office of Climate Change. Cambridge: Cambridge University Press. http://www.hm-treasury.gov.uk/sternreview_index.htm.

Telfer, L., and M. James. 2008. 'Tar sands action guide'. Sierra Club of Canada. http://www.sierraclub.ca/national/programs/atmosphere-energy/energy-onslaught/tar-sands-action-guide.pdf.

Toronto Conference. 1988. *World Conference on the Changing Atmosphere Proceedings*. Geneva: World Meteorological Organization.

Wilbanks, T.J., and J. Sathaye. 2007. 'Integrating mitigation and adaptation as responses to climate change: A synthesis'. *Mitigation and Adaptation Strategies for Global Change* 12: 957–62.

Williams, J.L. 2008. 'Oil price history and analysis'. WTRG Economics. http://www.wtrg.com/prices.htm.

Wilson, C., and T. McDaniels. 2007. 'Structured decision-making to link climate change and sustainable development'. *Climate Policy* 7 (4): 353–70.

Winfield, M.S. 2008. 'Climate change and Canadian energy policy: Policy contradiction and policy failure'. *Behind the Headlines* 65 (1): 1–19. Toronto: Canadian International Council.

World Energy Council. 2005. 'Photovoltaic and wind energy capacity data'. World Energy Council. http://www.worldenergy.org.

Zahran, S., et al. 2008. 'Risk, stress, and capacity: Explaining metropolitan commitment to climate protection'. *Urban Affairs Review* 43 (4): 447–74.

7

Water Security: Current and Emerging Challenges

Reid Kreutzwiser and Rob de Loë

Learning Objectives

- To gain an understanding of the nature and significance of water security at both the national and local scales.
- To appreciate the role of water allocation systems in minimizing conflict over the use of water resources.
- To understand the role of multiple levels of government, and of governance, in addressing local water quality problems.
- To demonstrate how sovereign nations can resolve disputes related to water quality and quantity.
- To recognize that challenges to successfully managing water resources are often issues of governance rather than limitations in technology or scientific understanding.

INTRODUCTION

An Ontario Ministry of the Environment permit to the Nova Group for tanker export of Lake Superior water to Asia, which was subsequently rescinded, led to an International Joint Commission study of Great Lakes diversions and consumptive uses (IJC 2000) and helped to secure a 'good-faith' 2005 agreement between Ontario, Quebec, and the eight Great Lakes states to ban large-scale water exports and diversions. Nevertheless, the possibility of bulk water exports from Canada under current free trade agreements remains a concern. In October 2005, many of the 1,700 residents of the remote northern

Ontario Cree community of Kashechewan were evacuated because of *E. coli* contamination of their drinking water. In 2007, the federal government agreed to contribute $200 million to address drinking water, flooding, and other concerns in the community (Brennan 2007). Water export and local drinking water contamination both illustrate the growing importance of water security.

'Water security' is a much-used term that means different things according to the context and the user's perspective. Most often it refers to the condition of the water resource— in other words, the extent to which it is threatened by overuse or contamination (Bruins 2000;

Falkenmark 2001). In this chapter, water security means access to adequate quantities of water, of acceptable quality, for human and environmental uses (Global Water Partnership 2000). Several aspects of water security pertaining to scale, types of water use, approaches to addressing water security challenges, and, importantly, conflict and uncertainty are worth noting.

Typically, concern for water security is at a national or global scale and relates to water scarcity (Postel 1996; Food and Agriculture Organization 2000; Coloumb 2002). Numerous threats to water security are evident, including population growth, urbanization, globalization, climate change, and, most recently, bioterrorism. Concern at this broader scale is certainly understandable, particularly in water-scarce countries where a significant portion of surface water flow originates outside the country and where arrangements to manage transboundary conflicts are weak or non-existent. For instance, in Egypt, 96 per cent of surface water flow originates outside the country (World Resources Institute 2000), and that country is almost entirely dependent on the Nile River for its water supply. A substantial increase in water withdrawals by upstream Nile River states, principally Ethiopia and Sudan, would have serious consequences for Egypt. In contrast, less than 2 per cent of Canada's flow originates externally. However, proposals for water export from Canada to the United States have, from time to time, sparked much concern and debate (Day and Quinn 1992).

Water security also has a local dimension. As communities grow, municipalities and local watershed management agencies are being challenged to meet increased water demands, to cope with increasing variability in supplies caused by droughts and climate change, and to meet stringent new drinking water standards imposed by senior governments. Even individual water users, such as homeowners who have to boil drinking water or farmers who may be unable to irrigate crops or water their livestock, may face challenges to the security of their water resources. Taken together, problems experienced by individual water users and local organizations can present major challenges. And cases such as the Nova Group water export proposal demonstrate that a local water development proposal can have national and international implications.

Human beings are by no means the only species facing water security concerns. World water withdrawals increased six-fold during the twentieth century, more than double the rate of growth in population (World Resources Institute 2000). While the benefits in terms of drinking water, agriculture, and other human uses are undeniable, there is a growing concern for the value of water in maintaining ecosystem functions and the impacts of human uses of water on the environment (Lemly, Kingsford, and Thompson 2000). Water security for the protection of wetlands, aquatic ecosystems, and biodiversity is critical for the well-being of these natural systems (Dyson, Bergkamp, and Scanlon 2003). This is an important lesson in the concept of sustainability (Mitchell and Shrubsole 1997), which recognizes that continued human prosperity depends on healthy natural ecosystems.

Appreciation of the need to ensure water security for the environment will necessitate greater attention to the implications of water withdrawals and trade-offs in water use. For example, many of the water allocation systems that exist today were not designed to accommodate environmental concerns such as ensuring that enough water is left in rivers and streams to support healthy aquatic environments. Thus, retroactively addressing the needs of aquatic habitats in water-short regions has meant that other uses—often agricultural irrigation— have had to be reduced (Lemly, Kingsford, and Thompson 2000).

There is growing recognition that the solutions to many water security issues transcend structural or engineering solutions. While adequate water infrastructure is critical, it is now well understood that water problems usually relate as much or more to social and governance factors (United Nations World Water Assessment Programme 2006). For example, many water shortage problems may best be resolved by adjusting human behaviour rather than by increasing supplies through building new dams. Experiences around the world demonstrate that eliminating wasteful and inefficient water use practices can delay, or entirely avoid, the need to find new water supplies (Gleick 2003).

Finally, as with other aspects of resource management, conflict and uncertainty are embedded in water security. Water allocation decisions, for instance, can generate conflict among users competing for supplies. The absence of clear allocation rules may produce even greater conflict. Contamination of water supplies can also generate much conflict as the various stakeholders attempt to apportion blame and responsibility for remedial action. Uncertainty is also apparent in the water security issue. Water managers have always had to deal with daily, seasonal, and annual changes in precipitation, stream flows, and lake levels, but these changes have fallen within a predictable range of variation. Climate change is upsetting these patterns, and water managers can no longer assume that the past will be a guide to the future (Milly et al. 2008). Thus, major decisions on infrastructure investments, which have long-term economic and other consequences, will have to be made in an environment of much greater uncertainty.

Our aims in this chapter are to examine water security concerns in Canada, and to explore how they are manifested in *water allocation* and *water quality management* across the country at scales ranging from national to local. Despite the fact that water quality and allocation issues are often interrelated and should be addressed together, we present examples that emphasize one more than the other to better highlight key challenges. To provide a foundation for the examples, the following section describes important aspects of Canada's water resources and human use of those resources.

WATER RESOURCES AND THEIR USE

Canada's Water Resource Base

Considering the size (9.2 million square kilometres) and physical diversity of its landmass and the importance of precipitation, evaporation, and plant transpiration in determining hydrological conditions, it is not surprising that Canada's water resources are both immense and variable. Precipitation averages about 600 millimetres annually, of which about 330 millimetres runs off in rivers and streams directly or after storage in wetlands or lakes or as groundwater. This represents 105,000 cubic metres per second (m^3/s) of river flow or discharge, or 7 per cent of the world's total *renewable* water resource. Fresh water in rivers, lakes, and wetlands covers almost 24 per cent of Canada's land area, and the volume of groundwater is many times that in rivers and lakes. Considering all sources of fresh water, Canada holds about one-fifth of the world's *total* freshwater water supply (Environment Canada 2002).

These figures suggest that Canada is water-wealthy, and one may wonder why water security should be an issue. Unfortunately, Canada's water wealth is more myth than reality (Sprague 2006). The commonly cited figure of one-fifth of the world's water supply includes water resources that, if used up, will not be renewed. For example, only a tiny portion of the vast water supply of the Great Lakes—approximately 1 per cent—is renewed on an annual basis; the remainder is a leftover of glaciation and thus must not be viewed as available for human uses. Furthermore, statistics portraying aggregate runoff and annual averages mask important spatial and temporal variability in water supplies. By and large, much of the water in Canada is found in sparsely populated areas. Moreover, Canada's water resources are threatened by a wide range of human stresses, including overuse, pollution, and climate change. Thus, water security is a very real concern in Canada.

Precipitation in Canada varies widely, from more than 3500 millimetres annually along the Pacific coast to less than 500 millimetres throughout the North and much of the Prairies. The wettest location, on average, is Ocean Falls, BC, with 4386 millimetres per year; the driest is Eureka, Nunavut, with 64 millimetres annually. This variability in precipitation, of course, translates into considerable variation in runoff across the country (Figure 7.1).

Temporal variation in precipitation and runoff is also very evident. More than a third of annual precipitation falls as snow, much of which is unavailable until spring melt. For instance, approximately 75 per cent of the stream flow in

Source: Adapted from Pearse et al. (1985)

Figure 7.1 Surface Runoff, Canada.
Note: Annual runoff, in mm, from annual precipitation.

the South Saskatchewan River Basin in southern Alberta comes from snowmelt, and most of this flow occurs between mid-May and mid-July (Alberta Water Resources Commission 1986). As a consequence of seasonal and annual variation in runoff, mean annual river discharge is an imperfect indicator of the water resource base. An alternative measure is reliable monthly discharge (Foster and Sewell 1981)—that is, flow equalled or exceeded, on average, in all but one month in 10 years. Reliable monthly flows are higher in Pacific coastal rivers, in the Great Lakes Basin, and in most northern rivers. In contrast, greater variation in precipitation and evapotranspiration contributes to less reliable flow in the Prairies. In the Assiniboine–Red Basin, reliable monthly flow is 3.9 m^3/s, or less than 8 per cent of mean annual flow. This certainly suggests drought susceptibility on the Prairies, although drought can occur in many parts of the country. In Ontario, for instance, four widespread droughts occurred between 1998 and 2007 (Environmental Commissioner of Ontario 2008).

At the other extreme, above-average discharges can cause flooding. Monthly, daily, and instantaneous flows can be many times the mean annual discharge. For example, during the 1964 flood on the Oldman River, instantaneous discharge near Lethbridge, AB, was 2,090 m^3/s on June 10, more than six times the June mean flow, almost 20 times the mean annual discharge, and close to 1,000 times the minimum daily flow recorded (Environment Canada 1973; 1989).

Water Use and Abuse

In discussing water use and its implications, it is helpful to distinguish between withdrawal and in-stream uses. The former involves the withdrawal of surface water or groundwater and often returns a portion of that water to surface flow. The portion not returned is consumed (e.g., lost to evapotranspiration in irrigation).

The latter, in-stream uses, does not involve withdrawal; rather, water serves as a medium for some use, such as hydropower generation or recreation.

Canadians are among the most prolific water users in the world. In terms of domestic use, Canadians use an average of 329 litres per day per person, twice the use in European households and nine times the use in East African households (International Institute for Environmental Development 2002; Environment Canada 2009a). For most Canadians, water used in and around homes is an obvious way in which they use water. However, water also is critical in a range of other sectors, including power generation, agriculture, industry, and recreation. Table 7.1 shows water withdrawals by region and sector. Water withdrawn for cooling during thermal power production is by far the largest use. However, very little of this water is consumed. In contrast, more than three-quarters of the water withdrawn for agricultural use is consumed. Thus, while Ontario is the largest water withdrawer in Canada for most uses, Alberta—where almost two-thirds of Canada's irrigation takes placed—is by far the largest *consumer* of water. Water withdrawals in Canada totalled 50,813 million cubic metres in 2005, an increase of almost 14 per cent over the previous decade.

Water withdrawals have important implications. Obviously, water is key to many economic activities. It is not surprising that agricultural water use is highest in Alberta and Saskatchewan where irrigation water demands are highest. However, water is also critical to ecosystems. Withdrawals of both surface and groundwater can adversely affect surface water flows, since the base flow in many streams and rivers is supported by groundwater discharge, particularly during the drier summer months (Dyson, Bergkamp, and Scanlon 2003). The quality of

Table 7.1 Water Intake in Canada (in Million m³) by Region and Sector, 2006

Region	Thermal Power	Manufacturing	Municipal*	Agriculture*	Mining	Rural*	Total
Atlantic	x	538	244	21	x	142	944
Quebec	x	1,833	1,582	113	24	220	3,772
Ontario	26,648	3,487	1,663	174	43	213	32,227
Prairies	x	675	624	3,592	x	129	5,021
British Columbia**	x	1,246	772	886	62	153	3,120
Total***	32,138	7,779	4,884	4,787	459	856	45,083
Per cent of total	63.1	15.3	9.6	9.4	0.9	1.7	100

Source: Environment Canada 2009b.

* Intake cannot be separated from total water use. Total water use reported here.
** Includes Yukon, Nunavut, and Northwest Territories.
*** Some values are not reported by region because of confidentiality and/or data issues; national totals may not equal the sum of the regions.
x Means the values are not available.

aquatic habitats may be impaired, flows may be insufficient for dilution of sewage treatment plant effluent, and less water may be available for other human uses. Problems are exacerbated when withdrawals increase during drought periods. Water withdrawals, thus, may contribute to conflicts over how water is allocated.

Major in-stream uses include hydropower generation, navigation, recreation, and waste disposal. The last of these involves the ability, or assimilative capacity, of receiving waters to accommodate a wide range of wastes. While in-stream uses do not involve water withdrawals, they are not without consequences. For example, hydropower production was 350,000 gigawatt-hours, or 59 per cent of total electric power production in 2006 (Minister of Industry 2008). Newfoundland, Quebec, and Manitoba produce virtually all their electricity through hydropower. While 'clean' in many respects, hydropower has been the major impetus for massive alteration of hydrological regimes. This alteration has involved construction of 650 large dams and 54 inter-basin diversions totalling 4,450 m³/s (Statistics Canada 2003). Much of this alteration has been in northern Canada, where loss of fish and wildlife habitat and social and

cultural impacts on Aboriginals' communities have been major consequences.

In addition to water use, Canada's water resources are subjected to numerous other stresses that threaten the security of water supplies. Water quality impairment of surface waters from point sources, such as sewage and industrial discharges, and from non-point sources, principally agricultural runoff, has been a concern for decades. Phosphorous and other nutrients, often from agricultural runoff, can accelerate eutrophication, or the biological aging, of surface water, thus reducing dissolved oxygen levels and degrading habitat quality for fish and other aquatic organisms. Industrial discharges and atmospheric deposition contribute a wide variety of toxic substances, including heavy metals and PCBs, to rivers and lakes. Hundreds of toxic substances have been detected in the Great Lakes.

The risk of large-scale waterborne disease outbreaks that killed thousands of Canadians in the nineteenth century has largely been eliminated through modern water treatment techniques. However, dangerous and sometimes deadly outbreaks do occur. The Walkerton, ON, case discussed below is perhaps the most famous recent Canadian example. However, it is not the only one. Cryptosporidium made thousands of people ill in Kitchener-Waterloo and Colling-wood, ON, in separate incidents in 1993 (Brod-sky 2001) and threatened the residents of North Battleford, SK, in 2001. Less dramatically but as potentially deadly, the Canadian Medical Asso-ciation in 2008 revealed that 1,766 boil-water advisories were in place in Canada, with many of them occurring in Aboriginal communities such as Kashechewan (Eggertson 2008).

Human activities also stress the quantity and quality of groundwater. Across Canada, ground-water withdrawals average only about 2 per cent of total water withdrawals (World Resources Institute 2000). However, about 30 per cent of all Canadians, and 82 per cent of the rural popula-tion, rely on groundwater for drinking water; groundwater also supplies 43 per cent and 14 per cent, respectively, of agricultural and industrial water use (Nolan 2005). Groundwater is contam-inated from a host of human activities, includ-ing leaking underground storage tanks, chemical spills, and urban and rural land-use practices. Rural residents dependent on groundwater for their drinking water are particularly vulnerable. In 1992, more than 1,200 farm domestic wells in Ontario were tested for nitrate, faecal coliform bacteria, and several herbicides, 34 per cent of wells were contaminated with bacteria, and 14 per cent exceeded Ontario Drinking Water Objectives for nitrate (Goss, Barry, and Rudolph 1998).

Not surprisingly, the intensification of live-stock farming is emerging as an issue, notably in parts of Quebec, Ontario, Manitoba, Alberta, and British Columbia. For instance, 35 per cent of Canada's hog production occurs in Quebec, and much of this production is concentrated in a few locations, such as the Yamaska River watershed (Coote and Gregorich 2000; Beau-lieu 2001). A single breeding sow produces three tonnes of solid manure per year. When not properly collected, stored, and disposed of, manure can contaminate surface water and groundwater (Kee 2001).

Climate change looms as another threat to the quantity and quality of water resources (Lemmen et al. 2008). Across Canada, climate change is resulting in increased air and surface water temperatures and increased precipitation and evapotranspiration, with a higher propor-tion of precipitation occurring as heavy rain-fall events (see Chapter 5). Despite increased precipitation, increases in evapotranspiration will mean reduced annual stream flow, lake lev-els, and groundwater recharge in much of the country. In coastal areas, reduced groundwater recharge and increased withdrawals may lead to saltwater intrusion of aquifers. In many areas, hydrologic variability will increase, meaning that more extreme flood and drought events will occur. Increased surface water temperature and less stream flow for dilution will degrade water quality in rivers and lakes and reduce the cap-acity of receiving waters to assimilate wastes.

Across Canada, recent years have been some of the driest on record. Significant economic and environmental consequences have resulted. For example, in 2001, drought-related crop insurance payouts exceeded $1 billion (Agricul-ture and Agri-Food Canada 2002). At the other extreme, the 1996 Saguenay River and 1997 Red River floods, with damages of $800 million and $500 million respectively, were among the costli-est floods ever. These extreme events may be a hint of what can be expected with climate warm-ing. They also highlight the need for effective water allocation systems.

WATER ALLOCATION

Deciding who gets to use scarce water resources has been a challenge for millennia. For instance, the Romans created a sophisticated legal system to allocate water (Wescoat 1997), and sophisticated community-level institutions for managing subterranean, gravity-driven filtration galleries known as 'qanats' have existed in the Middle East for centuries (Lightfoot 2000). Modern societies confront similar challenges. A strong allocation system that provides clear rules regarding who can use water and under what circumstances can help societies to avoid or reduce conflicts, promote efficient water use, and enhance water security. Protecting ecosystem functions is a contemporary concern. Unfortunately, many water allocation systems were established during a time when demands were less severe than, or quite different from, what they are now. Updating those systems to reflect modern circumstances and needs can be challenging.

All water allocation systems share the basic aim of deciding who gets water and under what circumstances. However, because water does not respect political boundaries, an important distinction must be drawn between water allocation that takes place *within* sovereign jurisdictions and allocation that takes place *between* sovereign jurisdictions (e.g., two neighbouring countries that share a river, such as Egypt and Ethiopia). Because the water allocation laws of one country do not apply in another, many countries that share water resources have formed treaties or agreements specifying how water will be allocated between them (Giordano and Wolf 2003). Canada and the US are world leaders in this respect. Within countries (or provinces or states with jurisdiction over water), water often is allocated to various competing users through a combination of common law (ancient customs and the cumulative decisions of judges) and statutes enacted by lawmakers. In Canada, provincial legislatures have enacted laws respecting water allocation, but some elements of common law still apply. For instance, the common law right to use water for domestic purposes still

exists in provinces despite the existence of statutory laws (de Loë et al. 2007).

In recent years, international trade agreements also have become important to water allocation in Canada (see Chapter 2). Key agreements include the North American Free Trade Agreement (NAFTA) among Canada, the United States, and Mexico and the General Agreement on Tariffs and Trade (GATT) administered through the World Trade Organization (WTO). Despite assurances to the contrary from the governments of Canada and the US, some commentators fear that these kinds of agreements are eroding Canada's sovereignty over water and will compromise the ability of governments to manage water resources (Elwell 2001; Nikiforuk 2007).

In this section, challenges and issues associated with water allocation in Canada are explored at two scales: international water allocation between Canada and the US and water allocation within Ontario. Each example illustrates major challenges societies face in determining how water should be shared among competing users. At the same time, these examples highlight some of the special challenges facing Canadians. They include working within a constitutional system that divides responsibility for water between the federal and provincial governments; adjusting to the implications of international trading agreements; dealing with the implications of climate change, which will alter hydrologic regimes across the country; and sharing water among uses and users within Canada as well as with a large and powerful neighbour to the south. The example of 'virtual water' is explored to highlight how the very nature of the water allocation problem is being redefined.

Water Allocation along the Canada–US Border

When rivers flow across international boundaries or when water bodies such as lakes are divided by these boundaries, special water allocation challenges exist. Questions of ownership, already perplexing within countries, become especially vexing when water is shared by two or more countries. Many rivers flow across the 8,900-kilometre Canada–US border, sometimes back and forth in different places, and the border also divides numerous lakes, wetlands, and aquifers. The Great Lakes are the largest shared water body between the two countries. Major transboundary rivers include the Columbia, which flows from British Columbia to the US; the Red, which flows from the US into Canada south of Winnipeg; and the St Lawrence, which drains the Great Lakes. Important water security issues exist for users on both sides of the boundary.

The Boundary Waters Treaty, signed in 1909, is a critical transboundary institutional arrangement that promotes water security for Canada and the US. Thus, in this section we explore its history and current role. However, any discussion of contemporary international water allocation also must consider the role of international trade agreements. While these arrangements were not created with water allocation in mind, they unquestionably play a role and may compromise water security in Canada.

The Boundary Waters Treaty of 1909

Major irrigation projects are expensive undertakings, and their success depends in part on a reliable water supply. Irrigation farmers need large quantities of water at key times during the growing season. The quantity of water available in the natural system is one key concern that affects their water security. Equally important are demands on the water from other users. If a water source is wholly within the jurisdiction of a particular province or country, then it can be managed according to internal rules. However, when the water source flows across an international boundary, water security is threatened, because users on one side of the boundary are not bound by the water allocation rules of the other country. This is the situation that existed in southern Alberta at the end of the nineteenth century.

Southern Alberta is semi-arid. Although the soils are good for agriculture, precipitation can be unreliable. Average annual precipitation in the region is 330 millimetres, considerably less than is received in other parts of Canada. In 1892, only 200 millimetres of precipitation fell during the year. Settlers, who had been led by the Canadian government to believe that southern Alberta was an agricultural paradise, became extremely concerned, and many pushed for irrigation (de Loë 2005). Consequently, during the late 1800s, several private companies attempted to construct irrigation projects in southern Alberta using water from the St Mary River, which rises in the state of Montana and flows into southern Alberta. Unfortunately, similar efforts were underway in Montana, and the American irrigators were not concerned about the needs of the Canadians. This situation became an ongoing irritant in relations between Canada and the US throughout the 1890s and early 1900s. Fortunately for Canada, an accident of geography provided the basis for a solution that led to a treaty.

The Milk River also rises in Montana, but after flowing across the international boundary into Alberta, it flows back into Montana. The Milk and St Mary watersheds are divided by a height of land known as the Milk River Ridge. When it became apparent that irrigators in the US planned to forge ahead with projects that would monopolize the waters of the St Mary River, the Canadian government authorized construction of a canal through the ridge. This canal, which was operational in 1903, allowed Canada to divert water from the Milk River into the St Mary River on the Canadian side of the boundary. This would have hurt US irrigation projects dependent on the waters of the Milk River that flowed back into the US (Mitchner 1973).

Recognition that Canada also could act unilaterally on boundary waters matters in western Canada brought the US to the bargaining table in 1904 (Mitchner 1973). These negotiations between the US and Britain (acting for Canada) led to the Boundary Waters Treaty of 1909. This treaty illustrates how some key water allocation problems that occur between countries can be resolved peacefully and effectively. It determined, among other things, how the flow of the St Mary and Milk rivers would be shared and established a bi-national body for addressing future boundary waters problems. That body, the International Joint Commission (IJC), was empowered to investigate boundary waters issues referred to it by the Canadian and US governments and to issue orders binding on both countries.

The Boundary Waters Treaty continues to guide transboundary water allocation between Canada and the US. Attention in the past few decades has focused on the waters of the Great Lakes. For instance, in its 2000 report, sparked by the Nova Group water export proposal, the IJC explored a number of pressing issues, including uncertainty over the potential impacts of climate change, the need for water conservation, and the significance of international trade agreements, especially as they relate to bulk water exports (IJC 2000). Reflecting concerns such as these, the provinces of Ontario and Quebec and the eight Great Lakes states signed in 2005 the Great Lakes–St Lawrence River Basin Sustainable Water Resources Agreement to protect and conserve the waters of the basin. In the next section, the issue of bulk water exports is examined to illustrate specific concerns that exist, especially related to trade agreements.

International Trade Agreements and the Issue of Bulk Water Exports

Canada is a trading nation. Much of its prosperity is due to access to the markets of other countries, especially those of its largest trading partner, the US. Consequently, the Canadian government during the past few decades has entered into agreements to liberalize trade and increase the flow of goods and services. These agreements have been accompanied by considerable controversy. As countries enter into these kinds of arrangements, they give up some measure of sovereignty (see also Chapter 2). For example, by becoming a signatory to the GATT—a global trade agreement—Canada has agreed to become subject to the binding arbitration mechanisms of an international body, the WTO.

In the context of water allocation, an especially important trade agreement that Canada has signed is the North American Free Trade Agreement, or NAFTA, (1994) with the US and Mexico. Critics of this agreement have raised the following concerns (Elwell 2001; Linton 2002; Nikiforuk 2007):

- Will international trade agreements prevent Canada from protecting and managing its water resources?
- Will Canada be forced to permit bulk water exports to the US and other countries?
- If bulk water exports do take place, will trade agreements make it impossible to 'turn off the tap'?

Much of the debate revolves around some highly technical, and to some extent hypothetical, questions. For instance, items that are 'goods' (such as automobiles and lumber) are specifically covered by NAFTA. Thus, water, when it enters commerce by becoming a good or product, such as an ingredient in a food or as bottled water, is covered by the agreement. However,

NAFTA does not specifically exclude water in its *natural state* (in other words, in rivers, lakes, reservoirs, wetlands, and aquifers). This left the question of bulk water exports open.

The Canadian government has taken steps to address this concern. The governments of Canada, the US, and Mexico issued a joint statement in 1993 that explained that 'NAFTA creates no rights to the natural water resources of any Party to the Agreement' and that water in any form is not covered by any trade agreement (including NAFTA) unless it enters commerce and becomes a good or product (Canadian Intergovernmental Conference Secretariat 1999). In 1999, as noted earlier, Canada and the US asked the IJC to explore issues relating to water quantity in the Great Lakes, including the impact of trade agreements on water management. In its report, the IJC (2000) recognized the concerns that critics raised but concluded that NAFTA and the WTO agreements would not force the government of Canada to permit bulk water exports and that if water exports took place, they could be terminated. Finally, in December 2001, an Act to amend the International Boundary Waters Treaty Act (the federal legislation through which the Boundary Waters Treaty of 1909 is implemented in Canada) received royal assent. The amendments in this Act will prohibit the bulk removal of water out of the Canadian portion of boundary water basins. Despite these changes, some critics are unconvinced and suggest that only amendment of NAFTA to explicitly state that water in its natural state is not subject to the agreement will be sufficient to dispel uncertainty (Linton 2002; Christensen and Lintner 2006).

International Trade in Virtual Water

While much of the focus in water allocation across international boundaries has been on water in its natural state, the concept of *virtual water* is garnering interest and concern. The concept refers to the volume of water required to produce a commodity—that is, the water 'embedded' in the commodity (Allan 1998). For instance, Hoekstra and Chapagain (2007) suggest that 200 millilitres of milk has a virtual water content of 200 litres, while a cotton T-shirt contains 2,000 litres. The import and export of various commodities can represent a substantial flow of virtual water between countries.

The concept and its relevance have been contested by some, including Merrett (2003), and the methodology involved in estimating virtual water flows requires various assumptions (e.g., regarding crop water requirements at the field level and crop yields). However, the comprehensive estimates produced by Chapagain and Hoekstra (2003; 2008) raise interesting questions pertinent to water allocation. They estimate that international virtual water flows between 1997 and 2001 averaged 1,625 billion m^3/year (Chapagain and Hoekstra 2008). Trade in crops and crop products accounted for 61 per cent of this flow across borders; livestock products and industrial products contributed 17 and 22 per cent, respectively. Chapagain and Hoekstra (2008) further estimate that 16 per cent of global total water use goes into the production of commodities for export.

Not surprisingly, Canada's net virtual water export of 59,888 million m^3/year makes it the world's second-largest net exporter of virtual water, behind Australia (Chapagain and Hoekstra 2008). The US imports a substantial volume of virtual water from Canada, and trade in livestock is an instructive example. Beef cattle is a water-intensive product; the virtual water content of beef produced in Canada is estimated to be 9,636 m^3/tonne (Chapagain and Hoekstra 2003). In 2006, Canada exported 206,430 more tonnes of beef cattle to the US than it imported (Agriculture and Agri-Food Canada 2009a), representing the export of some 1989 million m³ of virtual water. Almost 61 per cent of Canada's beef exports to the US that year came from Alberta (Agriculture and Agri-Food Canada 2009b).

Agriculture in semi-arid Alberta is heavily reliant on irrigation infrastructure that has involved substantial economic costs and has resulted in considerable stress on aquatic ecosystems (de Loë 1997). The social and economic benefits of irrigation development have been substantial for the region in general and for the livestock sector in particular, which benefits from a reliable feed source produced by irrigation. While the economic and social significance of Alberta's livestock industry must be acknowledged, the sustainability of current water use is in debate, and water allocation is a controversial issue in the province. Water resources in southern Alberta are over-allocated, as evidenced by Alberta Environment's 2006 moratorium on new water licences in three South Saskatchewan River watersheds through the Approved Water Management Plan for the South Saskatchewan River Basin (Alberta Environment 2009). In the context of limited water resources, uncertainty around climate change impact on those resources, and intense competition for water, the value-added of 'virtual water exports' through livestock exports dependent on water-intensive irrigated feed may become a concern in future.

Ontario's Permit to Take Water Program

Distinct water allocation systems have been developed within each province in Canada. Their roots can be traced back to one or more common legal heritages. For example, in Alberta, Saskatchewan, and Manitoba, water allocation takes place through licences issued by the respective provincial governments. Water allocation laws in these provinces have a common ancestor: the North-west Irrigation Act of 1894, enacted by the Canadian government, which was responsible for water in western Canada during this time. The 1894 statute is rooted in the *prior appropriation* model of water allocation developed in the western US. In this system, people who put water to use create a legal entitlement to that water, with the first person using the water having the strongest right. In Canada's adaptation of this model, the government issues licences to use water, with holders of earlier licences being entitled to receive their full measure of water before junior licensees receive any.

In Ontario, the water allocation system has its roots in a very different legal heritage. Two common law doctrines provide the foundation: the doctrine of *riparian rights* (for water that flows in known, defined channels—usually surface water) and the *rule of capture* (for groundwater) (Lucas 1990). Both doctrines are based on decisions by English judges in past centuries when water was relatively abundant, water uses were quite small (mostly small-scale agriculture), and understanding of hydrological systems was incomplete.

Under the legal doctrine of riparian rights, a person who owns land that borders on or is crossed by a watercourse with a known and defined channel is a 'riparian' and has certain rights to water. The riparian landowner is entitled to receive water undiminished in quantity or quality, may make reasonable use of the water, does not have to make use of the water to keep riparian rights, and can be sued for interfering with another riparian's rights to water. Unlike the prior appropriation system, the date of use is not important. Therefore, as new riparians use water, existing riparians must (in theory) adjust their use patterns. Regarding groundwater, completely different common law developed.

Riparian rights apply only to subsurface water when it flows in known and defined channels; this is very rare in Ontario. Therefore, the key common law doctrine that applies to groundwater is the 'rule of capture'. Under this doctrine, landowners may withdraw as much percolating water as they wish and are not liable for interferences resulting from the water-taking. Furthermore, landowners do not own the water under their properties until they 'capture' it (for instance, by pumping from a well).

When water supplies were relatively abundant, the shortcomings of these doctrines were easy to overlook. However, as demand for water increased in Ontario following the Second World War, problems increasingly became evident. For example, conflicts resulted when activities that consumed a lot of water often negatively affected downstream users. Also, not all activities needing a lot of water, such as industries, were located on land next to water bodies. This was a barrier to economic development.

Recognizing that a modern economy could not function well with this kind of water allocation system, the provincial government passed the Ontario Water Resources Act (OWRA) in 1961. This statute put in place a permit system, which requires that with a few exceptions, anyone taking more than 50,000 litres per day of surface or groundwater in Ontario must have a Permit to Take Water (PTTW). The OWRA supplanted—but did not completely extinguish—the common law rights flowing from the riparian doctrine and the rule of capture. Therefore, under the OWRA, no permit is ever required for people taking water for domestic water supply, livestock watering (when the water is not taken into storage), and firefighting. Permits also are not automatically required for water-takings that existed on or before 29 March 1961, but such users could be required to secure a permit. Small users (50,000 litres per day or less) normally do not

need a permit but may be required to secure one at the discretion of an official known as a director. Directors under the OWRA have considerable discretion to issue, refuse to issue, cancel, or impose terms and conditions on a permit after it is issued. This allows for much flexibility but can also foster uncertainty for water users.

Critics have pointed out numerous problems with Ontario's water allocation system under the PTTW program, including poor data collection, inadequate requirements for local consultation, and weak provisions for protecting environmental water requirements (e.g., Kreutzwiser et al. 2004). Many of these concerns were addressed through new regulations created under the OWRA in 2004 and 2007 and through an overhaul of the *Permit to Take Water Manual* that provides detailed guidance regarding the program (Ontario Ministry of the Environment 2005). Through a separate program, Ontario Low Water Response (OLWR), the provincial government established a system for responding to droughts and water shortages at the local scale (Ontario Ministry of Natural Resources et al. 2003). OLWR has brought a much needed *local*, watershed-based perspective to water allocation in Ontario, but implementation problems have occurred. For instance, there has been a marked unwillingness on the part of the provincial government to enforce restrictions during droughts on water-takings that will cause economic losses among water users.

Water allocation systems play a crucial role in shaping water security (de Loë et al. 2007). Whether or not Ontario's system, through the PTTW and OLWR, *currently* ensures that human needs for water are balanced against environmental requirements is an open question. The future is another matter entirely. Forecasts of climate change impacts on Ontario's water resources indicate that droughts, low stream flows, and reduced groundwater levels will become more common in future, with attendant impacts on the environment and human users of water (Chiotti and Lavender 2008). Both conflict and uncertainty can be expected to increase substantially. Thus, the critical link between water allocation and water security will need to be strengthened.

WATER QUALITY MANAGEMENT

Water quality is a function of its physical (e.g., temperature, turbidity), chemical (e.g., oxygen, nutrients, acidity, metals), and biological (e.g., bacteria, viruses) properties. As a result of natural processes, such as weathering of rock, runoff, and bank erosion, surface water may contain sediment, nutrients such as phosphorous and nitrogen, and heavy metals. Similarly, groundwater varies in its mineral content and may even contain nutrients leached from soil. Thus, water quality is naturally variable—but human impacts on the quality of surface water and groundwater are often more significant. Municipal sewage can introduce phosphorous and other nutrients, bacteria and viruses, and numerous chemicals into surface waters. Septic systems and agricultural activities can contaminate surface water and groundwater with nutrients and bacteria. Industrial facilities and waste disposal sites have been a major source of surface and groundwater contamination by a wide range of toxic substances.

Defining 'good' water quality is a challenge that involves considerable uncertainty. Ultimately, whether or not water quality is good depends on the intended use of the water and a determination of an acceptable level of risk to humans and aquatic systems. The *Canadian Environmental Quality Guidelines* (Canadian Council of Ministers of the Environment 2007) recommend levels of various substances that should not be exceeded in drinking water, in

freshwater aquatic environments, and in water used for irrigation and livestock watering. It is important to note that these guidelines are not legally enforceable. However, several provinces, including post-Walkerton Ontario, have used the guidelines as the basis for legal standards for drinking water quality (Christensen and Parfitt 2001).

Water quality management involves all levels of government and many water users. Historically, much of the concern has been with municipal and industrial point-source pollution. Considerable progress has been made in 'end of pipe' treatment of waste from these sources. In 1998, for instance, 97 per cent of the Canadian population served by sewers had some form of treatment, including 40 per cent with tertiary treatment (Statistics Canada 2003). Several highly toxic substances have been largely eliminated at source through bans in production or use (e.g., DDT and PCBs) or changes in industrial processes (e.g., elimination of chlor-alkali cells in pulp and paper mills). In contrast, non-point sources of pollution generally have been much more difficult to manage. Pollution from agricultural activities is certainly a major one. Promotion of voluntary adoption of agricultural best management practices has been successful up to a point, but recent incidents of contamination of municipal water supplies across Canada have drawn attention to the need for regulation of some agricultural activities. Atmospheric deposition of toxic substances is also a growing threat and one that is challenging to manage, particularly when some of the sources are outside of the country.

As with water allocation, the provinces have primary responsibility for water quality management in Canada. However, the federal government is also involved in some important respects. Notable, in a water quality context, are its responsibilities regarding boundary waters, interprovincial waters, and Aboriginal communities (see Chapter 4). The federal Fisheries Act conveys responsibility for maintaining the quality of fish habitat and prohibits the discharge of any substance deleterious to fish. The Canada Water Act provides for federal–provincial cooperation on management of water resources, including quality issues, and the Canadian Environmental Protection Act authorizes the federal government to regulate specific toxic substances. The federal and provincial governments share responsibility for public health, including the management of waterborne diseases. Municipalities are also very much involved in water quality management, most obviously through the provision of drinking water and sewage treatment services. It is worth noting, however, that because municipalities are creations of provincial governments, these local services are ultimately a provincial concern.

In this section, issues and challenges relating to water quality management are discussed at two scales: international water quality management in Great Lakes–St Lawrence River Basin and protection of the quality of municipal drinking water supplies, using the example of the contamination of the water supply of Walkerton, ON. These examples illustrate some major challenges for managing water quality, including building support for water quality management and integrating efforts among numerous stakeholders and levels of government and ensuring that local organizations have sufficient capacity to protect drinking water quality.

Bi-national Great Lakes Water Quality Management

The Great Lakes are great indeed, containing 22,800 cubic kilometres of water—one-fifth of the world's supply of fresh surface water (Environment Canada 2002). Over the past 300 years, the lakes have been subjected to various human stresses. Early disturbances were sediment and nutrient-loading from forest clearance and agriculture in the basin. Population growth and urbanization contributed nutrients and pathogens. By 1900, Lakes Erie and Ontario were sufficiently degraded that several fish species sensitive to eutrophic conditions, particularly reduced dissolved oxygen, had largely disappeared. Typhoid fever outbreaks in communities around all of the lakes were commonplace, with death rates 20 to 60 times higher than typical in northern European cities of the day. More recently, industrialization has contributed a variety of toxic substances threatening both aquatic organisms and human health.

Bi-national Great Lakes water quality management is significant because it reflects almost a century of collaboration between two countries. Motivated by the alarmingly high typhoid death rates, Canada and the US gave the newly created International Joint Commission one of its first references in 1912—to investigate the extent and causes of pollution in the Great Lakes and to recommend remedies. The commission's study was an impressive bacteriological investigation, involving the analysis of 18,000 water samples from 1500 locations around the lakes, which clearly identified untreated municipal sewage as the principal cause (Benidickson 2002). The report recommended that the two governments confer upon the IJC jurisdiction to regulate and prohibit such transboundary pollution. Despite nine years of negotiations, the two countries were unable to agree on implementation of this recommendation, primarily because the US was unwilling to transfer pollution regulatory jurisdiction to the IJC (Bilder 1972). Meanwhile, chlorination of drinking water, rather than sewage treatment, became the preferred solution to control waterborne diseases.

Following the 1912 reference, several other reference studies concerning Great Lakes pollution were completed, including the 1964 lower lakes reference, which formed the foundation for the 1972 United States–Canada Great Lakes Water Quality Agreement. While the IJC did not get implementation authority, as it had recommended in its reference report, the Water Quality Agreement did give the IJC new responsibilities, including monitoring implementation of the agreement. The establishment of specific water quality objectives was an important component in implementation. The focus of the 1972 agreement was on nutrient loadings to the lakes, and to the credit of both countries, very significant reductions in phosphorus concentrations have been achieved, largely through improvements in sewage treatment and limits on phosphates in detergents. By 2000, all of the lakes, with the exception of Lake Erie, met phosphorus concentration targets (Environment Canada and the United States Environmental Protection Agency 2001).

The agreement was renegotiated in 1978, placing more emphasis on toxic contaminants and an *ecosystem approach* to water quality management (see also Chapter 15). In the agreement, ecosystem refers to the interacting components of air, land, water, and living organisms, including humans, within the Great Lakes Basin. An ecosystem approach to water quality management is significant in several respects. It acknowledges that people are connected to the environment—their actions can affect water quality, and degraded water quality can impair numerous beneficial uses. It recognizes that the health of the Great Lakes is connected to the wider basin or watershed. The approach also recognizes that people are a critical part of the solution to water quality impairment and must be involved in managing the problem (see also Chapter 18).

In addition to emphasizing the ecosystem approach, the agreement called for the virtual elimination of discharges of persistent toxic substances. More than 360 chemicals have been detected in the Great Lakes, and 11 of them have been designated as critical pollutants because of their toxicity, persistence, ability to recycle in the environment, and tendency to bioaccumulate and biomagnify. A protocol, signed under the agreement in 1987, emphasized the importance of human and aquatic ecosystem health. It provided a mechanism for implementing an ecosystem approach through the development and implementation of Remedial Action Plans in a number of severely degraded harbours and other locations. Also, it required the development and implementation of Lakewide Management Plans, which focused on critical pollutants in open waters. Further, the protocol recognized the contribution to water quality degradation of non-point source contamination and airborne toxic substances.

Progress on toxic contaminants has been mixed (Environment Canada and the United States Environmental Protection Agency 2007). Levels of many chemicals in herring gull eggs have declined by 90 per cent. For instance, concentrations of dichlorophenylethylene (a breakdown product of DDT) in herring gull eggs in Toronto harbour have declined from more than 23 micrograms per gram (μg/g) in the early 1970s to less than 3 μg/g. These concentrations are still far above the federal tissue residue guidelines for the protection of wildlife of .014 μg/g of total DDT, including its breakdown products (Canadian Council of Ministers of the Environment 2007).

Atmospheric deposition of PCBs has also declined. Greater control over point sources of some contaminants has obviously helped to improve Great Lakes water quality. However, deformities in near-shore fish suggest that recycling of contaminants in sediments in harbours and other locations continues to be a problem, as does atmospheric deposition of many other chemicals and leaking of contaminants from waste disposal sites.

Some interesting and important shifts and accomplishments in Great Lakes water quality management are evident. The setting of specific water quality objectives and effluent discharge targets through the 1972 agreement was a significant accomplishment and a milestone in binational environmental management. The IJC's involvement in monitoring implementation progress has also been important, although in the absence of any regulatory control, the IJC must rely on moral suasion and fostering public pressure for enforcement. The initial attention to management of specific pollutants, such as phosphates, gave way to an approach that recognizes multiple contaminants and multiple sources and pathways of pollutants (e.g., atmospheric deposition, recycling, and point sources such as industrial discharges and waste sites). This growing recognition of the complexity of the water quality problem was evident in the 1978 agreement.

Remedial Action Plans

The Remedial Action Plans (RAPs), established for 43 of the most degraded locations around the Great Lakes, including 17 wholly or partly in Canada, are key mechanisms for implementing the ecosystem approach. The RAP process has encouraged stakeholders in designated Areas of Concern (AOCs) in the basin to develop and implement measures to restore and maintain specified beneficial uses, such as drinking water supply. An early review of the RAP program by Hartig et al. (1998) reported that more than half of the RAPs had demonstrated some success in ecosystem-based water quality management. Reflecting this success, two Canadian AOCs, Collingwood Harbour and Severn Sound, were 'delisted' in 1994 and 2003, respectively, based

on improvements in environmental conditions, and a third, Spanish Harbour, is currently in the recovery stage (IJC 2009).

Whether or not conditions in AOCs improve depends strongly on the willingness of stakeholders (individuals, industries, governments) to take actions and to change their behaviour. Sustained stakeholder involvement and support for the process depends on an ability to demonstrate progress. Cooperation and support are more easily obtained in smaller communities such as Collingwood, which are generally less complex politically and in which it may be easier to form partnerships among stakeholders and to reach consensus on issues (Hartig et al. 1998). In contrast, progress toward remediation of Hamilton Harbour has been very slow. The harbour's 500-square-kilometre watershed is home to 500,000 people and a number of large industries, including major steel producers (Stelco, Dofasco). Major problems are contaminated sediments, a legacy of decades of industrial and municipal discharges to the harbour, and degraded water quality from combined storm and sanitary sewer outflows. The costs of treating contaminated sediments alone could be as high as $1 billion, according to the International Joint Commission (IJC 1999).

Sustained public involvement and support is one critical precondition for success in efforts to clean up AOCs in the Great Lakes Basin (Mackenzie 1997). However, sustained leadership and funding from state, provincial, and national governments in the basin are also essential. In that respect, initiatives such as the Canada–Ontario Agreement Respecting the Great Lakes Basin Ecosystem are promising. This five-year agreement (2002–7) laid out environmental priorities and specific goals and actions. Among the many tangible results, releases of mercury, dioxins, and furans were reduced by 86 to 90 per cent compared to 1988 releases, more than 9,700 hectares of waterfowl habitat were protected and restored, and municipal waste-water pollution was reduced through a $500-million investment in treatment infrastructure by municipal, provincial, and federal governments (Canada and Ontario 2008). The agreement was renewed in 2007 and will remain in force until 2010.

Drinking Water Security: Lessons from Walkerton

Walkerton is a small community of approximately 4800 people located in southern Ontario. Like numerous other small towns in Ontario and across Canada, it has a municipal system that provides treated drinking water to its residents. The source for the town's water is groundwater, drawn from several nearby municipal wells. In May 2000, water contaminated by *Escherichia coli* O157:H7 and *Campylobacter jejuni* entered Walkerton's drinking water system through a municipal well located next to a livestock farm. Seven people died, and more than 2,300 became ill (O'Connor 2002a).

During the inquiry launched by the provincial government to investigate the tragedy, it became clear that the deaths and illnesses had been entirely preventable. The plant's operator was negligent and purposefully misled provincial authorities. At the same time, the inquiry concluded that the provincial government had contributed to the tragedy through cuts it had made to the Ministry of the Environment, the agency responsible for monitoring and regulating municipal water treatment systems. Between 1995 and 1997, the operating budget of the ministry had been cut by 44 percent (Kreutzwiser 1998). Interestingly, the inquiry concluded that the livestock farm that was the source of the contaminants was being operated properly. Had the municipal well been satisfactorily constructed and located, the contaminants likely would not have entered the drinking water system (O'Connor 2002a).

Justice O'Connor's Walkerton Inquiry examined both the causes of the tragedy and ways in which similar tragedies could be avoided in future. While the two reports together contain 121 recommendations, one important lesson that points the way toward enhanced drinking water security can be distilled from them: because activities on the land affect water quality and quantity, an integrated, multi-barrier approach to drinking water safety is necessary. As the tragedy demonstrated, the consequences of inadequate drinking water quality are extremely serious. Therefore, this multi-barrier approach must be based on legal standards and regulations. An additional lesson—which Justice O'Connor recognized but did not emphasize in his inquiry—is that not all local governments have the necessary skills, tools, and resources. In other words, some may not have the capacity to manage water effectively.

Land and water are closely interrelated systems. Some precipitation that falls on the land surface runs off into rivers and streams or is stored in wetlands and lakes. However, precipitation can also be absorbed by soil, and some of it will enter aquifers, where it will remain for varying periods of time. In some parts of the world, such as the Great Plains of the US, water in aquifers is, for all intents and purposes, 'fossil' water (de Villiers 1999). It entered the aquifers thousands of years ago and except for human pumping, does not interact with the hydrology of the surface. In contrast, in most of southern Ontario, groundwater and surface water are closely interrelated. Many rivers and streams depend on groundwater for a portion of their flow—sometimes their entire flow in dry summers. At the same time, contaminants on or under the land's surface, such as from buried wastes, road salt, or land application of manure, can be carried into many aquifers relatively quickly.

These facts were well known to water professionals long before the Walkerton incident. However, a study of the province's arrangements for water management and land-use planning up to May 2000 would lead one to conclude that they were two distinct and unrelated systems. Indeed, as noted earlier, this separation has also been a problem with water-taking. Water management was guided by provincial statutes such as the Ontario Water Resources Act, which regulated water-takings. This statute also regulated drinking water systems, but the focus was on treatment rather than source water protection. Land-use planning was guided primarily by the Planning Act and essentially is a local activity. Provincial policy statements briefly noted the importance of linking land and water in planning (Ontario 1997), but their effect on actual behaviour was minimal because they did not require source water protection or the integration of land and water management.

Along with numerous others in the water field (e.g., US EPA 1999; Conservation Ontario 2001; Christensen and Parfitt 2001), O'Connor (2002b) suggested a different approach. He recommended a multi-barrier approach to drinking water safety, which begins at the source and uses the watershed as an organizing framework. This approach draws on many of the principles of the ecosystem approach, which was introduced earlier in the context of RAPs on the Great Lakes (see also Chapter 15). For instance, it recognizes the critical role that water plays in ecosystems and that natural systems (in this case, the watershed) provide the most appropriate management unit. Additionally, it emphasizes that protection of water quality is not something that can be left to governments alone. Municipalities, conservation authorities (watershed-based water management organizations), and the provincial and federal governments play vital roles. However, water users—whether city dwellers or farmers—also have key roles. Of course, a watershed-based approach to source water protection also recognizes that it makes more sense to keep water clean at the source than it does to allow it to become degraded and then attempt to treat it in municipal systems.

GUEST STATEMENT

Integrated Watershed Planning Approaches for Addressing Uncertainty in Urbanizing Areas

Sonya Meek

Integrated watershed plans are becoming recognized for their role in guiding urban land-use planning decisions to ensure the continued protection of environmental systems and avoid conflict among the various water users. Many cities across Canada are experiencing rapid rates of urbanization in response to population growth. The conversion of natural or agricultural lands to urban uses, often coupled with increased demand for water supplies, can alter local watershed hydrology, thus posing threats to water security for human and environmental uses. Although many forms of uncertainty exist in growth planning, the process of integrated watershed planning offers several approaches for addressing this uncertainty while providing a sound basis for making informed decisions in a timely manner.

In general terms, integrated watershed planning offers all key stakeholders the opportunity to become engaged in the process of defining watershed goals and objectives and developing management strategies to achieve them, as appropriate for the given location. The resulting plan is based on an interdisciplinary and systems-based analysis. This approach establishes an understanding of water and its connections with other environmental, social, and economic systems.

The integrated watershed planning study completed for the Rouge River, located in the Toronto region, illustrates three typical forms of uncertainty in land-use planning and approaches used to deal with them (TRCA 2007):

1. **Possible futures:** To assess potential stresses on the watershed and determine effective management strategies, the Rouge River watershed study needed to anticipate future land use and climate. The precise extent and type of future land use in the watershed was not known at the time of the study, nor should it have been if in fact the plan was to guide these decisions. Although most global climate change models predict warmer and wetter conditions in the future for the Toronto area, there is a wide range of estimates of the magnitude and seasonal pattern of change. The study therefore defined *a series of potential future land-use scenarios*, representing progressively greater extents and different forms of urban growth. These scenarios included two climate scenarios that bracketed the range of climate change predictions. The careful choice of a reasonable number of scenarios allowed the study team to identify and focus on areas of relative sensitivity in the watershed.

2. **Limitations of science and predictive modelling capability:** An understanding of watershed systems and the order in which changes manifest themselves through these systems was available to guide an interdisciplinary analysis of the watershed's response to future scenarios. However, the degree to which relationships between systems can be quantified varies, as does our ability to define thresholds of integrity. Furthermore, there is a range of sophistication in the tools available to predict the effects of changes in one system on another, from computerized models to professional judgment. To help overcome the uncertainties associated with these limitations, the study team drew upon *multiple lines of evidence*, such as observations of similar patterns in results from different disciplines or neighbouring watershed studies, rather than relying only on modelling results from a single component study. Another important tactic was the *relative analysis* of scenarios rather than a reliance on absolute modelling predictions. In this way, any modelling errors were assumed to be treated equally across all scenarios.

3. **Effectiveness of technology:** The Rouge River watershed studies found that if urban development proceeds with the most innovative approaches to community design and stormwater management, it may be possible to maintain current environmental conditions, but some deterioration may still occur in parts of the watershed. Considering the heavy reliance on innovative and relatively untested technology, the multi-stakeholder team adopted a *precautionary approach* in its management strategy. It recommended that development should proceed at a pace that allows opportunities for necessary adjustments. It also called for active monitoring and evaluation of the performance of new methods and updated requirements for the next phases of development—a

form of *adaptive management*. A 'no regrets' aspect of the preferred management strategy involved active regeneration programs to increase the resilience of the natural systems to withstand the effects of human activities and climate change.

The multi-stakeholder and interdisciplinary characteristics of the integrated watershed planning process facilitated approaches for dealing with uncertainty in urban land-use planning. These approaches permit us to move forward while we continue to improve knowledge.

The term 'multi-barrier' is used to describe the management system advocated by O'Connor (2002b) because source water protection—keeping the water clean in its natural setting—is the first barrier. Subsequent barriers are put in place during treatment and distribution of water to consumers. For instance, in a watershed-based approach to source water protection, farmers would use best management practices to ensure that their activities minimized threats to water quality. Municipalities dependent on groundwater would develop wellhead and aquifer protection plans, which delineate areas where land-use activities could contaminate aquifers. These areas then would receive special protection. For instance, certain kinds of agricultural or industrial activities could be prohibited in critical zones, as would activities such as road salting and industries that used certain kinds of chemicals or practices (Witten and Horsley 1995). The cost-saving associated with keeping source waters clean can be substantial. For example, the cost of building new treatment plants for New York City was estimated at between US$3 and US$6 billion, while the cost of watershed management to keep the city's source waters clean is expected to be approximately US$1.5 billion dollars (Committee to Review the New York City Watershed Management Strategy 2000).

While source water protection should be the first barrier, appropriate technologies and legally enforceable standards remain key elements of drinking water security. For example,

contaminated water entered Walkerton's system through an improperly constructed and maintained well. This should have been detected by the treatment plant's operator but was not because he was poorly trained and negligent. Then, the contaminants should have been eliminated in the chlorination process, but the chlorinator was broken, and the water was allowed to enter the system untreated. None of these situations should have been allowed to occur.

Fortunately, the lessons of Walkerton were taken to heart in Ontario. In response to the recommendations of the inquiry into the tragedy, the province of Ontario strengthened provisions for drinking water safety in Ontario through its new Safe Drinking Water Act and instituted watershed-based source water protection under its new Clean Water Act. In 2009, local, watershed-based Source Protection Committees were completing inventories of their water resources, conducting risk assessments, and developing source protection plans that—*if implemented successfully*—will go a long way toward integrating land-use planning and water management and increasing drinking water safety in Ontario.

WATER SECURITY: THE WAY AHEAD

Water security is measured in terms of access to adequate quantities of water, of acceptable quality, for human and environmental systems. From this perspective, contamination and insufficient water for human uses and for the environment are threats to water security in Canada. This was demonstrated at various scales and in different contexts in the examples examined here:

- Irrigators and other water users in semi-arid southern Alberta depend on a limited surface water supply that they have to share among themselves and with their American neighbours. Unfortunately, the water allocation systems designed in the late 1800s and early 1900s to facilitate sharing of the water did not consider environmental needs. Thus, today's water managers and users are struggling to find ways to ensure environmental needs are met alongside human needs. This challenge will increase, because the rivers and streams of southern Alberta are fed from glaciers, and snow that melts in spring—both of which are being affected by climate change.
- Residents in Walkerton, ON, depend on an aquifer for their drinking water. The water security issue in this case relates to quality rather than supply. The aquifer provided an adequate quantity of water, but negligence and poor management led to contamination of that supply. Major steps have been taken to prevent another such incident in Ontario. However, implementation of this new approach will require a major, ongoing commitment on the part of the provincial government, local governments and watershed management agencies, and water users.

Two important lessons about water security emerge from the cases considered in this chapter. First, water security is enhanced when *integration* occurs—in other words, when key relationships are addressed. These relationships include the links between surface water and groundwater, between water and land, and, of course, between people and their environment. Second, it is clear that *governance* matters a great deal.

Water security is partly shaped by the physical availability and natural characteristics of the resource. However, an equally—if not more—important concern is the way by which people manage the resource. Attention to both integration and governance presents opportunities to address conflict and uncertainty.

Integration is Essential

The cases in this chapter illustrate that a critical factor shaping water security is the extent to which integration occurs. The term 'integration' is used throughout the water management literature to express this concern (Hooper, McDonald, and Mitchell 1999; Pahl-Wostl, Kabat, and Möltgen 2007; Ferreyra, de Loë, and Kreutzwiser 2008). Water management is *integrated* when important links and relationships are acknowledged in the decisions that resource managers make about water:

- Surface water and groundwater interact through the hydrologic cycle. Failure to recognize this relationship can compromise water security. For instance, the rule of capture that underlies water allocation among non-permit users in Ontario is fundamentally flawed from a hydrologic point of view. Over-pumping of groundwater can lead to reduced stream flows, with implications for fish habitat, recreation, waste disposal, and water users who depend on the stream for irrigation and other purposes. Water allocation systems that can better handle the surface water–groundwater relationships are better integrated, reduce conflict and uncertainty, and thus promote water security.
- Water interacts with land and other components of the biophysical environment. This was illustrated tragically in the case of Walkerton, where the municipal well was located in an area known to be susceptible to

contamination from nearby farms. Reflecting problems such as this, the Walkerton Inquiry stressed that the security of drinking water is enhanced by maintaining the purity of source waters and that the watershed provides the most appropriate focus for management activities. Similarly, a basic principle underlying the RAP process is that the quality of the water in the Great Lakes depends on a wide range of human activities in the Great Lakes Basin. A watershed-based approach to drinking water source protection, and an ecosystem approach to water quality management, can contribute to water security by more closely integrating land and water management.

Governance Matters

Another important lesson is that governance matters. Governance refers to the processes by which societies make decisions—in this case, decisions about water. Each case chosen for this chapter strongly illustrates that the quality of governance is a critical factor shaping water security. The quality of governance is determined by the design of institutional arrangements (such as treaties, laws, and organizations) and by the way in which decisions are made. Who makes those decisions also is a major factor shaping the quality of governance. Increasingly, it is being recognized that decisions that involve a broad range of stakeholders are more enduring than those made by a narrow set of interests.

- When confronted with the problem of sharing a scarce water resource across an international boundary, the governments of Canada and the US could have chosen many different approaches, including doing nothing (and living with an ongoing conflict). Instead, they worked together to design a treaty that has—for the most part—led to the equitable sharing of transboundary water resources for almost a century. This same treaty provided the basis for dealing with a range of problems not envisioned by its designers, most notably toxic contamination of the Great Lakes. At the same time, the system of governance developed to address water quality in the Great Lakes has accommodated a shift from decision-making that involved only governments to decision-making that involves all interested stakeholders through initiatives such as the RAP process. This has contributed to greater awareness of the problems among citizens as well as to effective solutions.

- While transboundary water allocation and water quality management between Canada and the US is not without its problems, it is an internationally recognized example of good governance. Unfortunately, uncertainty around NAFTA and its implications for bulk water exports and diversions continues to raise concern for transboundary water governance and Canada's water security.

- Residents of Ontario have also been confronted with evidence of weaknesses in the system of governance for water management in their province. The Walkerton case highlighted these weaknesses for drinking water supply. However, similar kinds of weaknesses exist in the realm of water allocation. Key challenges include finding ways to better consider land-use planning in water allocation decision-making (and vice versa) and to respond to current and future water shortages in ways that balance human and environmental needs more effectively. The watershed-based approach to drinking water source protection created by the Clean Water Act is a promising vehicle for addressing both water quality and water allocation issues. However, its success over the long term depends on continued commitment from landowners, municipalities, conservation authorities, and the provincial government.

FROM THE FIELD

Will we ever 'get it'?

The fundamental nature of the relationship of people with their biophysical environment and the significance of human behaviour for resource and environmental management have been recognized by scholars since the early 1950s. Yet, more than a half-century later, we still do not seem to 'get it', although we 'talk the talk' more than ever. A recent Conservation Ontario primer on what land-owners can do to protect water resources begins, 'Everything is connected through the water cycle and it is important to remember everyone lives downstream.' These are words to live by, but are people making this personal connection between water and their own behaviour? Are people doing all of the little things in their everyday lives that, collectively, can make a substantial difference in how water is used and managed? Ontario's post-Walkerton experience suggests that as a society we may still not be 'getting it'.

To be sure, there have been important changes in water governance, particularly in the regulatory regime for municipal water treatment. Recent provincial and federal support of water and waste-water infrastructure in Ontario has exceeded $2 billion. The multiple-barrier approach, promoted by Justice O'Connor in his Walkerton Inquiry report, has been embraced. Source water protection plans for municipal water systems will be in place across Ontario by 2013. The province has already provided municipalities and conservation authorities with more than $135 million in support of plan preparation.

To date, however, much emphasis has been on infrastructure investment, technical and legal concerns around water treatment, and in the case of source protection planning, technical dimensions of delineation of wellhead and surface water intake protection zones. However, are the keys to successful plan implementation well understood? Has a broader vision of watershed-based source water protection given way to a narrow focus on municipal water systems? While some $28 million in provincial stewardship assistance to landowners will have been expended between 2007 and 2011, it has been targeted to owners in proximity to municipal wellheads and intakes. The effort to promote water stewardship more broadly and to enhance people's appreciation of their connection to water has been led primarily by conservation authorities, public health units, municipalities, and several non-governmental organizations—with limited provincial financial support.

Importantly, very little attention has been given to the effectiveness of the messages being conveyed. Our sense, from several of our research projects during the past decade, is that we have a long way to go. A significant minority of urban dwellers do not know where their tap water comes from. Two-thirds of rural well-owners do not test their water for bacterial contamination at least annually. Well-owners' understanding of what they can do to protect the quality of the water in their wells is usually very limited. It is questionable if many people, whether urban or rural, have much of a grasp of what a 'watershed' is. We argue that a much better and broader public understanding of water and our intimate connection with it is a prerequisite to enhancing water security. That understanding alone, however, will not be sufficient. A creative, substantial, and sustained effort must be made to ensure that understanding is translated into enduring personal and collective behaviour that positively enhances security.

—Reid Kreutzwiser and Rob de Loë

REVIEW QUESTIONS

1. How might international trade agreements affect water allocation rules in Canada?
2. How does the constitutional division of powers between the federal and provincial governments in Canada affect how water resources are managed?
3. What does the Walkerton water contamination incident reveal about the importance of a multi-barrier approach to drinking water protection?
4. Why is the watershed often a relevant geographical unit in understanding and managing water resource problems?
5. Why does the International Joint Commission serve as an outstanding example of bi-national co-operation in managing shared water resources?

REFERENCES

Agriculture and Agri-Food Canada. 2002. 'Drought watch'. http://www.agr.gc.ca/drought/summ_e.html.

———. 2009a. 'Canada–United States cattle/beef comparison'. http://www.agr.gc.ca/redmeat/almrt 25cal_eng.htm.

———. 2009b. 'Livestock exported to the United States'. http://www.agr.gc.ca/redmeat/almrt4cal_eng.htm.

Alberta Environment. 2009. 'Approved water management plan for the South Saskatchewan River Basin (Alberta)'. http://www.environment.alberta. ca/documents/SSRB_Plan_Phase2.pdf.

Alberta Water Resources Commission. 1986. *Water Management in the South Saskatchewan River Basin: Report and Recommendations.* Edmonton: Alberta Water Resources Commission.

Allan, J.A. 1998. 'Virtual water: A strategic resource'. *Ground Water* 36 (4): 545–6.

Beaulieu, M.S. 2001. *Intensive Livestock Farming: Does Farm Size Matter?* Agriculture and Rural Working Paper Series, Working Paper no. 48. Ottawa: Statistics Canada.

Benidickson, J. 2002. 'Water supply and sewage infrastructure in Ontario, 1880–1990s: Legal and institutional aspects of public health and environmental history'. *The Walkerton Inquiry Commissioned Paper 1.* Toronto: Ontario Ministry of the Attorney General.

Bilder, R.B. 1972. 'Controlling Great Lakes pollution: A study in United States–Canadian environmental cooperation'. *Michigan Law Review* 70 (3): 469–556.

Brennan, R. 2007. 'Ottawa to rebuild troubled reserve'. *The Toronto Star.* 30 July. http://www.thestar.com/ News/article/241308.

Brodsky, M.H. 2001. 'A review of managing health risks from drinking water: A background paper for the Walkerton Inquiry [by D. Krewski, J.Balbus, D. Butler-Jones, C. Haas, J. Isaac-Renton, K. Roberts, and M. Sinclair] as it relates to the history of drinking water pollution outbreaks in Ontario'. Review prepared for Ontario Water Works Association and the Ontario Municipal Water Association.

Bruins, H.J. 2000. 'Proactive contingency planning vis-a-vis declining water security in the 21st century'. *Journal of Contingencies and Crisis Management* 8 (2): 63–72.

Canada and Ontario. 2008. *Canada–Ontario Agreement respecting the Great Lakes Basin Ecosystem. 2002–2007 Progress Report.* PIBS 6726e. Environment Canada and the Ontario Ministry of the Environment. Toronto: Queen's Printer.

Canadian Council of Ministers of the Environment. 2007. *Canadian Environmental Quality Guidelines.* Canadian Council of Ministers of the Environment. Updated.

Canadian Intergovernmental Conference Secretariat. 1999. 'Environment ministers meet at Kananaskis'. http://www.scics.gc.ca/cinfo99/83067000_e.html.

Chapagain, A.K., and A.Y. Hoekstra. 2003. *Virtual Water Flows between Nations in Relation to Trade in Livestock and Livestock Products.* Value of Water Research Report Series no. 13. Delft: UNESCO-IHE Institute for Water Education.

———. 2008. 'The global component of freshwater demand and supply: An assessment of virtual water flows between nations as a result of trade in agricultural and industrial products'. *Water International* 33(1): 19–32.

Chiotti, Q., and B. Lavender. 2008. 'Ontario'. In D.S. Lemmen et al., eds, *From Impacts to Adaptation: Canada in a Changing Climate 2007*, v. 6, 227–74. Ottawa: Government of Canada.

Christensen, R., and B. Parfitt. 2001. *Waterproof: Canada's Drinking Water Report Card*. Toronto: Sierra Legal Defense Fund.

Christensen, R., and A.M. Lintner. 2006. 'Trading our common heritage?' In K. Bakker, ed., *Eau Canada: The Future of Canadian Water Governance*, 219–39. Vancouver: University of British Columbia Press.

Coloumb, R. 2002. 'Water challenges for the 21st century'. *Water Science and Technology* 45 (8): 129–34.

Committee to Review the New York City Watershed Management Strategy. 2000. *Watershed Management for Potable Water Supply: Assessing the New York City Strategy*. Washington: National Academy Press.

Conservation Ontario. 2001. 'The importance of watershed management in protecting Ontario's drinking water supplies'. (Unpublished paper).

Coote, D.R., and L.J. Gregorich. 2000. *The Health of Our Water: Toward Sustainable Agriculture in Canada*. Publication 2020/E. Ottawa: Minister of Public Works and Government Services Canada.

Day, J.C., and F. Quinn. 1992. *Water Diversion and Export: Learning from Canadian Experience*. Geography Publication Series no. 36. Waterloo: Department of Geography, University of Waterloo.

de Loë, R.C. 1997. 'Practising the principles: Sustainability and the Oldman River Dam'. In D. Shrubsole and B. Mitchell, eds, *Practising Sustainable Water Management: Canadian and International Experiences*, 133–54. Cambridge: Canadian Water Resources Association.

———. 2005. 'In the Kingdom of Alfalfa: Water management and irrigation in southern Alberta'. In D. Shrubsole and N. Watson, eds, *Sustaining Our Futures: Reflections on Environment, Economy and Society*, 85–126. Geography Publication Series no. 60. Waterloo: Department of Geography, University of Waterloo.

de Loë, R., et al. 2007. *Water Allocation and Water Security in Canada: Initiating a Policy Dialogue for the 21st Century*. Guelph: Guelph Water Management Group, University of Guelph.

de Villiers, M. 1999. *Water*. Toronto: Stoddart Publishing.

Dyson, M., G. Bergkamp, and J. Scanlon. 2003. *Flow: The Essentials of Environmental Flows*. Gland, Switzerland: World Conservation Union.

Eggertson, L. 2008. 'Investigative report: 1766 boil-water advisories now in place across Canada'. *Canadian Medical Association Journal* 178 (10): 1261–3.

Elwell, C. 2001. NAFTA *Effects on Water: Testing for* NAFTA *Effects in the Great Lakes Basin*. Ottawa: Sierra Club of Canada.

Environment Canada. 1973. *Flood of June 1964 in the Oldman and Milk River Basins, Alberta*. Ottawa: Information Canada.

———. 1989. *Historical Streamflow Summary of Alberta to 1988*. Ottawa: Ministry of Supply and Services.

———. 2002. 'The nature of water'. http://www.ec.gc.ca/water/en/nature/e_nature.htm.

———. 2009a. 'Average daily domestic water use'. http://www.ec.gc.ca/water/images/manage/use/a4f4e.htm.

———. 2009b. 'Water intake in Canada, 2006'. http://www.ec.gc.ca/eau-water/default.asp?lang=Endn=851B096C-l.

Environment Canada and US Environmental Protection Agency. 2001. *State of the Great Lakes 2001*. Ottawa: Government of Canada and Government of the United States of America.

———. 2007. *State of the Great Lakes 2007*. Ottawa: Government of Canada and Government of the United States of America.

Environmental Commissioner of Ontario. 2008. *Getting to K(no)w. Annual Report 2007–2008*. Toronto: Environmental Commissioner of Ontario.

Falkenmark, M. 2001. 'The greatest water problem: The inability to link environmental security, water security and food security'. *Water Resources Development* 17 (4): 539–54.

Ferreyra, C., R. de Loë, and R. Kreutzwiser. 2008. 'Imagined communities, contested watersheds: Challenges to integrated water resources management in agricultural areas'. *Journal of Rural Studies* 24: 304–21.

Food and Agriculture Organization. 2000. *New Dimensions in Water Security—Water, Society and Ecosystem Services in the 21st Century*. Rome: Food and Agriculture Organization.

Foster, H.D., and W.R.D. Sewell. 1981. *Water: The Emerging Crisis in Canada*. Ottawa: Canadian Institute for Economic Policy.

Giordano, M.A., and A.T. Wolf. 2003. 'Sharing waters: Post-Rio international water management'. *Natural Resources Forum* 27 (2): 163–71.

Gleick, P.H. 2003. 'Global freshwater resources: Softpath solutions for the 21st century'. *Science* 302 (28): 1524–8.

Global Water Partnership. 2000. 'Towards water security: A framework for action'. In *Second World Water Forum*. 17 March. Stockholm: Global Water Partnership.

Goss, M.J., D.A.J. Barry, and D.L. Rudolph. 1998. 'Contamination in Ontario farmstead domestic wells and its association with agriculture 1: Results from drinking water wells'. *Journal of Contaminant Hydrology* 32: 267–93.

Hartig, J.H., et al. 1998. 'Implementing ecosystem-based management: Lessons from the Great Lakes'. *Journal of Environmental Planning and Management* 41 (1): 45–75.

Hoekstra, A.Y., and A.K. Chapagain. 2007. 'Water footprints of nations: Water use by people as a function of their consumption pattern'. *Water Resources Management* 21: 37–48.

Hooper, B.P., G.T. McDonald, and B. Mitchell. 1999. 'Facilitating integrated resource and environmental management: Australian and Canadian perspectives'. *Journal of Environmental Planning and Management* 42 (5): 747–66.

IJC (International Joint Commission). 1999. 'Hamilton Harbour Area of Concern status assessment'. http://www.ijc.org/comm/hamhar/hamharsa.html.

———. 2000. *Protection of the Waters of the Great Lakes: Final Report to the Governments of Canada and the United States*. Ottawa: International Joint Commission.

———. 2009. 'Great Lakes Areas of Concern'. http://www.ijc.org/rel/boards/annex2/rap_info.htm.

International Institute for Environmental Development. 2002. 'Drawers of water II: Thirty years of change in domestic water use and environmental health in East Africa'. http://www.iied.org/agri/dowrv-dowii.html.

Kee, V. 2001. *Intensive Livestock Operations in Quebec: A Discussion Paper*. Toronto: Sierra Club of Canada.

Kreutzwiser, R.D. 1998. 'Water resources management: The changing landscape in Ontario'. In R.D. Needham, ed., *Coping with the World around Us: Changing Approaches to Land Use, Resources and Environment*, 135–48. Geography Publication Series no. 50. Waterloo: Department of Geography, University of Waterloo.

Kreutzwiser, R.D., et al. 'Water allocation and the Permit to Take Water program in Ontario: Challenges and opportunities'. *Canadian Water Resources Journal* 29 (2): 135–46.

Lemly, A.D., R.T. Kingsford, and J.R. Thompson. 2000. 'Irrigated agriculture and wildlife conservation: Conflict on a global scale'. *Environmental Management* 25 (5): 485–512.

Lemmen, D.S., et al. 2008. *From Impacts to Adaptation: Canada in a Changing Climate 2007*. Ottawa: Government of Canada.

Lightfoot, D.R. 2000. 'The origin and diffusion of Qanats in Arabia: New evidence from the northern and southern peninsula'. *The Geographical Journal* 166 (3): 215–26.

Linton, J. 2002. *Canada on Tap: The Environmental Implications of Water Exports. A Report Commissioned by the Council of Canadians' Blue Planet Project*. Ottawa: Council of Canadians.

Lucas, A.R. 1990. *Security of Title in Canadian Water Rights*. Calgary: Canadian Institute of Resources Law, University of Calgary.

Mackenzie, S.H. 1997. 'Toward integrated resource management: Lessons about the ecosystem approach from the Laurentian Great Lakes'. *Environmental Management* 21 (2): 173–83.

Merrett, S. 2003. 'Virtual water and Occam's Razor'. *Water International* 28 (1): 103–5.

Milly, P., et al. 2008. 'Stationarity is dead: Whither water management?' *Science* 319: 573–4.

Minister of Industry. 2008. *Electric Power Generation, Transmission and Distribution 2006*. Statistics Canada catalogue no. 57-202-4. Ottawa: Ministry of Industry.

Mitchell, B., and D. Shrubsole. 1997. 'Practising sustainable water management: Principles, initiatives and implications'. In D. Shrubsole and B. Mitchell, eds, *Practising Sustainable Water Management: Canadian and International Experiences*,

1–25. Cambridge, ON: Canadian Water Resources Association.

Mitchner, E.A. 1973. *The Development of Western Waters, 1885–1930*. Edmonton: Department of History, University of Alberta.

Nikiforuk, A. 2007. 'On the table: Water, energy and North American integration'. Toronto: The Program on Water Issues, Munk Centre for International Studies, University of Toronto. http://www.powi.ca/pdfs/waterdiversion/waterdiversion_onthetable_new.pdf.

Nolan, L. 2005. *Buried Treasure: Groundwater Permitting and Pricing in Canada*. For the Walter and Duncan Gordon Foundation.

O'Connor, D.R. 2002a. *Report of the Walkerton Inquiry: Part One, The Events of May 2000 and Related Issues.*. Toronto: Ontario Ministry of the Attorney General, Queen's Printer for Ontario.

———. 2002b. *Report of the Walkerton Inquiry: Part Two, A Strategy for Safe Drinking Water*. Toronto: Ontario Ministry of the Attorney General, Queen's Printer for Ontario.

Ontario. 1997. *Provincial Policy Statement: Revised February 1, 1997*. Toronto: Queen's Printer for Ontario.

Ontario Ministry of Natural Resources, Ontario Ministry of the Environment, Ontario Ministry of Agriculture and Food, Ontario Ministry of Municipal Affairs and Housing, Ontario Ministry of Enterprise, Opportunity and Innovation, Association of Municipalities of Ontario, and Conservation Ontario. 2003. *Ontario Low Water Response: Revised*. Toronto: Province of Ontario.

Ontario Ministry of the Environment. 2005. *Permit to Take Water (PTTW) Manual*. Toronto: Queen's Printer for Ontario.

Pahl-Wostl, C., P. Kabat, and J. Möltgen. 2007. *Adaptive and Integrated Water Management: Coping with Complexity and Uncertainty*. Berlin: Springer.

Postel, S. 1996. *Dividing the Waters: Food Security, Ecosystem Health, and the New Politics of Scarcity*. Danvers: Worldwatch Institute.

Sprague, J. 2006. 'Great wet North?' In K. Bakker, ed., *Eau Canada: The Future of Canadian Water Governance*, 23–35. Vancouver: University of British Columbia Press.

Statistics Canada. 2003. *Human Activity and the Environment 2003*. Ottawa: Minister of Industry.

TRCA (Toronto and Region Conservation Authority). 2007. *Rouge River Watershed Plan—Towards a Healthy and Sustainable Future*. Report of the Rouge Watershed Task Force.

United Nations World Water Assessment Programme. 2006. *Water: A Shared Responsibility*. Barcelona: UNESCO and Berghahn Books.

US EPA (United States Environmental Protection Agency). 1999. *Safe Drinking Water Act, Section 1429 Ground Water Report to Congress*. EPA-816-R-99-016. Washington: United States Environmental Protection Agency, Office of Water.

Wescoat, J.L., Jr. 1997. 'Toward a modern map of Roman water law'. *Urban Geography* 18: 100–5.

Witten, J., and S. Horsley. 1995. *A Guide to Wellhead Protection*. ed. S. Jeer and E.K. Flanagan. Chicago: American Planning Association.

World Resources Institute. 2000. *World Resources 2000–2001: People and Ecosystems: The Fraying Web of Life*. World Resources Series. Washington: World Resources Institute.

PART 2

Enduring Concerns

Part 2 contains seven chapters related to fisheries, agriculture, forestry, wildlife, protected areas, mining, and urban environmental issues.

In Chapter 8, Dianne Draper examines the roles of different levels of Canadian governments, as well as the growing involvement of Aboriginal people, in marine and freshwater fisheries management. Subsequently, she reviews community efforts to participate in the governance of fishery resources. Ratana Chuenpagdee's guest statement then introduces the 'slow fish' concept, which emphasizes three actions: slow down fishing, scale down fisheries, and support small-scale fishing communities.

Iain Wallace and Mike Bklacich provide an overview in Chapter 9, another new chapter, of Canadian agriculture in a global context, the emergence of 'alternative' agriculture in Canada, and implications of climate change for the Canadian agricultural industry. The guest statement by Chris Bryant adds insights related to interconnections between agriculture and rural communities, with particular attention to climate change and regional development.

Chapter 10, focused on forestry, is also new. Kevin Hanna reviews Canada's forest types, then discusses the structure of the industry and the need for investment and innovation, with special attention to the role of tenure. In his guest statement, Roger Hayter argues that a key challenge is to be able to realize both economic and environmental values when managing the forest resource.

In Chapter 11, Graham Forbes observes that wildlife management is a broad field, extending from the interactions of species and habitat at scales of gene to community and to the ethics of killing for subsistence or sport. He notes that some conflicts result from competition with objectives for other resources, such as forestry, agriculture, and protected areas.

Phil Dearden then considers in the twelfth chapter the management of protected areas that often generates conflict and aspects of uncertainty involved in making management decisions. The national park system in Canada is the main focus of his analysis.

Mary Louise McAllister and Trish Fitzpatrick, in Chapter 13, examine minerals and mining. Their purpose is to review the challenges involved in moving the Canadian mineral industry in a more socio-ecologically sustainable direction. They review the role of different stakeholders and factors operating at all spatial and temporal scales related to the mineral industry's activities and consider their influence on shaping those activities.

In the final chapter of Part 2, Virginia Maclaren reminds us that with 80 per cent of Canadians living in urbanized areas, we need to pay attention to issues in our cities. Specifically, she considers opportunities and challenges related to waste management.

8

Marine and Freshwater Fisheries

Dianne Draper

Learning Objectives

- To recognize the vast extent of Canada's 'ocean estate' and freshwater fishery resources as well as some of Canada's responsibilities and efforts to seek their protection and sustainability.

- To appreciate the challenges (including uncertainty and conflict) that affect the highly interconnected marine and freshwater ecological systems and the human communities that depend on these systems.

- To understand some aspects of the nature and complexity of Canada's marine and freshwater management and potential changes possible in governance arrangements to achieve resource and community sustainability.

INTRODUCTION

Despite recent revolutionary discoveries about the workings of Earth's oceans, most of the sea remains mysterious, unknown, and unexplored. Yet, as the world's population has grown, people increasingly have exploited the 'deep frontier' (particularly the continental shelf areas) for food and resources—and left perceptible footprints in marine ecosystems as they have done so. Consider, for example, the extensive damage done by bottom trawling and other mobile fishing gear to the only known living siliceous sponge reefs in the world, located in the iceberg-scoured troughs of the continental shelf off British Columbia's coast.

First described by Canadian scientists in 1991, the four 9,000-year-old glass sponge reef complexes in Hecate Strait and Queen Charlotte Sound constitute a stable but structurally fragile ecosystem that provides three-dimensional cover (shelter from predators) and habitat for species such as young rockfish, lingcod, sea cucumbers, crabs, shrimp, and octopi (Conway et al. 1991; 2007). Areas of rich biodiversity, the 'living fossil' reefs in this 'seafloor Jurassic Park' contribute to the sustainability of local fish species (Geological Survey of Canada 2008).

Covering an area of about 1,000 square kilometres and consisting of interlocking glass-like structures that grow up to 21 metres tall in waters up to 255 metres deep, these reefs and the roles they play in maintaining healthy fish stocks and invertebrate communities remain a mystery, although research to determine their ecological and economic importance is ongoing (Department of Fisheries and Oceans 2000; Conway et al. 2007; Geological Survey of Canada 2008). Indeed, at the first Sponge Reef Symposium, held in October 2008, scientists shared their latest findings and focused on next steps in conservation and research (Sponge Reef Symposium 2008).

In 2002, after voluntary closures failed to protect the functional integrity of these internationally significant reefs from damage by the trawl fishery, the federal government (i.e., Fisheries and Oceans Canada, formerly the Department of Fisheries and Oceans and still referred to as DFO) closed the reef areas in Hecate Strait to bottom trawling, drawing the boundaries in close proximity to the reefs (Jamieson and Chew 2002). Scooping up bottom-dwelling species by dragging gear weighing more than one tonne along the ocean floor, bottom (ground) trawls cause the most severe ecological impacts of all fishing gear used in Canada (Fuller et al. 2009). Scientists estimate that more than 50 per cent of the glass sponge reefs have been damaged or destroyed by such fishing activity and equipment (Baker 2008; Hume 2008). In part, damage occurred because the size and shape of the reefs are different from what was estimated initially; after more precise mapping was undertaken, DFO responded by expanding the ground trawl closures to incorporate the reef areas that were not contained within the original closure boundaries.

Clearly, accurate mapping of sponge reefs and other sensitive marine habitats, and accurate establishment of boundaries, are vital steps in effective oceans and fisheries management. In this context, the February 2009 launch of Ocean in Google Earth (an initiative by Google, the National Geographic Society, the BBC, scientists, and other partners) is significant as a tool to map the oceans, as a means to understand more about human impacts on the Earth's life-giving oceans, and as a mechanism to generate more concern for the state of the oceans and the need to protect them.

With the 2005 discovery of additional, but smaller, glass sponge reefs in the shallower (95 to 140 metres deep) waters of the southern Strait of Georgia, there is an obvious need for action to ensure the long-term survival of these globally unique sponge reef complexes and to protect them from damaging human activities. Timely and decisive action is necessary, particularly since possible impacts from fishing operations, cable and pipeline installations, ocean dumping, gravel extraction, and potential oil and gas development have not been examined fully (Baker 2008; Canadian Parks and Wilderness Society 2008; Geological Survey of Canada 2008; Institute for Geology and Paleontology 2008). At the time of writing, the southern reefs remained unprotected, and no fishing closures were in effect. The long-standing efforts of the Canadian Parks and Wilderness Society (CPAWS) to secure permanent protection of the sponge reef complexes as Marine Protected Areas (MPAs) under Canada's Oceans Act, which would enable the reefs to attain World Heritage Site status, are important in protecting Canada's marine biodiversity, particularly in relation to expected climate change stresses on marine ecosystems (Department of Fisheries and Oceans 2000; Jamieson and Chew 2002; Institute for Geology and Paleontology 2008; Jessen and Patton 2008).

The glass sponge reefs example highlights some of the dimensions of uncertainty, complexity, and conflict that characterize the management of Canada's marine and freshwater fisheries. Scientific knowledge of the oceans and fresh waters that surround us, and the species and spaces within them, is incomplete; our ignorance of the existence of these sponge reefs reminds us that both known and unrecognized challenges may be associated with attempts to shape the future of the natural systems on which our survival depends. As well, conservation and fisheries interests (and other users) assess elements of our marine resources differently and hold different perspectives on the costs and benefits of using versus protecting global ecosystem values and services. Some interests might question whether fishing employment income justified structural and ecological damage to the reefs. Other interests might ask why the precautionary principle was not applied immediately to protect these globally unique reefs. (One of 27 principles contained in the 1992 Rio Declaration on Environment and Development, the precautionary principle indicates that when there are threats of serious or irreversible damage to the environment, a country should take protective action even if uncertainty exists). Twenty years after discovery of the first reef complexes, scientists, environmental non-governmental organizations (ENGOs), and citizens alike have wondered why the federal government has not yet instituted Marine Protected Area status under the Oceans Act to protect these invaluable environmental assets from damage by human activities.

Given that fisheries management really is people management (Birkes 2008) and given that our past decisions have set our present course toward the future, what innovative thinking and decisions must we now undertake to ensure the long-term well-being of the oceans, their fisheries, and fishers? These same questions are echoed with respect to freshwater fishery resources, since managers of both marine and freshwater environments and species face continuing conflict and uncertainty as they attempt to establish ecologically effective, economically efficient, and socially equitable solutions to the complex challenges they face.

Matters such as climate change, overexploitation of fishery resources, competing uses of aquatic environments, lifestyle changes, and globalization of trade and economies are among the challenges faced by fisheries managers, fishers, and local fishing communities around the world. In this chapter, the principal focus is on aspects of Fisheries and Oceans Canada's efforts to steward fishery resources and to safeguard

the interests of fishery-dependent communities, particularly in the marine context. It must be noted, however, that Canadian fisheries management is an extremely complex, multi-dimensional, and interdisciplinary subject; it is impossible to do it justice within one chapter.

First, a brief overview section identifies some of the threats facing Canada's marine and freshwater fishery resources and highlights the need to think differently about how fisheries are managed. While DFO typically has attempted to address the conflicts and uncertainties in managing fisheries through regulatory measures, shortcomings were recognized, and a new policy framework (Canada's Oceans Strategy) was developed, based on the principles of sustainable development, integrated management, and the precautionary approach (Fisheries and Oceans Canada 2002).

As resources have permitted, DFO has attempted to implement these three principles, but there is evidence of dissonance between strengthening the economic performance of fisheries and achieving sustainability. The relatively small contribution that marine and freshwater fisheries make to the Canadian economy is described, and the contrast with historic roles of fisheries as significant sources of community employment and sustainable livelihoods is noted. Reprising those roles for fisheries in the near future calls for proactive options that recognize 'fish and people' are an integrated system. Ratana Chuenpagdee's guest statement in this chapter, '"Slow Fish" for sustainability,' is a case in point.

DFO also faces complexities, conflicts, and uncertainties in fisheries management that derive from Canada's international treaty obligations as well as from its own operational deficiencies. Two examples are discussed briefly: extending Canada's exclusive economic zone beyond 200 nautical miles and protecting freshwater fishery habitat and resources. The chapter concludes by discussing the West Coast Vancouver Island Aquatic Management Board, a cooperative initiative designed to move the governance agenda forward in terms of integrated management activities on the Pacific coast. The need to manage fisheries as complex adaptive systems in order to achieve sustainability while reducing the incidence of uncertainty and conflict remains a challenging goal.

CANADIAN FISHERIES

Threats Facing Canada's Fishery Resources

Until both the east and west coasts of Canada experienced variable declines in the production and value of fish and sea products beginning in the early 1990s, Canada's apparently bountiful ocean and freshwater resources had provided consistent support for the lifestyles and livelihoods of Inuit, First Nations, and Métis cultures as well as other communities of fishers. Harvesting and processing of various fish and shellfish species, marine mammals, and seabirds for subsistence, ceremonial, and commercial purposes had contributed significantly to the life and economy of the nation. However, as DFO's fisheries management decisions in the 1970s and 1980s continued to emphasize production regimes, and as certain stocks and species came under increasingly intense and sometimes rapid exploitation, their 'inevitable' collapse pointed toward the need to think differently about fishery resources and their management.

These collapses brought the sustainability of Canada's fish stocks and fishing communities into question and revealed glaring gaps in our knowledge and understanding of the 'natural' and the human interactions that characterize our marine and freshwater fisheries environments and ecosystems. Physical scientists still

do not understand fully the nature and implications of environmental changes occurring in and affecting the ocean and freshwater ecosystems and their biodiversity; neither do social scientists understand, fully, the effects of such changes on human communities or the effects of human communities on ecological systems (Perry et al. 2008a, 2008b; Perry and Ommer 2008). The ways in which human communities have responded to ecological and social changes, over both the short and long term, have helped to focus attention on the need for citizen participation in adaptive fisheries management (see also Chapter 16) and the consequent drive toward new governance approaches (see also Chapter 2).

As research continues into natural uncertainties such as ocean–atmospheric interactions that result in ocean water temperature changes, shifts in ocean salinity and acidity, rising sea levels, and ozone depletion (all of which affect marine environments and marine life), human effects on ecosystem integrity and resource sustainability continue to compound the management challenges. For instance, human economic activities pose significant, ongoing threats to marine ecosystem integrity and biodiversity through the discharge of industrial and chemical effluents, municipal sewage, and agricultural runoff; oil and gas exploration and extraction; shipping activity; ocean dumping; logging and mining activities; and recreational and residential developments in coastal environments. Excessive harvesting of wild species has resulted in the depletion of Atlantic cod (*Gadus morhua*) and haddock (*Melanogrammus aeglefinus*) in the northwest Atlantic Ocean, while globally, 76 per cent of the 600 marine stocks monitored by the United Nations Food and Agriculture Organization are depleted, overexploited, or fully exploited—meaning they are not sustainable (United Nations Food and Agriculture

Organization 2007). Currently, the greatest threat to sustainability derives from undervaluing of ecosystems (Haggan 2008).

Are Canada's management processes and systems adequate and appropriate to protect (sustain) and manage the harvest of wild marine fish stocks? Do Canadians have confidence, particularly in the face of contemporary environmental change, that DFO's approach to biodiversity protection will secure the environmental health of these natural systems? Are social and economic 'extinction' the only options available to fishers and their rural coastal communities when fish stocks 'disappear'? Responding effectively to present and potential future fisheries management challenges requires Canadian governments not only to steward fishery resources but also to safeguard the interests of fishery-dependent communities (Senate of Canada 2005). Resolving the uncertainties and conflicts that exist and integrating both scientific questions and societal concerns in developing effective fisheries management approaches are necessities for sustainability of 'people and fish' in Canada.

Canada's inland (freshwater) fisheries resources are threatened too, principally by loss, degradation, and fragmentation of habitat, pollution, and alien invasive species (Rose 2005). Other sources of continued endangerment to fish include modifications of water flows within fish habitats because of construction of dams and other barriers related to water management activities, overharvesting of wild populations through commercial and recreational fishing, urban development activities that modify or convert habitat to alternative uses (e.g., construction in wetland or flood plain areas), and effects of climate change (Freese and Trauger 2000; Rose 2005). Agricultural, grazing, and ranching activities, extraction of resources in the logging and mining sectors, and the proliferation of infrastructure in urban areas (roads, housing,

light manufacturing plants) all contribute to losses in freshwater aquatic ecosystems and to the endangerment of fish species (Reynolds et al. 2006; Jelks et al. 2008). Uncertainties and conflicts associated with these threats require fishery managers to understand and enhance the roles of local fishers and fishery communities in strengthening management practices and improving management outcomes.

Addressing Conflict and Uncertainty in Canadian Fisheries

With coastlines on three of the world's oceans and thousands of fish-bearing lakes and rivers within our borders, Canadian efforts to address conflict and uncertainty in marine and freshwater fisheries management have been numerous and diverse. From the conduct of research on fish stocks and monitoring of catches, to establishment of fishing quotas and conservation areas closed to fishing, to affirmation of co-management agreements with Aboriginal peoples, to implementation of restructuring and adjustment programs, and to initiation of habitat protection and restoration projects, the federal government has played the key role in legislation, policies, and programs that govern Canada's marine and freshwater fish stocks. Aboriginal governance is also relevant, as is identified later in this chapter as well as in Chapter 4, and important management approaches and programs have been instituted as a result.

DFO's efforts to maintain stability in the complex biological, social, and economic system of Canada's fishery resources have occurred principally through the imposition of regulatory measures. However, the effectiveness of these measures in ensuring sustainability of Aboriginal, subsistence, recreational, and commercial fishery resources (i.e., wild fish stocks) and the communities that depend on those resources has continued to be challenged (Collins and

Lien 2002; Ommer 2002; Senate of Canada 2005; Perry et al. 2008a).

DFO recognized that a more effective approach was required to deal with the issues of failing ocean health, continued environmental degradation, growing user conflicts, and administrative and regulatory complexities (Fisheries and Oceans Canada 2007a). Thus, Canada's Oceans Strategy, called for under the Oceans Act and released in 2002, focused on oceans governance and the policy direction for Canadian oceans management.

The Oceans Strategy noted clearly that all Canadian citizens shared responsibility for achieving common objectives in oceans governance (Fisheries and Oceans Canada 2002). Within this strategy, DFO was committed to establishing institutional mechanisms (such as committees or management boards) in order to work collaboratively with associated federal government departments and with other levels of government. DFO also was to undertake integrated management planning to ensure ecosystems would be conserved and protected while generating wealth through ocean industries.

Significantly, the Oceans Strategy promoted stewardship and public awareness of the oceans and their resources and sought the involvement of all levels of government, Aboriginal organizations and communities, businesses, academia, non-governmental organizations, and citizens to promote oceans governance (Fisheries and Oceans Canada 2002). The strategy promised Canadians 'more direct involvement in policy and management decisions that affect their lives' and indicated that coastal communities would be actively involved in the development and implementation of sustainable oceans activities (Fisheries and Oceans Canada 2002). Given the broadening of management objectives that the strategy represents, given the expanded range of knowledge that the strategy implies will be used

in making decisions, and given that the participants and institutions involved will be linked in multiple ways, on multiple levels, as they attempt to deal with complexity and change, DFO's fisheries managers will need to learn to expect change and to manage for it. A shift to complex adaptive systems thinking would help to ensure productivity and resilience in the human and ecological systems DFO now co-manages (Birkes 2008; Brown 2008; see also Chapter 16).

According to the strategy, all ocean management decisions 'should' (not 'must') be guided by the three principles of sustainable development, integrated management, and the precautionary approach. The following discussion focuses on the first principle and the dissonance between achieving sustainability and strengthening the economic performance of fisheries: 'Not everything should be bought and sold on Bay Street. Fishing rights should be viewed as a heritage of coastal people not as another commodity like futures in pork bellies or something' (McCurdy, cited in Senate of Canada 2005, 4).

SUSTAINABILITY AND FISHERIES MANAGEMENT IN CANADA

Under the 1867 Constitution Act, the federal government manages fisheries in the public interest (because fishery resources belong to all Canadians). In 1868, the Canadian Parliament enacted the Fisheries Act, enshrining in law the conservation and protection of marine and freshwater fish and their habitat. Subsequent amendments to the Fisheries Act added specific habitat protection and pollution prevention provisions that, DFO claims, 'make the Act one of the strongest environmental laws in Canada' (Fisheries and Oceans Canada 2007b, 1).

Fisheries and Oceans Canada is entrusted by Parliament to administer all fisheries-related laws and to ensure the long-term sustainability of Canada's fisheries, fresh waters, and oceans, as well as to develop and implement policies and programs that support and renew both the governance and the economic viability of marine and freshwater fisheries (Treasury Board of Canada 2008). Three key statutes guiding DFO's activities are the Oceans Act (1997), the revised Fisheries Act (1996), and the Species at Risk Act (2002). The Oceans Act focuses DFO activities on oceans management, while the Fisheries Act relates to the management of fisheries, habitat, and aquaculture, and as one of three federal agencies charged with responsibility for the Species at Risk Act, DFO manages aquatic species at risk (see also Chapters 11 and 12). Other Acts and associated regulations and orders also confer various responsibilities to DFO in its management of more than 300 fish stocks (Senate of Canada 2005).

The current work that DFO undertakes, and reports to Parliament about, is based on its 2005–10 Strategic Plan, *Our Waters, Our Future*, which sets out the department's corporate (business) objectives and its new integrated approach to planning. In that plan, DFO's vision is to provide 'excellence in service to Canadians to ensure the *sustainable development* and safe use of Canadian waters' (Fisheries and Oceans Canada 2008a, emphasis added).

As DFO pursues the principle of 'sustainable development', its main focus is to 'support the building of a *strong economy* while protecting Canada's natural environment' (Fisheries and Oceans Canada 2008a, emphasis added). As an ongoing priority of the government of Canada, DFO is responsible to follow the directive that '*development is essential* to satisfy human needs and improve the quality of human life, but must be based on the efficient and environmentally responsible *use* of all of society's *scarce* resources' (Fisheries and Oceans Canada 2008a, emphasis added). While DFO recognizes the need to integrate environmental, economic, and social dimensions in planning for the future (to ensure benefits to this generation and those to come) and to involve Aboriginal people in decision-making about fisheries, and while DFO acknowledges the need to change the way it conducts its business to account for the impacts of its decisions on fisheries, its strategic plan directs the department to continue using ocean and freshwater resources to 'support *economic development activities* such as fishing, aquaculture, oil and gas development, navigation, ecotourism, forestry and urban development' (Fisheries and Oceans Canada 2008a, emphasis added).

Among the major impediments to thinking differently with regard to fishery resources and management is that continued economic growth has been a long-standing priority of the Canadian government. This priority was reiterated in terms of Canada's fisheries and oceans when, following delivery of the budget in 2009, Minister Gail Shea of Fisheries and Oceans noted that the federal government's plan to respond to challenges in the fishing industry was 'focused on supporting *development and growth*' and that the areas in which government investment would occur (including funds to boost the Coast Guard fleet, improve small-craft harbour infrastructural elements, and enhance access to credit for harvesters and processors) were those that would 'offer the most direct benefits to the Canadian *economy*' (Shea 2009, emphasis added). Since fish and fishers are an integrated system, has (or will) DFO's major focus on privatization of a common property resource for '*maximization of net income* from a specific quantity of fish' (Senate of Canada 2005, 5, emphasis added) result in sustainability of fishery resources and fishing communities? Not according to a witness at a 2005 Standing Senate Committee on Fisheries and Oceans hearing, who stated: 'The economistic thinking of DFO is very well established and its effects are very much negative effects on rural communities, on coastal communities' (MacInnes, cited in Senate of Canada 2005, 4).

DFO, like all federal government departments, reports to Parliament on the basis of its Program Activity Architecture (PAA). The PAA outlines how DFO's activities relate to the three strategic outcomes it has established and how DFO manages its resources to achieve those intended outcomes. DFO's three strategic outcomes are: (1) safe and accessible waterways; (2) sustainable fisheries and aquaculture; and (3) healthy and productive aquatic ecosystems (Treasury Board of Canada 2008). Each is carried out through 'program activities' and 'sub-activities'. For instance, under the strategic outcome of sustainable fisheries and aquaculture, DFO's PAA identifies three program activities: fisheries management, aquaculture, and science for sustainable fisheries and aquaculture. Activities and sub-activities of DFO's programs dealing with the strategic outcomes of sustainable fisheries and healthy ecosystems, as well as planned spending on these activities, are identified in Table 8.1.

DFO's 'sustainable fisheries and aquaculture' outcome is expected to deliver services that contribute to 'sustainable wealth for Canadians' and that 'create the conditions for improving the economic viability and performance of the fishing and aquatic sectors while ensuring sustainability' (Treasury Board of Canada 2008, emphasis added). The strategic outcome, 'healthy and productive aquatic ecosystems,' aims to provide 'sustainable development and integrated management' of fishery resources, in which sustainable development is defined as 'supporting a balanced approach to a wide range of economic opportunities while meeting important environmental protection needs and supporting the social needs of communities' (Treasury Board of Canada 2008, emphasis added). DFO's program activity, 'fisheries management', indicates DFO is

responsible for 'developing and implementing policies and programs to ensure the *sustainable use* of Canada's marine ecosystems' and to effect '*conservation* of Canada's fisheries resources *to assure sustainable resource utilization*' (Treasury Board of Canada 2008, emphasis added). Stated bluntly by the then minister of fisheries and oceans, Loyola Hearn, DFO's goal is to '*stimulate substantial growth in the industry's value* in an environmentally sustainable manner by *removing and/or reducing developmental constraints* and creating the necessary conditions for industry success' (Treasury Board of Canada 2008, emphasis added).

Not only does the preceding statement appear to carry an echo of the days prior to the groundfish collapse, when scientific monitoring and assessment advice were 'moderated' by politics and economics, but Hearn's statement also skirts or neglects key questions about what is meant by reduction and/or removal of developmental constraints. What constraints have been identified, what roles do they play, and how will they be removed? In what ways will sustainability of fish, fishers, and their communities be affected? Will conflicts between large- and small-scale fishers' enterprises (for access to locally adjacent resources, for example) be reduced as a result of easing developmental constraints? Will there be any less uncertainty about the long-term future of small-scale fishing communities and their capacity to promote employment and sustainable livelihoods (see the guest statement by Ratana Chuenpagdee in this chapter)? Will easing constraints help to enable the insights and specialized knowledge of local people to be employed in decision-making about the future of the fish stocks in their locales (see Neis and Lutz 2008)?

As Chuenpagdee (this chapter) and others (Birkes 2008; McCay 2008) have noted, different objectives (*viz.* economic growth and resource

Table 8.1 Fisheries and Oceans Canada: Selected Elements of Program Activity Architecture and Planned Spending

Strategic Outcome	Program Activity	Sub-activity	Planned Spending (on Program Activity: Millions of Dollars)

Sustainable fisheries and aquaculture	1. Fisheries management	• resource management
		• Aboriginal policy and governance
		  – Aboriginal fisheries strategy
		  – Aboriginal aquatic resource and oceans management
		  – policy and governance
		• salmon enhancement program
		• international fisheries conservation
		• conservation and protection
			• 2008–9: $385.8
			• 2009–10: $396.6
			• 2010–11: $388.6

	2. Aquaculture		• 2008–9: $5.0 • 2009–10: $5.0 • 2010–11: $4.9
	3. Science for sustainable fisheries and aquaculture	• fisheries resources • species at risk • aquatic invasive species • aquatic animal health • sustainable aquaculture science • genomics and biotechnology • science renewal	• 2008–9: $211.4 • 2009–10: $202.3 • 2010–11: $197.9
Healthy and productive aquatic ecosystems	1. Oceans management	• integrated oceans management • marine conservation tools	• 2008–9: $24.0 • 2009–10: $24.4 • 2010–11: $23.5
	2. Habitat management	• conservation and protection of fish habitat • environmental assessments • habitat program services • Aboriginal inland habitat program	• 2008–9: $109.3 • 2009–10: $103.8 • 2010–11: $100.2
	3. Science for healthy and productive aquatic ecosystems	• fish habitat • aquatic ecosystems • ocean climate	• 2008–9: $77.0 • 2009–10: $73.8 • 2010–11: $73.4

Source: Based on Treasury Board of Canada Secretariat. 2008. 'Fisheries and Oceans Canada 2008–9. Reports on plans and priorities'. http://www.tbs-sct.gc.ca/rpp/2008-2009/inst/dfo/dfo00-eng.asp.

sustainability) cannot be optimized simultaneously. Since environmental and sustainability concepts almost always seem to be expressed as the secondary components of DFO's outcomes (i.e., first economic growth, then sustainability of resources or communities, or conservation for the purpose of ensuring continued fish harvesting), the question arises: do DFO operations and processes assume that sustained exploitation is possible? Is a lack of conclusive scientific knowledge leading to confusion about how carefully any remaining fisheries must be 'developed'? Since many factors lie behind the collapse of groundfish stocks, their lack of recovery (after more than 15 years without fishing) may support the biological hypothesis that crustaceans, former prey species, may be feeding on larval-stage groundfish (see Choi 2008). However, since no consensus exists as to why this perturbed ecosystem seems to be 'hyper-stable' (almost ossified, too brittle to adapt) (Choi 2008), does the possibility exist that we have witnessed an ecosystem 'flip' to a new steady state? What are the implications of such a possibility?

Since we live in a world of endemic uncertainty (Myers 1990), it is critical that fisheries managers shift the ways they plan and adapt to expect the unexpected. Is such a shift possible in the 'over-organized' bureaucratic system within which DFO operates? Is the Canadian federal bureaucracy sufficiently 'open' to consider new governance perspectives and policy approaches (see Chuenpagdee, this chapter)? Will DFO be able to use science and local knowledge innovatively to help promote and build resilient governance systems capable of responding to global changes that affect marine ecosystems? Can human behaviour with respect to fisheries be managed effectively by DFO or any other agency?

While DFO labels itself as a 'sustainable development department [that] will integrate environment, economic and social perspectives to ensure Canada's oceans and freshwater resources benefit this generation and those to come', various elements of its mandate appear to confirm the precedence of economics over ecology (Fisheries and Oceans Canada 2008b). In identifying for Canadians what DFO does, for instance, the department indicated that it 'ensure[s] compliance with environmental standards and regulations *in support of economic development*' (Fisheries and Oceans Canada 2008b, emphasis added). One might have thought that ensuring compliance with environmental standards would be directed, in the first instance, toward protection of aquatic ecosystems.

While Minister Shea (2009) rightly noted that 'economic prosperity can't be achieved without

sustained and sustainable resources', a key question is whether the federal government actually recognizes the conflict between economic growth and sustainability. Since a strong correlation has been demonstrated to exist between economic growth and species endangerment (see, for example, Trauger et al. 2003; Lackey 2005), the stronger the human economy grows, the greater become the threats to fisheries resources (Rose 2005). Perhaps the 33 species of marine fish and 84 species of freshwater fish that were extinct, extirpated, endangered, threatened, or of special concern[1] in 2008 provide evidence of this correlation (Government of Canada 2008). Included among the endangered, formerly commercially harvested populations of marine fish on the federal 'list of wildlife species at risk' are specific populations of coho and sockeye salmon in British Columbia, Atlantic cod in Newfoundland and Labrador, and Atlantic salmon in New Brunswick and Nova Scotia (Government of Canada 2008).

As a result of the federal government's focus on economic growth, as well as a government-wide agenda to modernize and improve the way the public service and its programs are managed, accountability has become a prime expectation of all federal departments. While accountability for the stewardship of taxpayers' dollars is an important goal, the time, energy, and resources that federal departments spend in efforts to change their practices and to provide financial and human resources information to such agencies as the Treasury Board and Office of the Auditor General has increased. Fully one-third (five) of DFO's 15 priorities for 2008 to 2011 are management priorities that reflect the internal workings of the department. Conflict and uncertainty arise here too, as challenges are encountered in finding a meaningful balance between demonstration of fiscal accountability to government and social-ecological accountability to fishing communities, fish, and habitats. In a dynamic world, the ability of governance systems to be flexible and adaptable is vital; there is a danger that human control systems, like natural ecosystems, may become too rigid to adapt to rapidly changing conditions.

Chuenpagdee (in this chapter) confirms that the goals of fisheries management policies built on the dual elements of ecological sustainability and economic viability have been difficult to achieve. The major shift in governing perspectives she calls for was identified also by Chief Lucas of the Hesquiaht Tribe (west coast, Vancouver Island) who noted, 'All their [DFO] programs have titles like "renewal" or "revitalized" but under their management our communities are dying and all small-scale fishermen are facing extinction' (Lucas, cited in Senate of Canada 2005, 7).

The importance of incorporating basic principles and values in governance arrangements is indisputable, and slowing down and scaling down fisheries in support of small-scale fishing communities is among the alternative models of fisheries management that must be considered (Chuenpagdee, this chapter). New governance regimes that focus beyond economics to value ecosystems and their services and to support the development of local leaders and the collaborative rebuilding of trust and other social values would be part of proactive images for the future. What would sustainable outcomes be like if less emphasis were placed on maximizing economic development and use of fishery resources and greater prominence were given to ecological and social dimensions of the fishery? One example of the potential of such an approach is the West Coast Vancouver Island Aquatic Management Board. Their success is explored briefly in the section 'Toward Integrated Management'.

Since DFO's strategic plans emphasize the need to strengthen the economic performance of Canadian fisheries, the following section identifies the place of fisheries in the Canadian economy.

Fisheries in the Canadian Economy

Even prior to the 1990s declines in fish production, to ensure that future generations might still be able to consume ocean products, the federal government (DFO) had adopted restrictive measures such as licence buy-backs and quota programs. The effects of these measures, and other changes such as globalization of trade and climate change, have included restructuring of the workforce in the fishing industry (e.g., fewer small-scale independent operators), vertical integration in the industry, an increased emphasis on aquaculture, and shifts in the wild species harvested (Brown 2008; Kildow 2008; Perry et al. 2008b).

Various consequences of these changes on fishers and fishing communities have been identified, including fishers' reliance on government social welfare programs (such as the $1.9-billion Atlantic Groundfish Strategy from 1993 to 1997 and the $730-million Canadian Fisheries Adjustment and Restructuring Program in 1998); fishers retraining, retiring their licences, and exiting the industry; young people leaving their fishing communities because they have insufficient capital to enter the industry; rights-based fisheries, including individual transferable quotas (ITQs) and Aboriginal rights such as those obtained through the *Marshall* decision (see also Chapter 4); concentration of wealth in a smaller number of fishers; and political action, including organization of strong lobby groups, particularly in the wealthy fisheries, and use of blockades and strikes as a means of gaining access to and allocation of resources (Ommer 2002; Choi 2008).

Nevertheless, each of the commercial marine and freshwater, aquaculture, and recreational fisheries sectors contributes to the value of the fishing industry in Canada and the world. In 2005, 52,805 Canadian fishers harvested 1 per cent of the world's total volume of marine and freshwater fish (1.102 million metric tonnes, valued at $2.1 billion), while 3,920 aquaculturists produced 154,000 metric tonnes (less than 0.2 per cent of global production, valued at $715 million) (Department of Fisheries and Oceans 2007). More than 29,000 people were employed in fish-processing activities in 2005 (Office of the Auditor General 2009).

An estimated 85 per cent of Canada's catches and aquaculture production is exported to major markets in the United States, Europe (mainly the United Kingdom and Denmark), Japan, and China. In 2005, exports of fish products from marine, freshwater, and aquaculture sources reached $4.31 billion (Department of Fisheries and Oceans 2007). By 2007, however, the total production of ocean and freshwater fisheries continued to decline, to 1.019 million metric tonnes, and was valued at almost $1.9 billion (Fisheries and Oceans Canada 2008c).

Following the 1992 collapse of Atlantic cod stocks, and the subsequent moratorium on cod, crustaceans (lobster, snow crab, and shrimp) replaced groundfish (cod, flounder, halibut, sole) as the dominant commercial marine species harvested in Atlantic Canada. Although fishing is not the only factor controlling this system, because groundfish are natural predators of crab and lobster, reduced predation allowed rapid expansion of commercial fishing for major shellfish species.

By 2005, crustaceans accounted for 30 per cent of the total volume of landings but constituted almost 66 per cent of the total landed value in Canada (Department of Fisheries and Oceans 2007). Nova Scotia's marine fisheries (mainly scallop, herring, and shrimp) accounted for the largest proportions of the total volume (26 per cent) and the total value (35 per cent) of fish

landings in 2005, while Newfoundland and Labrador's volume of fish landings (mainly shrimp, queen crab, and mackerel) ranked second in Canada at 25 per cent of the total (Department of Fisheries and Oceans 2007). British Columbia ranked third in terms of volume of landings (mostly hake, salmon, and redfish species) and contributed 16 per cent of the total fishing value in Canada in 2005 (Department of Fisheries and Oceans 2007).

Major aquaculture production, particularly of salmon but also of mussels, trout, oysters, and clams, occurs in both British Columbia and New Brunswick. In 2005, the value of aquaculture production totalled almost $337 million in British Columbia and almost $231 million in New Brunswick (Department of Fisheries and Oceans 2007).

Contributing only 3 per cent of the commercial fishing value in Canada, freshwater fishing (mainly for walleye/pickerel, perch, whitefish, and trout) is most important in Ontario and Manitoba. In 2005, these two provinces accounted for 88 per cent of the overall $66-million landed value ($35 million and $23 million, respectively) of freshwater commercial species in Canada (Department of Fisheries and Oceans 2007).

Recreational fishing plays a larger role in the Canadian economy than commercial fisheries. In 2005, 3.2 million recreational anglers (78 per cent of whom were residents) harvested almost 72 million fish (including salmon, walleye, trout, perch, and bass) from the oceans, lakes, and rivers that flow within Canada. Recreational anglers contributed $7.5 billion to various local economies within all provinces and territories during 2005 (Fisheries and Oceans Canada 2007c).

Although the contribution of marine and freshwater fisheries to the national economy is relatively small,[2] marine commercial fisheries historically have been a vital source of employment in many isolated coastal communities. Prior to the collapse of groundfish stocks, for instance, half of Atlantic Canada's 1,300 fishing communities depended totally on Atlantic fisheries for their existence (Office of the Auditor General 1997). Between 1989 and 1996, when northern cod stocks off Newfoundland collapsed, fishing-related employment dropped more than 37 per cent, from 23,600 to 14,800 (Statistics Canada 2000).

Of the 16,547 fishing vessels active in 2005, 89 per cent were inshore vessels of less than 13.7 metres in length; they registered 54 per cent of the total Canadian landed value in 2005, while the 11 per cent of mid-shore and offshore vessels more than 13.7 metres long accounted for 46 percent of the landed value (Department of Fisheries and Oceans 2007). Traditionally, the use of inshore vessels provided significant community employment and sustainable livelihoods for fishers and their families. Employing the 'slow fish' for sustainability points raised by Chuenpagdee (this chapter), particularly the need to scale down fisheries and regain a balance between small-scale (inshore) and large-scale (mid-shore and offshore) fishing activities, would help to build viable, community-based fisheries for the future. In practice, will DFO's governance rhetoric support such a shift in perspective? Despite their small economic contribution, the politics and political economy of fisheries continue to be challenging (Pooley 2008), suggesting that fisheries managers would benefit from improved understanding of the values, interests, and practices of those who are attempting to maintain and rebuild local fisheries.

As a signatory to the 1982 United Nations Convention on the Law of the Sea (UNCLOS) and to the UN Food and Agricultural Organization's 1995 Code of Conduct for Responsible Fisheries, Canada is responsible for safeguarding the interests of its fishery-dependent communities through stewardship of fishery resources adjacent to these communities, providing preferential access to local fishers (particularly in subsistence, small-scale, and artisanal fisheries), and protecting the rights of fishers and fish workers to a secure livelihood. These requirements mean that DFO must address issues of uncertainty and conflict as the department strives toward sustainability to protect community interests. Doing so, however, remains a challenging task that requires consideration of alternative models of fisheries management, such as a shift to the 'slow fish' movement (Chuenpagdee, this chapter).

FISHERIES MANAGEMENT CHALLENGES IN CANADA

The nature, volume, and health of marine fish stocks available to Canadian fishers reflect the extent of the ocean area over which Canada has jurisdiction for harvesting and habitat protection, as well as environmental factors that affect the ocean environment and marine life. Similarly, the nature, volume, and health of freshwater fish stocks available to Canadian harvesters depend on the natural endowment of freshwater environments that support aquatic life as well as habitat and harvest management efforts. Given the complexity, uncertainty, and conflict that still surround the demarcation and possible extension of Canada's maritime boundaries, and the difficulties DFO has encountered in protecting freshwater fish habitat, these two issues are discussed as illustrations of fisheries management challenges in Canada.

Canada's Maritime Boundaries and Fishery Resources

At approximately 7.1 million square kilometres in area, Canada's 'ocean estate' represents a vast region within which Canada is permitted to exercise its sovereign rights under UNCLOS (Fisheries

GUEST STATEMENT

'Slow Fish' for Sustainability[3]

Ratana Chuenpagdee

Today we know much about marine and fisheries ecosystems and what has led to their deterioration. Numerous principles, approaches, and tools are employed at all levels to manage fisheries for sustainability. Why then are fisheries in peril worldwide? The question is no less relevant to Canada than to other countries where fisheries are the mainstay of its coastal communities.

Canada's Oceans Act corresponds well with the global efforts in addressing challenges and concerns in fisheries, including uncertainty and conflicts among resource users. In addition to incorporating co-management as part of the Oceans Strategy, interdisciplinary research methods, especially those aiming to integrate scientific knowledge with those of local and indigenous communities, are continuously employed. Yet achieving the goals of Canadian fisheries policies on building ecologically sustainable and economically viable fisheries has proven difficult. A major shift in the governing perspectives, including the creation of alternative images for the future, is required. 'Slow fish' is one such image.

For various lifestyle issues, such as food, dwelling, and travel, the 'slow movement' is being promoted to strike a balance between the right to enjoy, and respect for, the environment (Honoré 2004). The same can be attempted for the advancement of fisheries sustainability with the image of 'slow fish'. 'Slow fish' refers to three key actions: slow down fishing, scale down fisheries, and support small-scale fishing communities. 'Slow down fishing' means recognizing that the rate of growth and recovery of fisheries and ecosystems is incompatible with the exploitation rate. Slowing down the speed at which fishing takes place follows the precautionary principle in allowing time for managers and scientists to develop further understanding about ecosystems and how humans affect them. Importantly, it acknowledges long-term societal benefits (and not only those reflecting market values).

Next, 'scale down fisheries' deals with the need for reduction of fishing capacity, especially of fishing fleets that cause severe impacts on marine ecosystems, such as damage to habitats and by-catch. Studies show that ecological impacts from fishing using bottom-tending and mobile gear (trawls and dredges) are much more severe than other gear (hook and line, harpoon, dive) (Fuller et al. 2009). Although efforts have been made to minimize habitat and by-catch impacts, options of shifting from high-impact to low-impact gear need to be explored. Scaling down in fisheries also means correcting the imbalance between small- and large-scale fisheries by allocating more efforts to the sector that contributes to higher employment and sustainable livelihoods. According to some estimates (see Pauly 2006), about 70 per cent of global catches come from large-scale, industrial fisheries employing only a small fraction of people compared to small-scale fisheries.

In 'supporting small-scale fishing communities', the important contributions of the small-scale fishing sector to food security and viable livelihoods of local communities are acknowledged. As elsewhere, coastal fishing communities in Canada are faced with rapid transformation of their social environments caused by industrialization, coastal development, and globalization. These trends pose great challenges to maintaining their cultural identity and their ability to participate in resource management. Small-scale fishing communities are often marginalized partly because of the remoteness of their location, which weakens their marketing and political capacity. Appropriate programs to support small-scale fishing communities range from providing means to strengthen the market competitiveness of these communities to increasing education and enhancing options for livelihood opportunities in and outside the fisheries sector. Additionally, increasing the awareness of the general public about fisheries and ecosystem sustainability could create greater appreciation of the quality of the products from small-scale, locally owned and operated fishing enterprises.

The race for the last fish, the need to expand fisheries, and the neglect of the majority of fishers and their communities can be avoided by adopting a new 'interactive' governance approach focused on solving problems and creating opportunities through interactions among actors beyond governments (Kooiman et al. 2005). The emphasis should also be on the incorporation of basic principles and values to underpin desired governance arrangements. 'Slow fish' is an example of an image that could be promoted to help bring about fisheries sustainability. The actualization of 'slow fish', and hence the slowing down of the degradation of marine ecosystems and human communities that depend on them, will require bold proactive policies from decision-makers willing to make hard choices in the context of competing interests and values.

and Oceans Canada 2009). This ocean domain, embracing portions of the Pacific, Arctic, and Atlantic oceans, includes the longest coastline (almost 244,000 kilometres), the largest archipelago (the Arctic islands), and the second-largest continental shelf (about 3.7 million square kilometres) in the world. Other characteristics of Canada's three ocean environments are summarized in Table 8.2. Even in establishing the boundaries of Canada's ocean assets, however, complexity, conflict, and uncertainty exist.

As outlined under UNCLOS (the international regime giving legal powers and obligations to coastal states to manage and apply sound principles of resource management to the world's oceans) and as stated in Canadian law in the Oceans Act, six maritime zones make up our ocean estate: internal waters, territorial sea, contiguous zone, exclusive economic zone, continental shelf, and high seas (see Table 8.2 and Figure 8.1).

An additional maritime zone, the *area*, is the ocean floor beyond the continental shelf that UNCLOS has identified as the common heritage of humankind. No one state has sovereign rights to or jurisdiction over the area; rather, all states party to UNCLOS manage the resources of the area through the International Seabed Authority. The prime objective of the authority is to share benefits (royalties) derived from extraction of mineral resources in the area with developing nations. Royalties from exploitation of other non-living resources beyond 200 nautical miles on the continental shelf also are paid to and distributed by the authority (Fisheries and Oceans Canada 2009).

Normally, baselines used to make measurements of the sea are set at the low-water mark. Since Canada's coastline is highly irregular, an accepted practice is to use straight lines to join appropriate points on the coast. Canada has drawn its straight baselines using reference

Table 8.2 Characteristics of Canada's Three Ocean Environments

Characteristics	Pacific Ocean Environment	Arctic Ocean Environment	Atlantic Ocean Environment
Coastline length	• shortest coastline—approx. 27,000 km	• longest coastline—approx. 173,000 km	• coastline about 40,000 km
Continental shelf area	• narrowest shelf, 16–32 km wide • warmest waters (8–14° at surface) • very high biological productivity	• shelf area: >1 million km² • coldest ocean; lower organic biomass production • seasonal ice (1–2 m thick) • polynas (open water); important winter wildlife refuges	• shallow offshore 'banks' (<50 m deep); high biological productivity • relatively cold water • 10% of Grand Banks (250,000 km²) outside 200 n.m. EEZ
Major biological components	• 300 finfish species, including anadromous,[1] catadromous,[2] and other species • shellfish • seabirds • marine mammals • Aboriginal food fishery • aquaculture operations	• marine mammals (seals, whales) • seabirds • shellfish • large kelp biomass • whales hunted for food, cultural continuity • limited commercial potential	• groundfish,[3] pelagic species[4] • shellfish • marine mammals • seabirds • aquaculture operations
Commercial catches and values	Total Pacific catch (2007): 171,019 tonnes, live weight; value $293,940,000	Small commercial operations catch Arctic char, turbot; emerging fisheries	Total Atlantic catch (2007): 815,903 tonnes, live weight; value $1,593,618

Source: Based on Fisheries and Oceans Canada 2008c.

Notes:

1. Anadromous fish are hatched in rivers and streams, migrate to the ocean to live much of their lives there, and return to their birthplace to spawn and die (e.g., Pacific salmon).
2. Catadromous fish spend most of their lifecycle in freshwater but enter oceans to spawn (e.g., trout).
3. Groundfish normally occur on or close to the seabed (e.g., cod, flounder, halibut).
4. Pelagic fish or organisms such as plankton that swim or drift in the sea (e.g. herring, tuna, mackerel).

points permitted under UNCLOS (and set out under regulations in the Oceans Act). However, both the United States and the European Union dispute Canada's straight baselines that enclose the Arctic Archipelago, and Canada and Denmark disagree on the positioning of straight baselines in the Lincoln Sea (Nunavut) that relate to the sovereignty of Hans Island and surrounding waters. In the Beaufort Sea, Canada and the United States have a longstanding but 'managed' disagreement over the location of the maritime border between Yukon and Alaska; the dispute could escalate

if petroleum resources were to be found in the contested area.

Driven largely by the prospect of offshore petroleum wealth as well as access to fish, in early 2009 France indicated to the UN Commission on the Limits of the Continental Shelf that it would seek to extend its economic zone on the continental shelf off the French archipelago of St Pierre and Miquelon (located just south of the island of Newfoundland). In response, having previously contested the maritime boundary and accepted a 1992 International Court of Arbitration ruling, the Canadian government

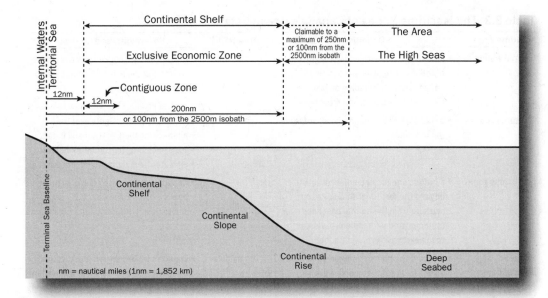

Figure 8.1 Schematic Drawing of Canada's Maritime Zones.

insisted that this boundary dispute already had been settled.

In the Bay of Fundy and the Gulf of St Lawrence, international legal considerations have meant that Canada has drawn 'fisheries closing lines' rather than baselines. Canada and the United States continue to 'manage' maritime boundary disputes off the British Columbia coast as well. These and other areas remain contentious.

As illustrated in Figure 8.2, Canada's continental shelf extends beyond 200 nautical miles (n.m.) from the baselines that enclose our internal waters. Article 76 of UNCLOS specifies the complex formula through which a coastal state may establish a new outer limit to its continental shelf and thereby confirm its sovereign rights in that area. Among other requirements, the thickness of sedimentary rocks in the shelf areas beyond 200 n.m. must be assessed to determine whether they form a natural extension of a state's land territory (United Nations Convention on

the Law of the Sea 2001). As a coastal state under UNCLOS, Canada has until 2013 (10 years from the time that UNCLOS was ratified by Canada) to prepare the scientific, technical, and legal details of its submission to the Commission on the Limits of the Continental Shelf.

Will the commission accept Canada's evidence that areas of the continental shelf located beyond the current 200 n.m. limit meet the criteria for extension of the outer limit of our exclusive economic zone (EEZ)? Uncertainty remains; other nations, such as Denmark, Norway, the Russian Federation, and the United States, are conducting their own scientific analyses of the seabed to determine their claims to sovereignty in the Arctic. While for some coastal states, gaining access to the potential energy resources located beyond the 200 n.m. limit appears to be a principal motivation for establishing sovereignty, extension of Canada's continental shelf would have fisheries implications as well. For instance, if Canada were able to claim an EEZ out

Table 8.3 The Maritime Zones of Canada's Ocean Estate

Maritime Zone	Definition	Area (km²)	Sovereignty and Control
Internal waters	• all waters landward of the straight baselines used in legally defining Canada's coast, including all lakes, rivers, and harbours, and most bays	2.5 million	• Canada has full sovereignty over these waters and the fish and habitat within them
Territorial sea	• 0–12 nautical miles (n.m.) seaward from the baselines	0.2 million	• Canada has sovereignty over the airspace, seabed, and subsoil in this zone, including fish and their habitat • all states have right of innocent passage
Contiguous zone	• a buffer zone located adjacent to and beyond the territorial sea, 12–24 n.m. seaward from the normal baselines (note: located within the first 12 n.m. of the exclusive economic zone)	—	• Canada may exercise control to prevent, or punish, infringement of its customs, sanitary, and other laws and regulations
Exclusive economic zone (EEZ)	• adjacent to and extending beyond the territorial sea, 12–200 n.m. beyond the baselines	2.9 million	• Canada's sovereign and jurisdictional rights include exploration, exploitation, conservation, and management of living and non-living resources in the waters above the seabed, in the seabed, and beneath the seabed (e.g., oil and gas)
Continental shelf	• the seabed and subsoil of the submarine areas that comprise the natural extension of Canada's land territory to the outer edge of the continental margin, or to a distance of 200 n.m. from the baselines, whichever is greater	—	• Canada may explore and exploit the mineral and other non-living resources of the seabed and subsoil, as well as living, sedentary organisms
High seas	• the area beyond the EEZ or the outer limit of Canada's continental shelf	—	• no state may claim sovereignty over any high seas area

Source: United Nations Convention on the Law of the Sea 2001; Fisheries and Oceans Canada 2009.

to the maximum 350 n.m. from the baselines, Canada would have jurisdictional rights to manage prime Atlantic Ocean fish habitat and harvest areas known as the 'Nose' and the 'Tail' of the Grand Bank (see Figure 8.2).

If Canada's case for extension of the continental margin were accepted, an incremental ocean area estimated at 1.5 million square kilometres would be included on the Atlantic and Arctic coasts (there is no extended continental shelf on the Pacific coast), giving Canada the total 7.1-million-square-kilometre ocean estate identified previously. Even with an enhanced

Canadian Coast Guard presence, how well would Canada be able to protect important ecosystems in our ocean estate from environmental degradation, particularly as new ocean resource uses are emerging? If our EEZ were to expand, how effectively would Canada be able to help combat the illegal, unreported, and unregulated international fishing that currently occurs on the high seas? Although DFO considers that Canada is 'an influential global leader' in international fisheries reform and oceans governance mechanisms, Canada has limited 'on the water' capacity to patrol the existing EEZ and even less capacity

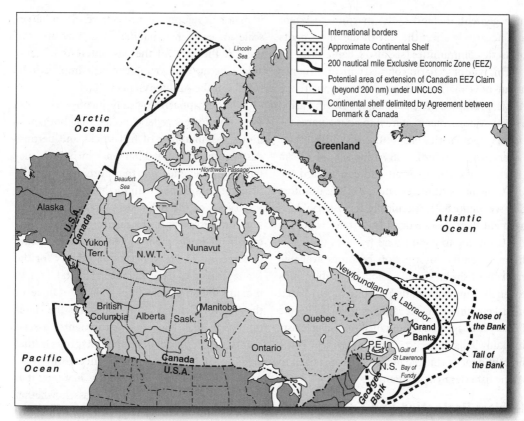

Figure 8.2 Canada's 'Ocean Estate' Showing Potential Outer Limit of the Continental Shelf.

to enforce national and international regulations within a potentially expanded EEZ. Canada must demonstrate capability as well as willingness to fulfill its international treaty obligations if our oceans leadership is to be fully credible and if sustainability of resources and fishery-dependent communities is to be achieved.

Of increasing consequence to Canada's jurisdiction in the Arctic is a specific provision known as the 'Arctic Clause' (Article 234 of UNCLOS) through which Canada enforces its laws and regulations (notably the Arctic Waters Pollution Prevention Act) to prevent, reduce, and control marine pollution from vessels operating in ice-covered waters. This Act allows Canada

to extend its pollution jurisdiction to 200 n.m. (from the current 100 n.m.).

Although commercial fisheries in the Arctic Ocean are limited (see Table 8.1), the ability to protect fisheries interests against potential effects of pollution is vital, particularly if climate change encourages greater shipping activity through the Northwest Passage and in other Arctic waters (see Figure 8.2). Furthermore, if climate change and the retreat of sea ice threaten to cause fundamental transformation of marine food webs in the Arctic (including the possibility that the range of commercial fish stocks could extend northward), commercial exploitation of marine resources could, without careful management, destabilize the marine ecosystem and negatively affect the way of life of residents of Arctic communities (see also Chapter 5).

This type of scenario, coupled with uncertainty about how marine ecosystems function in the Arctic, prompted the United States' North Pacific Fishery Management Council to adopt a precautionary approach to ensure the sustainability of fishery resources in the US Arctic EEZ off Alaska (north of the Bering Strait and east toward the US–Canada maritime boundary in the Beaufort Sea; see Figure 8.2). Specifically, its Fishery Management Plan closes federal (not state) waters in the US Arctic to commercial fisheries until sufficient scientific information becomes available regarding fish stocks and ecological relationships to enable implementation of ecosystem-based management of Arctic species (North Pacific Fishery Management Council 2009). Given existing knowledge gaps, how strongly evident is the precautionary principle in Canada's approach to Arctic fisheries management?

Habitat Protection and Freshwater Fishery Resources

Healthy habitat is a fundamental requirement for sustaining fish stocks (and also for human health, recreational use, and water quality) (see also Chapter 3). Yet, protecting fish habitat within the 9 per cent of our land surface that is covered by fresh water appears to have been a difficult and uncertain undertaking. Unfortunately, as the Commissioner of the Environment and Sustainable Development noted, 'the current state of Canada's fish habitat [in fresh water and estuaries] is unknown' (Office of the Auditor General 2009).

The objectives of the federal Fisheries Act include prohibiting the harmful alteration, disruption, or destruction of fish habitat and preventing pollution by prohibiting the deposition of harmful substances into our waters. Yet, even if major development projects result in 'large-scale losses of fish habitat', DFO may approve them if they are 'in the best interests of Canadians because of socio-economic implications' (Office of the Auditor General 2009).

While it appears logically inconsistent, DFO thus administers federal policy that both promotes sustainability of fish stocks and permits destruction of the habitat on which such sustainability depends. Furthermore, little monitoring of compliance with the habitat protection provisions of the Act takes place. For instance, the Commissioner of the Environment and Sustainable Development noted that after being in force for 23 years, DFO's 1986 Policy for the Management of Fish Habitat (Habitat Policy) remains incompletely implemented (Office of the Auditor General 2009). Since it is the Habitat Policy that triggers environmental assessments of proposed development projects that are known to have harmful effects on freshwater (and marine) fish-bearing habitats, it is significant that DFO has failed to assess whether its decisions on mitigating (damage reduction) measures and compensation (creation of habitat elsewhere) have been effective in meeting the principle of no-net-loss of habitat (Office of the Auditor General 2009).

Knowing that habitat loss and degradation are key threats to sustainability of fish stocks, why does DFO continue to authorize projects without accurate and timely information on the pre- and post-project state of fish habitat and without the ability to assess whether departmental actions are fulfilling the responsibility for fish habitat protection? Are approved projects causing more damage to fish habitat than the amount DFO authorized? Uncertainty continues about the effectiveness of the mitigating measures and compensation requirements to protect fish and their habitat (i.e., to achieve the purposes of the Act and policy). In turn, gaps in DFO's documentation standards and lack of compliance with its departmental control processes result in inadequate data by which effective evaluation of its own program decisions may be undertaken (Office of the Auditor General 2009).

Provinces, territories, and municipal governments have specific legislated roles and responsibilities related to fishery resources, while industries and conservation groups help to support the administration and enforcement of the Fisheries Act and its provisions, usually through agreements to undertake collaborative work on fish habitat protection and collection of habitat information. The Oceans Act contains explicit provision to ensure Aboriginal and treaty rights are neither abrogated nor derogated. But even as DFO has made progress in 'environmental modernization' (e.g., in streamlining environmental reviews of low-risk activities and strengthening its partnership arrangements), DFO's budget and personnel have been reduced, constraining improvements in habitat conservation and protection processes (see Table 8.2).

In Ontario, for example, 36 conservation authorities review low-risk project proposals and issue letters of advice on DFO's behalf. However, accountability mechanisms are lacking, so DFO could not ascertain whether assigned activities had been carried out in compliance with its policies and guidelines (Office of the Auditor General 2009). Although Aboriginal and stewardship groups, among others, collect habitat information, DFO lacks both the complete scientific data needed to establish a national baseline for the state of Canada's fish habitat and habitat indicators by which progress toward the Habitat Policy's long-term objective of a net gain in fish habitat may be assessed (Office of the Auditor General 2009).

That these inadequacies in DFO's management processes can negatively affect fish habitat is evident. For instance, in 1995 DFO approved the mining of 300,000 tonnes of gravel from Foster Bar in the Fraser River. Assuming that the removed gravel would be replaced naturally within one to three spring runoff periods, DFO's authorizations did not require compensation plans. By 2008, 'nature' still had not replaced the gravel.

In 2004, DFO and the province of British Columbia signed a five-year Fraser River Gravel Removal Plan Agreement. DFO, the province, local governments, and First Nations agreed to this gravel removal for flood control and erosion management purposes. However, the bars, islands, and secondary channels that are created as the shifting flows of the Fraser River move over these gravel deposits create high-quality habitat for at least 28 species of fish. DFO determined that gravel removal was harmful to fish habitat, yet ministerial authorizations for gravel removal were granted but without adequate information to assess possible impacts of gravel removal on fish stocks. In 2006, up to 2.25 million pink salmon were killed when the improper construction of a causeway to access one gravel removal site resulted in a downstream side channel drying up and exposing the salmon nests (Office of the Auditor General 2009).

Fisheries and Oceans Canada's inability to demonstrate adequate administrative control and enforcement of its habitat protection responsibilities under the Fisheries Act (Office of the Auditor General 2009) means that uncertainty exists regarding DFO's ability to protect fish habitat from the adverse effects of human activity. DFO needs to use these incidents, in combination with contemporary research, to adapt and improve its requirements for habitat compensation, to determine the effectiveness of its mitigation and compensation measures, and to ensure that its risk-based approach enables effective project monitoring and ascertains proponents' compliance with the Fisheries Act and all of DFO's project-specific terms and conditions.

TOWARD INTEGRATIVE MANAGEMENT

The Standing Senate Committee on Fisheries and Oceans reported in 2005 on DFO's plans to modernize Canada's marine coastal fisheries. The committee commented that part of DFO's strategic plan for fisheries management was to decrease its involvement in favour of greater industry involvement and co-management; DFO's role would evolve 'away from top-down management toward shared stewardship' (Senate of Canada 2005, 10). However, in early efforts to implement its new approach, DFO identified stakeholders as fishing licence- and quota-holders only, excluding the views of other stakeholders, including community interests.

The Senate Committee heard conflicting social and economic interests expressed by industry and community witnesses, with Chief Lucas indicating, 'We want a different kind of wealth; we want community wealth' (Lucas, cited in Senate of Canada 2005, 39). Since the Oceans Act specifies the importance of communities and stipulates that oceans must be managed collaboratively, coastal community interests in stewardship and conservation align with the ethic DFO says it promotes. Managing resources to benefit both present and future generations means that people in coastal communities adjacent to fishery resources must be involved in decision-making.

The Oceans Action Plan (2005) pointed DFO toward new oceans governance arrangements (specifically, 'integrated management' under the Oceans Act). DFO defined integrated management as a 'collaborative planning process that brings together interested stakeholders and regulators to reach general agreement on the best mix of conservation, sustainable use and economic development of marine areas for the benefit of all Canadians' (Fisheries and Oceans Canada 2006). Since fisheries management failures are due to institutional failure at least as much as to science failure (Jarre et al. 2008), a shift toward collaboration and inclusivity as mechanisms to help reverse the neglect of the fishing majority and their communities (Chuenpagdee, this chapter) is appropriate.

In 2002, in response to eight years of regional activism and demands for consideration of different approaches to aquatic resources management and for an enhanced role in decision-making in British Columbia—from coastal communities, First Nations (particularly the Nuu-chah-nulth), the provincial government, and public interest groups—the Aquatic Management Board (AMB) for the West Coast of Vancouver Island was established. DFO, the BC provincial government, Nuu-chah-nulth First Nation, and local governments collaboratively negotiated the terms of reference for the AMB.

Initially a three-year pilot project and the first board of its kind in Canada, the AMB provided a forum for coastal communities and others to 'participate more fully with governments in all aspects of the integrated management of aquatic resources' (West Coast Vancouver Island Aquatic Management Board, n.d.). The board consisted of eight non-governmental members representing fish harvesting and processing, tourism, environmental, labour, and aquaculture interests as well as eight governmental members, two each from federal, provincial, Nuu-chah-nulth, and local governments.

This diverse group of board members agreed to support the Nuu-chah-nulth principle of *hishukish ts'awalk* (everything is one), ecosystem management, and the Oceans Act principle of integrated management. They also agreed to adopt a second Nuu-chah-nulth principle, *isaak* (respect), and to work together in a fashion that respected their different perspectives as they integrated knowledge and expertise from both local and scientific sources (Pinkerton, Bedo, and Hanson 2005). Underlying use of these Nuu-chah-nulth principles is the belief that interactions with others should produce mutually beneficial outcomes (including ecosystem sustainability and biodiversity) that lead to continual improvement (West Coast Vancouver Island Aquatic Management Board, n.d.). The AMB's inclusion of these basic principles and values is noteworthy in the context of the 'slow fish' approach to governance (Chuenpagdee, this chapter).

A formal evaluation of the AMB's progress over its pilot period was conducted and concluded that the AMB is 'a unique and significant pilot effort in multi-party regional integrated aquatic management' (Pinkerton, Bedo, and Hanson 2005, i). In recommending continued support beyond its pilot stage, the evaluation team noted that the importance of the AMB's ambitious and complex work (see Table 8.4) stemmed from 'building a collaborative planning process with a wide spectrum of stakeholders and sectors' (Pinkerton, Bedo, and Hanson 2005, i).

The AMB's success in fostering effective working relationships between Aboriginal and non-Aboriginal fishing communities in the region derives from several sources. Among the 'intangible' success factors are the existence of goodwill and an ongoing commitment to management principles among board members and

Table 8.4 Selected Contributions of the West Coast Vancouver Island Aquatic Management Board to Integrated Fisheries Management, 2002–5

Projects	Social and Economic Benefits	Contributions to Integrated Fisheries Management
Gooseneck barnacle fishery *'blazes a path for innovative ways of rule-making, monitoring, and enforcement' (p. 7)*	• new experimental fishery • employed 32 harvesters, 12 jobs in monitoring, management, purchasing, transportation, processing (2004–5) • landed value of more than $180,000 • opening fishery required extensive research integrating local knowledge and natural science • raised $450,000 in funding to develop fishery	• revived a fishery that otherwise would be closed and probably poached • fishery is sustainable; monitoring of stock and habitat occurs; precautionary approach used • developed efficient electronic data collection on stock and habitat conditions, enabling rapid response to threshold warnings • AMB won harvesters' confidence by including their knowledge in generation of harvest rules
Web Atlas and documents database (www.westcoastaquatic.ca) *'[anyone] can instantly retrieve key information sources' (p. 7)*	• created Internet-accessible, integrated, geo-referenced database (>100 layers) in GIS format, describing terrestrial and marine environments, resource uses, and coastal communities • linked to a document database • allows the public to contribute its own digital mapping data • system has potential for significant contribution to future coastal zone planning and management	• enables integrated perspective on aquatic resources and habitats • supports government agency decision-making and integration • ultimately, AMB aims to include works by other regional agencies and groups, thereby reducing duplication, promoting co-ordination, and identifying gaps in service delivery • some First Nations groups want their use and occupancy research added to site

continued

Table 8.4 Continued

Projects	Social and Economic Benefits	Contributions to Integrated Fisheries Management
Economic development 'AMB *serves as a catalyst for other activities*' (p. 8)	• job creation: barnacle fishery plus 44 short-term and 7 permanent jobs • raised >$484,000 (plus barnacle fishery funds), leveraged ~$187,000, received >$193,000 contributions in kind, administered $210,000 in government-sponsored programs	• AMB activities encourage entrepreneurship, help groups form partnerships and launch fishery and management initiatives • charitable status enables diversification of funding sources, types
Applied ecological and social research and action [Board] '*contributes to the body of research . . . on wild fish*' (p. 8)	• applied research, information gathering, and restoration projects, including: – sea otter recovery – sea lice – mock oil spill response exercise – water quality impacts on shellfish – mapping (e.g. kelp, eelgrass, salmon stocks) – fish habitat restoration projects – First Nations training for snorkel surveys	• mapping projects help to protect local wild stocks; identify areas of high ecological, social, and economic values for proactive protection; provide information on productive capacity of habitats, data for treaty negotiations, and employment opportunities • developed oil spill training program available to all regional communities • research on how co-operative stewardship is developed, supported, and extended • habitat restoration works constructed
Administrative leadership 'AMB's *participation was acknowledged as resulting in a more truly integrated plan than was initially envisioned*' (p. 10)	• AMB facilitated events such as World Salmon Summit Satellite Conference– • oversees Area F Clam Management Board • facilitated Henderson Lake Fish Sustainability Plan • undertakes coastal zone planning	• economic benefits • develops new and innovative forms of resource management • brings wide variety of groups together to integrate information, knowledge, experience, perspectives, and interests • supplies background information; develops process options • undertakes comprehensive reviews of plans, gets feedback from all stakeholders at once, reminds people of their common vision to minimize partisan objections to plans, increasing likelihood of success

Source: Based on Pinkerton, Bedo, and Hanson 2005.

stakeholders. The long history of cooperation among the Nuu-chah-nulth people plays a role, as do the ongoing, collaborative planning processes that include all stakeholders in the board and in its decisions. By bringing together disparate sectors, avoiding 'separate management solitudes', and maintaining interpersonal relationships through the AMB, true integrated management takes place (Pinkerton, Bedo, and Hanson 2005).

The board's multiple leadership roles provide tangible evidence of its ability to make policy contributions to senior governments, reduce regional conflict, build capacity, and stimulate economic development. For instance, the AMB serves as a discussion and learning forum for the multiple sectors to voice their concerns and vent frustrations (and thus reduce conflict). Similarly, the board acts as a mediator in everyday interactions among local fishers, biologists, and DFO officials, reminding people of the benefits of working together toward long-term goals. The AMB's professional staff promote bottom-up rule-making by fishers such that harvest management problems are reviewed against principles and objectives and resolved, in turn promoting support of regulations and reducing enforcement costs (Pinkerton, Bedo, and Hanson 2005).

When Nuu-chah-nulth community concerns come to the board, it functions as a culture broker, promoting communication between parties with different perspectives and world views and merging local and scientific knowledge to promote understanding. The board has created new social, economic, and ecological protocols for managers; has educated and mobilized fishers, scientists, agencies, and the general public to think differently about linking fishery development with social sustainability and conservation; and, has promoted use of local knowledge in management decisions.

In gaining its credibility, the AMB has emphasized building relationships and healing rifts between stakeholders and communities, gaining stakeholders' trust in the board, contributing to economic development and ecosystem management, focusing on long-term sustainability of fishery resources, and developing the human and social capital necessary to generate new cooperative behaviours in the region that promote integrated aquatic management. Under its business plan (2007–12), the AMB continues to demonstrate that its form of governance will 'enable Canada to implement its obligations under the Oceans Act in a cost-effective manner' (West Coast Vancouver Island Aquatic Management Board 2007).

CONCLUSION

Resolving the conflict and uncertainty inherent in the demands of fishers and citizens for more effective approaches to the protection of marine and freshwater habitats and ecosystems requires collaboration among all levels of government (in their social and political roles as legislators and resource managers), scientists (as researchers and knowledge brokers/disseminators), communities (as knowledge sources and as the locus of coping and adaptive responses to change), and individuals (as participants in and agents of change).

In Canada there remains 'a long way to go' in fully developing the process of integrated fisheries management to collaboratively achieve sustainability (meaning progressive improvement in human and environmental affairs and not merely continued economic growth and expansion). The AMB has demonstrated to DFO that community-based people management promotes improved fisheries management. Effective fisheries management processes value ecosystems and their services and support development of local knowledge and leadership capacities and the advancement of new forms of governance that mature within and emanate from local and regional fishing communities. New governance models, such as those developed by the AMB, actualize the importance of values, principles, and trust as the basis for progress toward sustainability. Will other fishing communities in Canada be able to develop similarly effective governance regimes? Will DFO's policies help or hinder other communities in adjusting their governing approaches (perhaps to 'slow fish') and in becoming more resilient in the face of global changes? We need to remember: *hishukish ts'awalk*!

FROM THE FIELD

During the late summer of 2009, fishers eagerly were awaiting a bumper catch as 10.5 million sockeye salmon were expected to return to the Fraser River after spending two years in the open ocean. The anticipated run never materialized; about 9 million fish appear to have died at sea, making this the lowest return on record, and no one can explain definitively why. Ironically, pink salmon (about one-tenth the value of sockeye to fishers) migrated to the Fraser in higher than expected numbers. Perhaps nothing more clearly illustrates the challenges involved in conducting research on Canada's complex marine and freshwater fisheries resources and management issues than these apparently contradictory events.

Despite general belief that our science and predictive computer models have improved, the mysterious disappearance of the sockeye salmon reminds us vividly that we continue to base management decisions on an incomplete understanding of fishery resources and their environments. Natural scientists know almost nothing about what happens to juvenile sockeye salmon as they travel down the Fraser, past fish farms, and into the ocean; no one seems willing to fund the research required to answer these basic questions. While Canada's 2005 Wild Salmon Policy contains the tools to take precautionary actions to limit activities harmful to salmon, that policy has yet to be funded and implemented. Thus, the spectre of political considerations trumping science continues.

A bright spot in research is the integrative, interdisciplinary research collaborations among social scientists, humanists, and natural scientists that have attempted to offer solutions for institutional weaknesses in Canadian fisheries management. Research outcomes have included ethical, social, and environmental approaches, strategies, and procedures for fisheries management institutions that, if applied, could protect the legitimate interests of fish, fishers, and fishing communities. Yet, because these outcomes challenge the central power base of the federal government (DFO), they have been met with resistance (and limited research funding).

Community-based co-management systems strive to maintain or restore healthy fisheries ecosystems. Such systems also support sustainable communities, including small-scale fishers and their proven methods of sustainable production. Visible co-management success stories, such as the Aquatic Management Board (AMB) for the west coast of Vancouver Island, provide momentum to changing philosophical perspectives (that see fisheries as more than economic resources to be exploited for maximum gain; that promote 'slow fish' and other locally controlled, socially acceptable, and environmentally sustainable policy and management options) and shifting governance approaches (to ensure social justice). They represent positive trends in Canadian fisheries research and practice.

Society's tendency to label (fisheries) researchers as 'experts' brings with it significant responsibility to 'get it right'—to conduct research that yields accurate, timely, and relevant information and insight for innovative decision-making about marine and freshwater fishery resources, environments, and fishing-dependent communities. Remembering the concept *hishukish ts'awalk* (everything is one), the sockeye disappearance reminds us that humility is an important quality in those who are or would be leaders in research (even if academia does not reward them for working toward policy and management reforms). While not a common topic of discussion among researchers, humility embodies the notion of respect (*isaak*) and the hope that the work of scholars in the natural sciences, social sciences, and humanities will contribute to deeper understanding of fisheries issues in Canada.

—DIANNE DRAPER

Acknowledgement

I would like to thank Robin Poitras for preparing Figures 8.1 and 8.2.

Notes

1. COSEWIC (Committee on the Status of Endangered Wildlife in Canada) status categories used in the SARA (Species at Risk Act) public registry (with respect to marine and freshwater fish) are: extinct (wildlife species that no longer exist); extirpated (wildlife species that no longer exist in the wild in Canada but exist elsewhere); endangered (wildlife species facing imminent extirpation or extinction); threatened (wildlife species likely to become endangered if limiting factors are not reversed); and special concern (wildlife species that may become a threatened or an endangered species because of a combination of biological characteristics and identified threats) (Environment Canada 2009).

2. Statistics Canada does not include fish in its calculation of Canada's wealth because it 'lacks the necessary data and methods to assign a value to these resources' (Islam 2007). However, in conjunction with agriculture, forestry, and hunting, fishing industries contributed 2.2 per cent to Canada's Gross Domestic Product in 2005 (Statistics Canada 2008).

3. Based in part on: R. Chuenpagdee and D. Pauly, 2005. 'Slow fish: Creating new metaphors for sustainability'. In J. Swan and D. Gréboval, eds, *Overcoming Factors of Unsustainability and Overexploitation in Fisheries: Selected Papers on Issues and Approaches*, 69–82. International Workshop on the Implementation of International Fisheries Instruments and Factors of Unsustainability and Overexploitation in Fisheries. FAO Fisheries Report no. 782. Rome: Food and Agriculture Organization.

REVIEW QUESTIONS

1. Why is accurate mapping of Canada's ocean areas, and the resources they contain, of significance to Canada and Canadians? What kinds of conflicts and uncertainties have arisen (or might occur) if ocean boundaries were not clearly established or formally agreed upon? At which scale (local, regional, national, or international) do you think the most difficult conflicts relating to establishing boundaries might occur? What would those conflicts be about, and why would they be the most difficult ones to resolve?

2. If fisheries management really is people management, then what might be the roles of trust, respect, and other social and ecological values in: improving fisheries management? advancing the sustainability of Canada's fisheries? leading toward 'slow fish'? advancing the governance of our fishery resources? What factors could work against the potentially positive effects of trust, respect, and similar social-ecological values?

3. If you were a resident of a small, economically struggling fishing community and were given the opportunity to speak directly with the minister of Fisheries and Oceans Canada, what would you ask the minister to do to help change the future of your community? What factors would you need to take into account in making your request to the minister a 'balanced' one? What would your hoped-for outcomes include? What kinds of information might the minister need from you in order to respond effectively to your request?

4. Discuss the terms 'sustainable development' and 'sustainability' as they relate to DFO and its efforts to manage Canada's marine and freshwater fishery resources. What challenges do you think each term raises for DFO's responsibilities to plan and manage freshwater and marine fisheries in an integrated fashion for the benefit of all Canadians, both present and future?

5. If fisheries management failure results from institutional failure, in what ways might (or might not) the West Coast of Vancouver Island Aquatic Management Board's collaborative approach toward fisheries management help to overcome institutional failure? How does using an ecosystem approach help to advance integrated management of aquatic resources? Why is it important to integrate relevant local knowledge together with ecological, social, and economic information in adaptive management? Do you know of other communities or agencies that have achieved success using management principles similar to those the AMB used? What lessons about responsibility and accountability might DFO learn from the AMB (and perhaps others) that could be applied in other fishing communities?

REFERENCES

Baker, M. 2008. 'Undersea world of glass: Georgia Strait sponge reefs need protection'. *Wild at Heart* (CPAWS-BC newsletter) 19 (3): 6, 7.

Birkes, F. 2008. 'Coping with global change in marine systems: Exploring some conceptual issues'. http://web.pml.ac.uk/globec/structure/fwg/focus4/symposium/keynotes.htm. Forthcoming in R.E. Ommer, I.R. Perry, P. Cury, and K. Cochrane. 2010. *Resilience of Fisheries Systems to Global Change: A Social-Ecological Perspective*. Toronto: Wiley-Blackwell.

Brown, K. 2008. 'Vulnerability, adaptive capacity and resilience in marine and coastal social-ecological systems'. http://web.pml.ac.uk/globec/structure/fwg/focus4/symposium/keynotes.htm.

Canadian Parks and Wilderness Society. 2008. 'Sponge reef symposium: Scientists to highlight research on B.C.'s underwater "living dinosaurs"'. http://www.cpawsbc.org/press/10.27.08.hp.

Choi, J.S. 2008. 'State transitions of ecological systems as a driver of social change and how adaptive social change can help stabilize ecological systems'. http://web.pml.ac.uk/globec/structure/fwg/focus4/symposium/s1.htm.

Collins, R., and J. Lien. 2002. 'In our own hands: Community-based lobster conservation in Newfoundland (Canada)'. *Biodiversity* 3 (2): 11–14.

Conway, K.W., et al. 1991. 'Holocene sponge bioherms on the western Canadian continental shelf'. *Continental Shelf Research* 11 (8/10): 771–90.

———. 2007. 'Mapping sensitive benthic habitats in the Strait of Georgia, coastal British Columbia: Deep-water sponge and coral reefs'. *Current Research* A2: 1–6.

Department of Fisheries and Oceans. 2000. 'Hexactinellid sponge reefs on the British Columbia continental shelf: Geological and biological structure'. DFO *Pacific Region Habitat Status Report 2000/02*. http://www.pac.dfo-mpo.gc.ca/sci/psarc/HSRS/hab-02.pdf.

———. 2007. *Canadian Fisheries Statistics 2005*. Ottawa: Fisheries and Oceans Canada.

Environment Canada. 2009. 'The status of wild species in Canada'. (SARA General Status Report 2003–2008: Overview Document). http://www.sararegistry.gc.ca/document/default_e.cfm?documentID=1757.

Fisheries and Oceans Canada. 2002. *Canada's Oceans Strategy: Our Oceans, Our Future*. Ottawa: Fisheries and Oceans Canada.

———. 2006. Integrated management homepage. http://www.pac.dfo-mpo.gc.ca/oceans/im/default_e.htm.

———. 2007a. 'Canada's Ocean Action Plan: Foreword'. http://www.dfo-mpo.gc.ca/oceans-habitat/oceans/oap-pao/page01_e.asp.

———. 2007b. 'Oceans and fish habitat: Policies and legislation'. http://www.dfo-mpo.gc.ca/oceans-habitat/habitat/policies-politiques/index_e.asp.

———. 2007c. *Survey of Recreational Fishing in Canada 2005*. Ottawa: Fisheries and Oceans Canada.

———. 2008a. '2005–2010 strategic plan: Our waters, our future'. http://www.dfo-mpo.gc.ca/dfo-mpo/plan-eng.htm.

———. 2008b. 'Our organization: Vision, mission, mandate'. http://www.dfo-mpo.gc.ca/us-nous/vision-eng.htm.

———. 2008c. 'Summary of Canadian commercial catches and values'. http://www.dfo-mpo.gc.ca/communic/Statistics/commercial/landings/sum0407_e.htm.

———. 2009. 'Canada's ocean estate: A description of Canada's maritime zones'. http://www.dfo-mpo.

gc.ca/oceans/canadasoceans-oceansducanada/marinezones-zonesmarines-eng.htm.

Freese, C.H., and D.L. Trauger. 2000. 'Wildlife markets and biodiversity conservation in North America'. *Wildlife Society Bulletin* 28 (1): 42–51.

Fuller, S.D., et al. 2009. 'How we fish matters: Addressing the ecological impacts of Canadian fishing gear'. http://www.howwefish.ca.

Geological Survey of Canada. 2008. 'Sponge reefs on the continental shelf: A joint project between the Pacific Geoscience Centre and the University of Stuttgart'. http://gsc.nrcan.gc.ca/marine/sponge/index_e.php.

Government of Canada. 2008. 'Species at Risk Public Registry. (A to Z species index)'. http://www.sararegistry.gc.ca/sar/index/default_e.cfm.

Haggan, N. 2008. 'Including cultural and spiritual values in Pacific NW ecosystem management'. http://web.pml.ac.uk/globec/structure/fwg/focus4/symposium/s4.htm. Forthcoming in R.E. Ommer, I.R. Perry, P. Cury, and K. Cochrane. 2010. *Resilience of Fisheries Systems to Global Change: A Social-Ecological Perspective*. Toronto: Wiley-Blackwell.

Honoré, C. 2004. *In Praise of Slowness: How a Worldwide Movement Is Challenging the Cult of Speed*. New York: HarperCollins.

Hume, S. 2008. 'Rare creatures in danger'. *Vancouver Sun* 18 November. 081118.pdf.

Institute for Geology and Paleontology. 2008. 'Recent hexactinellid sponge reefs on the continental shelf of British Columbia, Canada'. http://www.porifera.org/a/ci.

Islam, K. 2007. 'Canada's natural resource wealth at a glance'. *EnviroStats* 1 (3): 3. dsp-psd.pwgsc.gc.ca/collection_2007/statcan/16-002-X/16-002-XIE2007003.pdf.

Jamieson, G., and L. Chew. 2002. 'Hexactinellid sponge reefs: Areas of interest as Marine Protected Areas in the north and central coast areas'. *Canadian Science Advisory Secretariat Research Document 2002/122*. http://www.dfo-mpo.gc.ca/csas.

Jarre, A., et al. 2008. 'Multi-criteria decision support: A toolbox for integration, communication and collaboration in marine social-ecological systems under global change'. http://web.pml.ac.uk/globec/structure/fwg/focus4/symposium/s6.htm.

Jelks, H., et al. 2008. 'Conservation status of imperiled North American freshwater and diadromous fishes'. *Fisheries* 33 (8) 372–407.

Jessen, S., and S. Patton. 2008. 'Protecting marine biodiversity in Canada: Adaptation options in the face of climate change'. *Biodiversity* 9 (3/4), 47–58.

Kildow, J.T. 2008. 'The social tipping point: Can policy catch up with the science of global climate change?' http://web.pml.ac.uk/globec/structure/fwg/focus4/symposium/keynotes.htm. Forthcoming in R.E. Ommer, I.R. Perry, P. Cury, and K. Cochrane. 2010. *Resilience of Fisheries Systems to Global Change: A Social-Ecological Perspective*. Toronto: Wiley-Blackwell.

Kooiman, J., M. Bavinck, S. Jentoft, and R. Pullin, eds. 2005. *Fish for Life: Interactive Governance for Fisheries*. Amsterdam: Amsterdam University Press.

Lackey, R.T. 2005. 'Economic growth and salmon recovery: An irreconcilable conflict?' *Fisheries* 30 (3) 30–2.

McCay, B. 2008. 'Signs & waves: Responses to change in three marine-social-ecological systems'. http://web.pml.ac.uk/globec/structure/fwg/focus4/symposium/keynotes.htm. Forthcoming in R.E. Ommer, I.R. Perry, P. Cury, and K. Cochrane. 2010. *Resilience of Fisheries Systems to Global Change: A Social-Ecological Perspective*. Toronto: Wiley-Blackwell.

Myers, N. 1990. *The Gaia Atlas of Future Worlds: Challenge and Opportunity in an Age of Change*. New York: Doubleday.

Neis, B., and J. Lutz. 2008. 'Making and moving knowledge: Interdisciplinary and community-based research in a world on the edge'. http://web.pml.ac.uk/globec/structure/fwg/focus4/symposium/s7.htm.

North Pacific Fishery Management Council. 2009. 'Arctic fishery management plan'. http://www.fakr.noaa.gov/npfmc/current_issues/Arctic/ARCTICflier209.pdf.

Office of the Auditor General. 1997. 'Report of the Auditor General of Canada', ch. 14. http://www.oag-bvg.gc.ca/internet/English/parl_oag_199710_e_1147.html.

———. 2009. '2009 Spring report of the Commissioner of the Environment and Sustainable Development. http://www.oag-bvg.gc.ca/internet/English/parl_cesd_200905_00_e_32510.html.

Ommer, R., ed. 2002. *The Resilient Outport: Ecology, Economy and Society in Rural Newfoundland*. St John's, NL: ISER Books.

Pauly, D. 2006. 'Major trends in small-scale marine fisheries, with emphasis on developing countries, and some implications for the social sciences'. *Maritime Studies (MAST)* 4 (2): 7–22.

Perry, R.I., et al. 2008a. 'Coping with global change in marine social-ecological systems'. *FishBytes* 14 (5): 1–2.

———. 2008b. 'Interactive responses of natural and human systems to marine ecosystem changes. http://web.pml.ac.uk/globec/structure/fwg/focus4/symposium/closing.htm.

Perry, R.I., and R. Ommer, convenors. 2008. 'Coping with global change in marine social-ecological systems'. Symposium held in Rome, July. http://www.peopleandfish.org. Forthcoming in R.E. Ommer, I.R. Perry, P. Cury, and K. Cochrane. 2010. *Resilience of Fisheries Systems to Global Change: A Social-Ecological Perspective*. Toronto: Wiley-Blackwell.

Pinkerton, E., A. Bedo, A. Hanson. 2005. 'Final evaluation report: West Coast Vancouver Island Aquatic Management Board (AMB)'. www.westcoastaquatic.ca/AMBFinalEvaluation22March05.pdf.

Pooley, S. 2008. 'FAO comments, July 11, 2008'. http://web.pml.ac.uk/globec/structure/fwg/focus4/symposium/closing.htm.

Reynolds, L., et al. 2006. 'Study report on economic growth and fish conservation'. http://www.fisheries.org/units/wqs/Study_Report_for_Internet_IV_(With_Hyperlinks).pdf.

Rose, C.A. 2005. 'Economic growth as a threat to fish conservation in Canada'. *Fisheries* 30 (8) 36–8.

Senate of Canada. Standing Committee on Fisheries and Oceans. 2005. 'Interim report on Canada's new and evolving policy framework for managing fisheries and oceans'. http://www.parl.gc.ca/38/1/parlbus/commbus/senate/com-e/fish-e/rep-e/repintmay05-e.htm.

Shea, G. 2009. Address to the Standing Committee on Fisheries and Oceans, 10 February. http://www.dfo-mpo.gc.ca/media/speeches-discours/2009/20090210-eng.htm.

Sponge Reef Symposium. 2008. 'Sponge Reef Symposium October 29, 2008 Workshop Notes'.

Statistics Canada. 2000. *Human activity and the environment 2000*. Ottawa: Minister of Industry.

———. 2008. 'Human activity and the environment: Annual statistics 2006'. Table 3.10. http://www.statcan.gc.ca/pub/16-201-x/2006000/4177446-eng.htm.

Trauger, D.L., et al. 2003. 'The relationship of economic growth to wildlife conservation'. *Wildlife Society Technical Review 03-1*. Bethesda, MD: Wildlife Society.

Treasury Board of Canada. 2008. '2008–2009 reports on plans and priorities, Fisheries and Oceans Canada. http://www.tbs-sct/rpp/2008_2009/inst/dfp/dfopr-eng.asp.

United Nations. 2001. United Nations Convention on the Law of the Sea of 10 December 1982. http://www.un.org/Depts/los/convention_agreements/texts/unclos/closindx.htm.

United Nations Food and Agriculture Organization. 2007. 'General situation of world fish stocks'. http://www.fao.org/newsroom/common/ecg/1000505/en/stocks.pdf.

West Coast Vancouver Island Aquatic Management Board. n.d. 'West Coast Vancouver Island Aquatic Management Board: About'. http://www.westcoastaquatic.ca/about.htm.

———. 2007. 'Coastal prosperity and health: A five year plan'. http://www.westcoastaquatic.ca/Aquatic_Mgmt_Board.htm.

9

Agriculture and Rural Resources

Iain Wallace and Mike Brklacich

Learning Objectives

- To understand the three major factors (globalization, the organic agriculture movement, and global climatic change) contributing to current and future uncertainties in Canadian agriculture.
- To appreciate the capacity of Canadian agriculture to manage these uncertainties.
- To understand how uncertainties might prompt or deter future conflicts in Canadian agriculture.

INTRODUCTION

In two important respects, Canadian farmers are more exposed to radical uncertainty than producers in other sectors of the economy. First, their prosperity depends significantly on the unpredictable climatic conditions to which they are exposed each season. Even those whose operations predominantly take place in buildings, such as poultry and mushroom producers, require inputs that originate in the fields that are the principal environment of agricultural production. And for most farmers, the possibility of drought, excessive moisture, hail damage, insect infestation, early or late frosts, or any other climate-related damage to crops or livestock is simply a fact of life. Insurance may be available to cover some of these potential losses, but it has a cost, and it may not be adequate.

A second source of uncertainty originates in the economic circumstances of most Canadian farmers. Although the average size (area) and capitalization of farms in Canada continues to grow (albeit with significant differences in mean size by region and sector), the average farmer remains a small player within the agri-food production system. The firms supplying purchased inputs that farmers use and the firms that buy farms' agricultural commodity outputs invariably have greater power to determine prices than do the farm operators, thus placing them in a 'cost–price squeeze'. This reflects the highly integrated structure of the modern agri-food industry and the presence of oligopolistic markets (dominated by a few large firms) both upstream and downstream of the farm itself. Costs and revenues are therefore more uncertain than in many other sectors of the economy. Given the amount of capital the average farmer needs and the loans necessary to meet short- and long-term production goals, financial uncertainty is another reality that shapes Canadian agriculture.

If uncertainty is endemic to Canadian farming, conflict is becoming equally so. Ever since the emergence of urbanized industrial societies, there has been a structural divergence of interest between farmers wanting a reasonable return on their capital and labour and city dwellers wanting cheap food. Throughout the twentieth century, the demographic and electoral balance between the farm-based population and the urban population shifted in Canada, and this became reflected in the priorities of politicians. Whereas governments have acted to support the farming community in ways described below, their overall priority has been to ensure an adequate and affordable food supply to the urban majority. Support for agricultural research and the development of advanced technologies to increase agricultural productivity have been ways of promoting this agenda.

Conflicts of interest between urban society and the agricultural community over food prices may be thought of as a national-scale phenomenon. Conflicts can be found at other geographical scales, however. Alongside the growing integration of the domestic agri-food sector is the increased globalization of food markets. Canadian farmers find themselves in a world in which other producers frequently enjoy more favourable resource endowments for agricultural production and lower labour costs. In a global policy environment shaped by neo-liberal commitments to freer trade and the abolition of national protection of domestic producers, this means that there is often conflict between the interests of Canadian farmers and those of competing farmers overseas, both in the temperate world and in more tropical environments (see also Chapters 1 and 2).

At the other end of the scale spectrum, the local impacts of agricultural production are a frequent source of conflict between farmers and their neighbours. The increasing size of agricultural operations and their increasingly discernable impact on rural environments with respect to local water quality, noxious smells, or the noise of mechanized operations has produced clashes over what is legitimate activity in rural areas where non-farm residents are recent arrivals in the local community. This has prompted some jurisdictions to introduce 'rights to farm' legislation to protect the interests of agricultural producers.

Uncertainty and conflict are thus themes that clearly feature in the appraisal and management of Canada's agricultural and rural resource base. In many respects, they are inseparable from the practice of farming in an industrialized—even post-industrialized—society. They therefore have a long history. But changes in the global context, both of accelerating environmental change and of increased economic integration, have substantially changed the priority issues over recent decades. In this chapter, we provide an overview of the shape these changes are taking. We focus on three elements of the emerging picture. Following this introduction, we situate Canadian agriculture in its global context, primarily from an economic geography perspective. Second, we focus on the emergence of 'alternative' agriculture in Canada, primarily associated with the growth of demand for 'organic' produce and the critique of conventional agriculture it embodies. Finally, we look at the implications of global climate change for the Canadian agricultural industry.

CANADIAN AGRICULTURE IN A GLOBAL WORLD

Historical Context

The emergence of a global market for food in the nineteenth century resulted from the demands of growing urban populations in industrializing Europe. Great Britain in particular needed to augment its limited domestic agricultural output with temperate foodstuffs from overseas, and Canada found its place as a food exporter primarily in supplying wheat to Great Britain and other foreign markets. The Canadian Prairies were settled on this basis, and the wheat staple played a major role in fostering economic development in Canada from the 1880s to the 1920s. It required the construction of an extensive transportation system to ship the grain overseas, stimulated the manufacturing sector, and sustained the immigration that contributed to a growing domestic economy. In contrast to the specialization and export-dependence of Prairie agriculture, farming elsewhere in Canada continued until the 1960s to be typified by mixed crop and livestock production oriented to the domestic market.

The past 50 years have seen dramatic changes in the global food market, and these changes have brought both opportunities and challenges to Canadian agricultural producers. The world's population has more than doubled, and per capita food consumption has increased as living standards have risen. At the same time, the area of land under cultivation has risen, and the productivity of agriculture has increased dramatically in many parts of the world. As a result, to date, global food demand and supply have kept approximately in balance, although this aggregate picture hides the persistence of regional food deficits and malnutrition, notably in Africa and parts of Asia.

The potential world market for Canadian food exports has thus grown both in size and variety, but at the same time competition from a growing list of other exporting nations has forced Canadian producers to be increasingly efficient and oriented to higher value-added foodstuffs. Moreover, Canadian farmers, together with their counterparts throughout the industrialized world, have experienced two powerful trends that have affected their livelihoods.

One has been the growing integration of the agri-food system in which they play an increasingly subordinate role to more powerful corporate actors 'upstream' and 'downstream' of the farm. The other has been shifting cultural and political attitudes toward agriculture. These attitudes have, on the one hand, prompted governments to reduce their intervention in agricultural markets, thereby allowing economic competitiveness alone to determine outcomes. But on the other hand, paradoxically, they have given rise to new expectations about what governments should promote as desirable agricultural policy, such as environmental sustainability. As a result, these sometimes opposing policy objectives can be found within the same institution, as when Agriculture and Agri-Food Canada encourages innovations in biotechnology while also funding support for organic farming.

Trade Policy Issues

Despite the enormous diversification of its economy over the past century, Canada remains a major player in global agri-food trade: in 2006 it was the world's fourth-largest exporter and fifth-largest importer (AAFC 2007). This significant two-way trade is comparable to that in energy (see Chapter 6) in that it creates challenges for national policy-makers because of the distinctive domestic geography of the two flows. The bulk of Canada's exports continue to come from the Prairies in the form of grains, oilseeds, cattle, and beef products. Its imports are much more diversified. They include a wide range of tropical and semi-tropical foodstuffs that cannot be produced commercially in Canada, but they also include temperate zone fruits and vegetables that replace domestic supplies in winter (and may compete with them in summer) and a wide range of processed and manufactured items such as New Zealand lamb and French cheeses. These products, which are purchased across the country, can represent a competitive threat to the commodities produced primarily by the farmers and food processors of southern Ontario and Quebec.

For Canada's agri-food exporters to prosper in world markets, federal policy needs to be aligned with the international moves to remove subsidies and other protective measures in this sector, and this is a matter of concern especially (though not exclusively) to western farmers. Simultaneously, however, there are political pressures to protect domestic agri-food producers, especially those most exposed to import competition, and this implies federal (and sometimes provincial) efforts to oppose moves to freer agricultural trade. These pressures come primarily from Quebec and Ontario. So it is no wonder that the federal approach to international agricultural trade negotiations has been 'ambivalent', if not outright schizophrenic (Wallace 2002, ch. 8)! Not that Canada is alone in this regard: the United States, long the proponent of reducing barriers to international trade, is just as guilty of trying to have its cake and eat it too by demanding open markets for its agricultural exports, but protecting by numerous particular subsidies (such as for peanut or cotton growers) many of its own farmers.

The signing of the Canada–United States Trade Agreement in 1989 and its subsequent expansion in 1994 into the North American Free Trade Agreement (NAFTA), which includes Mexico, was seen by many observers as a major source of uncertainty and potential disruption for Canadian agriculture (Troughton 1991). The greater agro-climatic potentials of the warmer agricultural lands to the south, together with the wage differentials between Canadian and Mexican farm workers and the economies of scale of large American farms and processing plants compared to those in Canada, were feared to be a recipe for disaster. Certainly, considerable

rationalization of small and underutilized processing plants such as those involving cheese or vegetables took place (Wallace 1992), but the alarmist scenarios have not, in general, come to pass. Conflict and uncertainty have originated much more in continuing political obstacles to free cross-border trade in agri-food products than from disparities in economic competitiveness. In particular, members of the US Senate have considerable power to invoke a wide range of protectionist measures in support of domestic farmers at the expense of Canadian agricultural exports. Even if these measures are eventually ruled to be inconsistent with NAFTA obligations, the process of challenging them is protracted and expensive, meaning that this sort of action serves its purpose of harassing Canadian producers and depriving them of sales (Skogstad 2008).

International comparisons of the support given to farmers through various forms of government action are measured in terms of the Producer Support Estimate (PSE) (OECD 2008). This support may involve anything from tariffs or import quotas to institutional support for farmers, such as subsidies to offset drought impacts or legislation to control the markets for agricultural produce. In Canada, the most significant form of market regulation is 'supply management' of the dairy and poultry sectors (Chiotti 1992; Bowler 1994; OMAFRA 2007). The main argument in favour of this arrangement is that with annual production set to match estimated domestic annual demand, farmers are guaranteed a steady market and consumers a reliable supply. Farmers purchase the right to supply a share of the market (a quota). The main arguments against the system are that by artificially restricting the market to domestic suppliers, consumers lose the benefits of competition, and the quota system makes the adjustment of the industry to technological change or economies of scale more difficult to achieve.

Froment (2008) provides data illustrating both sides of this argument. Whereas the average price of raw milk in Canada in 2006 was twice the average price in the United States, prices in Canada were markedly more stable over the decade 1998–2008 than those south of the border. So Canadian consumers pay more for milk and milk products than they might if there were cross-border trade, but Canadian dairy farmers are assured of a steady market, which helps to maintain the prosperity of the rural economy in the regions where they are concentrated.

The allocation of shares of the national milk market among the provinces significantly favours Quebec, especially with respect to 'industrial milk' used to manufacture dairy products (as opposed to 'fluid milk' sold to consumers). In 2008, Quebec, with 38 per cent of the Canadian dairy herd, had 45 per cent of the national milk quota, compared to 33 and 32 per cent, respectively, for Ontario and 15.5 and 12.3 per cent, respectively, for BC and Alberta combined (CDC 2008; CDIC 2009). Giving Quebec a disproportionate share of the industrial milk quota is one way of supporting dairy farms in remote regions of that province that are distant from metropolitan markets, and it enables the federal government to counteract the political support for separatism in 'le Québec profond' (Segal 2008). But the policy comes at the cost of hindering the growth of the dairy processing industry in the west and is thus a source of regional discontent in that part of the country.

Agricultural Specialization

Trade policy issues are in many respects reflections of more deep-seated forces that have been shaping the global agri-food sector for many decades. The mechanization of agricultural production started in the nineteenth century and saw, for instance, the Toronto-based Massey-Harris farm equipment manufacturer emerge as a major international firm (Winder 2002). But the thorough transformation of farm production and its incorporation into a fully industrialized agri-food system is a process that has greatly accelerated since the 1960s (Goodman and Redclift 1991). Farmers have been on a treadmill of technological and economic change that has required them to pursue continuous investments in new technology and increasing economies of scale if they are to survive. These pressures to adapt and the uncertainty of success are reflected in the increased specialization of farm enterprises and increasing average size of their operations, together with the continued exodus of conventional small farms from the industry. To be economically viable, small farms increasingly have to find a particular market niche that is not so exposed to the 'cost–price squeeze' of industrial-scale agricultural commodity production (see the discussion of organic farming below).

The number of farms in Canada has declined almost continuously since 1941, to a total of 230,000 in 2006 (AAFC 2007). By 2005, Canada's largest farms (defined by Statistics Canada as having more than $500,000 in gross annual revenues) comprised only 11 per cent of all farms but produced 55 per cent of total farm revenue. They were also the only size category in which farm income constituted the bulk of farm household income: below that threshold, off-farm income contributes a rising proportion of the income of farm households, with government support payments also being important to farms in the $250,000 to $500,000 range (Statistics Canada 2008). The push within 'productivist' agriculture to increasing size and specialization is evident both in crop and livestock production so that, for instance, pigs are now concentrated on farms with an average herd of 1,308 animals and the average size of a dairy herd increased by 44 per cent (to 57 cows) between 1996 and 2006 (AAFC 2007).

To better understand the dynamics of specialization and the cost–price squeeze and their spatially differentiated outcomes, it is helpful to draw on the theoretical work of Goodman, Sorj, and Wilkinson (1987). They identify two movements working to unify production in the agri-food industries, both of which significantly reduce the role and independence of the farmer: *appropriationism* and *substitutionism*. Appropriationism focuses on the transformations that have taken place in the agricultural supply industries (machinery, fertilizers, specialist services, and, not least, seeds and animal embryos themselves). These supply industries have expanded in scope but simultaneously have been consolidated in the hands of transnational corporations, almost none of them now Canadian-owned (Kneen 1995; 1999). They provide inputs on which the modern farmer is not only technologically dependent but in some sectors increasingly legally dependent, as illustrated by the active enforcement of corporate intellectual property rights with respect to genetically modified seeds (CBC News 2004; Moore 2008). This 'upstream' element of the agri-food sector has grown by taking over (appropriating) activities formerly carried out on the farm, as in the transition from horses to tractors in the mid-twentieth century and the displacement of retained seeds by purchased seeds more recently.

Substitutionism refers to the transformation of the agri-food sector 'downstream' of the farm. Its results are most evident in the greatly reduced share of consumer food spending that flows back to the farmer as the processing and retailing industries have massively increased their role in the food supply system. The urban and time-pinched population of contemporary Canada has largely substituted the supermarket for the farmers' market and processed foods and purchased meals for home-cooked fresh or basic ingredients. The purchasing power of the major

food chains and their control of access to shelf space give them the ability to put downward pressure on prices paid to producers, thus contributing to the cost–price squeeze.

The geographical consequences of these structural characteristics of the agri-food system are well summarized by Smith (1984), based on research in Quebec but applicable more widely. His modern adaptation of von-Thunen's analysis of the spatial structure of agriculture (see Chisholm 1979) recognizes the importance of location for both producers and consumers (Figure 9.1). Metropolitan markets, where 68 per cent of Canadians live (Statistics Canada 2009), are typically adjacent to productive agricultural land that supports relatively prosperous farms, either in supply-managed dairying or in field crops supplying the urban market. This inner zone also provides the greatest potential market for specialty producers, such as the organic farms reviewed later in this chapter. The concentration of consumer purchasing power makes this zone the most attractive one for the large food retailers whose weekly 'specials' help to keep down the cost of processed foods.

Beyond the metropolitan-dominated zone lies an intermediate one that offers reduced market options to the farmer. Lacking ready access to urban consumers, producers here are more dependent on selling their output to food processors, often under contract. This pattern characterizes the majority of dairy farmers in rural Quebec, forming the basis of cheese and other dairy product manufacturing, and is equally evident in the regional concentrations of potato producers in the upper Saint John River Valley in New Brunswick and in southwestern Manitoba that form the basis of the frozen french-fry industry (Glover and Kusterer 1990). Most Prairie farmers, whether engaged in grain, oilseeds, or beef production,

Inner zone of prosperous farmers. This coincides with the area of greatest consumer purchasing power and is the location of the large retail grocery chains and processing companies. Small independent retailers meet specialist consumer needs.

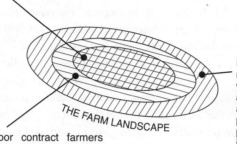

Extensive, isolated rural area with a rapidly declining farm population, widespread rural poverty and farm abandonment. No production contracts are available; there are no processing plants. The small, declining consumer population has low, shrinking purchasing power. There are no major chain store outlets, only small, expensive, independent retailers.

THE FARM LANDSCAPE

Fringe zone of poor contract farmers closely bound to the industrial market. This is an area with a low population density and weak purchasing power. The area is served mainly by independent stores and small processing firms; there are few chain store outlets or major processing firms.

Source: Derived from Smith 1984.

Figure 9.1 Zonation of the Contemporary Agri-food System around Major Metropolitan Markets.

fall into this category of depending on markets dominated by food processors or transnational marketing firms. And in this intermediate region, less attractive to the big supermarket chains, the food retailing sector is more in the hands of independent grocers.

Smith's (1984, 367) outer zone is one of remote rural areas usually characterized by a 'declining farm population, widespread rural poverty and farm abandonment'. Far from metropolitan markets and rarely of interest to food-processing firms, these regions are an ongoing challenge to policy-makers. Some entrepreneurial initiatives may prosper based on regionally specialized crops, such as blueberries in eastern Quebec, but for the most part, these are areas where the population and the economy are both in decline. The retail sector is small and fragmented, and its remoteness entails higher delivery costs, which all combine to make processed foods expensive for local consumers.

ORGANIC AGRICULTURE IN CANADA

Since the 1990s, there has been a noticeable increase in the size and scope of organic agriculture in Canada. This is part of a global phenomenon, for the amount of land farmed organically, primarily in developed nations, rose from 10 million to 24 million hectares between 2000 and 2004 (Mason and Spaner 2006). This section explains the growth of organic production as a linked set of responses to the perceived negative characteristics of 'productivist' agriculture on the part of both producers and consumers. For many producers, the environmental impacts of industrialized monoculture and ethical issues around intensive livestock operations have become increasingly disturbing. The opportunity offered by the growth of demand for organic products to pursue more environmentally appropriate practises, often on small or medium-sized farms, has been a major attraction. For consumers, driving forces have been worry over the health risks associated with industrialized farming, such as the presence of pesticide residues on fruits and vegetables, and ethical concerns about the welfare of farm animals within conventional agriculture. Arguments have also been made about the declining nutritional characteristics of increasingly processed foods and the superior taste of organic produce. The increasing scale of production that seems constantly required of farmers tied to the technological treadmill of industrialized agriculture has fostered a desire among many producers and consumers to return to an agricultural production system that rebuilds a sense of community linking farmers and those who consume their produce.

Yet, the path to achieving that goal has not been easy. Developing an organic production system involves numerous challenges beyond the simple transition from one production regime to another. The question of what constitutes 'organic' production has itself been an issue of dispute. The general principles are clear: a commitment to natural biological processes in place of reliance on industrial inputs, especially pesticides and fertilizers; a refusal to use genetically modified organisms (GMOs); and animal husbandry that avoids the use of antibiotics and the confinement associated with conventional production systems (Wunsch 2004; Canadian Organic Growers 2007). Developing an agreed-upon set of guidelines as to what qualifies as 'organic' has been a protracted process. Institutions to accredit agricultural producers as 'certified organic' have emerged under varying auspices, and only after a delay of some years has a Canadian standard that forms the basis for certification as an organic producer come into being (Canadian Organic Growers 2007). It has been developed to be compatible with emerging international standards of certification, which is important to Canadian organic farmers, notably of grains and oilseeds, seeking to supply growing foreign markets for their output (AAFC 2007). Equally, the question of what qualifies as an 'organic' manufactured food product has been contested, with different parties arguing for varying degrees of purity in terms of the presence of non-organic ingredients.

In many ways, the geography of organic production in Canada resembles that of conventional producers. The largest area under organic production is found on the Prairies, where it consists primarily of grain and oilseed crops (AAFC 2007). Organic production in the more urbanized parts of the country, principally southern Ontario and Quebec and the lower mainland of BC, is much more diversified and includes field crops, fruits, and livestock. In these regions, the opportunities open to organic farmers are, to a significant degree, determined by their accessibility to metropolitan markets, and in this they resemble other forms of niche production. For instance, within the metropolitan fringe, organic producers have access not only to farm-gate sales and urban farmers' markets but also to business purchasers such as restaurants and even a few public sector buyers such as the University of Guelph (Savour Ottawa 2007; George-Cosh 2007). All of this has increased opportunities for local economic development at the regional scale.

Factors Encouraging the Growth of Organics

Health and Food Quality— Happy Cows Don't Go Mad

The growth of the market for organics in Canada was given a major boost by a number of food scares in Europe in the 1980s and 1990s, of which bovine spongiform encephalopathy (BSE, popularly known as mad cow disease) in the UK was the most potent (Blay-Palmer 2005). This threat to human health was traced to the industrial use of diseased cows' brains as animal feed, which then allowed the disease to be communicated to humans. It highlighted for consumers the dangers inherent in profit-maximizing industrial agriculture and for many people became a reason to favour organically produced food—for instance, grain-fed cows—or to switch to a vegetarian (and preferably organic) diet. Canada subsequently had its first brush with mad cow disease in 1993 in Alberta, traced to cows imported from the UK (Health Canada 2005). A new case in 2003, with no identifiable foreign links, soon led to the closure of the US border to Canadian cattle and beef and to other international embargoes that were not fully lifted until 2006 (CBC News 2006; AAFC, 2007).

A further boost to organic producers has been associated with the attempt since the late 1980s by the multinational firm Monsanto to introduce milk from cows that had been given recombinant bovine somatropin (rBST, or bovine growth hormone) to increase their production. Popular opposition to this expression of corporate agriculture united many farmers (particularly farm women; see Mackenzie 1992) and consumers. Concern was expressed about the reported negative health effects on cows and the unknown health consequences should the artificial hormone be transmitted through the milk to human consumers. Despite pressure from Monsanto and the acceptance of rBST in the United States (but not in the European community or Australia), the Canadian government decided not to permit its use (Forge 1998). Finally, fears over the health effects of pesticide residues in fruits and vegetables have inclined many people to buy organic (Canadian Organic Growers 2007). Parents in particular expressed concern about residues in baby food and children's food (CFIA 2005). Issues such as those cited above have had a major impact in shaping consumer values and have created an environment in which the virtues of organic production have come to be seen by a much wider section of Canadian society than previously.

At the same time that health concerns have made organic produce more attractive, greater consumer interest in food quality has also worked in favour of this sector. The focus on cooking and food quality seen in the increased popularity of television programs devoted to chefs who have then become celebrities in their own right has encouraged consumers to purchase less-processed and fresher ingredients. This has worked in favour of suppliers close to major metropolitan markets and has created a niche in which organic producers have flourished (Beauchesne and Bryant 1999). Not all local produce is organic (just as not all organic produce is local; see below), but proximity to the purchaser allows for greater communication between farmer and consumer and enables organic producers to educate potential customers about the merits, both nutritional and ecological, of their produce (Miller 2008). Movements such as those inspired by the '100-mile diet' (Smith and Mackinnon 2007) have reinforced the creation of communities of interest between city dwellers and the farmers in their surrounding rural areas.

Environmental Sustainability— Happy Cows Don't Shit in the Water Table

Conventional productivist agriculture appears bound to continued growth in the size and technological intensity of farm units and to the creation of increasingly artificial environments. The creed of most Canadian organic farmers counters both these trends. The scale of production, particularly of non-grain crops, is generally restricted to traditional family-scale holdings.

And the form of land management reverts more closely to a pre-industrial model based on crop rotation and in which labour inputs are much more significant than those of technology. As a result, greater attention is given to working with ecological variety at the farm level. In these ways, organic producers claim that they are reversing the decline in natural fertility and soil structure typical of contemporary industrial agriculture (Canadian Organic Growers 2007). The smaller scale of organic farms also ensures that they avoid the negative consequences of concentrated pollutants, whether artificial or natural, as in the problems associated with feed-lot manure production and runoff that led to the fatal *E. coli* water contamination at Walkerton, ON, in 2000 (Prudham 2004) (see also Chapter 7).

Climate change mitigation policy has increased the competitive appeal of organic production in that it favours low-input agriculture. But the power of the conventional farming sector is such that rather than favouring substantially increased investment in organic production systems, federal policy has made expansion of corn-based ethanol production the major agricultural sector response to the challenge of climate change. This policy fosters expansion of conventional row-crop monoculture, which has been shown to have negative impacts on soil quality (nutrient content and soil erosion, particularly in times of drought on the Prairies and heavy rain in eastern Canada) and is the antithesis of organic soil management.

Rural Economic Sustainability— Happy Cow on the Plate

By 2006, the annual value of Canadian organic farm production had reached $986 million and involved 3,618 certified producers. To put this in perspective, these certified growers represented only 1.5 per cent of all Canadian farms, but more than 15,000 farms (6.8 per cent of all operators) reported at least one type of organic product (AAFC 2007). These non-certified farms are most associated with animal products, whereas field crops predominate on certified farms. We have seen that the organic sector tends to be more associated with fostering local and regional links between producer and consumer than is conventional agriculture and thus has a potentially greater role to play in fostering regional economic sustainability. The largest number of certified farms is found in Saskatchewan (1,230) where they constitute 2.5 per cent of all holdings. The highest proportion of organic farms is found in British Columbia (2.8 per cent, 484 farms) and in Quebec (2.7 per cent, 816 farms). In comparison, Ontario's 497 certified organic farms constituted only 0.9 per cent of the provincial total (Dorais 2007). Clearly, there are fewer opportunities for developing regional producer–consumer links on the Prairies than in the more densely populated regions of the country, but even in Saskatchewan, the more regionalized input profile of organic producers can contribute to rural sustainability.

Challenges and Opportunities
Economic

The small percentage of Canadian farms that have switched to at least some branches of organic agriculture means that there is still room for uncertainty about the long-term effects of that transition. Nevertheless, there are growing signs, in Canada and internationally, that organic producers enjoy economic, environmental, and social benefits compared to equivalent conventional producers. The principal economic advantage stems from lower input costs, which for cereals can be up to one-third below comparable farms relying on purchased agri-chemicals. This more than compensates for the fact that on average, plant yields in organic systems are 10 per cent lower (MacRae, Frick, and Martin 2007). A second source of economic advantage is the ability of organic farmers to charge premium prices compared to the market prices of commodities produced under conventional conditions, although this differential varies considerably between individual products and over time (AAFC 2007; Jowit 2008). A third factor that helps to improve net financial returns is the reliance on more direct marketing channels, which allows organic farmers to keep a greater share of the price consumers pay. Finally, the greater diversity of production on most organic farms and their land management practices can give them greater resilience to adverse climatic conditions and thus less exposure to income fluctuation (MacRae, Frick, and Martin 2007).

Nevertheless, together with these advantages there are a number of challenges to the economic competitiveness of organic producers. The first relates to higher labour costs compared to those of conventional producers, particularly for horticultural items. A second reflects aspects of government policy that favour conventional agriculture. For instance, in the early 2000s, organic agriculture was receiving only about 1 per cent of the federal research budget for agriculture, well below its share of the farm sector overall (Blay-Palmer 2005). On the Prairies, organic cereal farmers are charged premium rates for crop insurance, and as a result (perhaps also because they feel that their more resilient operations are less at risk) the majority do not take advantage of the insurance cover (MacRae, Frick, and Martin, 2007). The relatively small size of the organics sector, in terms of both the numbers and size of farms and the volume of sales to consumers, represents a third challenge. For, while producers seek to capitalize on premium prices and direct sales to a local community of buyers, the

sector finds it hard to benefit from economies of scale, particularly in processing and marketing, that would help it to be seen as an attractive and affordable alternative to a wider range of consumers (AAFC 2007).

Following trends in the United States and Europe where the market for organic food is much larger, when consumer demand does show a sustained increase, the sector becomes attractive to large firms within the conventional agrifood sector that capitalize on scale economies and other forms of market power (Dimitri and Greene 2002). Organic fruits and vegetables can now be found in most supermarkets, and large retail chains such as Loblaws and Shoppers Drug Mart have introduced own-brand organic foods, but their supply chains are international in scope and not necessarily structured to encourage local sourcing (Shaw 2008). As domestic organic growers become more numerous, however, they are becoming better equipped to supply the volume of produce demanded by the major chains, as is well illustrated by the emergence in Quebec of an association of horticultural producers, Les Producteurs biologiques Symbiosis (Dorais 2007). A member of this group that supplies Loblaws President's Choice brand and claims to be Canada's largest supplier of organic tomatoes, Serres Jardins-Nature, has an extensive distribution network in eastern Canada and the northeastern United States that it supplies from its peripheral location in New Richmond, Gaspésie (Jardins-Nature 2009).

Farm and Community

The economic structure of organic farming offers an identifiable strategy for small- to medium-scale operators who wish to secure the future of the family farm. Yet, any person or household setting out to become an organic producer faces a number of hurdles. For existing conventional farmers, there is an element of having to 'unlearn' familiar practices and become expert in organic. Government support for organic training has been limited, and post-secondary programs in organic agriculture have been slow to appear. The Nova Scotia Agricultural College in Truro was the first institution to receive federal research support, but there is now an infrastructure of programs spread across the country. The learning process can be accelerated through the network of organic non-governmental organizations that exist both nationally (such as Canadian Organic Growers) and internationally, with workshops and websites that offer relevant material (Blay-Palmer 2005).

It is not only farm personnel who may need reorientation, however; the land itself, if it has been in conventional agricultural production, requires a transition period to allow chemical residues to be eliminated from the soil. Only then can a producer obtain organic certification and begin to benefit from the premium prices associated with organic produce. At that stage, however, conflict can arise between organic farmers and neighbouring land managers, whether they are conventional farmers or the operators of golf courses, for example, because the integrity of organic production may be threatened by potential contamination from agri-chemicals in the local groundwater or the drift of chemicals from airborne crop-spraying or of pollen from GMO crops nearby (Blay-Palmer 2005).

GLOBAL CLIMATIC CHANGE AND CANADIAN AGRICULTURE

Much of Canada's agricultural land is located near the northern climatic frontier, and climatic variability has understandably been both a positive and negative influence on the agrifood sector from local through national scales. Climate uncertainty, including variability across the seasons as well as within a season, has without doubt been a key factor that has shaped the structure of Canadian agriculture, and in this sense, uncertainty is not new. For example, the drought of 2001 negatively affected agriculture in all regions of Canada, but the effects were highly variable, with the Prairie region suffering the greatest losses (Lemmen and Warren 2004). Should droughts become more intense and frequent in the future, it is possible that conflicts over water resources will become more common.

What is new are the uncertainties stemming from the impact of human activities on the Earth's climate system (see also Chapter 5). The Intergovernmental Panel on Climate Change (IPCC) reported in its third assessment that there was mounting evidence that human activities had been a major factor contributing to recent warming (IPCC 2001). In its most recent report, IPCC offered a stronger conclusion: 'most of the observed increase in global average temperatures since the mid-20th century is very likely due to the observed increase in anthropogenic greenhouse gas concentrations' (IPCC 2007, 10). Despite advances in climate change science, much uncertainty remains, and there is less consensus on how fast climate change might occur on a global scale and on regional variability within Canada (Weaver 2004). Collectively, anthropogenic climate change has added several new layers of uncertainty to Canadian agriculture, including:

- the pace at which temperature, precipitation, and other climate parameters will change in various regions across Canada;
- the impacts of these climates on regional agro-climatic potentials and crop productivity; and
- the capacity of Canadian farmers to adapt to unprecedented changes in climate.

GUEST STATEMENT

Climate Change and Regional Development: Challenges for Resolving Conflicts and Managing Uncertainty for Agricultural and Rural Communities in Canada

Christopher Bryant

Managers of 'rural' activities function within an increasingly open system and must deal with external forces often beyond their control—e.g., globalization, regional change processes, government actions, social movements in the environmental domain, and climate change. The vulnerability of Canada's rural communities to external forces has been a long-standing concern (Halseth, Markey, and Bruce 2009).

One major preoccupation concerns *relationships with major urban-metropolitan regions*. While rural communities in resource peripheries provided raw materials for the early development of the urban-industrial complex, their demographic and economic bases have become undermined by the continued growth of our large urban-metropolitan complexes. Rural communities in their urban fields face other pressures, including competition for labour and the land resource. Some have also experienced growth as a result either of exurban migrants seeking a place for retirement or simply a 'better' place to live or of interregional migrants. Migrants bring additional social stresses and conflicts but also opportunities (e.g., many place a high value on the natural environment) (SRQ 2008). The net result is a heterogeneous rural community, making the search for solutions difficult.

Climate change and its uncertainties constitute another major external stress (Bryant, Singh, and Thomassin 2008). How climate change affects agriculture, for instance, requires understanding farmers' roles and their variable adaptive capacities

in modifying relationships between climate change and farming.

These various stressors generate complex patterns of conflicts and opportunities. Canada's senior governments have devised interventions to sustain rural communities—e.g., the federal Community Futures program and Quebec's Local Development Centres (Bryant and Cofsky 2004). Some were truly innovative (Community Futures) and even relatively holistic—a critical prerequisite for addressing rural community futures successfully. But there are serious shortcomings because they have not been part of a truly holistic regional development program for a whole province, let alone nationally. Internationally, few instances exist of coherent national frameworks in which rural development is pursued while controlling continued concentration in the urban-industrial complex (a major exception is France). Furthermore, in Canada, innovative rural intervention must be contrasted with the continued direct and indirect support to major cities that dwarfs investments in rural areas.

Climate change and increasing variability in climate conditions have not yet been integrated effectively into rural development programs. Widespread public acceptance of its 'reality' exists, and various governments are taking it more seriously. But little concerted action has occurred in rural communities. British Columbia is one exception—for example, provincial support of the Cariboo-Chilcotin Pine Beetle Coalition regional initiative. In Quebec, with the creation of an Inter-ministerial Committee on

Climate Change, most provincial ministries have recognized the consequences of climate change. Further, in 2001, a unique not-for-profit consortium, Ouranos, was created, comprising several provincial ministries, Environment Canada, and Hydro-Québec, among other players. It supports research into climate change and adaptation, and builds awareness among decision-makers in Quebec and elsewhere.

Some limited actions are notable in Quebec. For instance, some farmers have begun to integrate climate change into their long-term planning, such as dealing with low snow accumulation and its damaging effects on blueberry production. Also, several groups of farmers have participated in research workshops, which during 2007–8 focused on the co-construction of collective intervention to build and enhance the farming community's (and other players') capacity to adapt to climate change. Other participants included regional representatives of provincial ministries and the Financière agricole, representatives of organizations such as watershed management committees, Clubs-conseils en Agro-Environnement, the Saguenay region Blueberry Producers Association, rural development officers from local development centres, and municipal governments. Their enthusiasm and constructive participation demonstrated recognition of climate change as an important stressor and their belief that collective intervention necessarily involves local and regional players.

Conflicts and uncertainties created for rural communities by these different processes vary regionally. Senior governments have often exacerbated problems by directly or indirectly favouring continued dominance of major urban areas. Confronting conflicts and uncertainties for rural areas requires the co-construction of appropriate collective intervention to build and enhance capacities to address them, identify and take advantage of opportunities, and plan strategically. Senior governments have important roles to play in creating an enabling framework. However, they must work closely with players with a permanent presence on the ground to ensure that the critical roles of participating with rural communities and entrepreneurs in their strategic planning and providing them with appropriate and timely information and advice are performed.

Climate Implications for Canada's Agro-climatic Potential

Less than 10 per cent of Canada's total land base has the physical potential to support commercial crop production on a sustained basis, and the agricultural potential of these lands is often constrained by either cooler temperatures or moisture deficits. Assessments of the climatic effects of human activities suggest considerable warming for most of Canada, with longer and warmer frost-free seasons (FFS) that would be more conducive to commercial agriculture than current thermal regimes. Estimates of extensions of the FFS are not uniform across Canada, ranging from a minimum of one week to a maximum of nine weeks. In the main agriculture regions in the Prairies, Ontario, and Quebec, most estimates suggest that the FFS will be extended by up to three weeks (Brklacich et al. 1997; Senate of Canada 2003; Gameda, Bootsma, and McKenney 2007; Sauchyn 2007). Most estimates of long-term global climate change indicate winter warming in the higher latitudes will exceed summer warming, and extensions of the FFS in the Peace River region are expected to be somewhat shorter than in other regions of Canada (Brklacich, Curran, and Brunt 1996).

Future moisture regimes, including the total amount of precipitation as well as its distribution during the extended FFS, are more difficult to estimate than temperature, thereby adding additional uncertainty for long-term planning for Canadian agriculture. There are indications that water shortages may lead to more competition and conflicts over this resource and thereby adversely affect Canadian agriculture to a greater extent in the future (Lemmen and Warren 2004; see also Chapter 7). This is of particular concern in the Prairie region where expected decreases in snow accumulation coupled with the retreat of glaciers may well reduce runoff and impose even greater pressures upon scarce water resources (Nyirfa and Harron 2004; Sauchyn 2007).

In addition, future climates are expected to create new opportunities for fruit production in eastern Canada. Warmer summer temperatures plus the milder winters are expected to reduce cold stress and improve prospects for fruit trees such as apple, but the expansion of areas within Ontario and Quebec that could support fruit trees could be tempered by additional freeze–thaw events in the December through March period (Rochette et al. 2004).

Overall, there is a strong consensus that global climatic change will result in longer and warmer FFS across Canada, but the potential benefits for commercial agriculture will in all likelihood be partially offset by increases in seasonal moisture deficits. These adjustments are expected to be highly variable across Canada and from year to year, thereby adding to the uncertainties that Canadian farmers will need to manage in the future.

A Crop Productivity Perspective

Climate change will have both positive and negative impacts on crop yields. Elevated CO_2 levels will increase water-use efficiency and productivity, and warmer and longer FFS will increase the range of crops that might be grown and allow farmers to use longer-season crop varieties. Potential negative impacts include exposure to a wider range of pests, potential crop damage from extreme weather events such as hailstorms and extreme heat, and more variable weather. Overall, the net impact of climatic change on crops produced by Canadian farmers is both speculative and uncertain (Brklacich et al. 1997; Lemmen and Warren 2004).

For the main agricultural regions within central Canada and on the Prairies, it is estimated that yields for most major field crops will improve (Brklacich et al. 1997). Yield increases of 20 per cent are anticipated for canola, corn, and wheat on the Prairies (McGinn et al. 1999), but these improvements hinge upon sufficient water supplies to sustain plant growth and potentially an increase in the use of agricultural chemicals to manage weeds and agricultural pests that will benefit from warmer temperatures (Boland et al. 2004; Sauchyn 2007). In Quebec, crop yield sensitivities to climate change are expected to be more mixed, with increases of 20 per cent estimated for corn but declines of up to 30 per cent projected for wheat and soybeans (Singh et al. 1998). In the Peace River region and further north, much of the expected warming is expected during the winter months. This represents a significant shift in regional climate, but potential increases in crop yields or the range of crop that might be grown near the current northern frontier for Canadian agriculture will be somewhat offset by expected declines in summer precipitation and concomitant increases in crop moisture stress (Brklacich, Curran, and Brunt 1996)

Overall, global climatic change is expected to improve yields for many crops in many regions of Canada, but these benefits are expected to be highly variable, and more erratic crop yields are likely to increase uncertainty in Canada's agricultural sector.

The Potential for Canadian Agriculture to Adapt to Climate Change

Canadian agriculture is a dynamic sector that is constantly exposed to multiple stresses and opportunities and constantly responding to these stimuli. 'Adaptation to climate change refers to adjustments in ecological, social and economic stimuli and their effects or impacts' (Smit and Pilifosova 2003, 9), and it involves measures taken at the farm level as well as adjustments in agricultural policies and programs at sub-national, national, and international levels (Burton and Lim 2005). Agricultural adaptation includes reactive responses to specific events as well as longer-term proactive or strategic responses to future threats (Bryant et al. 2000) and routinely involves a mixture of reactive and proactive actions implemented over several years (Brklacich et al. 2000). Overall, adaptation is routinely used by the Canadian agricultural sector to manage risks and reduce uncertainties (Wandell and Smit 2000).

Recent research has confirmed that climate variability and change does not always lead to adaptation measures in Canadian agriculture. In some cases, Canadian farmers are able to cope with the change without making adjustments to their operations, and in other cases institutional and/or economic barriers discourage adaptation (Burton and Lim 2005; Brklacich 2006). When agricultural adaptation does occur, farmers tend to apply numerous measures that are integrated into broader strategies to manage a wide range of threats that extend beyond climatic change (Smit and Pilifosova 2003).

Table 9.1 summarizes the range of adaptation options currently available to the Canadian agricultural sector, including producers, governmental organizations, and the agri-food industry, and that will continue to help them manage risks and future uncertainties. This extensive toolkit provides Canadian agriculture with multiple mechanisms to cope with climate uncertainties, but the economic losses and the personal hardships that have been triggered by recent extreme weather events suggest that

Table 9.1 Adaptation Options for Canadian Agriculture

1. **Technological Developments**
   - New crop varieties more suitable to altered climates
   - Long-term seasonal forecasts to assist with cropping decisions
   - Farm-level management systems to reduce climate-related risks

2. **Government Programs and Insurance**
   - Ad hoc compensation to reduce losses
   - Insurance to reduce losses from crop failures
   - Subsidies and incentive programs that stimulate farm-level adaptations

3. **Farm Production Practices**
   - Enterprise diversification in order to spread out risks
   - Irrigation and water conservation to improve water supply
   - Changing the timing of farm operations (e.g., seeding earlier to take advantage of warmer temperatures)

4. **Farm Financial Management**
   - Income stabilization to reduce year-to-year fluctuations in profits
   - Diversifying household income to include non-farm employment
   - Purchasing crop insurance

Source: Derived from Smit and Skinner 2002.

'[n]otwithstanding the technological and management adaptation measures available to producers, Canadian agriculture remains vulnerable to climatic variability and climate change' (Bryant et al. 2000, 48).

Recent work suggests that many Canadian farmers recognize opportunities and threats stemming from global climatic change, but their decisions on how to grapple with the uncertainties are not based on climatic stimuli in isolation (Bradshaw, Dolan, and Smit 2004). Agricultural adaptation routinely employs multiple strategies that are framed within the overall sustainability of agricultural systems and rural communities (Bryant et al. 2000; Stroh Consulting 2005; Brklacich 2006), and evidence is growing that there is a 'mutually supportive relationship between sustainable agriculture and climate change adaptation' (Wall and Smit 2005, 113).

It is important, however, to recognize that many of Canada's rural communities have been in decline for several decades and continue to be threatened by multiple stresses, ranging from urban intrusions into the countryside to the continued erosion of the social fabric that once defined rural Canada (Troughton 2004). Therefore, the capacity to adapt to climatic change is shaped by the overall health of these communities as well as by climatic change itself. Overall, 'The livelihoods of rural Canadians are already stressed by low commodity prices and trade conflicts . . . and climate change will bring additional challenges, which may aggravate the current situation' (Senate of Canada 2003, 56).

CONCLUSION

Overall, Canadian farmers and rural communities have been exposed to multiple stresses for many decades, and Canadian agri-food systems have responded to these challenges and opportunities. Globalization of agricultural systems, the emergence of organic agriculture, and climate change are three factors that are currently shaping and will continue to shape Canadian agricultural systems over the next few decades. The question is not whether Canadian agricultural systems and rural communities will be able to respond to any one of these forces, but rather it is about the capacity to manage all of these (and other) uncertainties and the conflicts that might arise as competition for scarcer agricultural resources, especially water, intensify. Canadian agricultural systems are supported by innovative farmers and relatively strong formal and informal agricultural institutions. Nevertheless, the adaptive capacity of Canadian agriculture to thrive among these new uncertainties continues to be debated, and there is an urgent need to improve programs to promote adaptation as a means of managing uncertainties within Canadian agriculture.

FROM THE FIELD

Definition of a farmer: an individual out standing in his/her field.

The above adage captures much of the recent reshaping of Canadian agriculture. For many years, ownership of high-quality farmland, coupled with robust rural communities, was the backbone of agriculture, and Canada's *outstanding* farmers were truly *out standing* in their fields. But much has changed. Agriculture has become more dynamic and more complex over the past half century. Today, Canadian farmers must also be sound business managers and entrepreneurs with expertise to manage multiple uncertainties and conflicts tied to economic and other forms of globalization, the changing position of the sector in Canadian society, and changing agriculture–environment relationships.

Agriculture remains among the top five sectors in both exports and imports in Canada, and it has been fully integrated into national and global economies for decades. The realities of farming in Canada are vastly different from the realities of the past when public policies were most often designed to nurture and, when necessary, protect the sector from external competition. These tendencies are deeply rooted in Canadian politics but clash with a global economy, which has given rise to the current conflicts between protectionism and free trade that have not been resolved and continue to be negotiated at local through global scales.

The changing context of agriculture within Canadian society is an ongoing tension that is also continuously reshaping the sector. As Canada becomes increasingly urban, fewer Canadians have direct on-farm experiences. There is a growing sense among many agricultural communities of a loss of position within Canada and that public policies are increasingly slanted toward urban needs. The agriculture sector is responding in many ways, and the recent growth of alternative food networks in near-urban areas is as much about the reconnecting of urban and rural areas as it is about less intensive production systems. Relationships between agriculture and Canadian society will continue to evolve, and the marketing of the importance of agriculture in an increasingly urban Canada will become more vital in the future.

And finally, anthropogenic global climatic change is adding new uncertainties and conflicts. Canadian farmers are used to coping with variable weather, so it should not be surprising that concerns go beyond more reliable weather forecasting. Mitigation measures to reduce human influences on climate systems and adaptation measures that will allow sectors to cope with more uncertainty represent two responses to climate change. Canadian agriculture plays important roles in both areas, ranging from restoring carbon in agricultural soils to the use of farm management and technologies to reduce production risks. Perhaps the most important lesson from recent Canadian research into climate–agriculture relationships is the debunking of mitigation and adaptation as potentially adversarial responses and a gradual shift toward identifying an appropriate blend of mitigation and adaptation responses.

These and other forces shaping Canadian agriculture intersect at many scales and form part of the reality faced by Canadian farmers and the agricultural sector. The capacity of Canadian farmers to continue to be *outstanding* as well as *out standing* in their field will rely heavily on research that starts with a recognition that addressing future uncertainties and conflicts hinges upon our collective capacity to understand and embrace the dynamics and complexity of the sector.

—Iain Wallace and Mike Brklacich

REVIEW QUESTIONS

1. What is the capacity of Canadian agriculture to cope with and manage uncertainties and conflicts stemming from globalization, the organic agriculture movement, and global climatic change?
2. What are the global through local connections that link Canadian agriculture to globalization, the organic agriculture movement, and global climatic change?

REFERENCES

AAFC (Agriculture and Agri-Food Canada). 2007. 'An overview of the Canadian agriculture and agri-food system'. Ottawa: AAFC. http://www4.agr.gc.ca/resources/prod/doc/pol/pub/sys/pdf/sys_2007_e.pdf.

Beauchesne, A., and C. Bryant. 1999. 'Agriculture and innovation in the urban fringe: The case of organic farming in Quebec, Canada'. *Tijdschrift voor economische en sociale geografie* 90: 320–8.

Blay-Palmer, A. 2005. 'Growing innovation policy: The case of organic agriculture in Ontario, Canada'. *Government and Policy* 23: 557–81.

Boland, G.J., et al. 2004. 'Climate change and plant diseases in Ontario'. *Canadian Journal of Plant Pathology* 26: 335–50.

Bowler, I.R. 1994. 'The institutional regulation of uneven development: The case of poultry production in the province of Ontario'. *Transactions of the Institute of British Geographers* NS 19: 346–58.

Bradshaw, B., H. Dolan, and B. Smit. 2004. 'Farm-level adaptation to climate variability and change: Crop diversification in the Canadian Prairies'. *Climatic Change* 16: 1–23.

Brklacich, M. 2006. 'Advancing our understanding of the vulnerability of farming to climate change'. *Die Erde* 137 (3): 181–98.

Brklacich, M., et al. 1997. 'Adaptability of agricultural systems to global climatic change: A Renfrew County, Ontario, Canada, pilot study'. In B. Ilbery, T. Rickard, and Q. Chiotti, eds, *Agricultural Restructuring and Sustainability: A Geographic Perspective*, 185–200. Wallingford, UK: CAB International.

———. 2000. 'Agricultural adaptation to climatic change: A comparative assessment of two types of farming in central Canada'. In H. Millward, K.B. Beesley, B. Ilbery, and L. Harrington, eds, *Agricultural and Environmental Sustainability in the New Countryside*, 40–51. Halifax: Saint Mary's University Press.

Brklacich, M., P. Curran, and D. Brunt. 1996. 'The application of agricultural land rating and crop models to CO$_2$ and climatic change issues in northern regions: The Mackenzie Basin case study'. *Agricultural and Food Science in Finland* 5: 351–65.

Bryant, C., et al. 2000. 'Adaptation in Canadian agriculture to climatic change and variability'. *Climatic Change* 45: 181–201.

Bryant, C., and S. Cofsky. 2004. *Central State Policies and Programs for Local Economic Development: What Do They Contribute to Sustainability?* Aberdeen: University of Aberdeen and the IGU Commission on the Sustainability of Rural Systems, 16–21.

Bryant, C., B. Singh, and P. Thomassin. 2008. *The Co-construction of New (Climate Change) Adaptation Planning Tools with Stakeholders and Farming Communities in the Saguenay–Lac-Saint-Jean and Montérégie Regions of Québec*. Ottawa: Research Report, Natural Resources Canada, CCIAP.

Burton, I., and B. Lim. 2005. 'Achieving adequate adaptation in agriculture'. *Climatic Change* 70: 191–200.

Canadian Organic Growers. 2007. 'Organic agriculture in Canada'. http://www.cog.ca/documents/Organic%20Agriculture%20in%20Canada.ppt.

CBC News. 2004. 'Percy Schmeiser's battle'. Ottawa: Canadian Broadcasting Corporation. http://www.cbc.ca/news/background/genetics_modification/percyschmeiser.html.

———. 2006. 'Timeline of BSE in Canada and the U.S.' Ottawa: Canadian Broadcasting Corporation. http://www.cbc.ca/news/background/madcow/timeline.html.

CDC (Canadian Dairy Commission). 2008. 'Market sharing quota'. Ottawa: CDC. http://www.cdc-ccl.gc.ca/cdc/index_en.asp?caId=812&pgId=2180.

CDIC (Canadian Dairy Information Centre). 2009. 'Dairy facts and figures'. Ottawa: CDIC. http://www.dairyinfo.gc.ca/pdf/dairy_cows_by_prov.pdf.

CFIA (Canadian Food Inspection Agency). 2005. 'Young children's food chemical residues project: report on agricultural pesticides residues 2003-2004'. Ottawa: CFIA. http://www.inspection.gc.ca/english/fssa/microchem/resid/2003-2004/todenfe.shtml#4.

Chiotti, Q. 1992. 'Sectoral adjustments in agriculture: Dairy and beef livestock industries in Canada'. In I. Bowler, C. Bryant, and M.D. Nellis, eds, *Contemporary Rural Systems in Transition. Volume 1: Agriculture and Environment*, 43–57. Wallingford, UK: CAB International.

Chisholm, M. 1979. *Rural Settlement and Land Use: An Essay in Location*. 3rd ed. London: Hutchinson.

Dorais, M. 2007. 'Organic production of vegetables: State of the art and challenges'. *Canadian Journal of Plant Science* 87: 1055–66.

Dimitri, C., and C. Greene. 2002. 'Recent growth patterns in the U.S. organic foods market'. *Agricultural Information Bulletin 777*. Washington: US Department of Agriculture, Economic Research Service. http://ers.usda.gov/publications/aib777/aib777c.pdf.

Forge, F. 1998. 'Recombinant bovine somatropin (rBST)'. PRB 98-1E. Ottawa: Parliamentary Research Branch. http://dsp-psd.communication.gc.ca/Collection-R/LoPBdP/BP/prb981-e.htm.

Froment, G. 2008. 'Overview of the Canadian dairy industry'. agrecon.mcgill.ca/courses/430/notejh/Froment-08.ppt.

Gameda, S., A. Bootsma, and D. McKenney. 2007. 'Potential impacts of climate change on agriculture in eastern Canada'. In E. Wall, B. Smit, and J. Wandeleds, eds, *Farming in a Changing Climate: Agricultural Adaptation in Canada*, 53–66. Vancouver: University of British Columbia Press.

George-Cosh, D. 2007. 'Matters of taste'. *The Globe and Mail* 16 October. http://www.globecampus.ca/in-the-news/globecampusreport/matters-of-taste.

Glover, D., and K. Kusterer. 1990. 'McCain Foods, Canada: The political economy of monopoly'. In D. Glover and K. Kusterer, eds, *Small Farmers, Big Business: Contract Farming and Rural Development*, 73–93. Basingstoke, UK: Macmillan.

Goodman, D., B. Sorj, and J. Wilkinson. 1987. *From Farming to Biotechnology*. Oxford: Blackwell.

Goodman, D., and M. Redclift. 1991. *Refashioning Nature: Food, Ecology and Culture*. London: Routledge.

Halseth, G., S. Markey, and D. Bruce, eds. 2009. *The Next Rural Economies: Constructing Rural Place in a Global Economy*. Oxfordshire, UK: CABI Press.

Health Canada. 2005. 'Overview of Canada's BSE safeguards'. http://www.hc-sc.gc.ca/fn-an/securit/animal/bse-esb/safe-prot-eng.php.

IPCC (Intergovernmental Panel on Climate Change). 2001. *Climate Science 2001: The Physical Science Basis*. Report from the IPCC Working Group I Report. Cambridge: Cambridge University Press.

———. 2007. *Climate Science 2007: The Physical Science Basis*. Report from the IPCC Working Group I Report. Cambridge: Cambridge University Press.

Jardins-Nature 2009. 'About our company'. http://jardinsnature.ca/english/distributeurs/a-propos-de-nous.

Jowit, J. 2008. 'Was the organic food revolution just a fad? Fear for farmers as shoppers tighten belts'. *The Guardian* 29 August. http://www.guardian.co.uk/environment/2008/aug/29/organic.food.

Kneen, B. 1995. *Invisible Giant: Cargill and Its Transnational Strategies*. Halifax: Fernwood.

———. 1999. *Farmageddon: Food and the Culture of Biotechnology*. Gabriola Island, BC: New Society Publishers.

Lemmen, D., and F. Warren. 2004. *Climate Change Impacts and Adaptation: A Canadian Perspective*. Ottawa: Natural Resources Canada.

McGinn, S.M., et al. 1999. 'Agroclimate and crop response to climate change in Alberta, Canada'. *Outlook on Agriculture* 28 (1): 19–28.

Mackenzie, F. 1992. '"The worse it got, the more we laughed": A discourse of resistance among farmers of eastern Ontario'. *Environment and Planning D: Society and Space* 10 (6): 691–713.

MacRae, R.J., B. Frick, and R.C. Martin. 2007. 'Economic and social impacts of organic production systems'. *Canadian Journal of Plant Science* 87: 1037–44.

Mason, H.E., and D. Spaner. 2006. 'Competitive ability of wheat in conventional and organic management systems: A review of the literature'. *Canadian Journal of Plant Science* 86: 333–43.

Miller, M. 2008. '"Not just about the vegetables": Community supported agriculture and discourses of the local and nature in the Ottawa area'. (Carleton University, Ottawa, MA thesis).

Moore, E. 2008. 'The new agriculture: Genetically engineered food in Canada'. In M. Howlett and K. Brownsey, eds, *Canada's Resource Economy in Transition*, 83–101. Toronto: Emond Montgomery.

Nielsen Canada. 2007. 'More than half of Canadian households purchased organics in 2006'. http://ca.acnielsen.com/news/20070514.shtml.

Nyirfa, W.N., and B. Harron. 2004. *Assessment of Climate Change on the Agricultural Resources of the Canadian Prairies*. Prairie Adaptation Research Collaborative (PARC). Summary Document no. 04-02.

OECD (Organisation for Economic Co-operation and Development). 2008. 'Agricultural support estimates'. *OECD Factbook 2008*. Paris: OECD. http://titania.sourceoecd.org/vl=3742173/cl=18/nw=1/rpsv/factbook/100301.htm.

OMAFRA (Ontario Ministry of Agriculture, Food and Rural Areas). 2007. 'Supply management systems fact sheet'. Toronto: OMAFRA. http://www.omafra.gov.on.ca/english/farmproducts/factsheets/supply.htm.

Prudham, Scott. 2004. 'Poisoning the well: Neoliberalism and the contamination of municipal water in Walkerton, Ontario'. *Geoforum* 35 (3): 343–59.

Rochette., P., et al. 2007. 'Climate change and winter damage to fruit trees in eastern Canada'. *Canadian Journal of Plant Science* 84: 1113–25.

Sauchyn, D. 2007. 'Climate change impacts on agriculture in the Prairies'. In E. Wall, B. Smit, and J. Wandel, eds, *Farming in a Changing Climate: Agricultural Adaptation in Canada*, 67–80. Vancouver: University of British Columbia Press.

Savour Ottawa. 2007. http://www.ottawatourism.ca/savourottawa.

Segal, H. 2008. 'From Champlain to Charest: The case for Quebec's leadership'. *Policy Options* July/August: 26–30.

Senate of Canada. Standing Committee on Agriculture and Forestry. 2003. *Climate Change: We Are at Risk*. Final Report. Ottawa: Senate of Canada.

Shaw, H. 2008. 'Shoppers Drug Mart Corp. steps up organic food fight against Loblaw'. *National Post* 6 March. http://network.nationalpost.com/np/blogs/fpposted/archive/2008/03/06/shoppers-drug-mart-corp-steps-up-organic-food-fight-against-loblaw.aspx.

Singh, B., et al. 1998. 'Impacts of a GHG-induced climate change on crop yields: Effects of acceleration in maturation, moisture stress and optimal temperature'. *Climatic Change* 38: 51–86.

Skogstad, G. 2008. 'The two faces of Canadian agriculture in a post-staples economy'. In M. Howlett and K. Brownsey, eds, *Canada's Resource Economy in Transition*, 63–82. Toronto: Emond Montgomery.

Smit, B., and O. Pilifosova. 2003. 'From adaptation to adaptive capacity and vulnerability reduction'. In J. Smith, R.T.J. Klein, and S. Huq, eds, *Climate Change, Adaptive Capacity and Development*, 9–28. London: Imperial College Press.

Smit, B., and M. Skinner. 2002. 'Adaptation options in agriculture to climate change: A typology'. *Mitigation and Adaptation Strategies for Global Change* 7: 85–114.

Smith, A., and J.B. Mackinnon. 2007. *The 100-Mile Diet: A Year of Local Eating*. Toronto: Vintage Canada.

Smith, W. 1984. 'The "vortex model" and the changing agricultural landscape of Quebec'. *The Canadian Geographer* 28: 358–72.

SRQ (Solidarité Rural du Québec). 2008. *Études de cas sur la néo-ruralité et les transformations de la communauté rurale*. Research Report. Nicolet: SRQ.

Statistics Canada. 2008. 'Canada's largest farms'. *The Daily* 25 July. Ottawa: Statistics Canada. http://www.statcan.gc.ca/daily-quotidien/080725/dq080725a-eng.htm.

———. 2009. '2006 Census: Portrait of the Canadian population in 2006: Findings'. Ottawa: Statistics Canada. http://www12.statcan.ca/census-recensement/2006/as-sa/97-550/p13-eng.cfm.

Stroh Consulting. 2005. *Agriculture Adaptation to Climate Change in Alberta: Focus Group Results*. Alberta Agriculture, Food and Rural Development, Conservation and Development Branch.

Troughton, M. 1991. 'An ill-considered pact: The Canada–U.S. Free Trade Agreement and the agricultural geography of North America'. In R.C. Rounds, ed., *Trade Liberalization and Rural Restructuring in Canada*. Working Paper no. 1. Brandon, MB: Rural Development Institute.

———. 2004. 'Agriculture and rural resources'. In B. Mitchell, ed., *Resource and Environmental Management in Canada: Addressing Conflict and Uncertainly*, 3rd edn, 233–64. Don Mills: Oxford University Press.

Wall, E., and B. Smit. 2005. 'Climate change adaptation in light of sustainable agriculture'. *Journal of Sustainable Agriculture* 27 (1): 113–23.

Wallace, I. 1992. 'International restructuring of the agri-food chain'. In I. Bowler, C. Bryant, and M.D. Nellis, eds, *Contemporary Rural Systems in Transition. Volume 1: Agriculture and Environment*, 15–28. Wallingford, UK: CAB International.

———. 2002. *A Geography of the Canadian Economy*. Toronto: Oxford University Press.

Wandel, J., and B. Smit. 2000. 'Agricultural risk management in light of climate variability and change'. In H. Millward, K. Beesley, B. Ilbery, and L. Harrington, eds, *Agricultural and Environmental Sustainability in the New Countryside*, 30–9. Winnipeg: Hignell Printing.

Weaver, A. 2004. 'The science of climate change'. In H. Coward and J. Weaver, eds, *Hard Choices: Climate Change in Canada*, 13–43. Waterloo, ON: Wilfrid Laurier University Press.

Winder, G. 2002. 'Following America into the second industrial revolution: New rules of competition and Ontario's farm machinery industry, 1850–1930'. *The Canadian Geographer* 46: 292–309.

Wunsch, P. 2004. 'There's more to organic farming than being pesticide-free'. *Canadian Agriculture at a Glance*, 179–88. Catalogue no. 96-325-XPB. Ottawa: Statistics Canada, Agriculture Division.

10

Transition and the Need for Innovation in Canada's Forest Sector

Kevin Hanna

Learning Objectives

- To develop a basic understanding of the diversity of forests in Canada and the importance of the forest sector to the Canadian economy.

- To understand the scale of the Canadian forest sector relative to other global competitors and the key challenges that affect the Canadian industry.

- To understand the importance of tenure arrangements to forest management and to recognize the important historical foundations of the present tenure and policy setting. Forests in various forms cover about 402 million hectares of Canada's landmass. This represents about 10 per cent of the world's total forest cover and 30 per cent of the world's boreal forest (NRCAN 2007a). Provincial governments own just over 90 per cent of forestland, making Canada distinct among developed nations—a country with overwhelming state ownership of productive forests.

While in Canada the forest resource is essentially government property, the private sector provides the means of exploitation and resource development. For more than a century, this government–industry compact had served business and the provinces well. Governments collected economic rents directly and realized other revenues indirectly. Forest resources were exploited to advance economic growth, and wood provided the basis for development of many of Canada's hinterland regions. Industry profited by turning the nation's wood fibre wealth into paper, construction products, and other wood-based goods. All the while, Canada's provinces provided a politically stable setting for

the creation of a profitable industry based on an extensive resource. This management regime has been described as public forest management for private timber production (Howlett 2001, 8).

Canada's forests, especially in BC, Quebec, and Ontario, have long been significant instruments for regional economic development, and while the industry has had its 'ups and downs', the forest sector remains integral to the economic well-being of many Canadian regions and a significant contributor to the nation's positive balance of trade. In recent decades, the industry has expanded greatly in Alberta and Saskatchewan and has become a dominant economic force in New Brunswick. Canada, in so many respects, owes much of its economic well-being to the hewing of wood.

The comfortable relationship between governments and industry has unravelled somewhat in recent decades, in no small part because many Canadians have increasingly viewed their nation's forest legacy as something other than fibre waiting to be exploited. There is an increasing desire to see some forestlands preserved, management practices changed, and more 'value-added products' from this significant public resource. This new view is manifest in years of conflict over how Canada's forests are managed, how they should be used—if at all—and who should benefit from public forest resources. In some areas, First Nations—Aboriginal peoples—have staked new claims to some forestlands, creating uncertainty about the future of the industry and established patterns of resource use but at the same time presenting new opportunities for co-management and regional economic development. First Nations have emerged as significant players in resource management in Canada, and in BC their role is now central to many forest management and planning decisions (also see Chapter 4).

In this chapter, I provide a brief characterization of Canada's forest types, an extended discussion of the structure of the industry and the need for investment and innovation, and a history of Canadian forest policy. Understanding the history of forest policy not only helps us to understand how the contemporary setting formed, it also illustrates the origins of some of the uncertainties that affect the forest sector. The importance of tenure is emphasized (conditions of forest use set by a provincial government). Tenure in many respects is at the heart of both industry survival and many of the forestry challenges facing Canadians. Conflict is noted throughout as a significant driver in recent forest policy initiatives, a point reinforced by Roger Hayter's guest statement.

The relationship among industry health, policy development, and environmental opportunity and performance is an integration challenge in which the economic well-being of forest communities and the industry that supports them are intertwined with resilience and sustainability objectives. Other authors in this book also address forestry issues. Reed (Chapter 19) considers the experience of forestry governance and transition and the ability of the discourse to inform other areas of environmental and resource management. Noble (Chapter 16) addresses adaptive management as a formal approach based on passive or active experimentation and draws on forestry and grazing case studies. The approach here is broad and intended to provide a snapshot of the forest sector and its place in Canada's economy and landscape.

CANADA'S FORESTS

Few nations can boast the scale and diversity of forest types that Canada contains. The forested area of Canada is about 310 million hectares, with an additional 92 million hectares of 'treed land' (lands not forested, which includes areas such as wetlands, muskeg, bog, swamp with some trees, and land with scattered trees) (NRCAN 2001). Some 89 per cent of Canada's forestland is considered to be stocked (currently supports trees). Of this land, 31 per cent is young, 37 per cent is mature or over-mature (standing wood that has reached the stage where it is near the end of its natural lifespan), and 32 per cent is uneven-aged (a forest stand with three or more distinct age classes) or unclassified as to maturity (NRCAN 2001). The stocked forest has a wood volume of 29 billion square metres, and about 77 per cent of this is dominated by coniferous wood species while the remaining 23 per cent is mostly broadleaved trees (NRCAN 2001).

GUEST STATEMENT

Forest Resources

Roger Hayter

*The time has come to dispel the notion that
forestry is mainly about tending woodyards.*
[Westoby 1989, 213]

In Canada, Westoby's claim of 20 years ago that
global forestry practices need to give greater priority to the non-industrial values of forests is
strikingly underlined by the highly publicized opposition of environmental non-government organizations (ENGOS) to industrial forestry. These conflicts
emerged in the 1970s, typically as locally organized
protests, and escalated rapidly in the 1980s as
global events orchestrated by multinational ENGOS
that involved attacks on markets as well as logging
sites and international criticism (and 'shaming')
in the media. In coastal British Columbia, logging
blockades by individual groups led to various
globally famous 'wars in the woods' in Clayoquot
Sound, the Carmanah Valley, and the Stein in the
1980s and 1990s, culminating on the mid-coast in
the Great Bear Rainforest campaign and multi-party
agreement of 2006, generally acknowledged as a
great victory for ENGOS.

Ironically, but not coincidently, as government
and corporate policies responded to a more serious
appreciation of the non-industrial values of forests, the economic performance of Canadian forest
industries has become deeply troubled. For almost
40 years, deepened environmental concerns have
been combined with increased industrial volatility, and in 2008 and 2009 the industry was *again*
characterized by widespread job losses, plant closures, and loss of profitability. Evidently, there are
serious challenges facing long-run sustainability

of both the industrial and non-industrial values of
Canadian forests.

During the long expansionary period of the 1950s,
1960s, and early 1970s, Canada's forest industries
experienced substantial growth and stability, but
if firms, communities, and labour prospered, the
environment did not. Industry (and government) dis-
dain for the environment is captured by two firmly
entrenched (and convenient) beliefs of the time:
Canadian forests would self-renew, and old-growth
wood is decadent and needs to be logged before it
becomes worthless.

The 1970s heralded fundamental changes,
however, in the industry's evolution and environmental beliefs. Growth levelled off, and industrial
performance became volatile; the energy crises and
strong recessions of the 1970s (1971, 1976) were followed by the remarkably long recession of the early
1980s and further significant downturns in the early
and late 1990s, while the 2009 crisis is enormously
threatening to forest industries across the country.
Roughly in tandem, the industry's environmental
credentials have improved. Since 1970, programs
of reforestation have expanded significantly across
Canada, old growth is no longer deemed decadent,
air and water pollution has been reduced, and forest conservation areas have expanded.

Environmental conflict did not cause forest industry volatility, but the two trends are connected
(Hayter 2003). The booming 1950s and 1960s greatly
extended forest exploitation by mass production and
pushed the industry rapidly to the mature end of the
resource lifecycle—which envisages resource exploitation passing through stages of early growth, boom,

maturity, and bust—with its concomitant increased costs and declining access to quality resources (Clapp 1998). The same trend, and fears for the complete elimination of old growth, sparked environmental conflict. Meanwhile, especially since the 1980s, forest industry volatility has been reinforced by the need to incorporate the implications of information and communication technology, notably with respect to micro-electronics, and to cope with remarkably enduring and punitive US tariff protectionism. Moreover, especially in British Columbia, ENGOS have allied themselves with US protectionists in attacking Canadian industrial forestry, while Aboriginal land claims have added cultural–political dimensions to economy–environment debates over the use of Canadian forests.

The central challenge facing Canadian forest policies now is to ensure that *both* economy and environmental values are realized. ENGOS are fond of saying, with legitimacy, that industrial forestry cannot be sustained by ignoring environmental priorities. But the environment also needs the economy. Conservation is costly, and caring approaches to forestry require achievement of the legitimate material needs of people and communities. In the past, technological change often had dire environmental consequences. Technological and institutional innovation, however, is the key for the future reconciliation of economic and environmental values. Despite frequent pleas, innovation has never been central to Canadian forest policies, within the provinces or federally, and Canadian forestry innovation has been adequate rather than stellar. Now is a good time to change this attitude. If so, an appropriate innovation system for Canadian forests requires greater emphasis on interdisciplinary research, including a stronger role for universities and a more cooperative (less adversarial) use of science between ENGOS and industry.

Forest Types

Canada's westernmost province, British Columbia, is arguably the most significant province in terms of wood production and forest diversity (Figure 10.1). High-elevation montane, subalpine and Columbia forests cover vast portions of the province's interior and mountain regions. The extensive coastal forests, with large tracts of temperate rainforest, were the historical anchor of the early BC forest industry and in recent decades have become the setting for acrimonious conflicts over land use (see Figure 10.2). The colossal Douglas fir, red cedar, and Sitka spruce found in these forests have assumed an iconic status among environmentalists and indeed the Canadian and global public. The province's interior forest systems, with large areas of lodgepole and ponderosa pine, are perhaps less evocative, but with the

spread of insect epidemics, they may be more endangered than their coastal brethren.

On the BC coast, the rotation lengths (the time needed for trees to mature to a harvest age) can be relatively short. Coastal forests that may appear pristine because of the presence of large trees may actually be second-, perhaps even third-growth (since being logged). A walk though many Vancouver Island forests will reveal not only large trees but hidden among the growth the stumps of giants cut long ago, many with springboard notches still evident (a notch cut to allow a board to be placed where a logger would stand while using a crosscut saw). Less resilient to human use are British Columbia's high-elevation forests—alpine and subalpine—containing Engelmann spruce and subalpine fir (Figure 10.3). These forests cover a substantial area, with harsh climates, but at the same time

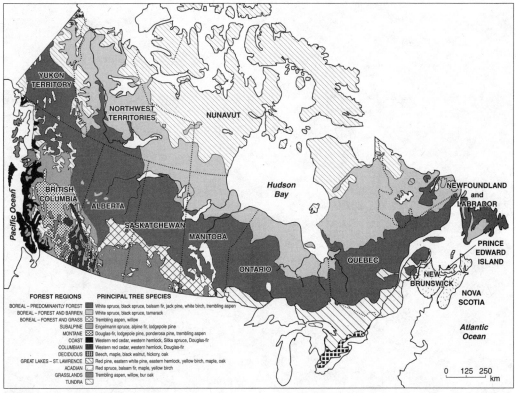

FOREST REGIONS	PRINCIPAL TREE SPECIES
BOREAL – PREDOMINANTLY FOREST	White spruce, black spruce, balsam fir, jack pine, white birch, trembling aspen
BOREAL – FOREST AND BARREN	White spruce, black spruce, tamarack
BOREAL – FOREST AND GRASS	Trembling aspen, willow
SUBALPINE	Engelmann spruce, alpine fir, lodgepole pine
MONTANE	Douglas-fir, lodgepole pine, ponderosa pine, trembling aspen
COAST	Western red cedar, western hemlock, Sitka spruce, Douglas-fir
COLUMBIAN	Western red cedar, western hemlock, Douglas-fir
DECIDUOUS	Beech, maple, black walnut, hickory, oak
GREAT LAKES – ST. LAWRENCE	Red pine, eastern white pine, eastern hemlock, yellow birch, maple, oak
ACADIAN	Red spruce, balsam fir, maple, yellow birch
GRASSLANDS	Trembling aspen, willow, bur oak
TUNDRA

NRCAN 2007b.

Figure 10.1 Forest Regions of Canada Based on Forest Composition.

they are delicate and susceptible to environmental change. They do not regenerate easily or quickly and are especially vulnerable to human activities. As timber supplies decline in other areas, the lower reaches of these forests may come under harvesting pressure.

The boreal forest extends across the nation from northern BC across northern Canada to Quebec and Newfoundland and Labrador (Figure 10.1). This is the largest forest type in Canada. The boreal forest is dominated by coniferous trees—black spruce, jack pine, and balsam fir. The colder and drier climates that characterize the boreal zone mean that rotation lengths are longer—even relatively small-diameter timber can take a long time to mature relative to wood

in warmer zones. In the northern boreal forest, notably in regions such as Yukon, trees grow very slowly, and logging can take on the characteristic of 'timber mining' (forests are managed as a non-renewable resource, or there is no expectation of regeneration within a reasonable time). Canada's boreal forest may well prove to be among the globe's great carbon sequestration resources. Extensive logging has threatened the boreal habitat for a range of wildlife such as the woodland caribou. While the boreal forest may seem impenetrable, and indeed infinite, in some regions it is at risk.

The Great Lakes–St Lawrence forest covers most of central Ontario and Quebec. This forest is vast and is composed of commercially

(Kevin Hanna).

Figure 10.2 Dense Coastal Second-Growth Forest on Western Vancouver Island, BC.

(Kevin Hanna).

Figure 10.3 High-Elevation Forest in the Columbia Selkirk Mountains, BC, mostly Engelmann Spruce and Subalpine Fir.

important conifers such as eastern white pine, red pine, and eastern hemlock. Deciduous tees—birch, maple, and red oak—are also found mixed within this forest. The deciduous forest covers southeastern Canada and contains mostly angiosperms such as oak, maple, beech, and birch. In southern Ontario, the Carolinian forest, historically always a very small area, is still found in small patches. The Carolinian is perhaps the most curious forest found in Canada, with its unique assemblages of trees and other plants—reminiscent of warmer southern climes. While the Carolinian is dominated by American beech and sugar maple—not particularly rare trees—it is the rare plants, such as the Kentucky coffee tree, black gum, sassafras, pawpaw, and tulip tree, that give this forest its unique qualities.

The Acadian forest is found in the Maritime provinces. This forest is composed primarily of sugar maple, American beech, yellow birch, red spruce, and eastern hemlock. The forest industry in the Maritime provinces is significant and centred to a great extent on private lands. The transition of the Acadian forest through the pressure of industrial forestry poses new challenges for forest management.

Management Challenges

Forestry practices have helped to change the characteristics of Canada's forest; fire suppression has been particularly effective. Many fire-climax forest systems (forest types reliant on natural fire to create conditions for the next succession stage or for tree regeneration, often through frequent burning of the understorey) now contain large volumes of over-mature timber and high ground-level fuel loads and have been in a climax (a final, relatively self-perpetuating stage [Kimmins 1995]) longer than would occur in a natural state (without suppression of natural fire). Harvesting operations have also changed forests, with regeneration geared toward replacing natural systems with tree species most valued by the market. Succession is the gradual replacement of one plant community of plants by another; it is a natural process but can be accelerated by silvicultural choices.

Silviculture decisions (management of the growth of trees from seedling to harvest, including harvest methods [Kimmins 1995]) affect nutrient levels, soil conditions, forest hydrological systems, plant community composition, and animal populations. The pursuit of sustained-yield forest management (a harvest rate that balances net growth with the amount harvested) has also helped to change Canada's forests. The fall-down effect (the wood supply gap between the end of old-growth wood supplies and the availability of second growth) observed in some regions would seem to indicate that sustained yield has not been achieved. Instead, what we have is more akin to a sustained cut—sustained for a while at least. However, far from seeing Canada's forestland wholly diminish in area, as some might assume from the language of environmentalism, what we see today (compared to the conditions at the time of European contact) is perhaps not so much a case of 'less forest' but rather 'transformed forests'. In south-central British Columbia, for example, fire suppression has allowed pine forests to become denser and larger in area. Early images of the Fraser Canyon in BC show mountainsides denuded by fire and logging, a marked contrast to the forests that cover the mountainsides today. In parts of eastern and Maritime Canada, forests have advanced and regenerated significantly since early settlement years.

Human settlement has also altered forest cover. Ontario's Carolinian forest occupies a tiny portion of its original area, mostly because of settlement and the logging of high-value hardwood species. The mixed woodlands of Ontario's Bruce Peninsula and Manitoulin Island contain few old-growth pines, and after more than a century of logging and fire, the forest cover is for the most part today a mixture of small-diameter cedar, spruce, and fir, with some hardwoods. Industry and provincial governments have built extensive road networks in order to access timber in remote areas. This brings in hunters and other recreationalists who have varying impacts on the backcountry. Roads also disturb biophysical systems and can weaken ecological connectivity. While Canada contains significant forested landscapes, resource use has changed the structure and composition of the nation's forests at both large and small scales.

In the past decade, a key challenge has emerged for Canadian forest management. The mountain pine beetle epidemic in BC highlights the interactions of forestry practices and climate change. Already, BC's interior pine forests have experienced devastation from the pine beetle, and the epidemic seems poised to spread across Canada wherever pines are found. This may have profound implications for Canada's boreal forest over the next few decades. The area of BC currently affected is about 13.5 million hectares, about four times the size of Vancouver Island (BC MFR 2008b). The extent of the epidemic stretches from the US border to as far north as Fort St John, but the beetle is found throughout the western US and even into Mexico (BC MFR 2008b).

The 'glut' of beetle-killed pine has skewed BC's forest industry. While BC has sought to manage the supply of dead wood by encouraging salvage operations, the stumpage system (the fee paid to a provincial government for cutting publicly owned trees) has not been sufficiently modified to encourage more substantial use of this wood. Thus, industry has been in essence high-grading (taking the most desirable and valuable wood while leaving other wood standing or on the ground) pine beetle–killed timber, and vast amounts of this wood have remained standing. A few years after death, these trees become unusable to industry. Timely harvesting is essential to reduce waste. The pine beetle has now spread into Alberta, and its move eastward seems inevitable. While the pine beetle has always been endemic in some Canadian pine forests, the current epidemic conditions point to a range of potential environmental challenges that may affect much larger portions of Canada's pine forests as environmental change occurs (see Box 10.1).

THE FOREST INDUSTRY

Employment and Economic Importance

The scale of Canada's forest resources, in terms of species, volume, and quality of wood, seems unrivalled. Only Russia has the 'real' potential to generate as much softwood volume and quality. Despite being the largest global exporter of forest products, Canada's forest sector is in many respects underdeveloped, undercapitalized, and lacking in innovation. The industry has not created the scale of firms or the competitive, indeed global, dominance that it could have with such a remarkable resource at hand.

Estimates of forest sector employment vary. One estimate by the Forest Products Association

Box 10.1 The Mountain Pine Beetle Epidemic: Factors and Biology

The mountain pine beetle, *Dendroctonus ponderosae*, has a lifespan of about one year. The insect is about the size of a grain of rice. Pine beetle larvae spend the winter under bark. They continue to feed in the spring and transform into pupae in June and July. Adult beetles leave infested trees during the summer and into the early fall. The beetle transmits a fungus that stains a tree's sapwood blue. Comprehensive testing has confirmed that the blue stain caused by the beetle has no effect on the wood's strength properties.

The mountain pine beetle prefers mature pine, although in epidemic conditions it will attach to other conifers. At 80 years, lodgepole pine trees are generally considered to be mature, and BC now has about three times more mature lodgepole pine than it did some 90 years ago, mainly because fire suppression has been so successful. Hot and dry summers also leave pine drought-stressed and more susceptible to attack by the mountain pine beetle.

Cold weather kills the mountain pine beetle. Mountain pine beetle eggs, pupae, and young larvae are the most susceptible to freezing temperatures. But temperatures must be consistently below −35°C or −40°C for several straight days to kill off large portions of mountain pine beetle populations. In the early fall or late spring, sustained temperatures of −25°C can freeze mountain pine beetle populations to death. A sudden cold snap is more lethal in the fall, before the mountain pine beetles are able to build up their natural anti-freeze (glycerol) levels. However, with the advance of climate change, we no longer see such cold conditions.

Source: BC MFR 2008b.

of Canada (FPAC 2005) puts the number of people directly or indirectly employed at about 864,000, while another estimate (NRCAN 2007a) places employment in the forest industry at about 750,000. Suffice it to say that the sector provides one of the largest single employment categories in the country and is one of the few commodity or manufacturing sectors to have a truly national presence. The same cannot be said of the energy sector (e.g., especially oil and gas), which is significant mostly in the western provinces, or for that matter of manufacturing, which is largely based in central Canada. All provinces and territories have forest-sector activities. Fibre production arguably provides the greatest direct and indirect economic benefits. Other benefits such as hunting or other recreational activities or non-timber forest products are also important, but they may be poorly accounted for in forest management and planning because they are not adequately measured and valued (see also Chapter 11).

Non-timber forest products (NTFs) are non-wood products of biological origin derived from forests, other wooded land, or trees outside forests (e.g., farms and wetlands) (UN FAO 1999, 63). Non-timber forest products include forest plants and mushroom products and services (BC MFR 2008a). By 1997, NTFs in BC were valued at $600 million per year, with more than 30,000 British Columbians earning all or part of their living from this part of the forest sector. More than 200 forest-based species are harvested, with mushrooms and floral greenery constituting the largest product categories, but there is increasing interest in the potential of the wild-harvested nutraceutical and bio-products (BC MFR 2008a). Salal, for example, a staple in floral arrangements across Canada, is harvested in BC's coastal forests.

Overall, the economic contributions of the forest sector are least in Prince Edward Island,

Nunavut, and the Northwest Territories (where there are no industrial forestry operations of national significance). All other Canadian regions have a notable forest industry. Even in Yukon where the forest industry is relatively small (compared to provincial forest industries), the sector is an important contributor to economic diversity and is seen by the Yukon territorial government as a key component of economic diversification efforts.

The forest industry is the economic base for about 300 Canadian communities, most of them rural or remote and many small (NRCAN 2007a). Jobs in the forest industries tend to be high-paying and increasingly require better skill sets than in past decades. As of 2002, the average forest sector wage was about $54,000; this is 70 per cent higher than the Canadian average (FPAC 2005). However, employment levels have been declining. The most notable declines in employment over the past four decades have occurred in the processing industries. This has been due to advances in the technology of wood processing and mill and plant closures. While employment in harvesting has also declined over the years, in part due to technological advances, it has tended to remain stable relative to processing; most declines in harvesting employment are attributable to downturns in market demand for pulp and paper and construction wood products.

From January 2003 to April 2008, largely because of mill closures and slowdowns, more than 38,000 jobs were lost in forest sector–dependent communities (NRCAN 2007a). The multiplier effect of the industry is significant. Each job in the forest sector supports another 1.5 indirect jobs, so when good-paying forestry jobs disappear, community incomes drop, and the revenues from other sectors can decline dramatically, leading to regional economic decline and instability (NRCAN 2007a). Many of these effects—community decline, a smaller tax base, and greater social impacts and costs—have been evolving slowly over recent decades. This transition has been the source of considerable debate, with some seeing it as inevitable as the forest-based economy gives way to competition, there is less demand for the products, and the cost of labour increases. Other factors such as the placement of forestland in parks or protected areas (as a result of environmentalism), decades of poor investment levels, business management problems, and federal government policies have also affected the sustainability of the forest industry.

An Export Industry

As I noted above, Canada is the world's largest exporter of forest products. Box 10.2 shows the value of total sales (domestic and export) at about $84 billion (in 2005), which was 60 per cent of the nation's merchandise trade surplus and 2 per cent of Canada's GDP. Forest products are now Canada's third largest export to the United States and Europe and the largest export product class to India, China, Japan, and South Korea (FPAC 2008).

The importance of the US and the habit of relying on this large and convenient market has been described as continentalism (Hayter 2000). Canadian federal and provincial governments have long encouraged continentalism. However, it means that the export market is reliant on the performance of the American economy. The US is a convenient market and has been a relatively stable importer of Canadian wood products. Nevertheless, the lack of a broader export market means that Canada's forest products companies are less resilient to economic downturns or trade disputes than their competitors in nations with more diverse export strategies. However, a recent and significant downturn in the US housing market (beginning in the last quarter of 2008 and extending into 2009) and the corresponding decline in US imports of Canadian wood products highlights the perils that come with dependence on the US market.

Canada also faces increasing competition from wood-producing nations in the less developed world, from emerging wood industries in Russia, and from a strong industry in Scandinavia. These regions have seen their share of global markets increase dramatically in recent decades, and their forest product firms have gained global prominence. The Russian government hopes to

---

Box 10.2 Value of Canadian Forest Products, 2005 (in Billions of Dollars)

Total revenue $84 billion
Forestry and logging activities $9 billion
Wood products $41 billion
Paper and pulp products $33 billion
Export sales $42 billion
Contribution to Canada's trade balance $32 billion
Contribution to Canada's trade surplus 60%
Export sales to US $35 billion
Annual research and development expenditures $506 million

Source: FPAC 2008.

increase the value of production in its forest industry from the present US$10 billion annually to some US$100 billion annually within the next two decades—in contrast, the present value of export shipments from Canada is US$40 billion (FPAC 2005).

Russia is seeking to expand its domestic production capacity. As part of this strategy, Russia has imposed stringent tariffs (export taxes) on raw log (unprocessed trees with limbs removed but no additional processing) exports, just at a time when it has become the world's largest log exporter (PwC 2008). The increase in tariffs will be substantial—from 6.5 per cent (a minimum of 4 euros per cubic metre) to new duties as high as 20 to 25 per cent for softwood, increasing to tariffs of 80 per cent (a minimum of 50 euros per cubic metre) by the end of 2009 (PwC 2008). For birch pulpwood, the tariff is delayed until 2011, when it rises from 0 to 80 per cent. The impact, as these policies are fully implemented, will likely be that harvesting timber in Russia will only be economic if the logs can be processed locally (PwC 2008). However, this is the aim—to develop greater domestic production and processing capacity. Conversely, in Canada there are no such tariffs on raw log exports. Indeed, in BC where there were once fairly strict prohibitions on exporting raw logs, such exports have increased in recent years.

Raw log exports are an emotional issue in Canada. In 2006, Canada ratified a softwood lumber agreement with the US after years of dispute over issues ranging from how Canadian provinces price public timber to US tariffs on specific products. The new agreement was a hurried affair brought forward by a federal Conservative minority government, and it may well open the door to increased raw log exports. Under the new agreement, taxes will be levied on coniferous wood products, sawn or chipped, sliced or peeled, planed, sanded,

or finger-jointed, of a thickness exceeding six millimetres, and on wood siding, flooring, and fencing (Parfitt 2006; 2007). However, as Parfitt (2006; 2007) writes, throughout the agreement's appendix, the word 'logs' is absent. Indeed, a key issue in the softwood lumber dispute has been the export of raw logs from Canada, particularly from British Columbia (BC Stats 2003).

The American argument has been that restricting exports of raw logs constitutes a subsidy to Canadian forest companies, since putting logs up for auction will supposedly bring a higher price than companies are currently paying through Canadian stumpage systems (BC Stats 2003). Some working in the forest sector say that there should be a complete ban on log exports, since Canadian timber is a public good that should confer benefits to Canadians and exporting raw logs is tantamount to exporting processing jobs (BC Stats 2003). Those working in harvesting are perhaps less likely to oppose raw log exports if it means harvesting jobs are retained. The recent softwood agreement indicates that the US has no interest in limiting the import of Canadian raw logs. There are significant portions of the western US sawmilling industry that rely on Canadian wood to support American jobs. However, in the US raw log exports from federal land are still prohibited.

There is no outright ban on raw log exports from BC. Section 127 of the BC Forest Act states that all timber harvested from Crown land must be either used in BC or manufactured within the province into other goods (BC Stats 2003). However, Section 128 provides for an exemption if the timber is surplus to the needs of BC processing facilities, if logs cannot be processed economically near the harvesting area and cannot be transported economically to another facility in BC, or if an exemption would prevent waste or improve the utilization of the wood (BC Stats 2003).

From 1998 to 2002, the quantity of BC logs shipped out of the country rose by more than 360 per cent to almost 4 million cubic metres, meaning a rise in export value from about $128 million to $515 million—most logs went to the US (BC Stats 2003). The portion of BC logs shipped to other countries had generally been less than 1 per cent of timber harvested in the province, but by 2002 the amount had grown to more than 5 per cent. The volume of logs exported is still small relative to BC's total harvest, but it is growing fast, and contrary to Russia, where the policy is to encourage domestic processing through aggressive export taxes, the recent Canadian direction is to make such exports easier.

There is an argument that supports raw log exports. This position holds that if the domestic processors cannot use the wood or if higher prices can be realized by selling raw logs to buyers outside Canada, then economic benefits will still be realized. Harvesting jobs will be created or maintained, and there will be income for the provincial government. The argument is extended to the state of the processing sector and the lack of innovation; thus, if Canada's sawmills and pulp and paper plants cannot innovate or make the necessary adjustments to compete in the global market, should harvesting jobs be sacrificed if there are buyers for the nation's logs? There is no easy answer, but the debate highlights the complexity of issues that at first glance may seem simple. An increase in log exports is indicative of a range of problems in the forest industry.

The Scandinavian experience is also cautionary for Canada. With a total forestland area considerably smaller than Canada's, Finland and Sweden still maintain export forest industries second and third to Canada's, and they do not need to export logs. Instead, they need to import wood. Even with domestic supply problems, increasing conflict over forestland use, and looming embargos or tariffs on logs from Russia, these nations arguably remain at the top of global forest-sector competitiveness.

Since 1990, Canada's forestry exports to the US have declined by 12 per cent, in part because of increasing competition from other nations (FPAC 2005). Despite the importance of growing manufacturing, technology, and service sectors, commodities remain the staple of the Canadian export economy, and forest products account for a 'lion's share'. Any real decline in the status and global export dominance of the Canadian forest industry will have profound impacts on the Canadian economy and the fiscal well-being of federal and provincial governments. Unlike the oil and gas sector, the forest industry is sustainable. Assuming that forest resources are managed appropriately and the industry regains its competitive advantage, it will make a major contribution to the Canadian economy as long as there is demand for wood.

Competitiveness and Innovation

Over the past decade, the profile of Canadian forest product sales has changed markedly (Figure 10.4). Paper and paper products sales have grown slowly compared to construction products. There is little indication that demand for paper products, especially newsprint, will rebound. Even in good economic times, the appetite for paper has not grown at the same rate as in previous decades. The decline in demand will be exacerbated by the present recession.

Between the late 1980s and 2000, North American demand for newsprint, a major Canadian paper product, declined by about 14 per cent (FPAC 2005). This all reflects complex changes in the information technology sectors, reading habits, and new advertising formats—especially the growth of digital communication and more advertising resources allocated to electronic formats.

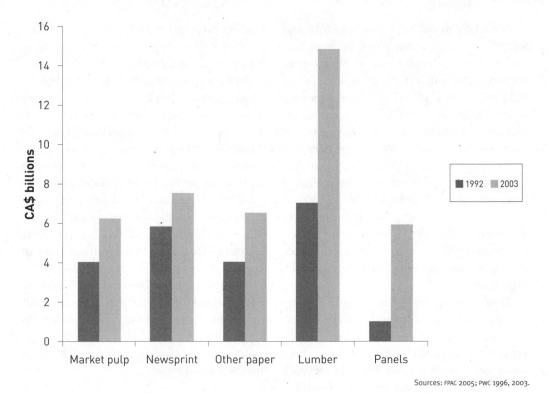

Sources: FPAC 2005; PWC 1996, 2003.

Figure 10.4 Forest Product Sales, 1992 and 2003.

The greatest growth in demand has been for lumber and panels. However, sales of these products are dependent in large part on housing demands. The recent mortgage crisis in the United States, followed by a recession that has hit not only the American economy but also Europe and Asia, will also have a substantial impact on demand for Canadian construction wood products. Since Canada's forest industry is in great part reliant on the American market, a slowdown in that economy will reverberate through Canada's forest-sector communities.

If the Canadian dollar declines relative to the euro and US dollar, as it did in the second half of 2008 and into 2009, Canadian forest products may regain a certain competitive advantage. A low-value currency cannot, however, sustain the industry in the long term, and the deep

structural issues that affect the competitiveness of Canada's forest industries will eventually have to be addressed.

In terms of company size, the ranking of global forest, paper, and packaging products firms is revealing (see Table 10.1, ranking based on PwC 2008). Not one Canadian firm is among the top 10 global companies ranked by sales. Domtar comes in at 15, making it the only Canadian company within the top 20 global firms. Domtar's sales in 2007 were just over half of those for tenth-ranked Smurfit Kappa, an Irish company, and Domtar's net income is a dismal third of Smurfit Kappa's. Indeed, it was not until 2007 that a Canadian firm made it into the top 20 (FPAC 2008). In comparison, three Finnish companies are among the top 10. In total, only 13 Canadian firms rank among the top 100 global

Table 10.1   Top Global Forest, Paper, and Packaging Companies (Ranked by Sales in Billions, 2007)

Rank	Company	Country	Sales	Net Income	ROCE (%)
1	International Paper	US	21,890	1,168	5.7
2	Stora Enso	Finland	18,322	(291)	3.5
3	Kimberly Clark	US	18,266	1,822	15.2
4	Svenska Cellulosa	Sweden	15,675	1,056	5.7
5	Weyerhaeuser	US	13,949	462	2.5
6	UPM	Finland	13,748	111	3.5
7	Oji Paper	Japan	10,758	146	2.3
8	Metsallito	Finland	10,507	(12)	1.4
9	Nippon Unipac	Japan	9,990	195	2.1
10	Smurfit Kappa	Ireland	9,963	202	8.0
15	*Domtar*	*Canada*	*5,947*	*70*	*4.5*

Source: Based on PWC 2008.

forest, paper, and packaging product companies (PwC 2008).

An alternative measure provides some indication of Canadian stock performance in the forest product sector. The Standard and Poors (S&P) Timber and Forestry Index places two Canadian companies among the top 10 firms by weighting—West Fraser at seventh and Canfor at eighth. Both are BC-based. Sino-Forest ranked sixth, but despite being incorporated in Canada, it is really a China-based company. The S&P Index is based on a modified market capitalization scheme, with an emphasis on stock weighting (S&P 2008). While providing an alternative view to sales-based ranking, the S&P Index does not really indicate the size of the firm with as much overall meaning as the PwC (2008) rankings. Ultimately, the largest firms in terms of employment, scale of global operations, and ability to invest in innovation are often going to be those with the highest sales, the most cost-efficient operations, and good records of product innovation.

The size of a forest company is not always a reliable indicator of performance; indeed, many small firms in Canada are successful and remain competitive (FPAC 2008). However, larger firms, and larger industrial operations, possess several basic advantages (FPAC 2008):

1. reduced harvesting and production costs through economies of scale;
2. greater capacity to invest in innovation and in forestry and product research;
3. economies of scale in distribution and potential access to lower cost of capital; and
4. access to greater resources to invest in environmental services and in production facilities.

Among the five global companies with the highest rates of return on capital invested (ROCE), not one is Canadian, but among the five global companies with the largest losses, the top two are Canadian (PwC 2008, 8). It should come as no surprise that among the top 10 global firms with an ROCE greater than 10 per cent, not one is Canadian. This performance extends to the reinvestment ratio (which measures the extent to which capital investment is replacing aging assets). From 2003 to 2007, Canadian reinvestment rates were below 1 per cent. In 2007, the rate was a remarkably low 0.4 per cent (PwC 2008). It represents the lowest forest-sector reinvestment ratio in the developed world.

Investment in research and development is the foundation for innovation (Hayter 2000). A study conducted for the Canadian Forest Service (Stargate Consultants 1999) highlights that investment in science and technology for Canada's forest sector is low in relationship to the importance of the sector to the Canadian economy. Research and development activities across Canadian industrial sectors are underfunded compared to those in other advanced economies (Hayter 2000, 357). Research and development investment in the forest sectors in Canada is no exception and is below the levels of most major competitors (Binkley 1995). As an example, we can look at comparisons with Sweden and the US. In Canada and the US in the late 1990s, more than 85 per cent of public investment in forest-sector science and technology was in forestry (forest management–related activities) rather than in processing or product innovation, while in Sweden, just over 50 per cent of the public research investment was in forestry— the remainder was in product and processing innovation (Stargate Consultants 1999). At the same time, Canada differs from both the US and Sweden in terms of private-sector investment in forest-sector research. Investment by the Canadian private sector was less than 40 per cent of that of Sweden and only 6 per cent of the US level (Stargate Consultants 1999).

In comparison to the US and Sweden, Canada's investment in forest product research is markedly lower. This has been the case for some time. By 1997, less than 50 per cent of Canadian investment in forest research, both private and public, was in forest products, but in the US and Sweden, around 80 per cent of the total research investment was devoted to forest products. In terms of the actual amount spent, public investment in forest products research is much lower in Canada—only about 15 per cent of that of the US and just 20 per cent of that of Sweden (Stargate Consultants 1999).

Binkley (1995) writes that the lack of research and development investment stems in part from three factors:

1. the problems associated with being a net exporter with a large share of many global markets;
2. the small size of Canadian firms when compared with global competitors; and
3. Canada's collective failure to articulate a meaningful forest-sector strategy with the capacity to guide policy and management decisions of governments, industry, and other stakeholders.

Binkley's (1995) observations still hold true. The culture of the industry may also be part of a larger innovation investment problem. Canadian firms have benefited from easy access to the US market, relatively supportive stumpage rates, and provincial governments with forest management regulations that some say favour the industry. This has likely contributed to an industry that is conservative in its innovation choices, since for some time there was little pressure to do things differently. It has also helped to weaken the backwards linkages that at one time supported a diverse Canada-based manufacturing and service sector based on serving the forest industry.

In the early 1900s, Vancouver was the centre of industrial activities such as equipment manufacturing, transportation services, warehousing, and shipbuilding that support the forest industry (Hayter 2007, 47). The Granville Island area in Vancouver was home to many small companies that made goods for the logging industry, such as cable, rope, engines, and forged goods. The forest industry for a time had a relatively strong spin-off effect, but it has greatly diminished. Today, the equipment used to harvest Canada's wood, everything from chainsaws to yarders, feller-bunchers, and skidders, are most likely to have been made in Sweden, Finland, the US, or Japan. In 2008, Madill Equipment, based in Nanaimo, went into bankruptcy. Madill was the last large manufacturer of harvesting equipment in Canada. Today, even the small axes used by timber cruisers and survey crews are likely to bear the hallmark 'made in Sweden'.

The existing problems in Canada's forest sector, rooted in policies that focus on the close and convenient US market, have helped to diminish backward linkages, leading to the importation of equipment for harvesting and processing, significant foreign ownership, poor resource pricing (low-cost resource sales), and the lack of a strong national forest industry policy. These issues have not come about in a short period, although in some respects they have become all the more troubling and immediate in recent decades.

CANADIAN FOREST POLICY

Conflict and Policy Evolution

Canadian forest policy has evolved over the past 140 years. Present-day policy regimes that affect the competitiveness and social responsiveness of the forest sector reflect incremental and often hesitant change. Kimmins's (1987, 14–15) comment more than 20 years ago that forest resource management tends to follow a predictable pattern through six stages continues to be relevant. In some places, the progression occurs over a long period, while in others it may be compressed:

1. unregulated exploitation with no thought about depletion or conservation;

2. a perception of real or threatened future shortages of forest resources;

3. exploitation of non-depleted forests at more remote locations, then a repeat of stages 1 and 2, or a move to stage 4;

4. institution of simple non-ecological regulations to control forest cutting and management;

5. realization that simple administrative procedures are not enough to ensure an adequate supply of timber and other resources; and

6. initiation of ecologically based forest resource management, with some increasing success in conservation, and timber production.

In Canada, we are firmly at the early stage 6, although shades of the fourth and fifth stage remain in some management practices. Since the 1940s, we have also seen the advance of new approaches. Multiple-use forest management emerged as a method by which resource managers seek to maximize product mixes for forestland by developing strategies for deriving products and services from forestland, often by separating them spatially or temporally. Later, integrated resource management (IRM) came about as a process aimed at integrating diverse interests and variable objectives as well as blending multi-sectoral perspectives into forest management, planning, and decision-making (Slocombe and Hanna 2007).

Integrated resource management remains a dominant paradigm in forest management agencies, although it has been greatly refined. More recently, ecosystem-based management (EBM) (see Chapter 15 for an expanded discussion of EBM) has advanced in some jurisdictions, notably in British Columbia where planning processes in the north and central coast regions have adopted it as the organizing foundation for forest resource management. However, even in BC, the explicit use of EBM as a management principle or policy approach is limited to specific regions. The concept has yet to gain broad currency across the province.

Quebec has also moved to adopt an ecosystem-based, or whole forest, approach to managing forest resources. As in BC, conflict played a role in this policy development. Beginning in 1996, the Quebec government undertook a review of its forestry regime—the Commission to Review Public Forest Management in Quebec (the Coulombe Commission), which culminated in the commission's 2004 report (Quebec 2004). Public controversy over forest management had sparked the creation of this independent inquiry into forest management practice and overall policy (Thiffault et al. 2007, 29). As a result of the inquiry, the Quebec government's forestry regime now identifies sustainable forest management as a key principle, embracing criteria such as biological diversity and social responsibility—ideas that were certainly not a priority 40 years ago (Thiffault et al. 2007, 29).

A universal shift is seen in the general move to enhance public participation, and efforts to incorporate non-timber values into management have also become prominent elements in forestry practice—but I would suggest that it has yet to move significantly beyond rhetoric. The efforts seen in BC and Quebec, for example, are in many respects still recent, and the extent to which they will really influence practices has yet to be determined.

Howlett and Raynor (2001) write that the move to IRM in the 1970s and 1980s was spurred in no small part by conflict between industry and environmentalists. As industry moved into more remote regions and environmental organizations and local people sought to preserve special places, provincial governments modified existing management regimes toward IRM (Howlett and Raynor 2006, 46). As I note above, recent moves in BC and Quebec to embrace ecosystem-based management and similar approaches also reflect the importance of conflict. Provincial governments have sought to develop inclusive processes to arrive at decisions about forest use. In BC, the acrimonious conflicts of the 1980s and 1990s led the government to embark on a series of broad planning initiatives. The most recent, the regionally focused Land Resource Management Plan process, is highly consultative and results in diverse management processes reflecting the characteristics of the communities and landscapes of each plan's respective area (Hanna et al. 2008).

In Ontario, conflict in the 1990s over logging in Temagami and other areas led the provincial government to initiate a consultation process called Lands for Life. The process was aimed at helping various stakeholders reach agreement on how Crown lands should be used. Lands for Life resulted in the Ontario Living Legacy, which expanded the province's parks system. Conflict has emerged as a major factor in forestry policy and has greatly contributed to parks and protected areas designation (Hanna et al. 2008; see also Chapter 12). The importance of conflict should not be underestimated in the Canadian natural resource policy discourse.

Conflict flowing from environmentalism has only erupted in recent decades, but the increasing complexity of forest management issues, and the move to Kimmins's sixth stage, have been developing for much longer. While Canadian forest policy sat contentedly at the fourth stage for a very long time, conflict has hastened shifts and introduced new social objectives and values into the mix. Many of the most 'advanced' management paradigms and policy innovations are actually quite recent, but the most enduring policy base—tenure—has its roots in events that occurred in the early 1900s. Despite recent changes, contemporary forest policy is still very much set on this foundation.

Antecedents: The Conservation Era and Provincial Control

As Canada's population and economy grew, wood became integral to the building of the nation's cities, farms, and railways. The growth of the Canadian economy, albeit at rates slower than in the US, meant that the domestic appetite for Canadian wood also grew—and it grew fast. The US would also import Canadian wood to help fuel the expansion of the American West just as American supplies from the Great Lakes states were in decline. This export relationship was to be the harbinger of the continentalism that permeates the Canadian industry. Canada sought to position itself as the natural supplier of wood to the US market from a relatively early date. By 1871, the Canadian forest industry rivalled agriculture in importance, and fully half of Canadian men were employed one way or another in the forest industries (Gillis and Roach 1986).

The initial stirrings of forest policy came about in 1880 when groups of lumbermen from

---

Box 10.3 The Forest Policy Legacy of the Conservation Era

The conservation era influenced forest resource use and ownership and resulted in far-reaching public policy decisions that remain influential today. Eight legacy points stand out. They are still very much present in the vision and mission statements and management ethics of forest management agencies across North America:

1. Forest resources could be depleted and indeed were being depleted.
2. Forest resources should remain in the public domain.
3. The state would take an active role in natural forest management, and public agencies would be established to manage this vast land base.
4. Professional resource managers should manage forest resources on a scientifically rational basis. The rise of resource professionalism sees its beginnings in the conservation era.
5. Forest resources should be managed on a basis that sees the resource renewed—a sustained-yield basis.
6. Public resources should benefit society and not just a privileged few; as Gifford Pinchot declared, forest resources should provide the 'greatest good for the greatest number of people'.
7. There should be an institutionalized role for preservation. (On this score, there was a split in the conservation movement between the utilitarian conservationists and the aesthetic conservationists. Regardless, the parks and wilderness systems of today owe their origin to the early conservation movement.)
8. Utilitarianism (usefulness of forests to humans) would guide resource management agencies' decision-making, albeit within a setting based on a conservation ethic and sustained yield, and where possible a utilitarian approach would extend to the realization of multiple products and uses from forestlands.

Ontario and Quebec began to be concerned about the decline in both the quality and the quantity of trees (Gillis and Roach 1986). This decline (akin to Kimmins's second and third stages) led those most closely tied to the forest resource to question the limits of what had been a seemingly perpetual forest. About the same time, the American conservation movement was gaining momentum, and soon there was a bilateral discourse on the future of forest use. Eventually, this discourse would yield the first suggestions of policy. The suggestions were modest: reserve forestlands should be established (not open to settlement or conversion to agriculture), these reserves should be patrolled, and elemental fire protection should be established. Basic regulation of cut rates would come later. However, the response from provincial governments was mixed. Quebec embraced the suggestions relatively early, but Ontario took three more years to provide a meaningful legislative response (Gillis and Roach 1986).

In many respects, this pattern became the harbinger of a policy setting we still observe in Canada. Provincial governments today pursue a varied set of forest policies that reflect different cultures of resource use and different perspectives on conservation and sustainability. As the provinces were formed, they took on responsibility for natural resource management and developed policies to encourage resource exploitation. Since their populations were small and their budgets modest, they looked on their forestlands as a lure to attract development and investment capital (Haley and Nelson 2007; Pearse 1976). Forest policy in this period was based mostly on creating a climate in which business could prosper based on resource use. Governments would realize greater revenues, and 'men' would be put to work.

The role of the federal government today is quite different from what it was just over a century ago. For many years, the federal government managed, or perhaps more accurately determined, the rate and location of woodcutting for Canada's western territories. The last remnant of this federal power exists in Yukon, Nunavut, and the Northwest Territories, but only Yukon has a forest industry of any note, and responsibility for day-to-day authority over Yukon's forests has essentially been transferred from the federal government to the Yukon government.

The last notable transfer of federal forest management responsibility happened during the tenure of William Lyon Mackenzie King in the early 1920s. His government moved to give Alberta, Saskatchewan, and Manitoba authority over the vast federal forestlands in those provinces and in doing so ended a long and, many would argue, productive period of dominion forest management. The remaining national forests and railway grants in British Columbia would also be transferred in the 1920s.

In reflection, King's motives might seem short-sighted. In moving natural resources to the provincial purview, he ceded most decision-making not only over forest management but also over energy resources to provincial governments. This was to have a profound impact—one felt to this day—on the ability of the federal government to guide resource management in Canada, to develop a national policy for any single natural resource, or indeed to advance national environmental policies. Others would argue that in doing so King had simply righted a provincial inequity, since the other provinces had already assumed authority over the natural resources within their borders. Regardless of one's perspective, if federal control over natural resources in Canada's West had been retained, as the American federal government had for large areas of its West, the face and strength of national resource policy in Canada would be quite different today—for better or worse.

The fundamental policy decisions made at that time would shape the nature of forest management in Canada for the next century. I argue that while the decision to transfer forestland from federal to provincial control was certainly important to policy development in Canada, an even more elemental choice had been made decades earlier—the decision to keep forestland in the public domain. This decision has had repercussions that resound to this day.

The decision to keep forestland in the public domain was a legacy of the conservation era. The conservation movement had its deepest roots in the US, but they would extend into Canada and would indeed be nourished by Canadian participation and a cross-border pollination of ideas and intellectual influence. Canadians contributed to the American conservation movement from the start, in no small part by collaborating with American colleagues in joint forestry congresses. These congresses helped to advance the discourse over forestland use in the US, spurred by transformation of the 'wide open' West of the American imagination into a settled landscape dotted with towns, logged forests, and rangelands dissected by fence lines. Such images, coupled with the growing influence of writings by people like George Perkins Marsh and John Muir, helped to build a gradual but influential movement that would seek in part to 'save' undeveloped lands in the West from rapacious development.

The conservation movement, driven by images of an end to the abundance of western forests and rivers, soon gained supporters in governments in Canada and the US. Eventually, the movement would evolve and become more complex in terms of its constituents' visions, some of which would be irreconcilable. A fracture emerged between utilitarian conservationists, who wanted public lands used, albeit wisely and to benefit the public at large, not just a few,

and the aesthetic conservationists, who wanted some lands preserved from development. The utilitarian perspective would guide resource management paradigms, while the aesthetic visionaries would see certain victories in the establishment of national parks.

If the 'closing of the American frontier' marked the end of the 'West', perhaps no politician was more moved by it than Theodore Roosevelt. He used his sudden ascension to the American presidency to preserve vast tracts of forestland, range, and river valleys within the domain of the US federal government. Several previous presidents had used the US General Land Law Revision Act to keep large areas of land within federal ownership, but Roosevelt used this authority with particular zeal. Under his term, the US Forest Service was created, and Gifford Pinchot—one of the chief theorists of utilitarian conservation—became its first chief. Pinchot was an advocate of public control of public lands. In his view, keeping land in the public domain and the scientifically rational management of public resources would yield the 'greatest good for the greatest number'.

Pinchot's ideas would extend to Canada, where his protégés would provide the backbone for forest management in British Columbia and the ideals of the conservation era would help to guide policy in the Dominion Forest Service. Pinchot spoke passionately and directly to Canadians about public ownership, encouraging Canada to keep forestland within the realm of public control. This he saw as providing the foundation for developing not only an industry but also equity of opportunity that would be the basis of nation-building—a message with which some politicians in the government of Wilfrid Laurier in the early twentieth century readily agreed.

The Canadian embrace of the conservation ideals had little to do with the imagery or romanticism of western landscapes that had been so influential in the US. There is little to suggest that Wilfrid Laurier shared Roosevelt's enthusiasm for the outdoors, but he did want Canada's West to grow, and he knew that the resource base was a key part of its attraction to settlers and investors. The federal government under his regime managed its western forest holdings on the basis of fundamental conservation ideals—a resource management agency was formed, regulations introduced, and the notion of rational management advanced. Indeed, conservationism benefited mostly from the support of Clifford Sifton, the minister responsible for the forestlands. When Sifton departed the cabinet, the influence of conservationism waned, but it did not disappear.

There were political stresses and pressure from western interests to promote development of forestlands, but for the most part there was a relatively cohesive approach to forestland management while the federal government maintained its role. With Mackenzie King's final transfer of forests, a disparate collection of policies emerged, and for Manitoba, Saskatchewan, and Alberta, there was indeed little initial interest in forest management issues—a fact that really did not change until after 1950.

The present role of the federal government in forestry is limited to trade issues and research. With respect to trade policies, the current federal Conservative government has been less aggressive than its predecessors in defending the Canadian forest industry against what Roger Hayter (in his guest statement in this chapter) aptly calls 'enduring and punitive' American protectionism. Nor have federal governments, past and present, sought to create substantive policies aimed at reducing the continental focus of the industry. While some federal funds are allocated, mostly in conjunction with the provinces, to support employment or industry adjustment

programs, overall Canada's federal government has little to no authority over the nation's forests and today offers no enforceable, comprehensive federal forest policy or particularly strong leadership in developing a national policy based on collaboration with the provinces and territories.

Policies that might be considered national are derived mostly from vision statements produced cooperatively by the provinces. Examples include the National Forest Strategy, the Canada Forest Accord, and the recent (2008) 'A Vision for Canada's Forests' by the Canadian Council of Forest Ministers. However, these are broad statements of good intent; they do not constitute 'hard policy'. The Forest Accord commits its signatories to specific policy actions, but there are no penalties for non-compliance, and measuring performance is difficult because many of the objectives and terms are very broad and lack detail. Nevertheless, Quebec and Alberta have refused to sign the accord. While such instruments represent recognition of issues and policy needs, these statements often lack implementation frameworks and are, all said and done, unenforceable.

National strategies in Canada are a negotiated product, which is not necessarily a bad thing, but they emerge merely as guidelines for policy development and management conduct. This stands in marked contrast to Canada's primary competitor nations—for example, Sweden, Finland, and the US—where extensive national forest policies exist: to advance a competitive industry, to support an industry that for each of these nations has a truly global presence, to support management of public and private forestlands, and to advance research both for growing trees and for processing wood. In Canada, the provinces set forest policy, and the flavour of guiding principles depends in large part on the conditions that they set with respect to Crown forest tenure.

TENURE

Tenure Systems

The contemporary importance of tenure in Canadian forest management is an enduring legacy of the conservation era. It was noted above that Canada has a unique dichotomy in its forest industry–public forestland ownership and a private-sector industry. The institutional arrangements under which any sector of the economy functions have major implications for its behaviour, performance, and competitiveness (Haley and Nelson 2007, 630). Tenure arrangements provide the structure under which forest companies operate. Tenure dictates responsibilities, outlines rights with respect to resource use, establishes the period of access (duration of tenure), and defines conditions to be met in order to gain access (such as a competitive bid or other application process).

Two basic forms of tenure exist in Canada. Volume-based tenure provides the right to cut a specific amount of timber within a defined time period. Area-based tenure provides the right to cut timber within a defined area of Crown land (Boyd 2003, 144). Most provinces allocate the larger part of their annual allowable cut (AAC) (amount of timber a provincial government allows to be harvested in a given year) through area-based tenures (Haley, Luckert, and Hoberg 2008). In Ontario and Quebec, all of the AAC is allocated through area-based tenures, while only BC and New Brunswick allocate less than half their AAC to area-based tenures.

The conditions of tenure have a major impact on the behaviour of firms. By establishing rights and setting conditions under which these rights are held, provincial governments have some control over the actions of forest companies and the extent to which these firms support broader social goals (Haley and Nelson 2007; Haley and Luckert 1990). Haley and Nelson (2007, 632) write that

the history of forest policy in Canada is largely the story of adjusting tenures to respond to changing public attitudes toward the use of forest resources and methods of forest management.

Tenure systems in BC were relatively unchanged from 1860 until the late 1970s when the six basic forms were grouped under one form known as Timber Licences. Today, there are 15 different timber tenures in BC. New forms, such as Community Forest Agreements, emerged in the late 1990s, and in many respects this was in response to conflict over forest use. The BC government created new tenure systems and adjusted existing ones to respond to acrimonious disagreements over the conditions of forest use, notably on the BC coast, and established approaches to forest management.

In Table 10.2, an outline of the basic elements of tenures in BC shows varied conditions, rights, and duration periods. Some forms have also become non-renewable. British Columbia has a diverse collection of tenure forms relative to other provinces. Most of the significant changes have occurred within the past three decades, and this has been largely in response to conflict. In contrast, Ontario has three basic tenure forms (see Table 10.3). They encompass a range of activities under a single tenure rubric. In comparison to the complex tenure options in BC, the Ontario system appears relatively simple in form but provides an effective framework for defining rights and obligations. Activities that in BC are allocated under a specific, or unique, tenure arrangement are covered by one of the three tenure forms in Ontario. What is also interesting is that they are relatively recent forms, brought about largely through Ontario's Crown Forest Sustainability Act. As in BC, the Ontario Act was part of a policy response to conflict over forest resource use.

Tenure stability has an impact on the willingness of firms to invest in forest management.

Table 10.2 Major Tenure Types in British Columbia (as of 2009)*

Tenure Name	Area- or Volume- Based	Duration	Rights	Responsibilities
Timber Licence	Area	No longer being issued. Existing licences have variable terms and may be extended.	Exclusive right to harvest merchantable timber in a specified area.	Operational planning, road-building, reforestation, stumpage payments.
Tree Farm Licence (TFL)	Area	25 years, replaceable every 5–10 years.	Virtually exclusive right to harvest timber and manage forests in a specified area. May include private land.	Operational planning, inventories, reforestation, stumpage payments, obligation to use logging contractors for a portion of the volume harvested each year, with some exceptions.
Forest Licence	Volume	Up to 20 years. May be replaced every 5–10 years or non-replaceable.	Right to harvest an AAC in specified Timber Supply Area or TFL area.	Operational planning, road-building, reforestation, stumpage payments. May be required to use logging contractors for all or part of the volume harvested.
Pulpwood Agreement	Volume	Up to 25 years. No longer issued.	Conditional right to harvest 'pulp quality timber' where other sources are insufficient or uneconomic.	Operational planning, obligation to maintain a pulp timber processing facility, obligation to purchase wood residue and pulp logs produced in the pulpwood area, reforestation, stumpage payments.
Timber Sale Licence (TSL)	Volume and area	New forms are up to 4 years, non-replaceable. Existing TSLS are replaceable.	Issued by competitive auction.	Operational planning in limited cases, stumpage payments. May be obligated to operate in accordance with certification organizations such as the Forest Stewardship Council of Canada or others.
Community Forest Agreement	Area	Probationary Agreements are for 5 years. Following an evaluation, may be extended or converted to long-term form of 25 to 99 years, replaceable every 10 years.	Exclusive right to a First Nation, municipality, or regional district to harvest an AAC in a specific area. May include right to harvest, manage, and charge fees for botanical forest products and other products. May be competitively or directly awarded.	Strategic and operational planning, inventories, reforestation, stumpage payments.

continued

Table 10.2 Continued

Tenure Name	Area- or Volume- Based	Duration	Rights	Responsibilities
Community Salvage Licence	Volume and area	Up to 5 years.	Provides communities the right to remove timber that is dead, damaged, diseased, windthrown, or left over from logging from a specified area	Operational planning, reforestation in some cases, stumpage payments.
Forestry Licence to Cut	Volume	Up to 5 years.	Right to harvest and/or remove timber from specified areas. Types are designed to meet different purposes, such as small-scale salvage, timber removal for scientific purposes, forest health, small commercial purposes (firewood, fence posts).	Operational planning (if a major licence or if issued under a Pulpwood Agreement), stumpage payments, reforestation where clear-cuts are larger than 1 hectare.

* Does not include small-area or small-volume tenures.

Source: Based on BC MFR 2006.

Table 10.3 Major Tenure Types in Ontario (as of 2009)*

Tenure Name	Area- or Volume-Based	Duration	Rights	Responsibilities
Wood Supply Commitments	Volume and/or location	10 to 20 years or legacy commitments that predate the current Crown Forest Sustainability Act but may or may not have termination dates.	Access to a supply of forest resources from Crown lands.	Do not provide a licence to harvest wood. Mills with wood commitments must enter into arrangements with forest resource licence-holders to obtain the committed wood fibre.
Sustainable Forest Licences (SFL)	Area	Up to 20 years. Reviewed every 5 years, providing that certain conditions are met. The licence-holder must comply with the terms and conditions of the licence. An independent forest audit must recommend renewal. Transferable. May not be sold.	Generally govern all of the area in a management unit of Crown forest but do not convey any rights to Crown land. Rights to harvest all species of trees found in a licensed area.	Forest management planning, gathering forest information for the Crown with compliance with the Crown Forest Sustainability Act (CFSA), Forest Operations and Silviculture Manual, Forest Information Manual, and Scaling Manual. Constructing forest roads that serve the public at large, regenerating the forest, compliance planning and monitoring. Payment of stumpage.

continued

Table 10.3 Continued

Tenure Name	Area- or Volume-Based	Duration	Rights	Responsibilities
Forest Resource Licences (FRL) Activities covered: Non-commercial uses, such as firewood and construction purposes. Harvest of forest resources reserved to the Crown on patent land. Personal use, salvage.	Area	Up to 5 years. Can be extended for 1 year under defined, special circumstances.	Typically cover only portions of management units and may overlap with an area covered by a SFL. While the FRL and SFL may share portions of the same area, each licence-holder will typically have the right to harvest different stands. FRLs may also overlap. Harvest rights are agreed upon by the licensees. May allow for the harvest of certain amounts and species of timber. Where FRLs overlap, the most recent licensee has preference over earlier licence-holders for the species and volumes outlined in the licences.	Compliance with CFSA, Forest Operations and Silviculture Manual, Forest Information Manual, and Scaling Manual. Payment of stumpage.

Does not include small-area or small-volume tenures

Source: Based on Ontario Ministry of Natural Resources 2009.

Some years ago, I spoke with an economist employed by a large BC-based forest products firm. He commented that their planning cycle was typically five years for financial and investment decisions because the firm considered tenure unstable. He noted that over the years, the BC government had made several changes to tenure conditions that had reduced or relocated cutting rights and this had compelled the firm to change its planning decisions, which was neither easy nor inexpensive. The perception of tenure instability also made the firm reluctant to make significant capital investments in new facilities or new product development. Instead, the firm adopted a short horizon for planning and held back some investments in its processing facilities.

Tenure Reform

Forest tenures have become increasingly complex across Canada. There are tenures of varying vintages and a diversity of rights, largely because they were established at different times and reflect evolving societal values (Pearse 1976). As provincial governments seek to respond to increasing conflict over forest resource management, they often resort to 'tinkering' with tenure, since it is the key policy tool available to them. Tenure reform has long been talked about in Canada, but we have seen few substantive changes. Haley and Nelson (2007) have argued that tenure reform is essential now, noting that existing systems are designed for past economies and social conditions. They propose seven criteria if tenure arrangements are to better reflect environmental, economic, and social sustainability objectives (Haley and Nelson 2007, 634–7):

1. Social legitimacy to incorporate social concerns into forest policy, responding to community values.
2. Flexibility to accommodate changing economic circumstances and social preferences.
3. Transparency so that tenures are simple and understandable to all.
4. Security to encourage new capital investment in the forest sector. Lack of security has a negative impact on competitiveness and sustainability. From a conservation perspective, tenure uncertainty may be a disincentive to invest in the renewal and management of forest resources.
5. Diversity in tenure systems allows for licences of different sizes and purposes held by a variety of industrial (e.g., firms of different sizes and ownership formats) and non-industrial (e.g., communities and Aboriginal) tenure-holders. This can result in more complex benefits and forest uses.
6. Minimum regulatory compliance costs commensurate with public objectives to help promote market competitiveness.
7. Timber pricing that promotes efficiency, provides incentives for socially responsible forest stewardship, and achieves equity.

If tenure reform is desirable, then how substantial does change need to be? Suggestions range from selling public forestland to private interests to a stronger state role, perhaps even provincial governments assuming a larger role in the sector by investing or owning forest companies. Haley and Nelson (2007, 637) outline three levels of tenure reform:

1. retain the basic structure of a tenure system and the allocation of harvesting rights, but redefine some rights and requirements to address specific problems;
2. retain many features of an existing tenure system but redefine rights and requirements; introduce significant systemic changes involving the reallocation of rights, the roles of government and firms, and the processes used to value and sell public timber; and
3. discard an existing tenure system, completely or in part, and replace it with alternative arrangements for allocating rights to public forestland and timber.

While the third option is the most sweeping, it also seems the most unlikely, even though few would advocate keeping the status quo. In practice, the pattern has been to follow the first and second. In BC, this has led to an increasingly complex collection of tenure arrangements, while in other provinces existing forms have been redefined in terms of new expectations and responsibilities for tenure-holders within existing or more simplified rubrics. Ontario, for example, has created a system of tenure that relies in large part on self-regulation. Companies are responsible for ensuring their own compliance with regulations, auditing their operations, and reporting the results to the province. While this provides a greater measure of autonomy and a degree of independence for tenure-holders, there is as yet no indication that it necessarily improves forest management or that it can maintain the level of accountability the public expects. Tenure reform should not mean the abrogation of provincial responsibility.

At various times, it has been suggested that Canada follow aspects of the Scandinavian model of predominantly private ownership with a combination of small-scale and large company forest holdings. An argument is that firms and individuals might be more willing to invest in innovative forestry practices if they owned the land, thus creating a 'culture of forest stewardship' based on the benefits of private property. However, in Canada, public lands are in some respects 'sacred territory' and there may be little public support for selling provincial forests, especially not to large companies. There may also be little interest among firms when it comes to buying forestland, especially cutover areas. The present system has served the industry well.

Another option is to increase the management role of communities through community-held tenures. Community forestry is one alternative model for tenure and forest management. However, despite the existence of community forests across Canada and active community forestry programs in most provinces, they account for a relatively small portion of industrial wood production. Nor is there any guarantee that community-held tenure systems would necessarily result in forest management all that different from established approaches or that innovation in processing and product development would necessarily be advanced by community-based tenures. To illustrate, on one hand, community-based tenures may result in a more stable, long-term vision of forest management with resource and community sustainability as the guiding ethics. Alternatively, some communities might support a short-term emphasis on harvesting as they seek more immediate opportunities for employment and prosperity, perhaps motivated by economic downturn or other hardships.

Tenure opportunities for small firms and individuals could be provided or expanded, perhaps by creating many small private forests, but would smallholdings or corporate ownership result in better forestry practices? Experience in the US and Canada suggests that some companies with private forests are not always the best forest managers. Some have logged their lands quickly for short-term profit, made few investments in regeneration, and moved to sell denuded forestland as 'development land'. However, others see their forests as the foundation of their long-term survival. Firms do not all behave in the same way. They possess different 'cultures' reflected in their approach to forest management and what they are willing to do to make a profit.

Small private holdings would require capacity-building and stable investment sources. Moreover, just as some large firms have, individuals may log their lands quickly for the sake of short-term profit. A Finnish colleague recently commented that a shift in the profile of smallholding ownership in his country had also changed the approach to forestry. At one time, people (often farmers) who lived 'on the land' owned most small forest holdings. This profile has changed, with most of these lands now owned by people who live in urban areas; many are retired people, and their approach to their forests is more likely to emphasize fibre production for quicker profits rather than managing forestland as part of their immediate social and physical landscape. In Scandinavia, governments act to dull some of these negative effects by regulating private forestry practices, requiring forest plans, guiding cutting rates, and financing forestry—all while supporting private forestland ownership. Tenure reform in Canada is needed, but it will have to respond to the unique conditions found in each province.

CONCLUSION

The forest sector in Canada is an important part of the national economy. Its contributions to the balance of trade, employment, and community stability are significant. However, despite being the largest global exporter of forest products, Canada has not created firms of a scale necessary to compete in the increasingly complex and competitive global forest products market. Investment in research and development remains low relative to other significant forest-product nations, and Canadian firms tend to be among those with the highest financial losses globally. Over-reliance on the American market will weaken the resiliency of the Canadian industry in times of economic downturn and makes the Canadian forest industry especially vulnerable to trade disputes with the US.

Conflict dynamics have also emerged in recent decades as many Canadians rethink the ways they want public forests to be managed. This has moved provincial governments to enact a range of policies in an attempt to change industrial forestry practices and respond to new social definitions of forests and desired uses. Much of the policy innovation has not only been conflict-driven. There have been moves toward stronger integrated resource management and ecosystem-based approaches, but the most substantial alterations have tended to involve changes to tenure systems.

In Canada, tenure reform will require a careful consideration of the lessons observed in other places. While we can look to jurisdictions such as Sweden and Finland for information and experience, Canada's forests, geography, social expectations, and forest industry cultures are distinct. The answer to tenure reform lies in creating tenure systems that provide appropriate opportunities for individuals, communities, and forest companies and establish conditions that will encourage stable, long-term business investments. Tenure will also have to better account for new and increasingly complex visions of the diverse benefits our forests provide and their importance to our economy. The criteria of Haley and Nelson (2007) point to the need not only for restructuring but also for stability and innovation. In other words, we need balance among regulation, state ownership, market mechanisms, and more private forestland. We must also seek policies that balance the economic, social, and ecological value of our forests—a point reinforced by Hayter in his guest statement.

Canada's forests are sustainable, and the industry can be resilient. This remarkable resource base can support communities, employment, and a vibrant and competitive industry if the right policy conditions are established. With the advance of new management approaches and updating of tenure systems, progress has been made. However, more work remains to be done, and difficult decisions will have to be made soon.

FROM THE FIELD

I spent this summer (2009) in south central British Columbia at my parents' ranch. It was a tough forest fire season. Most days in July and August, the valley was shrouded in a smoke haze thick enough to obscure the mountains. Smoke was flowing from fires hundreds of kilometres away, some small and some covering thousands of hectares. Some mornings the smoke was strong enough to make my clothes smell like a campfire. The hills nearby are covered with dead pines, killed by the pine beetle epidemic. Over the past few years, the progression of red, dead, beetle-killed forest has spread up the slopes of our valley like a rash. Logging cannot keep ahead of the dead wood, and firms, loggers will tell you, are high-grading (taking only the best) the dead pine.

I was standing in a supermarket line in Merritt, BC. The fellow behind me struck up a conversation; he had just been laid off by Canfor—he worked for them for more than 20 years. I know many people just like him—my first summer job was logging in mountains above Yale, BC. Colleagues in government confide that they wonder whether the industry has a future in BC. Many forest firms in Canada were having problems when the economy was strong, but now times are tough. AbitibiBowater has filed for bankruptcy protection, and so have more than a few smaller firms. Neither the state of the forests nor the industry that relies on them appears too healthy.

My research, whether about an integrated resource management perspective or examining how impact assessment has affected forestry practice, has a strong forestry thread. In recent years, I have increasingly worked with colleagues in Scandinavia, and this has exposed me to new issues, ideas, and challenges in forest management. There has been a tendency on the part of some in North America to look at Scandinavia and see a wholly positive model of forest policy, but the truth, as with many things, is a bit murkier.

While Canadians have seen some large and acrimonious conflicts over forest use, particularly in BC, conflict is also present in Finland and Sweden, where debate is occurring over the ecological qualities of what are industrial forests. And just as in Canada, the industry has been negatively affected by the recent recession, jobs have been lost, and communities are worried about the future. But it has also become apparent to me that there are things we can learn. While Canada's forests are different, and sometimes similar, the stronger willingness in Scandinavia to invest in greater product innovation and forestry research, to explore more diverse tenure arrangements, and to seek varied markets, as well as a broader social appreciation for the forest industry, might make it easier for the Scandinavians to adapt to difficult times.

As Canadians we have a global advantage in the quality and quantity of our wood, our transportation systems, access to more markets than we realize, and an increasing willingness to manage our forests with ecological values in mind. My research and experience have led me to believe that what is needed now is a shift in a forest sector culture that has relied for too long on easy access to a remarkable resource but has only produced mediocre results. Canada is not the global leader, but it should (and could) be.

—KEVIN HANNA

REVIEW QUESTIONS

1. Define the following terms: sustained yield, fall-down, succession, over-mature, fire-climax, even-aged stand.
2. What are raw log exports, and why are they controversial?
3. Why is tenure an important theme in Canadian forest management?
4. What is the difference between area-based and volume-based tenure?
5. How did the conservation era influence the present profile of forest ownership in Canada?
6. Why are Russia and Scandinavia important when considering the competitive position of the Canadian forest industry?
7. What are the conditions that have allowed the mountain pine beetle to reach an epidemic state?

REFERENCES

BC MFR (British Columbia Ministry of Forests and Range). 2006. *Timber Tenures in British Columbia: Managing Public Forests in the Public Interest.* Victoria: Ministry of Forests and Range.

———. 2008a. *Non-Timber Forest Products.* Victoria: Ministry of Forests and Range.

———. 2008b. *Facts About BC's Mountain Pine Beetle.* Victoria: Ministry of Forests and Range.

BC Stats. 2003. *Exports, February 2003.* Victoria: BC Stats, Queen's Printer.

Binkley, C.S. 1995. 'Designing an effective forest sector research strategy for Canada'. *The Forestry Chronicle* 71: 589–95.

Boyd, D. 2003. *Unnatural Law: Rethinking Canadian Environmental Law and Policy.* Vancouver: University of British Columbia Press.

Clapp, R.A. 1998. 'The resource cycle in forestry and fishing'. *The Canadian Geographer* 42: 129–44.

FPAC (Forest Products Association of Canada). 2005. *Competition and Consolidation in Canada's Forest Products Industry.* Ottawa: FPAC.

———. 2008. 'The industry, economic impact'. http://www.fpac.ca/en.

Gillis, P.R., and T.R. Roach. 1986. *Lost Initiatives: Canada's Forest Industries, Forest Policy and Forest Conservation.* New York: Greenwood Press.

Haley, D., M. Luckert, and G. Hoberg. 2008. *Interprovincial Comparison of Crown Forest Tenures.* Presentation to the Forest Tenure Conference, Vancouver, June.

Haley D., and H. Nelson. 2007. 'Has the time come to rethink Canada's Crown forest tenure systems?' *The Forestry Chronicle* 83 (5): 630–41.

Hanna, K., et al. 2008. 'Conflict and protected areas establishment: British Columbia's political parks'. In K. Hanna, D.S. Slocombe, and D. Clark, eds, *Transforming Parks and Protected Areas: Policy and Governance in a Changing World,* 137–53. Abingdon, UK: Routledge/Taylor and Francis.

Hayter, R. 2000. *Flexible Crossroads: The Restructuring of British Columbia's Forest Economy.* Vancouver: University of British Columbia Press.

———. 2003. '"The war in the woods": Post-Fordist restructuring, globalization and the contested remapping of British Columbia's forest economy'. *Annals of the Association of American Geographers* 93: 706–29.

Howlett, M. 2001. 'Policy regimes and policy change in the Canadian forest sector'. In M. Howlett, ed., *Canadian Forest Policy,* 3–20.Toronto: University of Toronto Press.

Howlett, M., and J. Rayner. 2001. 'The business and government nexus: Principal elements and dynamics of the Canadian forest policy regime'. In M. Howlett, ed., *Canadian Forest Policy,* 23–62. Toronto: University of Toronto Press.

———. 2006. 'Convergence and divergence in "new governance" arrangements: Evidence from European integrated natural resource strategies'. *Journal of Public Policy* 26 (2): 167–89.

Kimmins, J.P. 1985. *Forest Ecology.* New York: MacMillan.

NRCAN (Natural Resource Canada). 2001. *Canada's Forest Inventory (Canfi) 2001.* Ottawa: Natural Resources Canada, Canadian Forest Service.

———. 2007a. *State of Canada's Forests 2007.* Ottawa: Natural Resources Canada, Canadian Forest Service.

———. 2007b. *Canada's Forest Regions, Map.* Ottawa: Natural Resources Canada, Canadian Forest Service.

Ontario Ministry of Natural Resources. 2009. 'Ontario's tenure and licensing system'. Toronto: Queen's Printer for Ontario. http://www.mnr.gov.on.ca/en/Business/Forests/2ColumnSubPage/STEL02_167460.html.

Parfitt, B. 2006. 'Softwood deal will spur more raw log exports'. Editorial. Vancouver: Canadian Centre for Policy Alternatives.

———. 2007. 'Wood waste and log exports on the BC coast'. *Behind the Numbers: Economic Facts, Figures and Analysis* June. Vancouver: Canadian Centre for Policy Alternatives.

Pearse, P.H. 1976. *Timber Rights and Forest Policy in British Columbia: Report of the Royal Commission on Forest Resources.* Victoria: Queen's Printer.

PWC (Pricewaterhouse Coopers). 1996. *The Forest Industry in Canada.* Vancouver: Pricewaterhouse Coopers.

———. 2003. *The Forest Industry in Canada.* Vancouver: Pricewaterhouse Coopers.

———. 2008. *Global Forest, Paper and Packaging Survey, 2008 Edition, 2007 Results.* Vancouver: Pricewaterhouse Coopers.

Quebec. Commission d'étude sur la gestion de la forêt publique québécoise. 2004. *Rapport*. Québec : Bibliothèque nationale du Québec.

S&P (Standard and Poors). 2008. *S&P Global Timber and Forestry Index*. New York: Standard and Poors.

Slocombe, D.S., and K. Hanna. 2007. 'Integration in resource and environmental management: Towards a framework'. In K. Hanna and D.S. Slocombe, eds, *Integrated Resource and Environmental Management: Concepts and Practice*, 1–20. Toronto: Oxford University Press.

Stargate Consultants. 1999. *International Comparison of Forest S&T Funding and Management in Canada, the United States, Sweden and Brazil*. Prepared for the Canadian Forest Service. Nanaimo, BC: Stargate Consultants.

Thiffault, N., et al. 2007. 'Adaptive forest management in Québec: Bits of the big and small pictures'. *Canadian Silviculture* (May): 26–9.

UN FAO (United Nations Food and Agriculture Organization). 1999. 'FAO forestry towards a harmonized definition of non-wood forest products'. *Unasylva* 50 (198): 63–64.

Westoby, J. 1989. *Introduction to World Forestry*. Oxford: Basil Blackwell.

11

Managing for Wildlife in Canada

Graham Forbes

Learning Objectives

- To appreciate that wildlife is difficult to manage because personal values and not science dictate how important wildlife is.
- To realize that much research, resources, and management focus on species we hunt and fish rather than on all species.
- To recognize that there have been major successes in managing wildlife but the overall situation may be getting worse.
- To understand the range of tools that managers use to maintain wildlife in Canada.
- To recognize that there will be increased conflict within society on how much habitat we want for species that need large areas or live where society will want to extract resources.

Managing wildlife means managing people. This seems contradictory, but in fact no management of wildlife occurs separate from what society wants. And the hard part of management is that different people want different things. Conflict in wildlife management derives from human values; what I think animals should be used for likely conflicts with what you think those uses should be. To get at the core of wildlife management conflicts, take a look at your own values and answer this question: 'Why don't you hunt animals?' Or if you do hunt: 'Why isn't it cruel?' Opponents of hunting call hunting unethical, unnecessary, and damaging to nature. Proponents will tell you the exact opposite and that a strong hunting heritage is the best way to keep natural land safe from development (Manore and Miner 2006).

Questions such as these form the basis of conflict in the management of wildlife today. The values people hold regarding wildlife will dictate their actions toward animals, nature, and people with different values. In various surveys (summarized in Gilbert and Dodds 1992), the most common reported attitudes toward wildlife were an apathetic and neutral attitude (35 per cent) or one concerned with domesticated or captive objects such as pets and zoo animals (35 per cent). These opposing attitudes toward wild animals illustrate why conflict is common in wildlife management. A large number of people support the rights of animals, humane practices, and animal welfare (20 per cent), but an equal number (20 per cent) support consumptive activities such as hunting, trapping, and fishing (Gilbert and Dodds 1992).

Part of the conflict relates to changing societal values—a new majority will conflict with the legacy of those they displace. Many Canadians hunted 100 years ago, but as few as 7.3 per cent conducted the consumptive activities of hunting or trapping in 1991, declining to 5.1 per cent in 1996 (DuWors et al. 1999). In some areas, hunting is more common; for example, in Quebec 12 per cent of the population is interested in hunting (Government of Quebec 2009a). The province of Ontario passed legislation (Heritage Hunting and Fishing Act) in 2002 that maintains hunting as a right of heritage protected under law. Such a law is meant to counteract pressure from people promoting the rights and protection of animals, often at the expense of hunting and trapping. Wildlife is one of the few issues reflecting such a polarized range of values held by society: wildlife is both cute and dangerous, a cost to agriculture and forestry or a benefit to ecotourism, for use by people or for protection from people. With such opposing values, the opportunity for conflict is inevitable.

Another area of conflict in wildlife management is how to maintain threatened species. In abstract terms, most people support the protection of endangered wildlife. The conflict arises when the management becomes real. Do you believe a landowner should not build an apartment complex because the site contains a rare plant species? What if it were your land and you paid the taxes?

Uncertainty in wildlife management, as in other resource sciences, is mainly due to ignorance. Managing for the sustainability of many diverse species is difficult when we understand only small components but are forced to manage regardless. The recent emphasis on the maintenance of biodiversity (a term essentially meaning all life and, to some scientists, the processes that create it [Biodiversity Science Assessment Team 1994; Bocking 2000, although see Ghilarov

1996 for an alternate view]) is an even riskier endeavour. We have only the taxonomic name for most of the 71,309 species we know about in Canada (Mosquin, Whiting, and McAllister 1995), hardly enough knowledge to confidently say that we are maintaining them in today's changing environment. A mere 0.03 per cent of the species in Canada are vertebrates (Mosquin, Whiting, and McAllister 1995), the group on which we have spent most of our research effort (Freedman et al. 2001).

The definition of wildlife used by government agencies includes all living species, from bacteria to bison, including plants. The national wildlife policy defines wildlife as 'wild mammals, birds, reptiles, amphibians, fishes, invertebrates, plant, fungi, algae, bacteria, and other wild organisms' (Wildlife Ministers Council of Canada 1990). However, to the public, wildlife is usually 'animals', and often the image is more likely to be a mammal than a fish and more likely a game species such as moose than a non-game species such as a flying squirrel, and rarely would someone call mushrooms 'wildlife' (Figure 11.1).

Much of the broader definition of wildlife has been enveloped by the term 'biodiversity'. Government agencies were mandated to indicate progress toward goals outlined in the United Nations Convention on Biological Diversity, ratified by Canada in 1992. In practice, most government agencies responsible for wildlife have set larger goals and strategies for maintaining biodiversity and specific goals related to species that are hunted or are under threat (see Chapter 3). I recognize the broader definition but because the dominant public interest, and hence much of the management, is directed toward vertebrates, my focus in this chapter will be on the management of game species, threatened species, and species used as indicators of sustainability. Conflict in wildlife management revolves around two themes: the need for management and the ownership of wildlife.

(G. Forbes)

Figure 11.1 The public's concept of wildlife would relate more to the moose (left) than to the tree stump (right), although both are considered wildlife. This dichotomy is also expressed in management—strategies vary from the single-species approach such as moose management to a community approach that would maintain a diversity of species. The stump likely contains several thousand species of wildlife.

The field of wildlife management is broad, ranging from the interactions of species and habitat at spatio-temporal scales of gene to community to the ethics of killing for subsistence or sport. The objective of this chapter is to highlight issues of wildlife management that relate to aspects of conflict and uncertainty. Some conflicts result from competition with the objectives of managing for other resources, such as forestry, agriculture, and protected areas. Conflict exists because wildlife objectives often are a constraint on the amount of resources and profit that can be extracted from the land. The management

of these resources is discussed in other chapters and can serve as the context for my discussion on the objectives of wildlife management.

IS MANAGEMENT NECESSARY?

Wildlife managers are inclined to state that they manage people more than they manage wildlife. Societal values involving equitable access to resources, humane treatment of animals, and concepts of stewardship and sustainability drive the actions of wildlife managers. The extent of management action is often the catalyst for conflict. How proactive should wildlife management be? Most Canadians probably want, and many likely are even proud of having, grizzly bears in Canada, but few people want them too close or too abundant. Abundant animals can have indirect effects that must be considered. For example, it is possible to correlate the number of wildlife–vehicle collisions with the density of moose and deer in the forest; having many moose available for hunting and viewing opportunities is weighed against the probability of fatal collisions. Through hunting quotas and habitat manipulation, wildlife managers have the tools to set the density and demographic structure (i.e., sex ratio, age class distribution) of many game species.

Conflict arises over when and where these tools should be applied. Some people argue for allowing nature to take its course, to allow populations of species to rise and fall in density-dependent oscillations of predator and prey cycles or changing environmental conditions

that create drought or severe winters. An individual starving because of limited resources is 'nature's way'. This view contrasts sharply with the ethical basis for much game management, particularly furbearer management—many individual furbearers are considered surplus, above the annual capacity of the ecosystem, and trapping saves them from an inhumane death of starvation in winter (Caughley and Sinclair 1994).

Notwithstanding these ethical issues, there is historical evidence that at one time, wildlife management in Canada was in dire need of strengthening (Table 11.1). It is hard to imagine today, but the beaver, a common sight in much of rural and forested Canadian landscapes, was near extirpation (extinct) from Canada by 1900 (Novak 1987). Even harder to envision is the near-demise of white-tailed deer. A population of 13 million in the early 1800s was down to 350,000 by the end of the century, a 97 per cent decrease in 50 years over eastern North

Table 11.1 Terrestrial Bird and Mammal Species in Canada: Dramatic Population Increases due to Wildlife Management Activities

Species	Cause for Decline	Cause for Increase
Birds		
Canada Goose	Market hunting	Game regulations, refuges, restocking
Snow Goose	Market hunting	Game regulations, refuges, restocking
Osprey	Toxins, predator control	Ban on DDT, protection of nest and bird
Bald Eagle	Toxins, predator control	Ban on DDT, protection of nest and bird
Peregrine Falcon	Toxins, predator control	Ban on DDT, protection of nest and bird
Hudsonian Godwit	Market hunting	No hunting allowed
Trumpeter Swan	Market hunting	No hunting allowed, restocking
Wild Turkey	Market hunting, habitat loss	Game regulations, restocking
Mammals		
Musk Oxen	Market hunting, fur trade	Game regulations, restocking
Wood Bison	Market hunting	Restocking, refuges
Pronghorn Antelope	Market hunting	Game regulations, restocking
White-tailed Deer	Market hunting	Game regulations, refuges
Beaver	Fur trade	Game regulations, restocking
Sea Otter	Fur trade	International regulations on harvest
Fur Seal	Fur trade	International regulations on harvest
Polar Bear	Trophy hunting, bounty	International regulations on harvest

Sources: DeMarais and Krausman (2000); Mathiessen (1987); Stewart (1978).

America (McCabe and McCabe 1984). In the prairie regions of the continent, pronghorn antelope were reduced from approximately 30 million at the time of European colonization to 15,000 by 1915, but thanks to hunting restrictions and translocations, they now number approximately one million in total, with 30,000 in Canada (Yoakum and O'Gara 2000). However, the failures are equally impressive. The millions of plains bison and passenger pigeons did not recover; two species so abundant that their demise was considered impossible were gone after 100 years of intense hunting (Table 11. 2).

Unregulated hunting was a major mortality factor on the viability of game species in the 1800s. Large-scale market hunting in which wildlife could be sold for profit along with agricultural products was eventually made unprofitable in the United States with the Lacey Act of 1900 (Gilbert and Dodds 1992). The Act made transport of wildlife across state borders illegal, effectively cutting off the supply of wildlife that came from the hinterland to the markets of New York and Philadelphia. In Canada, the Conservation Act of 1909 established a similar set of regulations. Fur seals of the Pacific Coast, nearly exterminated for their pelts, were protected and then rebounded under an Act signed by Canada, the United States, Japan, and Russia in 1911.

The Migratory Bird Act, signed by Canada, the United States, and Mexico in 1918 (revised as the Migratory Bird Convention Act, 1994), ended the killing of all migratory birds except for certain game species such as waterfowl. Most shorebird species that staged on the mudflats of Prince Edward Island during migration, for example, could no longer be shot and sold at market—the occasional shorebird decoy can still be found in antique stores across the Maritimes.

Although federal legislation has been critical in protecting birds, the dramatic increase in other game species has been the result of many actions. Habitat creation projects, refuges, public education, and provincial regulations on the number, sex, and age that can be killed all contributed to the healthy populations we see today.

Table 11.2 Terrestrial Bird and Mammal Species in Canada: Once Abundant but Now Extirpated or Reduced to Very Low Numbers

Species	Cause for Decline	Cause for Failure to Recover
Birds		
Passenger Pigeon	Market/sport hunting	Extinct before regulations enacted
Eskimo Curlew	Market/sport hunting	Extinct before regulations enacted
Labrador Duck	Market/sport hunting	Extinct before regulations enacted
Great Auk	Sailor food source	Extinct before regulations enacted
Greater Prairie Chicken	Habitat loss	Habitat loss
Mammals		
Sea Mink	Fur trade	Extinct before regulations enacted
Black-tailed Ferret	Habitat loss, bounty	Habitat loss, loss of prairie dog colonies
Prairie Dog	Habitat loss, bounty	Habitat loss
Swift Fox	Habitat loss, bounty	Habitat loss
Grizzly Bear (Prairies)	Sport hunting, bounty	Extirpated before regulations enacted
Bison (Prairies)	Market/sport hunting	Habitat loss
Eastern Wolf	Bounty	Coyote competition, anti-wolf values
Woodland Caribou	Market/sport hunting	Habitat loss, parasites

Sources: DeMarais and Krausman (2000); Mathiessen (1987); Stewart (1978).

The conflict in management is more pronounced when one value is perceived to supersede other values (Figure 11.2). Government-sponsored wolf control on Vancouver Island and in Yukon during the 1990s illustrates this competition between values. Managers proposed that controlling populations of the main natural predator (the grey wolf) would allow populations of the prey (moose and caribou in Yukon, black-tailed deer on Vancouver Island) to be maintained at steady levels, an economic and cultural boon for subsistence and recreation hunting by local and visiting people (Boertje, Kelleyhouse, and Hayes 1995; Reid and Janz 1995). Not everyone appreciates such 'hands-on' management. The idea of subjecting nature, especially in such relatively wild regions, to such manipulation was not popular with people who valued wilderness and the role of top carnivores in ecosystems (e.g., Hummel 1995).

Conflicts also arise when there is a perception of too much wildlife. Hyper-abundant species have become a management issue in several parts of Canada, particularly in urban areas, where a large proportion of people demand control methods, such as sterilization instead of culling. Often, the abundance is related to lower mortality that increases populations, behavioural changes that become threats to people, and damage to other species. White-tailed deer and Canada geese in southern Ontario have been culled in attempts to lower population density. Mild winters and the absence of major predators (e.g., wolves, humans) near urban areas and in protected areas have allowed deer to thrive to levels at which driving safety and agricultural production are threatened. Canada geese along Lake Ontario have become year-round residents, thus avoiding a migration that once subjected them to considerable mortality pressure from hunting. Bull elk avoided wolves by moving into—and taking over—the golf courses and lawns in the town of Banff, Alberta. Males in breeding condition and females with young were a safety hazard; instances of goring and trampling made national headlines.

SPECIES OF CONCERN

Unlike the situation with respect to game species, most people, regardless of ideology, recognize the need for management when rare wildlife is at stake. Endangered species, by definition, are unlikely to persist without management intervention, and most people agree that hands-on, proactive management is required.

(G. Forbes)

Figure 11.2 Public values regarding wildlife often conflict. The thousands of tourists hoping to hear wolves howling in Algonquin Provincial Park, Ontario, perceive wolves differently from the person who snared wolves along the park boundary as a service to deer and farmers.

The United States established a nationwide law for the management of endangered wildlife in 1973. Although most Canadian provinces have endangered species legislation, a federal law in Canada was established only in 2003. The Species at Risk Act (SARA) is the legal framework for the protection of endangered species and their breeding habitat on federal land. The Act promotes a strategy of cooperation and stewardship among provincial governments and private landowners but can apply penalties and fines for gross violations of the Act.

By contrast, the American legislation protects listed species and their habitat on all land. As well, their legal system can be used to force assessments of species status. Critics of the American approach believe that the Act hinders cooperation between state and federal governments and that landowners may 'shoot-shovel-shut-up' rather than have government force protective measures on their land. Critics of SARA feel that stewardship models can only be successful if there is financial compensation to landowners (not in the Act) and that the Act will only have the required strength to protect species on a small portion of Canada—i.e., federal land south of the territories, or 6 per cent (Bocking 2001). Another controversy regarding SARA is that politicians have the final say on which species will be protected, raising concerns that protection for species that cause economic impacts will be denied. For further reading on the history and politics of endangered species legislation in Canada, much of which would apply to other environmental legislation, see Beazley and Boardman (2001).

At present, the Committee on the Status of Endangered Wildlife in Canada (COSEWIC) makes the decision on which species require protection. Consisting of government biologists and academics, COSEWIC was previously validated by the Canada Wildlife Act (1985) and now by SARA. Subcommittees review the status reports for select species and then recommend the status of the species to the full committee. The committee ranks rarity based on categories of risk of extinction, and the federal minister of the environment decides whether the species will be listed. As of 2008, 13 species were extinct in Canada, and 238 species were endangered. Most of the listed species are plants (Table 11.3).

The polar bear is a good example of the application of science and uncertainty in assigning the risk of extinction. Polar bears have become the 'poster child' and marketing tool for several wildlife conservation organizations, principally because they are impressive animals that are threatened by climate change. Pictures of a skinny bear floating on a little patch of ice in an unfrozen Arctic sea are powerful messages about global warming (Chapter 5). In May 2008, the United States designated the polar bear as 'threatened', and the expectation from many was that Canada would do the same. However, COSEWIC recommended retaining a status of 'species of concern', a status for which vigilance is required but immediate concern for the population is not warranted. This is the lowest level of threat within SARA. A status of 'endangered' means extirpation is likely without immediate action, and 'threatened' means that the species will likely become endangered if action is not taken.

The public response was varied: some thought that polar bears were abundant, others that they would be extinct in a few years. Newspapers noted political pressure from Arctic communities worried about losing lucrative income—as much as $3 million per year in the Arctic is derived from southern hunters paying for hunting part of the quota granted to the Inuit (CBC News 2008). Issues of race, culture, traditional knowledge, and perceived interference by 'condominium conservationists' on northern

Table 11.3 Number of Species in COSEWIC Categories with Corresponding Funding from Recovery Efforts, Indicating That Most Species Listed and Managed are Mammals and Birds

Status	Mammals	Birds	Reptiles	Amphibians	Fish	Arthropods	Molluscs	Plants	Mosses	Lichens	Total
Extinct	2	3	–	–	6	–	1	–	1	–	13
Extirpated	3	2	4	1	4	3	2	3	1	–	23
Endangered	21	27	14	7	37	17	17	89	7	2	238
Threatened	18	18	13	6	25	6	6	52	3	2	149
Special concern	26	23	9	7	41	5	5	33	4	5	158
Total (% of total)	70 (12%)	73 (13%)	40 (7%)	21 (4%)	113 (19%)	31 (5%)	31 (5%)	177 (30%)	16 (3%)	9 (2%)	581 (100)
% Eligible for RENEW funding*	12.7	13.2	7.4	4.6	17.5	3.8	4.1	33.5	1.9	1.4	100
% 2005–6 RENEW funding	34	28	8	4	12	3	4	6	1	n/a	100

Source: COSEWIC 2008a; RENEW 2006.

*RENEW funding is provided for species listed as extirpated, endangered, or threatened.
The % eligible is based on the number of listed species in 2004, the list that would influence funding in the following 2005–6 year.

realities became entwined with a listing process that attempts to rely on data and ignore political and economic issues when listing a species.

As a member of the Terrestrial Mammal Subcommittee, the group that recommended that the polar bear not receive increased protection, I believe it would be informative for readers to understand where the science fits into the decision. COSEWIC has strong numeric guidelines on when a species is endangered, or threatened, or of special concern. Endangered, for example, requires a population to have either declined to below 70 per cent or be projected to decline more than 50 per cent in the next three generations, have a distribution of less than 5,000 square kilometres, or have a population of less than 2,500 individuals. Rules also exist regarding what should be protected—a species is not always apparent. Black ducks, for example, appear as different from mallard ducks but genetically would not be recognized as distinct species (Ankney, Dennis, and Bailey 1987). There are subspecies, hybrids, types, forms, and disjunct populations.

In response, COSEWIC works with 'designated units' (DU), a unit identifiable both taxonomically and genetically. The DU is critical to the polar bear status. Because male bears can travel immense distances and interbreed, there is not enough genetic difference between polar bears from Yukon to Greenland to recognize separate populations (Paetkau et al. 1999). Thus, the status must be for one DU—all the polar bears in Canada. This mixing is fortunate for the viability of bears but unfortunate for listing, because bears are in serious decline in some parts of their range (e.g., in southern Hudson Bay, likely because of global warming, and adjacent to Greenland, likely because of overhunting). However, in much of the Arctic, data from researchers and wildlife managers indicate that the population is doing well or increasing. There are more than 15,000 bears in Canada, they live over a huge area

(5.6 x 10⁶ square kilometres), and population models based on demographics and mortality do not show significant enough population decline (McLoughlin, Taylor, and Dowsley 2008). Thus, at the scale of one DU, polar bears do not warrant threatened or endangered status.

But where is the risk? Do we know enough about genetic diversity to justify one DU as accurate? Thiemann, Derocher, and Stirling (2008), for example, promote five DUs. The models used to project future population levels did not have much input from the effects of melting sea ice because there is actually little data on how bears will respond. At the time of assessment, there were reports of emaciated bears dead in water (Monnett and Gleason 2006), but are these events common, and do we just notice them now with global warming in mind? Or are they evidence of a widespread event? When management requests the input of 'science', scientists will deliver advice based on existing knowledge. We did have evidence of population stability, but we did not have evidence of imminent population decline as a result of melting ice.

What if the science and management were wrong? The risk is that the polar bear will be extirpated because action will be delayed and the population will decline to the point that we cannot manage it back to viability. However, if global warming continues, polar bears will be extirpated regardless of their COSEWIC status. And this raises a truism in management: management should never be fixed, or reliant, on one tool. First, bears are a game species, and managers can close hunting completely if need be, regardless of a listing within SARA. Second, if evidence builds of a population decline for the DU, a new status report will be commissioned, and the status will be reassessed. And third, the big picture is critical—successful management of polar bears actually means 'managing' or resolving climate change.

Many years in the making, Canada's federal endangered species legislation is still controversial (Amos, Harrison, and Hoberg 2001). And although SARA was promoted as a model that would minimize a polarized environment for managing endangered species such as had developed in the United States, the involvement of politicians in deciding which species are listed resulted in a lawsuit within a few years of SARA being enacted. In British Columbia, nongovernment organizations launched a lawsuit to force the federal minister of the environment to require the provincial government to protect spotted owl habitat. Spotted owls in British Columbia had declined to 19 birds by 2007, mainly because of forest harvesting (COSEWIC 2008b). But the forestry industry and jobs would likely be affected if this old-growth forest specialist were to be recognized and given greater protection. In another example, the plains bison was recommended as threatened by COSEWIC but not placed on the SARA list by government because listing would hinder the bison farming industry. Lawsuits are being considered.

Notwithstanding the political aspects of listing, the preservation of many species will be very difficult—and expensive. Some species listed under SARA will be difficult to maintain because of the scale of the problems that caused the decline: Atlantic salmon, for example, are affected by a range of terrestrial and oceanic factors; woodland caribou use older forests that do not contain deer, but deer are expanding northward; the eastern population of wolverine needs a massive land base to be viable; and, the St Lawrence population of beluga whales lives at the exit-point for the industrial pollutants of the Great Lakes region.

One endangered species that will depend on Canada to save it is the Vancouver Island marmot, found nowhere in the world except in a few mountain meadows on Vancouver Island. The

population declined from a high of 300 to 350 individuals in 1984 (Bryant and Janz 1996) to approximately 30 in 2002 in the wild (Marmot Recovery Foundation 2002). The decline is possibly due to animals dispersing and establishing territories in forestry clear-cuts where over-winter mortality and increased predator densities have created 'mortality sinks' (Bryant 1996). The most recent recovery plan called for establishing a captive breeding population to safeguard the species (Janz et al. 2000). In response, a consortium of partners, including forestry companies, environmental groups, and the public, raised funds to build a breeding facility near suitable habitat for eventual restocking of marmots into the wild. Although many questions still remain about the effectiveness of planned recovery efforts based on captive breeding and reintroduction, captive breeding had increased the world population to more than 100 individuals by 2002. At the end of 2006, only 23 mature marmot remained in the wild (Bryant 2007); captive breeding now appears integral for this species' survival.

Reintroductions

A parallel committee to COSEWIC, the Recovery of Nationally Endangered Wildlife (RENEW), was established in 1988 and restructured in the 1996 Accord for the Protection of Species at Risk between provincial and federal governments. In 2005–6, RENEW reported spending of $41 million on the part of 170 organizations that assisted research and actions for the recovery of listed species (RENEW 2006). Most of the funding (62 per cent) went to recovery projects for mammals and birds, even though they make up less than a third of eligible species (see Table 11.3).

Several endangered species have been bred in captivity or captured and transported to vacant habitat in an attempt to recover endangered populations. Having most of the population of a rare species located in one area increases the risk of extirpation as a result of disease or other disturbance. The reintroduction of peregrine falcons from western Canada to eastern Canada in the 1980s is considered a great success, and the subspecies was down-listed from endangered to special concern (COSEWIC 2007). Similar success has occurred in bringing the swift fox back to Canada; 479 were released into southern Alberta and Saskatchewan between 1983 and 1996, and by 2008 an estimated 654 lived in Canada (Gedir 2008).

Reintroductions and introductions have been a common management practice for game species. Many jurisdictions have restocked areas that had lost a valued species because of past overexploitation or habitat change. Furbearers and ungulates are most frequently transferred across or within provinces. For example, Nova Scotia was repopulated with moose from Alberta and marten from New Brunswick; Newfoundland introduced moose from New Brunswick. Other species include wild turkey released into southern Ontario from Michigan, and numerous fish species are frequently moved by fisher and government agencies.

Illegal Trade

One way to maintain rare species is to limit their commercial value. Canada is a signatory to CITES, the Convention on International Trade in Endangered Species of Wild Fauna and Flora. In Canada, CITES is enforced through the Wild Animal and Plant Protection Act and Regulation of International and Interprovincial Trade Act (1996). Both serve to limit and control the traffic of endangered species used as food, pets, medicine, fashion, or culture. The importation of a range of species of trees, coral, insects, shells, plants, and vertebrates is controlled through a permit system. In Canada, species of value on international markets include gyrfalcon and peregrine falcons for falconry, sturgeon for

caviar, and whale and bear (polar and grizzly pelts, black bear gall bladders).

WHO OWNS WILDLIFE?

The ownership of wildlife has been a contentious issue for centuries. In medieval Europe, royalty controlled all resources, including wildlife. Game species were maintained for recreational hunts using hounds and falcons. The eventual replacement of aristocracy with democratic institutions led to the establishment of wildlife as a common good, and this idea has persisted over the past 100 years.

In Canada, the issue of ownership has gained recent prominence. Judicial rulings on Aboriginal rights, particularly in eastern Canada (see Chapter 4) and the emergence of game farming in agriculture have reignited a question that runs to the core of wildlife management. Is wildlife public property or the property of the landowner? Can I sell meat for profit? Do decisions on wildlife management first belong to Aboriginal people? Does wildlife have its own rights? The question of ownership is important because the mandate to manage—to develop policy and set regulations—derives from a legal basis of ownership.

Wildlife management in Canada today is a product of our colonial history (Loo 2006). Emigrants from Europe left a land where most wildlife was inaccessible, far from the city, or controlled by owners of vast estates. When they arrived in Canada, wildlife was dramatically accessible; pioneers lived off the game on their property and adjacent wilderness, and most of this wilderness was Crown land—free for travelling and hunting.

The concept that wildlife also was free became engrained in colonial society. The British North America Act (1867) and its replacement, the Constitution Act (1982), divided public ownership of resources, including wild-life, between two levels of government. The federal government has responsibility for wildlife on federal lands (i.e., national parks and refuges, military facilities), marine mammals, migratory birds, and migratory mammals crossing federal–provincial borders. Provincial governments are responsible for all other wildlife and have legislation in place to regulate the harvest, farming, and captivity of wildlife.

Hunter Access to Wildlife

Wildlife will cross ownership boundaries, and since ownership of unmarked animals is difficult to prove, many nations established laws establishing wildlife as common property. Today, the landowner has the authority to control access to the land, and this control is becoming a source of conflict. Proponents argue that landowners should be able to derive income from wildlife on their property—charging for access is a potential boon to rural economies. Opponents argue that such development will lead to the downfall of game management in Canada.

Under the present system, wildlife has more value when alive because living animals produce the surplus animals that are hunted. In pay-to-hunt management, there is increased value on the animal when it is dead. Opponents also contend that hunting will become economically segregated, citing the European-style situation in which fewer lower-income people hunt because they cannot afford to pay the access fees charged by landowners (Canadian Wildlife Federation 1992). The extent of the problem (or opportunity) varies across Canada. In regions such as the Maritimes, where private land ownership is above 50 per cent, or in southern Ontario and Quebec, access to hunting areas is declining compared to access in northern and western regions. In New Brunswick, the provincial government implemented a system

for limiting the creation of private hunting grounds. In response, some landowners charge fees for people to park on their land while hunting. A secondary benefit may be that more land is kept as productive habitat for game species because the cash incentive may offset the pressure to over-harvest woodlots.

Game Farming: Conflict between Farm and Forest

One of the largest areas of conflict in wildlife management is the domestication of indigenous species and the importation of exotic species. Game farming is a relatively new management approach in Canada, only becoming common in the 1980s. The main indigenous species farmed are Atlantic salmon, wapiti (elk), and plains bison, mainly in western Canada. Common non-native species include red deer (a subspecies of elk from Europe and New Zealand), fallow deer, wild boar, and various bird species (White 2000). Government regulations vary across Canada. Some provinces, such as Alberta, permit farming of indigenous species only, while some provinces only allow exotic species to be farmed. Transportation of exotic species across borders requires an environmental impact assessment for potential impact on native species.

Breeding wildlife in captivity is contentious because it invokes three major value sets—ideology, sustainability, and ethics. Opponents criticize game farming because (1) it removes the division of wildness and private ownership, a division seen by many conservationists since the days of market hunting as the foundation for healthy wildlife populations (ideology); (2) it may harm native populations through disease transmission and by making poaching unenforceable (sustainability); and (3) certain types of domestication, such as shooting reserves, are considered unethical (ethics) (Canadian Wildlife Federation 1992; CCWHC 1994; White 2000).

In contrast, proponents of game farming argue that (1) there is public demand for specialty products (e.g., antler velvet for medicine, lean meat for consumption) and if products are secured from domesticated herds, it will lessen impact on wild herds; and (2) disease transmission can be controlled with adequate resources. Fenced shooting reserves are not as popular in Canada as they are in the United States, but if the trend toward decreased access for hunters increases, the market for secure hunting grounds will increase.

Wildlife Disease

An area in wildlife management that combines elements of conflict and uncertainty is the transmission of disease from wildlife to humans or domesticated animals. Three types of wildlife disease are of interest: (1) those that occur in wild animals at low, insignificant levels but can increase to significant mortality levels as a result of overcrowding, such as occurs in captive industries (e.g., aquaculture, game farming); (2) those that can affect domesticated species and indirectly affect people; and (3) those that can directly harm people. The first group was discussed previously, the second group is an area of conflict, and the latter group is the cause of greatest uncertainty.

The management of wildlife diseases that affect domestic species and may harm people is laden with conflict. Farming operations and nations want disease-free labels so that they can export meat, milk, and breeding stock easily. An outbreak of tuberculosis in game farm elk in the early 1990s cost Canada its TB-free status for meat export, valued at $1 billion annually. The concern that diseases such as bovine tuberculosis and brucellosis in wild bison could infect free-ranging cattle led to a program to kill any bison that leave Yellowstone National Park, Montana, and to controversial proposals by agriculture interests to kill the bison in Wood Buffalo National Park, Canada (Aniskowicz 1990). Because the bison population is affected but can withstand the disease and transmission from bison to cattle is unlikely (Shaw and Meagher 2000), the issue is mainly an economic one. The conflict arises over the issue of the extent to which wildlife should be controlled by domesticated life.

The appearance of chronic wasting disease (CWD) in the elk farms of western North America has been of great concern. CWD is thought to be related to mad cow disease (bovine spongiform encephalopathy), a disease that can cause death in humans. Single infected elk from game farms were recorded in Saskatchewan in 1996 and 1998 (both had been imported from the United States) and in Alberta in 2002. Two wild mule deer were infected near the Alberta–Saskatchewan border in 2001, prompting massive surveillance programs that required the killing and testing of thousands of captive and wild hoofed mammals.

Disease in free-ranging deer makes containment much more difficult, and similar testing has been underway in the United States. In Wisconsin, 14 wild white-tailed deer had been infected by 2002. That summer, Wisconsin initiated a partial eradication program for wild deer; the goal is to kill all deer in the infected counties and reduce deer population by 50 per cent in the adjacent 10 counties, an estimated 15,000 deer. Approximately $27 million was spent between 2002 and 2007 in managing this disease (Wisconsin Department of Natural Resources 2008).

The outbreak of CWD may have major consequences for wildlife and game farming. Concerns over infected product has severely limited export markets for antler velvet (of medicinal value in Asia), breeding stock, and meat (a specialty product in Europe), and many game farmers report being near bankruptcy. In 2002, as a means of staying in business, the Alberta Elk Association requested, but was denied, government permission for fenced hunting reserves. The impact of this new disease on wild populations is not known, but the eradication programs for wild deer continue. Nor is it known what the impact on game farming and rural economies will be if hunters give up eating meat. However, the outbreak will be cited by opponents of game farming as a prime example of how captive wildlife negatively affects wildlife in the 'wild'.

The third category of disease includes wildlife diseases proven to be harmful to humans in a direct manner. These diseases create uncertainty because they seem to appear out of nowhere and have the potential for tragedy. People may pick up diseases if they go to certain locations. For example, ticks associated with white-tailed deer and certain rodents carry Lyme Disease, which can cause arthritic symptoms and blindness. People travelling or working in such areas can expect some degree of risk. Similarly, trappers know that they must be careful not to contract tularemia when skinning animals. However, society becomes much more concerned when wildlife diseases leave the wilderness and enter our secure urban domain.

Outbreaks can occur almost overnight. Diseases such as rabies spread rapidly, carried by infected dispersing mammals that transmit the virus in fights with any creature they come across. The first Canadian case of non-bat rabies recorded in Ontario, and likely most cases since, can be traced to a single Arctic fox that moved south from James Bay in 1954 and passed rabies into red fox populations (Johnston and Beauregard 1969). A virulent strain associated with the raccoon has been moving north from Florida over the past 15 years, bringing rabies to New Brunswick in 2000. Hanta virus, the cause of numerous deaths across North America, including 20 deaths in Canada by 2006, was only discovered in 1993. The disease is carried by small rodents, mainly deer mice, and is transmitted in aerosols of mouse urine and feces (CCWHC 1998). More recently, migrating birds spread the West Nile virus across North America within just three years. The response for both diseases is expensive eradication and public education programs.

Part of the uncertainty stems from lack of knowledge about how much of a problem these threats are. In addition to West Nile virus, six other arboviruses (mosquito- or tick-transmitted) have caused death in Canada (CCWHC 1999). Now that West Nile virus is here, will it lead to an occasional human fatality or to thousands? Should wetlands be drained and chemicals applied to wide areas to kill mosquitoes? What is the financial and ecological cost weighed against human life?

Conflict between Societies

A second major area of conflict relates to whether all people have equal access to wildlife. At present, wildlife is public property that cannot be sold for profit and is managed under specific rules empowered by federal and provincial laws. Aboriginal people counter that access to wildlife is guaranteed in various treaties and was never negotiated in regions where treaties were not made, such as the Arctic and in British Columbia. Recent rulings by the Supreme Court (e.g., the *Marshall* decision, 1999) interpreted treaties in the Maritime provinces as providing Mi'kmaq First Nations a share in fisheries resources and the opportunity to make a modest livelihood from selling wildlife (see Chapter 4). Although selling wildlife could be considered a partial return to the perils of the market hunting of the nineteenth century, the rulings stated that conservation objectives must supersede harvest. Much of the conflict to date, notably regarding lobster fishing, has been about whether Native harvest will be regulated by Aboriginal or non-Aboriginal governments or be subject to the same rules as non-Aboriginal harvest.

In terms of wildlife management, harvest quotas and seasons are based on models requiring accurate data on the age, gender, and location of hunted animals. Co-operation between Aboriginal hunters and government agencies becomes integral to sustaining wildlife populations. In many jurisdictions, the new arrangement is for Aboriginal harvest data to be provided to provincial agencies or, in the case of the territories, the establishment of co-management boards (Sandlos 2007). The hope is that conflicts can be resolved and that hunted wildlife is not negatively affected while the two societies develop a new relationship.

Wildlife and Recreation

The concept of sustainable development holds great appeal when compared to the environmental impacts of short-sighted practices. However, the concept also implies that any type of development can be sustainable, which raises concern that wildlife could be unduly exploited. The ecotourism industry promotes passive exploitation of wildlife, with people visiting their habitats and viewing the more spectacular aspects of their behaviour (Knight and Gutzwiller 1995). The majority of ecotourism activities probably have no or little detrimental effect on wildlife. The tens of thousands of birdwatchers walking the trails of Point Pelee National Park, ON, during spring migration are not a problem unless they leave the trails and trample the undergrowth.

However, the conflict between ecotourism and wildlife increases when animals are forced to accommodate visitors during important periods of their lifecycle. People will pay for spectacular sightings, often of rare species, but these events typically occur during short periods

or in limited areas. Polar bears feed on garbage for tourists while waiting for sea ice to form near Churchill, MB. Endangered right whales travel the coast of the Americas to breed in splashing and rolling spawning groups at the surface of the Bay of Fundy, typically surrounded by a flotilla of whale-watching boats. Killer whales off British Columbia and beluga whales in the St Lawrence River experience similar pressure. Most of the world's population of semipalmated sandpipers, sometimes 100,000 birds at a time, need to rest at high tide within metres of birdwatchers on the mudflats near Moncton, NB.

As ecotourism becomes more popular, government agencies are pressured into developing guidelines or regulations that will mitigate potential impacts. For example, industry associations and the federal Department of Fisheries and Oceans have developed guidelines for whale-watching: boats must avoid intercepting whales, travel slowly, and not encroach on the whales. Birdwatchers and dogs must stay behind marked lines when shorebirds are roosting along beaches in the upper Bay of Fundy. Sections of beaches are closed during piping plover nesting season.

In a survey of Canadian participation in nature, birdwatching and wildlife viewing away from home was one of the most common (18.6 per cent of the population over 15 years of age) outdoor activities in Canada. Visiting natural areas was involved in half of all outdoor activities. The economic value of wildlife and nature is considerable. Nature-related tourism was worth $5.6 billion in 1991, $11 billion in 1996, and is expected to continue increasing (DuWors et al. 1999). The income to government and businesses is considerable at provincial scales as well: for example Quebecers spend approximately $300 million annually on observing wildlife and birdwatching (Government of Quebec 2009b).

STRATEGIES ON HOW TO MANAGE

The management of wildlife in Canada has evolved from a period of unregulated market hunting, to grand legislative regulations and protected refuges, to data-intensive simulation modelling of habitats and populations (Burnett 2002; Colpitts 2003). The tools available to managers vary from protected areas to predictive models of population viability. Research and experience have shown that some species create habitat for other species and thus protecting one species may help many species, a cost-effective means of management. Wildlife managers have also learned that they must consider wildlife from the genetic to the landscape level. In this section, the main strategies for managing wildlife in Canada are explained.

Wildlife Refuges

An early example of wildlife management was the establishment of wildlife refuges, generally small pieces of land containing productive habitat. At a time when people were hunting with little regulation, the refuge was seen as a safe haven for populations of game species. As well, protected populations would reach high densities, and as offspring dispersed from the refuges, the refuge would act as a source of animals, replenishing hunted populations outside the refuge. The concept is well established in the public view of wildlife management. National wildlife refuges are common throughout the United States. In Canada, most provinces have established refuges offering similar functions as other protected areas, such as nature reserves and wilderness parks. By 2003, the Canadian Wildlife Service had established a network of 143 national wildlife areas and migratory bird areas. The refuges are usually for the protection of breeding and migrating waterfowl. At least

400 sites in Canada are dedicated to wildlife conservation (Rubec and Turner 2003).

The concept of the game reserve is much less popular today than it was 100 years ago. The confidence people have today in the way we manage a game species, enforce regulations, and manage habitat over much of the larger landscape has raised questions about having areas where hunting and resource extraction are not allowed. This view may conflict with the views of some hunters who benefit from hunting near refuges and protected areas and the views of people who value the idea of some land being off-limits to hunting.

Combining Population and Habitat

'Without habitat . . . there is no wildlife. It's that simple' (Wildlife Habitat Canada 2008, 1). An important development in wildlife management has been the integration of populations with habitat. It is obvious that habitat quality will influence the number and size of game species, but it was not until the 1980s that most government agencies developed specific objectives connecting the two variables. Previously, game managers would develop harvest quotas based on the location, number, and sex of a species that could be killed each season. Habitat was assumed to be variable but abundant enough that specific objectives were not required. As larger amounts of habitat (e.g., old-growth forests) were modified by forest harvesting, mining, urban development, and values besides hunting, the need to establish the amount of habitat needed by moose, deer, and ducks became a matter of concern.

This need to label the landscape relative to its ability to meet a population objective led to the concept of the Habitat Suitability Index (HSI) and Habitat Capacity (HC) models (Anderson and Gutzwiller 1996). The HSI is now a major tool in wildlife management, and HSIs have been developed for many game and non-game species. An HSI quantifies the value of a variable such as amount of large spruce trees or canopy cover, often assigning a break-point or threshold below which the species will decrease. Each forest stand or wetland can then be ranked for its suitability for a species, and the total available area, configuration, and spatial design can be assessed. As valuable as HSIs are, they are limited by our knowledge of which parameters are important to that species and the contents and accuracy of the database.

Viable Populations

The number of individuals for a viable population can be determined by several models. Most game species have been studied well enough that species-specific models are shared among jurisdictions and then adapted to local conditions. These models often estimate the population size and warn of destabilizing trends—for example, if too many breeding females are taken in a harvest.

For species for which specific models do not exist and for rare species, the number of individuals for a viable population can be estimated using population viability analysis (PVA) models (e.g., Vortex, Ramas, GAPPS, INMAT) (Brook, Burgman, and Frankham 2000). These models incorporate general knowledge on the effects of inbreeding and the demographic risk of small population size with local data on variable environments, mortality, and productivity to produce a probability of extinction or extirpation over a certain time period.

A limitation of PVA is that we usually lack the data on important variables in the model. For example, a PVA on wolverine using Vortex requires knowing the number of young born to one-, two-, three-year-old (and so on) female wolverines. Mortality and dispersal rates are equally unknown. The result is a choice between simpler models with data but lacking rigour and complex models with incomplete data for key variables. The accuracy of either is suspect in most cases, and PVA, at present, is best used to identify only broad levels of risk to a population (e.g., whether 50 or 500 individuals are required).

The application of habitat supply and viable populations of an indicator is well illustrated by the management of old forest in New Brunswick. The need for action resulted from projections that timber harvest rates would soon remove most spruce-balsam fir forest over 60 years old on Crown land (Erdle 1998). Provincial biologists identified 32 vertebrate species associated with this habitat type. Of these, one species, the American marten, requires the largest single patch (500 hectares home range for an adult male) of forest per individual. It was decided that the specific habitat requirements of this species would be maintained, thereby acting as an umbrella or surrogate indicator for the needs of the other 31 species. The amount of habitat needed was determined by multiplying the Minimum Viable Population by the home range, and 10 per cent of each Crown forest licence has to be maintained under specific structural and spatial characteristics. As these old areas succeed to young forest, new areas must be found to replace them.

This innovative program has been in place since 1992. Nonetheless, the program has its share of conflict and uncertainty. The 10 per cent objective resulted in a 12 per cent reduction in the annual harvest of trees (Erdle and Sullivan 1998), a significant impact on the timber-driven economy of the region. Trappers are concerned that the projected habitat amount will support much fewer marten compared to the population existing today. And do marten really indicate old-growth forest?

Indicator Species

A major aspect of uncertainty is associated with the use of indicators for planning. Indicators offer a means of managing complex relationships: understanding and protecting several species may be less expensive and faster than trying to plan and monitor many species (Croonquist and Brooks 1991; Mills, Soule, and Doake 1993). The risk is that the indicator may not indicate as accurately as expected (Broberg 1999; Lindenmayer et al. 2002). Indicators for use in resource management planning need to be specialized to a habitat type, exhibit consistent response to changing parameters, and be common enough to avoid population change due to chance events (Lindenmayer and Franklin 2002).

In the New Brunswick example, the choice of marten as an indicator of mature softwood forest was based on existing knowledge from western North America. Recent research in eastern North America (e.g., Payer and Harrison 2000) suggests that marten depend more on coarse woody debris, overhead canopy cover, and height of tree than on the softwood forest itself. These features typically occur in softwood stands, but because they may occur in other forest types, the use of the marten as an indicator for mature softwood is less definitive. Since the marten was a surrogate for 31 other species, and the present understanding of the needs of these species is met by the objective, the program still meets its goal and has been retained. It does illustrate, however, the risk of using indicators and the need for ongoing refinement (see also Chapter 3).

The Importance of Scale

Several recent paradigms are of interest to wildlife management, notably the concept of ecological integrity, the single-species approach, and landscape ecology. Ecological integrity is the concept that an ecosystem is composed of interacting units that can respond to stressors such as catastrophic disturbance from hurricanes or introduced species (see also Chapters 3 and 15). An ecosystem has integrity when it has the capacity to recover and function within some defined limits (Woodley, Kay, and Francis 1993). The term has become the foundation for protected areas management in Canada's national parks (see Chapter 12). Although it is difficult to define capacity levels, the concept provides an ecological framework that wildlife managers can consider when they are changing densities of certain predator or prey species or reintroducing species to the ecosystem.

The single-species approach dominated wildlife management activities until the advent of the biodiversity paradigm. Society valued game species such as moose and waterfowl, and wildlife managers developed habitat models for maintaining these species. In northern Ontario, forest planning on Crown land since the 1980s created a landscape of 80- to 130-hectare clear-cut patches in accordance with moose habitat guidelines (Thompson and Stewart 1997). The result benefited one species but possibly at the expense of species requiring larger or small patches of different-aged forest. Wildlife planners today may have habitat objectives for a wider range of species, the goal being that the entire community of plants and animals will persist.

One of the most important influences on wildlife management has been recognition of the importance of scale (spatial and temporal) in biology and management. Traditionally (and still today), most research and management was conducted at local-level spatial and temporal scales. For example, a review of 80 small mammal studies revealed that 70 per cent were conducted in less than a five-hectare area, the size of a city block (Bowman, Corkum, and Forbes 2001). The resolution (size of sample unit: 0.1 m^2 vs. 1 m^2, or 1-second vs. 1-day intervals) and extent (overall sampling area) of the research can lead to different conclusions. Management of white-tailed deer habitat in winter focused on the amount of canopy cover and tree species within a forest stand, but we now also realize that there need to be suitable stands across the landscape, because deer will migrate up to 90 kilometres to a winter refuge (Demarais, Miller, and Jacobson 2000). Moose, for example, may use a thick-canopy-cover habitat (less snow) within a stand in winter but be most abundant in regions with high amounts of open canopy (more food) (Forbes and Theberge 1993). Much of the recognition of larger-scale effects came about as a result of following tagged (i.e., radio-collared) animals and discovering how much interaction occurs between populations and how far individuals move between seasons.

Landscape ecology considers how structures facilitate or hinder movement of wildlife (and other features), how the land is dynamic, and how genes, species, and communities interact with abiotic features across scales. The result is a conceptual template that wildlife management should consider if it wants to maintain the full range of species and ecological processes (Schmiegelow, Machtans, and Hannon 1997; Venier et al. 1999; but see Lindenmayer et al. 2002).

DEALING WITH UNCERTAINTY

If a root of uncertainty is ignorance, then I suggest that greater effort is required on (1) long-term policies and strategies to co-ordinate activity; (2) research to improve understanding; (3) monitoring to warn of problems; and (4) adaptive management strategies to allow us to learn from experience.

Co-ordinated Strategies

The Wildlife Ministers Council, consisting of provincial, territorial, and federal ministers with responsibility for wildlife, developed a national wildlife policy in 1990. The goal of the policy is to maintain and enhance the health and diversity of Canada's wildlife for the benefit of wildlife and people today and in the future (Wildlife Ministers Council of Canada 1990). Three tasks are required to fulfill this goal: (1) maintain and restore ecological processes; (2) maintain and restore biodiversity; and (3) ensure that all use of wildlife is sustainable. Similar policies exist at the provincial level. Nova Scotia, for example, intends to meet its goals through an increased ability to determine population levels and distribution, intensifying research, and minimizing conflict between wildlife and other values by promoting integrated land use (Government of Nova Scotia 1987).

The funds required to manage wildlife originate from many sources. Government agencies receive the majority of their funds from general tax revenue. This system is different from the sources of funding in the United States. The Pittman-Robertson Federal Aid in Wildlife Restoration Act of 1937 directs taxes from the sale of ammunition and hunting equipment to wildlife agencies. The fund raises considerable operating dollars but also ties agency budgets to the popularity, and arguably the promotion, of hunting values (Patterson, Guynn, and Guynn 2001). The decline in hunting activity has affected agency budgets, and the promotion of hunting conflicts with the values of groups opposed to consumptive wildlife use (Gilbert and Dodds 1992).

Numerous programs raise funds and direct the money back into wildlife management. Waterfowl hunters in Canada are required to purchase an annual 'duck stamp' administered by a non-government group, Wildlife Habitat Canada. Since 1985, this program has directed a total of $33 million, mainly from selling stamps at $17 each, toward habitat improvement and research projects in Canada. Similarly, funds are raised on the sale of a range of consumptive and non-consumptive wildlife products in the provinces (e.g., Nova Scotia Habitat Conservation

Fund, Ontario Living Legacy, Buck for Wildlife in Alberta, and Habitat Conservation Trust Fund in British Columbia). The funds can be significant. For example, New Brunswick has a population of less than one million people but raises $1 million annually from small ($10 to $30 for fishing licences, $5 for specialty licence plates) surcharges on hunting and fishing licences and special vehicle licence plates. A council of interested citizens distributes monies from the New Brunswick Wildlife Trust Fund to clubs and researchers to rehabilitate streams, inventory rare species, restock salmon, and in general help wildlife.

A major source of funds for wildlife management comes from the North American Waterfowl Management Plan, an agreement between Canada, the United States. and Mexico that contributed $1.2 billion to wetland inventory, acquisition, improvement, and research in Canada between 1986 and 2008. More than half of the funding originated in the United States and was spent in Canada because habitat in Canada is the main source of ducks and geese; waterfowl hunted during winter in the United States typically breed in Canada. The lands support waterfowl but also support the many species that use wetlands as well as their ecological functions.

Numerous non-government agencies expend significant funds and effort on wildlife management. Groups such as Ducks Unlimited, Trout Unlimited, the Ruffed Grouse Society, the Canadian Wildlife Federation, and various provincial hunting organizations are very active in acquiring and maintaining wildlife habitat. Over a 15-year period (1984–99), hunters in Canada contributed $335 million in voluntary donations and involvement in habitat restoration and $600 million in mandatory fees (Wildlife Habitat Canada 2000). Groups associated with non-consumptive use undertake similar activities. Well-known groups such as the Canadian Nature Federation, the World Wildlife Fund Canada, the Nature Conservancy, and many provincial organizations work mainly toward land acquisition, but they also finance research and public education.

Research

The uncertainty associated with managing complex ecological processes is best addressed through research (Figure 11.3). Unfortunately, wildlife research is difficult to conduct well, partly because of the complexity of the system but also because of difficulties of scale and an inability to establish controlled experiments. Research is conducted by many of the organizations previously mentioned, often in partnership with each other. In the 1990s, the Canadian Wildlife Service helped to finance and establish Cooperative Wildlife Ecology Research Networks in Atlantic and western Canadian universities. In 1989, a cooperative was established between the New Brunswick Fish and Wildlife Branch of the provincial government and the University of New Brunswick. Other cooperative research groups exist among wildlife veterinarians and managers (University of Saskatchewan) and wildlife genetic laboratories (e.g., the Ontario government and Trent University). Such cooperative ventures seek to maximize the value of research effort and funds directed to information needed by managers.

Monitoring

The history of wildlife research and management is replete with stories of animals being introduced into unfamiliar areas and then forgotten. In the 1960s, it was common to build ungulate enclosures in order to measure plant growth free from browse pressure from deer or moose. These enclosures would be maintained for several years but then often forgotten and rediscovered years later in the middle of the forest by the

(J. Honderich)   (G. Forbes)

Figure 11.3 Management requires knowledge of an animal's habitat use, mortality factors, and movement. Radio-tagging, as in the case with this cougar in Alberta, is commonly used to track an individual's life history.

In the past decade, the general public has become involved with researchers in monitoring changes in wildlife and ecosystems—in this case, sparrow species in Nova Scotian meadows.

new biologist on staff. However, monitoring can be valuable if properly maintained. For example, an innovative new program involves setting lynx trapping quotas on the basis of the fluctuating abundance of snowshoe hare, a prime prey species monitored in the Gaspé Peninsula by provincial biologists.

Lack of appropriate monitoring is a common problem in the management of resources. People and priorities change, and budgets rarely allow for ongoing monitoring of old projects and creation of new projects. Since the 1990s, a major effort has been made to ensure that monitoring becomes a part of environmental management. The National Biodiversity Convention Office, established in 1991 as part of Canada's commitment to the Convention on Biological Diversity, has reported Canada's progress on maintaining diversity to the United Nations every four years. Each province has established the status of wildlife (plants, birds, mammals, herpetofauna, fish) and reports on changes in population trends. A second initiative is the creation of a national network of monitoring projects. The Environmental Monitoring and Assessment Network

(EMAN) supports research and volunteer-based monitoring across Canada on many aspects of the environment. EMAN, in association with many government and non-government groups, has initiated such programs as Frog Watch, Ice Watch, Plant Watch, and Worm Watch to measure changes in distribution and timing of the breeding or flowering of species thought to indicate environmental change.

Since the 1980s, a major effort has been directed toward involving the public in monitoring wildlife. In addition to the EMAN projects, Bird Studies Canada, with a variety of partners, co-ordinates volunteer efforts on programs such as Project Feeder Watch, Marsh Monitoring Program, Christmas Bird Counts, and Canadian Lakes Loon Survey.

The Breeding Bird Survey (BBS), administered in Canada by the Canadian Wildlife Service, has been conducted annually across North America since 1966. The survey is conducted over 450 permanent routes, each 40 kilometres long. Expert volunteers record the bird species seen or heard for three minutes over every 0.8 kilometre along the road. Results from the BBS formed the basis for reported declines in many species of birds, particularly birds that breed in Canada but over-winter in the tropics. Notably, it was these declines that led to much of the present concern over the effects of forest fragmentation and the need for landscape-level planning in wildlife management.

The strength of monitoring is (1) longevity, (2) standardized data collection and reporting protocols, and (3) systematic data records and management. Most monitoring projects are only a few years old and will require ongoing support. Standardized data collection and data that can confidently be interpreted are integral to comparing results over many years.

Adaptive Management

There are two approaches to gaining new knowledge. The first and more common approach is to 'try, try, and try again': implement an idea as a treatment, measure its success in terms of cost, replicability, and clarity, and then implement a modified version to improve it, if need be. The second approach is to implement several treatments at once, along with a control mechanism whereby change can be referenced independent of the manipulation. Each treatment is monitored and compared to the control site, and the most effective treatment is chosen. The second approach, termed adaptive management, is by its design a superior method of acquiring knowledge (Baskerville 1985; and see also Chapter 16). But it is also more expensive to establish and requires significantly more planning than the traditional approach.

Adaptive management has become very popular in certain industries such as forestry (see Chapters 10 and 19). The response of wildlife to forest management activities can be determined from an adaptive management framework. In game species management, the idea has

been slower to establish. Large-scale manipulation of animal populations is an expensive and long-term investment. Many game species are long-lived, and population-level response to manipulations will take 5 to 50 years depending on the species. For example, dramatic population fluctuations in predator–prey interactions of moose and wolves still were occurring on Isle Royale, Michigan, 30 years after disturbance (Peterson and Page 1988). The need for better understanding persists, and considerable foresight will be required today if society and wildlife are to benefit in the future.

CONCLUSION: IS THE SKY FALLING, OR IS THIS THE BEST OF TIMES?

The greatest area of uncertainty is whether predictions about the effects of global climate change are valid (see Chapter 5). If they are valid, wildlife and society will experience catastrophic impact within a mere 100 years, or more profoundly, within the lifetime of our children. Modelling exercises predict that the rate of adaptation required by species of plants will need to be 10 times faster than those of species indicated during the last post-glaciation period (Malcolm and Markham 2000). The same study suggests that 33 per cent of Canada's ecosystems will be subject to these required rates of adaptation movement; taiga/tundra, boreal forest, and temperate forest ecosystems will experience the greatest threats. Since so many species use these habitats, it is predicted that they will need to disperse quickly or somehow adapt to the new environment that overtakes them.

Another question that needs to be addressed is whether our successes and failures in wildlife management in Canada are a product of Canada's relatively limited exposure to human populations. How certain are we of our science in assuming that we can avoid the loss of the largest mammals as our population grows and our ecological footprint expands northward? Canada is unique among industrial nations, a massive land base and shoreline with a wealth of large mammals and wild herds and flocks. Compared to most of Europe, Canada has abundant moose, bear, wolves, caribou, freshwater fish, and waterfowl. Interestingly, Europe once had the same abundance.

Depending on one's perspective, wildlife today is either better off or on the edge of a precipice. Many species that we trap, hunt, or fish are at population levels higher than they were 100 years ago. However, many species that are sensitive to fragmentation, loss of unique habitat, or busy roads (e.g., reptiles) face an uncertain future. Managing woodland caribou, for example, will be very difficult and controversial; climate change, forestry access roads, Aboriginal rights to harvest, and wolf control issues will all come into play in the near future (Hummel and Ray 2008). I would offer the observation that both perspectives are accurate. Game species management may have been very successful, but the extent of change and our ignorance will limit our success in maintaining many non-vertebrate species.

As has always been the case with wildlife management, how and when we respond to these issues will ultimately be determined by societal values.

FROM THE FIELD

'A good wolf is a dead wolf, and that goes for wolf researchers.' And so began my conversation with a rather large man who wasn't fond of wolves, nor, apparently, of researchers. I wasn't particularly concerned about being threatened (I had just finished a third aerial telemetry flight of the day and was groggy from motion sickness pills), and when it became evident that I wouldn't be goaded into a fight, he switched to funding: 'I don't want my tax dollars spent on wolves. Who is funding you?!' I responded, 'we are fully supported by the Jehovah's Witnesses,' which wasn't true but was enough to deflect his anger, and he drove off muttering about 'bloody wolves'.

Fieldwork on wolves was a great experience —six years of finding, capturing, radio-collaring, and tracking a top predator in Algonquin Park, Ontario. We camped along logging roads, snowshoed beside tracks, and could hear wolves howling most nights. We flew in small airplanes to find the radio-collared members of a pack, landing the plane on a lake to snowshoe to a feeding site and try to ascertain causes of death, health of the prey, and the role of this small type of wolf in his ecosystem. Most of the time was spent picking up wolf scat—invaluable in determining what wolves were eating but also very smelly after a few weeks in the truck. When the research indicated that 'park' wolves were being killed by people outside the park, the research went from the field to the office—and the media. Should there be a ban on killing park wolves? What was the risk to livestock if more wolves lived? Conversely, what was the risk to the Algonquin wolf, now a recognized and threatened species, if too many were killed?

Working on wolves provided insight into the duality of wildlife management in Canada. We want our wildlife as long as they don't get in the way. The wolf seems either loved or hated by society. Wolves are valued as a symbol of wilderness and a source of pride because Canada has many, and thus Canada is still wild, still natural. By contrast, Europe and the United States have few, the big predators are gone, something is lacking. If you want wildlife fieldwork, radio-collaring is as good as it gets—and to work on wolves in Canada is the pinnacle. They are intelligent, rare, travel long distances, have advanced social behaviour, and are apex predators. For these same qualities, wolves are hated. They prey on cute baby deer or valuable farm animals. Their intelligence and pack structure make them hard to exterminate. And so if you have an affinity for wolves, you will be the envy of grad students at a wildlife conference, but you also run the risk of having your tires slashed or being punched by a wolf-hater. Our relationship with wildlife is based on values. My wife also worked on wildlife in Algonquin. People would invite us to their homes for interviews and tea. They would relate all sorts of stories and memories—and complain about wolves. Her thesis was on deer, so I kept quiet in the corner.

—GRAHAM FORBES

REVIEW QUESTIONS

1. Why is managing wildlife often about managing people?
2. How do wildlife managers avoid or minimize conflict with the values of a diverse society?
3. What are the risks and benefits of game farming?
4. What is the risk of relying on an indicator species to manage habitat?
5. Should politicians decide which species are placed on the endangered species list?

REFERENCES

Amos, W., K. Harrison, and G. Hoberg. 2001. 'In search of a minimum winning coalition: The politics of species at risk legislation in Canada'. In K. Beazley and R. Boardman, eds, *Politics of the Wild: Canada and Endangered Species*, 137–66. Toronto: Oxford University Press.

Anderson, S.H., and K.J. Gutzwiller. 1996. 'Habitat evaluation methods'. In T.E. Bookhout, ed., *Research and Management Techniques for Wildlife and Habitats*, 592–606. Bethesda, MD: Wildlife Society.

Aniskowicz, B. 1990. 'Life or death: A case for the defence of Wood Buffalo National Park's bison'. *Nature Canada* 19: 35–8.

Ankney, C., D.G. Dennis, and R.O. Bailey. 1987. 'Increasing mallards—Decreasing American black ducks: Coincidence or cause and effect?' *Journal of Wildlife Management* 51: 523–9.

Baskerville, G. 1985. 'Adaptive management: Wood availability and habitat availability'. *The Forestry Chronicle* 61: 171–5.

Beazley, K., and R. Boardman, eds. 2001. *Politics of the Wild: Canada and Endangered Species*. Toronto: Oxford University Press.

Biodiversity Science Assessment Team. 1994. *Biodiversity in Canada: A Science Assessment for Environment Canada*. Ottawa: Environment Canada.

Bocking, S., ed. 2000. *Biodiversity in Canada: Ecology, Ideas and Action*. Peterborough: Broadview Press.

———. 2001. 'Endangered species: A historical perspective'. In K. Beazley and R. Boardman, eds, *Politics of the Wild: Canada and Endangered Species*, 117–37. Toronto: Oxford University Press.

Boertje, R., D. Kelleyhouse, and R. Hayes. 1995. 'Methods for reducing natural wolf predation on moose in Alaska and Yukon: An evaluation'. In L. Carbyn,

S. Fritts, and D. Seip, eds, *Ecology and Conservation of Wolves in a Changing World*, 505–14. Edmonton: Canadian Circumpolar Institute.

Bowman, J., C. Corkum, and G. Forbes. 2001. 'Spatial scales of trapping in small-mammal research'. *Canadian Field-Naturalist* 115: 472–75.

Broberg. L. 1999. 'Will management of vulnerable species protect biodiversity?' *Journal of Forestry* 97: 12–18.

Brook, B.W., M. Burgman, and R. Frankham. 2000. 'Differences and congruencies between PVA packages: The importance of sex ratio for predictions of extinction risk'. *Conservation Ecology* 4: 6.

Bryant, A. 1996. 'Reproduction and persistence of Vancouver Island marmots in natural and logged habitats'. *Canadian Journal of Zoology* 74: 678–87.

———. 2007. *Update COSEWIC Status Report on Vancouver Island Marmot*. Prepared for Environment Canada.

Bryant, A.A., and D.W. Janz. 1996. 'Distribution and abundance of Vancouver Island marmots (*Marmota vancouverensis*)'. *Canadian Journal of Zoology* 74: 667–77.

Burnett, J.A. 2002. *A Passion for Wildlife: The History of the Canadian Wildlife Service*. Vancouver: University of British Columbia Press.

Canadian Wildlife Federation. 1992. *Game Farming in Canada: A Threat to Native Wildlife and its Habitat*. Ottawa: Canadian Wildlife Federation.

Caughley, G., and A.R. Sinclair. 1994. *Wildlife Ecology and Management*. Cambridge: Blackwell Science.

CBC News. 2008. 'US polar bear decision condemned in the North'. www.cbc.ca/canada/north/story/2008/05/15/bear-reax.html.

CCWHC (Canadian Cooperative Wildlife Health Centre). 1994. 'Disease transmission from game ranch animals: Risks to free-ranging wildlife'. *Canadian Cooperative Wildlife Health Centre Newsletter* 2 (3).

———. 1998. 'Chronic wasting disease in a game farm elk'. *Canadian Cooperative Wildlife Health Centre Newsletter* 5 (3).

———. 1999. 'West Nile virus and other zoonotic arboviruses'. *Canadian Cooperative Wildlife Health Centre Newsletter* 6 (2).

Colpitts, G.W. 2003. *Game in the Garden: A Human History of Wildlife in Western Canada to 1940*. Vancouver: University of British Columbia Press.

COSEWIC (Committee on the Status of Endangered Wildlife in Canada). 2007. *Assessment and Update Status Report on the Peregrine Falcon* (Falco peregrinus) *in Canada*. Ottawa: Canadian Wildlife Service.

———. 2008a. *Canadian Species at Risk: December 2008*. Ottawa: Canadian Wildlife Service.

———. 2008b. *COSEWIC Assessment and Update Status Report on the Spotted Owl*. Ottawa: Canadian Wildlife Service.

Croonquist, M., and R. Brooks. 1991. 'Use of avian and mammalian guilds as indicators of cumulative impacts in riparian-wetland areas'. *Environmental Management* 15: 701–14.

Demarais, S., and P. Krausman, eds. 2000. *Ecology and Management of Large Mammals in North America*. Upper Saddle River, NJ: Prentice-Hall.

Demarais, S., K. Miller, and H. Jacobson. 2000. 'White-tailed deer'. In S. Demarais and P. Krausman, eds, *Ecology and Management of Large Mammals in North America*, 601–28. Upper Saddle River, NJ: Prentice-Hall.

DuWors, E., et al. 1999. *The Importance of Wildlife to Canadians: Survey Highlights*. Ottawa: Canadian Wildlife Service.

Erdle, T. 1998. 'Progress toward sustainable forest management: Insight from the New Brunswick experience'. *Forestry Chronicle* 74: 378–84.

Erdle, T., and M. Sullivan. 1998. 'Forest management design for contemporary forestry'. *Forestry Chronicle* 74: 83–90.

Forbes, G., and J. Theberge. 1993. 'Multiple landscape scales and winter distribution of moose, *Alces alces*, in a forest ecotone'. *Canadian Field-Naturalist* 107: 201–7.

Freedman, B., et al. 2001. 'Species at risk in Canada'. In K. Beazley and R. Boardman, eds, *Politics of the Wild: Canada and Endangered Species*, 26–48. Toronto: Oxford University Press.

Gedir, J. 2008. *Update COSEWIC Status Report on Swift Fox*. Prepared for Environment Canada.

Ghilarov, A. 1996. 'What does biodiversity mean—Scientific problem or convenient myth?' *Trends in Ecology and Evolution* 11: 304–6.

Gilbert, G., and D.G. Dodds. 1992. *The Philosophy and Practice of Wildlife Management*. Malabar, FL: Krieger Publishing.

Government of Nova Scotia. 1987. *Wildlife: A New Policy for Nova Scotia*. Halifax: Government of Nova Scotia.

Government of Quebec. 2009a. 'Wildlife in Quebec life size'. www.mrnf.gouv.qc.ca/english/international/wildlife.jsp.

————. 2009b. 'Wildlife and nature count'. www. faunenatureenchiffres.gouv.qc.ca.

Hummel, M. 1995. 'A personal view on wolf conservation and threatened carnivores in North America'. In L. Carbyn, S. Fritts, and D. Seip, eds, *Ecology and Conservation of Wolves in a Changing World*, 549–51. Edmonton: Canadian Circumpolar Institute.

Hummel, M., and J. Ray. 2008. *Caribou and the North: A Shared Future*. Toronto: Dundurn Press.

Janz, D., et al. 2000. 'National recovery plan for the Vancouver Island marmot (*Marmota vancouverensis*) 2000 update'. RENEW Report no. 19. May.

Johnston, D., and M. Beauregard. 1969. 'Rabies epidemiology in Ontario'. *Bulletin of the Wildlife Disease Association* 5: 357–70.

Knight, R., and K. Gutzwiller. 1995. *Wildlife and Recreationists*. Washington: Island Press.

Lindenmayer, D., et al. 2002. 'On the use of landscape surrogates as ecological indicators in fragmented forests'. *Forest Ecology and Management* 159: 203–16.

Lindenmayer, D., and J. Franklin. 2002. *Conserving Forest Biodiversity*. Washington: Island Press.

Loo, T. 2006. *States of Nature: Conserving Canada's Wildlife in the Twentieth Century*. Vancouver: University of British Columbia Press.

McCabe, R.E., and T.R. McCabe. 1984. 'Of slings and arrows: A historical retrospection'. In L.K. Halls, ed., *White-Tailed Deer: Ecology and Management*, 19–72. Harrisburg, PA: Stackpole Books.

McLoughlin, P., M. Taylor, and M. Dowsley. 2008. *Update COSEWIC Status Report on Polar Bear*. Prepared for Environment Canada.

Malcolm, J., and A. Markham. 2000. *Global Warming and Terrestrial Biodiversity Decline*. Gland, Switzerland: World Wildlife Fund.

Manore, J., and D. Miner. 2006. *The Culture of Hunting in Canada*. Vancouver: University of British Columbia Press.

Marmot Recovery Foundation. 2002. *Marmoteer: Newsletter of the Marmot Recovery Foundation* no. 13 (fall).

Mathiessen, P. 1987. *Wildlife in America*. New York: Viking Penguin Books.

Mills, S., M. Soule, and D. Doak. 1993. 'The keystone-species concept in ecology and conservation'. *Bioscience* 43: 219–24.

Monnett, C., and J.S. Gleason. 2006. 'Observations of mortality associated with extended open-water swimming by polar bears in the Alaskan Beaufort Sea'. *Polar Biology*. DOI10.1007/s00300-005-0105-2.

Mosquin, T., P. Whiting, and D. McAllister. 1995. *Canada's Biodiversity: The Variety of Life, Its Status, Economic Benefits and Unmet Needs*. Ottawa: Canadian Museum of Nature.

Novak, M. 1987. 'Beaver'. In M. Novak, J. Baker, M. Obbard, and B. Malloch, eds, *Wild Furbearer Management and Conservation in North America*, 282–313. Toronto: Ontario Ministry of Natural Resources.

Paetkau, D., et al. 1999. 'Genetic structure of the world's polar bear populations'. *Molecular Ecology* 8: 1571–84.

Patterson, M., D.E. Guynn, and D.C. Guynn, Jr. 2001. 'Human dimensions and conflict resolution'. In S. Demarais and P. Krausman, eds, *Ecology and Management of Large Mammals in North America*, 214–32. Upper Saddle River, NJ: Prentice Hall.

Payer, D., and D. Harrison. 2000. *Influences of Timber Harvesting and Trapping on Habitat Selection and Demographic Characteristics of Marten*. Orono, ME: University of Maine.

Peterson, R., and R. Page. 1988. 'The rise and fall of Isle Royale wolves, 1975–1986'. *Journal of Mammalogy* 69: 89–99.

Reid, R., and D. Janz. 1995. 'Economic evaluation of Vancouver Island wolf control'. In L. Carbyn, S. Fritts, and D. Seip, eds, *Ecology and Conservation of Wolves in a Changing World*, 512–22. Edmonton: Canadian Circumpolar Institute.

RENEW. 2006. *2005–2006 Annual Report*. RENEW Report no. 17. Ottawa: Canadian Wildlife Service.

Rubec, C.D., and A. Turner. 2003. *Contributing to Ecological Integrity—Environment Canada's Protected Areas: A Discussion Paper*. Ottawa: Canadian Wildlife Service.

Sandlos, J. 2007. *Hunters at the Margin: Native People and Wildlife Conservation in the Northwest Territories*. Vancouver: University of British Columbia Press.

Schmiegelow, F., C. Machtans, and S. Hannon. 1997. 'Are boreal birds resilient to forest fragmentation? An experimental study of short-term community responses'. *Ecology* 78: 1914–32.

Shaw, J., and M. Meagher. 2000. 'Bison'. In S. Demarais and P. Krausman, eds, *Ecology and Management of Large Mammals in North America*, 447–66. Upper Saddle River, NJ: Prentice Hall.

Stewart, D. 1978. *From the Edge of Extinction.* Toronto: McClelland and Stewart.

Thiemann, G.W., A.E. Derocher, and I. Stirling. 2008. 'Polar bear conservation in Canada: An ecological basis for identifying designatable units'. *Oryx* 42 (4): 504–15.

Thompson, I., and R. Stewart. 1997. 'Management of moose habitat'. In A. Franzmann and C. Schwartz, eds, *Ecology and Management of North American Moose,* 377–402. Washington: Smithsonian Institute Press.

Venier, L., et al. 1999. 'Models of large-scale breeding bird distribution as a function of micro-climate in Ontario, Canada'. *Journal of Biogeography* 26: 315–28.

White, R.J. 2000. 'Big game ranching'. In S. Demarais and P. Krausman, eds, *Ecology and Management of Large Mammals in North America,* 260–76. Upper Saddle River, NJ: Prentice Hall.

Wildlife Habitat Canada. 2000. *Investors in Habitat: Hunter Contributions to Wildlife Habitat Conservation in Canada.* ed. J. Powers. Ottawa: Wildlife Habitat Canada.

———. 2008. Webpage. http://www.whc.org.

Wildlife Ministers Council of Canada. 1990. *A Wildlife Policy for Canada.* Ottawa: Canadian Wildlife Service.

Wisconsin Department Natural Resources. 2008. *A Plan for Managing Chronic Wasting Disease in Wisconsin: The Next Ten Years.* Madison: Wisconsin Department of Natural Resources.

Woodley, S., J. Kay, and G. Francis. 1993. *Ecological Integrity and the Management of Ecosystems.* Delray Beach, FL: St Lucie Press.

Yoakum, J.D., and B.W. O'Gara. 2000. 'Pronghorn'. In S. Demarais and P. Krausman, eds, *Ecology and Management of Large Mammals in North America,* 559–77. Upper Saddle River, NJ: Prentice Hall.

12

Parks and Protected Areas

Philip Dearden

Learning Objectives

- To understand why park and protected area management is prone to conflict and uncertainty.
- To appreciate the recent history of conflict and uncertainty in park and protected area management in Canada.
- To be familiar with national park policies and legislation.
- To understand the purpose of system planning, the current state of system completion in Canada, and the main sources of conflict and uncertainty.
- To appreciate the challenges involved in the designation of marine protected areas in Canada and why they are a main source of conflict.
- To understand some of the main sources of uncertainty in protected area management, including Aboriginal peoples, falling visitation, impacts of global change, and the need for active management.

INTRODUCTION

Conflict and uncertainty are endemic aspects of protected area (PA) planning and management. The focus may change over time, as does the balance between these aspects, but they are enduring features of PA management. For example, in the 1990s there was enormous conflict regarding the designation and management of PAs in Canada and headline news on the topic was common. In 1993, more than 800 people were arrested as they campaigned to have Clayoquot Sound on the west coast of Vancouver Island given a 'protective' designation. Similar protests had arisen all across the country, from the boreal forests of the Prairie provinces, to the Carolinian forests

of Ontario, to the Main River in Newfoundland. Since that time, the scale and frequency of these high-profile conflicts have diminished considerably. Nevertheless, levels of uncertainty seem to have increased on several fronts. One is global climate change. PAs are designated mostly to protect biodiversity. In the face of global climate change conditions, it is uncertain how these changes will affect park biodiversity and the management interventions that may be invoked. A second challenge relates to falling visitation in many park systems, the reasons behind this fall, and the possible implications.

In this chapter, I outline some of the main elements in management of PAs that generate conflict and discuss various aspects of the uncertainty involved in making management decisions. The discussion starts with an outline of the main values associated with PAs and why they are so prone to conflict. It continues with an overview of conflict in PA designation and management before discussing the role and nature of uncertainty. Most of the discussion is focused on the national park system. Although provincial parks and other forms of PA designation in Canada are important, they are also quite diverse and lie beyond the scope of this chapter.

PROTECTED AREA VALUES AND CONFLICT

The World Commission on Protected Areas (WCPA) of the International Union for Conservation of Nature (IUCN) defines PAs as 'A clearly defined geographical space, recognised, dedicated and managed, through legal or other effective means, to achieve the long-term conservation of nature with associated ecosystem services and cultural values' (Dudley 2008, 8). Thus, PAs are units of land or sea on which development restrictions are placed in order to protect various qualities of the natural environment. Commonly recognized PAs include ecological reserves and national and provincial parks. There are also many other categories, including wilderness areas, marine conservation areas, and wildlife sanctuaries.

There is a misconception that PAs are set aside from any 'use'. On the contrary, PAs have many uses, which are generally divided into two categories, ecological and social. Ecological uses include:

- maintaining the diversity and populations of wild animals and plants;
- providing individuals to repopulate areas from which species have been previously removed;
- maintaining natural environmental processes; and

- providing natural areas where the workings of relatively undisturbed environmental systems can be studied and understood.

Protected areas also play valuable social roles:

- maintaining areas of exceptional natural beauty;
- protecting landscapes and features from which many people derive spiritual fulfillment;
- providing areas for nature-based recreational activities;
- providing outdoor schoolrooms where people can learn about natural systems; and
- providing economic incentives for conservation activities through compatible tourism development in surrounding regions.

All these uses require the protection of a relatively natural environment, which is incompatible with the land-use changes associated with agriculture, forestry, mining, reservoir flooding, and other such manipulations. Hence, provision of PAs and other forms of resource development are usually mutually exclusive. This characteristic explains why PA designation is often such a conflict-ridden enterprise. Normal economic development processes are curtailed for the much less tangible returns associated with these other ecological and social uses.

CONFLICTS AND SYSTEM PLANNING

System planning is the development of networks of PAs such that a rational and effective distribution is manifest that will protect representative examples of ecosystems within protected areas. System planning helps to reduce uncertainty in that the desired distribution of PAs is established on scientific grounds, but it can also lead to conflict because areas identified may also be desired for other uses. In 1970, the Parks Canada System Plan was first drafted, giving the rationale for the establishment of parks: they would seek to represent the physiographic nature of Canada through identification of 39 terrestrial natural regions (see Figure 12.1). The goal was to have one or more national parks in each of these regions. Rapid growth occurred in the early 1970s but slowed in the 1980s (see Figure 12.2). By the time of the Parks Canada centennial celebration in 1985, less than 50 per cent of the target had been achieved (Dearden and Gardner 1987). A system plan was subsequently devised for marine areas in 1996, representing similar principles and with 29 areas identified.

The provinces adopted a similar approach, and in 1992 Canada's federal, provincial, and territorial ministers of environment, parks, and wildlife signed a Statement of Commitment to Complete Canada's Network of Protected Areas. Terrestrial systems were to be completed by 2000, whereas marine designation was to be 'accelerated'. There has been impressive growth in designating new PAs at all levels, but Canada is still far from meeting these commitments. Even Parks Canada has significant gaps (see Figure 12.1). In terms of overall protection of Canada's ecoregions, 29 per cent are provided a high level of protection (i.e., more than 12 per cent of their area), 12.4 per cent moderate protection (6 to 12 per cent), 41.9 per cent low protection (less than 6 per cent), and 16.6 per cent have no PAs (Environment Canada 2006).

Parks Canada's performance targets for creating and expanding new parks were to increase the number of represented terrestrial regions from 25 in March 2003 to 34 by March 2008 and to increase the number of represented marine regions from two in March 2003 to eight by March 2008 (Parks Canada 2007). Neither of these targets was met, and in 2009, 28 terrestrial

regions out of a total of 39 were represented by national parks.

As for the marine regions, in 2009 two of the 29 were represented, but they are from existing areas and not new acquisitions protected under the National Marine Conservation Areas Act. As a result, the goal was reduced to four of 29 in the 2007/8 corporate plan, but as yet no areas have been designated. Less than 0.5 per cent of Canada's marine area is set aside in protective designation, and Canada ranks seventieth globally in terms of the percentage of oceans protected (Environment Canada 2006). Projections suggest that optimistically, Canada will achieve 33 per cent of its international target goal by 2012 (Roff and Dearden 2007).

There have been major increases in the designation of protected areas at the provincial level. In 1968, Ontario had 90 PAs totalling 1.6 per cent of the province by area; 40 years later this had grown to 632 PAs totalling more than 9.4 million hectares, 8.7 per cent of the province. Nova Scotia in 2007 passed the Environmental Goals and Sustainable Prosperity Act requiring the province to protect 12 per cent of its terrestrial area by 2015 (Government of Nova Scotia 2007) and by 2009 had achieved 8 per cent. Quebec announced an additional 1.8 million hectares of protected area in late 2008 and is close to protecting 7 per cent of its total area. Wilderness protection became an election issue in the November 2008 provincial election, and both the Liberals and the Parti Québécois made commitments to achieve the 12 per cent target. The election winner, the Liberals, also promised that 50 per cent of northern Quebec would receive protective status. In BC, the area of parkland doubled between 1977 and 2005 and in 2009 totalled more than 12 million hectares. BC is the only jurisdiction to accomplish the internationally accepted 12 per cent target of land set aside in protected areas suggested by the World

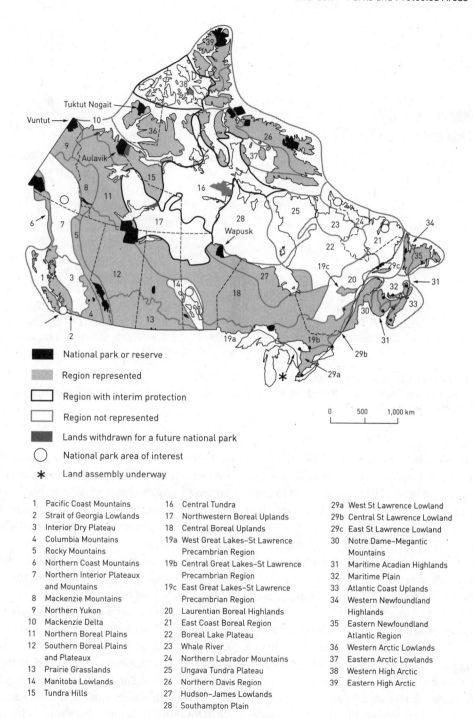

National park or reserve

Region represented

Region with interim protection

Region not represented

Lands withdrawn for a future national park

National park area of interest

∗ Land assembly underway

0   500   1,000 km

1 Pacific Coast Mountains	16 Central Tundra	29a West St Lawrence Lowland
2 Strait of Georgia Lowlands	17 Northwestern Boreal Uplands	29b Central St Lawrence Lowland
3 Interior Dry Plateau	18 Central Boreal Uplands	29c East St Lawrence Lowland
4 Columbia Mountains	19a West Great Lakes–St Lawrence	30 Notre Dame–Megantic
5 Rocky Mountains	Precambrian Region	Mountains
6 Northern Coast Mountains	19b Central Great Lakes–St Lawrence	31 Maritime Acadian Highlands
7 Northern Interior Plateaux	Precambrian Region	32 Maritime Plain
and Mountains	19c East Great Lakes–St Lawrence	33 Atlantic Coast Uplands
8 Mackenzie Mountains	Precambrian Region	34 Western Newfoundland
9 Northern Yukon	20 Laurentian Boreal Highlands	Highlands
10 Mackenzie Delta	21 East Coast Boreal Region	35 Eastern Newfoundland
11 Northern Boreal Plains	22 Boreal Lake Plateau	Atlantic Region
12 Southern Boreal Plains	23 Whale River	36 Western Arctic Lowlands
and Plateaux	24 Northern Labrador Mountains	37 Eastern Arctic Lowlands
13 Prairie Grasslands	25 Ungava Tundra Plateau	38 Western High Arctic
14 Manitoba Lowlands	26 Northern Davis Region	39 Eastern High Arctic
15 Tundra Hills	27 Hudson–James Lowlands	
	28 Southampton Plain	

Figure 12.1 Terrestrial System Plan for Parks Canada, Showing Current State of Representation.

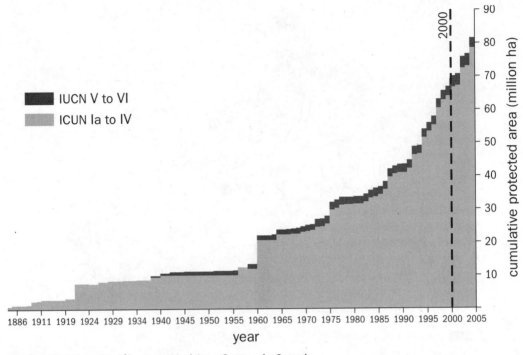

Figure 12.2 Growth of the Protected Area System in Canada.
Note that IUCN classifies PAS into different categories (Dudley 2008), with categories I to IV being the most highly protected categories.

Commission on Environment and Development (WCED 1987). However, no provincial government has fulfilled the 1992 commitment to complete a representative network of protected areas (Environment Canada 2006).

In 2009, about 10 per cent of Canada's terrestrial area had been awarded protective designation, well short of the average 14.6 per cent protected by OECD countries (Environment Canada 2006). However, 95 per cent of Canada's terrestrial protected areas fall within IUCN categories I to IV (Dudley 2008) and hence have a strong protective mandate. Among OECD countries, Canada ranks sixteenth out of 30 in terms of the proportion of land protected (the US, for example protects almost 25 per cent compared with our 10 per cent) yet ranks fourth in terms

of proportion of land with strong protection. Furthermore, Canada has some two-thirds of its protected area within a small number of sites, each more than 300,000 hectares in size. Few countries have the ability to preserve such large intact landscapes.

The PA system has grown enormously within the past 40 years. Uncertainty regarding the lands to be protected has been reduced, significant conflicts have been generated, trade-offs have been made, and although much remains to be done, Canada seems to be entering a more mature phase of PA management. Canada is signatory to the UN Convention on Biological Diversity, which calls for 'the establishment and maintenance by 2010 for terrestrial and by 2012 for marine areas of comprehensive, effectively managed, and ecologically representative national and regional systems of protected areas.' Canada will not be able to meet these international commitments for quite some time after the target dates. The marine system is in its infancy and getting off to a very slow and underfunded start. A significant number of the ecoregions of Canada still remain unprotected. Canada is not a world leader in term of the proportion of area set aside, yet we have some of the largest and wildest PAs on Earth. Progress has undoubtedly been made, but much remains to be done. This progress can be largely attributed to the conflicts in the 1990s over PA establishment and also to strong international support for PA establishment as a cornerstone of biodiversity protection.

Conflicts over the designation of PAs in Canada are by no means over, but since the year 2000 there has been a considerable reduction in the occurrence and strength of these conflicts. However, the establishment of marine protected areas (MPAs) is still a major source of conflict, which will continue into the future. Three federal agencies, Parks Canada, Fisheries and Oceans Canada (DFO), and Environment Canada, have mandates at the federal level to establish MPAs (Dearden and Canessa 2008). One of the main reasons for lack of progress is the uncertainty regarding the governance relations among these agencies, how they liaise with provincial initiatives, and the resulting conflict among the agencies (Gardner et al. 2008). As a result of these problems, the three federal agencies jointly developed a Federal Marine Protected Areas Strategy to co-ordinate their efforts to establish a network of MPAs of ecologically significant and representative areas (Government of Canada 2005). However, significant problems remain, and progress is slow.

Another area of significant conflict regarding governance relates to the role of Aboriginal peoples and their involvement in negotiations over MPA establishment and management. This concern is not unique to MPAs and includes terrestrial protected areas as well (Dearden and Langdon 2008), but it is a main point of conflict in the development of some MPAs, particularly on the west coast. In some cases, such as the attempts to establish greater protection around Haida Gwaii, the First Nations have been very supportive, and progress is being made. The greater challenge occurs in areas where many different First Nations are involved and there may be significant conflict among these nations as to rights and responsibilities. This is the case with the effort to establish a National Marine Conservation Area (NMCA) in the southern Gulf Islands where some 23 different First Nations jurisdictions are involved.

Other major sources of conflict relate to consultations with affected communities and particularly with primary resource users, mostly fishers, who often oppose protective designations. Successful community involvement is considered a cornerstone of effective MPA designation and management around the world (Charles and Wilson 2009). Sites that have been established in Canada, largely by DFO under the Oceans Act, are characterized by strong community support, and one of the main challenges is to build such support among a wider base, including Aboriginal peoples.

Conflict in the establishment of marine parks has occurred on both coasts. For example, local people successfully prevented establishment of a marine park by Parks Canada in the West Isles region of New Brunswick (Butler 1994). Perhaps the most serious case occurred in Newfoundland where in 1997 Parks Canada announced a half-million-dollar feasibility study on establishing a National Marine Conservation Area off the shores of Terra Nova National Park. The study generated an increasing degree of conflict as it proceeded and was abandoned two years later with very little to show for the effort, other than it acting as a focus for anti-government feelings in post-cod-crash Newfoundland. Lien (1999) offers several reasons for the failure:

- There was a lack of understanding of what an NMCA was and might be. The official mandate—to protect, conserve, and encourage public understanding, appreciation, and enjoyment of marine areas representative of ocean environments—held little resonance for economically depressed Newfoundland outports.
- Although ostensibly there was support from all federal and provincial government agencies, in reality there was little tangible support, adding to the overall uncertainty.
- Local timing was poor, with economically depressed communities and inadequate investment of time and money in creating greater awareness of the possible advantages of an NMCA creation.
- There was a deep distrust of government agencies built upon past experience and a perception of chronic mismanagement of the east coast fisheries on the part of the Department of Fisheries and Oceans.
- Parks Canada under-resourced the initiative and failed to entrust local managers and the advisory committee with sufficient budgetary control.
- There was a lack of technical support from qualified fisheries scientists who could provide serious evaluation of the possible conservation options being considered.
- The bill to create National Marine Conservation Areas (C-48) was under consideration by a parliamentary committee as the feasibility study was proceeding. Many of the provisions of the bill were seen to be in conflict with the more flexible approach being discussed in Newfoundland. This created a lot of concern about possible restriction of fishing rights if the feasibility study were to be approved.

Careful attention will have to be directed toward resolving these kinds of issues if Canada is going to show much progress in establishing MPAs. Since the first MPA evaluations more than 40 years ago, there has been virtually no improvement to the marine conservation estate, despite clear signs of ecological stress resulting from traditional ocean management approaches. There is increasing evidence of the positive role that MPA establishment can have on fisheries over the long term (e.g. see Wallace 1999; Committee on the Evaluation, Design and Monitoring of Marine Reserves and Protected Areas in the United States 2001; Lubchenco et al. 2003; Shears et al. 2006; Davis 2008). As with the terrestrial debate, conflict levels will fall as awareness of park values grows and as resource users reduce the uncertainty surrounding their resource extraction rights.

CONFLICTS AND MANAGEMENT PLANNING

Once PAs are established, conflicts often occur over how they should be managed. National parks have a specific mandate to ensure that the parks remain 'unimpaired for future generations'. However, many conservationists became increasingly concerned during the late 1980s and 1990s that such was not the case. Banff became a lightening rod for this concern. During the 1980s, the 'crown jewel' of the Canadian national park system became subject to seemingly endless commercial development, and conflict between pro-development interests and environmentalists heightened. In response, the minister placed a moratorium on development within the corridor and convened a task force—the Banff–Bow Valley Study (BBVS)—to provide informed direction to decision-making in the valley.

The task force explored the environmental, social, and economic aspects of development in the park. It found the grizzly bear populations declining rapidly and the aquatic ecosystems in ruin due to exotic introductions and dams. The recommendations to the minister warned of the 'serious, and irreversible, harm to Banff National Park's ecological integrity' (Banff–Bow Valley Study 1996, 4). The report called for stricter limits on growth and more effective methods of managing and limiting human use, as well as regional co-ordination outside park boundaries. It called for increased public involvement in decision-making and improvements in education, awareness, and interpretation to inform visitors of threats to ecological integrity. It also cited inadequate funding to meet requirements for ecological integrity. In short, 'while Parks Canada has clear and comprehensive legislation and policies, Banff National Park suffers from inconsistent application of the National Parks Act and Parks Canada's policy' (Banff–Bow Valley Study 1996, 14). The 1988 amendments to the National Parks Act explicitly stated that ecological integrity was the primary mandate of the national parks, yet management decisions were being made in violation of the mandate.

After the report's release, the minister, Sheila Copps, announced that no new parcels of land would be made available for development, the closure of the Banff airstrip, and the removal of the buffalo paddock and cadet camp—all of which block a wildlife corridor. She also limited the population of Banff to 10,000 residents and pledged to restore aquatic diversity and clean up sewage polluting waters in the park. In 1998, Copps unilaterally reduced the amount of commercially zoned land in Banff, allowing only 350,000 square feet of additional commercial development rather than the 850,000 square feet the town council proposed. Swinnerton (2002) provides a more detailed review of the follow-up to the report—how it subsequently influenced

the park management plan for Banff and other parks, the implications for park planning processes and tourism, and the relationship to planning for areas surrounding parks. Pavelka and Rollins (2008) also discuss the impacts on the community and tourism. The BBVS was significant because it demonstrated how the ecological integrity of national parks could suffer from overuse and development. It was the first time that such a substantial and public process had been used to study a national park in detail. The report received considerable media response, raising public awareness of the threats to national parks. The BBVS also acted as an alarm for the government—if Banff were under threat, then what about the other parks?

The Auditor General's 1996 Report

One month after the Banff–Bow Valley Report was released, the Auditor General publicly criticized Parks Canada's poor performance, particularly regarding ecological integrity and new park establishment (Auditor General of Canada 1996). The audit found that Parks Canada lacked the knowledge base for sound ecosystem-based management (see Chapter 15) and that in many cases, information and monitoring of natural resources were outdated, with few parks having research plans. The average age of park plans was 12 years, yet the National Parks Act states that plans must be reviewed every five years. Few management plans had clear ecological integrity statements, and links between business plans and management plans were also rare. The report recommended that Parks Canada improve the information coverage in its *State of the Parks Report* to include more detailed ecological integrity data. The Auditor General also expressed concern over initiatives aimed at increasing visitation in already stressed parks and the lack of clear direction and objectives in relation to tourism. Interpretation services, particularly for communicating ecological integrity issues, were also found to be lacking.

The report also decried the slow creation of new national parks and the lack of resources to support new parks. Noting the difficulties in securing land, the report recommended that Parks Canada pursue alternate forms of land ownership for national parks, a recommendation made some 10 years earlier by an Environment Canada task force (Environment Canada 1987). The audit also noted the lack of a plan for completing the National Marine Conservation Areas network. Like the Banff study, the Auditor General's report received extensive media coverage, placing the message of 'parks in peril' into the public mind.

In December 1998, the Auditor General completed a follow-up audit (Auditor General of Canada 1998). Since the first audit, no new parks had been created, 16 national parks still had outdated management plans, and many parks had still not completed ecological integrity statements. More positively, the follow-up report praised the BBVS and suggested that changes to the Banff management plan be mirrored in other parks. The follow-up report also noted positive changes in the 1997 *State of the Parks Report* and the increased importance placed on both monitoring and ecological integrity indicators (see Chapter 3). However, the audit report also noted that the changes '[have] not been able to prevent an increase in the threat, from both external and internal sources, to the ecological integrity of most national parks' (Auditor General of Canada 1998).

The 1997 *State of the Parks Report*

It was not just external reports that were documenting the challenges faced by Parks Canada in managing the conflicts between use and protection of ecological integrity in parks. Amendments to the National Parks Act in 1988 required the minister to report to Parliament every two years on the state of the national parks and the progress toward establishing new parks. Armed with improved ecological information, the 1997 *State of the Parks Report* marked a new phase in accountability and placed a greater focus on ecological integrity. Most of the report spoke to problems related to maintaining ecological integrity. It presented ecological indicators for biodiversity, ecosystem functioning, and stressors in the parks. Nineteen out of 36 parks reported stresses originating from park management practices, 26 reported stresses from visitor and tourism facilities, and 21 reported stresses from exotic vegetation. Other external stresses included forestry (18), agriculture (17), mining (15), and sport hunting (11).

The report also provided general recommendations for improving ecological integrity, including increased active management, regional integration, public participation and interpretation, restoration programs, and the creation of better partnerships with other interests both inside and outside the parks. The report supplied the government with another loud and clear warning of the serious nature of park problems and marked the first major internal acceptance of the problems within Parks Canada, using scientific evidence and indicators.

The Panel on the Ecological Integrity of Canada's National Parks

In response to the preceding reports and pressure from conservation groups, the minister formed the Panel on the Ecological Integrity of Canada's National Parks. The mandate of the panel was to 'assess the strengths and weaknesses of Parks Canada's approach to the maintenance of ecological integrity and provide advice and recommend how best to ensure that ecological integrity is maintained across the system of national parks' (Parks Canada Agency 2000, 1-2).

The panel's report (Parks Canada Agency 2000) made 127 recommendations, with a clear central theme that emphasized the continued ecological deterioration of the parks system. The panel suggested that an important contributing factor to the deterioration was the lack of an internal 'conservation culture' within Parks Canada—that is, in contrast to the priority given to ecological integrity in the National Parks Act, Parks Canada as an entity views ecological integrity as only one of many objectives. The panel was 'told that the major hurdles to achieving the mandate could be found within the organization' (Parks Canada Agency 2000, 2–4).

The panel also found that Parks Canada lacked the right tools for maintaining ecological integrity. Science and traditional knowledge were rarely used in decision-making; planning structures were linear and inadequate for fostering increased understanding of ecosystems and stakeholder values; basic inventories and monitoring were incomplete and insufficient; and laissez-faire management was causing harm to park ecosystems. To deal with incomplete knowledge of park ecosystems and to facilitate learning and understanding while actively pursuing restoration and ecological integrity, the panel recommended the adoption of adaptive active management (see also Chapter 16). This required, among other things, a change in planning mechanisms, the use of indicators and monitoring tied to goals and objectives (see Chapter 3), more frequent and detailed reporting on an individual-park level, and an increase of science capacity in the organization.

Another significant theme of the report was the need for increased cooperation across boundaries to deal with external impacts on park ecosystems. The panel reiterated earlier studies suggesting that parks seek partnerships and increased cooperation with landowners in the greater ecosystem (public and private) and jurisdictional cooperation between federal and provincial governments.

The panel also found that park messaging and interpretation failed to communicate the key purpose of national parks—to protect and represent the main ecoregions in Canada. In order to raise awareness of park stresses and to garner greater public support, the panel recommended that Parks Canada improve its communication and education programs to include all relevant information—'not just "good news" but also the hard realities and critical issues about stresses that affect national parks' (Parks Canada Agency 2000, 10-3). The report also commented on the need for external education efforts beyond the traditional scope of park interpretation in order to encourage an external culture of conservation.

Little in the panel report was new. The heavy emphasis on the conservation culture of Parks Canada was to some extent surprising and an aspect that had received little previous attention from academics (but see Payne 1997). However, the report served to legitimate the severity of the problems for the minister, other government departments, politicians, and the general public. It marked the institutional acceptance of the challenges faced by the parks and, as such, reduced the level of conflict that had become increasingly visible over the previous 20 years. With overall acceptance of the challenges, the problem became more one of working together to resolve them, rather than pitting the agency against conservationists, scientists, and the NGO community. Parks Canada addressed many of the points raised by the panel (see Parks Canada Agency 2001) and subsequently received additional funding to implement recommended changes.

The Auditor General's 2005 Report

In 2005, the Auditor General undertook a further audit of Parks Canada's activities, looking at 12 parks to see how Parks Canada plans and manages selected ecological monitoring and restoration activities and uses those activities to enhance public education and visitor experience. The Auditor General also examined the quality of the reports Parks Canada had produced on the state of national parks in 1997, 1999, and 2001. The findings were:

- Significant issues in ecological integrity, including ones related to biodiversity, ecosystem functions, and stressors, were being addressed through monitoring and restoration activities, but gaps in coverage existed. Gaps also existed regarding how these activities were planned and managed. For example, at the park level the central planning document is the park management plan. However, in 6 of the 12 parks, these plans were not up-to-date, and annual reports on the implementation of the plans were not being produced on a regular basis by all parks.
- Increasing understanding through public education is fundamental to maintaining and restoring ecological integrity. In this regard, objectives for enhancing public education through monitoring and restoration were lacking at the park level, and the results of monitoring and restoration projects were not used to full advantage in park-level communication materials.
- With new funding received in 2003 ($75 million over five years and $25 million annually thereafter), Parks Canada had implemented measures to improve monitoring and restoration and their use in enhancing public education and visitor experience.
- The 1997 *State of the Parks Report* was relatively good in terms of setting baselines on the state of the parks. However, the subsequent two reports did not make use of the potential offered by the 1997 report, making it difficult

to determine how the state of the parks had changed. Overall, these reports need to report more consistently on changes and trends in the state of parks over time. More information on the results of Parks Canada's actions is also needed (Auditor General of Canada 2005).

Again, the report talks of progress but with many challenges ahead. Parks Canada also realizes this. The most recent *State of Protected Heritage Areas* report was published in March 2005, but the report for 2007 was still unavailable in August 2009. However, more recent information is contained in Parks Canada's *Performance Report* (Parks Canada Agency 2007). The performance targets and progress reported for ecological integrity (EI) are shown in Table 12.1.

At the provincial level, all jurisdictions have recognized the importance of protecting EI either in legislation or policy (Environment Canada 2006). However, only four jurisdictions have specified objectives or indicators. Systematic measures to assess and report on the state of their PA systems exist only in Ontario.

UNCERTAINTY AND PROTECTED AREA POLICY AND LEGISLATION

An important component of reducing conflict is providing greater certainty about the way in which resources will be managed. Many conflicts in national park management, for example, can be attributed to the so-called dual mandate enshrined in the National Parks Act of 1930, dedicating the parks 'to the people of Canada for their benefit, education and enjoyment and such Parks shall be maintained and made use of so as to leave them unimpaired for the enjoyment of future generations' (Canada 1930). This broadly defined legislative mandate failed to provide administrators with a 'clearly defined park purpose to guide

Table 12.1 Performance Expectations and Status Related to Ecological Integrity

Performance Expectation	Status
National park and NMCA management plans will be on schedule and consistent with management plan guidelines by March 2010.	As of March 2007, 33 of 42 national parks had approved management plans consistent with the 2000 Guidelines for Management Planning. Three national parks operate under interim management guidelines, and the remaining six are engaged in the planning process.
Develop fully functioning EI monitoring and reporting systems for all national parks by March 2008.	Two national parks meet initial conditions for a fully functioning ecological integrity monitoring and reporting system, with the expectation that two-thirds of the parks will do so by March 2008.* The remaining one-third of parks will have most of the elements of an ecological monitoring and reporting system in place by March 2008.
Develop selected indicators and protocols for measuring NMCA ecological sustainability use by March 2009.	Minimal progress was made in 2006/7.
Improve aspects of the state of EI in each of Canada's 42 national parks by March 2014.	

By March 2008, about 32 of 42 national parks will have functioning monitoring systems (Woodley, personal communication, 2008).

Source: Parks Canada Agency 2007.

them' (Canada 1969). Over the years, decision-makers from the minister to individual park superintendents have had considerable discretion about whether they would emphasize the 'protection' or 'made use of' side of the equation. Parks Canada has tried to constrain this latitude through more explicit policy statements and new legislation. The national parks mandate was clarified in the 1964 and 1979 national park policies, both of which clearly prioritized ecological integrity and park protection over 'enjoyment' and tourism (Canada 1969; 1983), and there have been several more recent improvements in both policy and legislation for national parks. They will be reviewed in the next section, along with the other relevant legal and policy initiatives.

The 1988 Amendments to the 1930 National Parks Act

The 1988 amendments to the 1930 National Parks Act stated that the '[m]aintenance of ecological integrity through the protection of natural resources shall be the first priority when considering park zoning and visitor use in a management plan' (Canada 1988). This amendment legally clarified the purpose of the national parks and bound the government to protecting park ecosystems, with implications for planning, management, and operations both inside and outside park boundaries. The 1988 amendments also enabled the government to proclaim wilderness areas within the park, heightening the level of protection on these lands by prohibiting any activities 'likely to impair the wilderness character of the area' (Canada 1988).

Several other amendments were designed to strengthen accountability. Park management plans, while already required by policy, became a legislative requirement. The minister must table management plans in Parliament no later than five years after park establishment, and these plans must consider resource protection, zoning, and visitor use. Public participation was also strengthened through the amendments, requiring the minister, 'as appropriate', to provide opportunities for public participation in park matters. The amendments also required the minister to report to Parliament every two years on the state of the national parks and progress toward establishing new parks. Furthermore, the amendments increased fines and punishments for poaching and otherwise disturbing park ecosystems, legislated a ban on ski-hill development within park boundaries, enabled local government for park communities, and included marine parks in the mandate for the first time.

The 1995 Revised National Parks Policy

Policy provides central direction to government activities but is not enforceable by law. Nevertheless, policy has been important for park management since the early 1960s when the first comprehensive policy was issued to give coherent and consistent direction to the organization. The 1964 National Parks Policy (Canada 1969) was the first document clearly establishing the protection of the environment as the first priority for park decision-making. It also began to delineate acceptable and unacceptable uses and development in national parks. There have been two major revisions since then (1979 and 1995), placing successively greater importance on ecological integrity.

The 1995 revisions further defined how the national parks should be managed to maintain ecological integrity. Policies related to park establishment and the system plan now included ways to 'prevent the loss of ecological values during the feasibility assessment process' (Parks Canada 1995, 1.3.5) on future national parks. Sections related to management further emphasized the primacy of ecosystem health and recognized the important role of conservation biology and ecosystem-based management for maintaining health. For example, concepts like working

with park neighbours on land-use and pollution issues, controlling exotic species within the park, and maintaining long-term monitoring data were all espoused by 1995 policy. While previous policies also emphasized working with adjacent landowners, the focus was on economic rather than ecological integration. For example, a 1979 policy encouraged the development of tourist infrastructure outside park boundaries by providing financial assistance to municipal governments (Canada 1983). This policy did not appear in the 1995 revisions; instead, it was replaced by policies on ecological integrity and ecosystem-based management.

This emphasis could also be seen in zoning policies, which stated explicitly for the first time that 'Zones I and II will together constitute the majority of the area of all but the smallest national parks, and will make the greatest contribution toward the conservation of ecosystem integrity' (Parks Canada 1995, 2.2.3.2). There was also an entire section dedicated to policies related to wilderness areas, a new management tool created with the 1988 amendments to the National Parks Act. Another significant shift in policy reoriented interpretation and education toward ecological and environmental issues and 'on challenges to maintaining the ecological integrity of national parks in order to foster greater public understanding of the role that protected spaces play in a healthy environment' (Parks Canada 1995, 4.2.6). Additionally, visitor management processes were clearly laid out, and national park communities were limited to no new growth. More details and examples of current management practices that follow these policy directions can be found in Wright and Rollins (2008).

The 1998 Parks Canada Agency Act

Throughout the 1990s, the federal government placed increasing emphasis on 'fiscal responsibility' and debt reduction. This resulted in decreased public funding, public sector layoffs, and privatization of some Crown corporations and services. Parks Canada suffered a 25 per cent decline in revenue between 1995 and 2000. Thus, the Parks Canada Agency Act (PCAA) came into being, and Parks Canada became an operating agency (a distinct legal entity, a departmental corporation) with the same mandate under National Parks Act. Primarily, the PCAA pushed the organization toward operating, as Parks Canada CEO Tom Lee succinctly stated, 'in a more business-like manner' (Searle 2000, 123).

The Act affected the organization in four main areas: accountability, financing, human resources, and administration. For accountability, the agency still reported to the minister, but a new position of chief executive officer was created (to be appointed by cabinet) to head the agency and be responsible to the minister. The Act also called for three new reports: a five-year corporate plan, an annual report on the agency's operations, and a five-year human resource management report. Each must be tabled in Parliament. The PCAA also strengthened public participation by requiring the minister to provide feedback on the agency's performance.

Through this Act, Parks Canada gained greater financial flexibility and responsibility, such as a two-year rolling budget and full retention of, and reinvestment authority for, all revenues. The Act established a dedicated, continuing account to be used to fund new national parks, and the agency acquired authority to borrow funds for land acquisition. For human resources, the agency became an employer in its own right, separate from the federal government, giving the CEO bargaining authority with employees. Also, changes to administration enabled by the Act allowed for greater flexibility, with the agency now able to contract out day-to-day operations. In addition, the agency was permitted to accept gifts and donations of property.

The 2000 National Parks Act

In October 2000, a new National Parks Act was passed. The Act further clarified the Parks Canada mandate and legislated many recommendations made by conservation groups, the Ecological Integrity Panel, the Auditor General's report, and park academics alike. To start, 'ecological integrity', while used in the 1988 amendment, is clearly defined in the new Act: 'With respect to a park a condition that is determined to be characteristic of its natural region and likely to persist, including abiotic components and the composition and abundance of native species and biological communities, rates of change and supporting processes' (Canada 2000). The Act also stipulated greater detail in management plans, including a long-term ecological vision and ecological integrity objectives and indicators. It formally established seven new parks, raised poaching fines, capped commercial development in park communities, and simplified park establishment and enlargement processes. The new Act further refined the 1988 wilderness areas amendments to expedite their declaration by requiring the cabinet to declare wilderness areas one year from the management plan approval.

The 2002 National Marine Conservation Areas Act

In 2002, Bill C-10 was finally passed by Parliament, enabling Parks Canada to establish a system of MPAs known as National Marine Conservation Areas (NMCAs). The Act has a very different orientation from that of its terrestrial counterparts in that several agencies have jurisdiction within the MPAs and the focus is on conservation rather than protection per se, working in close co-operation with local interests. Total prohibitions, such as of hydrocarbon and mineral development, are minimal, and well-known highly destructive practices, such as bottom-trawling, are not excluded.

Judging the success of this legislation will take some time. First, some NMCAs have yet to be established. Gwaii Hanaas (BC) and Western Lake Superior (Ontario) are likely early candidates, although by early 2009, seven years after passage of the legislation, no area had been protected under the Act. Then it remains to be seen whether the close cooperation between different government agencies and local interests actually results in conservation benefits or just in parks that exist in name only. Key to the outcome will be the internal zoning of the NMCAs that will set aside some areas where extractive activities will be prohibited. The location and size of these zones will be crucial, for as Davis (2008) points out, research repeatedly shows that without complete cessation of fishing activities, marine biodiversity will not be conserved.

UNCERTAINTY AND PROTECTED AREA MANAGEMENT

Despite the changes outlined above that help reduce uncertainty in terms of park mandate, a high degree of uncertainty continues surrounding park management. This section outlines some main areas of uncertainty and potential responses.

Aboriginal Concerns

Legislation and policy in broader areas than parks can have a major effect on PAs. One of the most influential external influences over the past decade has been the role of Aboriginal peoples (see also Chapter 4). Aboriginal peoples in northern Canada have played a significant role in national park planning and development, while in southern Canada their role has varied from park to park. Aboriginal peoples have little or no involvement in some national

parks, while initial steps toward cooperation have been taken in others. Northern national parks have been established in conjunction with Aboriginal land claim settlements, while park reserves await claims settlement before attaining full park status. Overall, more than 50 per cent of the land area in Canada's national park system has been protected as a result of Aboriginal peoples' support for conservation of their lands (Parks Canada Agency 2000). Indeed, Dearden and Berg (1993) suggest that First Nations have emerged as the most dominant force influencing the establishment of national parks in Canada.

There is, however, great uncertainty associated with Aboriginals' involvement. A long line of Supreme Court decisions recognizing Aboriginal title in many resource management issues has set the overall context for greater involvement in park designation. Parks Canada policy and legislation have followed a similar trend. The National Parks Act extends harvesting rights to a larger number of parks, including all those established by agreement. The Act also makes provisions for the removal of non-renewable resources in the form of carving stone in order to support traditional economies.

The Act does not guarantee joint management for Aboriginal peoples whose traditional lands fall within national parks. Such joint management regimes are only specified in the policy and then only for Aboriginal groups that have successfully completed land claims settlements. Accordingly, a number of Aboriginal peoples whose traditional lands are affected by national park initiatives, but who have not concluded land claim settlements with the government, may or may not have an opportunity for joint management. However, on a policy basis, Parks Canada has been very active in developing not only a formalized consultative process but cooperative management arrangements as well. For example, for the Gulf Islands National Park

Reserve, Parks Canada, on a policy basis and in advance of treaty settlement, developed three co-operative management arrangements with Aboriginal groups to ensure consultation and input into major park decisions that affect the Aboriginal groups involved.

Parks Canada has tended to follow an ad hoc approach regarding Aboriginal peoples rather than working within a nationwide comprehensive policy. Land claim settlements themselves, rather than national park policy or legislation, determine the role of Aboriginal peoples in planning for and managing national parks. This has given rise to subtly different kinds of parks in northern and southern Canada, because a significant park planning and management role is given to Aboriginal peoples in northern Canada where parks are tied to settlement of land claims. Land claims in northern Canada will likely be completed sooner than those in southern Canada. In addition, many Aboriginal peoples in the south must look to old treaties and the National Parks Act and policy, rather than to comprehensive land claim settlements, to protect their interests. This has generated conflict in many locations. At Riding Mountain National Park in Manitoba, for example, many Aboriginal people remain frustrated over their exclusion from hunting or ceremonial activities in the park, and there are still unsettled land claims and grievances from earlier land expropriations. However, communication has increased in recent years, even though local Aboriginal people still do not have a formal management role. A specific land claim has been settled with the Keeseekowenin First Nation, and a parcel of land on the western shores of Clear Lake has been returned to the First Nation. A Senior Officials Forum with First Nations participation has been established to discuss park management issues, and a historical use study has been completed in co-operation with Aboriginal people.

Visitor Management

An important dimension of PA management is visitor management. To be effective, management must understand the nature of human use and whether and how it needs to be changed to meet PA objectives. Analytical social science capacities in the parks system are particularly deficient (Needham and Rollins 2008). Furthermore, the means of changing behaviour have concentrated heavily on one approach: regulation. Other aspects of behaviour modification need to be given more emphasis, including economic incentives, building civic action through local community support, and creating greater awareness among visitors of desirable and undesirable activities through interpretation (Hvenegaard, Shultis, and Butler 2008). Unfortunately, the latter has been curtailed over the past few years as financial exigencies required managers to cut costs. This lack of foresight has been recognized by Parks Canada, and more resources are now being directed toward interpretation. However, other park agencies have not learned from this experience and have curtailed educational services (see also Chapter 18 regarding 'social learning'). BC Parks, for example, announced in 2002 that all visitor centres and interpretation services would be cut immediately, and they have not been reinstated.

In the past, the major issue related to visitor management was how to design effective management interventions to reduce the impact of ever-growing numbers of visitors. In many parks and park systems, this is no longer a problem because visitation levels are declining. Visitation to national parks, for example, has declined, and even if changes in the way in which visitation is counted, implemented in 2000, are taken into account and the mountain parks excluded, visitation still shows a decline. Visitation to Banff was 3.2 million in 2006, for example, down from the 4.7 million of 2002. These trends are also seen in many provincial parks as well as internationally. Three main factors seem to explain this phenomenon:

- The baby boomers who were active in parks throughout their lives are getting older and not as capable of strenuous exercise. Even those who still engage in outdoor activities on a regular basis now prefer a gourmet dinner and a soft bed afterwards rather than sleeping in the pup tent they did when they were younger. In surveys conducted for Parks Canada, 63 per cent of respondents over the age of 65 said they used to visit parks but no longer do so.

- Surveys also show a drop in visitation among those in the 18 to 24 age group, historically one of the main users of parks. Similar results have been found in the US. This drop in visitation by younger people is being at least in part attributed to the so-called 'nature deficit disorder' (Louv 2005). Young people in this generation have grown up in a more urbanized world, cut off from everyday exposure to the natural environment and with a strong predilection for electronic gadgetry. They understand and feel more comfortable with their electronic world than the challenges posed by the outdoor world. This trend, apparent in many countries throughout the world, affects not only park visitation but also the entire gamut of society's interaction with the environment. What happens when the nature-deficit generation of today becomes the decision-makers of tomorrow?

- Surveys have also found that minorities and new immigrants are unlikely to be park visitors and supporters. Immigrants born outside of Canada make up 20 per cent of the population but only 10 per cent of visitors to national parks. Many immigrants are likely to have grown up in a large city rather than on a poor rural farm. They are used to an urban lifestyle, need to establish themselves in their new country, and may not have the necessary free time and money to get out to the parks. Many are also unaware of the vastness of Canada and what there is to see in the parks.

Will this trend of declining visitation continue? Some researchers working on the impacts of climate change on parks suggest that some parks, such as Banff, may experience increases in visitation as a result of warming (Lemieux and Scott 2005). However, if visitation continues to decline, should this trend generate concern, and if so, what could be done? On the one hand, parks could provide more commercialized activities and businesses and more motorized recreation and use these features to attract more visitors. Declining visitation results in declining park revenues, and the main goal is to return visitation figures to their historical high points. On the other hand, park policies cannot be driven simply by changing public tastes or declining revenues. Otherwise, they will become slaves to recreational fashion, with cellphone towers, touch-screen computers, and jet skis replacing the nature the parks were created to protect. An emerging challenge of the future will be to develop *appropriate* marketing for parks so that they can attract a growing number of visitors and to ensure that once they are there, these visitors will enjoy the experience, helping to build the parks' constituency in society.

Parks Canada is already showing some leadership in addressing this problem and is placing an increasing emphasis on understanding the visitor experience, as illustrated by the creation of the External Relations and Visitor Experience Directorate in 2005. One program targets schools. The Parks Canada in Schools program connects with teachers of history, social studies, geography, and natural science in Grades 4 to 12 in all provinces and territories. The 'Teacher's corner' on the Parks Canada website provides bilingual, curriculum-based learning resources for teachers across the country. Visits to the site grew from 378,079 in 2005/6 to 834,369 in 2006/7. Other aspects of Parks Canada's visitor experience program are detailed by Jager et al (2006).

Ecological Integrity and the Need for Active and Adaptive Management

There is still uncertainty over the precise nature of 'ecological integrity' and how it might be achieved. Each national park has to include an operational definition of ecological integrity in its management plan (an Ecological Integrity Statement, or EIS) and the means by which it will achieve the goal. This involves the establishment of various indicators, design and implementation of a monitoring program, and the resources, both financial and scientific, to carry out such programs. Furthermore, for most parks it will mean an increasingly active management program. Historically, many park managers have simply hoped to 'protect' the park environment and that ecological integrity would be maintained. However, many park environments have already been changed from their natural conditions through fire suppression, predator kills (see Chapter 11), and deliberate introduction of alien species, particularly fish. In addition, there have been numerous unplanned changes, such as the rapid infestation of alien species, in many parks.

In light of these changes, widespread agreement exists that management has to be more active in the parks in terms of influencing ecological variables. Activities include habitat restoration, creation of wildlife corridors, reintroduction of extirpated species, prescribed burning, and management of hyper-abundant species, such as white-tailed deer populations in Point Pelee National Park in Ontario (see also Chapter 11). However, due to a lack of knowledge of ecosystem processes, there is often considerable debate among scientists as to how such programs should be implemented (e.g., see Woodley 2008 on fire suppression).

Managers often have little means of reducing uncertainty surrounding these issues. As Woodley (2008, 114) points out, it is a big step for many managers 'to go from being a protector of "nature", where nature always knows best, to actively managing ecosystems for a defined ecological goal'. Often, there are significant knowledge gaps regarding the functioning of park ecosystems and hence few guidelines regarding the most appropriate active management interventions to initiate. One way that scientists deal with this kind of problem is through adaptive management (See Chapter 16).

Adaptive management is a formal process of testing the efficacy of different solutions to a problem. Management interventions are seen as hypotheses and result in a continual learning process. Adaptive management adopts a systems approach, recognizes the near ubiquity of change, uncertainty, and complexity in natural and human systems, and accepts the need to learn by doing. It calls for the use of science and modelling in determining management actions, experimental management that is carefully monitored to provide feedback, and regular evaluation and revision of management based on what is learned (Murray and Marmorek 2003). Adaptive management was one of the main recommendations endorsed by the Ecological Integrity Panel (Parks Canada Agency 2000).

Challenges from Without

Many of the problems that affect PA management originate outside the parks themselves, and there is a high degree of uncertainty as to the nature of the problem and how it may be addressed. The nature and scale of these problems vary considerably, from the impacts of forest harvesting on adjacent lands to international and global problems such as acid rain and global climatic change. These forces have prompted parks to become involved in land and water management issues outside their administrative boundaries. The generic term for such an approach is ecosystem-based management (see also Chapter 15).

Ecosystem-based management was formally recognized as a basic principle in the 1994 Parks Canada Policy and NMCA system plan (Parks Canada 1995). The Panel on the Ecological Integrity of Canada's National Parks made management of whole ecosystems, using the best scientific knowledge and involving local people and their concerns, a key part of its recommendations (Parks Canada Agency 2000). Grumbine (1994) identified 10 themes for ecosystem management: hierarchical context, ecological boundaries, ecological integrity, data collection, monitoring, adaptive management, interagency co-operation, organizational change, humans embedded in nature, and values. His definition, 'ecosystem management integrates scientific knowledge of ecological relationships within a complex sociopolitical and values framework toward the general goal of protecting native ecosystem integrity over the long term' (1994, 31), is widely cited, as is his list of goals:

- maintain viable populations of all native species in situ;
- represent native ecosystem types across natural range of variation;
- maintain evolutionary and ecological processes;
- manage over long enough periods of time to maintain the evolutionary potential of species and ecosystems; and
- accommodate human use and occupancy within these constraints.

Grumbine (1994) presents an ecological view of ecosystem-based management. Many ecologists and other natural scientists, who may not even agree with the use of 'ecosystem' in such a broad context or who dislike the vagueness of 'ecosystem management', are relatively content with species conservation as the basis for ecosystem management coupled with its scientific, rational character. This approach can be strengthened by placing greater emphasis on the contributions of conservation biology, landscape ecology, and systems ecology to land management, resulting in an approach based on integrated ecological assessment and adaptive management (Hanna and Slocombe 2007; Slocombe and Dearden 2008).

A fundamental premise of ecosystem management is moving the minds and actions of PA managers from the 'boundary thinking' that dominated actions for the first century of the national parks movement to consideration of the spheres of influence that affect parks beyond the administrative boundary. This new 'thought boundary' may be extremely large and will vary in size according to the particular threat being faced or resource being protected. Ecosystem-based boundaries are often based on natural units, particularly the home ranges of key species or watersheds, or a combination of factors. However, other dimensions of park values are also important and may not be covered by imposing another rigid boundary. Aesthetic threats, for example, may be defined in terms of the viewshed from the park (Dearden 1988). Park managers may also be able to wield powers of persuasion with neighbouring land-users when legislative powers do not exist. In other cases, however, when ecological values are being degraded by air pollution, for example, determination of the 'thought boundary' may be much more difficult, which is a challenge to the manager who must address the problem

(Welch 2002). Sparkes (2003), in an interesting application, examines the role of social capital in ecosystem-based management. Social capital is the supply of active connections between people, including the trust, mutual understanding, shared values, and behaviours that bind the members of human networks and communities together and make co-operative action possible.

Some of the best known examples of this kind of approach in practice are biosphere reserves. The Biosphere Reserve Program was started by the United Nations Educational, Scientific and Cultural Organisation (UNESCO) in 1968. Each reserve has three basic functions: a *conservation* function, to contribute to the conservation of landscapes, ecosystems, species, and genetic variation; a *development* function, to foster economic and human development that is socially and ecologically sustainable; and a *logistic* function, to provide support for research, monitoring, education, and information exchange related to local, national, and global issues of conservation and development. Typically, reserves have three zones:

- a *core* zone: strictly protected areas with very little human influence, used to monitor natural changes in representative ecosystems and serve as conservation areas for biodiversity;
- a *buffer* zone: areas surrounding the core zone where only low-impact activities are allowed, such as research, environmental education, and recreation; and
- a *transition* zone: the outer zone where sustainable use of resources by local communities is encouraged and the impacts can be compared to zones of greater protection.

National parks are often key components of the core zone. Waterton Lakes National Park in Alberta was the first biosphere reserve to be declared in Canada. NGOs with such diverse mandates as the Nature Conservancy of Canada and the Rocky Mountain Elk Foundation have worked with park staff and nearby landowners to address conservation, rancher, land development, and other issues. However, despite some indications of success, studies of landscape change in the Greater Waterton Ecosystem illustrate the scale of the ongoing challenge in maintaining ecological integrity (Stewart, Harries, and Stewart 2000). Other examples of biosphere reserves using Canadian PAs include Georgian Bay Islands National Park, the Niagara Escarpment Biosphere Reserve, the Pacific Rim and the Clayoquot Sound Biosphere Reserve, and the Greater Fundy Ecosystem.

Zorn and Stephenson (2002) report on an evaluation of the ecosystem management program in the Ontario region of Parks Canada. Evaluation was based on 11 components (Ecosystem Conservation Plan, greater park ecosystem inventory and analysis, greater park ecosystem, area of cooperation, ecological integrity indicators, scientific research program, ecological indicators, information network, stakeholder analysis, partnership group management guidelines, ecological integrity monitoring program, and communication strategy). The study found that the parks were strongest in terms of strategic planning and weakest in scientific research and monitoring. Not surprisingly, the latter requires the greatest amount of consistent funding.

At even larger scales, greater attention is now being given to bioregional approaches to incorporating PAs into landscape planning. No matter how well individual parks are planned, are managed, and liaise with their neighbours, they are still, in most cases, going to end up as isolated outposts of 'nature' in a highly developed landscape. The implications of this are serious. For example, a study by Landry, Thomas, and Nudds (2001) found that 'the small size, high visitation rates, and ecological isolation of southern parks mitigate against MVPs [Minimum Viable Populations] of wolves, black bears and grizzly bears being maintained there . . . that most of Canada's national parks cannot indefinitely sustain MVPs of these large carnivorous mammals' (19). Theberge and Theberge (2008) provide an overview of the application of ecological concepts to the management of protected areas in a Canadian context and emphasize the need to 'think big' and 'think connected' if park ecosystems are to be maintained relatively intact.

These concerns point to the need for more extensive networks that link PAs and can provide migration routes for wide-ranging species. This is also a sound strategy given the uncertainty associated with the likely impacts of global climatic change. As climatic conditions change, species will have to be able to move across the landscape to reach more favourable environmental conditions.

Work is only just beginning on the implications of climate change for PAs. Canada will be one of the most affected countries in the world (see Chapter 5). Scientists predict that each 1°C rise in temperature will cause biomes to migrate northwards some 300 kilometres. Given the predicted increase of 2 to 5°C in 70 to 100 years, this will translate into 600 to 1,500 metres in elevation and 300 to 750 kilometres in distance. Species must either be able to migrate fast enough to keep up with these changes, evolve to deal with them, or go extinct. Certain biomes, such as arctic-alpine and the boreal forest, will be very vulnerable to these changes.

For PAs, climate change has very serious implications. On the one hand, PAs will have a huge role to play in terms of helping to sequester carbon from the atmosphere. On the other hand, their role of providing refuge for natural populations will be vulnerable to the changes described above. PA networks must be made as resilient as possible against these changes.

One of the main mechanisms for achieving this is through large-scale bioregional planning that emphasizes connectivity, especially north–south connectivity among PAs. New PAs will be required to help facilitate migration, provide source populations, and provide suitable habitat for incoming populations. Private lands will also be important. Whitelaw and Eagles (2007) offer an example from Ontario that illustrates planning on private lands for the kind of long and wide conservation corridors that will be required in the future.

A survey of PA jurisdictions in Canada found that 80 per cent had not completed a comprehensive assessment of the potential impacts and implications of climate change on policy and management and have no adaptation strategy or action plan. Furthermore, 86 per cent of respondents felt that they did not have the capacity to deal with climate change issues (Lemieux, Beechey, and Scott 2007). Hannah (2003) suggests that one of the most important steps required to deal with global change in PAs is to improve existing management as soon as possible before climate change raises new challenges. Unfortunately, in almost all provincial jurisdictions in Canada, politicians have been driving things the other way by consistently cutting the funding available to park agencies.

Suffling and Scott (2002) have used climate change models to indicate some of the possible effects on the national parks system over the next century. They conclude that all regions and parks will be dramatically affected and suggest that Parks Canada is going to have to interfere increasingly in park ecosystems to maximize the capacities of ecosystems and species to adapt to climate change. Some parks are already showing significant changes, with Wood Buffalo, for example, having lower precipitation leading to changes in water quality and quantity and ecosystem health, an influx of invasive species, and a greater probability of fire (Wiersma et al. 2005). A study on Ontario parks was undertaken by Lemieux et al. (2008) to examine what types and forms of adaptation are possible, feasible, and likely, who would be involved in their implementation, and what is required to facilitate or encourage their development and implementation (i.e., commencing the 'mainstreaming' of climate change adaptation). All park agencies should be involved in this kind of study.

There is no doubt that bioregional-scale thinking will be required as an adaptive strategy for PAs in the context of climate change. Thinking on these bioregional approaches was stimulated by development of the Yellowstone to Yukon (Y2Y) initiative for a linked set of protected areas spanning the length of the Rocky Mountains (Locke 1997). Prompted by science that showed that certain species migrated all along the Rockies, the initiative attempts to see how different PAs along the way can be linked together (see Figure 12.3). Two designations by the BC government have been key in realizing the vision. The Muskwa-Kechika Management Area (MKMA) covers some 4.4 million hectares and since 1997 has been managed primarily for conservation, including 1.1 million hectares in park areas. Any development in the areas outside of the parks must, by law, allow for environment as a prime consideration. This area is adjacent to the latest designation, where after seven years of roundtable discussions, 475,000 hectares of new parks have been added to the existing 116,000 hectares, with an additional 410,000 hectares of special management zones and 900,000 hectares of no-logging. Together with the MKMA, these new areas create a conservation area of 6.3 million hectares connecting the ecosystems of the Rocky and Cassiar mountains with those of the Pacific coast. This area provides a home to the greatest combined abundance and diversity of large mammals in North America, among them thousands of moose, elk, and caribou and the continent's largest concentration of Stone sheep, all accompanied by a healthy population of predators. In fact, it was wolves from this area and Alberta that were used in the wolf reintroductions to Yellowstone.

CONCLUSION

There have been significant changes in PAs in Canada over the past 15 years. The amount of land protected has risen substantially as politicians have responded to public demands, independent reports, and outside pressures to increase the size of park systems. Overall, this has reduced the amount of highly visible conflict that characterized the earlier years of the past decade. Little progress, however, has been made on the marine front. The next 10 years will probably witness increased conflict here as the general public, NGOs, and conservation scientists push for establishment of MPAs, often against the short-term interests of local resource users.

Similarly, conflicts over management decisions in the parks raised by conservation scientists and the NGO community for many years were legitimized by several independent, government-appointed investigations that prompted more serious attention to be given to the protection of ecological integrity. However, there is often scientific uncertainty as to how to move forward on some of these challenges.

Legislation and policies governing PA agencies have undergone substantial revision. A new National Parks Act was passed in 2000, along with a new Act to define the operational characteristics of Parks Canada. An Act specifically dedicated to establishing MPAs was enacted in 2002, and national parks policy was also revised. These actions have helped to reduce uncertainty by constraining the latitude that decision-makers have to interpret park goals.

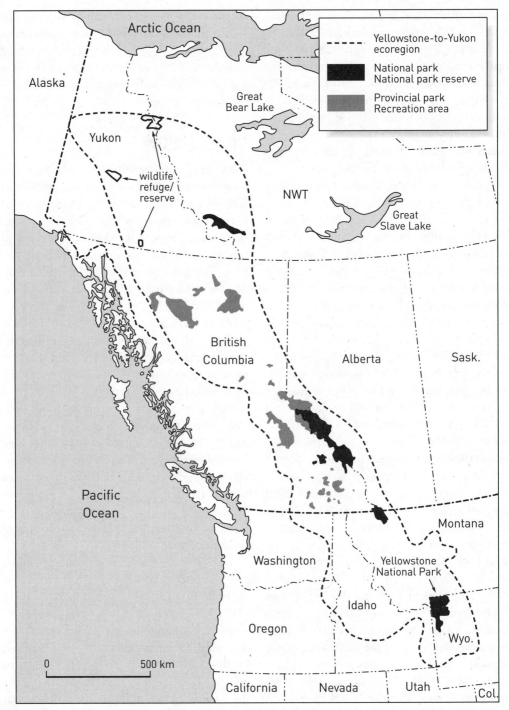

Figure 12.3 Some of the Main Protected Areas Linked through the Yellowstone to Yukon Initiative.

Overall, however, the management context has become more uncertain, largely because of external factors beyond the influence of park managers. How can a manager guarantee a park environment unimpaired for future generations when some of the most powerful factors that influence that environment are spawned continents away?

Conflict and uncertainty still remain as central themes in understanding PA management. Their nature and focus change over time, but they will not disappear. Effective management, therefore, will have to plan for it. Resource management approaches such as ecosystem-based management and adaptive management will be central to these efforts.

FROM THE FIELD

The good thing about working in the parks and protected areas field is that you get to visit and work in some of the most spectacular landscapes on Earth. This is especially true in Canada, with wilderness settings on par with any in the world. The bad thing is that often your work involves some aspect of monitoring, or gaining greater understanding of, the deterioration of that PA's environment. However it wasn't always like that; indeed, one of the main messages from parks is the speed of change that has occurred in global ecosystems.

My first research in a national park took place more than 35 years ago and would fall firmly under what I call the 'outdoor laboratory' category where we use PAS to understand how ecosystems function in the most pristine environments on Earth. However, the fundamental tenet—that we are studying 'untouched' ecosystems—is increasingly at variance with reality. Every raindrop falling on Earth now bears the imprint of industrial activity. The chemicals they carry have a profound effect on soils, water, and living organisms. Global climatic change driven by human activities penetrates into the most inaccessible parts of the world and has no respect for human boundaries, such as those around national parks.

The scale of changes now witnessed, encompassing not only terrestrial but also marine ecosystems,

engulfs all the land and seascape. Park environments play a major role in terms of monitoring these changes, but the idea of untouched environments is no longer tenable.

Yet much can be learned from parks. The changes mentioned above need to be communicated effectively to an ever more urbanized society. To the urbanite, the landscape of the Ice Fields Parkway linking Banff and Jasper National Parks looks as unfamiliar and remote as that of the moon. It looks wild and untouched. The visiting public will have no comprehension of how their daily activities at home, the chemicals they use in their gardens, in cleaning their houses, and on the foods they buy have a direct impact on the well-being of these seemingly remote environments—unless we tell them. Using parks as a main vehicle to raise environmental literacy and change public behaviours to more sustainable practices is one of the main opportunities and challenges for a new generation of park researchers.

The range of research challenges and skills required in PA research is increasing. My early research, and most park research in Canada today, was focused on the purely biophysical aspects of parks and often had little, if any, direct management implications for parks. Such research is still important, but we need considerably more interest

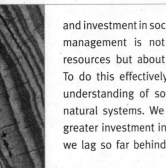

and investment in social science research. Resource management is not so much about managing resources but about managing human activities. To do this effectively requires the same detailed understanding of social systems as we have of natural systems. We also need a proportionately greater investment in marine understanding, since we lag so far behind in that area. These changes mirror my own career as I have sought to adapt my research to the most pressing questions. I began my career examining biophysical research in terrestrial environments in Canada; now my main focus is on social dimensions of marine conservation areas globally. It's been quite a journey.

—PHILIP DEARDEN

REVIEW QUESTIONS

1. What are some of the main uses of parks and protected areas?
2. Why are parks and protected areas particularly vulnerable to conflict in resource management?
3. What are the main elements of the policy and legislative framework for national parks in Canada?
4. Why are marine protected areas vulnerable to conflict?
5. What is the current status of PAs in Canada in terms of the amount of land or sea protected? Outline three main areas of uncertainty in PA management in Canada.
6. Why do you think that park visitation is falling?
7. What are some of the recommendations on how PAs can adapt to global climate change?
8. What is active management, and why is it increasingly necessary?
9. What is the role of Aboriginal peoples in PA establishment and management in Canada?

REFERENCES

Auditor General of Canada. 1996. *Report of the Auditor General of Canada to the House of Commons*, ch. 31: 'Canadian Heritage–Parks Canada: Preserving Canada's natural heritage'. Ottawa: Office of the Auditor General of Canada.

———. 1998. 'Follow-up of recommendations in previous reports'. In '1998 Report of the Auditor General of Canada'. http://www.oag-bvg.gc.ca/domino/reports.nsf/html/9631ce.html.

———. 2005 'Ecological integrity in Canada's national parks'. In '2005 Report of the Auditor General of Canada'. http://www.oag-bvg.gc.ca/internet/English/parl_cesd_200509_02_e_14949.html.

Banff–Bow Valley Study. 1996. *Banff–Bow Valley: At the Crossroads*. Summary Report of the Banff–Bow Valley Task Force. eds R. Page, S. Bayley, J.D. Cook, J.E. Green, and J.R. Brent Ritchie. Prepared for the Honourable Sheila Copps, Minister of Canadian Heritage. Ottawa.

Butler, M. 1994. 'When you can't build fences: The West Isles Marine Park proposal'. In C. Lamson, ed., *The Sea Has Many Voices*, 127–41. Montreal: McGill-Queen's University Press.

Canada. 1930. *National Parks Act*. Ottawa: Queen's Printer for Canada.

———. 1969. *National Parks Policy*. Ottawa: Queen's Printer for Canada.

———. 1983. *Parks Canada Policy*. Ottawa: Minister of Supply and Services Canada. Catalogue no. 62-109/1983E.

————. 1988. *An Act to Amend the National Parks Act—Bill C-30*. Ottawa: Minister of Supply and Services Canada.

————. 2000. 'National Parks Act'. http://laws.justice. gc.ca/en/N-14.01.

Charles, A., and L. Wilson. 2009. 'Human dimensions of marine protected areas'. *ICES Journal of Marine Science* 66: 6–15.

Committee on the Evaluation, Design and Monitoring of Marine Reserves and Protected Areas in the United States. 2001. *Marine Protected Areas: Tools for Sustaining Ocean Ecosystems*. Washington: National Academy Press.

Davis, G. 2008. 'Designing ocean parks for the next century'. *The George Wright Forum* 25: 7–22.

Dearden, P. 1988. 'Protected areas and the boundary model: Meares Island and Pacific Rim National Park'. *The Canadian Geographer* 32: 256–65.

Dearden, P., and L. Berg. 1993. 'Canadian national parks: A model of administrative penetration'. *The Canadian Geographer* 37: 194–211.

Dearden, P., and R. Canessa. 2008. 'Marine protected areas'. In P. Dearden and R. Rollins, eds, *Parks and Protected Areas in Canada: Planning and Management*, 3rd edn, 403–31. Toronto: Oxford University Press.

Dearden, P., and J.E. Gardner. 1987. 'Systems planning for protected areas in Canada: A review of caucus candidate areas and concepts, issues and prospects for further investigation'. In R.C. Scace and J.G. Nelson, eds, *Heritage for Tomorrow: Proceedings of the Canadian Assembly on National Parks and Protected Areas*, v. 2, 9–48. Ottawa: Minister of Supply and Services Canada. Catalogue no. R62-232/2-1987E.

Dearden, P., and S. Langdon. 2008. 'Aboriginal peoples and national parks'. In P. Dearden and R. Rollins, eds, *Parks and Protected Areas in Canada: Planning and Management*, 3rd edn, 373–402.

Dudley, N., ed. 2008. *Guidelines for Applying Protected Area Management Categories*. Gland, Switzerland: IUCN.

Environment Canada. 1987. *Our Parks—Vision for the 21st Century*. Report of the Minister of Environment's Task Force on Parks Establishment. Waterloo, ON: University of Waterloo, Heritage Resource Centre.

————. 2006. *Canadian Protected Areas Status Report 2000–2005*. Ottawa: Environment Canada.

Gardner, J., et al. 2008. *Challenges and Opportunities in Progress towards Canada's Commitment to a National Network of MPAs by 2012*. Vancouver: CPAWS-BC.

Government of Canada. 2005. *Canada's Federal Marine Protected Areas Strategy*. Ottawa: Fisheries and Oceans Canada.

Government of Nova Scotia. 2007. *Bill No. 146: Environmental Goals and Sustainable Prosperity Act*. Halifax: Ministry of Environment and Labour.

Grumbine, R.E. 1994. 'What is ecosystem management?' *Conservation Biology* 8: 27–38.

Hanna, K.S., and D.S. Slocombe, eds. 2007. *Integrated Resource and Environmental Management: Concepts and Practice*. Toronto: Oxford University Press.

Hannah, L. 2003. 'Protected areas management in a changing climate'. In N.W.P. Munro et al., eds, *Making Ecosystem-Based Management Work: Connecting Managers and Researchers*. Wolfville, NS: Science and Management of Protected Areas Association (SAMPAA). www.sampaa.org.

Hvenegaard, G., J. Shultis, and J.R. Butler. 2008. 'The role of interpretation'. In P. Dearden and R. Rollins, eds, *Parks and Protected Areas in Canada: Planning and Management*, 3rd edn, 202–34. Toronto: Oxford University Press.

Jager, E., et al. 2006 'Managing for visitor experiences in Canada's national heritage places'. *Parks* 16: 18–24.

Landry, M., V.G. Thomas, and T.D. Nudds. 2001 'Sizes of Canadian national parks and the viability of large mammal populations: Policy implications'. *The George Wright Forum* 18: 13–23.

Lemieux, C.J., et al. 2008 *Changing Climate, Challenging Choices: Ontario Parks and Climate Change Adaptation*. Waterloo, ON: University of Waterloo, Department of Geography Publication Series.

Lemieux, C.J., T. Beechey, and D. Scott. 2007. 'A survey on protected areas and climate change (PACC) in Canada: Survey update'. Canadian Council on Ecological Areas. *ECO* 16: 2–3.

Lemieux, C.J., and D.J. Scott. 2005. 'Climate change, biodiversity conservation and protected areas planning in Canada'. *The Canadian Geographer* 49: 384–99.

Lien, J. 1999. 'When marine conservation efforts sink: What can be learned from the abandoned effort to examine the feasibility of a national marine conservation area on the NE coast of Newfoundland?'

Paper presented at the 16th Annual Conference of the Canadian Council for Ecological Areas (CCEA), Ottawa, 4–6 October.

Locke, H. 1997. 'The role of Banff National Park as a protected area in the Yellowstone to Yukon mountain corridor of western North America'. In J.G. Nelson and R. Serafin, eds, *National Parks and Protected Areas: Keystones to Conservation and Sustainable Development*, 117–24. Berlin: Springer.

Louv, R. 2005. *Last Child in the Woods: Saving Our Children from Nature-Deficit-Disorder*. Chapel Hill, NC: Algonquin Books.

Lubchenco, J., et al. 2003. 'Plugging a hole in the ocean: The emerging science of marine reserves'. *Ecological Applications* 13 (supplement): S3–S7.

Murray, C., and D. Marmorek. 2003. 'Adaptive management: A science-based approach to managing ecosystems in the face of uncertainty'. In N.W.P. Munro et al., eds, *Making Ecosystem-Based Management Work: Connecting Managers and Researchers*. Wolfville, NS: Science and Management of Protected Areas Association (SAMPAA). www.sampaa.org.

Needham, M., and R. Rollins. 2008. 'Social science, conservation and protected areas' theory'. In P. Dearden and R. Rollins, eds, *Parks and Protected Areas in Canada: Planning and Management*, 3rd edn, 135–68. Toronto: Oxford University Press.

Parks Canada. 1995. 'Parks Canada guiding principles and operational policies'. http://parkscanada.pch.gc.ca/library/PC_Guiding_Principles/Park1_e.htm.

Parks Canada Agency. 2000. *Unimpaired for Future Generations? Protecting Ecological Integrity within Canada's National Parks*. Volume II: *Setting a New Direction for Canada's National Parks*. Report of the Panel on the Ecological Integrity of Canada's National Parks. Ottawa.

———. 2001. *First Priority: Progress Report on Implementation of the Recommendations of the Panel on the Ecological Integrity of Canada's National Parks*. Ottawa: Minster of Public Works and Supply. Catalogue no. R62-336/2001E.

———. 2007. *Performance Report for the Period Ending March 31st, 2007*. Ottawa: Parks Canada Agency.

Pavelka, J., and R. Rollins. 2008. 'Case study: Banff and Bow Valley'. In P. Dearden and R. Rollins, eds, *Parks and Protected Areas in Canada: Planning and Management*, 3rd edn, 272–93. Toronto: Oxford University Press.

Payne, R.J. 1997. 'The new alchemy: Values, benefits and business in protected area management'. In N. Munro, ed., *Protected Areas in Our Modern World*, 89–95. Proceedings of a Workshop held as part of the IUCN World Conservation Congress, Montreal, 1996. Parks Canada Ecosystem Science Review Reports 005.

Roff, J., and P. Dearden. 2007. 'Establishing a network of MPAs by 2012: The way forward'. Workshop Report. In *Ecosystem-Based Management: Beyond Boundaries. Proceedings of the Sixth International Conference on the Science and Management of Protected Areas*, 42. Wolfville, NS: Science and Management of Protected Areas Association (SAMPAA).

Searle, R. 2000. *Phantom Parks: The Struggle to Save Canada's National Parks*. Toronto: Key Porter Books.

Shears, N.T., et al. 2006. 'Long term trends in lobster populations in a partially protected vs no-take marine park'. *Biological Conservation* 132: 222–31.

Slocombe, D.S., and P. Dearden. 2008. 'Protected areas and ecosystem-based management'. In P. Dearden and R. Rollins, eds, *Parks and Protected Areas in Canada: Planning and Management*, 3rd edn, 342–70. Toronto: Oxford University Press.

Sparkes, J. 2003. 'Social capital as a dimension of ecosystem-based management'. In N.W.P. Munro et al., eds, *Making Ecosystem-Based Management Work: Connecting Managers and Researchers*. Wolfville, NS: Science and Management of Protected Areas Association (SAMPAA).

Stewart, A., A. Harries, and C. Stewart. 2000. 'Waterton Biosphere Reserve landscape change study'. In *Canada MAB 2000 Landscape Changes at Canada's Biosphere Reserves*, 13–20. Toronto: Environment Canada.

Suffling, R., and D. Scott. 2002. 'Assessment of climate change effects on Canada's national park system'. *Environmental Monitoring and Assessment* 74 (2) 117–39.

Swinnerton, G. 2002. 'Case study: Banff and Bow Valley'. In P. Dearden and R. Rollins, eds, *Parks and Protected Areas in Canada*, 2nd edn, 240–64. Toronto: Oxford University Press.

Theberge, J.C., and J.B. Theberge. 2008. 'Application of ecological concepts to the management of protected areas'. In P. Dearden and R. Rollins, eds,

Parks and Protected Areas in Canada: Planning and Management, 3rd edn, 84–109. Toronto: Oxford University Press.

Wallace, S.S. 1999. 'Evaluating the effects of three forms of marine reserve on northern abalone populations in British Columbia, Canada'. *Conservation Biology* 13: 882–7.

Welch, D. 2002. 'Acid deposition continues to threaten Canada's southern and eastern national parks'. In S. Bondrup-Nielsen, et al., eds, *Managing Protected Areas in a Changing World. Proceedings of the Fourth International Conference on Science and Management of Protected Areas*, 321–30. Wolfville, NS: Acadia University, Science and Management of Protected Areas Association (SAMPAA).

Whitelaw, G.S., and P.E.J. Eagles. 2007. 'Planning for long, wide conservation corridors on private lands in the Oak Ridges Moraine, Ontario, Canada'. *Conservation Biology* 21: 675–83.

Wiersma, Y.F., et al. 2005. *Protected Areas in Northern Canada: Designing for Ecological Integrity*. Gatineau, QC: Canadian Council on Ecological Areas.

Woodley, S. 2008. 'Planning and managing for ecological integrity in Canada's national parks'. In P. Dearden and R. Rollins, eds, *Parks and Protected Areas in Canada: Planning and Management*, 3rd edn, 111–32. Toronto: Oxford University Press.

WCED (World Commission on Environment and Development). 1987. *Our Common Future*. Toronto: Oxford University Press.

Wright, P., and R. Rollins. 2008 'Managing the national parks'. In P. Dearden and R. Rollins, eds, *Parks and Protected Areas in Canada: Planning and Management*, 3rd edn, 237–71. Toronto: Oxford University Press.

Zorn, P., and W. Stephenson. 2002. 'Ontario national parks' ecosystem management program and assessment process'. In S. Bondrup-Nielsen, et al., eds, *Managing Protected Areas in a Changing World. Proceedings of the Fourth International Conference on Science and Management of Protected Areas*, 398–413. Wolfville, NS: Acadia University, Science and Management of Protected Areas Association (SAMPAA).

13

Canadian Mineral Resource Development: A Sustainable Enterprise?

Mary Louise McAllister and Patricia Fitzpatrick

Learning Objectives

- To understand the inherent challenges involved in nudging the Canadian mineral industry, a global leader, in a more socio-ecologically sustainable direction.
- To appreciate the various actors and factors operating at all spatial and temporal scales related to the mineral industry's activities and to consider their influence in shaping those activities.
- To recognize the broad issues related to the growing complexities of decision-making for resource development, including a growing diversity of actors vying for policy influence, burgeoning environmental problems (from global climate change to local environmental contamination), and competing world views about which policy approaches would foster a sustainable way forward.

INTRODUCTION

Canadian mineral resource development is woven throughout the country's history, literature, landscapes, economy, politics, and culture. The quest for mineral wealth has led to the establishment of hundreds of communities and regional economies and stimulated the building of extensive transportation and communication links. Today, the industry is a world leader in mineral exploration and production, expanding well beyond national boundaries. The Canadian mineral sector has reached into the economies, societies, and environments of many countries.

While these developments generate economic wealth, they come with a price. The collective effects of past and present mining practices have left future generations with substantial environmental and socio-economic costs. Mining affects air, water, food chains and wildlife migration patterns, physical landscapes, and local cultures, particularly those of indigenous peoples. Abandoned mines are scattered throughout the country, often posing a threat to groundwater, rivers, and ecosystems. Mining also contributes to the cumulative impacts of all industrial activity, sometimes creating unanticipated and undesirable environmental consequences.

In addition to the growing complexity associated with attempts to maximize wealth while minimizing harm, the policy arenas governing resource management have become crowded. While at one time the main players were members of the industry and domestic governments, many more groups have claimed a stake in this sector. The policy community (that is, those groups that influence decisions related to mineral policy) now comprises a wide range of communities of interest. International, domestic, and local organizations, including governments, industries, non-governmental groups, indigenous peoples, unions, and communities, all vie to influence the use and management of resources. These 'stakeholders' in the mineral sector are now considering how the economic benefits of mining can be realized without accompanying long-term problems. While most players agree that mining should proceed in environmentally and socially responsible ways, few can reach consensus about how that should happen, who should pay, and what is meant by the term 'responsible'.

Conflict and uncertainty have always accompanied the 'boom-and-bust' industry. Yet today the problems are magnified. Decision-makers are expected to respond to complex problems with a balanced management approach that will recognize often competing interests. The goal is to find broadly acceptable solutions that will allow the industry to continue to produce revenues and profits. If this goal is not accompanied by environmentally sound approaches (in terms of both social and biophysical considerations), diverse interest groups will vocally remind them either directly or through the media. If this proves ineffective, the natural capital (the physical resources) will continue to degrade, inevitably undermining the health of desired ecosystems.

What is meant by a sustainable mineral industry? At one time, it may have referred to an enterprise able to sustain itself indefinitely into the future. It has, however, come to mean much more than that by encompassing environmental principles as well. Some might feel that sustainable mining is an oxymoron, given that the resource is classified as non-renewable. If a mining operation were to take place, however, a sustainable approach to that activity would integrate social, biophysical, and economic considerations into the planning processes from the first stages of exploration until post–mine closure. How such processes should take place is a matter of some debate between, and within, the various stakeholder groups. As the following section indicates, part of the challenge of reaching agreement about how to proceed in a sustainable manner lies in the diverse nature of the industry itself.

PROFILE OF THE MINERAL INDUSTRY

The Canadian mineral industry produces a wide range of minerals by employing a variety of techniques. It is a global industry, with more than 50 per cent of the world's mining companies based in Canada (NRCAN 2007a). The provinces with the overall highest levels of mineral production are Ontario, British Columbia, and Quebec. Mineral extraction includes, among others, the vast oil sands[1] operations in Fort McMurray, AB, the diamond mines in the Northwest Territories, the long-lived nickel and copper operations in the Sudbury region, gold and base metals in the famous Val-d'Or in Quebec, and the small rock and gravel quarries located near cities, used to build and maintain roads and infrastructure. In January 2008, 766 mines were in operation, including 63 metal and 703 non-metal developments (Mining Association of Canada 2008a).

Canada is one of the world's top exporters of metals and minerals (non-fuel but including coal). It has maintained that position for a long time and has responded to competitive and regulatory challenges with new mine management technologies that bear little resemblance to those of its early roots centuries ago. Its exports were worth $74.7 billion in 2006, with $49.1 billion of these exports going to the United States (NRCAN 2007b, 1.8).

The Old Myths and the New Realities: The Modern Mining Operation

> There's gold, and it's haunting and haunting;
> It's luring me on as of old;
> Yet it isn't the gold that I'm wanting
> So much as just finding the gold.
>
> *Robert Service, 'The Spell of the Yukon', 1907*

Mining is one of the oldest enterprises in the world. In British Columbia, one of the first known quarries was in use as early as 8000 BCE (Dickason 1992, 78). John Udd (2000) notes that the first record of mining by Europeans occurred in 998 CE when the Vikings mined ore in Newfoundland. In the seventeenth and eighteenth centuries, coal was mined in what is now known as the Maritime provinces and exported to Europe. By the nineteenth century, mining was an established enterprise in Canada. Today, the work of modern exploration companies has changed in many ways from Robert Service's poetic depictions of the gold rushes of more than 100 years ago. And the process of mining itself has come a long way from the days when miners used handheld explosives and picked away at a seam. While those pictures may once have reflected reality, these stereotypes are at odds with the operations of modern exploration and mining. Today the industry relies heavily on information technologies, automation, and even robotics.

The lifecycle of modern mineral development and production and its impact on Canada might be viewed in two primary ways. The first is in terms of the mineral product lifecycle. Traditionally, the mineral and metal processing sector has been viewed as having five main stages: (1) mining and quarrying, (2) smelting and refining (between smelting and refining, milling and/or flotation often takes place to produce a concentrate or intermediate product), (3) production of semi-fabricated parts, (4) production of fabricated parts and simple products, and (5) product assembly (NRCAN 1998, 3). This linear depiction has now become somewhat outdated as new technologies are introduced throughout the mining process. For example, it is now possible to produce metals from some ore types at the mine site using hydrometallurgical processes (referred to as hydromet) to refine metal, thereby eliminating the need for smelters (Doggett 2002).

A second way to view the lifecycle has to do with the cycles associated only with primary production, or the process of mining ore. This perspective (and the one that is applied in this chapter) considers the life of the mine as it progresses through the stages of exploration, mine development and operation, closure, and reclamation.

The Canadian exploration industry has concerned itself primarily with the search for base (e.g., nickel, copper, zinc, lead) and precious metals. Until the global economic turmoil that started in the last quarter of 2008, Canadian exploration had been increasingly robust since the previous economic downturn of 1999. Such robustness was attributed to a favourable economic climate and continuing high commodity prices as well as federal tax incentives (NRCAN 2007c). Provincial government support has also played a role. Quebec, for example, experienced a mining boom between 2004 and 2008 that saw exploration and development expenditures more than double to $476.3 million. The provincial government provides fiscal incentives through exploration allowances, tax credits, tax deductions through a flow-through shares system, and research and development assistance (Quebec 2009).

During exploration, modern geophysical methods assisted by satellites and computers are employed to identify potential mineral wealth. Despite all the advanced technological tools, however, the process of discovering a viable deposit is not an exact science. It requires considerable skill, knowledge, resources, and a measure of luck (or, some would say, intuition) to recognize and discover mineral potential. The exploration process requires access to very large tracts of land to enable the discovery of a single ore deposit that could be considered potentially economically viable. Moreover, from those prospects, only one in every 1,000 may actually become a mine. Access to land for exploration, coupled with some regulatory certainty regarding the rules governing land tenure, is a very important requirement of the exploration industry. Millions of dollars are invested in searches for that one deposit that can compensate the company for all the investments that did not pan out.

Before a potential prospect can be developed into a mine, two fundamental criteria must be considered: economic grade (concentration of mineral) and volume of ore. The deposit needs to be of a grade and size that will allow a company to pay its costs and make a profit (Placer Dome Inc. 1997, 28). The potential costs and profits are affected by many factors, including world mineral prices, energy, labour, and infrastructural costs, political and regulatory conditions, civil stability of a region, and socio-cultural and biophysical considerations. Some marginal mines (i.e., those with a lower economic grade or smaller ore size) may not be developed if the potential costs outweigh the benefits or revenues. Nevertheless,

through the use of modern mining technologies, it is now possible to mine low-grade deposits by extracting huge volumes of ore. It can cost many hundreds of millions or even billions of dollars to build a mine of this kind. Before that much capital is invested, a feasibility study is undertaken to assess anything that might influence the costs or benefits of a proposed mine in order to determine whether the mine will be economically viable.

Feasibility studies include assessment of existing environmental conditions and the measures required to limit the impact on the ecosystem. In an ideal situation, nearby local communities are consulted about how, and if, a mine should be developed. Without community support, mining operations can and do encounter difficulties that can lead to expensive delays. From its perspective, the community must weigh the potential economic and possible employment benefits of the mine against the potential adverse environmental impacts of the new operation. Communities, or more senior governments, in various parts of the world may be prepared to accept high environmental, social, or health and safety risks so that a marginal mine can become operational and provide much needed revenues.

In Canada, the mine development process is subject to numerous environmental laws and regulations (See Box 13.1 for an example of the legislation that applies in the province of British Columbia). Environmental assessments and plans must be approved, regulations followed, and permits and licences acquired. Once the decision is made to go ahead with the mine and all regulatory requirements have been fulfilled, mine construction can begin.

The process of mining—drilling, blasting, loading, and hauling—will take place in either open-pit (surface) or underground operations. The type of equipment used is a reflection of the mine plan. Open-pit mines often use very

---

Box 13.1 Mining in British Columbia:
Federal and Provincial Legislation and Regulation

Federal Legislation
- Canadian Environment Assessment Act
- Canadian Environmental Protection Act
- Fisheries Act
- Navigable Waters Protection Act
- Migratory Birds Act

Provincial Legislation
- Mines Act
- Environmental Assessment Act
- Environment and Land Use Act
- Environment Management Act
- Forest Act
- Health Act
- Waste Management Act
- Water Act
- Water Protection Act
- Wildlife Act
- Fisheries Act
- Fish Protection Act

Source: Mining Association of British Columbia 2008.

---

large pieces of equipment designed to move huge amounts of ore and waste in order to make lower-grade ore deposits economically viable. What is lacking in ore grade is compensated for in volume, thereby lowering unit costs through economy of scale. Underground mines require different equipment that can move in small, restricted places. Under these conditions, however, the equipment itself is not able to move such large tonnages as is the case with some surface mines, resulting in higher unit operating costs. A higher grade of ore is required for the operation to be profitable. Underground mines also can be very expensive to develop, especially on a per unit of capacity basis (Placer Dome Inc. 1997, 38–9).

Once begun, the operation of most modern mines in Canada is no longer labour-intensive. Miners now routinely guide drills, hoists, or vehicles from control consoles.

In remote regions of Canada (and in many other countries), companies are also engaged in 'fly-in' mining, or long-distance commuting to remotely located ore bodies. Modern fly-in mining operations bring in workers from regional centres for a one- or two-week shift.

The life of a mining operation can vary tremendously. It may range from the few years required to mine a small gold deposit to one that operates throughout the lives of several generations of miners, ending up as a deep, underground operation such as can be found in the nickel and copper mines of Sudbury.

A mine closure plan should be implemented when the ore body is depleted to the degree that it is no longer economic. Today in Canada, before a mine is permitted to open, companies are expected to demonstrate that they have a satisfactory closure plan and the financial ability to safely decommission and reclaim land after mine closure. This means that the mine tailings have been adequately contained, the environment returned to an acceptable state, and the future welfare of employees and the community satisfactorily considered.

Although mining is the oldest industrial activity in the world, it would be inaccurate to portray mining as a sunset industry. The modern enterprise is highly automated, using advanced technologies throughout its lifecycle. Unlike the situation in years gone by, mining is regulated by numerous statutes that must be satisfied before the proponent can move to the production stage. Most often, a government's ultimate decision to let a project proceed or not is more heavily influenced by factors of profits, revenues, and employment than anticipated environmental risk. While there are some recent exceptions to this statement (as discussed below), governments historically have attempted to regulate and manage the risks and benefits of resource development rather than to apply the 'precautionary principle' and stop a project from proceeding. To do the latter would have very immediate tangible economic and political costs for Canada, which is heavily dependent on staples production.

IMPORTANCE OF MINING TO CANADA

Throughout Canadian history, the natural resources sector has been an integral part of nation- and province-building, so much so that resource development was once considered synonymous with the public interest. The postwar period, in particular, saw concerted effort to stimulate the primary sector. The politics of the time, one commentator observed, 'manifested a philosophy of development at all costs' (Robson 1992, 67).

Mining constitutes one of the essential economic pillars of Canada, and despite its waning impact on the public agenda, it continues to play that role. That role goes beyond its economic importance; it is the foundation of numerous regional economies throughout Canada. In 2007,

approximately 51,000 people were employed in the industry, with an estimated 312,000 people employed in 'downstream' minerals industries (Mining Association of Canada 2008a). Thus, the mining industry suggests that one in every 46 employed Canadians works within its industry. Mining also provides economic opportunities for 1,200 Aboriginal communities located within 200 kilometres of minerals and metals activities. However, this can, and does, also lead to friction within communities when mining activities conflict with traditional Aboriginal lifestyles and values (NRCAN 2001b).

The mineral industry is a high-productivity/high-income sector (Dungan 1997, 7). The value of production of metals and minerals (including fuels) in 2006 exceeded $77 billion. If fuel production is excluded, the value of metals and minerals production was $33.6 billion. The largest provincial producers of metallic and non-metallic minerals are Quebec, Ontario, and British Columbia.

The industry is not only important for its direct contributions to the economy but is also directly linked 'upstream' (backward linkages) and 'downstream' (forward linkages) with the Canadian industrial structure (Dungan 1997, 174). Access to lower-cost raw resources led to the development of many manufacturing and service activities, such as those related to transportation or other businesses. The Mining Association of Canada (2008a) estimates that 70 per cent of Canada's port volumes and 55 per cent of rail volumes are generated by the industry. Canada's telecommunications industry—mobile communications and remote sensing, for example—was developed to service many resource-related activities. On a more mundane level, sand and gravel quarries were (and still are) essential for the building of roads and infrastructure (McAllister 1992). The metal recycling business is also growing. It tripled in volume between 1991 and 2005, reaching a value of \$4.8 billion (NRCAN 2006). Mining has contributed to the development of numerous regional economies, generating revenues as well as indirect wealth in related secondary and tertiary industries. It has also contributed, however, to social and environmental costs.

THE COSTS OF MINING

While a large new mineral discovery may make newspaper headlines, the industry has also attracted attention because of mine disasters (most notably the explosion of the Westray Coal mine in the early 1990s in Nova Scotia that killed 26 miners), the notorious Bre-X gold scam in which investors lost millions, and tailings dam failures contaminating watersheds in remote parts of the world. These impacts are not easily assessed using conventional cost–benefit analyses. Although various efforts have been made, it is difficult to calculate socio-economic costs to communities (particularly indigenous communities) when mining practices or mine closures seriously disrupt a way of life, or the biophysical costs of mining throughout all its stages, or the impact on Canada's reputation when its very high-profile international mining industry attracts negative global attention.

Socio-economic Impacts

In the early morning of 9 May 1992 a violent explosion rocked the tiny community of Plymouth, just east of Stellarton, in Pictou County, Nova Scotia. The explosion occurred in the depths of the Westray coal mine, instantly killing the 26 miners working there at the time.
[Richard 1997, summary]

The Westray coal mine disaster, generally thought to be preventable, highlights the human costs of mining when interest in economic development and profit supersedes regulations and procedures to ensure human health and safety. The report of the public inquiry after the disaster noted that there was a trade-off between productivity and human health and safety. The commission did not mince words: 'The Westray Story is a complex mosaic of actions, omissions, mistakes, incompetence, apathy, cynicism, stupidity, and neglect' (Richard 1997, summary). That said, many representatives of other mining companies would maintain that the irresponsible behaviour that led to the Westray disaster is the exception and not the rule in contemporary mines operating in Canada.

Social concerns go beyond the physical health and safety of the workers. The families of miners often live with the looming prospect of the inevitable temporary shutdown or permanent closure of a mine. As the well-known expression in the mineral industry goes, 'the day a mine opens, it begins to close.' While some communities, such as Sudbury, do diversify from the single economic base of a mine operation, others inevitably decline when there appear to be no other prospects for economic development. These kinds of problems contributed to the adoption of 'fly-in' mining operations, an alternative to the expense and problems associated with setting up a permanent town.

Aboriginal communities are inevitably disrupted with the introduction of mining in their vicinity, even though numerous economic advantages are often offered to communities located near a mining operation. For example, the diamond mines employ 975 Aboriginal people, almost one-quarter of their workforce (Mining Association of Canada 2008b). But those advantages can be a mixed blessing as elders and other community members struggle to maintain the community's traditions, cultural vitality, and health in a rapidly changing environment. As Susan Wismer (1996, 10) notes when discussing the impact of diamond development on northern communities living near Great Slave Lake:

> The importance of wildlife and of the land which supports it is clear and well-known. Centuries of life on the land have created a rich store of traditional knowledge about its care and well-being. Unfortunately, traditional knowledge has very few stories to tell about the desirability of diamonds and diamond mining. Although mining is a very old activity in human terms, it has not been a traditional activity during the 9,000 known years of Aboriginal habitation in Canada's Northwest Territories. Mineral development has been going on in Canada's North for about 100 years.

Nevertheless, exploration and development will continue as northern communities consider how best to defend their traditional ways while benefiting from the influx of capital from mining activities. The challenge is amplified because communities become dependent on a wage-based economy and mine-related employment, only to see the wealth eventually disappear from the local economy once a mine permanently shuts down. Social and environmental impact monitoring, impact benefit agreements, and other initiatives (discussed below) attempt to bring a measure of social justice, environmental mitigation, and economic development to Aboriginal peoples affected by exploration and mining.

Biophysical Costs

Mining, at all its stages of production, will generate various changes in affected ecosystems, depending on the mineral being developed, the location of the mine, the mining methods used, and possible cumulative effects that occur in conjunction with other nearby natural or industrial activities. In addition to its immediate impacts, mining can also have a 'shadow effect'. This means that the effects of mining can spread beyond the disturbance caused in the immediate region, such as with the development of roads that disrupt wildlife and migration patterns. Air, soil, water quality, and food webs can all be negatively affected by the mining process itself and from the contaminants associated with the industry, such as metals radionuclides and cyanide (Ripley, Redmann, and Crowder 1996, 110).

Acid drainage is generally considered the biggest physical environmental challenge faced by the industry. Even without mining, an ore body can naturally contaminate soils and water, leading to acidified lakes. When mining occurs, the waste piles or tailings, which can contain sulphide minerals, become exposed to air and water, leading to a chemical reaction referred to as acid generation, leaching metals into the environment. Once ore has been milled (the process by which minerals are extracted from the ore), substances containing trace metals and chemicals are discharged from the mill. To prevent possible environmental contamination, these substances must be contained and are often deposited in a tailings pond. Other times, tailings have been disposed on the ocean floor. Unfortunately, in various parts of the world, tailings dam failures have occurred or, at times, tailings have been discharged directly into water systems (Veiga, Scoble, and McAllister 2001, 193).

A generally accepted principle is that company practices in other countries should be held to the same standards as if the mine were operating in Canada. Large corporations are closely watched by an effective global network of environmental non-government organizations. Poorly run operations, however, are difficult to shut down because of the associated economic benefits, employment, and infrastructural assistance. Poverty-stricken communities and governments often see few alternatives and will accept the environmental costs in order to provide local people with a higher economic standard of living. Well-established mining companies, however, have been rethinking their policy of riverine disposal because of the international pressures they have faced as a result of this practice.

Canada, although a wealthy country, also has rural and remote regions in need of employment and economic opportunities. While environmental practices have vastly improved over the past century, serious concerns remain. One of the highest-profile issues revolves around looming environmental disasters as a result of abandoned mines—most of them in rural and remote areas—leaving massive environmental liabilities associated with acid rock drainage. While 10,000 identified mines in Canada have been orphaned (abandoned), there is no national inventory of orphaned mines or forgotten mine sites, leaving an unknown legacy (NRCAN 2001c).

When Royal Oak Mines went into receivership in 1999, it left behind 237,000 tonnes of arsenic trioxide dust located in the mine shaft and chambers of the Giant Mine in Yellowknife (SRK Consulting and SENES Consulting Ltd 2007). Thus, the Canadian public through the federal government was left responsible for the cleanup. Indian and Northern Affairs Canada (INAC), in consultation with an independent, peer-reviewed technical committee, proposed a remediation plan. The department determined that it would secure the dust in the current storage chambers through the 'frozen block method'. The department will freeze (and maintain in perpetuity) the chambers to restrict arsenic from leaching into the groundwater (and ultimately Great Slave Lake). Local residents, including the Yellowknives Dene, and the city of Yellowknife have expressed concerns about several aspects of the frozen block plan, including the fact that the contained site will have to be maintained in perpetuity, the potential danger posed by a flow of vehicular traffic over the current claims block, and how the site may be used in the future. At the time of writing, the proposed remediation plan for the site was undergoing an environmental assessment under the Mackenzie Valley Resource Management Act, 1998.

Modern, active mines have extensive closure and reclamation plans that did not exist when many of the now-abandoned mines were first opened. When the owner cannot be identified, however, liability reverts to the provincial or federal government (if it is in federal jurisdiction)—that is, the responsibility becomes that of all Canadians. Nevertheless, many firsthand observers of the Giant Mine are concerned that disasters can still happen in the future even with modern mines and regulatory structures (MiningWatch Canada 1999).

The environmental and human costs have only more recently become matters of urgent public concern. Resource decision-makers and managers are now confronted with the complex task of attempting to assess the relative benefits of development against the often uncertain environmental or human health implications.

CONFLICT AND UNCERTAINTY IN CANADIAN MINING: COMPETING PERSPECTIVES

Changing Policy Communities

The past century has seen the intensification of conflict and uncertainty in mineral resource management and decision-making. Concerns over the manner in which natural resources have been exploited have multiplied along with the number of active political players with diverse values and goals. A group of actors with a shared interest in a particular policy area such as mineral development might be referred to as a 'policy community', defined as follows:

A policy community is that part of a political system that—by virtue of its functional responsibilities, its vested interests, and its specialized knowledge—acquires a dominant voice in determining government decisions in a specific field of public activity, and is generally permitted by society at large and the public authorities in particular to determine public policy in their field [Pross 1986, 98].

In the early twentieth century, the mineral policy community consisted predominantly of governments wishing to promote resource development in order to raise revenues and the companies directly involved in the mining sector or benefiting from mineral development. Over time, unions became important players in setting employment, health, and safety standards. In recent decades, growing public concern about the adverse impacts of mining contributed to an uncertain policy environment as governments attempted to find a balance between often irreconcilable goals. Public interest groups proliferated, most notably environmental non-government organizations, demanding an opportunity to influence public decision-making. In addition, as Greg Poelzer (2002, 88–9) has suggested, Aboriginal peoples are no longer at the margins of environmental decision-making but are now one of the five major actors (see Chapter 4). For their part, mining-affected communities have achieved standing as an important actor in resource decision-making. Increasingly, communities are included in decision-making processes as more independent entities. This contrasts with the distant past when company-owned towns were directed and governed by the mine owners and managers. All these actors have competing agendas and different perspectives about how mineral resources might best be exploited, conserved, or protected.

The Mineral Industry

Mining for the Benefit of All Canadians
[cover of the 1998–9 Mining Association of Canada annual report]

Members of the industry and those involved in related enterprises are primary actors in the mineral policy community. Members of the industry are proud of their contribution to Canada and of the mining traditions that often extend back several family generations. It has, however, had its share of challenges. Always subject to the volatility of world prices, the industry now has to cope with a rapidly shifting global political and economic environment. These challenges can be roughly subdivided into two categories: those intrinsic to the industry and those associated with emerging social and political trends.

Endemic or intrinsic challenges include:

- the cyclical market environment;
- the economic implications of commodity production;
- the location and manner in which mining takes place, leading to conflict with other stakeholders such as Aboriginal peoples or advocates for the environment;
- the constitutional division of responsibilities for resources, which contributes to an uncertain regulatory climate; and
- the nature and composition of the industry itself.

The mineral industry and other export-oriented primary enterprises are subject to the vagaries of the international markets and the fluctuating prices offered for commodities. This was particularly the situation when the global economic turbulence that started in late 2008 made the tricky business of commodity price forecasting virtually impossible. Mining must be cost-competitive through efficient production of its commodities. Global markets are unpredictable, and the industry goes through periods of 'boom and bust' when prices and demand for mineral commodities surge and fall. Efforts to keep costs down are often hindered by what mining representatives see as an uncertain regulatory environment, worsened by overlapping federal–provincial jurisdiction over primary resources.

Moreover, mineral development often takes place in areas where it can come into conflict with the interests and demands of other resource users such as Aboriginal communities or advocates for the environment. This is the case with the industry's requirement for access to large tracts of land for exploration purposes. Land is increasingly subject to multiple claims, including the creation of parks and protected areas, making it off-limits to exploration.

Finally, the industry itself has difficulty in operating in a cohesive manner, although it has seen significant improvements in this regard during the past couple of decades. Part of the challenge is the number of diverse industry associations that reflect the varied activities employed in a sector that differs widely in the type of minerals produced and processed and the scale on which the activity takes place. The huge Fort McMurray oil sands development in Alberta is a very different proposition from a small sand and gravel pit. Mineral production is a very different type of enterprise from exploration or metal production. Organizations include prospectors and developers associations, mining associations, coal associations, base metal groups, equipment suppliers, and professional technical societies. The composition of associations also reflects the jurisdictional and political divisions of the country. Associations may be national, provincial, territorial, or regional in scope.

It is difficult to relay a consistent voice to the public when the industry itself is so varied and some of its maverick individuals are not easily brought onside to deliver a consistent message. The national Mining Association of Canada (MAC) represents the most unified voice of the industry, with members spanning many aspects of exploration, mining, smelting, refining, and semi-fabrication, accounting for the majority of Canada's output of metals and major industrial materials. Its mandate is to promote the development of the mineral industry. It has worked assiduously to develop a more cohesive industry, its first major step in that direction being the 'Keep Mining in Canada' campaign in the 1990s aimed at helping to raise the profile of the industry with decision-makers. Despite codes of environmental ethics and socially responsible behaviour required of members of MAC, it is not always easy to develop a consensus about policy directions among its varied membership. For their part, individual companies have difficulty changing their own internal corporate culture when the sustainability policies of the senior members of the company do not, in practice, automatically filter down effectively to the mine-site operations.

Members of the industry also have disparate ideas about the nature of their broader societal responsibility. Some point out that their primary fiduciary or legal responsibility is to maximize profits for their shareholders and that the industry should only be responsible for running an efficient, law-abiding operation and to pay the required fees and taxes. Others recognize that additional voluntary initiatives are important to maintaining an image of good corporate citizenship and avoiding unwanted and perhaps ineffective blanket regulations imposed by a 'command and control' governing regime.

The Canadian mineral industry has actively engaged in voluntary initiatives. These measures have ranged from voluntary pollution control by individual companies and reduction of greenhouse gas emissions, participation in government–industry initiatives, and adoption of and adherence to association voluntary codes of best available environmental practices. The Mining Association of Canada instituted the first multi-stakeholder initiative of its kind in the world, the Whitehorse Mining Initiative (WMI). The final accord of the WMI stated its commitment to foster a 'socially, economically and environmentally sustainable, and prosperous mining industry, underpinned by political and community consensus' (Whitehorse Mining Initiative 1994, 5). More recently, the Mining Association of Canada (2006) unveiled its corporate responsibility plan addressing social, economic, and environmental performance (Towards Sustainable Mining, discussed below). This approach is similar to that taken by the International Council on Minerals and Metals.

Despite such efforts, the industry has historically struggled to reach out effectively to other stakeholders and to cultivate contacts over a period of time. The very people it was hoping to engage in a productive dialogue—the environmental community—viewed with great suspicion broad efforts such the Global Mining Initiative (discussed below). It takes a long time to build trust, particularly for an industry trying to live down errors of the past.

The industry has tended to engage in short-term, high-profile public relations approaches rather than to build long-term societal support. Industry has some distance to travel before it can begin to manage under the holistic and integrated template more characteristic of an integrated resource and environmental management paradigm (Clausen and McAllister 2001b). Recent efforts have addressed this difficulty through mining associations at the domestic level and internationally through the International Council on Minerals and Metals. Nevertheless, as discussed below, the challenges of integrating these values are enormous when one

considers that it is being attempted at a global level by a highly diverse industry often viewed as non-renewable and a heavy user of environmental goods and services. The socio-ecological impacts of one corporation alone are tremendously diverse, depending on the context in which its operations take place. This challenge is magnified by the many stakeholders and companies involved.

All these developments are taking place while the industry itself has been rapidly reshaping, with the emergence of new global players and many mergers and acquisitions (M&As) creating a climate of uncertainty for all involved. To cite just two examples, in late 2006 the Canadian nickel giant Inco was acquired by the Brazilian Companhia Vale do Rio Doce (CVRD). Falconbridge Inc. was acquired by the Swiss company Xstrada. Globalization was cited as one of the reasons for these M&As, the argument being that it has become economically necessary for operations to take place at a scale sufficiently large and diversified to respond to customers also operating on a global scale (Humphries 2006). Other rationales included cost reduction, productivity improvement and capital efficiency, and rapid industry growth (Humphries 2006). The boom-and-bust nature of the mineral industry was cast in sharp relief when the global economic crisis came to a head in the fall of 2008, following on the heels of a worldwide surge in demand for mineral commodities.

Economic considerations aside, the global mineral industry as a whole faces a growing public requirement to secure the approval or co-operation of civil society. The mining sector is dealing with its historical legacy in terms of the cumulative environmental problems caused by earlier mining activities. These incidents, highlighted by the media, concerned citizen groups, and non-governmental organizations, make it difficult for the industry to maintain a positive profile.

Mining also must compete with many other interest groups for attention on government and public agendas, reflecting the issue-attention cycle described in the introductory chapter. It has had to work hard to overcome a dated 'pick and shovel' image of a sunset industry and to emphasize that it is a high-tech, safe industry with many spin-off benefits and is therefore a good investment.

With the development of the world wide web, companies have become aware that the influence of critical non-government organizations does not stop at national borders and that the industry will be held accountable for its activities throughout the globe. During the past decade, international agencies and conferences have been exploring ways to ensure that international mining companies behave responsibly wherever they have operations. For example, between 1998 and 2006, the International Development Research Centre (IDRC) in Ottawa set up a new program called the Mining Policy Research Initiative (MPRI) to support community-based research in Latin America, where Canadian mining operations take place, to help communities benefit rather than suffer from the effects of mining operations. Other Canadian organizations such as the North-South Institute have similar agendas.

As a response to these challenges, international mining associations have begun directing their attention to global strategies of change. The most recent and ambitious of these was a two-year, industry-driven initiative entitled the Mining, Minerals, and Sustainable Development Project (MMSD), which included regional and national partners throughout the world. In advance of the 2002 World Summit on Sustainable Development in Johannesburg (commonly referred to as Rio +10), the MMSD was set up, proclaiming itself to be an initiative that would allow the industry to move forward. The

industry, faced with criticism from non-governmental groups throughout the world, launched the MMSD initiative in order to build bridges. It also could be viewed as a way of reassuring bankers that mining was still a good investment in the face of civil disturbances related to mine development.

MMSD documents state that it was an 'independent process of participatory analysis aimed at identifying how mining and minerals can best contribute to the global transition to sustainable development' (Mining, Minerals, and Sustainable Development Project 2002, 8). The International Institute for Environment and Development managed the project under contract to the World Business Council for Sustainable Development. MMSD was part of a broader process called the Global Mining Initiative. Many environmental and community-based organizations around the world reacted to the introduction of the short-term initiative with suspicion and dissatisfaction with its goals. Nevertheless, the MMSD forged ahead and released its report, *Breaking New Ground*, in May 2002. The goal of the initiative was to 'unearth the most controversial problems of mineral development and its impacts on poverty, human well-being, the environment, and other factors key to sustainable development' (Environmental News Services 2002, para. 5). On the basis of the research, the report called for the following actions:

- an industry protocol for sustainable development;
- a commitment to address the negative legacy of the past;
- supporting legislation for artisanal and small-scale mining;
- integrated management of the full mineral chain;
- more effective government management of mineral investment; and

- a more equitable international trade regime for minerals (International Institute for Environment and Development and World Business Council for Sustainable Development 2002).

In order to implement the proposed actions, the International Council on Mining and Metals (ICMM) was formed in 2001 to foster the sustainable development of mining. It represented the world's leading minerals and metals companies, adopting 10 guiding principles in 2003 to inform the organization's sustainable development framework. In May 2008, ICMM adopted an assurance procedure, with the stated goal of providing independent assurance that those guiding principles and commitments would be met by each company (ICMM 2008).

Attempts at sustainability frameworks in a global industry such as mining are plagued by a host of problems ranging from definitional issues of sustainability and the desired ecosystems to be protected (and their associated spatial and temporal boundaries), to questions of enforcement and transparency, to consistency of meaning and practice between corporate head office and the mine site. It is also unknown whether these initiatives will serve as a useful long-term communication bridge to other members of the policy community. Partnerships between groups with vastly diverging value systems normally take a long time to establish. The political and cultural dynamics of the main players make it difficult to find common ground. At a minimum, however, a sustainability strategy will have to be sold to an industry that itself is internally divided about the best way to proceed. Only when the mineral sector can agree on a reasonably cohesive vision and translate the words into practice will it be in a position to begin persuading other stakeholders that they can effectively work together toward sustainability.

Governments: Competing Agendas, Conflicting Mandates

If it can't be grown, it has to be mined.
from Natural Resources Canada,
Minerals and Metals Sector
website, 2002, 'Mining facts'.

Federal, provincial, municipal, and territorial governments have always been involved in the active promotion of mineral development. The provincial governments own the resources and gain valuable revenues from their exploitation, and municipalities constitutionally fall under the jurisdiction of the provinces. As discussed in the introductory chapter and Chapter 2, however, the federal government has many responsibilities that overlap with those of the provinces. At both the federal and provincial levels, specific departments have been given the mandate to regulate and promote mineral resource development, even though other departments or ministries also have responsibilities related to mining. Further, the requirement to both promote and regulate development has led to an often conflicting mandate within a single department.

At the federal level, for example, Natural Resources Canada is divided into several sectors, one of which is Minerals and Metals (MMS). This sector, although renamed a number of times, has historically been viewed as responsible for maximizing the economic efficiency of mineral development. Environmental protection was the responsibility of other departments, such as Environment Canada. This 'silo' approach, while helpful in keeping lines of accountability clear, did little to ensure that biophysical health and social considerations were adequately integrated into decision-making processes. In recent years, governments have responded to this problem by developing policy guidelines that reflect the principle of 'sustainable development'.

However, as illustrated by two gold–copper mine projects in British Columbia, the federal government's commitment to sustainable development can be constrained. The Kemess North copper–gold project involved the expansion of an existing mine north of Prince Rupert by Northgate Mineral Corporation (Joint Panel Review 2007). The proposed development included the construction of a new open-pit mine and modification of an existing mill and related infrastructure such that the life of the mine would expand by 13 years. The joint federal–provincial panel convened to review the project concluded in its 2007 report that the project would not have a net contribution to sustainable development. Of particular importance to the panel was the destruction of Duncan (Amazay) Lake, which has local traditional use and spiritual significance for the Tse Keh Nay and the Gitxsan First Nations. Furthermore, the panel expressed significant concern about the need for long-term monitoring and mitigation (thousands of years) post-closure. Thus, the panel recommended that the project not proceed, a decision accepted by the federal government.

A second copper–gold mine in British Columbia had different implications for sustainability. The Red Chris development, 80 kilometres south of Dease Lake, triggered environmental assessments (EA) by both the provincial and federal governments. Although the federal authorities originally determined that the project would be subject to a comprehensive study, based on the volume of ore extraction projected, they revised the scope to focus only on those aspects directly under their jurisdiction (e.g., fish and fish habitat and explosives). As result, they conducted a less rigorous environmental screening. The project was subsequently approved in May 2006.

MiningWatch Canada successfully appealed the EA decision on the grounds that there had been a lack of public consultation surrounding the revised assessment track. In reaching its decision, the lower court found that the responsible federal authority had 'side-stepped its statutory requirements . . . in a guise of a decision to re-scope the Project, the . . . [federal departments, specifically the Department of Fisheries and Oceans and Natural Resources Canada] acted beyond the ambit of their statutory power' (2008 FCA 209).

The proponent, as well as the federal and provincial governments, appealed this decision to the federal courts, which this time found in favour of the federal government, affirming the government's right to utilize discretion when scoping a project. The outcome of this court challenge will have significant implications for the mineral industry and indeed for all large developments in Canada. In this case, the position of the federal government is not consistent with the concept of one project, one assessment, which recognizes that without the mine, the stream crossings and explosive permits *would not be necessary*. If the federal government continues to assess only those components for which it has regulatory authority, the principle of taking a holistic approach to project planning necessary for sustainable development will be lost. In late 2008, MiningWatch got leave to appeal this decision to the Supreme Court of Canada.

The ability of the federal government to implement policies that allow mining companies to continue to produce and contribute revenues and jobs to the economy, while at the same time formulating ecologically sound regulatory policies, is a challenge. Moreover, while it is relatively straightforward to develop tools for economic productivity, such as those used to calculate ore yields and grade and processing efficiency, it is another thing to develop widely supported standards and assessment methods appropriate for long-term sustainability (Clausen and McAllister 2001a). Institutional cultures also take a long time to change.

The Minerals and Metals Sector of Natural Resources Canada is not structurally well-equipped to regulate the industry when it is actively engaged in promoting economic exploitation and use. It is a difficult balancing act to achieve. The mission of the MMS is to:

advise the Government on—and to advance its agenda for—the economic, social, environmental, scientific and technological spheres through the development and use of minerals and metals. An important role for MMS is to generate and share knowledge as a basis for sound decision-making that affects the sustainable development of minerals and metals for the benefit of Canadians [NRCAN 2001a, 9].

The department has adopted the conventional definition of sustainable development: 'Development that meets the needs of the present without compromising the ability of future generations' (NRCAN 2001a, 5). MMS has drawn on the results of a number of heavily promoted multi-stakeholder consultative initiatives aimed at seeking a sustainable path forward. These efforts include the Whitehorse Mining Initiative (1994), the Minerals and Metals Policy (1996), the development of criteria and indicators for sustainable development in the mineral sector (1998–2000), and a strategic vision for the sector for the years 2001–6 that would help Canada become 'a role model for the world in applying sustainable development through good stewardship of its minerals and metals resources to enhance the quality of life of Canadians and their communities' (NRCAN 2001a, 9).

In the first decade of the twenty-first century, the federal government increasingly turned its attention to corporate social responsibility (CSR), an approach that places more of an onus on industry to take voluntary measures to improve its performance with respect to the social and environmental costs and benefits of its activities. The Canadian mineral industry includes many companies that work internationally and are not subject to the same principles and practices in their offshore operations that are required in Canada. Negative coverage regarding various environmental and socially unacceptable practices internationally resulted in a proposed strategy to regulate activities of Canadian mining companies operating abroad (Campbell 2008). In consultation with industry, the National Round Table on the Environment and Economy proposed the International Finance Corporation's Performance Standards and Voluntary Principles on Security and Human Rights as a preliminary framework for evaluating CSR. However, as noted by Campbell (2008, 393), there has been no official response to the proposal, although there are signs that industry associations have been lobbying to ensure that it is not adopted.

Various provincial government mining departments have initiated or participated in multi-stakeholder initiatives. The degree to which they adopt 'sustainability' mandates, rather than the traditional pro-development approach, to policy and regulation varies throughout the country and is influenced by the ideological orientation of the governing party.

Aboriginal Peoples

Mineral exploration and development has occurred in many parts of Canada on lands where Aboriginal peoples have maintained a use, affinity for the land, and occupation for thousands of years. Aboriginal communities have depended on the land for a warehouse of natural resources for survival. In the past, mineral activity has disrupted the traditional lifestyle and has left few traditional economic opportunities for Aboriginal peoples.

[*Whitehorse Mining Initiative 1994, 35*].

Aboriginal peoples throughout Canada have interacted with the environment and harvested and extracted resources since long before contact with Europeans. Knowledge (commonly referred to as traditional ecological knowledge) was passed down through generations. Much of this knowledge emphasized a holistic approach, which valued humans as part of (not distinct from) the broader environment—a perspective that fits well with an integrated approach to resource management. With the colonization of North America and the imposition of European-based values, indigenous peoples were usually marginalized in resource decision-making until quite recently.

In the past few decades, Aboriginal peoples have achieved modern treaty settlements in various parts of the country, gained international recognition of their human and indigenous rights, received several favourable and important Supreme Court decisions reaffirming their resource and land rights, and have been included in various types of co-management agreements, compensation packages, and consultative initiatives (see also Chapter 4). As a result, Aboriginal peoples are increasingly empowered and in a much stronger position to assert their own values and needs with respect to resource decision-making.

Nevertheless, complications abound. Many Aboriginal peoples and communities have different views and ideas about how to blend traditional and contemporary values concerning how natural resources should be used. Moreover, their alliances with other interest groups, such as industry, government, or environmental groups, often vary from group to group, sometimes pursuing environmental agendas while at other times advancing their productive interests (as resource developers). Greg Poelzer (2002, 104) has noted that

[t]he role that First Nations play in shaping environmental policy is diverse. In some instances, such as James Bay, First Nations pursue non-productive interests. In other instances, such as Clayoquot Sound and the Stoney Reserve, productive interests are at the fore. Moreover, as the cases demonstrate, First Nations and environmentalists are not inherently natural allies. In each of these cases, however, First Nation state actors proved central to determining policy outcomes.

In the mineral sector, Aboriginal people have had many historical reasons to develop a distrust of the industry. Industrial activities frequently brought serious disruption to isolated communities not equipped to deal with large-scale extractive enterprises. Those who wished to participate in the wage economy offered by the industry often lacked the skills, education, and assistance needed to be able to hold anything but the most menial of jobs. Moreover, mining activities usually have an impact on the traditional way of life.

Some attention is now being given to social and biophysical monitoring in order to develop indicators to help assess the impact of mining on Aboriginal communities. Brenda Parlee, in her work undertaken with the Lutsel K'e Dene First Nation on Great Slave Lake, adopted the Chipewyan terminology for monitoring, which translated as '[w]atching, listening, learning and understanding about changes in the community' (Parlee 1998, 11).

Assuming that all these changes have been effectively mapped (which is not often the case, because the changes can take place over many years and be manifested in complex ways), attempts are now being made to compensate Aboriginal communities. Contemporary impact and benefit agreements (IBAs) between mining companies and Aboriginal peoples are private contractual agreements devised to provide Aboriginal peoples (and the mining companies) with some measure of security (see discussion by Galbraith, Bradshaw, and Rutherford 2007; Fidler and Hitch 2007). As the title implies, the ostensible goal of these negotiated agreements is to offer some benefits to Aboriginal peoples. The possibilities are wide-ranging and could include a mix of job training and employment, monetary compensation or revenue-sharing, joint venture agreements, health and education facilities, and so on. At the same time, these agreements are also aimed at minimizing the adverse impacts of mineral development. Fidler and Hitch (2007, 5) suggest that IBAs can also serve to address broader policy issues, including 'corporate social responsibility, EA deficiencies, and Crown consultation deficiencies'.

Unfortunately, despite increasing uptake by the mineral sector, it is difficult to evaluate the success of a specific IBA. Most IBAs include 'confidentiality provisions that prevent parties to the Agreement from divulging their contents' (O'Faircheallaigh 2005, 631). This poses a methodological quandary for evaluating the success of any one agreement in achieving its objectives. Research thus relies on a generalized list of topics potentially covered by IBAs. Initial studies suggest that Aboriginal communities that have entered into IBAs have improved socio-economic conditions, particularly when contrasted against Aboriginal communities across Canada (e.g., Prono 2007). Nonetheless, a number of issues remain. For example, lack of specific knowledge about IBA content impedes the ability of Aboriginal groups entering into negotiations to learn from past experience (Fidler and Hitch 2007). Second, although IBAs have increasing uptake, they are only required by law in a few jurisdictions, primarily those

in the Canadian North with settled modern land claims (Klein, Donihee, and Stewart 2004). Thus, it is premature to consider IBAs as part of the development framework across Canada. Finally, given the lack of public knowledge or independent oversight surrounding this tool, it is troubling to think that it may be used to address public policy issues, such as EA deficiencies and inadequate Crown consultation.

As is normal with any clustering of a diverse group of peoples and interests, there is usually no consensus among the different Aboriginal communities about the best way to proceed when confronted with a new resource development in their midst. Communities themselves are torn about the best way to achieve a balance among a multiplicity of traditional and modern values as well as ecological, social, and economic concerns. These debates take place at a time when rapid changes can happen on a daily basis when a mine is being developed: tides of strangers flow and in and out of the area, construction equipment begins to reshape the physical environment, and the community itself experiences internal turmoil. Such situations do not allow the community much room for careful reflection or consideration of all the possible implications of choices that could have far-reaching consequences.

Although faced with enormous challenges, many indigenous peoples throughout the world are learning how to develop cooperative strategies in which their interests intersect regionally, nationally, and internationally. In Canada, one of the largest associations, the Assembly of First Nations (AFN), speaks to the broader issues concerning Aboriginal peoples.

Other groups, such as the Canadian Aboriginal Minerals Association (CASMA), have a distinctly different focus. CAMA helps to establish contacts and affiliations between industry and Aboriginal peoples. The organization states that

[CAMA] was formed out of a need expressed by Aboriginal communities. Their priorities are the environment, employment and training, and economic development. Establishing relations with mineral companies to explore and develop mineral resources is seen as a way to achieve economic self-sufficiency [Canadian Aboriginal Minerals Association 2002, n.p.].

Given the diverse numbers of Aboriginal peoples and cultures in Canada, it is not surprising that different groups develop distinctive allegiances and associations. What most (if not all) of these groups do have in common, however, are fundamental beliefs about the following principles: that indigenous peoples should have a right to speak for themselves; that decisions affecting them should recognize their long-overlooked rights and concerns; and that age-old traditions and knowledge about resources and the environment should be acknowledged. Yet, the number and diversity of grassroots organizations, as with the environmental organizations, make it difficult for industry, government, and others to find commonality with the often shifting, and elusive, coalitions of interest.

Environmental Non-government Organizations

The very real legacy of mining includes an estimated twenty-seven thousand abandoned mines across Canada, billions of dollars of remediation liability for acid mine drainage contamination, extensive disruption of critical habitat areas, profound social impacts in many mining communities, and the boom and bust upheaval of local economies. The cost of Canadian mining operations in other parts of the world has been no less dramatic.

[*MiningWatch Canada 2002*].

Environmental non-governmental organizations (ENGOs) have developed effective networks that often coalesce around certain issues. These organizations represent a powerful policy community. Larger ENGOs have annual budgets in excess of $1 million. While this figure is small compared to what industry may be able to spend on public relations campaigns, it does allow environmental organizations to become established institutional players with their own staffed offices and well-developed lobbying techniques. These groups have also made most effective use of e-mail lists and websites. As Jeremy Wilson (2002, 62) has noted,

The movement continues to do a better job of mobilizing and capitalizing on the political energy and talents of women and young people than do most other political organizations. It has skilfully exploited the potential for alliances with soulmates in other parts of the world and other parts of the Canadian political landscape. It has responded consistently and sure-footedly to both new technology and the changes in political constraints and opportunities associated with globalization.

The agenda-setting influence of ENGOs on the mineral sector was first made powerfully clear when they went head-to-head with the Canadian mineral industry in the early 1990s over a large mineral deposit in Windy Craggy mountain at the confluence of the Tatshenshini and Alsek Rivers in northwestern British Columbia. The deposit was said to contain a huge copper deposit (with by-product cobalt, gold, and silver). This was an exciting discovery for an industry that was looking to renew its depleted reserves. Environmental groups were very worried about the potential impacts of the mine on wildlife, particularly on the Chikat Bald Eagle

Reserve and the grizzly bear population. They were not persuaded that the project's proponent would be able to mitigate adverse environmental effects. The issue had international dimensions, since the site was near an American national park in Alaska. An extensive North American environmental network, including the Western Canada Wilderness Committee, the Sierra Club, the US National Parks Conservation Association, and the World Wildlife Fund, lobbied extensively (McAllister 1992, 14–15). As a result of many factors, the area became the Tatshenshini Alsek Wilderness Park in 1993 and subsequently was designated a World Heritage Site by the United Nations Educational, Scientific and Cultural Organization (UNESCO)—an important achievement for many environmental groups but a bitter blow to the industry.

Many environmental organizations are actively involved in mining issues. Some are local or grassroots 'issue-specific' groups concerned about a particular issue, such as industrial contamination of a particular site or the location of a sand and gravel pit. Others are 'sector-specific', such as the BC Environmental Mining Coalition and MiningWatch Canada. Still others are broad-based environmental groups, such as Friends of the Environment or the World Wildlife Fund, which include mining development as one of a number of causes.

In this era of globalization, environmental groups often network internationally. Mining-Watch Canada, for example, signed the London Declaration in May 2001, proclaiming that the MMSD and the broader Global Mining Initiative (GMI) was '[t]he latest in the series of corporate-led propaganda offensives' (Mines and Communities 2001). Another international coalition of interests referred to the GMI as a 'global corporate greenwash' (CORPWatch 2002). While many of the mainstream groups state that they are not opposed to mining as such, they often

do not accept the way in which the mineral industry pursues its agenda as part of a global, market-driven sector where other considerations take second place. Concerns include the impact of exploration and development in environmentally sensitive or protected areas, biophysical degradation, inadequate remediation measures, community dislocation, and limited local control.

Mining Communities and Labour

Toronto? That's just the place you go to get the train to Cobalt

[popular saying during the Cobalt (silver) boom; Angus and Griffin 1996, 17].

The agendas of labour unions and local communities are not identical, but the kinds of issues that preoccupy them in a mining town are often similar. Since the workers often comprise a significant portion of the citizenry, many of a resource town's issues are also those of the union. They include workplace safety, employment opportunities, skills training, community development, and mine closure.

Rural and remote communities have their own sets of dilemmas and needs related to resource issues. If they have come to depend on mining for employment in their region, they are often anxious to maintain the industry even under adverse conditions. Since their fortunes are often tied to international markets, resource-based towns experience the worst of the boom-and-bust economy, fuelled by high salaries and community growth when world demand for the resource is high and lay-offs or even closure when demand falls or the ore body becomes depleted.

Given the level of uncertainty, communities and workers now expect to be consulted on environmental mitigation, local economic development, and in particular, mine closure and rehabilitation issues. One example of a consultation process that began far in advance of closure occurred in Kimberly in southeastern British Columbia, a mining community established more than 100 years ago. TeckCominco's Sullivan mine in British Columbia was closed in 2001. A number of employees and local residents of Kimberly were fourth-generation mine employees. In November 1990, the company announced that the mine would close in 10 years In addition to a provincially approved mine closure and reclamation plan, a public liaison committee was formed to review the plan under the guidance of the provincial government. Participants included industry, government officials, environmental groups, and members of the community. Discussions related to employment and economic diversification as well as long-term environmental impacts. While one review of the process found local attendance was low, members of the community were able to ensure that their views were expressed at the meetings of the committee over a 10-year period (Anderson 2002).

Communities are becoming more assertive and proactive when dealing with mining companies and are now negotiating for their long-term futures, as can be seen in the proactive example offered by some northern Manitoba mining communities. In 2002, the northern mining communities (Flin Flon, Lynn Lake, Leaf Rapids, Thompson, and Snow Lake) initiated a collective strategy to diversify by developing a regional economic development plan. Their efforts included cooperative attempts to attract new investment in mineral development and other enterprises and to establish themselves as a regional tourist destination based on their history as mining communities and as good locations for recreational sports (Team Manitoba 2000). Members of the provincial mineral policy community, including First Nations, the Métis

Nation, the mineral industry, the provincial government, and northern community councils also worked together to sign a long-term sustainable strategy to improve educational and business opportunities, develop joint ventures, establish cross-cultural training, and maintain diverse ecosystems (Manitoba 2000).

The Mineral Policy Community

The last part of the twentieth century experienced the rise of diverse communities of interest, all claiming a stake in the future direction of mining. Despite differences among and within them, all the actors in the mineral policy sector seem to agree on a few main points.

- The industry and the issues that surround it have become globalized to an extent that broader strategies for change in the mineral sector cannot be readily confined within national borders.
- Biophysical and social issues must be adequately considered in all mining-related activities.
- Consultation with affected parties (however defined) is an essential part of any long-term strategy toward a sustainable future.
- Communities affected by mining are now seen as important stakeholders whose needs and interests must be considered through inclusive decision-making processes.

CONCLUSION: TAKING THE INITIATIVE AND FINDING THE BALANCE

Resource management is a term that implies, to some, that the environment can be effectively controlled. Dissatisfaction with that term has led to the suggestion that it would be more appropriate to attempt to manage human interactions with the environment. Yet this too is fraught with difficulties. The allocation, use, or protection of a resource is based on decisions informed by a particular set of ethics or values. It has to do with the views of decision-makers and broader societal actors about how best to advance certain goals as interpreted by a framework of beliefs that they use to make sense of their environment.

The history of mineral development (and resource development in general) has been heavily imbued with the values of individualism, economic development, science, and technology. In Canada, those goals have been tempered by the recognition of collective rights and Crown ownership. The general public interest and the goal of nation-building have been served in many ways through intensive resource development. In the twenty-first century, other values emphasizing social or biophysical sustainability are now becoming politically salient. These influences have led decision-makers to reassess traditional approaches to mining and governance. Multi-stakeholder consultations or attempts to find common ground are as much discussions about values and ethics as they are about land use, economic development, or co-management agreements.

The Canadian mineral industry has been forced to adapt to many changes over the past century. It has had to become technologically innovative and highly efficient in order to respond to international competitive challenges. Its exploration industry has branched out into all corners of the world. When confronted with growing environmental regulations, the industry developed new technologies and standards to mitigate the adverse impacts of mining. Well-suited to those kinds of challenges, the industry has retained a leading position as a world producer of minerals.

The industry is now working its way through less familiar territory. It is learning how to negotiate in political and social arenas in order to develop partnerships internationally, regionally, and locally. In the third edition of this book, published in 2004, McAllister observed that many of its members had yet to accept that not all biophysical and social issues were resolvable with technological fixes or financial compensation. Rather, she suggested that an integrated approach was needed, one designed by a coalition of interests and incorporating concerns for social equity and ecological diversity. Government mining departments, like the industry, have also had to change the way in which they interpreted and pursued their mandates, incorporating a broader spectrum of interests into decision-making processes and engaging the public.

In the intervening five years, a different trend has emerged. On one hand, the industry is developing and contributing to policies designed to ensure that projects respect environmental standards. Furthermore, with interest in CSR guidelines for industrial activities abroad, there is some promise that the industry is making efforts to improve its environmental approach. On the other hand, remnants of past activities, and recent federal decisions and actions related to new and revised mineral developments projects, highlight the challenge of effectively regulating industry activities in Canada, at least from a federal perspective.

For their part, many non-government organizations, Aboriginal communities, and mining communities have yet to be persuaded that the global, capitalist mineral sector is capable of developing an ecologically and socially sustainable path. They continue to monitor its actions closely and press for accountability. The mineral industry may indeed be a sustainable enterprise, at least in the immediate, economic sense of the term. What remains to be seen is whether the conflict and uncertainty that surround mineral activities today will direct the sector toward a more ecologically sensitive and sustainable future.

FROM THE FIELD

A generation ago, research in the Canadian mineral policy sector focused heavily on public institutions and their relationship to the mining companies. Such investigations would often involve wading through dusty library shelves loaded with boxes bulging with legislative, budgetary, regulatory, and taxation papers as well as weighty tomes containing discussions about resource economics, federal–provincial relations, and annual reviews of mineral exploration, development, and production. Research findings and recommendations were strongly imbued with a scientific management approach whereby researchers valiantly struggled to maintain the impossible position of being 'objective' in order to provide the best policy analysis and advice of the times, which often focused on fostering the revenue-generating activities of the resource sector.

Fast-forward to the 2000s. The orientation has changed dramatically. While the same documents might still be perused, albeit in electronic form, research approaches have expanded to encompass a wider appreciation of grounded theory and case study research. The mineral policy community has diversified to include many competing groups with an even larger number of perspectives and ways of valuing and assessing the full social and ecological costs of extracting, producing, and exporting mineral wealth. The old top-down, economically driven decision-making approaches are losing some ground to public pressures for participatory, cooperative, and integrated approaches to resource policy-making.

What does this mean for the researcher? Resource policy and management questions today are much more all-encompassing and complex—namely, how to foster desired, sustainable, resilient socio-ecological systems. How do you get there? Well, rigorous research and careful methodological approaches are still as important as ever. But today's researcher must be much more informed about the social and biophysical setting in which mineral development takes place, recognizing that each case is unique with its own set of imperatives. Moreover, expectations for researchers in the resource field these days are such that their results should have not only academic but practical value as well.

So, a note to the future mineral policy researcher: do your literature reviews and analysis, and develop criteria for assessment. Then take time to *really* understand the dynamics of resource decision-making on the ground. Go to affected mining communities in which the First Nations peoples, government officials, mineworkers, health care workers, and business owners are engaging in hotly contested debates about the role of mineral development in their region. Live there. Listen. Observe. Participate in the community. Be useful where you can. Be patient. Different cultures have diverse ways of using time and sharing perspectives. Effective communication is as much about learning to appreciate and understand diverse world views and local knowledge as it is about sharing your own perspectives.

And when you are ready to leave the community, make sure that you leave something useful behind with that community—something more than policy recommendations buried in a university research paper. It could be as basic as a summary of a focus group session that you convened, one that presents the community's collective vision of their valued places, spaces, traditions, and dreams for the future. Remember, the members of this locality who shared their time with you are co-participants in your quest to make a worthwhile contribution to a more sustainable world.

—Mary Louise McAllister and
Patricia Fitzpatrick

Acknowledgements

The authors would like to thank Bruce Mitchell (University of Waterloo). We also thank Michael Doggett (Queen's University) and Robert Irvine (Natural Resources Canada) for their thoughtful and insightful comments on the first draft of this chapter. Brittany Shuwera completed preliminary research for the revised draft. Any errors of fact or interpretation are the authors' own.

Note

1. The term 'mineral industry' tends to refer to all mineral extraction with the exception of oil and gas. Oil sands, however, are mined and therefore are considered part of the mineral industry.

REVIEW QUESTIONS

1. Who are the different stakeholders in the Canadian mineral sector? What are their different perspectives, concerns, and world views with respect to sustainability and mineral activities?
2. Do you think that these world views could be integrated in a way that would foster a mineral industry that contributes to a more sustainable world? What steps would have to be taken in order to see the industry move in the direction you believe is sustainable?
3. What are the federal and provincial governments' responsibilities for ensuring that mining is conducted sustainably in Canada? How should these responsibilities be implemented?

REFERENCES

Anderson, R.R. 2002. 'Healthy mining communities: The interdependency of companies and communities'. (University of British Columbia, Department of Mining and Mineral Process Engineering, MA thesis).

Angus, C., and B. Griffen. 1996. *We Lived a Life and Then Some: The Life, Death, and Life of a Mining Town*. Toronto: Between the Lines.

Campbell, B. 2008. 'Regulation and legitimacy in the mining industry in Africa: Where does Canada stand?' *Review of African Political Economy* 117: 365–85.

Canadian Aboriginal Minerals Association. 2002. Webpage. http://www.Aboriginalminerals.com.

Clausen, S., and M.L. McAllister. 2001a. 'An integrated approach to mineral policy'. *Journal of Environmental Planning and Management* 44 (2): 227–44.

———. 2001b. 'A comparative analysis of voluntary environmental initiatives in the Canadian mineral industry'. *Minerals and Energy: Rare Material Report* 16 (3): 27–41.

CORPWatch. 2002. 'Campaign: Alliance for a Corporate-Free UN, "Girona Declaration" Corporate Europe Observatory (CEO), May 27, 2002'. http://www.corpwatch.org/campaigns/PCD.jsp?articleid=261.

Dickason, O.P. 1992. *Canada's First Nations*. Toronto: McClelland and Stewart.

Doggett, M. 2002. Personal communication. Department of Geology, Queen's University, Kingston, 3 July.

Dungan, P. 1997. *Rock Solid: The Impact of the Mining and Primary Metals Industries on the Canadian Economy*. Toronto: University of Toronto Institute for Policy Analysis.

Environmental News Service. 2002. 'World's biggest mining firms confront their bad reputation'. 6 May. http://www.ens-newswire.com/ens/may2002/2002-05-06-05.asp.

Fidler, C., and M. Hitch 2007. 'Impact and benefit agreements: A contentious issue for environmental and aboriginal justice'. *Environments* 35 (2): 49–69.

Galbraith, L., B. Bradshaw, and M.B. Rutherford. 2007. 'Towards a new supraregulatory approach to environmental assessment in northern Canada'. *Impact Assessment and Project Appraisal* 25 (1): 27–41.

Humphries, D. 2006. 'Industry consolidation and integration: Implications for the base metals sector'. GFMS Precious and Base Metals Seminar, 14 September. http://www.nornik.ru/_upload/presentation/Humphreys-GFMS.pdf.

ICMM (International Council on Mining and Minerals). 2008. 'Assurance development framework: Assurance procedure'. London: ICMM. http://www.icmm.com/our-work/sustainable-development-framework/assurance. See also ICMM 2008, 'Our history'. http://www.icmm.com/about-us/icmm-history.

International Institute for Environment and Development and World Business Council for Sustainable Development. 2002. 'Breaking new ground: What can minerals do for development'. News release, 1 May.

Joint Panel Review. 2007. *Kemess North Copper–Gold Mine Project*. Ottawa: Canadian Environmental Assessment Agency.

Klein, H., J. Donihee, and G. Stewart. 2004. 'Environmental impact assessment and impact and benefit agreements: Creative tension or conflict?' Paper presented at the 24th Annual Conference of the International Association for Impact Assessment, Vancouver. http://impactandbenefit.com/Kleinetal_2004_Creativetensionorconflict.pdf.pdf.

McAllister, M.L. 1992. *Prospects for the Mineral Industry: Exploring Public Perceptions and Developing Political Agendas*. Working Paper no. 50. Kingston: Centre for Resource Studies.

Manitoba. 2000. *Manitoba's Minerals Guideline: Building Relationships and Creating Opportunities: Guiding Principles for Success*. See also Prospectors and Developers Association of Canada, spring 2000, *PDAC in Brief*.

Mines and Communities. 2001. 'The London Declaration'. 20 September. http://www.minesandcommunities.org/Charter/londondec.htm.

Mining Association of British Columbia. 2008. 'Environmental laws and regulations'. http://www.mining.bc.ca/environmental_laws_and_regulations.htm.

Mining Association of Canada. 2006. 'Towards sustainable mining: Guiding principles'. http://www.mining.ca/www/Towards_Sustaining_Mining/Guiding_Principles/Guiding_Principles.php.

———. 2008a. 'Facts and figures: A report on the state of the industry'. http://www.mining.ca/www/media_lib/MAC_Documents/Publications/2008/08_FFeng.pdf.

———. 2008b. 'Some facts about Canada's diamond mining sector'. http://www.mining.ca/www/media_lib/MAC_Documents/Diamond_Affairs/2008/Diamond_Fact_Sheet.April.08.pdf.

Mining, Minerals, and Sustainable Development Project. 2002. *Project News Bulletin* 21 (April): 11.

MiningWatch Canada. 1999. 'Taxpayers to pay for massive cleanup at northern mines'. Joint news release of MiningWatch Canada and the Yukon Conservation Society. 2 September.

———. 2002. 'Aims and objectives'. http://www.miningwatch.ca/MWC_profile_short.html#anchor28160192.

NRCAN. 1998. *From Mineral Resources to Manufactured Products*. Ottawa: Minister of Public Works and Government Services Canada.

———. 2001a. *Focus 2006: A Strategic Vision for 2001–2006. Minerals and Metals Sector*. Ottawa: Minister of Public Works and Government Services Canada.

———. 2001b. 'Communities'. In *Facts on Mining*. Ottawa: Minister of Public Works and Government Services Canada.

———. 2001c. 'Orphaned and abandoned mines'. In *Canada's Natural Resources: Now and for the Future*. http://www.nrcan-rncan.gc.ca:80/mms/pdf/orphan-e.pdf.

———. 2006. 'Recycling metals'. In *Canadian minerals yearbook 2006*. Ottawa: Minister of Public Works and Government Services Canada. http://www.nrcan.gc.ca/mms/cmy/content/2005/72.pdf.

———. 2007a. 'Canada's international presence'. In *Canadian Minerals Yearbook 2006*. Ottawa: Minister of Public Works and Government Services Canada.

———. 2007b. 'General review'. In *Canadian minerals yearbook 2006*. Ottawa: Minister of Public Works and Government Services Canada. http://www.nrcan.gc.ca/mms/cmy/content/2006/01e.pdf and http://www.nrcan.gc.ca/mms/cmy/content/2006/08.pdf.

———. 2007c. 'Canadian exploration scene'. In *Canadian minerals yearbook 2006*. Ottawa: Minister of Public Works and Government Services Canada. http://www.nrcan.gc.ca/mms/cmy/content/2006/01.pdf.

O'Faircheallaigh, C. 2005. 'Indigenous participation in environmental management of mining projects: The role of negotiated agreements'. *Environmental Politics* 14 (2): 629–47.

Parlee, B. 1998. *A Guide to Community-Based Monitoring for Northern Communities*. Working Paper no. 5. Yellowknife: Canadian Arctic Resources Committee, Northern Minerals Program.

Placer Dome Inc. 1997. *The Mine Development Process*. Vancouver: Placer Dome Inc.

Poelzer, G. 2002. 'Aboriginal peoples and environmental policy in Canada: No longer at the margins'. In D.L. Vannijnatten and R. Boardman, eds, *Canadian Environmental Policy: Context and Cases*, 2nd edn, 87–106. Toronto: Oxford University Press.

Prono, J. 2007. 'Assessing the effectiveness of impact and benefit agreements from the perspective of their Aboriginal signatories'. (University of Guelph, Department of Geography, MA thesis).

Pross, P. 1986. *Group Politics and Public Policy*. Toronto: Oxford University Press.

Quebec. 2009. 'Québec mineral exploration and development highlights'. http://www.mrnf.gouv.qc.ca/english/mines/quebec-mines/2009-02/highlights.asp.

Richard, L.P. 1997. *Report of the Westray Mine Public Inquiry*. Executive Summary. Halifax: Published on the authority of the Lieutenant Governor in Council by the Westray Mine Public Inquiry.

Ripley, E.A, R.E. Redmann, and A.A. Crowder. 1996. *Environmental Effects of Mining*. Delray Bach, FL: St Lucie Press.

Robson, R. 1992. 'Building resource towns: Government intervention in Ontario in the 1950s'. In M. Bray and A. Thomson, eds, *At the End of the Shift*, 97–119. Toronto: Dundurn Press.

SRK Consulting and SENES Consulting Ltd. 2007. 'Giant Mine remediation plan'. http://reviewboard.ca/upload/project_document/1207685649_Giant%20Mine%20Remediation%20Plan.pdf.

Team Manitoba. 2000. 'Prosperity and commitment to the future: Come share the wealth: History. Cities of Lynn Lake, Leaf Rapids, Thompson, Snow Lake, Flin Flon'. Brochure. See also 'Speaking notes of the mayor of Leaf Rapids', Manitoba Mining and Minerals Convention, 16–18 Nov. 2000, Winnipeg.

Udd, J. 2000. 'A chronology of minerals development in Canada'. In Canadian Institute of Mining and Metallurgy, *A Century of Achievement—The Development of Canada's Minerals Industries*, 52: Appendix. See also Natural Resources Canada, Minerals and Metals Sector, 2001, http://www.nrcan.gc.ca/mms/chrono-e.htm.

Veiga, M.M., M. Scoble, and M.L. McAllister. 2001. 'Mining with communities'. *Natural Resources Forum* 25 (3): 191–202.

Whitehorse Mining Initiative. 1994. *Leadership Council Accord: Final Report*. November. Available from Natural Resources Canada, Minerals and Metals Sector.

Wilson, J. 2002. 'Continuity and change in the Canadian environmental movement: Assessing the effects of institutionalization'. In D.L. Vannijnatten and R. Boardman, eds, *Canadian Environmental Policy: Context and Cases*, 2nd edn, 46–65. Toronto: Oxford University Press.

Wismer, S. 1996. 'The nasty game'. *Alternatives* 11 (4): 10–17.

14

Waste Management: Moving up the Hierarchy

Virginia W. Maclaren

Learning Objectives

- To understand that waste is both a valuable resource and a source of conflict.
- To understand the composition of waste, how it is produced, and strategies for reducing the amount of waste sent to landfill.
- To become familiar with some of the key debates in waste management.

INTRODUCTION

Current waste management practices in Canada emphasize techniques and approaches that avoid or minimize the need for waste disposal. This is a significant reversal from practices that existed during most of the twentieth century, when the main role of the waste management professional was to collect and then dispose of waste as efficiently as possible by burning or burying it.

Rising concern for the environment during the 1970s and growing public opposition to the siting of landfills and incinerators changed our perspective on waste, and we now view it as a resource that contains valuable economic commodities. This perspective on waste management is not a new one in Canada, however. During the world wars, the level of resource extraction from waste materials was much higher than it is today because of the scarcity of virgin materials and restrictions on imports. However, once

these periods of national crisis ended, the value of waste declined again, and the emphasis on disposal returned. The evidence from Canada therefore suggests that the definition of what types of waste are resources and how valuable they are varies with time. The evidence from other countries suggests that it also varies with place. For example, in developing countries, many materials that Canadians send to landfills are considered valuable resources and are reused or recycled.

Waste differs in two important ways from other types of resources. First, unlike most other resources, the resource base for waste has been steadily increasing. Second, management of the waste resource differs in an important way from the management of other resources. One of the goals of modern waste management practice is to reduce rather than to increase or sustain the resource base. This is because reducing the amount of waste that ends up in incinerators or landfills helps to conserve the resource base of other renewable or non-renewable resources such as forests and coal. It also has a net beneficial impact on the environment.

Concern for the environment, rising levels of waste generation, and the challenges of siting waste management facilities have forced waste managers to consider a new approach to waste management that includes a much broader range of options than just disposal. This new approach, known as integrated waste management, adds source reduction, reuse, recycling, and energy recovery to the traditional options of incineration and landfill. The design of an integrated waste management system takes into account the relative environmental impacts and economic costs of the different options, as well as the views of different stakeholders, such as the public, government, industry, and non-government organizations.

This chapter contains four major sections. After first describing the nature and origin of waste, I discuss the key elements of an integrated waste management system, examine some recent policy innovations in waste reduction, and then discuss some of the latest developments in waste management facility siting. I conclude with a brief commentary on some of the major challenges facing the field of waste management.

A TYPOLOGY OF WASTES

Three distinct waste streams are, as much as possible, kept separate for the purposes of disposal. They are: (1) non-hazardous municipal and industrial-commercial-institutional (ICI) waste, (2) hazardous waste, and (3) radioactive waste. Municipal solid waste includes waste generated by households, municipal governments, and small commercial and retail activities. It excludes industrial waste, waste from large commercial, retail, and institutional generators, and

construction and demolition waste. These latter types of wastes are usually managed by the generators themselves rather than by municipalities. Much of the data on waste generation in various countries or cities refer to municipal waste because these data are publicly available and well tracked by municipal governments. However, the data may sometimes include ICI waste, so caution should be taken when comparing data across jurisdictions.

Hazardous wastes can come from a number of sources, but most come from the industrial sector. Non-industrial generators include households, institutions, and commercial establishments. Radioactive wastes can be subdivided into two categories: high-level and low-level wastes. High-level radioactive waste consists primarily of the spent fuel from nuclear reactors. Low-level radioactive wastes include uranium mining and refinery wastes and contaminated materials from nuclear reactors (not including spent fuel), institutions (medical and academic), and industry.

Another important distinction in a typology of wastes is that between liquid and solid waste. Municipal waste, by definition, is a solid waste but may include small amounts of liquid waste in unemptied containers. Both hazardous and radioactive wastes come in liquid and solid form. Gaseous wastes are a third type of waste, but they are normally discussed separately because their management strategies differ distinctly from those used for solid and liquid wastes.

WASTE GENERATION AND WASTE COMPOSITION

An understanding of where waste comes from, how much is being produced, and what it consists of is essential for good waste management planning. In particular, this knowledge is useful for identifying where efforts can be focused most successfully in order to promote reduction, reuse, or recycling of waste. Waste generation rates and waste composition vary spatially within a community, over time, and across economic sectors. Waste management strategies for restaurants may be quite different from strategies for office buildings. An understanding of the factors that influence waste generation is also beneficial in providing guidance on how to reduce generation rates.

The amount of waste produced can be measured in two ways: by volume or by weight. Although volume is a useful measure for evaluating landfill needs and remaining capacity, most waste statistics are in units of weight, since it is easier to obtain an accurate estimate of weight at disposal sites (by means of on-site scales) than the volume of waste in a truck. The discussion of waste generation and waste composition that follows therefore refers to weight measures in all instances.

In 2006, Canadians generated 1,072 kilograms of non-hazardous waste per capita and disposed of 835 kilograms per capita (Statistics Canada 2008a). This makes Canada one of the highest waste generators per capita in the world. Despite the efforts that individual Canadians and Canadian industry are making to reduce, reuse, and recycle their wastes, non-hazardous waste disposal per capita was 12 per cent higher in 2006 than in 2000. These per capita figures refer to waste requiring disposal from all sources, not just residential sources. Residential waste generation per capita was 397 kilograms in 2006, and disposal from residential sources was 283 kilograms per capita.

Of the total amount of non-hazardous waste sent for disposal in 2006, more than one-third came from residential sources and two-thirds from non-residential sources consisting of the two major categories of ICI waste (construction and demolition waste). The two sources diverted approximately the same proportion of their waste from disposal, producing an average diversion rate for Canada of 22 per cent.[1]

Municipal Waste

Waste generation per capita has been rising steadily in North America for many years. Waste composition has also been changing. Although there are no historical accounts of changes in waste composition in Canada, data from the United States indicate that the greatest change has been in the growth of plastics in the waste stream. Between 1960 and 2007, plastics found in discarded products and packaging rose from 0.4 per cent of the American municipal waste stream to 12.1 per cent (United States Environmental Protection Agency 2008). At the same time, the use of ferrous metals and glass declined as lighter plastic materials and non-ferrous metals (aluminum), particularly in packaging, replaced them.

During the late 1980s and the 1990s, our knowledge of the composition of solid waste in Canadian municipalities increased substantially as numerous municipal governments undertook waste composition studies as part of their efforts to plan for new waste disposal facilities. Figure 14.1 shows the average composition of residential waste sent for disposal in the Greater Vancouver Regional District (now known as Metro Vancouver) in 2004.[2] It does not include waste recovered for recycling. The waste composition profile is fairly typical of urban areas in Canada, with paper and organic wastes predominating. Paper and paperboard wastes include newspaper, fine paper, boxboard, cardboard, magazines, phone books, tissue paper, and other mixed papers. Organic wastes constitute about one-third of the waste stream and include food wastes and yard or garden wastes. A small but very important component of the waste stream in Figure 14.1 is household hazardous waste (HHW). It is important because even small amounts of hazardous waste can cause great harm to the environment

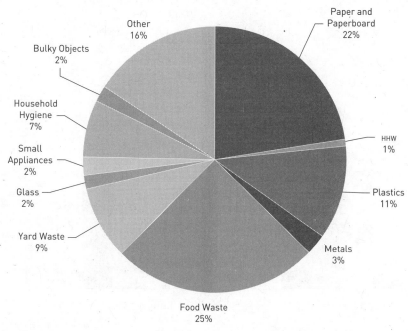

Source: Based on Technology Resource Inc. 2005.

Figure 14.1 Residential Waste Composition, Metro Vancouver, 2004.

and human health. Figure 14.2 illustrates the choices that Canadians made in 2006 for disposal of two types of HHW—namely, leftover paint and leftover medicine. A distressingly high percentage (40 per cent) of leftover medicine is dumped down the drain or put in the garbage. Much less leftover paint finds its way into the garbage, but the need for improved collection systems is apparent from the large percentage of waste that is left in storage rather than being taken to hazardous waste depots or returned to suppliers.

A number of factors affect the total amount, per capita amounts, and types of waste generated in different sectors. In the residential sector, the most important factors are household size, age structure of the household, annual household income, type of dwelling unit, geographical location, and time of the year. In the commercial sector, the major influencing factor is number of employees and type of commercial activity. In the industrial sector, the size of the firm and the type of industrial activity both have an important effect. The proportion of residential, commercial, institutional, and industrial waste in the municipal waste stream varies across urban areas and depends on the size of the community and the structure of economic activity within the community.

Household size affects residential waste generation because the more people in a household, the greater the amount of waste produced by that household. However, there are typically economies of scale in the production of household waste in that the per capita generation of waste is lower in a large household than in a small household. One example of scalar economies is in the purchase of daily newspapers. A three-person household only needs to buy one newspaper because it can be read several times. The per

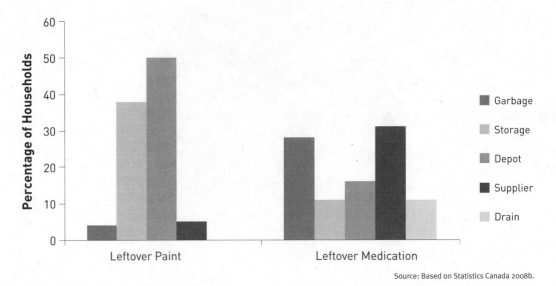

Source: Based on Statistics Canada 2008b.

Figure 14.2 Disposal Choice for Selected Household Hazardous Wastes, Canada, 2006.

capita generation of newspaper waste is therefore one third of a newspaper per day. However, the purchase of a daily newspaper by a single person household would result in a per capita waste generation rate of one newspaper per day. Another example of economies of scale is in the production of food packaging waste. Larger households can buy in bulk, while smaller households may rely more heavily on single-serving packages.

Age structure of the household can affect waste generation for a number of reasons. If infants are in the household, then disposable diapers can add a considerable load to the daily waste output. On an annual basis, children tend to have more material discards than adults, since they quickly grow out of clothes and toys.

Higher-income households normally produce more waste than lower-income households because their incomes give them the ability to purchase more material goods. They also tend

to have larger properties and thus more garden wastes. This is partially offset by the fact that members of higher-income households may eat out more and generate fewer food wastes.

The presence or absence of yard wastes has a significant influence on the amount and composition of waste generated in different types of dwelling units, at different times of the year, and at different locations across the country. The amount of yard waste produced per capita in apartment buildings is lower than that produced in single-family dwellings because individual apartment units do not have backyards but rather share a relatively small common area surrounding the apartment building. The amount of yard waste produced in the spring and fall months is greater than that produced in the summer and winter because of garden planting in the spring and fallen leaves in the fall. Finally, regions of Canada with a shorter growing season will produce less yard waste and will have waste composition profiles that reflect this difference.

The commercial waste stream differs from the residential waste stream in that the sources of waste are more heterogeneous. In the residential sector, there is a single producer—namely, the household—whereas in the commercial sector, waste comes from three general categories of producer: office buildings, retail businesses, and restaurants. Each sub-sector has a dominant waste component: mixed paper in offices, food waste in restaurants, and packaging in retail establishments. Another difference between the residential and the commercial sectors is that the standard per capita measure of waste production in the former refers to waste generation per household resident, while in the latter it is usually waste production per employee. Waste production levels per employee will vary with the type of commercial activity and the number of employees.

As with the commercial sector, the institutional sector covers a fairly broad range of activities, such as libraries, schools, and cultural institutions. It also includes non-hazardous wastes from hospitals. Paper is the main component of the waste stream in the first two sources, while the latter two have more diverse waste streams.

The industrial waste stream is even more heterogeneous than the commercial or institutional waste streams. According to the two-digit Standard Industrial Code (SIC) used by Statistics Canada, the industrial sector can be broken down into about 40 different industrial categories. Each category will tend to differ in the amount and type of waste produced, but within each category the generation and composition figures for comparable-size firms should be fairly similar.

Hazardous Waste

Hazardous wastes are those that cause or have the potential to cause harm to human beings or to other organisms because the wastes are toxic, corrosive, flammable, explosive, reactive, or pathological. Some of the properties of hazardous wastes are self-explanatory, but others deserve further discussion. Toxic wastes can lead to death or serious injury when inhaled, ingested, or absorbed. The chemical properties of corrosive wastes can cause deterioration of materials and body tissues at point of contact. Pathological wastes are hazardous because of their potential for spreading disease. By definition, radioactive wastes are hazardous wastes, but they are usually classified in a category of their own because of their unique handling characteristics and slow deterioration rates.

No agencies in Canada collect comprehensive data on the generation of hazardous waste. The best estimates are from the National Pollutant Release Inventory (NPRI) (http://www.ec.gc.ca/pdb/npri/npri_home_e.cfm), which tracks pollutant releases to air, land, and water from industrial and non-industrial sources on an annual basis. However, the NPRI database contains information on only a limited number of hazardous waste categories and excludes hazardous wastes produced by small generators.

Radioactive Waste

Nuclear fuel wastes and low-level radioactive wastes have different levels of radioactivity and require different disposal methods. Nuclear fuel wastes are high-level radioactive wastes that require shielding for both radioactivity and heat. Low-level wastes are low in radioactivity but still require some shielding.

An important characteristic of radioactive wastes is the half-life of the radionuclide. A half-life is the amount of time that it takes the activity of a radionuclide to decay by half. Radionuclides are generally harmless after they have decayed for 10 half-lives. The half-lives of some of the most common radionuclides range from about eight days for Iodine-131 to 4.5 billion years for

Uranium-238. Iodine-131 is used for medical diagnostics and radiotherapy, while the main sources of Uranium-238 are uranium mine tailings and uranium processing residues. The long half-lives of radionuclides in most radioactive waste means that such wastes can be a threat to human health for generations to come unless they are properly managed.

Almost 90 per cent of the high-level nuclear fuel waste in Canada comes from nuclear generating plants in Ontario. Ontario relies heavily on nuclear energy for electricity production. It accounts for about 50 per cent of the province's electrical generating capacity. Over the past 35 years, more than two million spent fuel bundles from nuclear power plants in Ontario, Quebec, and New Brunswick have been stored on-site at the nuclear facilities. The Canadian government is currently designing a site selection process for identifying a permanent, deep geological repository for high-level radioactive waste.

A major source of low-level radioactive waste in Canada is a refinery in Port Hope, ON, which processed radium in the 1930s and 1940s and uranium during World War II (Gunderson and Rabe 2000). The radioactive processing wastes were deposited in two unsecured sites close to Port Hope and in smaller dumps and ravines throughout the town. Soils from the dumpsites containing very small amounts of these processing wastes were inadvertently used for construction and landscaping of private residences in the town during the 1950s and 1960s, leading to widespread radioactive contamination (Gunderson and Rabe 2000). The federal government cleaned up the contaminated sites, beginning in 1975, and now stores about two million cubic metres of low-level radioactive waste and contaminated soils in short-term storage facilities in Port Hope and two surrounding communities. After an unsuccessful effort to find a permanent disposal site elsewhere in Ontario for this waste (described later in this chapter), the federal government signed an agreement with the affected communities in 2001 to build two long-term, low-level radioactive waste management facilities in the Port Hope area. The proposed sites are currently undergoing an environmental assessment review.

WASTE MANAGEMENT OPTIONS

Integrated waste management systems seek to divert as much waste as possible from disposal by following the waste management hierarchy. At the top of the hierarchy is *source reduction*, followed by *reuse*, *recycling* and *biological treatment*, *thermal treatment* (typically with energy recovery), and *landfill*. The precedence of a waste management option within the hierarchy is determined by its environmental performance.

One way to assess the environmental performance of waste management options toward the bottom of the hierarchy is through *lifecycle analysis* (LCA) of municipal waste management systems. LCA enumerates the energy, water, and material inputs and the outputs of airborne emissions, water effluents, and solid wastes at each stage of a waste material or waste product's lifecycle, from 'cradle' to 'grave'. The cradle of a waste management system is typically the point at which municipal solid waste is set out for collection, and the grave is disposal in landfill. Recycling, biological treatment, and thermal treatment occur in between. LCA studies confirm that recycling of materials in the waste stream is better for the environment overall than incineration with energy recovery and that incineration with energy recovery is better than landfilling (Finnveden et al. 2005). The one exception is plastics, which a few studies have shown have higher acidification effects and energy use when they are recycled rather than burned. Another cautionary note in the use of the waste management hierarchy is that it does not consider the social or financial costs and benefits of each option. For example, although incineration with energy recovery may be better for the environment than landfilling, it may also be far more costly and therefore unattractive to municipal governments.

The terms 'source reduction' and 'waste reduction' are sometimes used interchangeably to refer to the concept of reducing wastes at their source of generation. More often, however, 'waste reduction' is defined in a broader sense and, along with the term 'waste minimization', is used to refer to the combined activities of source reduction, waste reuse, and waste recycling. Most provinces and many municipalities have established waste reduction targets to be achieved within a given period of time. For example, half of the regional districts in British Columbia have adopted 'zero waste' as their long-term waste reduction goal or strategy. Zero waste focuses on encouraging the redesign of products and processes so that no waste is produced during manufacture and so that products can be more easily reused and recycled. It also involves building infrastructure, developing local policies, and delivering education programs that will support waste reduction at the household level. In some jurisdictions, recovery of energy from waste may be a part of a zero waste strategy.

Waste reduction strategies focus on the management of both post-consumer and post-production wastes. Post-consumer wastes are those generated by the consumer, such as the household or the retailer. Post-production wastes are those generated by the primary or secondary producer of consumer products and their inputs.

Source reduction occurs in the household when the householder makes purchasing decisions that reduce the amount or toxicity of products and materials entering the household over a given period of time. Examples of household source reduction include purchasing products that are more durable or have less packaging than those normally purchased. Backyard composting is also a form of source reduction, since it reduces the amount of waste collected, at source, from households. At the industrial plant, source reduction might encompass the use of more efficient manufacturing techniques that produce less scrap or generate fewer toxic wastes per unit of output.

Reuse occurs when materials or products can be reused in their original form or for their original purpose with no need to apply physical or chemical treatments. Some minor physical treatment may be required, such as washing or making repairs. In contrast, recycling of waste materials or products requires significant physical or chemical treatment and often results in a product that differs in form or use from the original. Biological treatment refers to the treatment of organic wastes by means of centralized composting.

Source Reduction Strategies

Less has been accomplished by municipalities in the area of source reduction of post-consumer wastes than with the other waste diversion options because the barriers inhibiting reduction are more difficult to overcome. For example, the responsibility for waste disposal rests with municipalities, yet control over packaging regulations lies primarily at the federal and provincial levels where trade-offs must be made between national or provincial and local interests. One of the concerns raised by industry is that imposing packaging controls on Canadian manufacturers and retailers will reduce profits and give a competitive advantage to foreign producers. Source reduction strategies at the local level have relied primarily on educational tools, such as consumer awareness campaigns encouraging consumers to avoid products that are over-packaged or disposable. Some municipalities are going further. In 2007, Leaf Rapids, MB, became the first municipality in Canada to ban single-use plastic shopping bags, and as of 2009 the City of Toronto required retailers to impose a charge of 5 cents per plastic bag. The proliferation of plastic water bottles is a concern to many municipalities, and in 2009 the Federation of Canadian Municipalities passed a resolution calling on all municipalities to phase out the sale and purchase of bottled water at their own facilities where appropriate and where tap water is available.

A stronger driving force toward source reduction of post-consumer wastes may be consumer awareness. Consumer awareness of the packaging and durability issues is compelling companies to re-evaluate the type of products that they produce and how they deliver them. In the 1980s, the emphasis on marketing was to make products as convenient to use as possible. Starting in the 1990s and beyond, marketing the environmental benefits of a product has become increasingly important.

During the past two decades, there has been a notable increase in the application of source reduction measures to the management of hazardous and non-hazardous industrial waste. The four main techniques for source reduction in industry are good operating practices, technology changes, input material changes, and product changes (Shah 2000). Good operating practices focus on administrative, procedural, and institutional approaches for minimizing waste and reducing waste toxicity during production. Examples range from very simple tasks such as fixing leaky valves to retraining employees to make them more aware of source reduction practices. Good operating practices are relatively inexpensive to implement and tend to be the first type of source reduction measure adopted by industry. Technology changes involve changes in production process, such as introducing automation or installing equipment that produces less waste. Input material changes include purification of existing input materials to remove toxins or substitution of input materials for new inputs that produce less waste or have a lower toxic content (e.g., substituting water-based solvents for organic solvents). As the name implies, product change refers to a change in the nature of the product so that it will produce less waste or release fewer contaminants into the environment when it is used. It does not refer to a direct reduction of the weight of a product's packaging (e.g., by substituting plastic packaging for paperboard packaging) but can denote a change in the product that will ultimately require less packaging.

Reuse Strategies

In the industrial sector, reuse strategies can have substantial economic savings. For example, the electroplating industry can reduce its liquid wastes by reusing its waste waters in the rinsing process. This will result in savings in jurisdictions where companies have to pay sewer surcharges for excessive waste water discharges. The reuse of waste water also means that the company's requirements for fresh water will decrease, as will its water bills.

Non-profit agencies play an important reuse role in the residential sector. Agencies such as Goodwill and the Salvation Army collect, repair, and redistribute old clothing, furniture, and numerous other items for reuse. Most cities also have thriving second-hand markets for a wide variety of products, such as books and sporting equipment. A significant, informal sector outlet for household waste reuse is the yard sale. Finally, consumer purchasing decisions can lead to reuse. For example, households can purchase reusable shopping bags or returnable bottles.

Recycling

There are several ways of sorting and collecting recyclable materials in the residential waste stream. One of the easiest ways to classify these methods is by the number of streams into which householders must sort their wastes. The practice of separating wastes at the point of generation is known as *source separation*. Table 14.1 shows the diversion rates being achieved by a selection of cities with source separation systems. All except Calgary provide curbside collection of recyclables. Note that these diversion rates include not only diversion by means of curbside collection of source-separated materials but diversion through other programs as well, such as backyard composting and deposit-return systems for selected beverage containers.

Edmonton and Calgary are examples of two-stream separation systems, consisting of a stream for recyclables and a stream for garbage. After collection, recyclables are sent to a materials recovery facility, or MRF, where a variety of automated and manual systems separate the mixed recyclables into different material sub-streams, including plastics, glass, mixed paper, cardboard, and ferrous and non-ferrous metals. Two-stream systems normally achieve moderate diversion rates. Calgary's two-stream system has lower diversion rates because of its reliance on recycling depots, which are less convenient for residents than curbside collection of recyclables. On the other hand, Edmonton has a very high diversion rate because instead of sending its garbage directly to a landfill, the city first sends it for processing at a co-composting facility. The facility has mixing drums that combine the household waste with biosolids (sewage sludge)

Table 14.1 Curbside Collection of Source-Separated Materials in Selected Cities

Municipality	Residential Diversion Rate	Source Separation Systems
Edmonton[a]	60%	Recyclables Garbage
Halifax[b]	55%	Recyclables Organics ('green cart') Garbage
Guelph[c]	44%	Recyclables (dry stream) Organics (wet stream) Garbage
Waterloo[c]	44%	Recyclables Organics ('green bin') Garbage
Greater Montreal[d]	32%	Recyclables Garbage
Calgary[e f]	16%	Drop-off recycling depots Garbage

[a] Source: City of Edmonton 2007.

[b] Source: Halifax Regional Municipality 2007.

[c] Source: Waste Diversion Ontario 2007.

[d] Source: Communauté métropolitaine de Montréal 2008.

[e] Source: City of Calgary 2006.

[f] In 2006, the City of Calgary did not provide curbside collection for recyclables but instead operated 50 recycling depots. In April 2009, the city introduced blue cart recycling services for all single family households.

before sending the mixture to aeration bays for composting.

The most popular source-separation system in Canada is the three-stream system. Householders separate out recyclables and organic waste from their garbage. Halifax introduced a three-stream system in 1999 and now has a residential diversion rate of close to 60 per cent. The advantage of a three-stream system is that it produces a high diversion rate and the organic wastes are relatively clean because they have not been mixed with garbage. Clean organics have little or no contamination from items such as broken glass or plastics. The compost produced from clean organics fetches a higher price on the market and brings in more revenue to the municipality. Although two-stream systems that only require separation of recyclables achieve lower diversion rates than three-stream systems, their popularity stems from the fact that they have lower capital and operating costs than either of the other two systems. Source-separation of organics in three-stream systems also requires significant behavioural changes on the part of households, meaning that participation rates can be low unless there is an extensive education campaign. Even after such an educational program, some householders may simply find it too difficult or inconvenient to separate out their organics.

Household *participation rates* in waste diversion programs vary widely and depend on a number of factors, including the way in which participation is measured and the convenience of the program. If we define recycling participation as occurring when a household sets out recyclables at least once during a given interval of time, then a participation rate based on a weekly time interval will almost certainly be lower than a participation rate based on a monthly time interval. Weekly participation rates provide an indication of how many collection vehicles are needed per week. Monthly participation rates are helpful in

evaluating the success of a program. A weakness of the participation rate is that it does not measure what percentage of recyclables in the home is set out, nor does it measure how much waste is actually diverted from incineration or landfill. The *diversion rate*, which measures the percentage of waste generated that is diverted from landfill, is actually a more useful measure for program evaluation, since a community may have a high participation rate but a fairly low diversion rate.

Residential diversion rates vary considerably across the provinces (Figure 14.3)[3] depending on the number of waste diversion policies in place, residential access to recycling and organics diversion programs, and the availability of markets for recyclables. Nova Scotia has Canada's highest diversion rate because it introduced several innovative waste reduction policies during the 1990s (Wagner and Arnold 2008). First, it has landfill disposal bans on a wide range of wastes, including food wastes, yard waste, selected plastics, beverage containers, newsprint, corrugated cardboard, and metal and glass food containers. All refillable and non-refillable beverage containers have deposits on them. To encourage greater use of refillable containers, consumers receive the full deposit back when they return refillables but only half the deposit when they return non-refillable containers. Finally, all residents of Nova Scotia have curbside collection of inorganic recyclables, and about 90 per cent of them have curbside collection and centralized composting of organic waste, including food waste, leaf and yard waste, and non-marketable paper products (such as soiled paper towels and napkins) (Nova Scotia Environment 2008).

Along with participation and diversion rates, an important measure of a municipal recycling program's success is the *capture rate*, or the percentage of recyclable materials recovered from households with recycling program access. High capture rates occur when participation rates are

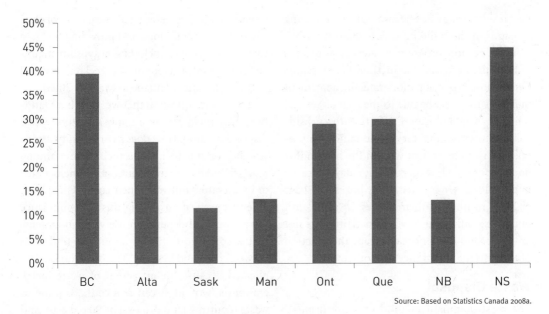

Source: Based on Statistics Canada 2008a.

Figure 14.3 Residential Diversion Rates by Province, 2006.

high and when the households participating recycle all of the materials that they are supposed to recycle. Some households may not recycle certain materials because they are not aware that the materials should be recycled or because it is less convenient to recycle them. Low capture rates are a sign that the municipality needs better public education programs to improve awareness among householders of what materials to recycle, how to prepare them, and why it is important to recycle.

One area in which waste diversion programs have lagged is in multi-family dwellings (typically defined as those with six or more units). In Toronto, with the second-highest percentage of multi-family dwellings (MFDs) in its building stock after Vancouver, residents of single-family dwellings (SFDs) diverted 59 per cent of their waste in 2008, while residents of MFDs diverted only 15 per cent of their waste (City of Toronto, n.d.). Part of the explanation for this difference is that all SFDs had green bin programs for diversion of organic waste while the MFD green bin

program is not due to be rolled out fully in the city until 2010. However, this explains only part of the difference in the diversion rates, because data also show that recyclables, collected from all SFDs and MFDs in the city, have a significantly lower capture rate in MFDs than they do in SFDs. For example, the capture rates for paper, aluminum cans, and glass are all close to 80 per cent in SFDs, while they range between 25 and 40 per cent in MFDs.

One reason for the difference noted above is that recycling in multi-family dwellings tends to be less convenient for residents because they may have limited storage space and most residents must carry their recyclables to containers in the basement or outside the building.[4] For single-family households, setting out recyclables for collection means carrying them only a few steps to the curbside. Another reason for the lower capture rates is that occupancy turnover tends to be higher in multi-unit buildings than in single-family households. This means that new residents may not be aware of the recycling program in the building unless there is an on-going education program to alert new tenants of both the need and the method for recycling. Finally, lower participation rates in apartment buildings may also be due to the lack of a social norm or 'neighbourhood' effect. With curbside collection programs for single-family house-holds, it is quite evident who in the neighbour-hood is not recycling on collection days, because householders set out recycling boxes (or bags or bins) in front of their homes. In multi-unit buildings, individual recycling activities are not as visible, and residents can escape the sanction of their neighbours if they do not recycle.

Waste Disposal

The most dominant method of waste disposal used in Canada today is landfilling. A few juris-dictions also have incinerators (with energy recovery). Incineration is an intermediate rather than a final disposal method, because some wastes, in the form of ash, remain even after incineration and must be sent to landfill.

There is widespread controversy over whether it is less environmentally harmful to burn waste or to bury it. To varying degrees, all waste manage-ment facilities produce negative environmental impacts in the form of noise, litter, odours, dust, and increased vehicle traffic. In addition, inciner-ation releases pollutants in the form of airborne emissions, ash, and quench waters. Landfill pro-duces leachate and gas. The main components of the latter are carbon dioxide and methane, both of which contribute to the problem of global warming (see Chapter 5). Methane may also pose a danger if it migrates underground to the base-ments of adjacent housing developments where it will explode if ignited. A beneficial impact of landfills is that methane gas has the potential for recovery and use as an energy source. Twenty-one landfills captured methane for energy recovery in Canada in 2005, generating 67 megawatts of elec-tricity, heating buildings, and providing fuel to a variety of facilities, including a gypsum manu-facturing plant, a greenhouse, a steel refinery, and a recycling plant (Methane to Markets 2008).

There is no question that we must have land-fills, since incinerators are only an intermediate disposal method. How does a community decide whether or not to include incinerators among its waste management facilities? Do the benefits of incinerators outweigh their costs? There is no objective way of answering this question, since the answer will depend on the value judgments that a community makes about the trade-offs between risks and benefits. The question is com-plicated because the attraction of waste incin-eration is twofold: it can reduce the volume of waste requiring land disposal by 90 per cent, and it can be used to produce energy for heating or electricity generation. Incineration accompan-ied by energy recovery from the heat generated during burning of the waste is usually referred to as energy-from-waste (EFW). The pressure to construct incinerators is greatest in municipal-ities and regions with little landfill capacity and competing land-use pressures. Incineration with EFW also contributes to community self-reliance by exploiting a local energy resource and reduc-ing the need for energy imports.

Those who oppose incinerators are con-cerned about the environmental impacts of incinerators and the possibility that the pres-ence of EFW incinerators may conflict with com-munity reduction and recycling programs. EFW plants require a minimum amount of waste each day, and if this waste is not available, then the plant must cease or reduce operations. To avoid these problems, most EFW plant operators require that the municipality supplying waste to the plant guarantee a minimum daily tonnage of waste and, sometimes, minimum energy con-tent. This type of guarantee has the potential to be a source of conflict between EFW plants and reduction–recycling programs. It may prevent a community from increasing its reduction–recycling efforts beyond the levels determined by a waste management facility that is lower down on the hierarchy of waste management options.

INNOVATIONS IN POLICY

Two major waste management policy initiatives adopted by many jurisdictions across Canada during the past 10 to 15 years to encourage waste reduction are user-pay programs and extended product responsibility policies. User-pay programs at the municipal level make households pay separately for the collection of some or all of the bags (or containers) of waste set out for collection. Households normally pay for waste collection services as part of their general property taxes and never see how much they are paying for waste collection and disposal. By assigning a cost to each bag of waste set out for collection, the expectation with user-pay programs is that this economic incentive will encourage householders to recycle more and to reduce the amount of waste that they produce.

Extended producer responsibility policies expand the traditional environmental responsibilities of producers beyond the production stage (i.e., beyond air pollution control and wastewater treatment) to subsequent stages in the lifecycle of a product, particularly to management of wastes at the post-consumer stage (Environment Canada 2001). They represent a step beyond a similar type of policy known as Product Stewardship. Product Stewardship is based on the concept of shared responsibility for a product's lifecycle impacts, allocating responsibility not only to the producer or manufacturer but also to the retailer, the consumer, and the municipality that collects waste. The paragraphs below describe these new policy initiatives in more detail.

User-Pay Programs

There are several different types of user-pay programs. With full user-pay systems, households pay for every bag set out, while partial user-pay systems allow households to set out one or more bags for free before they are charged for collection. To receive collection, households must purchase tags or stickers for their garbage bags, or purchase specially labelled garbage bags, or set out their waste in containers of set sizes specified by the municipality. The first system is by far the most popular one in Canada because of its relatively low cost and convenience for households. Research has shown that introduction of a user-pay system can reduce the amount of household waste sent to landfills by 15 to 45 per cent, of which up to 5 to 7 per cent may be attributed to source reduction and backyard composting and the remainder to increased recycling (EnvirosRIS 2001). Larger user fees produce greater amounts of diversion from landfill. To be successful, user-pay programs must be accompanied by access to a wide suite of diversion options such as recycling and yard-waste collection. One drawback of user-pay programs is that it is difficult to adapt them to multi-family dwellings. If a user fee is imposed on a multi-family dwelling, it is usually charged to the owner of a rental building or the board of a condominium rather than to individuals within the building. All tenants or unit-owners then usually share the fee equally, and the economic incentive to individuals for reducing waste is lost. To reduce management costs, the onus is then on the building manager to publicize waste diversion programs widely within the building and to make it as convenient as possible for residents to recycle and divert organics.

Although user-pay programs have gained popularity across Canada over the past decade, they have also faced fierce opposition in some municipalities. One of the arguments raised by opponents of user-pay programs is that their introduction encourages illegal dumping, burning, or littering as householders search for a way to avoid paying the user fee. In practice, although these problems may arise initially, they tend to decline or disappear over time as residents change their recycling or source-reduction behaviours or become accustomed to the fee (Miranda and Bynum 2002). The most serious opposition to user-pay programs comes from those who see them as a tax grab by the municipality. Opponents argue that residents already pay for waste management services in their tax bills so the user fee is simply a way of raising more taxes under the guise of a program that is already funded.

Extended Producer Responsibility

The first extended producer responsibility (EPR) policies in North America appeared in the 1970s when state and provincial jurisdictions became concerned about litter problems and in response introduced 'bottle bills'. The 'bottle bill' legislation required consumers to pay a deposit on beverage containers and retailers or brand-owners of certain beverages (typically soft drinks in glass bottles) to refund the deposit when the container was returned. All provinces in Canada now have beverage container legislation. In the 1990s, a second generation of EPR policies emerged, motivated by a quite different set of concerns. Among them were (Bury 1999) concerns over resource scarcity, concerns over the environmental impacts of production and product disposal, government and taxpayer fatigue with paying for the costs of disposal, recycling, and environmental remediation, and concerns over how to change consumer purchasing habits.

Internationally, the year 1991 marked the beginning of the modern era of extended producer responsibility with the introduction of the German Packaging Ordinance. The ordinance made industry in Germany responsible for its packaging waste by requiring producers of all types of packaging to either take back their packaging waste or contribute to the funding of the Duales System Deutschland (DSD), a packaging industry organization that manages an extensive packaging waste collection system for the country. Variations of the German Packaging Ordinance quickly spread to other European countries, and in 1994 the European Commission introduced its Packaging Directive with the objective of harmonizing packaging EPR policies in Europe. Like other EPR packaging policies, it sets overall recycling targets for materials contained in packaging waste and individual targets for different materials (60 per cent by weight for glass, paper, and paperboard; 50 per cent for metals; 22.5 per cent for plastics; and 15 per cent for wood), all of which had to be met by 2008.

Compared to Europe, North America has been slow in introducing modern EPR regulations. For example, the Canadian government chose to use a voluntary rather than a regulatory approach for encouraging packaging waste reduction in the early 1990s. It created the National Task Force on Packaging with a mandate to develop packaging waste reduction policies and a monitoring system for tracking the success of these policies. After a multi-stakeholder consultation process with industry, all levels of government, and non-governmental organizations, the task force agreed on an overall goal of reducing packaging waste in Canada by 50 per cent by the year 2000 compared to 1988 levels. This goal became part of a strategy for reducing packaging waste known as the National Packaging Protocol (NaPP). The reduction goal was met four years early in 1996 (Canadian Council of Ministers for the Environment 2000). There has been no detailed analysis of the specific types of packaging modifications made by industry to achieve this goal, but they probably included innovations such as light-weighting (substituting lighter packaging materials for heavier ones), reducing the volume of the product (e.g., by concentration), and removing unnecessary packaging used primarily for advertising. Although NaPP met its target, experience has shown that the most effective EPR programs are those that are mandatory (Quinn and Sinclair 2006).

British Columbia has been a leader among Canadian provinces in its commitment to mandatory EPR (see Table 14.2). The province passed North America's first beverage container deposit-refund legislation in 1971 and currently has more EPR programs than any other province in Canada, although Ontario is catching up quickly. All of British Columbia's programs are for household hazardous waste, with the exception of its beverage container legislation.

Pre-EPR policy for household hazardous waste in the province placed the onus on government and taxpayers for recovery. In 1990, British Columbia opened a small network of eight household hazardous waste depots that operated at an annual total cost of $1.4 million per year and served less than 0.5 per cent of the population (British Columbia Ministry of Water, Land and Air Protection 2002). Because of the high cost and low service coverage, the province closed the depots in 1994 and turned to EPR as a more appropriate policy approach.

In 1994, British Columbia mandated paint recovery by brand-owners. Product Care, a non-profit industry-sponsored organization, manages 104 permanent paint collection depots and offers a number of one-day collection events across the province, providing a much larger service coverage than the previous program. Product Care's industry members pay an 'eco-fee' to the organization to pay for the costs of collecting and managing the leftover paint. Members typically recover these costs by passing the eco-fee along to their customers at point of sale.

Since local governments have no regulatory powers for EPR, some have been using a voluntary approach. In 1997, the City of Ottawa decided that it needed a better system for recovering household hazardous waste than its single hazardous waste depot where recovery rates were very low. Recognizing that some retailers already took back certain used items from their customers (e.g., tires and used motor oil), the city approached these retailers to form a product stewardship partnership. The city's contribution to the partnership includes free communications materials about the program and the publication of a directory, in print and on the web (http://app01.ottawa.ca/takeitback/Welcome.do?lang=en), which lists the retailers participating and the products that they accept. The number of retailers in the program grew from 16 in 1997, taking back three products, to 547 retailers and charitable organizations in 2008, taking back 131 products (Reimer 2008). Retailers benefit from the program because it encourages repeat business; householders benefit because the large number of return facilities makes it easier for them to return used products than to dispose of them; and the city benefits because fewer household hazardous waste products end up in its landfill.

Several provinces are looking beyond existing household hazardous waste and beverage container EPR programs to consider a wider range of problematic goods in the waste stream. These are goods that are either bulky, are made of composite materials that make recycling difficult, or contain hazardous materials. Most electronic goods, particularly computers, have all of these characteristics and are a potential target for EPR. Old electronic goods requiring disposal, including computers and peripherals, telecommunications equipment, and video and audio equipment, are often referred to as 'e-waste', waste electrical and electronic equipment, or WEEE. The amount of

Table 14.2 Extended Producer Responsibility and Product Stewardship Programs in Canada, 2008

	Packaging					Automotive Products				Other HHW						Electronics		
	Milk	Beverage containers	Other	Printed Materials	Mercury Lamps	Used oil	Lead-acid batteries	Tires	Other	Paint	Solvents	Batteries	Fertilizers and pesticides	Pharma-ceuticals	Mercury products	TVs and computers	Telephones, VCRs, etc.	Other
NF		P				P		P										
PE		E				P	P	P										
NS	EP	P		P		E		P		P				P		E	E	
NB	P	P				P		P		E								
QC	EP	E	EP	EP		E		P		E								
ON	EP	EP	EP	EP	E				E	E	E	E	E	E	E	E	E	
MA	E	E	E	E		E		E	E		E	E	E	E	E			
SK		P				E		P		E						E		
AB	E	E				E		P		P						P		
BC	E	E			E	E	P	E		E	E	E	E	E	E	E	E	E

Source: Canadian Council of Ministers of the Environment 2009.

E = Extended producer responsibility
P = Product stewardship

e-waste is rising rapidly because of the growing number of people purchasing electronic goods and technological changes that quickly make electronic equipment obsolete.

One goal of EPR is to encourage producers to make product design changes that will reduce waste management costs. Such changes could include making a product easier to recycle or reuse, reducing packaging, and a variety of similar changes collectively known as Design for Environment (DfE). Although EPR has been successful in transferring waste management costs to producers, it has not been as successful in encouraging DfE. Various reasons explain this lack of success, prime among them being that current EPR policies allow producers to meet diversion targets collectively, for the industry as a whole, rather than ensuring that individual producers meet waste diversion targets for their own products. Future generations of EPR policy may be more effective in achieving the DfE goal and reducing waste at source.

WASTE MANAGEMENT FACILITY SITING

Several types of facilities may be included in an integrated waste management system. Among the most common are landfills, incinerators, transfer stations (where waste is transferred from small to large collection vehicles), composting facilities, and recycling plants. Each can have negative impacts on the biophysical and socio-economic environments. The impacts of incinerators and landfills on the environment have already been discussed. The remaining facilities can also have negative impacts, but with the exception of composting facilities, they are generally recognized as being less severe compared to incinerators and landfills. Little is known yet about the environmental impacts of emerging technologies for the thermal treatment of waste management, such as pyrolysis and gasification,[5] but they are being examined carefully as a way of increasing diversion of mixed waste and source-separated waste.

Siting waste management facilities has become a conflict-ridden process characterized by massive public opposition and disagreement over the environmental impacts of the facilities. Many characterize this type of opposition as NIMBY (not in my backyard) opposition. Dissatisfaction with traditional methods of facility siting has led to the emergence of a new approach that emphasizes co-operation over conflict. This new approach is known as the voluntary, 'open', or willing-host approach. The major difference between the traditional and the voluntary approaches is that in the former, the aim is to undertake a wide geographical search and choose the site that best meets a set of technical criteria, regardless of whether the local community in which the site is located has expressed willingness to accept the site; the latter process focuses on finding a willing host community in which there is at least one technically satisfactory site.

The first step for both the traditional and voluntary approach is to define a candidate area or general region within which the site search will be conducted. This is usually determined by considerations such as where the waste is generated, whether there are limits on how far the waste can be transported, and whether there are any political boundaries (national, provincial, or local) across which the waste cannot or should not be transported. The second step of the traditional process is to undertake an area screening by means of a technique known as constraint mapping. This technique maps information on environmental protection criteria across the candidate area and screens out the parts of the candidate area that do not meet minimum requirements for each criterion. The portions of the candidate area remaining after area screening are known as site areas. Standard area screening criteria for the siting of landfills include hydrogeological, soils, and land-use criteria. For example, any area with sandy soils would be considered a poor location for a land-fill because it does not provide natural contain-ment for leachate control.

The next step in the traditional approach is detailed analysis of data within the site areas and application of a second level of screening cri-teria in order to delineate potential sites. Typical screening criteria for landfill sites might include a minimum land area criterion or a criterion that stipulates avoidance of environmentally sensitive areas. The final step in the traditional approach is to undertake a detailed compara-tive evaluation of the potential sites and select the site that is best overall across a set of bio-physical, economic, and social criteria. This usu-ally involves identifying and weighing criteria to reflect trade-offs among them and then combin-ing the weighted scores of each alternative site into a single value. The potential site with the best score is considered to be the best site.

Critics of the traditional approach have noted a number of flaws. First, site selection relies heav-ily on scientific and technical criteria measurable on an 'objective scale'. This means that socio-psychological impacts, such as stress and lifestyle disruption, are frequently ignored because they are difficult to measure. Second, scientists and engineers usually make many of the decisions in the siting process concerning which criteria to include and how to measure and weight them. This can leave members of the general public frustrated and fuels opposition to a process over which they have little control. Research has shown that a community's concerns about a proposed landfill derive as much from inadequacies in the siting process as from the perceived impacts of the landfill (Wakefield and Elliott 2000). These con-cerns can endure even after a facility is approved

and has been operating for a number of years (Elliott and McClure 2009). However, the most likely outcome of a siting process without mean-ingful public involvement is not approval but rather an unhappy host community and rejec-tion of the site (see, for example, Baxter, Eyles, and Elliott 1999 for a case study of a failed siting attempt in the Region of Peel, Ontario). How-ever, even extensive public involvement from the beginning of the siting process is no guarantee that all stakeholders will be able to agree on an acceptable site because of conflicting value judg-ments over trade-offs among economic, environ-mental, and social criteria (see, for example, Ali 1999 for a case study of landfill siting in Guelph, ON, as well as Chapter 18).

In contrast to the traditional approach, the voluntary approach seeks cooperation with the public and attempts to locate a site in a will-ing rather than a resisting host community. A key element of the voluntary approach is that communities may opt out of the siting process at any stage. The first step in the process is to hold regional information meetings that inform local communities about the characteristics of the proposed facility and the nature of the sit-ing process. At this stage, communities have the opportunity to express an initial interest in hear-ing more about the facility. More detailed infor-mation meetings are then held in communities expressing interest. If the community politicians vote to continue with the process after this stage, then the search will begin for a potentially suit-able site within the community. Communities with no suitable sites must drop out of the sit-ing process. If one or more potentially suitable sites can be found, then the next step is to obtain community approval—for example, by holding a referendum or public meetings. Communities that approve the proposed site become poten-tial host communities. When the voluntary process produces more than a single potential host community, then the level of government responsible for finding a site must decide which of the sites being offered is the best.

In order for a community to volunteer as a host, it must perceive that the benefits of the facility will outweigh the costs. The most significant costs are the potential for harm to the natural environment and potential for negative social and health impacts. The primary benefits associated with hosting a facility are economic. Construction and operation of the facility will create employment and benefit the tax base of the community. In addition, monetary compensation packages may be offered to the community as an incentive for hosting the facility. In communities that have a declining economic base, the attraction of such economic benefits can be substantial and may outweigh concerns about social and environmental costs. An example of a potential host community where this type of trade-off might occur is a resource-based community in which resource extraction activities have closed down or are about to close down.

In principle, a siting approach that places a waste management facility in a willing host community would seem to be superior over one that imposes a facility on an unwilling host community. However, various problems with voluntary siting restrict its application in practice. First, and probably most important, there is no guarantee that any community will volunteer to host the facility. If there is no willing host, then the siting process must start again, after considerable time and money have been spent. Second, the dominant principle behind the voluntary approach is finding a site that is socially acceptable. There is a danger that this will be accomplished at the expense of protection of the natural environment. Third, some residents within the host community will experience more negative impacts from the facility than others because the facility will end up in their 'backyard'. This

group may be opposed to hosting the facility, but the rest of the community will be in favour if the perceived positive impacts on the community as a whole outweigh the negative impacts. Fourth, residents from communities close to the willing host may oppose the facility because they receive none of the benefits of the host community yet may be exposed (or perceive that they will be exposed) to increased environmental risks from proximity to the facility (Kuhn 1998). The fifth and final concern is a matter of ethics. It involves the types of communities most likely to volunteer to host waste management facilities. Only communities that have the greatest need for the economic benefits of these facilities are likely to consider volunteering. Ultimately, therefore, the poorest communities may end up bearing the greatest burden for the consequences of activities that take place elsewhere, such as nuclear power generation and industrial production.

The voluntary approach has been applied successfully in Canada to site hazardous waste facilities in Alberta and in Manitoba (Rabe, Becker, and Levine 2000). However, it has also had its failures. For example, British Columbia was unsuccessful in finding a site for a provincial integrated hazardous waste management facility (including waste treatment, incineration, and landfilling) using the voluntary approach. Kuhn and Ballard (1998) attribute this failure to weaknesses in the public consultation program and subsequent loss of trust in the process by the two willing host communities, Cache Creek and Ashcroft.

Another example of a voluntary approach that was not entirely successful occurred during the search for a low-level radioactive management site in Ontario. After a 10-year willing-host search, Deep River reached an agreement-in-principle with the federal government in 1995–6 to host the site. The agreement came after extensive public consultation and a community referendum in which 72 per cent of the residents of

Deep River voted to host the facility (Gunderson and Rabe 2000). Deep River is a town of 14,600 people that has been the site of the primary research and development facility for Atomic Energy of Canada Ltd since the end of World War II. Threatened by a declining economy, its motive for volunteering was to secure employment for a workforce that had extensive experience in the nuclear industry. However, in December 1996, the town council of Deep River voided the agreement, claiming that the federal government had refused to provide job guarantees that the town had thought would be part of the benefits package (Gunderson and Rabe 2000).

A final example of an unsuccessful application of the voluntary siting approach comes from Toronto. The City of Toronto began looking for a new landfill site in 1988 using a variety of traditional approaches but was not successful with any of them. Switching to a voluntary approach, Toronto finally identified a potential willing host for the city's new solid waste disposal site in 1998. The site was an abandoned open-pit mine in Kirkland Lake, ON, a small resource-based community located about 600 kilometres to the north of Toronto. Toronto's waste was to be transported there by rail and processed for recovery of recyclables before disposal of the remaining waste.

Although a majority of town council members in Kirkland Lake volunteered their community, there was considerable local opposition to the proposal among residents of the town and by the mayor of Kirkland Lake. Nearby communities were also strongly opposed because they were concerned about potential contamination of their groundwater. Despite vehement protests by northern Ontario residents and environmental groups at meetings of Toronto City Council, the city appeared ready to accept the Kirkland Lake offer. It was only a last-minute rereading of the proposed disposal contract that uncovered financial details that were unacceptable to the city. Toronto rejected the proposal and chose instead to start its search again, this time calling for proposals from private-sector operators of existing landfill sites.

Facing the closure of its last remaining landfill in 2002, Toronto finally awarded a contract for disposal of the city's garbage to a private landfill operator in the United States. Starting in 2003, 142 truckloads of waste travelled 375 kilometres every weekday from Toronto to a landfill site near Detroit, Michigan. That number has since fallen to 74 truckloads per day as Toronto has been successful in diverting more and more waste from landfill with a variety of programs, most notably its 'green bin' organics diversion program. Toronto finally found its own landfill in 2007 when it purchased an existing private landfill near London, ON. The capacity of the landfill for waste disposal should extend 20 to 30 years, depending on the level of waste diversion, and it will start taking all of Toronto's waste in 2011.

CONCLUSION

Conflict and uncertainty are present in many aspects of waste management. For example, as the focus of integrated waste management has turned to waste diversion, uncertainty has developed over the most effective ways to achieve high diversion rates. Since the conventional blue box system in communities across Canada can only achieve moderate diversion rates, it is being replaced by various high-diversion, two- and three-stream systems, some of which are so new that their long-term advantages and disadvantages are still unknown. Uncertainty in the public's mind about the existence and consequences of harmful emissions from incineration and its potential conflict with recycling has removed incineration as an integrated waste management option in many communities. Great uncertainty also exists surrounding the environmental and financial performance of emerging technologies such as pyrolysis and gasification. On a positive note, uncertainty has been decreasing in a few areas. For example, we now have much better municipal waste generation and waste composition data than we did 10 years ago. Progress is also being made toward standardizing these data across the country.

One apparently intractable conflict in waste management is community opposition to waste management facilities, particularly landfills. The emergence of the voluntary siting approach has not been as effective as many had hoped in resolving this conflict. Another point of contention in waste management is the user-pay system. Although it is a proven policy instrument for increasing waste diversion, the introduction of user-pay systems faces vehement opposition in many communities.

Challenges still lie ahead in the field of waste management. A principal challenge is keeping pace with the rapid evolution of new ideas for moving management of wastes up the hierarchy toward source reduction. Another challenge lies in resolving some of the key debates in waste management. For example, is it right for a community to impose its waste disposal problems on another community? Is widespread adoption of user fees the only way to make households aware of their wasteful habits? How close can we get to 100 per cent waste diversion or zero waste? Should all manufacturers be held accountable for the environmental impacts of their products throughout the product's lifecycle? Although the ethical nature of some of these questions and the environmental, social, and economic trade-offs that they imply place any definitive answers beyond the scope of this chapter, they are vitally important questions, the answers to which will likely shape the future of waste management planning in Canada.

FROM THE FIELD

When I was a young assistant professor in the early 1980s, newly arrived at the University of Toronto, I had never heard of waste management as a field of study. My interests then were more broadly in environmental conflicts, decision-making, and environmental assessment. In my third year, a colleague in the Department of Geography suggested that I apply for a seat on the City of Toronto's Recycling Action Committee. He had just stepped down from the committee and knew that there was an opening. At the time, a main focus of the committee was to bring the blue box to Toronto, following the lead of Kitchener, ON, the community that had introduced the blue box to North America in 1981. I had been looking for an opportunity to engage in community service related to the environment, and this sounded just about right. My application was successful, and influenced by my initial exposure to waste issues on the committee, I have subsequently devoted much of my research efforts to the 'garbage' field.

When I started conducting research on waste management, I discovered that three academic journals were available on the subject but all emphasized engineering and scientific aspects. Very few social scientists were writing about waste management at the time. As noted in this chapter, the way that we view waste has changed considerably over the past several decades, and these changing perspectives have produced a flourishing academic literature on the social, economic, and policy aspects of waste management.

To understand how much the field has grown since the 1980s, I searched all articles in the Social Science Citation Index database using the keywords 'waste management', 'waste reduction', 'waste reuse', 'recycling', and 'landfills'. Between 1980 and 1989, only 147 journal articles were published on these topics, between 1990 and 1999 there were 677, and between 2000 and 2009 there were 1052. This is just a crude indicator of the significant growth in academic writing on waste management, since it does not include all possible keywords that might be relevant. Nevertheless, it does reflect the growing importance that society places on managing waste as a resource rather than merely something to be sent for disposal—'out-of-sight, out-of-mind'.

Waste management research is still evolving and growing because there is still so much we don't know. One example is waste diversion behaviour, about which much has been written but that changes as source-separation programs collect new materials that place more and more demands on our ability to separate wastes. In many communities, there are long lists of what can and cannot be put in the blue box and green bin. In Toronto, the green bin program accepts disposable diapers but not baby wipes, which is difficult for many people to understand. We can put some types of plastics in the blue box but not all, and trying to remember and identify which ones belong is not an easy task. Although recycling has become a social norm, meaning almost everybody does it, many are not doing it right. Research can help us to determine the limits of pro-environmental behaviour, the impact on behaviour of better communication strategies, and the potential of other options for diverting and reducing waste, such as Design for Environment. Regarding waste management issues, therefore, the research opportunities are plentiful and address pressing environmental concerns.

—Virginia Maclaren

Notes

1. This diversion rate is lower than the actual rate because the Statistics Canada data do not include reuse activities (such as the sale of used clothing, the reuse of bottles in deposit-return programs, or the refurbishment of computers) or backyard composting.

2. In Figure 14.1, small appliances include waste electronic and electrical equipment; bulky objects include white goods (i.e., large appliances such as refrigerators and stoves), carpets, and furniture; household hygiene includes diapers, animal litter, and sanitary products; other includes textiles, rubber, leather, wood, and fine particles.

3. Note that Statistics Canada does not provide residential diversion quantities for Prince Edward Island and Newfoundland and Labrador for reasons of confidentiality (required when numbers are small). The overall diversion rates for these two provinces, including waste from both residential and non-residential sources, are 37.8 per cent and 6.9 per cent, respectively.

4. A relatively small number of multi-family dwellings have more convenient systems, such as collection of recyclables by maintenance staff from the chute room on each floor or specialized chute systems that direct recyclables to recycling bins at the bottom of the chute. In smaller multi-family dwellings (e.g., three to four storeys), residents may have curbside collection.

5. Pyrolysis and gasification are thermal treatments that use high temperatures to break down carbon-based materials, such as paper, plastics, and food waste, into gas, solid, and liquid residues. Unlike incineration, pyrolysis uses no oxygen, and gasification uses very little oxygen. Both produce syngas, which can be used as a fuel to generate electricity or steam.

REVIEW QUESTIONS

1.   Why is it important to understand waste-stream composition when planning a waste management system?
2.   What factors affect the generation of waste in the residential and ICI sectors?
3.   What types of activities make up the waste management hierarchy, from top (i.e., best for the environment) to bottom? How does lifecycle analysis help in measuring environmental performance within the hierarchy?
4.   What is waste diversion? What makes Nova Scotia a leader in waste diversion?
5.   Why are waste diversion rates so much lower in multi-family dwellings than in single-family dwellings?
6.   Why are user-pay systems and extended producer responsibility becoming popular as waste management policy tools? How effective are they, and can they be improved?
7.   What are the strengths and weaknesses of the voluntary versus the traditional approach to waste management facility siting?

REFERENCES

Ali, S.H. 1999. 'The search for a landfill site in the risk society'. *The Canadian Review of Sociology and Anthropology* 36 (1): 1–12.

Baxter, J.W., S.D. Eyles, and S.J. Elliott. 1999. 'From siting principles to siting practices: A case study of discord among trust, equity and community participation'. *Journal of Environmental Planning and Management* 42 (4): 501–25.

British Columbia Ministry of Water, Land and Air Protection. 2002. 'Overview of product stewardship in British Columbia'. http://wlapwww.gov.bc.ca/epd/epdpa/ips/index.html.

Bury, D. 1999. 'The Canadian approach to EPR'. *At the Source* fall: 2. http://c2p2.sarnia.com.

Canadian Council of Ministers for the Environment. 2000. 'National packaging protocol 2000, final report'. Ottawa: Canadian Council for Ministers of the Environment. http://www.ccme.ca.

———. 2009. 'Canada-Wide Action Plan for Extended Producer Responsibility'. Ottawa: Canadian Council of Ministers of the Environment. http://www.ccme.ca/assets/pdf/epr_cap.pdf.

City of Calgary. 2006. *Waste and Recycling Services Annual Report*. Calgary: City of Calgary.

City of Edmonton. 2007. 'EcoVision annual report 2007'. Edmonton: City of Edmonton. http://www.edmonton.ca/environmental/documents/EcoVisionAnnualReport2007.pdf.

City of Toronto. n.d. 'Residential diversion rate'. http://www.toronto.ca/garbage/residential-diversion.htm.

Communauté métropolitaine de Montréal. 2008. 'La gestion des matières résiduelles: un défi prioritaire pour la Communauté métropolitain de Montréal'. Report presented to the Commission of Transportation and Environment, National Assembly of Quebec. http://www.cmm.qc.ca/fileadmin/user_upload/pmgmr_doc/GMR/audition_transport-Memoire.pdf.

Elliott, S., and J. McClure. 2009. '"There's just hope that no one's health is at risk": Residents' reappraisal of a landfill siting'. *Journal of Environmental Planning and Management* 52 (2): 237–55.

Environment Canada. 2001. 'Extended producer responsibility and stewardship'. http://www.ec.gc.ca/epr/en/index.cfm.

EnvirosRIS. 2001. 'The waste diversion impact of bag limits and PAYT (pay-as-you-throw) systems in North America'. Toronto: City of Toronto, Works and Emergency Services Department, Policy and Planning. http://www.wdo.on.ca/wdo_reports3.htm.

Finnveden, G., et al. 2005. 'Life cycle assessment of energy from sold waste, part 1: General methodology and results'. *Journal of Cleaner Production* 13: 241–52.

Gunderson, W.C., and B.G. Rabe. 2000. 'Voluntarism and its limits: Canada's search for radioactive waste-siting candidates'. *Canadian Public Administration* 42: 193–214.

Halifax Regional Municipality. 2007. *Solid Waste/Resource Management System—Diversion Opportunities*. Report to Council, 1 February.

Hanisch, C. 2000. 'Is extended producer responsibility effective?' *Environmental Science and Technology* April: 170–5A.

Kuhn, R.G. 1998. 'Social and political issues in siting a nuclear-fuel waste disposal facility in Ontario, Canada'. *The Canadian Geographer* 42 (1): 14–28.

Kuhn, R.G., and K.R. Ballard. 1998. 'Canadian innovations in siting hazardous waste management facilities'. *Environmental Management* 22 (4): 533–45.

Methane to Markets. 2008. 'Methane to Markets Partnership Landfill Subcommittee country specific profile and strategic plan for Canada'. http://www.methanetomarkets.org/landfills/index.htm#profiles.

Miranda, M.L., and D.Z. Bynum. 2002. 'Unit-based pricing and undesirable diversion: Market prices and community characteristics'. *Society and Natural Resources* 15 (1) : 1–15.

Nova Scotia Environment. 2008. 'Not going to waste: Final report on Nova Scotia's 1995 Solid Waste Resource Management Strategy'. Halifax: Nova Scotia Department of the Environment. http://www.gov.ns.ca/nse/waste/docs/SolidWasteStrategyFinalReport2008.pdf.

Quinn, L., and A.J. Sinclair. 2006. 'Policy challenges to implementing extended producer responsibility for packaging'. *Canadian Public Administration* 49 (1): 60–79.

Rabe, B.G., J. Becker, and R. Levine. 2000. 'Beyond siting: Implementing voluntary hazardous waste siting agreements in Canada'. *American Review of Canadian Studies* 30 (4): 470–96.

Recycling Council of British Columbia. 2000. 'The "dirt" on composting in British Columbia'. Vancouver: Recycling Council of British Columbia, Organics Working Group. http://www.rcbc.bc.ca/publications/policy_reports/organics_recycling_reports.htm.

Recycling Council of Ontario. 2000. 'Assessment of multi-unit recycling in Ontario'. Toronto: Recycling Council of Ontario. http://www.wdo.on.ca/wdo_reports3.htm.

Reimer, G. 2008. 'City of Ottawa Take it Back! program'. Presentation at the Packaging Waste Reduction Forum, 10 September, Toronto. http://www.toronto.ca/garbage/packagingforum/pdf/4_reimer_coo_tib.pdf.

Shah, K. 2000. *Basics of Solid and Hazardous Waste Management Technology*. Columbus, OH: Prentice-Hall.

Statistics Canada. 2002. *Waste Management Industry Survey: Business and Government Sectors, 2000*. Ottawa: Statistics Canada.

———. 2008a. *Waste Management Industry Survey: Business and Government Sectors, 2006*. Ottawa: Statistics Canada.

———. 2008b. *Households and the Environment 2006*. Ottawa: Statistics Canada.

Technology Resources Inc. 2005. *Solid Waste Composition Study*. Burnaby: Greater Vancouver Regional District.

United States Environmental Protection Agency. 2008. 'Municipal solid waste in the United States: 2007 facts and figures'. Washington: US Environmental Protection Agency, Office of Solid Waste and Emergency Response. http://www.epa.gov/epawaste/nonhaz/municipal/pubs/msw07-rpt.pdf.

Wakefield, S., and S.J. Elliott. 2000. 'Environmental risk perception and well-being: Effects of the landfill siting process in two southern Ontario communities'. *Social Science and Medicine* 50: 1139–54.

Wagner, T., and P. Arnold. 2008. 'A new model for solid waste management: An analysis of the Nova Scotia MSW strategy'. *Journal of Cleaner Production* 16: 410–21.

Waste Diversion Ontario. 2007. 'Residential GAP diversion rates'. Toronto: Waste Diversion Ontario. http://www.wdo.ca/content/?path=page82+item35931.

PART 3

Responses

Part 3 contains six chapters, each of which considers a particular approach or perspective that can enable society to respond to conflict and uncertainty: an ecosystem approach, adaptive management, environmental assessment, social learning and participation, gender, and governance.

In Chapter 15, Scott Slocombe considers attributes of an *ecosystem approach*. He emphasizes that while no universally agreed-upon definition exists, all definitions include certain core elements. He examines how different interpretations have been applied to various management situations. In his guest statement, Michael Fox explores the challenges of moving from theory to practice.

In Chapter 16, Bram Noble explores the basic characteristics and principles of adaptive environmental management, developed to recognize that our knowledge is often incomplete and that uncertainty is normal. He also considers experience in Canada with adaptive management in situations including forestry, fisheries, minerals, wildlife, and parks.

John Sinclair and Meinhard Doelle co-authored Chapter 17, a new chapter dealing with environmental assessment, including initiatives for improvement: sustainability assessment, strategic environmental assessment, community-based assessment, and multi-jurisdictional assessment. In their guest statement, Graham Whitelaw and Dan McCarthy examine how the federal environmental assessment process was applied to the Victor Diamond Mine in northern Ontario.

Alan Diduck, in Chapter 18, examines participatory and social learning approaches. He outlines their advantages and disadvantages and concludes by reviewing examples of best practice applied in environmental assessment, participatory research, forestry, and, co-management.

Gender and governance are the twin foci of Chapter 19. Maureen Reed seeks to understand how aspects of gender in Canada affect institutions, capacities, and implementation. Her main example relates to understanding how 'gender in forestry' can inform institutions for forestry governance. Alison Gill's guest statement provides another perspective regarding gender when she considers implications of policies for resource-dependent regions in British Columbia.

In Chapter 20, Tony Dorcey explores how innovations in environmental governance have been applied in the Vancouver metropolitan area. He argues that participation in Canadian governance processes has evolved in three waves of innovation over the past 50 years: (1) experiments with public participation; (2) multi-stakeholder, conflict resolution, and consensus-building processes; (3) participatory approaches, including new multi-stakeholder processes and electoral reform.

15

Applying an Ecosystem Approach

D. Scott Slocombe

Learning Objectives

- To understand the history of ecosystem approaches.
- To recognize the common elements of ecosystem approaches.
- To understand the influence of ecosystem approaches in resource and environmental management.
- To be familiar with some examples of the applications of ecosystem approaches in resource and environmental management.
- To appreciate the challenges of applying ecosystem approaches.
- To be aware of emerging trends in the application of ecosystem approaches.

An ecosystem approach is difficult to define precisely. Indeed, there is no single, universally agreed upon ecosystem approach. There are many ecosystem approaches—variations around common concepts, goals, and methods—developed for application in a wide range of resource and environmental management, and other, problem contexts. The concepts and perspectives embodied in ecosystem approaches provide a framework for thinking about and intervening in resource and environmental problems that seeks to overcome common shortcomings of the way we as individuals, organizations, and societies tend to interact with nature and resources. The ideas and methods of ecosystem approaches are found in integrated resource management and watershed management, among others.

Ecosystem approaches address common problems that have been identified time and again in resource and environmental management over the past several decades:

- viewing people and their activities as separate from nature;
- fragmented knowledge or disciplines, ecosystems, jurisdictions, and management responsibilities;
- emphasizing single resource uses or economic sectors and avoiding or ignoring conflict over possible alternate uses;
- failing to recognize the myriad ways in which ecological and socio-economic systems are interconnected;
- ignoring the propensity of biophysical and socio-economic systems to change, sometimes rapidly and unexpectedly; and

• reacting to rather than anticipating change and problems, leading to attempts to eliminate uncertainty by controlling complex, dynamic systems instead of adapting to them.

The diverse origins of ecosystem approaches are a part of their strength, as I will explore below. Although clearly rooted in the ideas of 'ecology' and 'ecosystem', they were greatly expanded and developed, beginning in the late 1960s and 1970s, by researchers and practitioners in many disciplines: environmental studies, to be sure, but also political science, anthropology, sociology, human ecology, geography, and psychology. Current writers and environmental professionals often draw on several of these disciplines, depending on their specific background and working environment.

An implication is that ecosystem approaches have a range of central components, variously influenced by the several disciplines and professions that have shaped the development of ecosystem approaches and resource and environmental management in particular. As a hint of what is to come, recurring core elements of ecosystem approaches include systems concepts and analysis, ethical perspectives, stakeholder and public participation, a bioregional or place-based focus, efforts to identify and develop common goals, and developing detailed understanding of the (eco)system of interest.

The development of ecosystem approaches in the past 40 years has been strongly linked to several other approaches also represented by chapters in this book, especially adaptive management (Chapter 16), sustainability and environmental assessment (Chapter 17), and participatory approaches (Chapter 18), as well as ecosystem health and integrity (Chapter 3). The reader may want to be alert to links while reading this chapter and the others ones mentioned. Issues of governance have emerged as central

in many of these areas and are now being given more attention (Chapters 2 and 19).

In turn, ecosystem approaches have been discussed and influenced in Canada in many different resource and environmental management contexts. Among the more important of these are watershed management and restoration, especially in the Great Lakes Basin; forest management; comprehensive regional land-use planning; parks and protected areas; coastal management and ocean management; and environmental studies and monitoring. This chapter examines each of these areas, with specific examples, to illustrate ecosystem approaches.

The rest of the chapter follows up on the dimensions of ecosystem approaches introduced here, beginning with their origins, moving on to their common components and dimensions, and including current developments. The chapter then describes their elaboration and application in several aspects of resource and environmental management. A final section discusses the challenges and opportunities in wider use of ecosystem approaches and provides conclusions and reflections on current trends.

THE ROOTS OF ECOSYSTEM APPROACHES

Ecology is commonly defined in terms of the study of the distribution and abundance, and at least implicitly the interrelationships, of organisms. In this sense, it goes back to the late nineteenth century and has been influential beyond the original study of plants and non-human animals in encouraging studies of interrelationships and interactions of cultures, organizations, individuals, companies, and others. In the early twentieth century, human ecology was perhaps the first such expansion. It originated in efforts to better understand the interactions, health, economic status, and future prospects of people in the poorer, more densely populated parts of such cities as Chicago, New York, and London.

Originally, and still in the strict sense, ecology focuses on organisms and their interactions and limiting factors. Arthur Tansley coined the term 'ecosystem' in 1935 to provide a reasonably precise and comprehensive (or holistic) term for the set of biological and physical factors that affect an organism—going beyond other organisms to include things like soils, water, weather, and climate. Today, in strict natural science usage, an ecosystem is a locally distinct and coherent ecological community of organisms and the physical environment with which they interact. In the mid-twentieth century, 'ecosystem' was mostly used to expand understanding of the dynamics of particular organisms or communities of organisms (see Bocking 1994 and Likens 1992 for more detailed histories of the ecosystem concept, and Slocombe 1993a on ecosystem approaches).

By the 1960s, ecosystem science had come into its own, catalyzed by studies that sought to integrate knowledge of a wide range of processes (e.g., hydrological, soils, meteorological) and flows (e.g., energy, nutrients) with knowledge of organisms. This systems ecology was fostered by large international collaborative research programs such as the International Biological Program (IBP) in the 1960s and the Man and the Biosphere Program (MAB) in the 1970s. The latter had explicitly conservation-oriented aims. This research and program development significantly broadened ideas of ecosystems, contributing substantially to the development of the environmental movement in, for example, the new understanding it provided of the pathways of pollutants in the environment. The roots of many critical conservation approaches today, such as conservation biology, ecosystem integrity, and landscape ecology, lie in this research, which continues today—perhaps most

obviously in the work of large international projects on climate change such as the International Geosphere-Biosphere (IGBP), Human Dimensions of Global Change, and Millennium Ecosystem Assessment projects.

By the 1970s, as the environmental and conservation movements were developing along many routes and fronts, the ecosystem idea began to be adopted by researchers and professionals in many fields, as ecology had been earlier. The word 'ecosystem' is used analogously to its use in ecosystem science, but the concept has been broadened conceptually and is often not tied so firmly to a clearly definable place. In the broadest sense, an ecosystem approach studies something (a system of interest), its environment, and the interactions between them. One starts by defining an ecosystem of interest as the unit of study and then looks at the processes, flows, and interconnections within the ecosystem and between it and its environment. So the 'ecosystem' might be a park, a city, a family, a company, a culture, not just a specific, local wetland or forest or coral reef.

Examples of applications of ecosystem approaches go back to the 1940s and 1950s in cultural ecology efforts to understand cultural evolution and particular environments and more recently in ecological anthropology studies of the adaptation of societies to their natural environments (Moran 2000). In psychology, ecosystem approaches had a strong influence in the 1970s and 1980s among counsellors and therapists, who used them to focus on the way people perceive and relate to their environment. The individual's interactions with others, perception of the ecosystem, and processes of adaptation—all of which are strongly influenced by positive and negative feedbacks and mutually reinforcing (often dysfunctional) behaviours and cycles—play a part in how functional or effective an individual is in his or her environment. Such approaches have had considerable influence in management in the 1990s and beyond, most famously in Peter Senge's (1990) systems-based learning organization ideas.

During this period, ecosystem approaches also had a strong influence in the emerging fields of environmental studies and planning. For example, they were used to highlight ecological constraints and contexts for human societies (Darling and Dasmann 1969) and to broaden the context and highlight biophysical–socio-economic interconnections in planning for land use, urban areas, or protected areas (Caldwell 1970; Dorney 1973). Human ecology drew on similar ideas in this period, and some classic studies were completed (e.g., Boyden et al. 1981).

Perhaps the most influential area of development of ecosystem approaches, however, was in the management of the Great Lakes Basin and especially in revision of the original 1972 Great Lakes Water Quality Agreement (GLWQA) in the 1970s and early 1980s. The approach remains at the centre of the activities of the International Joint Commission in this region and of the management and science boards established under the GLWQA. It underlies such long-term goals as zero discharge of persistent toxic contaminants and the development of Remedial Action Plans for areas of concern (discussed below and in Chapter 7). As developed and applied in the Great Lakes Basin, the ecosystem approach is holistic, interdisciplinary (bridging and integrating traditional disciplines), goal-oriented, and participatory and tries to get people to recognize and act as though they are part of the ecosystem, not separate from it (see US National Research Council and Royal Society of Canada 1985; Hartig and Vallentyne 1989). This experience has probably most strongly influenced current thinking about ecosystem approaches as I will define and explore them in the next two sections.

In the past 15 years, ecosystem-based management (or, more commonly, ecosystem management) has become a major subject of academic and professional interest and development (see Slocombe 1993b, 1998; Grumbine 1994; Meffe et al. 2002; Layzer 2008). The development of ecosystem-based management has certainly been influenced by ecosystem approaches but also, again especially in the US, by protected areas planning, integrated resource management, regional planning, and the natural science ideas of ecology and conservation biology. Strictly speaking, an ecosystem approach is a philosophy or framework; ecosystem management is a practice with much narrower goals and more specific applications—although many writers blur the distinctions.

THE ELEMENTS OF ECOSYSTEM APPROACHES

Ecosystem approaches, like systems approaches, tend to be different things to different people. This is both a strength and a weakness—a strength to those seeking flexible, innovative, systemic approaches; a weakness to those attached to the rigour, repeatability, and idealized objectivity of traditional scientific approaches. To an ecologist, an ecosystem approach means either study of an entire local ecosystem in all its biophysical complexity or study of an organism in its ecosystemic context. Environmental studies and management use analogous approaches but with different foci (often larger areas, or people or organizations) and with a broader range of processes and elements included. It is this spatial and disciplinary broadening that is most controversial to some, yet most useful to others.

Broadly speaking, ecosystem approaches in the various disciplines that have drawn on them share a number of characteristics and differ in others. The biggest differences are probably in the basic dimensions of empirical/

theoretical, quantitative/qualitative, and equilibrial/dynamic. There are common and fundamental similarities, however, in the near-universal inclusion of some form of systems analysis; a focus on interactions, processes, and flows (matter, energy, information); an interest in both the biophysical and socio-economic aspects of the system and their interactions; an attention to institutions and management options; an integrative, holistic, interdisciplinary perspective; an orientation toward anticipation and adaptive responses; and, somewhat less often, an ethical perspective on action and decision-making.

More specifically, I have derived a series of characteristics of ecosystem approaches (Slocombe 1993a). Although few initiatives would embody all of them, most have several, and they are presented roughly in order of commonness. These characteristics are:

- describing parts, systems, and environments and their interactions;
- holistic, comprehensive, interdisciplinary description and analysis;
- including people and their activities in the ecosystem;
- describing system dynamics—e.g., with concepts of homeostasis (self-regulation), feedbacks, cause-and-effect relationships, self-organization, and so on;
- defining the ecosystem naturally—e.g., in biophysical and/or cultural terms—instead of arbitrarily;
- looking at different levels and scales of system structure, process, and function;
- recognizing goals and taking an active, management orientation;
- including understanding of actor and stakeholder relationships and interactions and institutional factors in analysis;
- using an anticipatory, flexible research and planning process;

- entailing an implicit or explicit ethic of quality, well-being, and integrity; and
- recognizing systemic limits to action—defining and seeking sustainability.

This list is presented as a set of goals or desired characteristics. How these characteristics can be achieved or implemented is a somewhat different question. There are probably several core theories and practices that tend to be used, most of which have already been mentioned. They would include systems approaches, comprehensive and/or systematic resource and environmental description, boundary-setting, adaptive, hierarchical complexity-based analysis, institutional analysis, public participation and conflict resolution approaches, and efforts to set common goals and/or develop a vision.

Systems approaches are varied and range from the highly quantitative, as in engineering and computer science, to the highly qualitative, as in sociology and organization science. Systems approaches in environmental studies are often somewhere in the middle. What they have in common are systematic efforts to define a core focus—the system of interest and the boundaries between it and its environment—and to identify the often hierarchical levels of subsystems that make it up. Special effort goes into identifying flows and other interactions between and within a system and its environment and subsystems. Further elaborations seek to understand what determines the state of the system, its tendencies toward stability or change, and feedback or self-organizing mechanisms that stabilize or destabilize the system (see Meadows (2008) for a fine introduction and Marten (2001) for much relevant elaboration in a human ecology/ ecosystem approach context).

Comprehensive resource description is an umbrella term for efforts to describe a system, or ecosystem, in terms of many different characteristics—e.g., ecology, geology, economy, or culture. Its roots lie in environmental impact assessment, comprehensive regional planning, and protected areas planning in the 1970s, when it was recognized that such a comprehensive knowledge base was critical to making good plans, decisions, and boundaries.

The key challenges are what information to include, where to find it, and how to organize and, ideally, integrate it. McHarg (1969) addressed these questions well and pioneered cartographic and geographic information systems (GIS)-based efforts to integrate and analyze information; Bastedo, Nelson, and Theberge (1984) outlined a simple, hierarchical, integrative, and multidisciplinary approach that has had considerable influence in Canadian ecosystem and protected areas management; and Jensen and Bourgeron (2001) provided a state-of-the-art description of current best practice using information technologies, quantitative analyses, and large-scale co-ordination and participation processes. The most recent approaches emphasize multi-scale connections, strengthening socio-economic dimensions, and increasing public participation in gathering data and access to data products, influenced by systems and complexity ideas (e.g., Danby and Slocombe 2005; Silver 2008; Bingham 2007; McLain et al. 2008).

Adaptive management builds on scientific and descriptive methodologies as well as systems dynamics ideas to argue for a new way of designing and learning from management actions (Chapter 16). It is a response to complexity and uncertainty in human–environmental systems and is addressed in the next chapter. Other significant practices include detailed analyses of the structure, processes, and interactions of institutions and organizations and how they do or do not effectively devise and implement solutions to problems or ways to achieve goals. It and methods for involving the public and seeking to resolve conflicts are addressed in Chapters 4 and 18.

DEVELOPMENT AND APPLICATION OF ECOSYSTEM APPROACHES

In this section, several examples of the application of ecosystem approaches in Canada are examined. The applications are watershed management and restoration, especially in the Great Lakes Basin; forest management; comprehensive regional land-use planning; parks and protected areas; coastal management and ocean management; and environmental studies and monitoring. Some use an ecosystem approach more explicitly than others, and some are more developed than others, but taken together they provide an illustration of ecosystem approaches in practice, as well as a range of ideas for applications of ecosystem approaches to other problems and places. Several of these areas of application are also treated in other chapters of this book.

Watershed Management and the Great Lakes Basin

As noted earlier, ecosystem approaches have a long history in the Great Lakes Basin on both sides of the Canada–US border. They emerged in response to the complex environmental problems of a very large, international lake system with small watersheds, a population well over 40 million, and some of the largest population and industrial concentrations in both Canada and the US (see Environment Canada and United States Environmental Protection Agency 1995; Grady 2007). Some of the defining environmental issues and controversies of the environmental movement—eutrophication of the lakes and toxic-pollutant landfills such as Love Canal in Niagara Falls, NY—took place in the basin.

Great Lakes Basin issues and approaches emerged in the context of a long history of watershed management approaches in many parts of the world. And as water quantity and quality issues continue to increase in number and significance, watershed management is also gaining greater attention (see Hooper 2005; Global Water Partnership and INBO 2009).

The size, complexity, and significance of the lakes and their economies have given rise to many different approaches, to such an extent, in fact, that only a few central ones can be mentioned here. Lee, Regier, and Rapport (1982) reviewed 10 early ecosystem approaches to the planning and management of the Great Lakes. They looked at initiatives at four levels—policy, strategy, tactics, and tools—and found that most shared several common features, which could serve as ecological criteria for an ecosystem approach. Their criteria primarily emphasized ecological characteristics of the system; system boundaries that reflect ecological integrity; use of mapping, monitoring, and modelling to assess ecological conditions; and consideration of system (ecological) responsiveness and self-regulation. Several of the initiatives discussed then evolved into or influence current management activities, notably Remedial Action Plans (RAPs), Lakewide Management Plans (LaMPs), State of the Lake Reporting, and the IJC's ecosystem approach itself. The first of these offers a well-documented example of good, specific implementation of an ecosystem approach (MacKenzie [1996] provides the most in-depth review), while LaMPs have seen considerable development as vehicles for implementation of an ecosystem approach over the past five years (GLC 2004; 2008).

Remedial Action Plans were recommended in 1985 by the Great Lakes Water Quality Board of the International Joint Commission and were included in the 1987 GLWQA revisions with the aim of reducing the impacts of pollution and toxic contamination on the water quality, fish and wildlife, and use and aesthetic value in the

43 most degraded sites around the basin—officially called Areas of Concern. Among the main Canadian Areas of Concern are Hamilton harbour, Toronto harbour, the St Clair River, the Bay of Quinte, the Spanish River mouth, Nipigon Bay, and Thunder Bay, all in Ontario. The RAP process was explicitly designed to bring institutions and stakeholders together to work collaboratively toward common, explicitly defined ecosystem goals (Hartig and Vallentyne 1989).

The RAP program requires a three-stage process for each Area of Concern. At the end of each stage, a report is prepared for provincial and state management agencies and the IJC to approve. The *first stage* involves investigation and documentation of environmental problems. *Stage 2* outlines appropriate ecosystem objectives, remedial measures, monitoring, and a timetable for completion. The *third stage* details proof of the restoration of beneficial uses and outlines continued monitoring. When monitoring has demonstrated a satisfactory level of restoration, the area can be removed from the list of Areas of Concern. Most areas are now in Stage 2 or 3; Collingwood harbour (in 1994) and Severn Sound (in 2003) are the only areas in Canada that have been removed from the list so far. As more areas move toward Stage 3, attention is shifting to monitoring and research needs to attest to remediation (Hall, O'Connor, and Ranieri 2006; Grapentine 2009).

The details of RAPs vary from place to place, but at their heart is a focus on bringing stakeholders together, with good information on impairments, to work on developing feasible solutions. The more successful RAPs have brought together environmental non-governmental organizations (ENGOs, industry, and local and senior levels of government) to identify and jointly implement and fund actions to improve the environment. This has rarely been a smooth process, and for some of the Areas of Concern it will be a very long-term process. A review of RAPs highlighted data compilation, defining guiding principles and a long-term vision, building partnerships, and using market forces as key contributors to success (Hartig et al. 1998), along with the following more specific principles:

- broad-based stakeholder involvement;
- commitment of top leaders;
- agreement on information needs and interpretation;
- action planning within a strategic framework;
- human resource development;
- results and indicators to measure progress;
- systematic review and feedback; and
- stakeholder satisfaction.

These are largely process characteristics and prerequisite actions for an ecosystem approach and RAPs. Other studies have sought to identify the specific actions taken and tools used to implement an ecosystem approach in RAPs (Hartig 1995). Examples include inventorying ecosystem features and developing a GIS database; developing partnership agreements for watershed planning and management; incorporating lifecycle analysis into pollution regulations and assessments; using stewardship techniques such as easements to protect habitat; establishing citizen stewardship programs; using economic instruments to develop mutually beneficial environmental and economic solutions; and ensuring environmental education and human resource development in government and the public to foster an ecosystem approach.

Mitchell (1998, 39–42) provides a lengthy list of lessons and principles for integrated watershed management and community-based approaches. Examples that complement the above lists include recognizing local context and conditions, the importance of having a vision and a champion, willingness to share or distribute power, the need to make decisions by consensus, the potential toll and burnout, especially on the part of public volunteers, and encouraging mutual teaching and learning.

There have been growing calls to revisit the Great Lakes Water Quality Agreement, and on 13 June 2009 the US and Canadian federal governments announced that they would review the agreement. Key issues include reflecting new understanding of the lakes, new threats such as climate change, invasive species, and new contaminants, and new political challenges and opportunities (see Babbage 2009). It will be a long and complex process, with numerous issues to address (cf. Krantzberg 2009).

Forest Management

Forest management has long been an economically critical and a practically and intellectually influential component of resource and environmental management (see Chapters 10, 16, and 19). Many important concepts, such as multiple use and integrated resource management, have emerged from forest management theory and practice. Ecosystem approaches and management have also had strong links to forestry at the national level in the US and, to a lesser degree, in Canada.

The influence of ecosystem approaches in Canadian forest management has been fairly subtle until recently, with few examples of highly explicit adoption of the ideas, even in the form of ecosystem management (but see McAfee and Malouin 2008). Nevertheless, some time ago Stan Rowe (1992) presented an explanation of and argument for an ecosystem approach in forestry. He saw an ecosystem approach in terms of focusing on wholes rather than parts, seeking sustainable resource use in terms of using the interest and not the capital, and focusing on the multi-dimensional landscapes and waterscapes in which the related activities of forestry, wildlife

management, recreation, and water management occur. This led him to suggest three central
questions: (1) To what extent can we maintain
natural, semi-natural, and artificial ecosystems
on public lands? (2) Since semi-natural ecosystems are the main part of our forested ecosystems, what are the amounts and dynamics of
their natural capital and interest? and (3) How
can we use and recognize scale and the need for a
regional approach to management? Ecoforestry
advocates draw on and endorse many of the
same ideas, seeing forests as ecological communities, seeking long-term sustainability, using an
ecosystem model, accepting nature's design or
limits, and fostering natural diversity and complexity, while recognizing local traditions and
needs (e.g., Drengson and Taylor 1997).

In Canada, one of the best examples of an
integrated, ecosystem approach to forestry is the
national Model Forest program. It was established in 1992 by the Canadian Forest Service of
Natural Resources Canada, initially for five years
and then extended in 1996 and again in 2002 for
another five years. The Model Forest network
includes 14 sites across Canada, three in both
Quebec and Ontario, two in British Columbia,
and one in each of the other provinces except
PEI.

Model forests were meant as sites for exploring and demonstrating approaches that move
beyond a narrow focus on sustained yield to
ecosystem management for multiple uses.
Model forests work by bringing together and
building partnerships among government levels
and agencies, local communities, First Nations,
ENGOs, and forest companies to 'gain a greater
understanding of conflicting views, share their
knowledge, and combine their expertise and
resources to develop innovative, region-spe-
cific approaches to sustainable forest management' (Model Forest Network 2009). Model
forests are large laboratories in which scientific,

practical, and process-oriented innovations can
be explored in an actual, working landscape.
Recent projects have extended their activities
into climate change, non-timber forest products,
and ecosystem-based management initiatives.
Several examples illustrate the range of these
initiatives.

The Prince Albert Model Forest includes
more than 367,000 hectares in central Saskatchewan, and a significant part of the region is
within the area of Weyerhaeuser Canada's forest management licence agreement. Other parts
are Prince Albert National Park and Candle Lake
Provincial Park. Ten partners from local community groups, First Nations, industry, and federal
and provincial resource management agencies
are involved. Projects have focused on ecosystem
health and local-level indicators for sustainable
forest management, integrated resource management, and communications and knowledge
exchange (see also Chapter 3). There has also
been a collaborative elk reintroduction project
(Bouman, Langen, and Bouman 1996; Model
Forest Network 2009). An Ecosystem-Based
Integrated Resource Management Plan, released
in 2000, focused on applying and integrating the
results of the projects.

The Fundy Model Forest covers 420,000 hectares in southeastern New Brunswick, of which
63 per cent comprises private woodlots, almost
15 per cent Crown land, 17 per cent industrial
freehold owned by JD Irving Ltd, and 5 per cent
Fundy National Park. The partnership involves
31 groups and is managed by partnership and
management committees, with six working
groups that guide the more than 300 projects
that have been and continue to be carried out.
Projects have addressed topics such as wildlife,
water quality, biodiversity, wood supply, socio-
economics, recreation, soils, valuing ecological
goods and services, planning for tough times,
education, and communications (Model Forest

Network 2009). Scientific research and wildlife–forestry protected area interactions have been a big part of the Fundy work (see Woodley and Freedman 1995; Forbes, Woodley, and Freedman 1999).

One dimension on which these two examples differ is the extent to which they directly involve other than essential stakeholders and partners, with Fundy being inclusive and Prince Albert relatively exclusive and using an advisory structure. A detailed analysis of the two suggested that the changes they introduced were incremental, emphasizing some operational procedures rather than challenging timber extraction paradigms, the structural relationships of power inequalities, and business as usual, as ecoforestry and deeper critics of industrial resource management would suggest is necessary to implement true ecosystem approaches in forest management in Canada (Beyers 2001).

Comprehensive Regional Land-Use Planning

Comprehensive regional land-use planning has a long history and in its early connections to human ecology and ecology in the work of such planners as Patrick Geddes and Benton Mackaye (see Weaver 1984) often overlapped with integrated resource management and ecosystem-based management activities. The distinction is perhaps in its emphasis on zoning and assigning land uses to particular plots of land. Key emphases are on data collection and mapping of both resource potentials and capabilities and actual use, the spatial representation in maps (going back to McHarg 1969), and the use of explicitly participatory processes to involve the public and achieve consensus on objectives and zoning. Plan effectiveness from an ecosystem approach perspective is variable, and the durability and implementation of plans, particularly for large, non-urban areas, is relatively low and brief.

Although there have been well-known, long-running initiatives in Alberta's East Slopes (Kennett 2002, 2009; CPAWS 2007) and in Ontario (from Strategic Land Use Planning in the 1970s to Lands for Life in the late 1990s), probably the best known have been in BC. In the early 1990s, the Commission on Resources and Environment (CORE) was established to resolve ongoing resource development versus conservation disputes throughout the province, of which Clayoquot Sound was the classic case. CORE lasted about four years and worked in four very large regions (Vancouver Island, Cariboo–Chilcotin, West Kootenay–Boundary, and East Kootenay).

The processes were often highly controversial and faced numerous challenges in the competition between resource development and environmental protection and in terms of the differing values and priorities of people in different parts of the province. There were also logistical challenges in identifying stakeholder groups at such scales and fostering equity in these groups' ability to participate, as well as the obvious difficulty of gathering and organizing information. Further, the commission faced institutional problems rooted in the destabilization of existing management and planning relationships (Owen 1998; Penrose, Day, and Roseland 1998). Gunton, Day, and Williams (2003) have reviewed the effectiveness of several of the succeeding Land and Resource Management Plan processes, and in Chapter 19, Reed makes the point that CORE was gender-blind, often ignoring and disadvantaging the concerns and conditions of women in forest-based communities in BC.

The CORE process worked by combining independent, expert technical support with professional facilitation and planning, and a structured consultation process that involved identifying representative sectors and limiting the representatives at the table. The successor to CORE was the Land and Resource Management

Plans (LRMPs). They are now carried out under the rubric of strategic land and resource planning and have been completed or are underway for almost all of BC, generally on smaller areas than the CORE processes (ILMB 2009).

In Ontario, the provincial government has recently undertaken extensive actions to protect significant natural and cultural features around the Greater Toronto area through the Oak Ridges Moraine Plan and Act of 2001–2 and the Greenbelt Plan and Act of 2005. They are different in spatial focus from the BC experience—urban and peri-urban—but have been innovative in that they combine ecological and socio-economic goals and use top-down and bottom-up planning approaches (Hanna, Webber, and Slocombe 2007; Whitelaw and Eagles 2007; Whitelaw et al. 2008). The number of comparative and systems-influenced studies of urban and regional sustainability is growing elsewhere as well (e.g., Forman 2008; Newman and Jennings 2008).

Parks and Protected Areas

Parks and protected areas management is and has been a major area of application of ecosystem approaches and ecosystem-based management (see also Chapter 12). This is in large part a reflection of the dual trends since the late 1970s toward emphasizing ecology and science in park management and recognizing that parks cannot be managed in isolation from their surroundings. Some of the earliest systematic explorations of ecosystem management focused on protected areas (Agee and Johnson 1988). The literature is large and is reviewed elsewhere (e.g., Slocombe and Dearden 2008); here I highlight the main points. As in forest management, ecosystem approaches in protected areas management have two core themes. The first is development of natural science information to support management of the protected area in the context of

interactions with the surrounding areas; the second is development of processes, institutions, and social science for the same purposes.

On the one hand, protected areas management is increasingly based on development of comprehensive information about parks and their surroundings, often drawing on the new ecosystem sciences such as conservation biology, landscape ecology, and population viability analysis. There is a strong emphasis on data collection and management, analysis, ongoing monitoring and assessment, and integrating information from a range of sources and disciplines (e.g., Forbes, Woodley, and Freedman 1999; Jensen and Bourgeron 2001).

On the other hand, at the same time, and not always easily, protected areas management seeks to better understand the institutions, the local communities, and the attitudes and values of people near parks and far from them and ultimately to foster collaborative management, develop goals, and in some places develop various degrees of collaborative management of parks. Sometimes this approach is premised on legal requirements such as land claims settlements, but even more often it recognizes that a protected area is embedded in a complex landscape of varied uses and perceptions that cannot be ignored. Several good examples of efforts to implement ecosystem approaches in protected areas management are biosphere reserves and the Ontario national parks ecosystem management process.

Biosphere reserves emerged from the UNESCO Man and the Biosphere Program mentioned earlier. The biosphere reserve system developed in the early 1980s as an attempt to move away from the traditional strictly protected, wilderness national park ideal and devise a designation suited to (the globally more common) complex landscapes, often modified over centuries by relatively low levels of human use (e.g., traditional pastoralism or hunting and gathering). It envisions a strictly protected core area, a buffer zone where traditional sustainable activities are allowed, and even further from the core a zone of transition where more intensive activities are permitted. Monitoring is performed throughout the reserve, while education and research are conducted in the buffer zone and more intensive tourism and recreation and other activities are permitted in the transition zone. UNESCO gives biosphere reserve designation to existing core protected areas that develop regulatory and/ or voluntary means for implementing buffer and transition zones. The program is popular, with more than 550 reserves around the world today. They vary tremendously in detail, but Canada has been a leader with 13 such reserves, from Waterton Lakes National Park in Alberta

(Figure 15.1) to Mont St Hilaire and Charlevoix in Quebec. The emphasis on research, monitoring, links between the core and surroundings, and education and other outreach to nearby residents is very close to an ecosystem approach. The reserves' role in fostering sustainability is also being explored (Jamieson et al. 2008), and, indeed, a fine publication by UNESCO (2000) has highlighted and elaborated on this aspect (see Box 15.1).

More broadly, there is resurgent interest in the regional context of protected areas and the connections between protected areas and regional conservation and development (Hanna, Clark, and Slocombe 2008). The Crown of the Continent region of southwest Alberta, southeast BC, and northwest Montana has long been a case study of this, with several national parks,

(Photo: S. Slocombe)

Figure 15.1 Looking west from the prairie to the mountains over the Waterton Biosphere Reserve sign with its recognition of public and private partnership.

Box 15.1 Ecosystem Approach Principles and Guidelines

On the direction of the Fifth Conference of the Parties to the Convention on Biological Diversity (CBD), the Subsidiary Body on Scientific, Technical and Technological Advice prepared principles and guidance on the ecosystem approach in 2000. UNESCO's Man and the Biosphere (MAB) program has taken the resulting principles and guidelines and illustrated them with examples from biosphere reserves around the world (UNESCO 2000). The 12-ecosystem approach principles and five operational guidelines for application of the ecosystem approach are listed below. This is a superb publication, available for download in Adobe Acrobat .pdf format at http://www.unesco.org/mab/publications/publications.htm.

Principles

- Land, water, and renewable resource objectives are matters for societal choice.
- Management should be decentralized to the lowest appropriate level.
- The effects of actions on adjacent and other ecosystems should be considered.
- Ecosystems must be understood in an economic context to recognize potential gains from management.
- Conserving ecosystem structure and function to maintain ecosystem services should be a priority.
- Ecosystems should be managed within the limits of their functioning.
- The ecosystem approach should be undertaken at appropriate spatial and temporal scales.
- Ecosystem management objectives should be set for the long term.
- Change is inevitable.
- An appropriate balance should be sought between biodiversity conservation and use, and an integration of these two goals should be pursued.
- All forms of relevant information—scientific, indigenous, and local—should be considered.
- All relevant sectors of society and scientific disciplines should be involved.

Guidelines

- Focus on functional processes and relationships within ecosystems.
- Promote fair and equitable access to the benefits of ecosystem functions and biodiversity and their products.
- Use adaptive management processes.
- Manage at scales appropriate to the problem and as decentralized as appropriate.
- Ensure co-operation between agencies and governments.

biosphere reserves, and international peace parks. Recent efforts to link science, policy, and management there have been well documented (Prato and Fagre 2007), along with historical perspectives on the politics and inter-actor dynamics of conservation (Sax and Keiter 2007) and recommendations for transboundary co-operation and cumulative effects assessment to foster more integrated ecosystem management (Pedynowski 2003).

Coastal, Ocean, and Fisheries Management

As a country with one of the world's longest coastlines and some of its major fisheries, Canada has long faced criticism from within and without for the absence of national, or even provincial, coastal zone and oceans management programs. And the errors made in management of at least our marine fisheries in the past 40 years are well known (see also Chapter 8). The absence of integrated coastal and ocean management has long been attributed to the division of ownership and management responsibility between the provinces and the federal government. The inadequacy of this state of affairs has been highlighted by fisheries problems and further emphasized by growing offshore development pressures on all three marine coasts from oil and gas development (Chapter 6), tourism and recreation, demands for marine protected areas (Chapter 12), and fishing (Chapter 8).

Recognizing some of these issues, and arguably seeking to pre-empt more protective measures that could emanate from Parks Canada, among others, the federal Department of Fisheries and Oceans (DFO) saw the passage of the Oceans Act in 1997. The Oceans Act enabled DFO to create marine protected areas (MPAs) and integrated coastal zone management plans, as well as to prepare guidelines and objectives for marine environmental quality and to promote an ecosystem approach. DFO is also part of the federal government's effort to explore the applicability and utility of adopting the precautionary principle, albeit in a fairly weak form (Government of Canada 2001).

Five years later, DFO released the Oceans Strategy (DFO, Oceans Directorate 2002) to facilitate implementation, since the Oceans Act was merely enabling legislation. The strategy aims to achieve its results through three main initiatives:

improving institutional governance mechanisms, especially through co-ordination and collaboration; implementing a program of integrated management planning with partners (pilot projects are underway in the BC central coast, the Eastern Scotian Shelf [described below] and the Arctic [see Siron et al. 2008]); and promoting stewardship and public awareness. Integrated management is to be guided by a number of principles: ecosystem-based management, sustainable development, precautionary approach, conservation, sharing responsibility for planning and protection, flexibility, and inclusiveness.

The Eastern Scotian Shelf Integrated Management Initiative (ESSIM) developed a collaborative management and planning process for the Eastern Scotian Shelf Large Oceans Management Area (about 325,000 square kilometres, with boundaries based on ecological, oceans use, and political and administrative criteria). The initiative was announced in December 1998 by the minister of fisheries and oceans. Its goals are:

- to integrate the management of all activities in the Eastern Scotian Shelf Oceans Management Area;
- to encourage the conservation, effective management, and responsible use of marine resources;
- to support the maintenance of natural biological diversity and productivity; and
- to foster opportunities for economic diversification and sustainable wealth generation for coastal communities and stakeholders (ESSIM Working Group 2001).

The ESSIM Forum (see Figure 15.2) is envisioned as the structure for moving forward. DFO's Oceans and Coastal Management Division serves as the secretariat, with the multi-stakeholder Oceans Management and Planning Group (OMPG) as the core vehicle for facilitating multi-party collaboration and co-ordination.

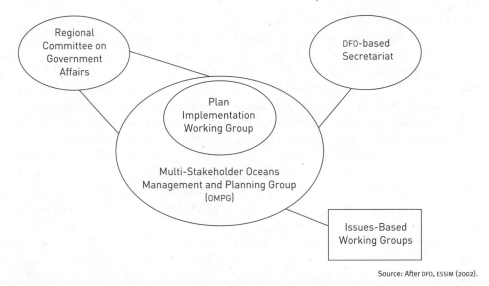

Source: After DFO, ESSIM (2002).

Figure 15.2 The Structure of the East Scotian Shelf Integrated Management Initiative.

Two intergovernmental committees engage governments: one for executive-level decision-making (the Regional Committee on Government Affairs) and another to foster operational activity as a subgroup of the OMPG (the Plan Implementation Working Group) (ESSIM Working Group 2008).

A first ESSIM Forum workshop was held in February 2002 with about 150 stakeholders and others. Conference proceedings were to be published later, and an initial co-ordinating group was to be established. Several technical reports and synthesis studies have also been prepared to support issue description, management area definition, and identification of ecosystem objectives. The ESSIM strategic plan applies to offshore areas of the shelf and was released in 2007. It has a broad intent to link common goals and objectives with sectoral plans and activities (ESSIM Planning Office 2007). Near-shore plans may be developed in collaboration with provincial and other governments and actors (e.g. Bras d'Or Lakes Collaborative Environmental Planning Initiative [CEPI] 2009).

There are many elements of ecosystem approaches and potential improvement here, but there are also grounds for scepticism (e.g., Anderson 2002). Most notably, critics point to DFO's poor track record in managing and protecting even traditional fisheries resources; the divisions within the department over protection versus exploitation and development of fisheries; very little experience or success at involving the public, communities, or fishers in fishery, coastal, or ocean management; and the federal government's general reluctance to do anything that might offend a province.

Environmental Studies and Monitoring

Not least because of the commonalities between adaptive management and ecosystem approaches, there have also been significant applications of the ecosystem approach to the research and study of human–natural systems and to ongoing monitoring of them. A basic step here is to develop ecosystem goals and objectives (Chapter 3). The

Great Lakes RAPs and LaMPs did this initially in a fairly simple, quite human-focused way and in considerably more detail over the years (e.g., Indicators for Evaluation Task Force 1996). Marmorek et al. (1993) provide another good example in their report on an ecosystem approach for British Columbia, providing lengthy examples of goals of different kinds and discussing the monitoring process, the distinctions between goals, objectives, and indicators, and the procedures and questions for setting goals and objectives. The starting point is to ask: What do we have now? What do we want in future? What can we get? How do we get there? And are management activities working?

An evaluation of the implications of the ecosystem approach for northern Canada (Keith 1994; see also Boyle 2004) drew explicitly on dynamic systems ideas to argue for the need to support self-organizing ecosystem processes, to understand whole ecosystems, to make full use of all available information rather than depending on any single type of knowledge, and to use the precautionary principle. Some of these ideas certainly found their way into Environment Canada's Northern Ecosystem Initiative, which included research on climate change, contaminants, monitoring, and capacity building (Environment Canada 2005; Hardi and Roy 2005). Cumulative effects assessment and environmental monitoring, including traditional ecological knowledge programs, are good examples of this approach (e.g., Baxter, Ross, and Spaling 2001; Duinker and Greig 2006; Dubé et al. 2006).

At the cutting edge of ecosystem approaches and complex systems theory are efforts to develop descriptive methodologies that link biophysical processes with socio-economic and governance structures and processes to better address the dynamic, complex character of linked human–natural systems (e.g., Kay et al. 1999; Waltner-Toews, Key, and Lister 2008).

A concrete manifestation of this is the development of a detailed set of indicators and a monitoring program for Ontario provincial parks. Boyle, Kay, and Pond (2001) started philosophically from an ecosystem approach and post-normal views of science as open and contested—not limited to experts and single, consensus answers—and recognized the need for participatory, diverse understandings and research in the face of complex, uncertain problems. In practice, this leads to indicator identification via a detailed analysis of the hierarchical-systems nature and interactions of the relevant natural and societal systems, as well as to a consultative process for review and discussion. The adoption of an adaptive ecosystem approach underscores the co-evolving nature of societal and ecological systems. In recent years, there has been extensive development of methodologies for assessing management effectiveness of protected areas (Hockings et al. 2006), although assessing conservation effectiveness remains less studied.

Such an approach requires the use of management strategies to foster system self-organization in desired ways, consultants as facilitators in a decision-making process informed by science that involves all stakeholders, and scientists with a focus on adaptation and learning (see Chapters 16 and 18). Monitoring, then, is an ongoing process and the primary tool for learning, and it addresses system context, flows across boundaries, and societal influences on the system (Boyle, Kay, and Pond 2001; Boyle 2004).

CHALLENGES AND OPPORTUNITIES FOR ECOSYSTEM APPROACHES

Ecosystem approaches are by no means a panacea for the problems of resource and environmental management. Much depends on their implementation, the ways the ecosystem notion is made manifest, the preferences and priorities of the researchers and managers involved, and, it is to be hoped, the characteristics of the problem. The site-specificity and problem-specificity of ecosystem approaches offer both a challenge and an opportunity. The challenge exists because there are costs to doing things at least slightly differently each time. The opportunity rests in the fact that there are enough commonalities and common challenges that lessons can be extracted and basic ideas reinforced (Slocombe 1998; Layzer 2008). And certainly, for ecosystem approaches as for resource and environmental management generally, there is much need for comparative studies and sharing of experience.

There have also been a range of specific criticisms of ecosystem approaches. Like systems approaches generally, they can be seen as too vague—what is an (eco)system?—and as tending to produce results that vary depending on who is doing the study. Others believe that such a widely applicable approach must have little that is unique and essential to it and therefore merely overlaps or duplicates work that could better be done within traditional disciplines such as ecology, economics, or engineering, to mention perhaps just the most common. Ecosystem approaches have also been specifically criticized for commonly assuming that systems tend toward stability or equilibrium, for generalizing from ecology to socio-economic systems, for ecological determinism, and for functionalist emphases. In addition, treating ecosystems and other complex systems as equivalent to organisms in terms of ecosystem health has also been open to criticism (see Slocombe 1993a). Finally, some ecosystem approaches and management regimes, like much resource and environmental management, tend to ignore power, gender, and related issues (see Chapter 19).

Ecosystem approaches in the form of ecosystem management are also subject to criticism, particularly in the United States. Critics question the very need for such management (i.e., ecosystems are not in bad shape), the presumption of the need for and benefits of a strong government role in management, and the potential benefits and role of science in management (see Lackey 1998).

Many of these criticisms are considerably less common now than they were for applications 20 or more years ago. Ecosystem approaches have certainly changed and evolved over the years. Perhaps most obviously, they have substantially incorporated complex, dynamic systems ideas, become increasingly linked with adaptive and ecosystem-based management and science-based approaches (see Bocking 1994, 2009; Kay and Schneider 1994; Kay et al. 1999), and contributed to such new, synthetic concepts and goals as ecosystem health and integrity (see Chapter 3). One area that remains a largely unfulfilled opportunity is increasing the investigation and variation of power, politics, and equity in implementing an ecosystem approach, although some progress has been made (see Chapters 3 and 19). Connecting with and learning from political ecology may be useful here (e.g., Neumann 2005).

Conflict and uncertainty in resource and environmental management emerge from many sources, some in the natural environment, some in the human, and more in the interaction of the two. Ecosystem approaches seek to reduce them through a combination of better information and understanding, changed interactions between and among people and organizations, changed expectations, and reduced demands of society on nature. Ecosystem approaches have made real contributions to improvements in the first two, but we have not made much progress on the last two, perhaps because they require too fundamental and philosophical or ethical a change in ourselves and our societies.

To the extent that ecosystem approaches are manifest in resource and environmental management, there are fundamental ethical issues of overconfidence to ponder. It is certainly arguable whether we can, or should, manage ecosystems rather than simply human activities in them—a distinction that underlies the difference in phraseology between 'ecosystem management' and 'ecosystem-based management' (cf. Slocombe 1993b; Woodley and Freedman 1995). Beyond that, deeper questions can be raised in terms of the great difficulty of defining any non-utilitarian, non-managerial relationship to non-human nature that ultimately implies a need for deep philosophical and structural changes in society, such as away from growth and strong property rights toward a partnership ethic (Merchant 2003).

While ecosystem approaches and their manifestations have been successful in bridging some boundaries—for example, between protected areas and their surroundings and between sectors and interests in at least some comprehensive watershed, regional, and forest management—other boundaries have not been so well addressed. There are those of gender, power, and race, as noted above. Others that deserve more attention than they have received in recent years include the links between terrestrial and marine systems, between commercial and non-commercial (formal/non-formal economy) activities, and among urban, rural, and wilderness systems. There is also a need for greater exploration of ethical issues and of the existence of limits and constraints.

There are also practical challenges to implementing an ecosystem approach (see the guest statement by Michael Fox):

- bringing institutions together;
- developing, managing, and disseminating the necessary information base so that all stakeholders have similar information;
- better using and integrating new governance and conflict resolution ideas; and
- balancing the substantive, research aspects with the consultative, participatory, goal-setting, and deliberative processes.

At least partly in response to these observed challenges, there is a growing focus on the bridging of boundaries, or integration, in resource and environmental management. Some stress human dimensions (Ewert, Baker, and Bissix 2004), some the diverse dimensions of and approaches to integration (Hanna and Slocombe 2007), some simulation modelling approaches to foster integration (Policy Research Initiative 2007), others the legal and institutional dimensions (Kennett 2006), and yet others adaptive and collaborative approaches to governance (Brunner et al. 2005; Armitage, Berkes, and Doubleday 2007).

Doing resource and environmental management in accordance with an ecosystem approach is expensive in terms of expertise, people, and dollars. It requires research, fieldwork, and monitoring, and it also requires often lengthy, complex processes of consultation, visioning, goal-setting, and sometimes conflict resolution. Even when governments commit to the idea, they tend not to follow through. Nowhere has this been more obvious in the past decade than with Parks Canada's efforts to implement the recommendations of the Panel on the Ecological Integrity of Canada's National Parks (Parks Canada Agency 2000; see also Chapter 12). Ultimately, implementing ecosystem approaches requires their elevation in the eyes of decision-makers and budgeters at the highest levels of government and business in order to ensure that good plans and ideas and processes are not made pointless by changing priorities and resources.

Theory and Practice in the Ecosystem Approach

Michael Fox

'This course covers too much material . . . global warming, deforestation, acid rain, toxic waste disposal . . . the list goes on and on. But why does nothing ever seem to happen in terms of government policies and actions? If so many people are outraged by these environmental problems, why is so little being done to solve them? You have pointed out all of these issues, but you never provide any solutions!'

I found the above comment written on the back of a student's examination booklet some 25 years ago. It was a stark reminder of the extreme frustration of some university students as they listened and learned about environmental issues. It struck at the root of what many accused academe of doing: identifying problems and critical issues, then developing theoretical approaches and solutions unrelated to the 'real world'. Statements such as this one helped to shape a new framework for thinking about—and intervening in—society's shortcomings when dealing with environmental issues and conflicts. The experience with one-off case studies of urgent issues and conflicts caused many academic disciplines to move toward a more comprehensive way of dealing with these issues through an ecosystem approach.

Over the past 25 years, much has changed in the way in which university programs have evolved in reflecting the power of an ecosystem approach. There was a gap between the 'theory' developed and taught within the university and the 'practice' of the communities and regions in which we lived. If university programs were to equip students to deal with the complex and conflict-filled world of human–environment interaction, the ecosystem approach appeared to be one of the best we could use.

A generation later, I have been pleased to see that today's students readily embrace the ecosystem approach and are quite capable of balancing theory and practice in their learning. With the rapid growth of co-operative education, service learning, international exchanges, practicum placements in public and private policy and research positions, as well as the interdisciplinary approach to environmental issues, academic programs assist our students in understanding their role in reducing the conflict and uncertainty between humans and the natural environment. I have seen dozens of students go to various policy and planning environments over the years with a clear understanding of the ecosystem approach. They were much better equipped to deal with resource and environmental conflict than the students of a generation ago. That was my conclusion until I joined a 'real-world' regional planning commission some years ago in which conflict and uncertainty persisted despite best efforts within the academy.

The development of an ecosystem approach has certainly been a resounding success within our academic approach to resource and environmental issues. There also is growing evidence of an ecosystem approach in much scientific and government policy analysis and planning at the local, regional, national, and international levels. We also can see the approach reflected in political policy platforms. Real progress has occurred in a number of jurisdictions, such as the Remedial

Action Plans, various conservation authorities, and regional land-use plans across the country.

And yet, if one sits at a planning commission table, in town council chambers, or at kitchen tables across this vast country, there continues to be an ever-present, perhaps heightened, sense of conflict and uncertainty among citizens and decision-makers when addressing environmental issues over the long term. In spite of a generation of applying an ecosystem approach to resource, environmental, and land-use issues, my experience suggests that a huge gap still exists between 'theory' and 'practice' in many plans, policies, and legislated practices.

The reality of having a three- or four-year mandate as a political decision-maker, surrounded by day-to-day issues and a large degree of uncertainty, often causes age-old human and economic reactions to dominate over long-term, ecosystem thinking. The gap between theory and practice at times becomes a chasm, with planners and environmental scientists only able to design and propose the most sustainable route to follow and politicians often opting to react to problems and controlling complex environmental systems by separating themselves from nature.

Rather than falling prey to frustration, however, it is important to continue to address conflict and uncertainty in environmental systems management through the educational process. This chapter certainly provides many examples of the success of the ecosystem approach in planning and environmental management. The key point is that it took a generation within the academy, so it may take another before political leaders and decision-makers will systematically inform their decisions through an ecosystem perspective.

The bridge over the theory–practice chasm can be a two-way link. Thus, we must continue to push the practice of everyday decision-making toward an ecosystem way of thinking and behaving—'practice into theory', so to speak. Ecosystem education outside of the academy will be a key to bridging this gap.

FROM THE FIELD

Applying an Ecosystem Approach

Experience in the field, as a researcher and as a consultant, has long been an important part of my work. One of my goals has always been to foster the intersection and mutual development of theory and practice, which I believe is particularly important in resource and environmental management. I have had the benefit of conducting field research in many rural and wilderness parts of Canada and elsewhere. I also have been able to speak formally and informally with resource and environmental managers, resource users, politicians, NGO leaders, and others about local and wider challenges and opportunities in resource and environmental management. This sort of ongoing consultation, in at least a few places regularly over 20 years, has been essential to understanding local concerns and their origins and how they change, or don't, over time, as well as how resource and environmental management really works: who's involved, what the key issues are, and what policies and actions could make a tangible difference. It's always a pleasure to find how willing the vast majority of people are to talk about such things and how much specific, dedicated people in every place know.

There are a few recurring issues, I think, although they may be phrased differently in different places. There is nearly always concern about maintaining local ecosystems and resources, usually in order to maintain current livelihoods or in the hope of better (safer, more secure, more viable) livelihoods. There are concerns with how local people's and governments' knowledge, perspectives, and priorities are taken into account by higher-level governments, whether provincial/territorial or national. There are concerns about the future and the ability of future generations to stay in the area. There are concerns about the sustainability of resources, often especially soils, water, and wildlife, in different combinations in different places. Concern about overarching issues such as economic development, climate change, or resource exploitation is increasingly present.

Numerous lessons emerge from this experience: the potential for change in local priorities and in local resources and ecosystems over time; the difficulties of seeing change at any one, particular point in time and place; the real difficulties of achieving effective collaboration, knowledge integration, and cross-cultural connection; the importance of thinking about connections across spatial scales; the need to take enough time for collaboration, integration, and learning and to ensure effective leadership and adequate resources in resource and environmental planning processes—always a challenge but especially so in times of fiscal restraint, whether politically or economically motivated.

There has been steady progress in much of Canada toward more collaborative models and in some cases the development of excellent plans and frameworks. However, there is consistently limited will and resources for monitoring and evaluation of progress. I look forward to continuing to track the evolution of Canadian resource and environmental management, in terms of both ideas and the issues and practices on the ground.

—SCOTT SLOCOMBE

Acknowledgements

I am grateful to the Social Sciences and Humanities Research Council of Canada and Wilfrid Laurier University for support over many years for the research on which this chapter is based. Dean Bavington and Rob Brown provided research assistance on an earlier project, which helped to inform this chapter.

REVIEW QUESTIONS

1. What is the core contribution of an ecosystem approach?
2. What are some common elements of an ecosystem approach?
3. Give two examples of the incorporation of an ecosystem approach into resource and environmental policy.
4. What are some of the challenges involved with implementing ecosystem approaches?
5. How might application of ecosystem approaches develop in the next five years?

SUGGESTED READING

Hanna, K.S., and D.S. Slocombe, eds. 2007. *Integrated Resource and Environmental Management: Concepts and Practice.* Toronto: Oxford University Press.

McAfee, B., and C. Malouin, eds. 2008. *Implementing Ecosystem-Based Approaches in Canada's Forests: A Science-Policy Dialogue.* Ottawa: Natural Resources Canada, Canadian Forest Service, Science and Programs Branch.

UNESCO. 2000. *Solving the Puzzle: The Ecosystem Approach and Biosphere Reserves.* Paris: UNESCO.

REFERENCES

Agee, J.K., and D.R. Johnson, eds. 1988. *Ecosystem Management for Parks and Wilderness.* Seattle: University of Washington Press.

Anderson, M. 2002. 'There is a way to save our oceans—But is there a will?' *The Globe and Mail* 14 August: A11.

Armitage, D., F. Berkes, and N. Doubleday, eds. 2007. *Adaptive Co-management: Collaboration, Learning and Multi-level Governance.* Vancouver: University of British Columbia Press.

Babbage, M. 2009. 'U.S. will look at ways to alleviate Buy American concerns: Clinton'. *The Globe and Mail* 13 June. http://www.theglobeandmail.com/news/politics/us-will-look-at-ways-to-alleviate-buy-american-concerns-clinton/article1181284.

Bastedo, J.D., J.G. Nelson, and J.B. Theberge. 1984. 'Ecological approach to resource survey and planning for environmentally significant areas: The ABC method'. *Environmental Management* 8 (2): 125–34.

Baxter, W., W.A. Ross, and H. Spaling. 2001. 'Improving the practice of cumulative effects assessment in Canada'. *Impact Assessment and Project Appraisal* 19 (4): 253–62.

Beyers, J.M. 2001. 'Model forests as process reform: Alternative dispute resolution and multistakeholder planning'. In M. Howlett, ed., *Canadian Forest Policy: Adapting to Change*, 172–202. Toronto: University of Toronto Press.

Bingham, B. 2007. 'Information management: Barrier or bridge to integrating natural resources science and management?' *George Wright Forum* 24 (2): 41–7.

Bocking, S. 1994. 'Visions of nature and society: A history of the ecosystem concept'. *Alternatives Journal* 20 (3): 12–18.

———. 2009. 'Leopold's challenge'. *Alternatives Journal* 35 (4): 12–17.

Bouman, O.T., G. Langen, and C.E. Bouman. 1996. 'Sustainable use of the boreal Prince Albert Model Forest in Saskatchewan'. *Forestry Chronicle* 72 (1): 63–72.

Boyden, S., et al. 1981. *The Ecology of a City and Its People: The Case of Hong Kong.* Canberra: Australian National University Press.

Boyle, M. 2004. 'Learning from history: Lessons for cumulative effects assessment and planning'. *Meridian* (Canadian Polar Commission) fall/winter: 6–12.

Boyle, M., J. Kay, and B. Pond. 2001. 'Monitoring in support of policy: An adaptive ecosystem approach'. In T. Munn, editor-in-chief, *Encyclopedia of Global Environmental Change*, v. 4, 116–37. London: John Wiley and Sons.

Bras d'Or Lakes Collaborative Environmental Planning Initiative (CEPI). 2009. Project website. http://www.brasdorcepi.ca.

Brunner, R.D., et al. 2005. *Adaptive Governance: Integrating Science, Policy, and Decision Making.* New York: Columbia University Press.

Caldwell, L.K. 1970. 'The ecosystem as a criterion for public land policy'. *Natural Resources Journal* 10 (2): 203–21.

———, ed. 1982. *Perspectives on Ecosystem Management for the Great Lakes: A Reader.* Albany: State University of New York Press.

CPAWS. 2007. 'The southern east slopes of the Rocky Mountains'. *Green Notes Newsletter* (December) 1–19.

Danby, R.K., and D.S. Slocombe. 2005. 'Regional ecology, ecosystem geography, and transboundary protected areas in the St Elias mountains'. *Ecological Applications* 15 (2): 405–22.

Darling, F.F., and R.F. Dasmann. 1969. 'The ecosystem view of human society'. *Impact of Science on Society* 19 (2): 109–21.

DFO (Department of Fisheries and Oceans), Oceans Directorate. 2002. *Canada's Oceans Strategy: Our Oceans, Our Future*. Ottawa: DFO.

Dorney, R.S. 1973. 'Role of ecologists as consultants in urban planning and design'. *Human Ecology* 1 (3): 183–200.

Drengson, A., and D. Taylor. 1997. 'An overview of ecoforestry'. In A. Drengson and D. Taylor, eds, *Ecoforestry: The Art and Science of Sustainable Forestry*, 17–33. Gabriola Island, BC: New Society Publishers.

Dubé, M., et al. 2006. 'Development of a new approach to cumulative effects assessment: A northern river ecosystem example'. *Environmental Monitoring and Assessment* 113 (1–3): 87–115.

Duinker, P.N., and L.A. Greig. 2006. 'The impotence of cumulative effects assessment in Canada: Ailments and ideas for redeployment'. *Environmental Management* 37 (2): 153–61.

Environment Canada. 2005. Northern Ecosystem Initiative website. http://www.pnr-rpn.ec.gc.ca/nature/ecosystems/nei-ien/index.en.html.

Environment Canada and US Environmental Protection Agency. 1995. *The Great Lakes: An Environmental Atlas and Resource Book*. 3rd edn. Toronto: Government of Canada; Chicago: US Environmental Protection Agency.

ESSIM Planning Office. 2007. 'Eastern Scotian Shelf Integrated Ocean Management Plan'. Dartmouth, NS: Department of Fisheries and Oceans, Oceans and Habitat Branch. http://www.dfo-mpo.gc.ca/Library/333115.pdf.

ESSIM Working Group. November 2001. 'The Eastern Scotian Shelf Integrated Management (ESSIM) Initiative: Development of a collaborative management and planning process'. Halifax: Department of Fisheries and Oceans, Oceans and Coastal Management Division. http://www.mar.dfo-mpo.gc.ca/oceans/e/essim/essim-reports-planning process-e.html.

ESSIM Working Group. 2008. 'Introduction to the ESSIM Initiative'. http://www.mar.dfo-mpo.gc.ca/oceans/e/essim/essim-intro-e.html.

Ewert, A.W., D.C. Baker, and G.C. Bissix. 2004. *Integrated Resource and Environmental Management: The Human Dimension*. Wallingford, UK: CABI Publishing.

Forbes, G., S. Woodley, and B. Freedman. 1999. 'Making ecosystem-based science into guidelines for ecosystem-based management: The Greater Fundy Ecosystem experience'. *Environments* 27 (3): 15–23.

Forman, R.T.T. 2008. *Urban Regions: Ecology and Planning beyond the City*. Cambridge: Cambridge University Press.

GLC (Great Lakes Commission). 2004. *LaMPs: Lakewide Management Plans and Initiatives*. Ann Arbor, MI: Great Lakes Commission.

———. 2008. *LaMPs: Lakewide Management Plan Updates for the Great Lakes*. Ann Arbor, MI: Great Lakes Commission.

Global Water Partnership and INBO. 2009. 'A handbook for integrated water resources management in basins'. http://www.inbo-news.org.

Government of Canada. 2001. *A Canadian Perspective on the Precautionary Approach/Principle: Proposed Guiding Principles*. Ottawa: Government of Canada.

Grady, W. 2007. *The Great Lakes: The Natural History of a Changing Region*. Toronto: Greystone Books.

Grapentine, L.C. 2009. 'Determining degradation and restoration of benthic conditions for Great Lakes areas of concern'. *Journal of Great Lakes Research* 35 (1): 36–44.

Grumbine, R.E. 1994. 'What is ecosystem management?' *Conservation Biology* 8 (1): 27–38.

Gunton, T.I., J.C. Day, and P.W. Williams, eds. 2003. 'Evaluating collaborative planning: The British Columbia experience'. *Environments* 31 (3).

Hall, J.D., K. O'Connor, and J. Ranieri. 2006. 'Progress toward delisting a Great Lakes Area of Concern: The role of integrated research and monitoring in the Hamilton Harbour Remedial Action Plan'. *Environmental Monitoring and Assessment* 113: 227–43.

Hanna, K.S., D. Clark, and D.S. Slocombe, eds. 2008. *Transforming Parks and Protected Areas: Policy and Governance in a Changing World*. London: Routledge.

Hanna, K.S., and D.S. Slocombe, eds. 2007. *Integrated Resource and Environmental Management: Concepts and Practice*. Toronto: Oxford University Press.

Hanna, K.S., S.M. Webber, and D.S. Slocombe. 2007. 'Integrated ecological and regional planning in a rapid-growth setting'. *Environmental Management* 40 (3): 339–48.

Hardi, P., and M. Roy. 2005. *Inventory of Ecosystem Indicators in Canada's North for the Northern Ecosystem Initiative*. Winnipeg: IISD for Environment Canada.

Hartig, J., ed. 1995. *Practical Steps to Implement an Ecosystem Approach in Great Lakes Management*. Windsor and Detroit: US Environmental

Protection Agency, Environment Canada, International Joint Commission, Wayne State University.

Hartig, J., et al. 1998. 'Implementing ecosystem-based management: Lessons from the Great Lakes'. *Journal of Environmental Planning and Management* 41 (1): 45–75.

Hartig, J., and J. Vallentyne. 1989. 'Use of an ecosystem approach to restore degraded areas of the Great Lakes'. *Ambio* 18 (8): 423–8.

Hockings, M., et al. 2006. *Evaluating Effectiveness: A Framework for Assessing the Management of Protected Areas.* 2nd edn. Gland, Switzerland and Cambridge: IUCN (World Conservation Union).

Hooper, B. 2005. *Integrated River Basin Management: Learning from International Experience.* London: IWA Publishing.

ILMB (Integrated Land Management Bureau). 2009. Land-use planning website. http://ilmbwww.gov.bc.ca/slrp/index.html.

Indicators for Evaluation Task Force. 1996. *Indicators to Evaluate Progress under the Great Lakes Water Quality Agreement.* Windsor, ON: International Joint Commission.

Jamieson, G., et al. 2008. 'Canadian biosphere reserve approaches to the achievement of sustainable development'. *International Journal of Environment and Sustainable Development* 7 (2): 132–44.

Jensen, M.E., and P.S. Bourgeron, eds. 2001. *A Guidebook for Integrated Ecological Assessments.* New York: Springer Verlag.

Kay, J.J., H.A. Regier, M. Boyle, and G. Francis. 1999. 'An ecosystem approach for sustainability: Addressing the challenge of complexity'. *Futures* 31 (7): 721–42.

Kay, J.J., and E. Schneider. 1994. 'Embracing complexity: The challenge of the ecosystem approach'. *Alternatives* 20 (3): 32–9.

Keith, R.F. 1994. 'The ecosystem approach: Implications for the North'. *Northern Perspectives* 22 (1): 3–6.

Kennett, S.A. 2002. 'Reinventing integrated resource management in Alberta: Bold new initiative or "déjà vu all over again"?' *Resources* 77: 1–7.

———. 2006. *Integrated Landscape Management in Canada: Getting from Here to There.* CIRL Occasional Paper 17. Calgary: University of Calgary.

———. 2009. 'Change to believe in: A legal check-list for Alberta's land-use framework'. *Resources* 104: 1–6.

Krantzberg, G. 2009. 'Renegotiating the Great Lakes Water Quality Agreement: The process for a sustainable outcome'. *Sustainability* 1: 254–67. www.mdpi.com/journal/sustainability.

Lackey, R.T. 1998. 'Seven pillars of ecosystem management'. *Landscape and Urban Planning* 40 (1): 21–30.

Layzer, Judith A. 2008. *Natural Experiments: Ecosystem-Based Management and the Environment.* Cambridge, MA: MIT Press.

Lee, B.J., H.A. Regier, and D.J. Rapport. 1982. 'Ten ecosystem approaches to the planning and management of the Great Lakes'. *Journal of Great Lakes Research* 8 (3): 505–19.

Likens, G.E. 1992. *The Ecosystem Approach: Its Use and Abuse.* Oldendorf/Luhe: Ecology Institute.

McAfee, B., and C. Malouin, eds. 2008. *Implementing Ecosystem-Based Approaches in Canada's Forests: A Science-Policy Dialogue.* Ottawa: Natural Resources Canada, Canadian Forest Service, Science and Programs Branch.

McHarg, I.L. 1969. *Design with Nature.* New York: Doubleday/Natural History Press.

MacKenzie, S.H. 1996. *Integrated Resource Planning and Management: The Ecosystem Approach in the Great Lakes Basin.* Washington: Island Press.

McLain, R.J., E.M. Donoghue, J. Kusel, L. Buttolp, and S. Charnley. 2008. 'Multiscale socioeconomic assessment across large ecosystems: Lessons from practice'. *Society and Natural Resources* 21: 719–28.

Marmorek, D.R., et al. 1993. *Towards an Ecosystem Approach in British Columbia.* Fraser River Action Plan 1993–16. Vancouver: ESSA Ltd for Environment Canada, BC Ministry of Environment, Lands and Parks, and BC Ministry of Forests.

Marten, G.G. 2001. *Human Ecology: Basic Concepts for Sustainable Development.* London: Earthscan.

Meadows, D. 2008. *Thinking in Systems: A Primer.* White River Junction, VT: Chelsea Green Publishing.

Meffe, G., et al. 2002. *Ecosystem Management: Adaptive, Community-Based Conservation.* Washington: Island Press.

Merchant, C. 2003. *Reinventing Eden: The Fate of Nature in Western Culture.* New York: Routledge.

Mitchell, B. 1998. *Sustainability: A Search for Balance.* Waterloo, ON: University of Waterloo, Department of Geography.

Model Forest Network. 2009. The Canadian Model Forest Network website. http://www.modelforest.ca/cmfn/en.

Moran, E.F. 2000. *Human Adaptability: An Introduction to Ecological Anthropology.* 2nd edn. Boulder, CO: Westview Press.

Neumann, R.P. 2005. *Making Political Ecology*. London: Hodder Arnold.

Newman, P., and I. Jennings. 2008. *Cities as Sustainable Ecosystems: Principles and Practices*. Washington: Island Press.

Owen, S. 1998. 'Land use planning in the nineties: CORE lessons'. *Environments* 25 (2–3): 14–25.

Parks Canada Agency. 2000. *Unimpaired for Future Generations? Protecting Ecological Integrity with Canada's National Parks*, v. 1, *A Call to Action*, v. 2, *Setting a New Direction for Canada's National Parks*. Report of the Panel on the Ecological Integrity of Canada's National Parks. Ottawa.

Pedynowski, D. 2003. 'Prospects for ecosystem management in the Crown of the Continent ecosystem, Canada: Prospects and recommendations'. *Conservation Biology* 17 (5): 1261–9.

Penrose, R.W., J.C. Day, and M. Roseland. 1998. 'Shared decision making in public land planning: An evaluation of the Cariboo-Chilcotin CORE process'. *Environments* 25 (2–3): 28–47.

Policy Research Initiative. 2007. *Integrated Landscape Management Models for Sustainable Development Policy Making*. Ottawa: PRI.

Prato, Tony, and D. Fagre, eds. 2007. *Sustaining Rocky Mountain Landscapes: Science, Policy and Management for the Crown of the Continent Ecosystem*. Washington: Resources for the Future.

Rowe, S. 1992. 'The ecosystem approach to forestland management'. *Forestry Chronicle* 68 (1): 222–4.

Sax, J.L., and R.B. Keiter. 2007. 'Glacier National Park and its neighbours: A twenty-year assessment of regional resource management'. *George Wright Forum* 24 (1): 23–40.

Senge, P.M. 1990. *The Fifth Discipline: The Art and Practice of the Learning Organization*. New York: Doubleday.

Silver, J.L. 2008. 'Weighing in on scale: Synthesizing disciplinary approaches to scale in the context of building interdisciplinary resource management'. *Society and Natural Resources* 21: 921–9.

Siron, R., K. Sherman, H.R. Skjodal, and E. Hiltz. 'Ecosystem-based management in the Arctic Ocean: A multi-level spatial approach'. *Arctic* 61 (supplement 1): 86–102.

Slocombe, D.S. 1993a. 'Environmental planning, ecosystem science, and ecosystem approaches for integrating environment and development'. *Environmental Management* 17 (3): 289–303.

———. 1993b. 'Implementing ecosystem-based management'. *BioScience* 43 (9): 612–22.

———. 1998. 'Lessons from experience with ecosystem management'. *Landscape and Urban Planning* 40 (1–3): 31–9.

Slocombe, D.S., and P. Dearden. 2008. 'Ecosystem-based management and park planning'. In P. Dearden and R. Rollins, eds, *Parks and Protected Areas in Canada*, 3rd edn, 342–70. Toronto: Oxford University Press.

UNESCO. 2000. *Solving the Puzzle: The Ecosystem Approach and Biosphere Reserves*. Paris: UNESCO.

US National Research Council (US NRC) and Royal Society of Canada (RSC). 1985. *The Great Lakes Water Quality Agreement: An Evolving Instrument for Ecosystem Management*. Washington and Ottawa: US NRC and RSC.

Vallentyne, J.R., and A.M. Beeton. 1988. 'The "ecosystem" approach to managing human uses and abuses of natural resources in the Great Lakes Basin'. *Environmental Conservation* 15 (1): 58–62.

Waltner-Toews, D., J.J. Kay, and N.-M. Lister, eds. 2008. *The Ecosystem Approach: Complexity, Uncertainty and Managing for Sustainability*. New York: Columbia University Press.

Weaver, C. 1984. *Regional Development and the Local Community: Planning, Politics and Social Context*. Chichester: John Wiley and Sons.

Whitelaw, G.S., P.F.J. Eagles, R.B. Gibson, and M.L. Seasons. 2008. 'Roles of environmental movement organizations in land use planning: Case studies of the Niagara Escarpment and Oak Ridges Moraine, Ontario, Canada'. *Environmental Planning and Management* 51 (1): 801–16.

Whitelaw, G., and P.F.J. Eagles. 2007. 'Planning for long, wide corridors on private lands in the Oak Ridges Moraine, Ontario, Canada'. *Conservation Biology* 21 (3): 675–83.

Wilson, J. 2001. 'Talking the talk and walking the walk: Reflections on the early influence of ecosystem management ideas'. In M. Howlett, ed., *Canadian Forest Policy: Adapting to Change*, 94–126. Toronto: University of Toronto Press.

Woodley, S., and B. Freedman. 1995. 'The Greater Fundy Ecosystem Project: Towards ecosystem management'. *George Wright Forum* 12 (1): 7–14.

16

Applying Adaptive Environmental Management

Bram F. Noble

Learning Objectives

- To understand the fundamental principles of adaptive environmental management.
- To be able to distinguish among the different types of adaptive environmental management.
- To understand the requirements for applying adaptive environmental management.
- To become familiar with a range of Canadian applications of adaptive environmental management and the lessons emerging from practice.

INTRODUCTION

Adaptive environmental management, or adaptive management for short, is perhaps the most misused and misunderstood concept in resource and environmental management. Adaptive management has become a catch phrase and is often erroneously equated with *managing adaptively* in response to changing conditions (see Murray and Marmorek 2004; Gregory, Olson, and Arvai 2006). Resource managers often think that they are doing adaptive management if they are involving the public on an ongoing basis and monitoring and adapting to expectations, changing conditions, and unexpected outcomes. However, while useful and important, managing adaptively and the capacity to adapt are not the same as adaptive management. Adaptive management is more than monitoring, adapting, and responding to the unexpected (Walters 1997); it

is a clearly articulated approach to environmental management that treats policies or management prescriptions as experiments to test hypotheses, monitors the outcomes in order to refine those hypotheses, and subsequently adapts policies and actions as new knowledge and understanding are gained (Taylor, Kremsater, and Ellis 1997). Adaptive management is a structured approach for proceeding with management actions despite uncertainties about the best course of action (Schreiber et al. 2004); it is a formal, systematic approach to learning from the outcomes of planned management interventions.

Adaptive management presents an alternative approach to environmental management and assessment for those who are dissatisfied with conventional principles and practices that adopt a 'learn-as-you-go' philosophy or that attempt to control environmental systems. Attempts to prescribe policies and management strategies that seek to control environmental systems are common in practice; however, resource and environmental systems are not static but are in constant change, and often even the most carefully crafted management programs, plans, and policies are, or become, inappropriate. Adaptive management encourages planners and managers to approach their work with the expectation that they may be incorrect but that the experience and lessons learned from lesser successes will allow them to benefit and improve future environmental management policies and practices. Expect the unexpected, and learn from it—this is the underlying principle of adaptive management.

In the 30-plus years since its inception, adaptive management has evolved considerably. First envisaged as an approach to simulation-based modelling of ecological systems, adaptive management is now often closely linked with adaptive capacity, socio-ecological systems, and social learning. This chapter provides an overview of the fundamental principles and concepts of adaptive environmental management. Adaptive management is approached as a scientific process and as one component at play in a variety of larger-scale processes such as ecosystem management (Chapter 15) and social learning (Chapter 18). Several thumbnail sketches of Canadian experiences with adaptive management are presented, and the implications for adaptive management implementation are discussed.

ORIGINS AND CHARACTERISTICS OF ADAPTIVE MANAGEMENT

The concept of an adaptive approach to resource and environmental management is not new. Dearden and Mitchell (2009) note that people in many parts of the world who depend upon subsistence agriculture have practised adaptive management for thousands of years. It was not until the 1970s, however, that adaptive management, as

a formal approach to resource and environmental management, really took form. In Canada, the development of an adaptive approach to resource and environmental management was a response to Environment Canada's initiatives to examine the potential usefulness of simulation modelling as a tool for environmental management. Adaptive management was pioneered during the early 1970s by C.S. Holling and C.J. Walters (and associates) at the Institute of Animal Resource Ecology, University of British Columbia, and at the International Institute for Applied Systems Analysis in Vienna.

Referred to in its original context as Adaptive Environmental Assessment and Management, or AEAM, adaptive management emerged as a systematic approach for the design and testing of creative management and policy alternatives. AEAM relies upon teams of scientists, managers, and policy-makers jointly to identify and define resource and environmental management problems in quantifiable terms and model alternative policy and management solutions (Holling 1978; Walters 1986). While simulation modelling was the initial focus of adaptive management, the scope became much broader when various common features of most environmental problems and analyses began to emerge. Holling and his colleagues concluded that the shortcomings of most environmental analyses and management systems stem from how one deals with the uncertain and how lessons learned from management experiences are communicated and incorporated into future management policies and practices.

The premise of AEAM is that environmental policy and management practices must be designed with uncertainty in mind. This means that resource and environmental managers must attempt to manage uncertainty by designing policies and management practices so that they are capable of not only dealing with unexpected events but also benefiting from them. This requires an experimental approach to management design and intervention, treating policies and management prescriptions as experiments and then testing and adapting them as new knowledge is gained. The initial developments in adaptive management in Canada, which included an examination of a range of issues from fisheries management to economic development, were summarized in a book, *Adaptive Environmental Assessment and Management*, edited by Holling (1978), which has become a benchmark in the evolution of how we approach uncertainty in resource and environmental management problems. The central theme of the book was, in short, how to plan in the face of the unknown.

While most of the attention in adaptive management has focused on addressing uncertainty, emphasizing 'how' to manage (Jacobson et al. 2009), the scope of adaptive management is much broader. In the process of managing uncertainty, adaptive management also presents the opportunity to develop multi-stakeholder relationships to address resource and environmental conflict (Walker et al. 2002). This is what Buck et al. (2001) refer to as the 'who' and 'what' of adaptive management—directly involving institutions and the communities affected by management decisions in the development of management goals and in the evaluation of the implications of management actions. In this way, adaptive management enables a mutual understanding among participants and facilitates movement toward consensus and insight into how, in the midst of uncertainty, resources and environmental systems can be managed to sustain resilience and support learning (Walkerden 2006).

Core Ideas of Adaptive Management

The term 'adaptive management' has been used in various ways over the years, and often rather loosely, simply to refer to 'learning from management mistakes'. However, adaptive management is more than a *haphazard* approach to learning by doing. Nyberg and Taylor (1995, 241) define adaptive management as 'a formal process for continually improving management policies and practices, by learning from the outcomes of operational programs'. In other words, adaptive management involves more than picking up the pieces from failed or poorly conceived policies and management practices—scientific methodology is a key ingredient (Schreiber et al. 2004). Adaptive management is a multi-step, deliberative process that involves exploring alternative management actions and making explicit forecasts about their outcomes, carefully designing monitoring programs to provide reliable feedback and understanding of the reasons underlying actual outcomes, and then adjusting objectives or management actions based on this new understanding. Adaptive management is an experimental framework that articulates policy and management interventions as experiments from the outset and monitors outcomes to refine hypotheses in order to build knowledge (Taylor, Kremsater, and Ellis 1997).

While there is no single best definition of adaptive management, fundamental and defining characteristics of adaptive management can be identified. They are summarized in Table 16.1 and described in the following subsections.

Favours Action

First, adaptive management favours action, since experience is the key to learning. Much of what is characterized as adaptive management borrows significantly from 'adaptive control process theory', which addresses how decision-making devices can be structured in order to facilitate learning from experience (McLain and Lee 1996). One of the underlying principles of adaptive management is that policy and management strategies will be best facilitated through an approach that

Table 16.1 Key Characteristics of Adaptive Management

Adaptive management:

1. Favours action
   a. experimental
   b. exploratory

2. Accepts and benefits from uncertainty
   a. accepts change as a positive occurrence
   b. benefits from the unexpected

3. Allows discretion
   a. non-mechanistic
   b. flexible to local needs and circumstances

4. Seeks resilience
   a. does not aim at short-term solutions
   b. vagueness or generality of goals
   c. seeks resilience not stability

5. Provides feedback
   a. monitors and evaluates system feedback
   b. integrative holistic approach

6. Facilitates learning and integration
   a. promotes institutional learning
   b. integrates scientific and policy information

allows adjustments to changing events, decisions, and circumstances. Thus, adaptive management is question-oriented and presents an experimental approach to environmental policy, planning, management, and implementation. An adaptive management process is designed from the outset to test clearly expressed ideas or hypotheses about system behaviour. Lee (1993) explains that these ideas or hypotheses usually represent various predictions of how particular systems might respond as a result of modifications in management practice or policy implementation. When the management practice or policy is tested, the adaptive approach is designed so that learning occurs and adjustments can be made based on this new understanding. Errors provide the information required to generate new hypotheses and to modify and adjust management policies and practices and thus create the experience and information upon which new knowledge is built (Holling 1978, 8).

Accepts and Benefits from Uncertainty

Second, adaptive management explicitly accepts that resource and environmental systems may contain surprises. An adaptive approach is based on the premise that by planning for and expecting the unexpected, policy-makers, planners, and managers are better equipped not only to adjust but also to improve policy and management practices as changes occur and as new information arises. Uncertainty is inevitable in resource and environmental management. Rather than trying to eliminate the uncertain or to suppress it, adaptive management recognizes uncertainty as an inherent part of the management and learning process. Adaptive management permits resource and environmental management practices to proceed regardless of uncertainty and tries to benefit from it (Hamouda et al. 2004). Unexpected events and system surprises provide opportunity for learning and improvement—the opportunity to learn what it is that we do not know.

Allows Discretion

Third, adaptive management is discretionary. As discussed previously, conventional blueprint planning approaches consist of a fixed set of actions that will not involve extensive modification, experimentation, or revision. Compliance and crediting are key. However, 'too much direction and control stifle creativity and enthusiasm, and lead to standardized approaches that may not fit local conditions' (Mitchell 1997, 142). Program and policy goals often change, and unexpected events can trigger corrective actions that result in progressively greater commitments to making management corrections and adjustments. Present decisions have future decision consequences as well as environmental ones, and subsequent induced decisions may often generate greater environmental concern than seemed possible at the outset. The solution is to learn and adjust, giving local implementers discretion to make minor management and policy modifications relative to local needs and particular conditions as they emerge. However, there is also a danger that too much discretion may allow management actions or implementation to stray too far from the intended design, making it difficult to test management hypotheses, isolate 'background noise', and learn anything meaningful from the results (Murray and Marmorek 2004).

Seeks Resilience

Fourth, adaptive management does not aim at a specific system state. A common goal of environmental management programs is to generate policies and management practices that result in stable socio-economic, cultural, political, and environmental behaviour. In other words, the goal of adaptive management is not to maintain an optimal state or stable condition but to develop optimal management capacity (Johnson 1999). What we do know about the behaviour of environmental systems is much less than what we do not know. In many instances, we do not even know what it is that we do not know. Environmental systems are dynamic, and precise system states can be difficult to predict. Problems often arise when management and policy actions are designed with such specificity in mind. Thus, adaptive management encourages generalizations, or even vagueness, concerning future system states.

Noble (2000) and Mitchell (2002) note that vagueness and ambiguity provide scope to custom-design policy and management implementation to suit differing and dynamic conditions. Rather than managing for a single, optimal state, emphasis is on managing within a range of acceptable outcomes while avoiding catastrophic and irreversible negative outcomes (Johnson 1999). This does not mean that specific hypotheses are not formulated and predictions not made but rather that there is no fixed endpoint. The goal, instead, is resilience in the face of surprise (Lee 1993, 63). Resilience refers to the ability of environmental systems not only to absorb but also to benefit from change. One of the underlying goals of adaptive management, then, is to understand resiliency or the magnitude of disturbance that can be experienced by a system before it 'flips' to another system state (Holling and Meffe 1996). This requires the formulation of hypotheses, the testing of multiple policy and management prescriptions, and policy and management goals and objectives designed at the outset with change in mind.

Provides Feedback

Fifth, adaptive management is an iterative process that provides continuous and important feedback concerning the validity of hypotheses, the state of the environmental system being managed, and the appropriateness of the management style, goals, and objectives. While monitoring itself is not adaptive management, monitoring plays a key role in this feedback process. For programmed or blueprint-style implementation, monitoring is used to determine whether specified objectives are being met and whether expected management or policy outcomes are occurring. In contrast, the role of monitoring in adaptive management and implementation is to test hypotheses, determine whether and when adjustments are required in policy and management practices, and whether goals and objectives need refinement based on new information gained from system feedback. By incorporating such information feedback loops into the management and policy process, environmental decision-makers are able to accelerate the rate at which they learn from management experience.

Facilitates Learning

Sixth, adaptive management is learning to manage by managing to learn (Bormann et al. 1994). Adaptive management experimentally compares policies or practices by evaluating alternative hypotheses about the environmental system and then seeks to integrate the results of management experiments into management decisions, policies, and institutional learning environments. However, learning in adaptive management can also adopt a more inductive approach in which insights are derived from new information and dynamic hypotheses guide reasoning and structured argumentation. This, explains Pahl-Wostl (2007), sharpens the awareness of managers of the need to be prepared for the unexpected.

Integrative

Finally, adaptive management is an integrative process. Adaptive management addresses problems that are almost always multidisciplinary and involve many interests, ranging from political decision-makers to disciplinary specialists.

As such, adaptive management requires an integrated, interdisciplinary approach, enabling decision-makers and experts to identify and address complex environmental issues and to appreciate where, and in what form, their information is useful to others. One of the limiting factors of most environmental analyses, for example, is the lack of communication, information feedback, and information integration into future practice. Adaptive management seeks to overcome this limitation by integrating an unfamiliar mix of interests, bringing together different approaches and views and drawing upon experiences in other areas and from other issues. Bringing policy decision-makers, disciplinary specialists, and concerned interests together in one arena enables the formation of interdisciplinary relationships and focuses policy and management on practical options and on important system indicators (Grayson, Doolan, and Blake 1994). This mix of interests enables researchers to better appreciate where, and in what form, their knowledge is useful to others.

APPLYING ADAPTIVE ENVIRONMENTAL MANAGEMENT

Francis and Regier (1995, 289) suggest that 'the chronology of resource management has been a history of crisis and unitary responses'. Managers often rush from one outbreak to the next, never having the time or resources to test, monitor, evaluate, and subsequently learn from management actions and experiences. One only has to look to the management history of the northern cod fishery on the Grand Banks of Newfoundland, for example, to see evidence of this (see also Chapter 8). Management of the northern cod fishery has been characterized by one 'band-aid solution' after another, including the Atlantic Groundfish Strategy (TAGS) implemented in

1992 to cradle the fishing industry following the collapse of the northern cod stocks. While TAGS did provide temporary income relief for those affected by the decline of the fishing industry, it contributed little to understanding the socio-ecological system or to longer-term, improved fisheries resource management. To date, fish stocks have failed to recover. There is an urgent need to replace such *react-and-control* strategies in resource and environmental management with *anticipate-and-prevent* ones.

The Adaptive Management Process

The basic tenets of adaptive management all deal with the unpredictable interactions between people and their environments and the need to design management experiences and practices with the intention of learning—from both the successes and the failures. Holling (1978) and Walters (1986) maintain that no single best approach exists for implementing adaptive management; however, common components characterize the adaptive management process. Nyberg (1999), for example, identifies several model components adopted by British Columbia's Forest Service, which can be translated into a generic guide for the adaptive management process. Adaptive management can be portrayed as an iterative process of problem assessment, followed by the design of policy or management experiments, the implementation of management actions, and evaluation and adjustment, all of which is informed by an ongoing system of monitoring and assessment. These steps are depicted in Figure 16.1 and summarized below. There is no single, best adaptive management design process. The specific nature and style of adaptive management implementation, including the timing, methods, and techniques employed, depend on the nature of the issue at hand. Arguably, however, the best adaptive management design processes ensure the ongoing engagement of the

institutions and communities directly affected by management actions (Jacobson et al. 2009).

Problem assessment: Problem assessment in adaptive management is similar to the scoping phase of environmental impact assessment (Chapter 17). Problem assessment may be done in a workshop setting, or through a series of workshops, to define the scope of the resource or environmental management issue, identify measurable management objectives and potential management actions, define hypotheses concerning alternative management actions, establish indicators for each objective, explore the potential effects of alternative actions, and identify and assess key gaps in understanding. There are a variety of tools and techniques to facilitate this, including *conceptual models, simulation models,* and AEAM *workshops.*

Monitoring and assessment[1]*:* Parallel to, or as part of, problem assessment, it is necessary to identify key system variables and to monitor and assess the system dynamics in the absence of any policy or management experimentation or in the absence of any adjustment to the current baseline condition. The objective is to develop a conceptual model of the system, outlining linkages and functional relationships between actions and indicators. Depending on the nature of the application, simulation or other types of decision-support models and frameworks may be constructed to facilitate a degree of understanding of the system at hand and the complexities involved and of how the effects of management actions will be assessed. Monitoring refers to the collection and analysis of repeated observations to evaluate condition changes or progress toward meeting specified management objectives.

Designing policy or management options or experiments: Based on knowledge gathered from the previous phases, alternative policy or management options or experiments are identified for implementation, and a monitoring program

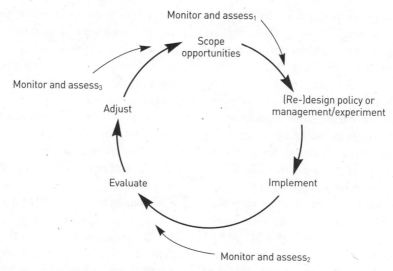

Figure 16.1 The Generic Adaptive Environmental Management Process.

is developed to provide feedback for learning. A decision is made as to whether adaptive management will adopt a *passive* or *active* approach (see below). Based on the objectives identified during the problem assessment stage and the chosen approach to adaptive management, alternative management options or experiments are evaluated and a single option or set of options chosen for implementation and monitoring.

Implementation: The management practice, policy adjustment, or experimental alternative(s) is implemented.

Monitoring and assessment[2]: Key indicator responses to the imposed management or policy action(s) are monitored and hypothesized relationships tested. Monitoring allows managers to gauge the *actual* effects or success of the management or policy action(s) and can also provide early warning of unanticipated change or surprise events. Monitoring should focus on validation of the model and relationships, the effectiveness of the management plan, and compliance with the original plan design.

Evaluation: The impacts, both intended and unintended, of the management action(s) are reviewed and evaluated with regard to hypothesized impacts and key questions, concerns, and issues identified. Here, actual results of the management actions are compared to the hypothesized results or forecasts made during problem assessment. The results are documented and communicated to managers, decision-makers, and other stakeholders and interests.

Adjustment: Goals, objectives, and experiments are revised as necessary. The most appropriate policy or management action is identified and the results communicated for integration into policy and practice.

Monitoring and assessment[3]: System dynamics are monitored and assessed over time, as well as the longer-term outcomes and system responses to the policy or management adjustments, alongside the suitability of revised goals and objectives.

Implementing Adaptive Management

The traditional way of dealing with the uncertain or the unknown from a scientific standpoint has been experimental approaches. In order for experimental approaches such as adaptive environmental management to succeed, Holling (1978) suggests that at least three minimum conditions must be met. First, the experiment itself must not destroy the experimenter. While experimental errors or lesser successes do provide new information and understanding necessary for learning, someone or some institution must remain who is capable of learning from such failures. Second, the experiment should not cause long-term, irreversible change to the environmental system. This may prevent any future learning or the opportunity to apply the knowledge gained from experimentation. Third, the experimenter or management organization must be willing to learn from mistakes and to start over again, armed with new information and understanding of the resource and environmental system. Failure cannot be treated as defeat.

Adaptive management may not always be the most appropriate strategy for resource and environmental management. Building on the above three minimum conditions, Murray and Nelitz (2008, 3) outline six questions to help determine whether adaptive management is appropriate to the resource or environmental problem at hand. If the answer to any of these questions is 'no', then adaptive management may not be the most appropriate choice of management tool or framework.

- Is there significant uncertainty about which management actions will best achieve the desired outcomes? If there is little or no uncertainty, then there is little need for adaptive management. This, however, does not negate the need for managing adaptively.
- Is a management experiment the best way to address or reduce this uncertainty? If trends analysis, experience from elsewhere, or previously collected data are sufficient, experimentation is not needed.

- Can a powerful enough management experiment be designed to discern the effects of the prescribed management action(s)? If it is not possible to confidently detect cause–effect or association, an experimental approach is of limited value.
- Is it feasible to undertake sufficient monitoring? There must be an opportunity to monitor long enough or over a broad enough scale to sufficiently detect the effects of the management experiment.
- Can there be safe failures? The results of failed management experiments must not be detrimental to the sustainability of the system.
- Is there sufficient support for implementation? Community, stakeholder, and institutional support are required for experimentation and over the long term to ensure the opportunity to learn from the results.

Implementation Styles

Despite the many real difficulties in meeting the above six questions, Holling (1978) argues for the need for innovative solutions to deal with uncertainty in resource and environmental management problems and that various experimental approaches are required. The objective is to develop interactive processes that recognize and reduce uncertainty and also benefit from it (Holling 1978). How adaptive management occurs in practice, however, varies considerably from one environmental problem to the next. Generally speaking, there are two broad classifications of adaptive management—active and passive (see Figure 16.2).

Active Adaptive Management

In active adaptive management, managers implement more than one policy or management practice as concurrent experiments to determine which best meets specified objectives. Active adaptive management is much more resource-intensive than passive adaptive management, but learning can occur much more quickly. Active adaptive management is the most powerful form of adaptive management, in which the management programs themselves are designed as experiments to test hypotheses about their effects on the particular resource or environmental system being managed (Nyberg and Taylor 1995). Central to an active adaptive approach is the design of policies and practices that not only recognize the unexpected but also benefit from it. There are two approaches to active adaptive management, prospective and retrospective.

Prospective adaptive management is the most intense and most ambitious form of active adaptive management. Prospective adaptive management involves active experimentation in which the environmental system is intentionally disturbed under controlled conditions in order to test hypotheses, monitor, and learn from a range of policy and management interventions. In other words, the resource planner or manager purposefully implements a series of changes or disturbances to the system as a management experiment and learns from the system's response in order to make more certain management and policy recommendations. Prospective adaptive management does not involve arbitrary perturbation of the environmental system just to see how it might respond. Specific and well-planned adjustments or experiments

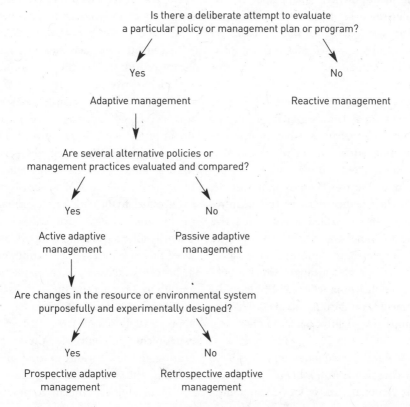

Source: Based on the Ontario Ministry of Natural Resources 1998. Adaptive Management Forum: Linking management and science to achieve ecological sustainability.

Figure 16.2 Adaptive Management Implementation Styles.

are initiated to address specific questions about change in order to improve policy and practice. Prospective, active adaptive management is certainly not the best type of management approach for all resource and environmental management situations. In cases where rare or endangered species, habitats, or ecosystems are involved, for example, prospective adaptive management can be risky business.

Active adaptive management can also be implemented less ambitiously through *retrospective* adaptive management. Underlying this second approach to active adaptive management is retrospective experimentation to provide insight and information about policy or alternative management practices. Changes, or experiments, in the resource or environmental system are not *purposefully* designed but are based on information and knowledge gained from previous natural or human-induced disturbances or conditions.

Retrospective adaptive management is particularly useful when data are available from past experiments or disturbances, when prospective experimentation is too risky, or when resources to design and implement an adaptive management system are scarce. There is some debate, however, as to whether retrospective adaptive management can be labelled as active adaptive management. Some argue that retrospective adaptive management is no more than retrospective analysis and simply a means by which a 'best' management policy or practice can be identified and implemented for the purpose of establishing a system of passive adaptive management.

Prospective Adaptive Management of Grazing in Grasslands National Park

Grasslands National Park is located in southwestern Saskatchewan, near the Saskatchewan–Montana border, and covers approximately 906 square kilometres of mixed-grass prairie. The park is a semi-arid ecosystem and was once subject to migratory and sedentary bison grazing, a disturbance considered to have contributed to the heterogeneity and ecological integrity of the landscape. Bison have been extirpated from the grasslands area since the turn of the twentieth century, and most of the region outside the park has since been cultivated. A policy of grazing exclusion on lands acquired for the park has been in place since 1987, consistent with the National Parks Domestic Animal Regulation. Recent monitoring in the park, however, has identified an increase in alien perennial grass and standing crop biomass, as well as a decrease in species of concern. These changes in grassland ecology are believed to be the outcome of the long-term exclusion of grazing activity.

In 2002, the Grasslands National Park Management Plan identified grazing as a necessary ecosystem function to restore the ecological integrity of the mixed-grass prairie. A prospective adaptive management experiment was developed whereby a series of grazing prescriptions would be implemented in the park to represent disturbance regimes consistent with historic patterns of migratory and sedentary bison herds. The experiment involves introducing livestock to the park for the primary purpose of determining how various levels of grazing intensity alter spatial and temporal heterogeneity of the mixed-grass prairie community. Up to 33 per cent of the park will be moderately grazed and up to 2 per cent of the park heavily grazed in nine large-scale experimental units, each approximately 300 hectares in size. The grazing experiment will control for the effects of non-grazing and test various grazing intensities on uplands, valleys, and riparian zones.

The grazing experiment commenced in the summer of 2006, with pre-treatment monitoring and assessment of indicators of ecosystem composition, structure, and function in each of the experimental units. In June 2008, livestock were introduced into six of the experimental units. The remaining three units will remain ungrazed as control sites. Compositional, structural, and functional indicators of ecological integrity will be monitored throughout the life of the experiment, to 2017, to determine how the restoration of grazing affects landscape heterogeneity and biodiversity. Data generated from the experiment will be integrated with a broader Grassland Ecosystem Management Support project to generate spatially explicit models of grazing effects in support of grazing management decisions. Adaptive management is being applied in Grasslands National Park in a way that no other national park has experienced—with the intent to cause disturbance to the ecological system and monitor its response. Information on Grasslands National Park's experimental grazing and adaptive management program, including annual progress reports, is available at http://www.grazingbiodiversity.org.

Retrospective Adaptive Management of Ontario's Moose Habitat Guidelines

In the early to mid-1970s, resource managers in Ontario recognized a problem with the sustainability of the province's moose population. An increasing hunter population with greater access to hunting areas, attributed to the mechanization of forestry practices, resulted in higher hunter success rates and a decline in the province's moose population (Timmermann, Gollat, and Whitelaw 2002). In response, several management control measures were implemented to address the problem, including reductions in moose harvests, delays in the opening date of the hunting season, and the establishment of wildlife management units to organize population data based on forest type and habitat potential. None of these react-and-control measures implemented in the 1970s were effective, and the

province's moose population failed to increase to a sustainable level.

In 1980, the Ontario government introduced a new policy for moose population management, accompanied by specific goals and objectives, including the objective to increase the population from 80,000 to 160,000 animals by 2000. Research on habitat management was a major focus of the new policy. In 1988, Timber Management Guidelines for the Provision of Moose Habitat were formally introduced in the province for the purpose of assisting resource managers in planning timber management activities to support moose population recovery (see Bottan, Euler, and Rempel 2002). The principal goal of the guidelines was to help increase Ontario's moose population by enhancing habitat through timber management processes, particularly modified clear-cut operations that include managing forest access, regeneration, and forest maintenance (Ontario Ministry of Natural Resources 1988). The Moose Guidelines Evaluation Program (MGEP) was established shortly thereafter to study moose population dynamics, habitat use, and productivity, feeding areas, and genetics. The MGEP was expanded in 1994 to include a long-term study of the efficacy of the guidelines.

In 1997, as part of the MGEP initiative, a retrospective adaptive management framework was developed and implemented near the Quetico Great Lakes–St Lawrence Forest Region to evaluate the suitability of the guidelines (see Rempel et al. 1997). The retrospective experiment was designed to compare alternative management practices and evaluate the effectiveness of the guidelines for providing suitable moose habitat, with the overall intent to provide the information necessary to improve the guidelines. Five different types of historical forest disturbances were compared—current moose guidelines modified clear-cut, progressive clear-cut landscape, wildfire burn area, mature uncut landscape without road access, and mature uncut landscape with road access. The objective was to determine whether landscape disturbance, hunter access, or the interaction of these factors explained the differences in moose density among landscapes and whether landscapes logged by modified clear-cutting, as prescribed by the guidelines, provided more suitable habitat than unmodified clear-cutting.

Sixteen years of moose population data and 19 years of Landsat images of the five different landscapes, totalling 240,000 hectares, were used to evaluate the guidelines. Results of the adaptive management experiment did not support the original hypothesis that the current guidelines, consisting of modified cutting in small clear-cut blocks, led to higher moose densities than conventional progressive clear-cutting. Using retrospective comparisons, current guidelines were found to be ineffective in mimicking natural disturbances, such as broad-scale burn or gap disturbance, and were insufficient for increasing moose density. The researchers concluded that if landscapes were managed to emulate natural burns combined with restrictions on hunter access, then the policy objective of increased moose density could be achieved. Rempel et al. demonstrated that co-ordinated harvest and habitat management is necessary to successfully manage moose populations and to achieve the 1980 policy objective of 160,000 animals.

Timmermann, Gollat, and Whitelaw (2002), however, note that the moose population has increased significantly above 1970 levels under the guidelines and that the management approach adopted by the guidelines might have been an appropriate one. They also note that the original policy objective of 160,000 animals may not be possible, given the constraints of current patterns of land use, hunting management, and other ecological factors in the province.

Recommendations by Rempel et al. (1997), and recommendations emerging from numerous other moose habitat and population studies conducted throughout the 1990s and early 2000s, are now being considered as part of a provincial moose management program review. The review was announced in 2007 and is still ongoing. As of April 2009, four draft moose management policies and procedural guidelines had been developed—namely, the cervid ecological framework, moose management policy, population objectives-setting guidelines, and harvest management guidelines. Information on the guidelines and on the review process can be found at www.mnr.gov.on.ca under the topics 'fish and wildlife' and 'forests'.

Passive Adaptive Management

Active experimentation is not always possible or feasible. In many cases, resource managers do not have the luxury of actively introducing disturbance to the environmental system or looking to past disturbances and data to simultaneously compare and experiment with alternative policy or management actions. In passive adaptive management, a single management policy or prescription that managers think is 'best' is implemented and then monitored to see if they were right.

Walters and Holling (1990) describe passive adaptive management as the use of historical data available at each time to construct a single best estimate or model for response, and the policy or management decision is based on *assuming* that the model is correct. The 'experiment'—that is, the policy or management prescription—and hypotheses are adjusted accordingly, and new knowledge is gained through the monitoring process. Changes in policy and management result from a recognized deficiency or failure in the existing management system to address changing resource or environmental conditions.

In other words, there is a real expectation that the policy or management practice will need to be adapted. Passive adaptive management is less resource-intensive than active adaptive management, but learning happens more slowly.

Passive Adaptive Management of Algonquin Provincial Park's Wolf Population, Ontario

Algonquin Provincial Park is the largest protected area in North America for the eastern wolf. In 1993, in response to concerns over a negative wolf population growth rate in the park and the impact of human-caused mortality from shooting and snaring outside the park, the Ontario minister of natural resources implemented seasonal protection of the wolf in three townships adjacent to Algonquin Provincial Park. The minister also appointed an Algonquin Wolf Advisory Group and commissioned a scientific review of the status of Algonquin wolves. The seasonal ban on hunting was intended to protect Algonquin Park wolves during the December to March period when wolves were believed to leave the park boundaries to prey on deer in the adjacent townships.

In 2001, following decades of wolf population research by scientists and government agencies, the Algonquin Wolf Advisory Group recommended that wolves be protected in additional townships adjacent to Algonquin Provincial Park and that a long-term monitoring plan be implemented to determine the effect of the management prescription. In response to these recommendations, the minister invoked a single management prescription as an experiment in wolf population management. The adaptive management plan would involve a temporary, year-round ban for 30 months on the harvesting of wolves in the 40 townships surrounding Algonquin Provincial Park. The plan would be subject to ongoing monitoring and evaluation to determine whether the management experiment had the desired effect on wolf population status.

Theberge et al. (2006) set out to test the adaptive management prescription, evaluating the probability of being able to test the null hypothesis that protecting wolves adjacent to Algonquin park for 30 months will not result in a significant positive population response. Key to passive adaptive management is maintaining focus on a few key variables that can be measured efficiently as conditions change and using the results to make decisions about management practices. Too many system variables identified in only a general fashion confound management and learning. Using average population growth rates and variances in growth rates, Theberge et al. applied a discrete population viability model considering population density, recruitment rates, mortality due to natural causes, and mortality due to human causes. Their analysis showed the testability and falsifiability of the hypotheses to be low, attributed to the stochastic nature of ecological systems and also to the short time horizon of the experimental treatment. Theberge et al. concluded that the null hypothesis could not be rejected and recommended that a precautionary principle should be adopted whereby the ban on wolf hunting would remain, even if it could not be proven effective in the time scale of the experiment. Theberge et al. argued that it would be better to err on the side of caution and retain the ban in the interests of conservation, even if in the longer term it proved ineffective, than to remove the ban when it might have been necessary.

The temporary ban on wolf harvesting in Algonquin continues, currently well beyond its initial 30-month experimental period. However, not all scientists agree with the management decision. Patterson and Murray (2008) revisited the issue and argued that population viability analyses conducted by Theberge et al. were statistically flawed. They demonstrated that recovery following the 30-month ban was a reasonable expectation and that change in the wolf population due to the ban could be detected. Patterson and Murray went on to note that the temporary ban on wolf harvesting outside the park was reasonable based on data and analyses available at the time, but they also argued that the perceived threat to the long-term viability of Algonquin wolves 'was greatly exaggerated' (2008, 84). Like many other applications of passive adaptive management, the Algonquin case study illustrates that the results of management experiments are sometimes uncertain, may conflict with stakeholder values, or do not necessarily conform to social and political time frames. Additional information on wolf management and science research in Algonquin Provincial Park can be found on the website 'The science behind Algonquin's animals' at www.sbaa.ca.

One Size Does Not Fit All

Adaptive environmental management embodies a simple imperative: policies, plans, and programs are experiments—learn from them. No particular style of adaptive management, however, is best suited to all situations. Furthermore, notwithstanding the benefits of adaptive management, it is not always necessarily the most desirable approach (Lee and Lawrence 1986; Noble 2000).

With regard to passive adaptive management, for example, Walters and Holling (1990) note two fundamental limitations. First, the passive approach often fails to accept change as a positive phenomenon. The objective is often to maintain the system in a steady state and attempt to restore it to its original condition through management and policy adjustments when changes occur. While this allows us to monitor and adjust as new information is gained, it provides less opportunity to actively design management and policy opportunities that are more resilient to changing circumstances and events.

Second, passive adaptive approaches may fail to detect *real* opportunities for improving system performance in that the *right* policy model and the *wrong* policy model may result in the same response pattern. For many ecological systems, there is considerable debate as to whether changes in system status are management-related or a product of natural environmental cycles. At the same time, however, control and experimentation are difficult and perhaps even risky in such large-scale environmental systems.

The limitations of the passive approach do not necessarily mean that an experimental approach to adaptive management is always preferred. Active experimentation may increase the rate of learning, but it may also increase the risk associated with potentially suboptimal management outcomes (McCarthy and Possingham 2007). Experimental approaches may not be appropriate when the risk to the environment or society is high, such as the risk of oil spills or threats to health and safety in offshore environments, or when the potential damage caused by management or experimental failure is irreversible. An active experimental approach may be too risky in situations in which experimentation takes place over large geographic scales, policy failure may endanger human health, or error may cause species extinction or irreversible environmental damage. But as Holling (1978, 8) argues, in the absence of an active adaptive approach, 'the search for a solution should not replace trial-and-error with some attempt to eliminate the uncertain or the unknown.' While efforts to reduce uncertainty are admirable, the goal is to develop more resilient policies and management practices.

In addition, conditions exist under which any style of adaptive management might be completely undesirable. Lee and Lawrence (1986) suggest that an adaptive approach is not appropriate when the problem is curable, the remedy is unique, and managers are held accountable for policy, plan, or project failure. First, adaptive management is continuous management. While there are advantages to managing over an extended period of time, adaptive management approaches are often inappropriate for situations for which a one-time, curable solution exists, such as the cleanup of an oil spill or a contaminated site. Second, if there is no learning curve—that is, the solution is unique to a particular problem—then a 'learning-by-doing' philosophy may be less useful, since the experience gained may not be transferable. Third, if managers are held responsible for errors and for policy, plan, or project failure, then they are unlikely to take the risk of adaptive management when the opportunity presents itself. Consequently, managers are drawn to more reactive management strategies.

LESSONS LEARNED FROM CANADIAN EXPERIENCES WITH ADAPTIVE MANAGEMENT

Although the adaptive approach to resource and environmental management has been advocated since the early 1970s, its success in practice has been mixed. The previous sections highlighted the various styles of adaptive management. Attention now turns to Canada's experiences with adaptive management implementation and the lessons learned from these experiences. Five case studies are presented, the first being a classic in Canadian adaptive management. The case study discussions are not comprehensive; rather, they serve to illustrate a range of experiences with adaptive management within the Canadian context from past to present.

New Brunswick Budworm Management

The case of spruce budworm management in New Brunswick is one of the earliest, and most commonly referenced, applications of adaptive management. In the early 1970s, the New Brunswick Department of Natural Resources introduced an adaptive management approach in response to a spruce budworm epidemic that threatened tree species of prime economic importance to the province's forest industry (Holling 1978). The final outcome of the management experiment, however, proved to be less than successful.

The spruce budworm feeds on the foliage of balsam fir and spruce trees, causing up to 80 per cent mortality. Although the budworm has occupied New Brunswick forests for centuries, it was not until the 1950s, when fir and spruce forest stands increased in economic importance in the pulpwood industry, that the budworm and its relationship to the dynamics of that forest were recognized (Baskerville 1995). In the 1950s, widespread evidence of damage to fir and spruce forests in New Brunswick emerged as a result of budworm defoliation. This damage was perceived as a threat to the forest industry and hence the economic base of the province. In response, the New Brunswick government joined with the forest industry in large-scale, aerial insecticide spray programs in 1952 to attempt to control stand defoliation. The large-scale spray program, however, proved to be less than successful. What started out as a short-term operation turned into a lengthy one, since protective spraying programs maintained forest stands that were regularly reinvaded after treatment (Baskerville 1995).

During the late 1960s and early 1970s, concern grew over the effects of long-term intensive spraying campaigns on both forest and human health. In 1972, the Canadian government contracted a team of researchers from the University of British Columbia to work with local scientists in developing and implementing a model that would lead to increased understanding of the budworm population and its interactions with forest and human health and provide insight on alternatives to the existing large-scale spray campaign (McLain and Lee 1996).

The team developed a complex computer-based model to simulate the interaction of budworms with the forest ecosystem and human population. Two major options were presented: an early low-dose spraying program for budworm population management and a change in current forest harvesting regimes (Baskerville 1995). The New Brunswick–appointed Task Force for the Evaluation of Budworm Control Alternatives selected the low-dose spraying option as the preferred policy on the grounds that changes in harvesting regimes would have severe economic repercussions for the forest industry. A committee was then established to implement, monitor, and evaluate the results of the policy experiment.

Implementation of the new policy, however, proved to be the most difficult stage of the management process and raised a storm of controversy among environmentalists and many foresters. The 'black box' nature of the modelling exercise was disconnected from the 'who' and 'what' of adaptive management (see Buck et al. 2001). The process had failed to effectively engage the institutions and interests directly affected by the management prescription in assessment and decision-making. The result was considerable conflict about how the budworm problem should be managed. Whereas the scientists involved in the adaptive management experiment viewed the problem as one of insect control, environmentalists perceived the problem as one of resource allocation and identified solutions in terms of change that would shift New Brunswick forest management away from pulpwood production. Environmentalists accused the government of failing to adequately assess the ecological impacts of the spraying program and argued that the proposed monitoring

program was biased in favour of the spraying alternative. Many foresters were also sceptical about the proposed policy approach and felt that the modelling exercise captured only a narrow view of how forest ecosystems function.

By the mid-1980s, the simulation model had been replaced. Managers and planners had greatly underestimated the complexity of the decision-making process. The management and modelling team assumed that decisions were made and implemented by levels of higher authority, notably the federal government of Canada. However, the truth was that a multitude of stakeholders at many levels, who influenced budworm management through less formal and indirect mechanisms, were involved in forest policy and management decision-making (McLain and Lee 1996). Adaptive management proved to be less than successful in addressing the budworm problem. It was not so much that the adaptive management 'experiment' itself failed but that the assumptions surrounding management practice and policy implementation underestimated the complexity of policy and management decision-making in the forest industry. Natural scientists and modellers, who likely expected 'rational' responses to their management prescriptions, dominated the team. The modelling-based approach was not sensitive enough to the probability of conflict and how that conflict might be addressed.

Sockeye Salmon Stock Rebuilding

Building on the modelling experiences from the New Brunswick budworm problem, an adaptive environmental management approach was applied to British Columbia's fisheries management sector in the mid-1970s. The purpose was to enhance salmon planning facilities and to rebuild declining sockeye salmon stocks in Rivers Inlet on the central coast of British Columbia. The experiment, however, proved to be completely unsuccessful in meeting its objectives.

The Rivers Inlet fishery was historically the third-largest producer of sockeye salmon in British Columbia, following the Fraser and Skeena rivers. Until the early 1970s, fisheries scientists believed that sockeye river exploitation rates at around 60 to 80 per cent per year would allow for sufficient escapement, averaging 400,000 to 600,000 spawning fish per year for Rivers Inlet (Walters, Goruk, and Radford 1993). However, suspicions of overfishing, based on low spawning returns in the mid-1970s, prompted several assessments of the Rivers Inlet stock and recruitment data. Assessments confirmed these suspicions and further added that, contrary to previous studies, optimum escapement levels of sockeye in order to facilitate healthy returns might be as high as one million spawning fish.

Concern within both government and the fishing industry was growing over declining stocks. Using this window of opportunity, the Department of Fisheries and Oceans drastically reduced the Rivers Inlet fishery during 1979 to 1984 in hopes that higher spawning escapements would lead to increased adult returns and hence a more profitable sockeye fishery. The Department of Fisheries and Oceans explicitly labelled their management efforts as an adaptive or experimental policy approach (Walters, Goruk, and Radford 1993).

Enhancement planning models were developed based on the lessons learned from the New Brunswick budworm experience and were therefore more sensitive to conflict. For example, the objectives of various interest groups and more sophisticated assumptions about real, implementable policy decisions were built into the stock simulation model (McLain and Lee 1996). One of the objectives of the model was to determine the effects of increased escapement on spawning returns. A multi-year experiment of increased escapements was implemented to determine whether larger spawning stocks would produce larger adult salmon returns.

Stock size did not increase as expected. Walters, Goruk, and Radford (1993) highlight the insufficient consideration given to whether the modelling system would be sensitive only to minor responses in sockeye populations to increased escapement. Also, little consideration was given to how many years of experimental data would be necessary to determine, with any degree of certainty, whether salmon recruitments had increased under the new policy regime and what variables should be monitored and at what resolution. It was only assumed that increases in stock size over the experimental period would be large enough to be reflected by escapement numbers.

Government and commercial fishers concluded that the adaptive management experiment had been a complete failure. Commercial fishers experienced a drastic loss in catch, and the experiment was unsuccessful in increasing recruitment levels, at least to the extent that the model could detect them. Too much was expected from the experiment in too short a time. The result was frustration and mistrust between commercial fishers and fisheries managers in the enhancement program. While the Rivers Inlet sockeye salmon adaptive management experimentation did provide managers with considerable insight for improving future fisheries policies and practices, in terms of direct environmental and socio-economic gain it was a failure.

British Columbia's Berries for Bears Project

The active management of grizzly bear habitat in British Columbia, referred to as the 'berries for bears' project, implemented in 1992 by the British Columbia Forest Service, is an example of prospective or experimental adaptive management. Grizzly bears are considered to be a vulnerable species in British Columbia—at risk because of low numbers and limited range, and in some low-lying coastal areas of the province they are considered to be threatened—likely to become endangered if the factors affecting their vulnerability are not addressed. The rich valley bottoms of the coastal areas of British Columbia, prime grizzly bear habitat, are also highly valued for commercial timber productivity, creating conflict between bear habitat and land use for commercial timber harvesting. The British Columbia government, through the Grizzly Bear Conservation Strategy and the Berries for Bears project, is actively taking steps to conserve and manage critical grizzly bear habitat in these areas (BC Forest Service 2001).

Current forestry practices in coastal valley bottoms limit the availability of adequate forage for grizzly bears. McLennan and Johnson (1993) explain that forage species necessary for supporting grizzly bear populations compete for growing space with regenerating seedlings. Short-term chemical control and harvesting operations reduce the availability of forage, which over the long-term results in dense forest stands with thick canopies and limited forage species. The problem is exacerbated because most forest operations occur in the lower elevations, particularly coastal areas, in prime grizzly habitat.

An active adaptive approach was implemented in 1992 by the Forest Service to promote the growth of grizzly bear forage, while, at the same time, to produce high quality timber. Guidelines for promoting the growth of forage species necessary to support grizzly bear populations were required immediately, notwithstanding the uncertainty surrounding their effectiveness. Relatively small experimental stands, ranging from 4 to 10 hectares, were established in the Coastal Western Hemlock biogeoclimatic zone to evaluate the effects of various silviculture treatments, spacing, and tree cluster arrangements on forage growth and timber quality. Results generated from these trials are now being used to develop or to fine-tune harvesting operations to support grizzly bear habitat. Based on knowledge gained from experimentation, guidelines have also been developed for British Columbia's silviculture practices. The guidelines identify the high-priority forested ecosystems associated with grizzly bear forage production, describe the recommended stocking standards for those ecosystems, suggest alternative silviculture treatments, identify main plant species considered to be preferred grizzly bear forage, and suggest parameters for conducting and monitoring field operations (BC Forest Service 2001).

Donna Creek Forestry and Biodiversity Project

The Donna Creek Forestry and Biodiversity Project (DCFBP) was initiated in 1991 by the Peace/Williston Fish and Wildlife Compensation Program, a joint initiative of BC Hydro, BC Environment, and TimberWest Forest Ltd. The objective of the Peace/Williston Wildlife Compensation Program is to protect and enhance fish and wildlife affected by hydroelectric dam construction and the creation of the Williston Reservoir. The DCFBP, located in the northern half of the province in the Williston watershed at the foot of the Wolverine mountain range, was initiated specifically to develop and test alternative forest harvesting practices designed to sustain habitat elements required by wildlife, particularly cavity-nesting birds and small fur-bearers such as the American marten (*Martes Americana*) and fisher (*Martes pennanti*) (Corbould and Hengeveld 1998).

Approximately 16 per cent of British Columbia's native wildlife can be classified as 'tree users', many of which are tree cavity–nesting birds and small mammals (Backhouse 1993). Birds and small mammals that rely on tree cavities are at risk in British Columbia's managed forests because of conventional clear-cutting harvest methods that remove all standing trees and stumps. The DCFBP tests for differences in use by cavity-nesters and by other breeding birds and furbearers among conventionally clear-cut blocks, old-growth control areas, and blocks cut with three modified harvesting techniques. The project was developed to determine whether creating three-metre high stumps and retaining small residual tree islands during clear-cutting would benefit cavity-nesting animals and to examine the characteristics and locations of these features that make them useful to the animals.

Three alternative experimental cutting blocks were established, each measuring approximately 100 hectares in size and divided into four equal sections. In each block, four harvest treatments were applied, including clear-cut leaving no stumps or tree islands, stub-cut leaving three-metre-high stumps but no tree islands, island-cut leaving tree islands but no stumps, and stub-and-island-cut leaving both tree islands and three-metre-high stumps (Corbould and Hengeveld 1998). The design is experimental, using small-scale treatment areas as the basis for scientific evaluation and learning. In order to determine the effects of each harvest method, several programs have been implemented to monitor both the quality of forest regeneration and the potential benefits for cavity-nesting birds and mammals, and several adjustments to harvest treatments have been made to the experiment and management prescriptions as a result of information gained from ongoing monitoring programs.

In a situation where uncertainty is high, controlled active experimentation is being used to address and reduce the uncertainty of management actions at the local, experimental plot level before implementing policy and management practices in the forest sector at large. The experiment also serves as a monitoring standard by which previously implemented forest management policies and practices can be evaluated and adjusted as new knowledge is gained. The DCFBP is an example of the long-term commitment required to undertake prospective adaptive management and to monitor and measure the effects of management prescriptions on the natural environment. The knowledge gained from smaller, controlled policy experimentations must then be passed on to the forest sector and to provincial biodiversity policy and forest management practices. As with most active experimental approaches to adaptive management, however, a major challenge is controlling for actual management and biophysical processes using the knowledge gained from policy experimentations implemented at relatively local scales on a number of experimental units (Walters and Holling 1990).

Diavik Diamond Mine's Aquatic Effects Monitoring Program Adaptive Management Plan

The Diavik Diamond Mine (Diavik), an unincorporated joint venture between Diavik Diamond Mines Inc. and Aber Diamond Limited Partnership, is located on a 20-square-kilometre island in Lac de Gras, Northwest Territories, approximately 300 kilometres northeast of Yellowknife. Lac de Gras has a 4,000-square-kilometre drainage area. With low water temperatures, limited light during the winter, and approximately eight months of ice cover, both lake nutrients and productivity are low. However, water in Lake de Gras is relatively pure and is important for the sustainability of wildlife in the project region.

Mineral claims were first staked at Diavik in the early 1990s, followed by several years of test drilling and site planning. A comprehensive environmental assessment review was implemented under the Canadian Environmental Assessment Act, with the project receiving approval by the federal government for permitting and licensing in 1999. Mine production commenced in January 2003. As part of the project licensing and approvals process, water licences for the mine operation required that Diavik include an adaptive management strategy as part of the aquatic effects monitoring program.

The Diavik Diamond Mine Adaptive Management Plan for Aquatic Effects (Diavik Diamond Mines Inc. 2007) provides a framework for how data in the mine's aquatic effects monitoring program will be used to identify additional mitigation strategies to minimize the impact of project operations on the aquatic environment. Adaptive management is defined in the plan as 'a systematic process for continually improving mine operation practices by learning from the outcomes of performance monitoring and review programs . . . a cyclical process of plan, monitor, review, revise plan, monitor, etc.'.

While the definition reflects a passive approach to adaptive management, at least in principle, a 2008 independent review of Diavik's adaptive management plan by ESSA Technologies suggests that in practice, adaptive management 'is being viewed with much less rigor than required to be done properly, or is being misunderstood as managing adaptively' (Murray and Nelitz 2008, 5). Monitoring and mitigation assessment at the Diavik mine does not conform to the experimental design of adaptive management, and the aquatic effects monitoring program itself was never intended to be undertaken for the purpose of improving management goals or objectives. Rather, the program was established primarily to manage impacts—both known and unexpected—through monitoring for management. Neither the mine itself nor the perturbations caused by its development and operations were designed as an experiment but rather as development activities with environmental effects to be monitored and managed.

As Murray and Nelitz (2008) explain, this does not mean that Diavik's aquatic effects monitoring program is not effective or that monitoring for management is not a worthwhile activity. Rather, it means that monitoring for management is not the same as adaptive management. At Diavik, as in many other resource development projects and monitoring plans, adaptive management has become a common phrase for managing adaptively.

When Does Adaptive Management Work Best?

Attempts to foster the development and implementation of adaptive management have enjoyed mixed success. But what lessons can be learned from Canada's 30-plus years of experience with adaptive management? Under what conditions does adaptive management work best? If one were to reflect on Canada's history of adaptive environmental management, both the successes and the failures, several lessons for future practice emerge. Notably, adaptive management seems to work best when the following minimal conditions are met.

A Mandate Exists to Take Action in the Face of Uncertainty

Adaptive management requires a mandate for action, since as mentioned previously, action is the key to learning. Managing under uncertainty requires that we do more than simply acquire knowledge over time by learning as we go; we must plan for uncertainty, explicitly test management practices, and adapt policy and management as new knowledge is gained.

Managers in British Columbia's forest sector, for example, recognized the uncertainties and the lack of understanding of the relationships between harvesting operations and grizzly bear habitat. However, notwithstanding the uncertainties involved, a clear mandate was provided to take action and to integrate the knowledge gained from localized and controlled adaptive experimentation into broader forestry policy and practices as the adaptive policy experimentation unfolded.

Successful implementation requires that managers are explicit, but not rigid, about their policy and management objectives. Specification of objectives facilitates the development and implementation of monitoring methods and techniques most appropriate for the circumstances and generates the most useful information to be carried forward.

Management and Policy Prescriptions Are Treated as Experiments

Good management does not necessarily begin with good science, but it rarely can occur without it. Treating resource and environmental policy and management interventions as experiments, and expecting to learn from them, is central to adaptive management. Adaptive management works best when managers and decision-makers acknowledge that uncertainty exists and accept the possibility that policy and management prescriptions may be incorrect. When managers and decision-makers treat policies and practices as experiments, there is willingness and an expectation to adjust along the way as new information is gained and to start over in cases where policies and practices turn out to be less than successful.

Continuous monitoring and evaluation of the Donna Creek project, for example, has revealed that the spatial scale of the adaptive management experiment was insufficient for determining the effects of various harvest treatments on small furbearers. Thus, recommendations were made to refine management goals and objectives and the spatial extent of the management experiment. Learning works best when managers and

practitioners have discretion to make modifications in policy and practice in response to local needs, conditions, and changing events and circumstances.

Sufficient Stability Is Present to Measure Long-Term Outcomes

Any agency adopting an adaptive approach must be prepared to devote more planning resources, management staff, and financial resources than those practising reactive management strategies. Adaptive management focuses on longer-term policy and management goals and not on short-term gain. Experimental learning takes time. Adaptive management is most successful when managers and experimenters are patient about the acquisition of results and the management institutions support the time and financial investments in a policy or practice about which it is uncertain in terms of at least part of the outcome.

The Donna Creek Forestry Biodiversity project has been ongoing since the early 1990s and was recognized at the outset as a long-term management experiment. Almost 20 years later, the project is still considered to be in its infancy. In sharp contrast, fisheries management organizations and the fishing industry in British Columbia were overly optimistic about the time required to see improvements in sockeye salmon recruitment levels. The expected benefits of adaptive management were not realized, and 10 years into the policy experimentation, it was considered an adaptive management failure. Similar challenges were evident with Ontario's adaptive management plan for the eastern wolf near Algonquin Provincial Park, mentioned earlier in this chapter. In this case, the management prescription of 30 months was insufficient to determine whether the experimental treatment had a significant effect on wolf population growth.

Mechanisms Are Available to Transfer the Results of Adaptive Management to Broader Policy and Management Practice

Various methods and techniques are available for implementing adaptive management, notably baseline research and monitoring; public participation, consultation, and multidisciplinary team workshops; technological forecasting and system simulation modelling; and geographic information systems. No particular set of methods and techniques will do all that is required of adaptive management in all situations. The key is to use methods and techniques most suited to the problem at hand. However, one must also have the appropriate methods and techniques to facilitate the actual implementation of the product(s) of an adaptive management process. There is more to adaptive management than modelling and simulation.

Learning from the New Brunswick experience, in which the complexity of policy implementation was severely overlooked by the technical modelling process, those involved in the Rivers Inlet experiment considered policy implementation mechanisms at the outset. McLain and Lee (1996, 141) explain that rather than assuming a value-free decision-making process in which fisheries managers were free to make decisions based solely on the modelling exercise, those in charge of the adaptive management program attempted to incorporate the values and objectives of various interest groups into the modelling process. Adaptive management practitioners must not only collect information; it must be evaluated and used as the basis for comparing outcomes and impacts with management expectations. This new understanding must then be integrated into future policies and management practices. In the case of experimental grazing in Grasslands National Park, for example, lessons learned from the adaptive management are being used to inform broader grassland management and policy decisions about grazing activities in the park.

Managers and Planners Work with Interests in a Co-operative Environment

As demonstrated by the shortcomings of the New Brunswick budworm management program, planners and managers must maintain the continued involvement of those with a vested interest in the outcome. Critics of the New Brunswick budworm management program felt that the budworm modelling exercise was too narrowly focused and biased in favour of spraying interests and that the model itself was overly optimistic and failed to define the problem from the perspective of all those affected by the decision outcome.

Similarities can be drawn between the contextual factors that make alternative dispute resolution work, outlined by Mitchell (2002), and adaptive management. In order for adaptive management to work effectively, managers must: (1) recognize that a problem or potential problem exists; (2) be motivated to address the problem or to manage it through experimentation; (3) represent all affected parties; (4) find consensus in the management approach to experimental design; (5) accept the need for challenging constructively; (6) focus on interests; and (7) invent options for mutual gain.

There is a growing recognition that given the multi-stakeholder nature of most environmental situations and the multiplicity of actors involved, the more immediate success factors in adaptive management application are organizational and social rather than technical. Included among these factors is the need for adaptive management to anticipate and manage conflict, to accommodate non-scientific forms of knowledge, and to promote a shared understanding among those involved with and affected by the management program.

IMPLEMENTATION CHALLENGES AND OPPORTUNITIES

The term 'adaptive management' has become familiar among resource managers and policy decision-makers, but we appear to be moving at a glacial pace toward its widespread application. Why does it seem that adaptive management has not received the acceptance and experienced the success that it should have? Why has adaptive management not lived up to its potential? Several key reasons can be suggested.

Inconsistency in Definition and Principles

First, adaptive management has suffered from the lack of a consistent and widespread *operational* definition. While numerous definitions are presented in the literature, Nyberg (2001) argues there is very little operational understanding among policy and management practitioners. Definitions of adaptive management emphasize a 'learning by doing' philosophy and treating policies as experiments and learning from them, but there is much confusion and misinterpretation of adaptive management as managing adaptively. There is also limited guidance concerning *how* managers can and should operationalize adaptive management under different situations to address pressing resource and environmental problems.

The Need to Move beyond Environmental Monitoring

Second, environmental monitoring is important to adaptive management, but environmental monitoring is not the same as adaptive management. It is widely believed that adding a monitoring program to almost any existing policy or project will lead to valuable insights and so-called adaptive management. While monitoring is an important component of adaptive management,

it is only one part of a much larger and integrative process. So-called adaptive management at the post-policy or project implementation stage is often no more than an add-on approach to problem management and is hardly a starting point for learning. Adaptive management must be recognized at the outset if it is to be effective.

Lack of Participative Approaches

Third, there is no single approach to adaptive management that will work best in all situations. Many of the adaptive management approaches, such as the case of New Brunswick budworm management, involve extensive computer simulation and modelling. Simulation and modelling, however, can be expensive and time-consuming. Furthermore, it does little to contribute to the identification of socially acceptable alternatives and their implementation when considerable mistrust exists among multiple interest groups, all with something to gain or lose from the policy or project outcome. In such cases, time is best spent on building multi-stakeholder management teams, identifying and managing conflict among stakeholders, and strengthening organizational relationships in order to identify the questions to be investigated and to define mutually agreeable alternative strategies. As Mitchell (1997, 142) notes, 'the adaptive approach seeks active participation of relevant participants . . . in belief that more participants will bring more information and perspectives to help define issues and develop solutions.' At the same time, however, one should not mistake a participatory process for resource and environmental decision-making as the same as adaptive management.

Focus on Short-term Results

Fourth, the most powerful form for learning by doing is active adaptive management. The problem, explains Nyberg (2001), is that many managers struggle to find a way to *do* active adaptive management. Social and political expectations do not necessarily operate on the same time scale as adaptive management, with expectation often focused on relatively short-term results. But the benefits of active adaptive management may take years or decades to surface. Thus, a major challenge for adaptive management is that although scientists may insist that revisions in management prescriptions not be made without proper falsifiable test results or that experiments continue in the absence of short-term indication of effectiveness, the necessary longer time frame is not always consistent with urgent management and policy requirements (Theberge et al. 2006).

Institutional Rigidity

Finally, adaptive management will not succeed simply because of the recognition that uncertainty exists, nor will it succeed simply with good experimentation and monitoring programs and by adopting the 'learning by doing' philosophy. Adaptive management requires that resource and environmental management institutions adopt a different style of management and policy decision-making. Management institutions must be willing to accept the fact that their actions and policies may be incorrect or less than sufficient and that by acknowledging their mistakes, they can learn from the lesser successes and improve future management practices. Those involved in

adaptive management experiments should not be held directly accountable for those lesser successes, unless provision is made for learning and improving from such experiences. The key is to recognize that failure is not always a manifestation of the lack of skill or ability and that both successes and failures can be instructive if there is a willingness to learn from them (Noble 2000).

CONCLUSION

Adaptive environmental management presents an innovative approach to addressing uncertainties in resource and environmental policy and management. The goal of adaptive management is not to eliminate the uncertain but to benefit from it through the development of more resilient policies and management practices. Learning from our management experiences—both successes and failures—is key, as is communicating this new knowledge into future practice. While the principles of adaptive environmental management offer many new possibilities for improving the way in which we approach environmental management, Canada's 30-plus years of mixed success with adaptive management reminds us that it remains a work in progress. Each time adaptive environmental management is applied, however, lessons can be learned for future applications—even from lesser successes.

FROM THE FIELD

Adaptive Management in Canadian Environmental Assessment:
Policy versus Practice

Evaluating the potential impacts of large-scale resource development projects is fundamental to the environmental assessment (EA) process. However, the likelihood of accurately predicting all

impacts associated with a proposed development, such as a mining operation or hydroelectric project, is virtually nil. Impact predictions in EA often are wrong!

In recognition of the uncertainties associated with predicting and managing the impacts of resource development, adaptive management (AM) is included under the Canadian Environmental Assessment Act as an important component of the EA process. Based on experience from the field, however, the policy and practice of AM in EA do not always align—and often for good reason.

Recognizing uncertainty is core to the AM process, but explicit recognition of uncertainty is typically *discouraged* in the practice of EA. The public and environmental regulators typically have little tolerance for uncertainty in resource development proposals. Consider a proposal for a large-scale mining operation near your home community. Would you be willing to accept uncertainty in the proponent's prediction of environmental impacts? Would you be willing to accept an experimental approach to impact management? Why then would a project proponent openly acknowledge uncertainty when predicting the impacts of a development initiative and propose impact management practices as experiments, when it is because of such uncertainty that the project might not be granted approval? A degree of confidence is often presented in the practice of EA that simply does not exist!

Under the federal EA process, AM is directly linked to mandatory follow-up and monitoring programs. In practice, however, scientists and project proponents sometimes have very different understandings and mandates concerning follow-up and monitoring—and thus AM. This was evident regarding the Voisey's Bay Nickel Mine in Newfoundland. Both federal scientists and the project's proponent agreed that monitoring programs should be scientifically defensible, with verifiable hypotheses.

However, federal scientists advocated a learning approach to facilitate a greater understanding of the way aquatic ecosystems function. The proponent, in contrast, was more interested in managing adaptively and focusing attention on the early detection and avoidance of project-related problems for which the proponent would be held responsible, rather than understanding ecosystem dynamics and learning from experimental design to improve future EAs.

Adaptive management is a sound scientific principle, and it makes good policy sense. However, when put to the test in EA, many practical challenges hinder its application. First, AM can be problematic when the project undertaking is unique, such as development in remote regions, and when local experience and understanding are lacking. Second, a project effect is sometimes curable rather than chronic. Third, a proponent's explicit recognition of uncertainty is limited in practice when such uncertainty is considered a reason for not allowing a proposed development to proceed. Finally, neither resource development projects nor the perturbations caused by project development and operations are designed as experiments but rather as development activities with environmental effects to be monitored and managed. In many cases, AM may not be the best approach, and managing adaptively may be entirely appropriate. Knowing when to use AM in EA, and how to deal with uncertainties in resource development, continues to be a major challenge for both proponents and regulators.

—BRAM NOBLE

REVIEW QUESTIONS

1. What is the difference between adaptive environmental management and managing adaptively?
2. What are the different components of the adaptive environmental management process?
3. Under what conditions is adaptive environmental management not appropriate or perhaps less useful than managing adaptively?
4. What are the different implementation styles of adaptive environmental management, and what are their relative strengths and limitations?
5. Based on lessons from Canadian experiences with adaptive management, under what conditions does adaptive management work best?
6. What are the main challenges to adaptive environmental management?

REFERENCES

Backhouse, F. 1993. *Wildlife Tree Management in British Columbia.* Victoria: BC Environment.

Baskerville, G.L. 1995. 'The forestry problem: Adaptive lurches of renewal'. In L.H. Gunderson, C.S. Holling, and S.S. Light, eds, *Barriers and Bridges to the Renewal of Ecosystems and Institutions,* 38–102. New York: Columbia University Press.

BC Forest Service. 2001. *Grizzly Bear Habitat in Managed Forests: Silviculture Treatments to Meet Habitat and Timber Objectives.* Victoria: Production Resources Branch, Government of British Columbia.

Bormann, B.T., et al. 1994. *Adaptive Ecosystem Management in the Pacific Northwest.* USDA Forest Service General Technical Report. PNW-GTR-341.

Bottan, B., D. Euler, and R. Rempel. 2002. 'Adaptive management of moose in Ontario'. *Alces* 38: 1–10.

Buck, L.E., et al. 2001. *Biological Diversity: Balancing Interests through Adaptive Collaborative Management.* Boca Raton, FL: CRC Press.

Corbould, F.B., and P.E. Hengeveld. 1998. *Donna Creek Winter Furbearer Surveys.* Peace/Williston Fish and Wildlife Compensation Program Report no. 179.

Dearden, P., and B. Mitchell. 2009. *Environmental Change and Challenge: A Canadian Perspective.* 3rd edn. Toronto: Oxford University Press.

Diavik Diamond Mines Inc. 2007. *Diavik Diamond Mine Adaptive Management Plan for Aquatic Effects.* Prepared for Wek'èezhii Land and Water Board, NT.

Francis, G.R., and H.A. Regier. 1995. 'Barriers and bridges to the restoration of the Great Lakes Basin ecosystem'. In L.H. Gunderson, C.S. Holling, and S.S. Light, eds, *Barriers and Bridges to the Renewal of Ecosystems and Institutions,* 239–89. New York: Columbia University Press.

Grayson, R.B., J.M. Doolan, and T. Blake. 1994. 'Application of AEAM (adaptive environmental assessment and management) to water quality in the Latrobe River catchment'. *Journal of Environmental Management* 41 (3): 245–58.

Gregory, R.D., D. Olson, and J. Arvai. 2006. 'Deconstructing adaptive management: Criteria for applications to environmental management'. *Ecological Applications* 16 (6): 2411–25.

Hamouda, L., et al. 2004. 'Adaptive management of salmon aquaculture in British Columbia'. Paper presented at the 2004 IEEE International Conference on Systems, Man, and Cybernetics, Las Vegas.

Holling, C.S. 1978. *Adaptive Environmental Assessment and Management.* Chichester: John Wiley.

Holling, C.S., and G.K. Meffe. 1996. 'Command and control and the pathology of natural resource management'. *Conservation Biology* 10 (2): 328–37.

Jacobson, C., et al. 2009. 'Toward more reflexive use of adaptive management'. *Society and Natural Resources* 22 (5): 484–95.

Johnson, B.L. 1999. 'Introduction to the special feature: Adaptive management—Scientifically sound, socially challenged?' *Conservation Ecology* 3 (1): 10.

Lee, K.N. 1993. *Compass and Gyroscope: Integrating Science and Politics for the Environment.* Washington: Island Press.

Lee, K.N., and J. Lawrence. 1986. 'Adaptive management: Learning from the Columbia River Basin fish and wildlife program'. *Environmental Law* 16 (3): 431–60.

McCarthy, M.A., and H.P. Possingham. 2007. 'Active adaptive management for conservation'. *Conservation Biology* 21: 956–63.

McLain, R.J., and R.G. Lee. 1996. 'Adaptive management: Promises and pitfalls'. *Environmental Management* 20 (4): 437–48.

McLennan, D.S., and T. Johnson. 1993. *An Adaptive Management Approach for Integrating Grizzly Bear Habitat Requirements and Silvicultural Practices in Coastal B.C.: Working Plan.* Smithers, BC: Oikos Ecological Consultants for BC Ministry of Environment, Wildlife Branch.

Mitchell, B. 1997. *Resource and Environmental Management.* Harlow, UK: Longman.

———. 2002. *Resource and Environmental Management.* 2nd edn. Harlow, UK: Prentice Hall.

Murray, C., and D.R. Marmoreck. 2004. 'Adaptive management: A spoonful of rigour helps the uncertainty go down'. Paper presented at the 16th International Annual Meeting of the Society for Ecological Restoration, 23–7 August, Victoria, BC.

Murray, C., and M. Nelitz. 2008. *Review of Diavik and EKATI Adaptive Management Plans.* Vancouver: ESSA Technologies Ltd for Fisheries and Oceans Canada, Western Arctic Area, Central and Arctic Region, Yellowknife, NT.

Noble, B.F. 2000. 'Strengthening EIA through adaptive management: A systems perspective'. *Environmental Impact Assessment Review* 20 (1): 97–111.

Nyberg, B. 1999. *An Introductory Guide to Adaptive Management for Project Leaders and Participants.* Victoria: BC Forest Service, Forest Practices Branch.

———. 2001. 'Adaptive management: Moving from theory to practice'. *Horizons* 4 (3): 12–14.

Nyberg, J.B., and B. Taylor. 1995. 'Applying adaptive management in British Columbia's Forests'. In *Proceedings of the FAO/ECE/ILO International Forestry Seminar,* Prince George, BC, 1–15 September, 239–45. Ottawa: Natural Resources Canada, Canadian Forest Service.

Ontario Ministry of Natural Resources. 1988. *Timber Management Guidelines for the Provision of Moose Habitat.* Toronto: Ontario Ministry of Natural Resources.

———. 1998. 'Adaptive management forum: Linking management and science to achieve ecological sustainability. http://www.mnr.gov.on.ca.

Pahl-Wostl, C. 2007. 'Transitions towards adaptive management of water facing climate and global change'. *Water Resources Management* 21: 49–62.

Patterson, B.R., and D.L. Murray. 2008. 'Flawed population viability analysis can result in misleading population assessment: A case study for wolves in Algonquin Park, Canada'. *Biological Conservation* 141 (3): 669–80.

Rempel, R.S., et al. 1997. 'Timber management and natural disturbance effects on moose habitat: Landscape evaluation'. *Journal of Wildlife Management* 61 (2): 517–24.

Schreiber, E.S.G., et al. 2004. 'Adaptive management: A synthesis of current understanding and effective application'. *Ecological Management and Restoration* 5 (3): 177–82.

Taylor, B., L. Kremsater, and R. Ellis. 1997. *Adaptive Management of Forests in British Columbia.* Victoria: BC Ministry of Forests.

Theberge, J.B., et al. 2006. 'Pitfalls of applying adaptive management to a wolf population in Algonquin Provincial Park, Ontario'. *Environmental Management* 37 (4): 451–60.

Timmermann, H.R., R. Gollat, and H.A. Whitelaw. 2002. 'Reviewing Ontario's moose management policy, 1980–2000: Targets achieved, lessons learned'. *Alces* 38: 11–45.

Walker, B., et al. 2002. 'Resilience management in socio-ecological systems: A working hypothesis for a participatory approach'. *Conservation Ecology* 6 (1): 14.

Walkerden, G. 2006. 'Adaptive management planning projects as conflict resolution processes'. *Ecology and Society* 11 (1): 48.

Walters, C.J. 1986. *Adaptive Management of Renewable Resources.* New York: Macmillan.

———. 1997. 'Challenges in adaptive management of riparian and coastal ecosystems'. *Conservation Ecology* 1 (2): 1.

Walters, C.J., R.D. Goruk, and D. Radford. 1993. 'Rivers Inlet sockeye salmon: An experiment in adaptive management'. *North American Journal of Fisheries Management* 13: 253–62.

Walters, C.J., and C.S. Holling. 1990. 'Large-scale management experiments and learning by doing'. *Ecology* 7 (16): 2060–8.

17

Environmental Assessment in Canada: Encouraging Decisions for Sustainability

A. John Sinclair and Meinhard Doelle

Learning Objectives

- To understand the collective experience with environmental assessment across Canada.
- To recognize the key components of an effective environmental impact process.
- To appreciate uncertainties and conflicts surrounding the practice of environmental assessment.
- To understand the main issues related to environmental assessment.
- To know what initiatives are being taken to improve environmental assessment.

GETTING TO KNOW EA

As you read the morning paper, you notice an article about a new waste incinerator being proposed on the edge of the green belt area that surrounds your town. The article indicates that the project proponents note the burgeoning waste problem in the region, with waste quantities increasing monthly, the lack of an environmentally secure landfill, and the growing energy demands of your community that the burning of waste could help to satisfy. The paper indicates further that the proponent has been searching for a willing host community and is now in discussions about the regulatory approvals needed for the project with both local and provincial governments. You are taken aback by the proposed location of the development, questioning how anyone could possibly think this is a suitable site given the ecological and social benefits of the neighbouring green belt; you are worried about the air pollutants and greenhouse gas emissions from the stack; you worry about effects on waste reduction efforts in your community; you decide that you need to find a way to intervene in the decision-making process.

It is not that long ago that people in the community of Selkirk, MB, found out about such a proposal for their community. They, like many other Canadians who decide to question and even challenge what someone else—an individual, organization, or institution—thinks is a great idea, had to learn quickly about environmental assessment (EA). At its core, EA is a decision-making process that helps to ensure what Beanlands and Duinker (1983) defined as 'minimum regret planning'. Through assessment, we attempt to ensure that externalities are identified, evaluated, and incorporated into the planning and decision-making processes. Furthermore, Meredith (2004, 469) observes that EA is 'in principle, no more than a process by which common sense concerns about community futures are incorporated into decisions—public or private—that will affect the future.' As such, EA is a tool available to governments to achieve the societal objectives of environmental protection and sustainable development (Lawrence 2003; Doelle 2008). In the case of the proposed incinerator, an EA process would attempt to ensure that the project proposal is environmentally and socially sound through activities such as considering alternatives, determining environmental and social impacts, and including public participation.

David Lawrence (2003, 7) offers a definition of EA that provides a reference point for our discussions in this chapter. He defines EA as a systematic process of:

determining and managing (identifying, describing, measuring, predicting, interpreting, integrating, communicating, involving, and controlling) the potential (and real) impacts (direct and indirect, individual and cumulative, likelihood of occurrence) of proposed (or existing) human actions (projects, plans, programs, legislation, activities) and their alternatives on the environment (physical, chemical, biological, ecological, human health, cultural, social, economic, built, and their interrelations).

Lawrence favours a broad definition of environment, captures a wide range of activities that create impacts, demonstrates that EA processes blend many activities, and establishes that a considerable amount of research and planning is needed to make an informed decision about whether an undertaking should proceed or not—the 'go or no go' decision.

EA is now carried out in more than 100 countries worldwide and has deep roots in many nations. In most jurisdictions, EA has evolved from being voluntary and discretionary to more mandatory in process and substance. EA also has evolved in other ways. As Meredith (2004) notes, the evolution of EA is in part due to the maturing of the environmental crises and the reciprocal understanding that continued human activity without limitations and without thorough consideration of the environmental implications of such activity is no longer possible. This view is captured by Gibson and Hanna's (2009) four stages in the evolution of EA processes (see Table 17.1).

EA processes, including those in Canada, are still in transition through these four stages and will continue to advance as we collectively learn more about the environment, the synergies between undertakings, and the impact of new types of physical projects we had not imagined before, such as shipping liquefied natural gas (LNG). This evolution has required, and continues to require, practitioners, decision-makers, and academics to stretch their thinking about the uncertainties inherent in moving forward toward achieving a more sustainable society. Canada has contributed significantly in this regard; in fact, all of the authors we have referenced so far are Canadian, and the early work of institutions such as the Canadian Environmental Assessment Research Council set the stage for Canadians to contribute to virtually every aspect of EA design, from empowering the public to strategic impact assessment. We will draw on these contributions throughout this chapter.

Table 17.1 Stages in the Evolution of EA Processes

Stage 1: Reactive pollution control responding to locally identified problems (most often air, water, and soil pollution) with technical solutions and issues addressed in often closed negotiations between government and polluters.

Stage 2: Proactive impact identification and mitigation through relatively formal impact assessment and project licensing but focused on biophysical concerns in the environment with no serious public role.

Stage 3: Integration of broader environmental considerations in project selection and planning through EA but in the context of the individual activities proposed. This involves consideration of a full range of factors such as cultural, historical, and economic impacts, the examination of alternatives, and public reviews.

Stage 4: Integrated planning and decision-making for sustainability, addressing policies and programs as well as projects and cumulative local, regional, and global effects, with decision processes that empower the public, recognize uncertainties, and favour precaution.

Source: Gibson and Hanna 2009.

Key Elements of Best Practice for EA

What constitutes the key elements of 'best practice' for EA has been the subject of considerable debate. This is due in part to the fact that what we expect from EA has evolved and will continue to mature. For example, during the formational period of EA, the Bruntland Commission had not yet popularized the term 'sustainable development', but today sustainability or sustainable development is a key outcome goal of many EA processes. Three frequently cited sources in both the national and international literature, which address the issue of key EA elements, are the Canadian Environmental Network (CEN) (1988), Gibson (1993), and Sadler (1996). The CEN's Environmental Planning and Assessment Caucus identified what it coined the 'core elements' of EA in the lead-up to discussions for a legislated EA process in Canada. They have been modified slightly over the years and are now called the 'basics principles' of EA, which can be viewed at www.cen-rce.org. Gibson based his contribution on a review of 20 years of experience with EA in Canada at the federal level as well as in Ontario. Sadler's principles were gleaned from evaluating EA practice and performance in an international context.

Eight characteristics of best practice for EA emerge:

1. a strong legislative foundation that establishes EA as a mandatory and enforceable process and that provides clarity, certainty, fairness, and consistency;

2. a broad definition of the environment and a process that sets out requirements that ensure that EA is applied to all environmentally significant undertakings;

3. a process aimed at identifying the best option rather than merely acceptable proposals, requiring critical examination of purposes, comparative evaluation of alternatives to the proposal, and alternative means of undertaking the proposal;

4. a process that limits ministerial discretion;

5. an open and fair process that provides a significant role for the public and contains provisions related to public notice, comment, access to information, and participant funding;

6. a process with enforceable terms and conditions for the approval of an activity;

7. a process that specifically addresses monitoring and other post-approval follow-up activities to ensure that terms and conditions have been properly met; and

8. a process with provisions for linking assessment work into a larger context, including the setting of overall biophysical and socio-economic impacts.

As Doelle (2008, 3) has summarized it, the basic principles of EA 'suggest an orientation toward problem solving, a clear link between the EA process and decision making, and a monitoring and feedback capability to ensure compliance and to evaluate the EA process and the project'. Implementation of these principles also ensures that we move beyond a narrowly focused planning process aimed mainly at mitigating biophysical impacts.

EA and the Administrative State

As Lawrence (2003, 13) suggests, 'at a most basic level a process is a series of actions directed toward an end.' Figure 17.1 outlines how government-directed EA processes are typically arranged. The start of the process is the development of an activity, project, or undertaking, with the desired end to be an enlightened decision about the environmental acceptability and sustainability of the initiative. The activities, projects, and policies subject to EA vary by jurisdiction. It is probably easiest to think about projects in your community, region, or province to get a sense of this—projects like new highways, road widening, dams, landfills, sewage systems, factories, mines, pipelines, wind turbines, bridges, LNG facilities. For each of these projects or activities, the proponent (individual, organization, institution, or corporation) develops a plan or proposal. This normally occurs outside of the administrative sphere of EA, since the proposal development process is not regulated by government. Ideally, the proponent will start to discuss the proposed activity with government regulators and the public as early as possible in the planning cycle in order to maximize the potential of such consultations resulting in modifications to the proposal. In the case of the incinerator example above, the proponent would start to detail its plan and identify alternatives and different ways of undertaking the work. It

would document testing at the preferred sites for soil and wind conditions and so on.

Once the proponent has developed the proposal and has submitted it to a regulatory agency, the EA process outlined in Figure 17.1 formally begins. Of course, it is likely that by this point the proponent will have had fairly extensive discussions about the proposed activity with government approval agencies. It is the public and non-regulatory government departments and agencies that will be more formally and widely involved through the EA process. Thus, EA amounts to opening up a political decision-making process—in our example, should we have an incinerator?—to public and expert scrutiny. This process of opening up (see Figure 17.1; see also Lawrence 2003) often includes the following:

Public participation is the foundation and cornerstone of effective and fair EA. The public should be involved in ways they (the public) deem appropriate throughout the EA process.

Process decision, the first step in the process, seeks to determine whether a proposal should be subject to EA and, if so, what type of EA process should apply.

Scoping requires consideration of what is included as part of the project and what the focus of the assessment will be. This involves determining which aspects to include, such as defining the need for the project, alternatives to the project, establishing environmental components, impacts to be considered, and other related matters.

Alternatives analysis involves assessing the choices for satisfying a particular need or opportunity. This step normally ends with the selection of a preferred alternative, which is then studied in greater detail. The analysis of alternatives should normally include the 'do nothing' alternative.

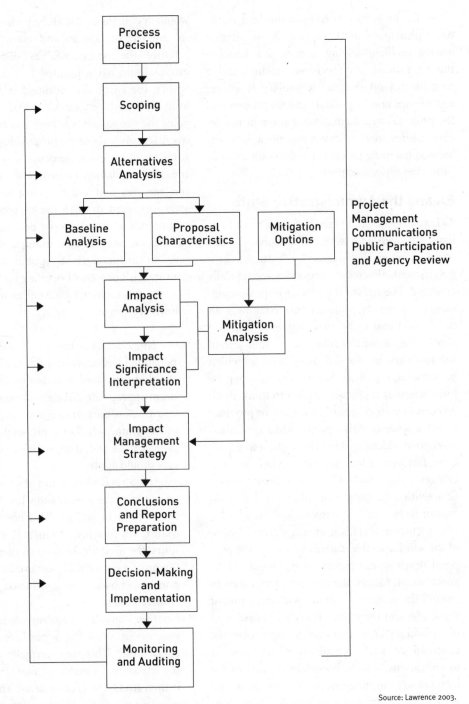

Figure 17.1 Example of a Conventional EA Process.

Source: Lawrence 2003.

Baseline analysis/proposal characteristics/mitigation options is a series of iterative activities that establish details of the environment within which an undertaking might be placed, develop a detailed project description, identify likely impacts, and consider ways of mitigating those impacts. This is when the actual assessment of the proposed project begins. The baseline data are used to predict likely impacts of the proposed undertaking and to determine the best mitigation strategies.

Impact analysis/impact significance/mitigation analysis involves characterizing the impact dimensions (e.g., intensity, duration, frequency, reversibility, cumulative effects), scope (including transboundary effects), and final prediction (likely outcomes). Mitigation and enhancement techniques are also analyzed to determine the extent of prevention, amelioration, rehabilitation, or compensation each provides.

Report preparation and submission occurs once the proponent has all this information in hand and has completed the final submission, often called an environmental impact statement or report. This final EA report documents all activities and outcomes of the assessment and is submitted to the final decision-makers.

Review and final decision is largely in the hands of the regulating agency and should be transparent, open, and accountable. The regulating agency may determine that further review by the public is warranted and call for mediation sessions or a public hearing. Once any further input has been collected, the responsible minister will report the government's decision in writing to the proponent and the public. This decision document, which in some jurisdictions is a licence, specifies the terms and conditions under which the project can or cannot proceed.

Follow-up and monitoring is the final step in the process. Monitoring (usually by the proponent) of the actual impacts of the undertaking and assessment of its contribution to sustainability are important components. Compliance enforcement by the regulating agency ensures that this step is taken seriously, in part to make certain that all parties collectively learn from the process. The EA process itself should also be audited to ensure that it was efficient, effective, and equitable.

A considerable literature focuses on each of these process steps. For example, Beanlands and Duinker (1983) have set the standard for identifying methods of documenting and understanding the receiving environment and for making predictions about interactions among ecological components of the environment and the project. There is considerable literature, much of it Canadian, around public involvement in EA, led by practitioners like Des Connor (2001; see also www.connor.bc.ca) and Richard Roberts (1998; see also www.praxis.ca). Numerous texts also devote chapters to each step of the EA process as well as to issues that have evolved (e.g., Petts 1999a; Wood 2003; Lawrence 2003; Noble 2006; Doelle 2008; Hanna 2009). We will explore the EA literature in greater detail as we review legislative approaches and discuss conflicts and uncertainties in the process.

LEGAL DIRECTION FOR EA IN CANADA

As Table 17.2 indicates, the primary EA processes in Canada are guided by legislative frameworks that outline the steps to be followed leading to a decision about the environmental impacts of a project. These legislative frameworks are discussed in detail below.

Table 17.2 A List of Primary Canadian EA Legislation, along with Government Websites Providing Access to the Legislation

Jurisdiction	Primary Legislation and Internet Web Address
Alberta	Environmental Protection and Enhancement Act, R.S.A. 2000, c. E-12 http://www.qp.gov.ab.ca/documents/Acts/E12.cfm?frm_isbn=9780779729241
British Columbia	Environmental Assessment Act, S.B.C. 2002, c. 43 http://www.qp.gov.bc.ca/statreg/stat/E/02043_01.htm
Canada	Canadian Environmental Assessment Act, S.C. 1992, c. 37, C-15.2 http://www.canlii.org/en/ca/laws/stat/sc-1992-c-37/latest/sc-1992-c-37.html
Inuvialuit Settlement Region	Inuvialuit Final Agreement, as implemented by the Western Arctic (Inuvialuit) Claims Settlement Act, S.C. 1984, c. 24, W-6.7 http://www.canlii.org/en/ca/laws/stat/sc-1984-c-24/latest
Manitoba	The Environment Act, C.C.S.M. c. E-125 http://web2.gov.mb.ca/laws/statutes/ccsm/e125e.php
New Brunswick	Clean Environment Act, R.S.N.B. 1973, c. C-6 http://www.gnb.ca/0062/PDF-acts/c-06.pdf
Newfoundland and Labrador	Environmental Protection Act, S.N.L, 2002, c. E-14.2 http://assembly.nl.ca/Legislation/sr/statutes/e14-2.htm
Northwest Territories	Mackenzie Valley Resource Management Act, S.C. 1998, c. 25, M-0.2 http://www.canlii.org/en/ca/laws/stat/sc-1998-c-25/latest
Nova Scotia	Environment Act, S.N.S. 1994-95, c. 1 http://www.gov.ns.ca/legislature/legc/statutes/envromnt.htm
Nunavut	Nunavut Land Claims Agreement, Article 12 Part 5 http://www.canlii.org/en/ca/laws/stat/sc-1993-c-29/latest
Ontario	Environmental Assessment Act, R.S.O. 1990, c. E-18 http://www.e-laws.gov.on.ca/html/statutes/english/elaws_statutes_90e18_e.htm
Prince Edward Island	Environmental Protection Act, R.S.P.E.I. 1988, c. E-9 http://www.gov.pe.ca/law/statutes/pdf/e-09.pdf
Quebec	Environment Quality Act, R.S.Q, c. Q-2 http://www.canlii.org/qc/laws/sta/q-2/20080515/whole.html
Saskatchewan	Environmental Assessment Act, S.S. 1979-80, c. E-10.1 http://www.qp.gov.sk.ca/documents/English/Statutes/Statutes/E10-1.pdf
Yukon	Yukon Environmental and Socio-economic Assessment Act, S.C. 2003, c. 7, Y-2.2 http://www.canlii.org/en/ca/laws/stat/sc-2003-c-7/latest

Source: Sinclair and Diduck 2009.

Federal EA

Since its enactment in 1995, the Canadian Environmental Assessment Act (CEAA) has been the cornerstone of federal EA. In general terms, an EA is required under CEAA before federal decisions are made about whether to allow a proposed project to proceed. If an EA is required, the assessment can involve a screening level assessment, a comprehensive study, a panel review, or mediation. At the end of the EA process, the results are presented to the decision-maker for a final project decision. Each basic element of CEAA is briefly discussed below, but please also access the CEA Agency website—www.ceaa.gc.ca —to obtain the Act, regulations, policy, guidance material, and current EA cases. Also, consult Doelle (2008) and Hazel (1999) for more detailed reviews of the CEAA process.

Is an EA Required?

Numerous provisions together determine whether a given activity triggers the Act (Figure 17.2). The basic approach is to require an assessment under CEAA before a federal authority makes a decision as specified in section 5 of the Act with respect to a project. The three main pillars on which this approach rests are the definition of 'project', the definition of 'federal authorities', and the list of federal decisions in section 5 of CEAA.

The first pillar for the application of the Act is represented by the definition of 'project'. The principal role of the definition is to limit the application of the Act mainly to physical works, such as stream crossings and road development. Decisions on policies, plans, and programs are not covered under the Act.

The second pillar for the application of the Act is the definition of 'federal authority'. 'Federal authority' is defined to include a minister, an agency, or other body established under federal legislation that is responsible to a minister,

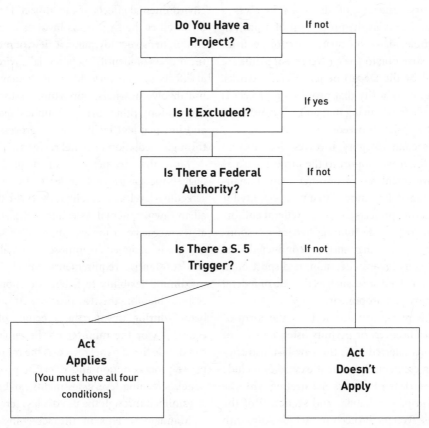

Figure 17.2 Does CEAA Apply?

a department, or departmental corporation and any other body prescribed by regulations.

The third pillar is section 5. If a decision specified in section 5 is required by a federal authority with respect to a proposed project, an EA will have to be done before that federal decision can be made. Federal authorities with section 5 decision-making responsibilities for a given project become responsible authorities (RAs) for the EA. Federal decisions under section 5 that trigger an EA fall into four categories: (1) decisions in which a federal authority is a proponent of a project; (2) decisions to financially support a project; (3) decisions to grant an interest in federal land to a project; and (4) federal 'regulatory' decisions.

The first category of decisions involves a federal authority as the proponent of a project. For example, many infrastructure projects from road construction to sewage treatment plants are proposed by the Department of Public Works. Any federal authority that proposes a project is responsible for ensuring that an EA is conducted before the project proceeds.

The second category involves decisions to provide financial support to the project, such as direct investment, a loan, or other form of financial assistance (e.g., money to a municipality for infrastructure projects). Again, a federal authority asked to provide funding becomes responsible for ensuring that an EA is conducted. The third category involves decisions to dispose of an interest in federal land, such as the sale of federal land to a project proponent.

Finally, regulatory decisions in the form of approvals, licences, or permits listed in a set of regulations referred to as the Law List also trigger an assessment. Prominent examples include section 35 of the Fisheries Act dealing with the protection of fish habitat and section 5 of the Navigable Waters Protection Act dealing with impacts on navigation.

The EA Process

Once a determination is made that CEAA applies, a decision has to be made as to whether the EA will be a screening, a comprehensive study, a panel review, or mediation (see Figure 17.3). Some aspects of this determination are set in law, while others are subject to criteria and discretion.

The process decision is linked to the scoping stage (see Figure 17.1). Whether the scope of a proposed marine terminal project includes the facility (e.g., oil terminal, grain elevator) it is intended to service, for example, may affect whether a screening or a comprehensive study is required. Once the process and scope have been determined, the EA process usually turns to the identification and assessment of potential environmental effects of the project. Depending on whether the EA process involves a screening, comprehensive study, panel review, or mediation, the process of identifying potential environmental effects can be very different. It may or may not involve hearings, opportunities for written submissions, other forms of public engagement, and independent facilitation or process control through mediation or panel reviews.

More than 99 per cent of all projects that trigger the EA process under CEAA undergo a screening-level assessment only (visit the Canadian Environmental Assessment Registry index at www.ceaa.gc.ca for examples). The screening process is designed to impose minimal process and substantive requirements, thereby offering maximum flexibility to federal decision-makers responsible for the EA process. At any time before, during, or after the screening of a project, an RA or the minister of the environment can decide that a panel review is the more appropriate process option and refer the project to a panel. Screening-level assessments can be further streamlined through the use of class screenings.

Mandatory steps in the screening process include a public notice of commencement of

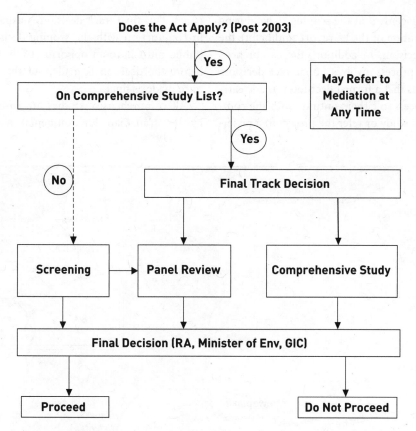

Figure 17.3 CEAA Track Decisions.

the assessment, determination of the scope of the project and assessment, preparation of the screening report, and the final project decision. Co-ordination among federal decision-makers and with other jurisdictions depends on the nature and extent of the involvement of multiple decision-makers and jurisdictions. Transparency and public engagement obligations are limited to certain notice requirements and minimum waiting periods before decision-making. Active public engagement in scoping, the preparation of the screening report, and the final decision is discretionary.

Five to 10 comprehensive studies are carried out under CEAA in an average year. Projects subject to comprehensive studies are generally large-scale projects not referred to panels. Comprehensive studies are required as the minimum level of assessment for all projects that meet the description of projects listed in the comprehensive study regulations. Examples include oil and gas developments, major infrastructure projects, and power plants. A comprehensive study, while still considered a form of self-assessment, is a hybrid between a screening and a panel review.

Required steps in the comprehensive study include and build on those required for a screening. After the notice of commencement of the EA, a comprehensive study includes mandatory public engagement at the scoping stage, during the preparation of the EA report, and before the project decision. In addition, the comprehensive study process involves a final track decision. This means that a formal decision is made early in the process to either continue with the comprehensive study or refer the project to a review panel. This final track decision is usually made in conjunction with the scoping decision. The public must have an opportunity to comment before the decision is made, but the decision, once made, is final.

All comprehensive studies are co-ordinated by the Canadian Environmental Assessment

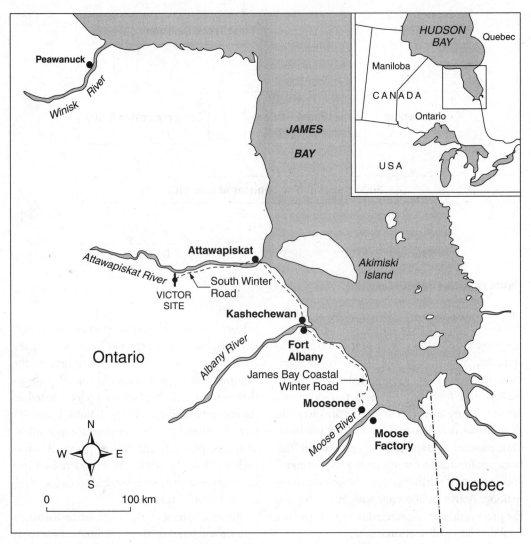

Figure 17.4 Map of Victor Diamond Mine Area (find from Whitelaw and McCarthy).

(CEA) Agency. The CEA Agency co-ordinates among various RAs and with other jurisdictions interested in the federal EA. Co-ordination includes the identification of RAs, the scoping decisions, preparation and review of the EA report, and the project decision. However, the power of the agency as co-ordinator is generally limited to process. The CEA Agency cannot make scoping decisions or final track decisions, and it cannot prescribe final project decisions for RAs.

The minister of the environment, who is not involved in a screening other than to decide whether a panel review is warranted, has an enhanced role in a comprehensive study. Most notably, the minister is required to review the comprehensive study report, seek public input, and determine whether the process is ready for RAs to make final project decisions. RAs are then required to make final decisions in light of the outcomes of the EA process and, in particular, by considering whether the project is likely to cause significant adverse environmental effects. As for screenings, RAs determine whether the project should proceed and if so, under what conditions. Follow-up is mandatory for all comprehensive studies.

GUEST STATEMENT

Learning from the Victor Diamond Mine Comprehensive Environmental Assessment, Ontario

Graham Whitelaw and Daniel McCarthy

The quality of Environmental Assessment (EA) implementation in Canada varies. Within the context of northern environments characterized by great uncertainty and potential conflict, some assessments have resulted in decisions that have benefited society and protected the environment (e.g., the first Mackenzie Valley gas pipeline EA [Gamble 1978]) and more recently also addressed aspects of sustainability (e.g., Voisey's Bay Mine and Mill Environmental Assessment [Gibson 2002]). Unfortunately, not all assessments are of such quality.

The Victor Diamond Mine Comprehensive Environmental Assessment (VDMCEA) (federal comprehensive study process) highlights serious potential deficiencies in assessment implementation and provides important lessons related to scoping and the need for regional land-use planning to complement EA. The diamond mine is located within the traditional homeland of the western James Bay Cree that includes the coastal First Nations (FNs) of Attawapiskat, Kashechewan, Fort Albany, and Moose Factory (see Figure 17.4).

The proponent, DeBeers Canada Exploration Inc., designed and carried out the assessment based on guidelines finalized in 2004 by the Canadian federal government (Government of Canada 2004). The assessment was approved in 2005. Numerous permits allowing for mine operation were subsequently issued, and mine production started in 2008. Although the assessment guidelines focused on all FN communities along the western James Bay coast, the proponent mainly worked with Attawapiskat FN. The other FNs were largely scoped out of the process because of the proponent's and Attawapiskat's view that the mine was located solely on the territorial lands of Attawapiskat and the mine would mainly affect Attawapiskat; this viewpoint has been disputed by other FN organizations. The territorial issues are complex and have yet to be resolved.

The proponent, supported by the Canadian government, also restricted the scope of the EA based on a strict interpretation of the Canadian Environmental Assessment Act that requires only the socio-economic impacts that have a direct biophysical link to be assessed. This led to an inadequate social assessment. The proponent-driven EA process, in this case, disadvantaged a number of FN communities (Whitelaw, McCarthy, and Tsuji 2009). Nishnawbe Aski Nation, the supra-regional political body consisting of 49 FNs within the territory of Treaty No. 9 and the Ontario portions of Treaty No. 5, requested that the EA process be elevated to a panel review process, but this request was denied. The panel review is the most thorough EA process at the federal level, led by a neutral appointed body well positioned to ensure

that all stakeholders are included in assessment processes.

The EA has had some 'positive' outcomes related to land-use planning. Many community members have learned about EA and the need for regional and community-based land-use planning to inform such processes (Minkin 2008). The mine was approved in the absence of any land-use plans for the FN communities and their homelands. The benefits of land-use planning in other jurisdictions have been well recognized (e.g., in the Northwest Territories [Armitage 2005]). The Mushkegowuk Council, the western James Bay FN governing body, in 2008 passed three resolutions indicating that Mushkegowuk FNs will not accept any further resource development, including mining, unless there is full consent by FNs, community land-use plans prepared by the communities are completed, and resource mapping of their territories is finalized (Mushkegowuk Council 2008). These developments suggest that there may be some short-term conflict, but they also hold promise that the Mushkegowuk FNs will achieve their vision for future development.

Panel reviews generally number between one and five assessments initiated under the Act in a given year. Examples include the Mackenzie Valley Pipeline Project, various LNG projects, and resource extraction projects. The important process decisions in a panel review are made by the minister, while the implementation of the process is through an independent panel. Once the decision to refer a project to a panel is made, the minister determines the scope of the project and assessment, sets the terms of reference of the panel, and appoints panel members. In practice, the panel is often involved in the scoping process and frequently holds scoping

hearings. Once the scoping determination is made, the review panel takes over the process in accordance with the terms of reference issued by the minister. It establishes procedures, holds hearings, reviews oral and written submissions, and prepares recommendations for decision-makers. Section 34 of the Act outlines the key responsibilities of the panel:

- ensure that the required information is obtained and made available to the public;
- hold hearings in a manner that offers the public an opportunity to participate in the assessment;

- prepare a report that includes conclusions, recommendations, and a summary of comments received from the public; and
- submit the report to the minister and the RA.

A review panel may summon any person to give evidence and produce documents considered necessary for the assessment. The panel has the same powers as a court, and its orders can be enforced in the federal court.

The panel prepares a final report. The report identifies whether the project is likely to cause significant adverse environmental effects and whether it should be allowed to proceed. If the panel recommends that the project be permitted to proceed, it will usually propose conditions and make other recommendations on how to minimize adverse effects and maximize expected benefits. Panels commonly also comment on the contribution the project is expected to make to sustainable development. The minister has the responsibility to make the panel report available to the public.

Public participation and transparency are significant features of panel reviews. Included are public notices of important steps in the process, direct access to the panel through hearings, and access to relevant information and documentation through electronic and paper registries. Intervener funding and the ability to adjust the nature of the hearing to the cultural norms or preferences of those interested in participating are other strengths. CEAA also provides for joint panel reviews with other jurisdictions, such as provinces, to allow for co-operation between them and the federal government.

CEAA also allows for mediation to resolve a particular issue or assess an entire project. The discretion to formally refer a project or issue to mediation rests with the minister of the environment. The minister can exercise this discretion if the interested parties have been identified and

are willing to participate in the mediation. Mediation has been used only rarely.

Decision-Making and Follow-up

The information gathering and assessment stage of the EA process is followed by the decision-making stage. This requires federal decision-makers to draw conclusions from the results of the EA process. This generally involves determining whether the project is likely to cause significant adverse environmental effects. This step should also track additional comments and recommendations from the EA that will assist decision-makers in making integrated decisions consistent with the purposes and preamble of the Act.

Once the results of the EA are considered, federal decision-makers can exercise their regulatory or other powers with respect to the project. Depending on the nature of the power, this can result in a refusal to permit the project to proceed in spite of a finding that the project is not likely to cause significant adverse environmental effects. Numerous related but separate considerations influence the final project decision. Is the project likely to cause significant adverse environmental effects? Will it contribute to sustainability? Are there other relevant factors that affect whether the section 5 power, duty, or function should be exercised to allow the project to proceed?

The final phase of the EA is the follow-up, monitoring, and compliance stage. This phase can include monitoring and follow-up requirements imposed on the proponent as well as efforts to ensure that decisions were consistent with the requirements of the Act. Monitoring and compliance should therefore involve proponents, federal decision-makers, the CEA Agency, and members of the public.

The Future of Federal EA

In March 2009, the federal cabinet registered an amendment to the Exclusions List Regulation under CEAA and introduced the Infrastructure Projects Environmental Assessment Adaptation Regulation. Both of these changes relate to projects funded, at least in part, through the federal government's 'plan' called Building Canada: Modern Infrastructure for a Strong Canada. In short, the amendment to the Exclusions List Regulation excludes many projects contemplated by the 'plan' from federal EA under CEAA over a two-year period. The Regulatory Advisory Statement published along with the amendment indicates that the change is meant to reduce the number of federal EAs by up to 2000 by excluding from federal EA projects such as the construction of rapid transit systems, the building of roads and public highways, the construction or modification of buildings to be used for various purposes, and potable and waste water treatment systems. The Adaptation Regulation is meant to apply to 'plan' projects not captured by the Exclusion List Regulation amendments. It removes the possibility of bumping up a screening to a panel review, eliminates the public consultation component requirements regarding the scope of a comprehensive study, and removes the possibility of bumping up a comprehensive study to a panel review. The regulation also authorizes the substitution of provincial EA processes for those of CEAA when the minister of environment deems it 'appropriate'.

In passing these regulations, the federal government took the unusual step of publishing them directly into the *Canada Gazette Part II*, eliminating any chance for public comment. The changes were introduced in this fashion, citing the national economic emergency. The regulatory modifications have caused concern among many groups and organizations because they undermine the years of work dedicated to moving CEAA toward best practice. Questions have also been raised as to why the government did not wait until the legislated seven-year review

of the Act, due to be carried out by Parliament in 2010, to introduce such substantive changes, particularly those relating to substitution (see section below). The level of public concern led Ecojustice, on behalf of the Sierra Club of Canada, to launch a lawsuit claiming that the federal government acted unlawfully in issuing the two regulations. Ecojustice (2009) lawyer Justin Duncan stated that '[n]one of the US, China or India have gotten rid of EA oversight as part of their economic stimulus plans.' For many, this move is a strong signal that the current Conservative government, at least, wants the federal government out of EA for all but a few projects.

Provincial and Territorial EA

Each province and territory in Canada also has its own legislated EA process (see Table 17.2). As observed by Fitzpatrick and Sinclair (2009), several jurisdictions believe they have the best EA process in Canada. The Environment Act (C.C.S.M. c. E-125) of Manitoba characterizes the standard provincial approach to EA in that it uses a list to identify the types of projects that trigger EA, has some flexibility in terms of process options for proceeding with the EA itself, and includes considerable discretion. Manitoba's Act sets a broad purpose:

> the intent of this Act is to develop and maintain an environmental management system in Manitoba which will ensure that the environment is maintained in such a manner as to sustain a high quality of life, including social and economic development, recreation and leisure for this and future generations.

The language of sustaining a high quality of life and development, along with the reference to future generations, suggests that the Act is intended to address sustainability concerns and promote sustainable development. The EA process is triggered when any person (individual, corporation, or government) wishes to 'construct, alter, operate or set into operation' a class 1, 2, or 3 development. Class 1 developments generally have the effect of discharging pollutants, class 2 developments have effects primarily unrelated to pollution or in addition to pollution, and class 3 developments have effects which are 'of such a magnitude or which generate such a number of environment issues that it is as an exceptional project.' The Classes of Development Regulation lists specific examples of which activities or projects fall within each class. Certain developments within the following sectors are listed: agriculture, energy production, fisheries, forestry, manufacturing, transportation, waste disposal and treatment, habitat modification, mining, recreation, and water development and control. Therefore, the Act applies to both private undertakings and projects by the government.

Other jurisdictions also use some form of physical work, such as any enterprise, activity, project, structure, modification, extension, or work program, as the trigger for their EA process. With the exception of Saskatchewan and Ontario, all use a list of projects that require assessment, normally along with ministerial discretion to include other projects. While Manitoba and Nova Scotia are the only jurisdictions to refer specifically to classes of development, the law and project lists normally distinguish between larger and smaller projects. In Saskatchewan, the process is triggered by the requirement for EA approval before proceeding with a 'development', with development being defined by the Act. Triggering in Ontario is based on the definition of an 'undertaking' provided in the Ontario Environmental Assessment Act, with all undertakings requiring EA approval.

The provincial website in Manitoba indicates that once triggered, the EA process consists of five steps: (1) the proponent files a proposal; (2) the proposal is screened by a technical advisory committee, and terms of reference for the EA may be set; (3) the proponent may be required to provide further information; (4) opportunity may be provided for public hearings; and (5) a licensing decision is made. Each step is described in detail in the Act and associated guidance material (see also Lobe 2009). The Act and regulations are not specific about which impacts must be assessed; however, the Licensing Procedures Regulation provides that a development proposal for any class of development must include:

- type, quantity, and concentration of pollutants to be released into the air or water or on land;
- impact on wildlife;
- impact on fisheries;
- impact on surface water and groundwater;
- forestry-related impacts;
- impact on heritage resources; and
- socio-economic implications resulting from the environmental impacts.

The director or minister may require further information and can also require a proponent to conduct specific studies for all three classes of development as part of the assessment process.

The EA process in other jurisdictions is also driven by a project proposal. This is usually followed by preliminary screening to make a final decision on the track of the assessment (class 1, 2, or 3 in the case of Manitoba, comprehensive EA or other form of EA in most other jurisdictions). For larger projects, the terms of reference for the assessment are then set before the final assessment and reporting are undertaken. The requirements for considering the environmental impacts of a project are generally flexible and quite broad.

The minister of conservation in Manitoba may require the proponent to provide funding to participants in the assessment process for class 1, 2, or 3 developments or for panel hearings. This provision is generally used to provide funding for participation in hearings. As in most other provinces, if hearings are called, the hearing panel's report is set in the context of recommendations to the minister. As such, the hearing panel does not make the final decision on whether the project proceeds. In Manitoba, the final decision rests with the director of approvals, who makes the final determination on whether to approve or reject a class 1 or 2 development. He or she must issue reasons to the proponent and the public if a decision is made not to issue a licence. For a class 3 development, the minister makes the final determination, but the approval of cabinet is required for refusal to issue a licence. Other jurisdictions also leave the approval of large projects to the ministerial level.

As Lobe (2009) observes, numerous weaknesses exist with Manitoba's Environment Act. Prime among them are: the lack of any definition of 'significant effects' despite the fact that the word 'significant' appears at many crucial points; the discretionary powers of the director of approvals and minister, apparent in the Act through the frequent use of 'Minister may' clauses (e.g., 'The Minister may request public hearings'); and the provision for staging developments that allow proponents to break their projects into component parts, thereby obtaining EA licence approval for each stage or part (e.g., approval for land clearing; approval for waste water treatment upgrades; approval for the development), leading to fragmentation of the EA process.

While most jurisdictions in Canada follow this standard EA approach, which closely resembles the conventional EA process (see Figure 17.1), each has unique characteristics. As with the federal process in Canada, the laws that underpin EA in each of these jurisdictions have

evolved and continue to evolve. Some important deviations exist compared to the basic approach outlined above. EA in Ontario and British Columbia, for example, has a narrow range of application. In Ontario, only projects or undertakings of the government are normally subject to EA. In both provinces, many projects or undertakings are excluded, and there is ample discretion for the minister to exclude others. While many provinces have some form of provision for participant funding, only Manitoba provides such funding in practice. In Quebec, differing EA provisions have been established for the James Bay and northern Quebec region and the rest of the province, making procedures in the north much different from those in the south. For example, in the north, time limits have been set for process steps for all but hearings activities, and advisory committees established for project review include the public. The territories and Quebec provide for the active participation of Aboriginal governments in the decision process and place an emphasis on Aboriginal participation within their EA processes. Consequently, careful consideration of the EA law, and what has been written about it, in the jurisdiction of interest is very important to understanding how the process will be undertaken.

Such consideration is particularly important when working with the EA processes in Yukon, the Northwest Territories, and Nunavut. Each of these northern jurisdictions has recently introduced EA, in part because of the intense pressure associated with mineral extraction and oil and gas development, and land claims play an important role in each. As Armitage (2009) outlines, this has resulted in new opportunities for more collaborative approaches to EA.

Nunavut has the newest EA process as established through the Nunavut Impact Review Board (NIRB). The EA process is not unlike that of provincial jurisdictions. The assessment is undertaken by the NIRB, and the scope of the assessment is usually broad in that the NIRB is to gauge and define the extent of regional impacts of a project and to take regional interests into account. The NIRB consists of nine members appointed by different levels of government: four appointed by the federal minister responsible for northern affairs upon nomination by a designated Inuit organization, two appointed by federal ministers, and two by ministers of the territorial government. A chairperson is appointed by the federal minister responsible for northern affairs from nominations agreed to and provided by board members. The nature of the land claims agreement on which the EA process rests means that Aboriginal communities play a much stronger role in the EA process than in federal and provincial EAs. For example, in public hearings, the NIRB must give due regard and weight to the tradition of Inuit oral communication and decision-making. Additionally, designated Inuit organizations have a special role in nominating panel members if a panel is required. The EA processes in Yukon and the Northwest Territories also establish a strong role for Aboriginal communities, with both requiring that decision-makers must consider and make use of traditional knowledge.

CONFLICTS, UNCERTAINTY, AND POSSIBLE WAYS FORWARD

As observed by Lawrence (2003, 10), 'the realization of EA objectives has been, at best, a mixed endeavor.' There have been advances in EA, but there continue to be significant challenges. In this section, we initially consider challenges to the EA process and outline how they might be addressed. We then consider possible alternative ways that EA might be practised to encourage more sustainable outcomes.

EA Process Challenges

As experience with EA in Canada matures, process challenges become more evident and often more complex. We cannot review all of the issues with EA process, so we have chosen a selection that highlights the ones most actively being considered by EA practitioners and policy-makers.

Use of Discretion

Discretion is a part of all EA processes in Canada and is probably inevitable. Many EA processes, including the federal EA process, have gradually evolved from a completely discretionary process to one that operates under a combination of legal obligations and discretion granted to government decision-makers. Only practical experience will determine whether decision-makers exercise discretion wisely or whether the exercise of discretion tends to reduce the effectiveness of the EA process. Doelle (2008) suggests that in most EA processes, the exercise of discretion has been particularly critical and controversial in four key areas: the trigger or application of the EA process, the choice among process options, the substantive scope of the assessment, and the decision that flows from the findings of the EA process (see also Green 2002).

Issues around discretion are difficult to resolve, since governments like to have the ability to modify individual EA processes and because discretionary provisions in EA law are generally immune to oversight through judicial review. The nature and extent of discretion will change over time as more experience is gained with EA. One approach to helping this process along is the development of policy documents and ministerial guidance on discretion. For example, public groups pushed for clarity on how discretion related to the 'need for' and 'alternatives to' components of CEAA should be implemented. This eventually led to an operational policy statement titled 'Addressing "need for", "purpose of", "alternatives to" and "alternatives means" under the Canadian Environmental Assessment Act', accessible on the CEA Agency website. In Manitoba, public participants also requested further clarity around discretionary decisions regarding the allocation of participant funding. This led to the establishment of a Participants Assistance Review Committee and may lead to further policy direction.

Effective or Efficient EAs?

One enduring challenge of most EA processes is the perceived trade-off between the effectiveness and the efficiency of the process. This perceived trade-off has dominated the implementation of EA since its inception. To what extent the perception is accurate is the subject of ongoing debate (see Doelle 2008, 31).

The main efficiency concerns in any EA process are the time and cost involved. These two concerns surface repeatedly, limiting the ability to incorporate requirements, measures, or steps into the EA process that would otherwise improve its effectiveness. To illustrate, consider the concern that lack of baseline data will hinder the ability of the EA process to make accurate predictions about the future consequences of a proposed activity. A simple answer would be to require that the baseline data be collected before the EA process is concluded. The reason most EAs do not require this is presumably that the cost and time involved are considered prohibitive.

A similar issue arises regarding the consideration of alternatives, something identified in the literature over the past 30 years as critical for effective EA. Nevertheless, many EA processes do not require consideration of alternatives or do so only for certain projects. When consideration of alternatives is required, the same detailed information about these alternatives is generally not available. The reason the full information on

alternatives is not gathered in EAs is that the cost and time involved are seen as undermining the efficiency of the process.

Both efficiency and effectiveness are important design criteria for EA. The idea, however, that we can either generically or on a case-by-case basis decide when to trade off effectiveness for efficiency is a fallacy. If all parties knew upfront the benefits of a thorough EA, we could avoid the trouble of doing the EA and just implement the project in a manner that maximizes benefits and minimizes risks and harm. The value of EA cannot, however, be determined by looking at the size or nature of individual projects; the only way to get the full benefit of EA is to provide the opportunity to identify unforeseen environmental, social, cultural, and economic consequences of a proposed project. To do otherwise can only result in an inefficient EA.

In 2007, the federal government established the Major Projects Management Office (MPMO) (www.mpmo-bggp.gc.ca/index-eng.php) to better co-ordinate federal regulatory approval for large resource projects, such as mines, dams, and oil sands development. This initiative was presented as an attempt to improve the efficiency of EA processes by reducing the time the assessment takes through improved co-ordination among government departments. In moving forward with initiatives like this, assessors need to be sure to focus on developing an EA process in which efficiency is not taken to a point where it starts to compromise effectiveness, since the MPMO is another pressure point pushing RAs toward efficiency.

The Relationship between EA and Regulatory Processes

As noted in the introduction, EA is widely viewed as a planning tool. Thus, projects and undertakings are considered from a future-oriented perspective—look before you leap. The notion of

EA as a planning tool is, however, presented in a much different manner in the literature than in practice. In the literature, the debate about EA as a planning tool is mainly about the limits of project EAs and the need for integration of project EAs with strategic EA, planning processes, and integrated decision-making. In practice, however, the debate revolves around how best to harmonize EA with regulatory processes. The result has often led to EA processes that tend to focus on regulatory approvals (e.g., should the proponent be allowed to disturb a watercourse?) rather than considering the broader planning implications of the project and alternatives to it (e.g., is the project the best way to meet the identified need?).

Joint assessments and substitution are examples of harmonized processes that have eroded the use of project EAs as a planning tool. An example is the regulatory process under the National Energy Board (NEB), which has more and more frequently entered into the EA process under CEAA. Most recently, the regulatory process of the NEB was used as a substitute process for CEAA in the case of the Emera Pipeline Project in Saint John, NB (see the section Multi-jurisdictional Assessment below).

Public Participation

While public participation is viewed by many as a cornerstone of EA (e.g., Petts 1999b, 2003; Sinclair and Diduck 2001; Wood 2003; Devlin, Yap, and Wier 2005), whether the benefits of participation are realized in practice depends to a large extent on the legislation and policy applicable to the particular EA. Despite the years of experience with public participation and volumes documenting how best to undertake it, meaningful participation has proven elusive (Blaug 1993; Petts 1999b; Diduck and Sinclair 2002; Sinclair and Fitzpatrick 2003; Meredith 2004).

Sinclair and Diduck (2009) have documented key concerns with EA public participation

processes, including lack of shared decision-making, lack of participation at normative and strategic levels of planning, information and communication deficiencies, insufficient resources for participants, accelerated decision processes, and weak participation in follow-up and monitoring programs. We highlight the first two points, since they might not be as clear as the others.

The lack of participation in decision-making is a serious shortcoming of Canadian EA processes for citizens and environmental non-governmental organizations (ENGOs). A helpful tool for understanding this problem is Arnstein's (1969) classic ladder of citizen participation (also discussed in Chapter 18; Dorcey, Doney, and Ruggeberg 1994; Connor 2001). Arnstein identified eight levels of public participation and associated degrees of power-sharing. Activists and ENGOs are often highly critical of processes that only consist of lower levels of involvement and argue that public participation should entail redistribution of decision-making power such that it is shared by the public.

Another serious issue for participants arises when they are actually involved in the assessment process. Participation can occur at the normative level of planning (decisions about what should be done), the strategic level (decisions to determine what can be done), or the operational level (decisions to determine what will be done) (Smith 1982). As noted above, the earlier participation occurs in the EA process, the more influence the public will likely have on important issues such as project need, purposes, and alternatives.

A good example of normative level involvement is the Mackenzie Valley Pipeline Inquiry of the 1970s, which was groundbreaking in the way it engaged potentially affected communities early in the planning process (Berger 1977). As that case showed, through early participation, basic choices can be considered (including whether the project should proceed at all) before political momentum builds for a project and substantial amounts of time and money are invested.

However, in Canada participation is not required legally until well into the planning process and sometimes only at the operational stage of planning. Participation at earlier junctures is left to the discretion of the proponent, and early involvement is sporadic at best. The Maple Leaf hog-processing EA in Manitoba is a good example. In that case, many members of the public as well as ENGO activists wanted to discuss large normative questions (e.g., should we have an industrial hog-processing plant in our community?) rather than detailed operational issues (e.g., how thick should the liner be for the anaerobic lagoon in the waste treatment facility?) (Diduck and Sinclair 2002).

Sinclair and Diduck (2009) also note that while EA practitioners and policy-makers face formidable challenges in creating more meaningful roles for the public in EA, some promising avenues are being considered, including early and ongoing participation, mutual learning, alternative dispute resolution, community-based assessment, and legal requirements. For example, any individual in Quebec can request a public hearing, and since it would be impossible to hold hearings for every request, the minister can ask that the parties consider mediation to resolve any dispute. As a result, Quebec has one of the only active EA mediation tracks and is gaining valuable experience in this approach to EA decision-making. Further, Sinclair, Diduck, and Fitzpatrick (2008) view taking greater advantage of the learning opportunities resulting from participation as a key future direction as it creates important opportunities for learning related to sustainability at individual and various organizational levels (Fitzpatrick and Sinclair 2003).

Consideration of Cumulative Effects

The EA literature offers many definitions of cumulative effects (CE) (see Spaling 1997; Ross 1998; Noble 2006). Assessment of CE is about determining the effects of a proposed activity in light of other existing, planned, and possible future activities and their interaction with anything also affected by the proposed activity. Doelle (2008) indicates there is a general recognition that in order for project-based EA processes to deliver an effective mechanism for predicting and evaluating future consequences of proposed actions and their alternatives, they have to be able to make meaningful predictions about CE. While there were high expectations when EA policy and legislation started to require consideration of CE, the practice of assessing CE issues such as space-crowding (multiple oil sands projects in a watershed) or multiple effects (adding pollutants to a river already subject to liquid effluent) has proven very difficult, divisive, and generally disappointing.

Duinker and Greig (2006) identify six major problems with CE assessment in Canada:

1. the application of CE assessment at a project level;
2. an EA process focused on project approval rather than on sustainability assessment;
3. a general lack of understanding of ecological impact thresholds;
4. the separation of CE from project-specific impacts;
5. a weak interpretation of cumulative effects by practitioners;
6. the inappropriate handling of potential future development.

They propose improvements to project EA to address these concerns, but in their view, the major difficulties with CE assessment can only be addressed through a shift toward regional

EA frameworks or land-use planning. Some lessons about taking a regional approach to CE have been learned in the case of the Great Sand Hills region in Saskatchewan. In reviewing this case, Noble (2008) concludes that success can be attributed to aspects of the assessment such as the strategic framework used, the future-oriented approach to CE considerations, and the valued ecosystems components (VEC)–based multi-scale analysis.

Sustainability Assessment

Sustainability assessment (SA) (or integrated or triple bottom-line assessment) is 'an umbrella term that embraces a range of processes that all have as their broad aim to integrate sustainability concepts into decision-making' (Pope 2006, v). While this may seem no different from the aim of best-practice EA, by taking the concept of sustainability as its foundation, SA strives for greater breadth of coverage and integration of concerns than EA's traditional emphasis on the biophysical environment. Sustainability minimally includes the social, economic, and biophysical environments and emphasizes their interconnections and interdependencies. Hence, SA is conceived as a fundamentally integrative process (Gibson 2006).

As seen in the stages of EA evolution described at the beginning of this chapter, EA has trended toward being a more comprehensive consideration of diverse factors, and as such SA can be seen as a matured (stage 4) version of EA that fully accounts for the social, economic, and environmental pillars of sustainability (Gibson et al. 2005). However, SA explicitly seeks to do more than assess the acceptability of a project or plan, and instead, '[s]ustainability assessments have the double role of vehicles for the general pursuit of sustainability and contributors to defining the specifics of sustainability in particular circumstances' (Gibson et al. 2005,

62). As well, SA can be applied at the project level, to programs, plans, or policies, or to existing activities (Pope, Annandale, and Morrison-Saunders 2004).

Many existing assessment procedures focus on single dimensions of sustainability (e.g., health impact assessment, social impact assessment), but most often these assessments are conducted independently without regard to their inherent interrelations and without an attempt to integrate them into a larger vision of sustainability (Hacking and Guthrie 2008). The same is often the case with separate social, economic, and biophysical assessments conducted as part of a decision-making process. This tendency is exacerbated by the momentum of established assessment practice in which relevant experts are trained in one of the three fields, information is collected and categorized separately, and government agencies typically have mandates tied to only one sustainability dimension (Gibson 2006). This can lead to seeing SA as primarily about balancing dimensions against one another via trade-offs.

Given prevailing alignments of power, it is feared that SA practice could end up sacrificing environmental protection in favour of economic growth, giving up the hard won gains of decades of EA advocacy (Morrison-Saunders and Fischer 2006). But this would be a distortion of a fundamental assumption of SA—namely, that the sustainability of any one dimension is necessarily dependent on that of the others. Instead, SA practice seeks mutually beneficial outcomes among dimensions that together further the cause of sustainability. This requires considering social, economic, and environmental factors concurrently and attending to the ways that they are mutually reinforcing or detrimental (Gibson 2006). This is not easy. The techniques described in the literature for conducting effective integrated assessments are generally more sophisticated, time-consuming, and costly than those used in typical EA regimes. They also tend to be unfamiliar to practitioners (Lee 2006).

A SA starts with an articulation of sustainability criteria to guide decision-making. While there are numerous broad articulations of sustainability, practical application requires they be translated into specific decision criteria appropriate to the particular context. In this way, the assessment is not measuring against a baseline but rather is determining the direction and distance to the target of sustainability for the specific matter under consideration (Pope, Annandale, and Morrison-Saunders 2004). This necessitates a highly participatory approach that engages stakeholders in order to generate a contextually appropriate vision of sustainability. It also enhances the SA's integrated character since 'public issue identification and priority setting processes often identify secure livelihoods, safety, health, vibrant and attractive communities, new opportunities and choice, and influence in decisions as key objectives. None of these is a purely social, economic or ecological matter' (Gibson 2006, 263). Further, a highly inclusive and participatory approach to SA can act as a safeguard against decision-makers discounting or neglecting factors identified as vital to the overall sustainability of the context inhabited by the participants.

There have been relatively few applications of SA. It is more common in industrial nations but is starting to evolve in developing nations, including countries in Africa (Kaliba and Norman 2004; Alshuwaikhat 2005; Dalal-Clayton and Sadler 2005; Govender, Hounsome, and Weaver 2006; Onyango and Schmidt 2007). Thus, SA is in its infancy, with many practical difficulties yet to be overcome, but it still stands as a potentially powerful tool for directing us toward sustainability.

Strategic Environmental Assessment

Strategic environmental assessment (SEA) is a formalized, systematic, and comprehensive process of evaluating the environmental impacts of a policy, plan, or program and its alternatives. As Partidário (1999, 64) notes,

> [It] is a systematic, ongoing process for evaluating, at the earliest possible stage of publicly accountable decision-making, the environmental quality, and consequences, of alternative visions and development intentions incorporated in policy, planning, or program initiatives, ensuring full integration of relevant biophysical, economic, social and political considerations.

SEA is thus an EA that goes beyond a single project to consider an industry sector, a region, or a particular policy, plan, or program (Dalal-Clayton and Sadler 2005). A SEA can be reactive in response to the proposal of a particular project, such as the first proposal to introduce a new technology or industry (e.g., an LNG facility or tidal energy technology) if it extends beyond the individual project to look at the whole technology, industry sector, or region. A SEA can also be proactive by seeking to address an identified sustainable development or environmental challenge. For example, a SEA could be initiated to develop an energy policy encompassing a range of environmental, social, and economic concerns related to climate change, air pollution, and energy security.

SEA's link to sustainability interests EA practitioners and academics most, since some feel that sustainability goals, such as those held by governments and donor agencies, are more likely to be advanced and realized if SEA is implemented (Lawrence 1997; Partidário 1999; Petts 1999b; Thérivel and Brown 1999; Noble 2002).

Noble (2006, 9) comments in this regard that '[a]dvancing the sustainability initiative will require increasing the application of EA principles beyond the project level . . . SEA is based on the notion that the benefits of sustainable development trickle down from policy decisions to plans, programs, and eventually to individual projects' (see also Partidário 1999; Vicente and Partidário 2006).

In practice, however, SEA is 'still quite new and relatively limited in terms of its adoption' (Noble 2006, 196), despite the perceived benefits. Various reasons have been offered. They include the lack of political will to subject policy, plan, and program decisions to such an invasive public process; lack of a common understanding of the roles SEA can play and should play in decision-making; challenges in meaningfully engaging the public in such a forward-looking process, especially when there is no immediate project at hand; lack of approaches and methodologies for how to do SEA; financial constraints; lack of agreement around the need for and benefits of SEA; and problems finding a government authority willing to co-ordinate and take responsibility (Partidário 1996; Petts 1999b; Noble 2006, 2008; Noble and Harriman-Gunn 2009).

In Canada, the assessment of policies, plans, and programs was originally included in the EARP (Environmental Assessment Review Process) Guidelines Order at the federal level. In return for limiting the application of CEAA to project assessments, a *Cabinet Directive on Environmental Assessment of Policies, Plan and Program Proposals* (Privy Council Office and Canadian Environmental Assessment Agency 1999) was introduced in 1999 and updated in 2004 (Canadian Environmental Assessment Agency 2004). Under the most current version, a SEA is expected whenever a proposal that may result in important environmental effects is submitted to a minister or cabinet for approval.

When appropriate, SEAs are encouraged to consider public concern about the possible consequences of a proposed policy, plan, or program. SEAs are also encouraged to help implement sustainable development goals.

The federal Commissioner of the Environment and Sustainable Development last carried out an audit of the implementation of the SEA cabinet directive in 2004. The overall conclusion was that the level of commitment to and compliance with the cabinet directive was low. On the basis of four case studies involving LNG facilities in Canada, Doelle (2008) concludes that the need for better use of SEA is apparent. SEA at the federal level is far from fulfilling its promise to help overcome the limitations of project EAs and deliver integrated decision-making for sustainability.

Efforts have been made to improve the use of the SEA process. The environment minister's Regulatory Advisory Committee (RAC) on CEAA identified SEA as a priority as a result of its review of CEAA in 2000. RAC is currently considering proposals on how to make better use of SEA within the federal EA process. In addition, there have been ad hoc efforts to make use of SEA, such as the SEA on tidal energy in the Bay of Fundy (Doelle 2009).

As Noble (2009, 119) indicates, 'Canadian provinces do not have any formal system of SEA comparable to that of the federal 1999 Cabinet Directive.' Like the review by Noble (2006), our own review of provincial legislation found that Alberta, the Northwest Territories, Manitoba, Newfoundland, Quebec, and Prince Edward Island had no requirement for higher-order SEA of policy, plans, or programs. Legislation in British Columbia, Ontario, New Brunswick, Nova Scotia, and Nunavut provide for the assessment of policy, plans, and programs but do not set out a formal process. Saskatchewan does have formal requirements for regional considerations, but they are not called SEA. Noble (2008) provides an example of a provincial SEA related to the Great Sand Hills region of Saskatchewan.

Community-Based Assessment

Community-based approaches to EA have emerged as important for development planning (Neefjes 2001; Spaling 2003; CIDA 2005). In contrast to the assessment approach often used for large projects, community-based EA (CBEA) 'has been adapted in an innovative way to smaller, community-based projects that utilize natural resources for basic livelihood needs' (Spaling 2003, 151; see also Pallen 1996; Neefjes 2001). Typical projects include boreholes, gravity water systems, small reservoirs, agro-forestry, fish ponds, construction of latrines, clinics, schools, and small bridges.

In community-based approaches to EA, a participatory forum facilitates a process of communal dialogue and collective decision-making that includes the development of goals, the sharing of knowledge, negotiation and compromise, problem-posing and problem-solving, evaluation of needs, definition of goals, and research and discussion usually around questions of justice and equity (Ameyaw 1992; Meredith 1992; Neefjes 2001; Spaling 2003). This process helps communities to clarify values, be more adaptive and proactive, respond to change, develop an appreciation for the human–ecological interface, set personal and communal goals, and participate in a process where they are heard (Meredith 1992; Keen and Mahanty 2006).

With the continued success of CBEA approaches in relation to development planning for projects, academics and EA practitioners are testing it in different contexts. Sinclair, Sims, and Spaling (2009) completed a community-based strategic EA of the Instituto Costarricense de Electricidad's (ICE—Costa Rica's publicly owned electrical and telecommunications company)

watershed management agricultural program that is to address erosion and contamination problems caused by conventional farming practices (i.e., planting homogeneous crops, a heavy reliance on chemicals, and regular tilling) in watersheds where ICE has hydro projects. Nine lessons were learned in recommending further testing of the approach for SEA. Sinclair and Diduck (2009) also note that many reasons for using the CBEA approach apply to small projects in Canada—the desire to ensure that projects are assessed, to do so cost-effectively and efficiently, and to build local capacity for participation in EA. For that reason, they identify CBEA as a promising future direction for EA.

Multi-jurisdictional Assessment

The widespread adoption of EA by different governments and institutions throughout the world has created a context in which any one project may trigger more than one assessment (Fitzpatrick and Sinclair 2009). As discussed above, the federal, provincial, and territorial governments all have their own EA processes, as do many major municipalities. For example, the Wuskwatim project involves the construction of a low head dam and three 230-kV transmission line segments in northern Manitoba. The proposal triggered reviews under the Canadian Environmental Assessment Act (S.C. 1992, c.37) and the Manitoba Environment Act (S.M. 1987-88, c. 26). Rather than duplicate efforts to evaluate a project's potential impacts on the environment, there is a growing demand for inter-jurisdictional co-ordination of EA practices. In this way, a proposed development undergoes one EA that addresses the requirement of two or more different processes.

Given the complexity of jurisdictional responsibilities between the federal and provincial governments, Canada has a growing history of inter-jurisdictional co-ordination. Three approaches to inter-jurisdictional co-ordination have been considered in Canada (Canadian Council of Ministers of the Environment 1998), including 'standardization', 'harmonization', and 'substitution'. Fitzpatrick and Sinclair (2009) discuss each of them in some detail, from which we have drawn the following overview.

Standardization involves devising one common EA process to be used across different jurisdictions. However, as shown above, process triggers vary by region. For example, some EAs are initiated by both public and private developments (e.g., the Manitoba EA process), while others are triggered only by public developments (e.g., the federal EA process). Some EAs focus on physical works (e.g., the federal EA process), while others consider projects, policies, and programs (e.g., the Ontario EA process). This made standardization efforts in Canada very complex, and they were quickly abandoned (Kennett 2000).

Harmonization involves rationalizing EA so that the requirements of all applicable legislation are met through one process. There are two traditions of harmonization in Canada: *bilateral agreements* between governments about how their processes will be harmonized and *project-specific agreements* between governments about how to proceed with an assessment for a specific project. Bilateral agreements are negotiated under the Canada-Wide Accord on Environmental Harmonization. Figure 17.5 illustrates the system of bilateral agreements across Canada. Bilateral agreements are negotiated between the federal government and provincial or territorial

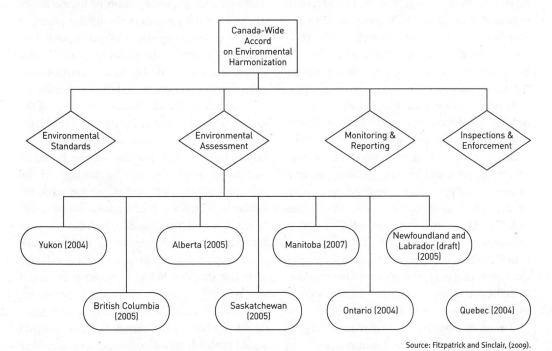

Source: Fitzpatrick and Sinclair, (2009).

Figure 17.5 The system of EA coordination negotiated between the federal and provincial governments.

Note: Quebec is not a signatory to the Canada Wide Accord on Environmental Harmonization, and, as such, the bilateral agreement negotiated between the federal government and Quebec is outside this system.

governments, usually at the initiation of the CEA Agency. A project-specific approach is employed when the federal/provincial/territorial governments have not established a bilateral agreement. As Fitzpatrick and Sinclair (2009) conclude, these agreements vary widely, even in relation to specific aspects of EA such as public participation. The Sable Gas project on the east coast is an example of a project-specific agreement.

The third approach to inter-jurisdictional co-ordination is *substitution*. According to the Canadian federal EA process, if a project under the Act is subject to a review by a federal authority that is deemed to be an appropriate substitute, that review may replace the federal EA process. The Emera Brunswick Pipeline Project, which involved the construction of a natural gas pipeline in New Brunswick, is the first and so far only case of EA substitution under the CEAA (see Schneider, Sinclair, and Mitchell 2007; 2009). The project triggered both the CEAA for EA purposes and the National Energy Board Act for regulatory purposes.

In their review of multi-jurisdictional co-ordination, Fitzpatrick and Sinclair (2009) conclude that the two approaches to harmonization (bilateral agreements and project-specific agreements) have the greatest potential to minimize duplication, avoid process uncertainty, and increase efficiency and effectiveness in EA, which are the stated objectives of inter-jurisdictional co-ordination. They see great benefit, in fact, in government agencies sharing their expertise, engaging, and working together. In 2008, the Canadian Council of Ministers of the Environment mandated an initiative to find ways of ensuring that inter-jurisdictional assessment is more efficient and effective. The working group will no doubt consider experience with standardization, harmonization, and substitution.

CONCLUSION

The 35 years since the inception of EA in Canada have been transformative (Meredith 2004). Legislated EA processes have been created in every jurisdiction throughout the country, satisfying one of the core principles of best-practice EA. The EA concept has moved well beyond federal and provincial levels of government and is practised by Aboriginal peoples, incorporated regions, municipalities, cities, and even institutions. There are decision-makers, EA practitioners, and academics who now have considerable experience with EA, which has generally improved implementation. There has been recognition and serious consideration of the many challenges to achieving effective, efficient, and equitable implementation of EA, necessary to ensure that EA results in thoughtful planning about the environmental implications and overall sustainability of the projects we undertake. Canadians have made significant contributions to EA practice, nationally and internationally.

Where does Canada place in the stages of the evolution of EA noted at the outset of the chapter? The answer is not an easy one, because each jurisdiction tends to practise its own brand of EA. The EA legislation may be similar, and the basic process steps followed are the same, but the guidance, policy, and practice are often quite different. Nevertheless, Canadian jurisdictions have collectively moved beyond stage 1, the reactive stage, in part because each EA process calls for proactive thinking before a project is approved and requires some involvement of the public, not just affected industries and government, in the process. Most jurisdictions in Canada would probably consider themselves somewhere between stages 3 and 4. In other words, they do proactively identify impacts in a formal way and integrate this information in a planning context that considers alternatives, a variety of impact types, and notions of sustainability.

A closer look at practice, however, raises doubts about how far we really have evolved beyond stage 1. The many conflicts and uncertainties we have outlined above underscore this, as do the recent regulatory changes to CEAA. For example, the move toward a greater emphasis on issues of regulatory approvals rather than planning greatly affects the extent to which sustainability can be considered, as do the limited consideration of cumulative effects, the marginalization of the public's voice in the process of decision-making, and disagreements over scope.

Moving forward to embrace concepts such as community-based assessment, sustainability assessment, and strategic assessment, while limiting discretion, will require significant political will—no matter how much closer these changes bring us toward a sustainable society. EA can open up political decisions to public scrutiny in a way that makes some politicians uncomfortable. This is especially problematic when the projects involved may provide the much-needed jobs, tax revenue, and other economic spin-offs that large projects often offer. This may in part be what stalled discussions in Manitoba regarding the transformation of EA into sustainability assessment. Status quo thinking or, worse, thinking that undermines the basic foundations of EA will not move us to the next generation of assessment. In fact, the public has often had to turn to the courts to try to ensure that basic EA principles, such as public participation, are upheld. The Red Chris case (*MiningWatch Canada v. Minister of Fisheries and Oceans et al.* 2008 FCA 209) recently proceeded to the Supreme Court of Canada to consider issues of discretion as they relate to public involvement in federal EA, with a decision expected in March 2010. The courts have become a last resort for the public when those responsible for implementation refuse to embrace the value of EA and instead at best turn it into a paper exercise and at worst violate legal requirements. At the end of the day, courts cannot move EA forward: they can only prevent backsliding. Moving EA forward will require enlightened decision-makers and proponents—and especially public support.

FROM THE FIELD

We have had the good fortune to work on the development and implementation of EA in Canada through the Environmental Planning and Assessment Caucus of the Canadian Environmental Network (CEN). The caucus, which we have both chaired at various times, includes representatives of member groups of the CEN from across the country. Long-time active participants come from groups such as the Bow Valley Naturalists Society, the Environmental Coalition of PEI, Clean Nova Scotia, Resource Conservation Manitoba, and West Coast Environmental Law (the websites of these groups highlight the issues that each addresses).

Through face-to-face meetings and other regular communication, the caucus allows its member groups to bring forward a strong public interest analysis of the many facets of EA. The caucus has a long history of guiding the development of EA law and policy in Canada, starting with development of the first federal EA legislation. Since then, we have prepared numerous briefs and position papers and have participated in many consultations. Recently, we worked on the issue of EA substitution, as reported in this chapter, and we are gearing up for the seven-year review of CEAA due to start in 2010. Our caucus plan, as well as related briefs and materials, can be viewed at www.cen-rce.org.

Through our association with the caucus, we have also served as members of the CEAA's Regulatory Advisory Committee (RAC). The RAC is a multi-stakeholder committee representing various industry groups, Aboriginal organizations, environmental groups, provincial governments, and federal departments. The role of RAC is to provide input to the federal minister of environment on regulatory and policy matters arising under CEAA. This is done through face-to-face meetings, conference calls, and the work of subcommittees.

An example of RAC subcommittee work we have participated in involves development of guidance materials for public participation in EA. Subcommittee members representing various sectors identified key components for the guidance material and an approach to writing the material, as well as reviewing the material created. The final May 2008 guidance materials can be viewed at www.ceaa.gc.ca under the 'Policy and Guidance' header.

As you might anticipate, subcommittee and full RAC committee discussions can be intense and colourful, since each stakeholder organization has its own view on how the EA process should unfold.

What makes the RAC work is the focus on substance rather than stakeholder positions, the range of experiences and expertise around the table, an effort to find common ground, and a certain level of trust that has developed from time to time when members consistently adhere to the principles of RAC.

RAC has strengths and limitations. Its strength has been the ability to influence government policy when there is common ground among industry, environmental groups, and Aboriginal organizations. RAC's greatest weakness is that it perpetuates inaction in areas for which common ground is not or cannot be found. As a result, many opportunities to strengthen the federal EA process were dismissed on the basis of a lack of consensus at RAC rather than on their merits. In spite of its limitations, RAC has stood the test of time as one of the most effective processes available for civil society input into law and policy-making in the environmental field in Canada. Unfortunately, the future of RAC is in question under the current government as the minister of environment has not called a meeting in over a year and during that time changes were made to CEAA regulations without consultation.

Whether through EA caucus, the RAC, or other forums, in our experience it is through collaboration that we tend to make progress on the many EA challenges we face.

—JOHN SINCLAIR and MEINHARD DOELLE

REVIEW QUESTIONS

1. What are some of the potential benefits of carrying out an EA of a project, plan, or policy?
2. What are the steps that an EA process normally follows? What are the types of information that decision-makers might require to move from step to step?
3. What tools or methods could be used to support the identification and classification of impacts?
4. Visit the websites listed in Table 17.2, and compare the provincial or territorial EA process in your area to that of the federal EA process. What are the similarities and differences (e.g., how is the environment

defined, and what types of projects are included?)? What stage in the evolution of EA do you think best characterizes your jurisdiction?

5. What modifications would be required to the EA law in your region to ensure that more sustainable decisions are taken?

6. What are the key differences between federal screenings, comprehensive studies, and panel reviews under CEAA?

7. Does your municipality have its own EA process? If so, what sorts of projects has it been applied to?

8. What would be the potential cumulative effects of a project under discussion in your jurisdiction? What other activities are already being undertaken in the region of the proposed project? Should a regional cumulative effects assessment be done, and if so, who should be responsible for doing it?

9. What EA activities highlight the efficiency versus effectiveness debate related to EA?

10. What are the main components of meaningful participation?

11. What should the role of public participation be in environmental decision-making, and what are the barriers to implementing meaningful public participation?

12. What are the benefits and challenges of applying EA to policy?

13. What approach would you use to do an EA of a new project on campus (e.g., a new building, sports field, parking garage), and what types of information would you need to collect to complete the assessment?

REFERENCES

Alshuwaikhat, H. 2005. 'Strategic environmental assessment can help solve environmental impact assessment failures in developing countries'. *Environmental Impact Assessment Review* 25 (2): 307–17.

Ameyaw, S. 1992. 'Sustainable development and the community: Lessons from the Kasha Project, Botswana'. *The Environmentalist* 12 (4): 267–75.

Armitage, D. 2005. 'Collaborative environmental assessment in the Northwest Territories, Canada'. *Environmental Impact Assessment Review* 25 (3): 239–58.

———. 2009. 'Environmental impact assessment in Canada's Northwest Territories: Integration, collaboration, and the Mackenzie Valley Resource Management Act'. In K.S. Hanna, ed., *Environmental Impact Assessment: Practice and Participation*, 2nd edn, 211–37. Toronto: Oxford University Press.

Arnstein, S. 1969. 'A ladder of citizen participation'. *Journal of the American Institute of Planners* 35 (4): 216–24.

Beanlands, G.E., and P.N. Duinker. 1983. *An Ecological Framework for Environmental Impact Assessment in Canada*. Halifax: Institute for Resource and Environmental Studies, Dalhousie University.

Berger, T.R. 1977. *Northern Frontier, Northern Homeland: The Report of the Mackenzie Valley Pipeline Inquiry*. Ottawa: Minister of Supply and Services.

Blaug, E.A. 1993. 'Use of environmental assessment by federal agencies in NEPA implementation'. *Environmental Professional* 15 (1): 57–65.

Canadian Council of Ministers of the Environment. 1998. 'A Canada-Wide Accord on Environmental Harmonization'. http://www.ccme.ca/assets/pdf/accord_harmonization_e.pdf.

Canadian Environmental Assessment Agency. 2004. *Guidelines for Implementing the Cabinet Directive on SEA*. Ottawa: Canadian Environmental Assessment Agency.

Canadian Environmental Network. 1988. *A Federal Environmental Assessment Process: The Core Elements*. Ottawa: Canadian Environmental Network, Environmental Planning and Assessment Caucus.

CIDA (Canadian International Development Agency). 2005. 'Environment handbook for community development initiatives: Second edition of the handbook on environmental assessment of non-governmental organizations and institutions programs and projects'. Gatineau, QC: CIDA. http://www.acdi-cida.gc.ca/CIDAWEB/acdicida.nsf/En/JUD-47134825-NVT.

Commissioner of the Environment and Sustainable Development. 2004. *Assessing the Environmental Impact of Policies, Plans and Programs*. Ottawa: Office of the Auditor General of Canada.

Connor, D.M. 2001. *Constructive Citizen Participation: A Resource Book*. Victoria, BC: Connor Development Services Ltd.

———. 2005. *Strategic Environmental Assessment*. London: Earthscan.

Devlin, J.F., N.T. Yap, and R. Wier. 2005. 'Public participation in environmental assessment: Case studies on EA legislation and practice'. *Canadian Journal of Development Studies* 26 (3): 487–500.

Diduck, A.P., and A.J. Sinclair. 2002. 'Public participation in environmental assessment: The case of the nonparticipant'. *Environmental Management* 29 (4): 578–88.

Doelle, M. 2008. *The Federal Environmental Assessment Process: A Guide and Critique*. Markham, ON: LexisNexis.

———. 2009. 'The role of strategic environmental assessments (SEAs) in energy governance: A case study of tidal energy in Nova Scotia's Bay of Fundy'. *Journal of Energy and Natural Resources Law* 27 (2): 112–44.

Dorcey, A., L. Doney, and H. Ruggeberg. 1994. *Public Involvement in Government Decision-Making: Choosing the Right Model*. Victoria, BC: Round Table on the Environment and the Economy.

Duinker, P.N., and L.A. Greig. 2006. 'The impotence of cumulative effects assessment in Canada: Ailments and ideas for redeployment'. *Environmental Management* 37 (2): 153–61.

Ecojustice. 2009. 'Regulations unlawfully gut key federal environmental law'. Media release, 20 April. http://www.ecojustice.ca.

Fitzpatrick, P., and A.J. Sinclair. 2003. 'Learning through public involvement in environmental assessment hearings'. *Journal of Environmental Management* 67 (2): 161–74.

———. 2009. 'Multi-jurisdictional environmental impact assessment: Canadian experiences'. *Environmental Impact Assessment Review* 29 (4): 252–60.

Gamble, D.J. 1978. 'The Berger Inquiry: An impact assessment process'. *Science* 99 (4332): 946–52.

Gibson, R.B. 1993. 'Environmental assessment design: Lessons from the Canadian experience'. *The Environmental Professional* 15 (1): 12–24.

———. 2002. 'From Wreck Cove to Voisey's Bay: The evolution of federal environmental assessment in Canada'. *Impact Assessment and Project Appraisal* 20 (3): 151–9.

———. 2006. 'Beyond the pillars: Sustainability assessment as a framework for effective integration of social, economic and ecological considerations in significant decision-making'. *Journal of Environmental Assessment Policy and Management* 8 (3): 259–80.

Gibson, R.B., et al. 2005. *Sustainability Assessment: Criteria, Processes and Applications*. London: Earthscan.

Gibson, R.B., and K.S. Hanna. 2009. 'Progress and uncertainty: The evolution of federal environmental assessment in Canada'. In K.S. Hanna, ed., *Environmental Impact Assessment: Practice and Participation*, 2nd edn, 18–36. Toronto: Oxford University Press.

Govender, K., R. Hounsome, and A. Weaver. 2006. 'Sustainability assessment: Dressing up SEA? Experiences from South Africa'. *Journal of Environmental Assessment Policy and Management* 8 (3): 321–430.

Government of Canada. 2004. *Victor Diamond Project DeBeers Canada Exploration Inc.: Guidelines for the Conduct of a Comprehensive Study and the Preparation of a Draft Comprehensive Study Report*. Ottawa: Natural Resources Canada, Environment Canada, Department of Fisheries and Oceans, Indian and Northern Affairs Canada.

Green, A. 2002. 'Discretion, judicial review, and the Canadian Environmental Assessment Act'. *Queen's Law Journal* 27 (2): 785–96.

Hacking, T., and P. Guthrie. 2008. 'A framework for clarifying the meaning of triple bottom-line, integrated, and sustainability assessment'. *Environmental Impact Assessment Review* 28: 73–89.

Hanna, Kevin S., ed. 2009. *Environmental Impact Assessment: Practice and Participation*. 2nd edn. Toronto: Oxford University Press.

Hazel, S. 1999. *Canada v. the Environment: Federal Environmental Assessment 1984–1998*. Toronto: Canadian Environmental Defence Fund.

Kaliba, A.R.M., and D.W. Norman. 2004. 'Assessing sustainability of community-based water utility projects in central Tanzania with the help of canon-ical correlation analysis'. *Journal of Environmental Assessment Policy and Management* 6 (1): 73–90.

Keen, M., and S. Mahanty. 2006. 'Learning in sustain-able natural resource management: Challenges and opportunities in the Pacific'. *Society and Nat-ural Resources* 19 (6): 497–513.

Kennett, S.A. 2000. 'Meeting the intergovernmental challenge of environmental assessment'. In P.C. Fafard and K. Harrison, eds, *Managing the Environ-mental Union: Intergovernmental Relations and Environmental Policy in Canada*, 107–31. Kingston, ON: School of Policy Studies, Queen's University.

Lawrence, David P. 1997. 'Integrating sustainability and environmental impact assessment'. *Environ-mental Management* 21 (1): 23–42.

———. 2003. *Environmental Impact Assessment: Practical Solutions to Recurrent Problems*. Hobo-ken, NJ: Wiley-Interscience.

Lee, N. 2006. 'Bridging the gap between theory and practice in integrated assessment'. *Environmental Impact Assessment Review* 26: 57–78.

Lobe, K. 2009. 'Environmental assessment: Manitoba approaches'. In K.S. Hanna, ed., *Environmental Impact Assessment: Practice and Participation*, 2nd edn, 346–60. Toronto: Oxford University Press.

Meredith, T. 1992. 'Environmental impact assessment, cultural diversity, and sustainable rural develop-ment'. *Environmental Impact Assessment Review* 12: 125–38.

———. 2004. 'Assessing environmental impacts in Canada'. In B. Mitchell, ed., *Resource and Environ-mental Management in Canada: Addressing Con-flict and Uncertainty*, 467–96. Toronto: Oxford University Press.

Minkin, D.P. 2008. 'Cultural Preservation and Self-Determination through Land Use Planning: A Framework for the Fort Albany First Nation'. (Queen's University, School of Urban and Regional Planning, Kingston, ON, Master of Urban and Regional Planning Thesis).

Morrison-Saunders, A., and T.B. Fischer. 2006. 'What is wrong with EIA and SEA anyway? A sceptic's per-spective on sustainability assessment'. *Journal of Environmental Assessment Policy and Management* 8 (1): 19–39.

Mushkegowuk Council. 2008. 'Mushkegowuk Coun-cil Resolution No. 2008-11-13 Territorial Map-ping and Land Use Planning; Resolution No.

2008-11-25 Mining Activities in Mushkegowuk First Nation Homelands; Resolution No. 2008-11-29 Resource Development Activities in the Mush-kegowuk First Nation Homelands'. Timmins, ON: Mushkegowuk Council.

Neefjes, K. 2001. 'Learning from participatory environmental impact assessment of community-centered development: The Oxfam experience'. In B. Vira and R. Jeffery, eds, *Analytical Issues in Natural Resources Management*, 111–25. New York: Palgrave.

Noble, B.F. 2002. 'The Canadian experience with SEA and sustainability'. *Environmental Impact Assess-ment Review* 22 (1): 3–16.

———. 2006. *Introduction to Environmental Impact Assessment: A Guide to Principles and Practice*. Toronto: Oxford University Press.

———. 2008. 'Strategic approaches to regional cumu-lative effects assessment: A case study of the Great Sand Hills, Canada'. *Impact Assessment and Project Appraisal* 26 (2): 78–90.

———. 2009. 'Promise and dismay: The state of stra-tegic environmental assessment systems and prac-tices in Canada'. *Environmental Impact Assessment Review* 29 (1): 66–75.

Noble, B.F., and J. Harriman-Gunn. 2009. 'Strategic environmental assessment'. In K.S. Hanna, ed., *Environmental Impact Assessment Process and Practices in Canada*, 2nd edn, 103–30. Toronto: Oxford University Press.

Onyango, V., and M. Schmidt. 2007. 'Towards a strate-gic environment assessment framework in Kenya: Highlighting areas for further scrutiny'. *Manage-ment of Environmental Quality: An International Journal* 18: 309–28.

Pallen, D. 1996. *Environmental Assessment Manual for Community Development Projects*. Ottawa: Can-adian International Development Agency, Asia Branch.

Partidário, M.R. 1996. 'Strategic environmental assessment: Key issues emerging from recent practice'. *Environmental Impact Assessment Review* 16 (1): 31–55.

———. 1999. 'Strategic environmental assessment—Principles and potential'. In J. Petts ed., *Hand-book of Environmental Impact Assessment*, 60–73. Oxford: Blackwell Science.

Petts, J., ed. 1999a. *Handbook of Environmental Impact Assessment: Environmental Impact Assessment:*

Process, Methods and Practice. Oxford: Blackwell Science.

———. 1999b. 'Public participation and environmental impact assessment'. In J. Petts, ed., *Handbook of Environmental Impact Assessment*, 145–77. Oxford: Blackwell Science.

———. 2003. 'Barriers to deliberative participation in EIA: Learning from waste policies, plans and projects'. *Journal of Environmental Assessment Policy and Management* 5: 269–93.

Pope, J. 2006. 'What's so special about sustainability assessment?' *Journal of Environmental Assessment Policy and Management* 8 (3): v–x.

Pope, J., D. Annandale, and A. Morrison-Saunders. 2004. 'Conceptualising sustainability assessment'. *Environmental Impact Assessment Review* 24: 595–616.

Privy Council Office and Canadian Environmental Assessment Agency. 1999. *The Cabinet Directive on the Environmental Assessment of Policy, Plan and Program Proposals.* Ottawa: Minister of Public Works and Government Services Canada.

Roberts, R. 1998. 'Public involvement in environmental impact assessment: Moving to a "new-think"'. *Interact* 4 (1), 39–62.

Ross, W.A. 1998. 'Cumulative effects assessment: Learning from Canadian case studies'. *Impact Assessment and Project Appraisal* 16 (4): 267–76.

Sadler, B. 1996. *Environmental Assessment in a Changing World: Evaluating Practice to Improve Performance.* Canadian Environmental Assessment Agency and International Association for Impact Assessment. Ottawa: Ministry of Supply and Services.

Schneider, G., A.J. Sinclair, and L. Mitchell. 2007. 'Environmental assessment process substitution: A participant's view'. Ottawa: Canadian Environmental Network. www.cen-rce.org/eng/caucuses/assessment/index.html.

———. 2009. 'Environmental assessment process substitution: Is meaningful public participation possible?' Ottawa: Canadian Environmental Network. www.cen-rce.org/eng/caucuses/assessment/index.html.

Sinclair, A.J., and A.P. Diduck. 2001. 'Public involvement in EA in Canada: A transformative learning perspective'. *Environmental Impact Assessment Review* 21 (2): 113–36.

———. 2009. 'Public participation in Canadian environmental assessment: Enduring challenges and future directions'. In K.S. Hanna, ed., *Environmental Impact Assessment Process and Practices in Canada*, 2nd edn, 56–82. Toronto: Oxford University Press.

Sinclair, A.J., A.P. Diduck, and P.J. Fitzpatrick. 2008. 'Conceptualizing learning for sustainability through environmental assessment: Critical reflections on 15 years of research'. *Environmental Impact Assessment Review* 28 (7): 415–522.

Sinclair, A.J., and P. Fitzpatrick. 2003. 'Provisions for more meaningful public participation still elusive in new Canadian EA bill'. *Impact Assessment and Project Appraisal* 20 (3): 161–76.

Sinclair, A.J., L. Sims, and H. Spaling. 2009 (in press). 'Community-based approaches to strategic environmental assessment: Lessons from Costa Rica'. *Environmental Impact Assessment Review*.

Smith, L.G. 1982. 'Mechanisms for public participation at a normative planning level in Canada'. *Canadian Public Policy* 8 (4): 561–72.

Spaling, H. 1997. 'Cumulative impacts and EIA: Concepts and approaches'. *EIA Newsletter 14.* www.art.man.ac.uk/EIA/publications/index.htm.

———. 2003. 'Innovations in environmental assessment of community-based projects in Africa'. *The Canadian Geographer* 47 (2): 151–68.

Thérivel, R., and A.L. Brown. 1999. 'Methods of strategic environmental assessment'. In J. Petts, ed., *Handbook of Environmental Impact Assessment*, 441–64. Oxford: Blackwell Science.

Vicente, G., and M.R. Partidário. 2006. 'SEA—Enhancing communication for better environmental decisions'. *Environmental Impact Assessment Review* 26 (8): 696–706.

Whitelaw, G., D. McCarthy, and L. Tsuji. 2009. 'The Victor Diamond Mine environmental assessment process: A critical First Nations perspective'. *Impact Assessment and Project Appraisal* 27 (3): 205–15.

Wood, C.W. 2003. *Environmental Impact Assessment: A Comparative Review.* 2nd edn. London: Prentice Hall.

18

Incorporating Participatory Approaches and Social Learning

Alan Diduck

Learning Objectives

- To be able to explain how participatory and social learning approaches help to deal with complexity, uncertainty, and conflict in resource and environmental management.
- To understand basic characteristics of participatory approaches and the key processes and elements of social learning.
- To appreciate the advantages and disadvantages of participatory and social learning approaches, including practical opportunities and challenges.
- To be able to identify examples of best practice regarding participatory and social learning approaches in resource and environmental management in Canada.

INTRODUCTION

In response to the seeming intractability of many resource and environmental problems, theorists and practitioners are relying increasingly on participatory approaches and social learning orientations to resource and environmental management. In this chapter, I explore that emerging perspective. The first section examines why such a view is being adopted, grounding the analysis in a discussion of uncertainty, conflict, and change. The second section discusses the nature of participatory approaches and social learning, centring on four key aspects of participation and three models of social learning. The third section reviews advantages and disadvantages of participatory and social learning approaches, identifying practical opportunities and challenges. Finally, the fourth section presents applications in selected Canadian resource sectors, management processes, and environmental problems.

WHY THE EMERGING PERSPECTIVE?

As discussed throughout this book, uncertainty and conflict—often related to the complexity of social–ecological systems—are prevalent conditions in resource and environmental management. The rationale for participatory approaches and social learning orientations is often grounded in implications related to one or all of these themes (e.g., Keen, Brown, and Dyball

2005; Ison, Röling, and Watson et al. 2007; Pahl-Wostl et al. 2007). The implications of uncertainty have received considerable attention in the literature (Holling 1978; Christenson 1985; Wynne 1992; Ludwig 2001; Folke et al. 2005). For example, Christenson's (1985) prototypical conditions of planning problems included 'chaotic conditions', in which uncertainty exists over both ends and means, and 'bargaining conditions', in which uncertainty exists over ends but means are known. Appropriate solutions under both sets of conditions include participatory processes such as group facilitation for problem identification, mediation of competing goals, and collaborative learning workshops. A point of convergence is that participatory and social learning approaches, emphasizing multiple knowledge sources, dialogue, mutual learning, and the continual evolution of ideas, are appropriate for circumstances characterized by high uncertainty. Only these types of approaches enable individuals, organizations, and communities to construct legitimate endpoints, identify appropriate technologies for reaching those endpoints, and navigate through complex social–ecological dynamics.

The implications of conflict for approaches to resource and environmental management have also received attention (Dorcey 1986; Lee 1993; O'Leary and Bingham 2003). Again, participatory and social learning approaches emerge as a key to resolving or limiting some forms of environmental conflict. For example, Lee (1993) synthesized collaborative conflict resolution ('bounded conflict') with adaptive management to offer a general model of social learning for sustainable development (which he called civic science) (see also Chapter 16). Further issues related to conflict are discussed later in the chapter.

Building on the work of Funtowicz and Ravetz (1993), Cardinall and Day (1998) addressed the implications of both uncertainty and conflict. They focused on decision stakes (interpreted here as representing potential for conflict) and management uncertainty. Decision stakes include individual preferences, social interests, and cultural beliefs, with beliefs having the highest potential for conflict. Management uncertainty comprises substantive uncertainty, subjective doubt, and procedural confusion, with procedural confusion having the highest uncertainty. When stakes and uncertainty are high, decision-making will affect core beliefs, problems will have diverse implications, and the relevance of information will depend on values and knowledge. For Cardinall and Day, such issues and problems are too complex for programmed, top-down management. Such an approach will be ineffective and potentially harmful because it will reinforce the biases that created the problems in the first place. They argued, therefore, that when stakes and uncertainty are high, social learning or 'civic exploration' is the best approach, relying on participatory ideals and methods.

THE NATURE OF PARTICIPATORY APPROACHES AND SOCIAL LEARNING

Given the relevance of participatory and social learning approaches to resolving complex, uncertain, and conflict-ridden resource and environmental problems, it becomes important to understand the nature of such approaches. The ensuing discussion describes characteristics of participatory approaches, focusing on four aspects: breadth, degrees (or extent), timing, and techniques. The subsequent section introduces social learning and provides linkages to fundamental characteristics of participatory approaches.

Characteristics of Participatory Approaches

Breadth of Participation

The issue of breadth relates to who is involved in any given resource and environmental management function. As discussed in Chapter 19, this is an important question that has social justice and political implications. That chapter reveals how gender analysis can illuminate systemic impediments to equitable participation by women. Further concepts pertaining to breadth, discussed below, include multiple publics, active and inactive publics, stakeholders, and barriers to participation. As well, important concerns for resource managers, particularly in the public sector, include the legitimacy of participation and accountability for social welfare.

A basic idea is that focusing on the general public as such is problematic. It creates the risk that the plurality of values, interests, and issues affected by most resource and environmental management decisions will be ignored. For this reason, managers have generally come to accept the notion of multiple publics in which the general public is viewed as a shifting multiplicity of organizations, individuals, interests, and coalitions. That is, the general public should be segmented and treated as many publics rather than *the* public. A related idea is the distinction between active and inactive publics (Mitchell 2002). The former includes industry associations (e.g., the Chemical Producers Association, the Mining Association of Canada), environmental non-governmental organizations (e.g., Greenpeace, Pollution Probe), and other organized interest groups. Many of these groups are devoted to resource and environmental issues and participate actively in diverse management functions, such as planning, assessment, monitoring, research, and policy development. Inactive

publics, on the other hand, are individuals and groups who do not typically become involved in resource and environmental or other civic issues.

Active publics can make important contributions to resource and environmental management. However, to take full advantage of participatory approaches, managers must often find ways to engage members of inactive publics (Larson and Lach 2007). In doing so, it is helpful to identify and deal with barriers to participation, which can include inadequate notice, incomplete or inaccessible information, insufficient resources, lack of opportunity, fragmented decision processes, ingrained technical or cultural perspectives, and lack of impact on decisions taken (Lowndes, Pratchett, and Stoker 2001; Diduck and Sinclair 2002; Petts 2004). A difficult problem is that some members of inactive publics simply do not have the time to participate in environmental or other community issues. Their time is spent on other priorities, including livelihood activities and, for some, meeting basic needs.

In working with active publics and seeking to engage inactive publics, managers often focus on stakeholders, or those individuals and organizations with an interest in, concern about, or legal responsibility for a decision or issue. Such a focus raises the challenge of establishing a legitimate basis for participation, particularly in situations characterized by high uncertainty and conflict. It also implies a need to avoid reducing the broad public interest to narrow bargaining among a small number of stakeholders (Parson 2000). To help respond to these challenges, Dorcey and McDaniels (2001) suggested that managers should avoid overrepresentation of the powerful and ensure that important interests not involved are still represented effectively. They further argued that it is essential for public involvement processes to have distinct terms of reference and clear lines of accountability.

Several desirable characteristics of participatory approaches to resource and environmental management can be identified respecting breadth of participation. They include involving active publics, engaging inactive publics, reducing barriers to involvement, identifying key stakeholders, establishing a legitimate basis for participation, and maintaining clear lines of accountability for broad public interests.

Degrees of Participation

Degrees of participation refer to the extent to which managers and their key publics share decision-making power. Arnstein's (1969) classic ladder of citizen participation is an important model touching on this issue. The model identified eight levels of involvement and associated degrees of power sharing (see Table 18.1). The bottom rungs of the ladder—manipulation and therapy (characterized as non-participation)—describe public relations exercises designed to educate or 'cure' citizens and gain their support. The middle rungs—forms of tokenism—include informing, which involves one-way flows of information from managers to citizens; consultation, in which citizens are given a voice but not necessarily heeded; and placation, in which citizens provide advice but do not participate in decision-making. The top rungs—referred to as forms of citizen power—are partnership (in which trade-offs are negotiated), delegated power, and citizen control, the last of which involves the highest degree of 'decision-making clout' (Arnstein 1969, 217).

Arnstein was highly critical of lower levels of involvement and argued that citizen participation should involve redistribution of power over political, economic, and environmental decision-making. In a slightly different approach, Dorcey, Doney, and Ruggeberg (1994) advocated a strategic and adaptive model. Building on the Arnstein ladder, they described a continuum of public involvement with increasing levels of public interaction, influence, and commitment. The lowest levels were *inform* and *educate*, with the highest levels being *seek consensus* and *ongoing involvement*. In this model, each level in the continuum could be an appropriate degree of participation, depending on the decision-making context. Further, lower levels of participation were viewed as important and often necessary supports for reaching higher levels of interaction.

Additional variations on the ladder of participation can be found in the literature, but using the Arnstein model, a participatory approach to resource and environmental management implies high levels of citizen control over decision-making. Supplementing this view, in the Dorcey model, participatory approaches

Table 18.1 Degrees of Citizen Control and Forms of Participation

Degree of citizen control	Form of participation
Citizen power	Citizen control
	Delegated power
	Partnership
Tokenism	Placation
	Consultation
	Informing
Non-participation	Therapy
	Manipulation

Source: Arnstein (1969).

are characterized by multiple levels of involvement, including shared decision-making and ongoing involvement.

Timing of Participation

Timing of participation examines at what point key publics are brought into planning and decision-making processes. Smith (1982) suggested that involvement could occur at normative levels of planning (i.e., when decisions are made as to what should be done), strategic levels (i.e., when decisions are made to determine what can be done), and operational levels (i.e., when decisions are made to determine what will be done). He concluded that public participation at normative and strategic levels of planning was rare, and more recent literature suggests that this continues to be the case. For example, Sinclair and Diduck (2009) reviewed Canadian environmental impact assessment (EIA) processes and found that involvement was typically not required legally until operational stages of planning, while involvement at earlier junctures was left to the discretion of project proponents. And although government officials encouraged project proponents to engage in early involvement, this occurred only sporadically. An implication is that legal requirements might be necessary to ensure participation at normative and strategic levels of EIA planning and decision-making (see also Chapter 17).

Employing Smith's framework, desirable participatory approaches to resource and environmental management are characterized by involvement at normative and strategic stages of planning and decision-making. The earlier the involvement occurs, the more influence interested publics can have on fundamental issues such as defining problems or opportunities, identifying goals, and analyzing alternative means of achieving the goals. Although early involvement is rare, it is crucial for ensuring that key publics

have an opportunity to influence 'big-picture' decisions with long-term implications.

Participation Techniques

Numerous public participation techniques exist, and several authors have presented helpful compilations and classifications. Mitchell's (2002) approach was to analyze techniques in terms of the basic functions of participation programs, including:

- Information-out, which should serve to meet the information needs and wants of key publics in a fair, systematic, and timely fashion. Techniques suitable for this function include advertisements, technical reports, and public meetings.
- Information-in, which should give interested publics opportunities to present their views on the issue or problem being considered. Appropriate techniques include public hearings, workshops, and interviews.
- Continuous exchange, which refers to mechanisms or systems for establishing ongoing dialogue among managers and key publics. Effective techniques for this function include advisory committees, task forces, and community boards.
- Facilitating consensus, a long-term function, requires techniques such as mediation and non-adversarial negotiation.

The International Association for Public Participation (IAP2) (2006) compiled a comprehensive list of techniques and classified them into six major groupings: passive public information, active public information, small group public input, large group public input, small group problem-solving, and large group problem-solving (see Table 18.2).

The methods in the first four classes are obviously suited to the information-out and

Table 18.2 Public Participation Techniques

Passive Public Information Techniques

Advertisements	Feature stories	Information repositories
News conferences	Newspaper inserts	Press releases
Print materials	Technical reports	Television
Websites		

Active Public Information Techniques

Briefings	Central contact person	Community fairs
Expert panels	Field offices	Field trips
Information hotline	Open houses	Technical assistance
Simulation games		

Small Group Public Input Techniques

Informal meetings	In-person surveys	Interviews
Small-format meetings		

Large Group Public Input Techniques

Public hearings	Response sheets
Telephone, mail, and Internet surveys	

Small Group Problem-Solving Techniques

Advisory committees	Citizen juries	Community facilitation
Consensus building	Mediation and negotiation	Panels
Role-playing	Task forces	

Large Group Problem-Solving Techniques

Workshops	Interactive polling	Sharing circles
Websites and chat rooms		
Future search conference (see Chapter 14)		

Source: International Association for Public Participation 2006.

information-in functions described above, and many of the problem-solving techniques are appropriate for continuous exchange and facilitating consensus. Generally, the problem-solving techniques are highly interactive or deliberative and involve communication, understanding, collaboration, and mutual learning. These are not only key dynamics in social learning, discussed below. They are also often associated with high degrees of participation (e.g., partnerships and shared decision-making). It is fair to say, therefore, that desirable participatory approaches to resource and environmental management are characterized by emphases on the continuous exchange and facilitating consensus functions of public involvement, as well as on a liberal use of interactive, problem-solving techniques.

Models of Social Learning

Social learning is an emerging framework in resource and environmental management. At its heart is the suggestion that learning is an idea that applies not only to individuals but also to social collectives, such as groups, formal organizations, and communities. Core concepts link individual and collective learning and suggest that participatory approaches (with characteristics such as those noted above) are central to learning by social collectives. This section describes three models of social learning with relevance to resource and environmental management. The first is the civics approach, founded on interactive and adaptive decision-making processes underlying mutual and collective learning. The second involves the development of collective cognition and the group dynamics needed to develop such cognition. The third includes similar process variables but is noted here for its elaboration of the types of learning outcomes required for achieving sustainability. A theme common to each of the models is the need for transdisciplinary knowledge—i.e., understanding that spans both the natural and social sciences plus local and traditional knowledge. Another consistent theme is that community self-organization must be at the heart of planning, management, and decision-making. See Debbe Crandall's guest statement for an account of community self-organization in relation to conservation of the Oak Ridges Moraine in southern Ontario.

GUEST STATEMENT

Community Self-Organization: The Seeds of Change for Sustainability
Debbe Crandall

Much of the literature on environment and resource management is written for the benefit of the resource manager. This is appropriate, given that ultimately the final decisions rest with those charged to manage. However, management issues facing our communities are increasingly complex and solutions more uncertain. Diduck describes the benefits of participatory approaches for managers to deal with uncertainty, integrating a diversity of perspectives and resolving conflict while fostering collective learning. Social learning outcomes can also create conditions for an emergent community structure that sets its own agenda rather than leaving the decisions solely to the discretion of the resource manager.

This alternative structure is community self-organization. In this chapter, Diduck describes three different models of social learning, all of which rely to some degree on this alternative vision. Community self-organization involves the community (a group of people organized around a collective issue) defining the vision, determining who the stakeholders are, establishing the process, and setting the agenda for achieving outcomes.

Is this easy? No. Is it worthwhile? Yes. Are there principles that should underlie such a model? Absolutely.

Social learning through action and participation is a transformative process and provides key community members with the requisite knowledge and skills to manoeuvre confidently within the political power hierarchy. Bonding and bridging social capital among the broad spectrum of actors evolves through years of participation in resource management issues.

While conflict is usually the entry point into community activism for most people, it is not conflict itself that leads to learning but rather how community groups respond. It is through individual and collective reflection on the failure to communicate or to achieve a desired outcome that one's world view changes. And for the curious folk, it is the thirst to understand foundational motivations more fully that teaches important lessons. Having a diversity of knowledge and skills available to the group as a whole is essential, but when this knowledge and skills are present within the membership, each meeting becomes a learning platform.

One of the biggest challenges to community activism is successfully moving back and forth between thinking locally, regionally, and globally. Getting stuck in the 'not in my backyard' mindset can be limiting, not only for the group's credibility but also for its ability to work effectively with other communities of practice or to collaborate with managers or industry members. Groups must move with confidence beyond their borders and engage with the broader community to forge alliances, seek different perspectives, and attract funding.

The Oak Ridges Moraine in southern Ontario is a fascinating case study in which community self-organization is emerging. In 2001, the provincial government, forced by daily barrages in the media to take action to resolve long-standing land-use conflicts, responded with the Oak Ridges Moraine Conservation Act and land-use regulation. The initial euphoria of the conservation community of having 'saved' the moraine and having greater certainty in planning policy was soon replaced by recognition that there was only the illusion of certainty.

The inherent limitations of rules and regulations, coupled with electoral timeframes and short attention spans, confirm that government on its own could not and would not deliver the kind of future for the moraine for which activists had fought so hard for many years. In response, a number of non-governmental organizations joined forces in 2005 with the goal of fundamentally influencing the way decisions are made to protect the moraine.

Organized around a multi-party community-based monitoring model, the Monitoring the Moraine (MTM) project is a collaboration among the Save the Oak Ridges Moraine Coalition, Citizens' Environment Watch, and the Centre for Community Mapping. MTM is guided by the vision that a broad governance arrangement is needed to champion the moraine and its policy framework. Drawing upon experiences, skills, and knowledge from a diverse range of people, MTM is grounded in the principles of collaboration, participation, social learning, and community volunteerism. Parallel initiatives at several universities are exploring social–ecological integrity and resilience through a sustainability framework, and work is underway to have the moraine designated as a UNESCO World Biosphere Reserve. Biosphere reserves are models of sustainability organized around goals of sustainable development, conservation, and building local capacity for research, monitoring, and education. Building capacity to achieve these goals requires cooperation among a plurality of stakeholders, so the conservation community has started to reach out beyond its borders to foster discourse with farmers, artists, health care and other service providers, local government, and academia about issues of community well-being and social innovation.

Years ago, at the beginning of the campaign on the part of a coalition of volunteer community activists to save the Oak Ridges Moraine, the end goal was provincial legislation. Little did we know that in the process of getting to what we thought was the solution, the true seeds for protecting the moraine's future were being sown: the community itself.

The Civics Approach

Sharing features with Lee's (1993) notion of civic science (mentioned earlier), the civics approach to resource and environmental management offers a holistic vision for understanding linkages between social and natural systems (Nelson 1995; Nelson and Serafin 1995, 1996; Lawrence and Nelson 1999; Sinclair, Diduck, and Fitzpatrick 2002). The approach recognizes cultural differences in how knowledge is validated and in how reality is understood and investigated. The civics world view is holistic, or integrated, focusing on goals such as sustainability, equity, and heritage diversity. It emphasizes transdisciplinary knowledge and focuses on interactive models of planning, which centre on broad participation, dialogue, negotiation, co-operation, and mutual learning. Further, it relies on adaptive strategies to learn from experience by correcting errors and profiting from successes (see Chapter 16). Civics advocates suggest these key features make the model suitable for resource and environmental problems characterized by high degrees of complexity and uncertainty. These features also help to respond to the conflict often associated with such problems, because the social learning envisioned in the model focuses not only on cognitive differences but also on conflict over norms and values:

. . . the learning should stress a humanistic as much as a scientific approach. It should stress the capacity to assess and make value judgments as much as technical learning. The new learning approach probably should search not only for the best overall alternative or goal but also for an array of acceptable ones reflecting the differing interests and values of concerned groups [Nelson and Serafin 1996, 12].

In addition to the general characteristics described above, the civics approach entails an interactive and adaptive decision process that underlies the social learning needed to address complexity, uncertainty, and conflict. The process includes seven key elements linked through nonlinear, dynamic interplay (see Figure 18.1).

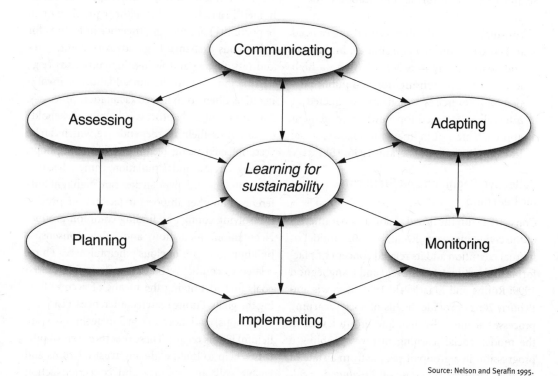

Source: Nelson and Serafin 1995.

Figure 18.1 Interactive and Iterative Decision Processes in the Civics Approach.

The process applies to individuals, groups, and organizations and envisions interplay across resource sectors and at multiple spatial and temporal scales. At the centre of the process is understanding, made possible by 'communication in its various forms; assessment activities of a regular or irregular type; strategic and other types of planning; implementation involving research and experiment; monitoring of different kinds of information; and adapting' (Nelson and Serafin 1995, 8).

An important implication of civics decision-making is that information—the basic currency of the interactive processes envisioned here—must be understandable to all participants. Moreover, it must be 'comprehensive enough to cover all facets of the situation and to reflect the interests of all concerned parties' (Nelson and Serafin 1996, 18). A further implication is that

> consultation and involvement processes need to be carried on in a continuous, interactive and adaptive way in order to achieve a high level of success. Consultation and public participation exercises which are conducted at intervals in a staged top down management approach are not in line with ongoing social learning [Nelson and Serafin 1996, 18].

Collective Cognition and Platforms for Learning

Consistent with the broad vision and parameters of the civics approach, Röling's (2002) model of mutual cognition and its related concept of platforms for learning (Maarleveld and Dangbégnon 1999; Röling and Maarleveld 1999; Leeuwis and Pyburn 2002) provide details of social learning processes at the individual and group levels. In the model, social learning involves enabling a progression in a group of people from a state of *multiple cognition* to a state of *distributed or collective cognition*. Cognition is treated broadly; it

embraces mental processes of acquiring knowledge, perceptions and theories of one's environment, and values, emotions, and goals respecting the environment. Multiple cognition means that people have multiple perspectives. Distributed cognition means that their perspectives have enough shared attributes to permit concerted action. Collective cognition involves shared attributes plus concerted action. The model also postulates that correspondence among the elements of cognition is a basic driver of a person's cognitive growth. Another driver is consistency between a person's cognitive elements and his or her environment.

The model suggests that intentionally designed platforms for learning can facilitate fundamental cognitive growth and progression from multiple to collective cognitive states. Platforms are deliberate interventions in which interdependent stakeholders in a resource issue (either a problem or an opportunity) are brought together in forums for continuous exchange (e.g., advisory committees and task forces) and facilitating consensus (e.g., negotiation and mediation). Platforms involve iterative, often complicated facilitation processes. An important early process is helping stakeholders recognize their interdependence, which can be accomplished using techniques like workshops, sharing circles, and simulation games. Second, given that interdependencies can highlight differences, another important facilitation process is resolving conflict by finding common ground. Negotiation, mediation, and other consensus-building tools are obviously helpful here. Third, related to conflict resolution is designing shared goals and identifying the means of accomplishing the goals. Future search conferences, visioning workshops, and nominal group techniques are helpful in this regard. These first three sets of processes help to build enduring trust relations and enable collective inquiry and concerted action toward a common goal.

Applied initially in the context of rural international development projects, aspects of the model have proved relevant to resource management problems in more developed countries (e.g., Maarleveld and Dangbégnon's [1999] investigation of fisheries in Benin and water resources in Holland). More recently, ideas from the model informed the theoretical and methodological bases of a multi-site study of integrated watershed management in the Netherlands, France, Italy, and the United Kingdom. The SLIM project (social learning for the integrated management and sustainable use of water at catchment scale) (from 2001 to 2004) investigated the application of social learning as a conceptual framework, an operational principle, a policy instrument, and a process of systemic change. The results of the project appeared in a special issue of *Environmental Science and Policy* (2007, v. 10, no. 6), on the project website (slim.open.ac.uk/page. cfm), and in other academic publications (e.g., Ison and Watson 2007).

Among the important results of the project was a new model—'a diagnostic framework'—for bringing stakeholders together to gain insight into their roles in complex natural resource management situations. The framework invites participants to engage in an iterative process of critical reflection and inquiry aimed at provoking transformational changes in understanding and practice (Steyaert and Jiggins 2007). The framework also identifies four key interacting variables that influence such changes (and which are themselves altered by the changes):

- stakeholders and stake-holding processes (in which interests and positions change over time and new stakes and stakeholder positions emerge);
- ecological components of ecosystems (i.e., what people understand about ecological

components and processes of ecosystems, as framed and structured by social relationships);
- institutional and policy frameworks (i.e., the organizations, laws, and norms that constrain and enable deliberative processes, concerted action, and adaptability); and
- facilitating learning processes and coordination among stakeholders (involving methods for using scientific knowledge and researcher interventions to enable concerted action by stakeholders).

Sustainability Learning

The third model of social learning encompasses similar outlooks and process factors as those described above and outlines the types of learning outcomes required for achieving sustainability. (See also the edited volumes by Keen, Brown, and Dyball [2005] and Wals [2007], which offer insights into social learning for sustainability.) The model was developed in the context of the HarmoniCOP project (harmonizing collaborative planning) (from 2002 to 2005), a wide-ranging investigation of participatory river basin management and planning in nine European countries. The project's purpose was to generate practical information about participation processes, taking an integrated, cross-sectoral approach. The results of the project are reported in a special issue of *Ecology and Society* (2007, v. 12/13), on the project website (http://www. harmonicop.uos.de/index.php), and in several journal papers (e.g., Tippett et al. 2005; Pahl-Wostl et al. 2007). What I focus on here is an aspect of the project that sheds light on whether social learning outcomes are consistent with sustainability criteria.

Tàbara and Pahl-Wostl (2007) developed a model explaining what is required for social learning to also be sustainability learning. Using what they label the SEIC framework, they suggested that for any given social–ecological system, social learning needs to support adaptive governance structures (S) which:

- assure that energy and resource (E) use is consistent with long-range needs and goals;
- generate diverse information and knowledge (I) about the system to ensure adaptability; and
- lead and manage change (C) using the three preceding factors so that the change does not exceed the system's size, thresholds, and connections.

In brief, they argued that sustainability learning would increase the capacity of people, organizations, and systems to anticipate and respond to the unintended, undesired, and irreversible negative impacts of development. They recognized that such learning would need to transform mainstream presuppositions, values, norms, and institutions respecting the role of humans in social–ecological systems. Further, it would need to be holistic (or at least integrative) in its perspective on social–ecological systems. Moreover, sustainability learning would recognize the limits of social–ecological systems, the inherent complexity and uncertainty of such systems, and that multiple scales of analysis are valuable in understanding system dynamics. Additionally, they argued that transdisciplinary understanding of social–ecological systems should be the foundation of sustainability learning.

The practical implications for public participation stemming from social learning orientations to resource and environmental management, such as the three models described above, are numerous. They include imperatives for breadth and diversity of participants, access to complete and understandable information pertinent to the problem, early and ongoing involvement, and high degrees of participation. They also include a need for deliberative

participation mechanisms, emphasizing conflict resolution, future visioning, and opportunities for concerted action. Moreover, social learning orientations imply adaptive and flexible institutional arrangements concerning participation as well as institutional support for fostering transdisciplinary perspectives on resource and environmental issues.

OPPORTUNITIES FOR AND LIMITATIONS OF PARTICIPATORYAND SOCIAL LEARNING APPROACHES

To further understand participatory approaches and social learning orientations to resource and environmental management, it is helpful to examine their advantages or benefits as well as their limitations or disadvantages. The next section summarizes the practical benefits, including types of learning outcomes, as well as benefits related to conflict resolution. The subsequent section reviews limitations and challenges, focusing on efficiency concerns and power imbalances.

Advantages or Benefits

A rich and diverse literature describes the benefits of public participation. Grima and Mason's (1983) early synthesis identified four basic positions:

- political–philosophical, which states that participation is consistent with government in the public interest and citizen involvement in the governing process and therefore helps to strengthen the democratic fabric of society;
- improved planning, a pragmatic view suggesting that participation is an effective means of improving decision-making;
- political market, which holds that in representative democracies, public participation

helps to meet a demand by citizens for direct engagement in the political system; and

- political conflict resolution, in which participation is viewed as a means of helping to resolve conflict or make difficult political decisions more acceptable.

In more practical terms, participatory and social learning approaches to resource and environmental management:

- provide access to local and traditional knowledge from diverse sources;
- bring alternative ethical perspectives into the decision-making process;
- help to define problems and identify solutions;
- broaden the range of potential solutions considered;
- enhance the sense of public ownership, legitimacy, and fairness of decisions taken;
- spread the responsibilities and risks associated with management actions;
- furnish access to new financial, human, and in-kind resources;
- prevent 'capture' of management agencies by the industry being regulated;
- encourage more balanced decision-making;
- increase accountability for decisions made;
- facilitate challenges to illegal or invalid decisions before they are implemented; and
- enable innovative governance and institutional reform (Smith and McDonough 2001; Mitchell 2002; McGurk, Sinclair, and Diduck 2006; Olsson, Folke, and Hughes 2008).

Conflict Resolution

In addition, participatory approaches have a range of practical benefits pertaining to conflict resolution. This is noteworthy, because as discussed earlier, conflict resolution is a key process in social learning. These benefits are particularly evident when emphasis is placed on interactive

techniques serving the continuous exchange and facilitating consensus functions of involvement programs. Moreover, they are most likely to arise when all key publics are involved and have realistic expectations regarding timing and degrees of participation. In such circumstances, participatory approaches to resource and environmental management:

- provide structured forums that can assist in resolving long-standing behavioural conflict (or historical differences);
- emphasize interests (needs, desires, fears, aspirations), central to overcoming entrenched positional bargaining;
- illuminate goals and objectives, necessary for working through value or normative conflict;
- furnish venues for clarifying cognitive conflict (or different ways of understanding a resource problem or situation);
- emphasize trade-offs, compromises, and consensus-based decisions;
- facilitate the building of trust and the creation of collaborative solutions;
- help to avoid costly and time-consuming litigation; and
- reduce the level of controversy associated with a problem or issue (Todd 2001; O'Leary and Bingham 2003).

Learning Outcomes

Evaluation frameworks for public involvement have long included consideration of learning-related benefits. For example, Godschalk and Stiftel (1981) proposed seven criteria for evaluation of public involvement. They included a public-awareness criterion (how knowledgeable did public participants become about the planning program?) and a staff-awareness criterion (how much did agency staff members learn from the information and views provided by participants?). More recently, as noted above, mutual and social learning outcomes have received considerable attention (e.g., Webler, Kastenholz, and Renn 1995; Daniels and Walker 1996; Alexander 1999; Keen, Brown, and Dyball 2005; Ison, Röling, and Watson 2007; Pahl-Wostl et al. 2007). An important aspect of this growing body of research is an emphasis on learning among all participants (including policy-makers, technical experts, and professional managers) in a given resource and environmental management activity. Another is the weight given to non-technical learning outcomes, including clarification of goals, identification of interests, and other aspects of conflict resolution. A third, as mentioned earlier with regard to the HarmoniCOP project, is a focus on the types of outcomes consistent with sustainability criteria.

Webler, Kastenholz, and Renn's (1995) early case study featured these aspects and illustrated the types of learning outcomes that could be derived from a participatory approach to EIA. Examining public involvement in the assessment of a waste disposal facility in the Swiss canton of Aargau, the authors found evidence of both intellectual and moral development. The former included learning about technical, interdisciplinary issues (e.g., waste stream composition and incineration technology), identifying self-interests, recognizing the interests of other citizens, practising integrative thinking, and learning about probabilistic risk analysis (e.g., risk cannot be zero but can be acceptable in certain circumstances). Moral development included developing a sense of self-respect and personal responsibility, being able to empathize with or adopt the perspective of others, developing problem-solving and conflict resolution skills, developing social rationality or a sense of solidarity, and learning how to co-operate with others in solving collective problems.

Limitations and Challenges

Despite the benefits for conflict resolution reviewed above, in some circumstances participatory approaches to resource and environmental management can exacerbate rather than resolve conflict (Parson 2000). As well, if divergent goals are deeply contested and interests are irreconcilable, collaboration and consensus are not realistic or even desirable objectives (Mostert et al. 2007). In such cases, it is more appropriate to focus on formal dispute resolution mechanisms (such as litigation and majority decision-making) and to use other governance tools (such as penalties or incentives).

Another limitation or challenge is that consensus decision-making, a process that seeks general agreement and is often used in participatory approaches, requires a high degree of trust, goodwill, and mutual respect among the parties. Behavioural conflict, or historical differences, can easily derail consensus-based decision processes. Cognitive conflict, as well, can be a significant barrier to consensus-building (Lubell 2000). In addition, since consensus decision-making seeks accommodation and compromise, it may not result in the best decision (measured against some external criteria). It simply results in a decision that all the parties can live with; better solutions are sometimes found through conflict and competition rather than through consensus and compromise (Muro and Jeffrey 2008). A further criticism is that consensus decision processes tend to reinforce the status quo or focus on incremental change, since radical steps are unlikely to be unanimously accepted (Beder 1994).

Power Imbalances

The discussion of breadth of participation referred to the importance of involving key stakeholders and raised two related challenges: the need to establish a legitimate basis for participation and the importance of not reducing broad social welfare concerns to negotiations among stakeholders. The earlier discussion offered responses to these challenges—namely, avoiding overrepresentation of social and economic elites, ensuring that the views of important inactive publics are represented, taking an inclusive approach to identifying stakeholders, and defining clear terms of reference and lines of accountability. In a political sense, these suggestions are means of addressing power imbalances among the stakeholders or key publics in any given resource and environmental management activity. A further suggestion is developing institutional arrangements (e.g., legal or constitutional mechanisms) to ensure procedural fairness and substantive equality for marginalized and disadvantaged groups in society.

As Buchy and Race (2001) and others have argued, addressing power relations should be paramount in developing and analyzing public involvement programs. Doing so can help to prevent public involvement processes from reproducing inequalities and injustices found in larger segments of society or from legitimating forms of governance that privilege narrowly defined economic goals at the expense of public goods and values (Masuda, McGee, and Garvin 2008; Raik, Wilson, and Decker 2008). It can also avert the development of social learning outcomes that do not fairly reflect the knowledge, values, and aspirations of 'laypeople' or non-experts as well as marginalized segments of heterogeneous communities (Armitage, Marschke, and Plummer 2008; Muro and Jeffrey 2008; Healy 2009).

Cost, Time, and Inefficiency

Another challenge to advocates of the participatory or social learning approach is responding to arguments that it is too costly, time-consuming, and inefficient. For example, critics argue that members of the public often expect to discuss questions or issues that have already

been resolved and therefore conclude that public involvement creates unnecessary costs and delays (Mitchell 2002). This is a valid concern but is likely a symptom of poorly designed public involvement programs rather than an outcome of involvement itself (Buchy and Race 2001). Mismatched expectations respecting degrees of involvement, inappropriate timing, badly chosen techniques, and other design errors are sources of inefficiency, not of weaknesses in public involvement itself.

An outcome that does relate more to the very nature of participatory approaches is that they tend to be more time-consuming and costly during initial stages of planning and analysis in comparison to top-down approaches with little or no public engagement (Rosenberg and Korsmo 2001; Rist et al. 2007). However, these initial investments are often returned later in the process because they help to avoid or minimize conflict. One example is when participatory approaches help to forestall formal legal challenges to regulatory approvals. Initial investments of time and money are also returned because they facilitate effective implementation, evaluation, and monitoring through co-operation with local publics. Hence, there could actually be efficiency gains from taking participatory approaches, contrary to the common view that such approaches are inefficient. However, the realization of such gains hinges on responding successfully to the particular challenges associated with implementing co-operative plans and decisions. These challenges include lack of strategic direction, lack of commitment to implementation, and failure to recognize that many objectives require shared decision-making during implementation (Margerum 1999).

Lack of Capacity

Another challenge lies in enhancing individual, organizational, and institutional capacity to conduct participatory resource and environmental management (Kapoor 2001; Sinclair, Diduck, and Fitzpatrick 2008). Stakeholders and other members of key publics sometimes do not have adequate skills, knowledge, and resources to participate in technical analyses of resource allocation problems. Similarly, technical experts and professional managers may lack the communicative skills and grounded knowledge to engage in discussions of values, interests, and local concerns. As well, management agencies can be resistant to participatory approaches because they do not have a vision or supportive culture for participation. Similarly, institutional arrangements present barriers if, for example, they codify restrictions on delegating planning and decision-making authority.

CANADIAN APPLICATIONS

Numerous applications of participatory and social learning approaches to resource and environmental management have been made in Canada. The Mackenzie Valley Pipeline Inquiry of the 1970s was groundbreaking in the extent to which it engaged potentially affected communities and took into account local values, concerns, and knowledge (Berger 1977). In British Columbia, the Commission on Resources and Environment processes of the 1990s involved several innovative experiments with multi-stakeholder planning and shared decision-making (Owen 1998)—although as Chapter 19 reveals, some of these processes still had serious shortcomings in terms of reducing gender-related, systemic barriers to participation. Manitoba's Consultation on Sustainable Development Initiative was a creative 18-month process of public participation at the normative phase of policy development (Sinclair 2003). A sustained civics approach to conservation in the Grand River Valley of southern Ontario helped

to reverse degradation of the river and protect heritage resources, resulting in the Grand being declared a Canadian Heritage River (Nelson and O'Neill 1990). Experiences in Nova Scotia show how community-based ecological monitoring can yield scientifically valid data for informing local planning and decision-making (Sharpe and Conrad 2006). More examples could be cited, but rather than present a comprehensive overview, this section discusses a small number of examples in more depth. The first part summarizes important issues in public involvement in EIA, and ensuing parts examine applications of participatory or social learning approaches in resource and environmental research, forest management, and co-management.

Environmental Impact Assessment

Public involvement has always been an important concern in EIA in Canada and abroad (see Chapter 17). Reflecting this concern, significant advances in public involvement thinking have come in the context of EIA. That said, few would argue that highly participatory approaches have been the norm in EIA in Canada. Exceptions can be found (such as the Mackenzie Valley Pipeline Inquiry, mentioned above), but relatively few cases have involved high degrees of participation (e.g., citizen control, to use Arnstein's language), involvement at normative levels of planning (e.g., during consideration of project need, purposes, and alternatives), or use of alternative dispute resolution. What follows outlines the basic building blocks of public involvement in EIA in Canada, summarizes continuing concerns in the field, and highlights research at the interface of involvement and social learning.

Sinclair and Diduck (2009) identified five fundamental elements of public involvement in EIA and reviewed the extent to which Canadian legislation included provisions for these elements. Their basic building blocks were adequate notice, access to information, participant assistance, opportunities for public comment, and public hearings. *Adequate notice* refers to letting members of active and inactive publics know in a timely manner (i.e., before approval decisions are made) that a proponent has submitted a project proposal. *Access to information* means sharing with interested parties information about the proposal provided by the project proponent, the EIA regulators, and members of the public. *Participant assistance* deals with the provision of funding from the project proponent or the EIA regulators to help active publics improve the effectiveness of their interventions in EIA approval processes. *Reasonable opportunities for comment* enable members of the public to comment on the project proposal, respond to the government's position on the proposal, and react to input from other participants. Finally, *public hearings*, often reserved for large or contentious projects, support decision-making and add certainty and transparency to the EIA process. (The federal and several provincial EIA processes—for example, those in Manitoba, Ontario, Nova Scotia, and Quebec—permit the use of mediation as an alternative to public hearings and at other decision points when there is conflict among the parties; see Box 18.1.)

Sinclair and Diduck (2009) also summarized six enduring concerns with public involvement in EIA in Canada: lack of shared decision-making, lack of involvement at normative and strategic levels of planning, information and communication deficiencies, insufficient resources for participants, accelerated decision processes, and lack of opportunity for involvement in follow-up. With regard to Arnstein's (1969) ladder of citizen participation, activists and environmental non-governmental organizations are often critical of EIA processes that consist only of lower levels of participation and argue that the effectiveness and equity of EIA would improve if public involvement entailed shared decision-making. Regarding the second enduring concern, critics argue that lack of involvement at normative and strategic levels of planning deprives members of the public from having influence on important issues such as project need, purposes, and alternatives.

Box 18.1 The Use of Mediation in EIA in Quebec

Among the provinces, Quebec's EIA process makes the most liberal use of mediation. In the process, any person may make a request for the Bureau d'audiences publiques sur l'environnement (BAPE) to conduct a public hearing. The BAPE is an independent agency in charge of EIA hearings in the province. In reviewing the request and other public input, the BAPE may ask the minister of sustainable development, environment and parks to permit it to use mediation rather than hold a public hearing. The objectives of the mediation are to resolve disagreements over the impacts of proposed projects and to agree on the outstanding issues among the parties, while respecting environmental values and community well-being. In 2004, the BAPE published formal rules of procedure for its mediations, which are available on its website (bape.gouv.qc.ca/sections/lois_reglements/eng_lois_ind.htm).

Source: Bureau d'audiences publiques sur l'environnement n.d.; Sinclair and Diduck 2009.

Information and communication deficiencies, a bundle of issues covering inadequate notice, inaccessibility of project documents, and lack of deliberative mechanisms, hamper informed discussion and decision-making and have an adverse impact on the overall quality of an EIA. Similarly, devoting insufficient resources to participants reduces the likelihood of having critical yet valid alternative points of view represented during the assessment. The fifth persistent concern relates to accelerated decision processes, manifested in short timelines for reviewing EIA documents and making approval decisions. Implemented to improve efficiency in EIA processes, the timelines, critics charge, hamper a careful, critical review of documents by both EIA officials and members of the public. Finally, lack of opportunity for involvement in EIA follow-up procedures is a barrier to members of the public becoming involved in mitigation efforts. It is also a hurdle to learning about the actual impacts of a project, the accuracy of predictions in the original assessment, and the efficacy of the overall EIA process.

The learning implications of public involvement in EIA have been the subject of research in Canada over many years. Studies have:

- highlighted the importance of non-formal education methods in the overall mix of public involvement techniques (Sinclair and Diduck 1995);
- used the principles of critical education to analyze opportunities for involvement and learning (Diduck and Sinclair 1997; Fitzpatrick and Sinclair 2003);
- pursued the transformative learning potential of involvement in EIA (Sinclair and Diduck 2001; Diduck and Mitchell 2003); and
- examined the organizational learning potential of involvement (Fitzpatrick 2006; Hayward, Diduck, and Mitchell 2007).

Sinclair, Diduck, and Fitzpatrick (2008) have offered a conceptual framework showing relationships among non-formal critical education, individual and social learning, public involvement, essential EIA functions, and sustainability criteria (see Figure 18.2). Fundamental contextual variables in the framework are the need for large-scale social change to achieve sustainability, the prevalence of conflict in resource and environmental management, and the inherent uncertainty and complexity of social–ecological systems. The framework views EIA as a valuable resource management function, especially when guided by sustainability principles, objectives, and tools (see Chapter 17). Post hoc monitoring establishes feedback loops for learning from experience, while strategic and cumulative assessments (again see Chapter 17) inform collective understanding of long-term effects and influence future decisions. Opportunities for meaningful involvement (see the features of participatory approaches discussed earlier) enable learning by people and social organizations. Active publics can use non-formal critical education to facilitate their involvement, affect approval decisions, and influence individual and social learning outcomes. These outcomes can be transformative, but it is essential that they be in line with sustainability principles, objectives, and methods.

Resource and Environmental Management Research

Research is an important management function in support of planning, decision-making, assessment, monitoring, evaluation, and problem-solving in general. Participatory approaches to research are gaining prominence in resource and environmental management, particularly for questions dealing with social systems and coupled social–ecological systems (Nelson 1991; Chambers 1997; Wilmsen et al. 2008). Rooted in

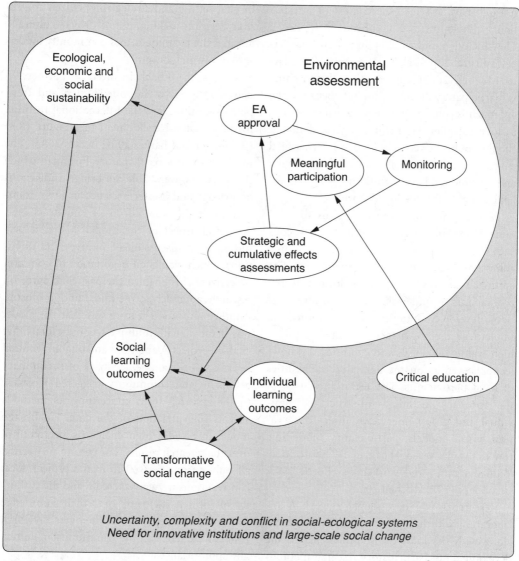

Figure 18.2 A Conceptual Framework of Learning for Sustainability through Participation in Environmental Impact Assessment.

Source: Sinclair et al. 2008.

diverse disciplines, including public health, sociology, planning, adult education, and anthropology, participatory research is often linked to political purposes such as capacity enhancement, socio-political empowerment, and environmental justice (Chambers 1997; Checkland and Holwell 1998; Kemmis and McTaggart 2005; Park et al. 1993).

For resource and environmental management, important streams of participatory research have arisen in the context of international development (Chambers 1994; 1997). They include rapid rural appraisal and participatory rural appraisal. With respect to the latter, Mitchell (2002) suggested that a better name is participatory 'local' appraisal, because the approach is equally applicable to urban settings as to rural ones. He further suggested that it should not be viewed as limited to international development, because its purposes, strategies, and techniques have significant potential for application in the developed world.

Participatory local appraisal aims for sustainable livelihoods and institutions. It is concerned with empowerment and developing community awareness and capabilities. The approach centres on enabling local research participants to undertake their own investigations, develop community-based solutions, and implement their own action plans. It places high value on local and traditional knowledge as well as on the analytical capabilities of local participants. The techniques used in the approach include semi-structured individual and group interviews, direct observation, visual models, participatory modelling, transect walks, seasonal calendars, institutional Venn diagrams, and workshops (Beebe 1995; Chambers 1997; Mitchell 2002).

Researchers who have sought to work in collaboration with indigenous communities have, along with the local communities, led in the development of participatory research methodologies (Jackson 1993; Smith 2005). A proliferation of work in the Canadian North has provided insight into models for participatory research on environmental change (Berkes 2002). Emphasizing local observations and place-based research, these studies have sought the integration of scientific with local and traditional knowledge. Truly community-based, several models are characterized by broad local participation

in goal-setting and strategizing, citizen control over decision-making, use of diverse culturally appropriate techniques, and community ownership of research outputs.

One such model involved a partnership between the Sachs Harbour Hunters and Trappers Committee and the International Institute for Sustainable Development (see Figure 18.3) (Reidlinger and Berkes 2001; Berkes 2002). The model created a participatory forum in which the partners were able to bridge indigenous knowledge and Western science in their examination of the local effects of climate change. A central aspect of the model is that it permitted open discussion of motives and intentions, which facilitated negotiation of common objectives. This enabled the parties to agree on research approaches, methods, and protocols. Another important feature was that the model was iterative, with multiple feedback loops that enabled project adaptations and ongoing mutual learning. For example, feedback of preliminary results to the community triggered feedback to the research team concerning its methods and its interpretation of the data. Finally, the model involved a high degree of local control over the research results. The community vetted all publications derived from the project, local contributions were given credit, and the authority of local participants over their knowledge was acknowledged. As well, the overall research results were left with the community in culturally appropriate ways.

To be sure, 'there is no magic way of constructing new models of community-based research, and there certainly is more than one way to do it' (Berkes 2002, 343). However, the model described above is a good one because it enabled the participants to work successfully through the political and philosophical challenges of sharing ideas, concepts, and goals across knowledge systems.

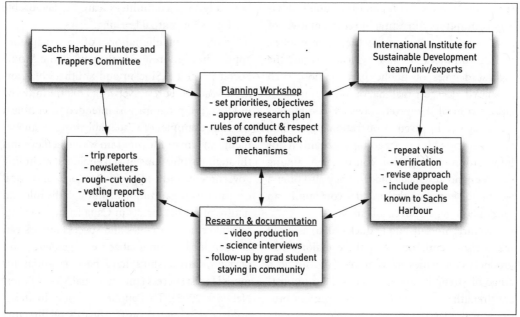

Source: Berkes 2002.

Figure 18.3 A Partnership Model for Research That Can Combine Western Science and Indigenous Knowledge.

Forest Management

As discussed in Chapters 10, 11, and 19, new perspectives and emphases have emerged in Canadian forest management over the past 15 to 20 years. Concepts such as 'multiple-use sustained yield' and practices such as clear-cutting are changing in response to the theory and practice of sustainable forest management. Sustainable forest management involves maintaining and enhancing the long-term health of forest ecosystems for the benefit of all living things while providing environmental, economic, social, and cultural opportunities for present and future generations (Canadian Council of Forest Ministers 1998). An implication is that an integrated, ecosystem approach to management is required. Moreover, given ecosystem and social complexities (and related uncertainties), an adaptive approach is needed (Chapter 16). This further

suggests a need for monitoring and local criteria as well as indicators that characterize sustainable forest management (Duinker 2001; also see Chapter 3).

The implications for institutional arrangements are that they should be decentralized and flexible, involving networks of organizations linked across spatial scales. This necessitates partnerships in the forest community and broad involvement from diverse sectors of society (Parkins 2006). In fact, after some initial growing pains (e.g., ineffective implementation, resistance to high degrees of involvement, and lack of breadth among the participating publics [Duinker 1998; Robinson, Robson, and Rollins 2001]), partnership or participatory approaches have become increasingly common.

A specific manifestation of participatory approaches is increased use by forest licensees (i.e., forest products companies) of public advisory committees consisting of representatives of local stakeholder groups. As the name implies, such committees do not typically entail the highest degrees of public participation, such as shared decision-making or other forms of citizen control. However, advisory committees often become involved in operational planning, and they provide an ongoing mechanism for information-sharing and consultation among licensees and key stakeholders. They also offer a vehicle for engaging the broader community in forest planning and decision-making.

McGurk, Sinclair, and Diduck (2006) assessed the advisory committees of three major forest products companies in Manitoba and found a range of strengths and weaknesses. An important strength with respect to the committees' processes was that they used multiple techniques to involve their members. Another was that the committees were open and transparent when it came to information-sharing and communication. A notable process weakness, however, was that committee members were not sufficiently accountable to their organizations for their committee activities. A second was that stakeholder committees did not adequately accommodate the status of First Nations as unique governments.

An important strength regarding the outcomes of the committees' work was the breadth of learning that occurred among the committee members. On a related note, a second outcome strength was the development of new, trusting relationships. Important outcome weaknesses were lack of influence on normative and strategic management decisions and failure to foster extensive involvement among members of the public beyond the stakeholder groups. While the results were mixed, the study confirmed that advisory committees hold promise as a method of actively involving a select group of stakeholders in forest management. It also demonstrated

how advisory committees can be favourable platforms for mutual learning.

Another manifestation of participatory approaches in forest management is the Model Forest Program (Sinclair and Smith 1999; Canadian Model Forest Network 2009). Initiated in 1992, the program was intended to establish working examples of sustainable forest management. An aim of the program was to achieve this through partnerships of stakeholders, including government, industry, First Nations, and local community organizations. Although the role and nature of model forests in Canada are evolving, several critical variables and success factors can be identified. From a citizen engagement point of view, partnerships have been essential for model forest success (International Model Forest Network 2005). This applies not only to stakeholders represented on the boards of directors but also to relations among model forests and broader publics and resource communities. Factors underpinning partnership success include having a broad range of stakeholders (including non-traditional partners such as religious institutions and the military), using participatory, transparent, and accountable decision-making structures, identifying collective shared benefits, resolving conflict through building consensus, engaging in dialogue and mutual learning, and being oriented toward concrete action.

Co-management

Important applications of participatory approaches to resource and environmental management have developed under the umbrella of co-management. Various definitions of co-management can be found, but a good starting point is offered by the International Union for the Conservation of Nature (1996, 1): co-management is 'a partnership in which government agencies, local communities and resource users, nongovernmental organizations and other stakeholders negotiate, as appropriate to each context, the authority and responsibility for the management of a specific area or set of resources.'

Although the term co-management has been used to describe everything from purely consultative arrangements to treaty rights recognizing co-equal status (Berkes 2007; 2009), general characteristics can be identified. Co-management arrangements typically include decision-makers other than state or industry managers, encourage participation of local resource-users, stress negotiation rather than litigation in situations of conflict, and try to combine Western science with traditional and local knowledge. A central feature of co-management is an agreement that specifies the rights and responsibilities of the parties with respect to an explicit management plan. Ideally, the agreement should be flexible, with mechanisms for ongoing review and amendment (Berkes et al. 2001).

Consistent with much of the discussion throughout this chapter, an important issue in co-management relates to degrees of participation, or the extent to which decision-making power is shared (Borrini-Feyerabend et al. 2004). With the Arnstein ladder of participation as a model, Berkes (1995) presented seven levels of co-management arrangements, ranging from weak systems that consist of token government consultations with local-resource users to full community control. At the bottom level (informing), local users are informed about rules and regulations. Level two (consultation) involves face-to-face contact as communities are consulted about projects. At level three (co-operation), managers begin to make some use of traditional and local knowledge. At level four (communication), real dialogue starts, and local concerns begin to enter management agendas. Level five (advisory committees) involves shared decision-making, including a search for common objectives. With level six (management

boards), local communities are actively involved in policy-making as well as decision-making. Finally, at the top level (community control and partnership), joint decision-making is fully institutionalized.

Echoing another of the basic concerns raised earlier, Houde (2007) argued that to fully account for the richness, depth, and complexity of traditional ecological knowledge, co-management arrangements in Canada need to be designed in such a way that Aboriginal communities are involved from the initial stages of decision-making—i.e., at the normative and strategic levels of planning. Such early involvement would facilitate increased Aboriginal control of traditional ecological knowledge and enable a greater sense of Aboriginal empowerment with respect to resource and environmental issues (see also Chapter 4). To achieve early involvement and see the full range of benefits from traditional ecological knowledge, flexible co-management frameworks are required that emphasize adaptation in response to lessons learned about the social–ecological systems in question and about each of the partners' needs and values.

Examples of co-management arrangements can be found throughout the country and in relation to various resource sectors. The majority of Canadian experiences, however, have been with respect to fish and wildlife management. The Beverly and Qamanirjuaq Caribou Management Board (BQCMB) is a longstanding Canadian co-management organization (Kendrick 2000; Beverly and Qamanirjuaq Caribou Management Board 2009). Founded in 1982, the BQCMB helps to manage the Beverly and Qamanirjuaq herds of barren-ground caribou, whose migratory routes include parts of Nunavut, the Northwest Territories, Saskatchewan, Manitoba, and Alberta. The range used by the Beverly herd over the past 60 years extends almost 1,000 kilometres from north to south and at least 600 kilometres from east to west. The extent of the range of the Qamanirjuaq herd has been similar: more than 1,000 kilometres from north to south and about 500 kilometres from east to west.

In 1994, population estimates numbered the herds at up to 776,000 animals. Use of the herds is very important for sustaining the livelihoods, culture, and lifestyles of about 20 Inuit, Dene, and Métis communities. The BQCMB is composed of members of communities from across the ranges of both herds and from the governments of Manitoba, Saskatchewan, the Northwest Territories, and Nunavut. The board is advisory and makes recommendations to government as well as conducting education, conservation, and management projects. Projects undertaken include development of comprehensive management plans, research into caribou habitat, production of educational materials for use in schools, publication of a newsletter, development of a hunters' code of ethics, and monitoring of caribou habitat and community use.

From the BQCMB and other co-management experiences, numerous factors that contribute to the success of co-management arrangements can be identified. A subset has implications respecting social learning. Dale (1989) identified factors contributing to shifts in mental frameworks that enable negotiation of co-management arrangements. They include having consistent advocates of co-operative approaches to resolving resource problems, having opportunities to reflect on and adapt to existing problems, and being presented with an external crisis outside the control of the principal actors (e.g., divisive court battles or threats to the resource from habitat loss). Berkes (2009, 1699) viewed social learning as crucial both for collaboration among co-management partners and as an outcome of that collaboration, and he suggested that 'co-management that does not learn often becomes a failed experiment.' Borrini-Feyerabend et al. (2004)

identified elements and processes that support learning by co-management organizations:

- having all key stakeholders involved in the initial negotiation process;
- producing and institutionalizing an organizational vision of a desirable future;
- developing objectives, strategies, and action plans for achieving the vision;
- identifying expected results and indicators with respect to each plan;
- implementing the plan and then monitoring and evaluating actions taken;
- modifying actions, plans, and agreements based on evaluation results; and
- engaging with resource-users throughout these processes in a participatory and ongoing manner.

The learning implications of co-management have come to the fore with the emerging concept of adaptive co-management. This concept, which integrates ideas from the adaptive management (see Chapter 16) and co-management literatures, emphasizes experiential learning (at both individual and social levels) through place-based, cross-scale, participatory governance (Olsson, Folke, and Berkes 2004; Armitage, Berkes, and Doubleday 2007). Reflecting many of the characteristics of participatory approaches discussed earlier, the conditions for successful adaptive co-management include having:

- stakeholders with shared interests;
- a long-term commitment to shared governance;
- resources for enhancing stakeholder capacity;
- participants who draw upon multiple knowledge systems and sources; and
- an overarching policy framework supportive of collaborative governance (Armitage et al. 2009).

CONCLUSION

Participatory approaches and social learning orientations are emerging as responses to complex resource and environmental problems rife with uncertainty and conflict. Participatory approaches are characterized by inclusive definitions of stakeholders and key publics, early and ongoing participation, citizen control over decision-making, and use of multiple involvement techniques, including interactive mechanisms. Social learning orientations emphasize collective learning by groups, organizations, and communities as a means of adapting to and driving social–ecological change. In general, highly participatory approaches provide the conditions for social learning. A range of practical benefits is associated with participatory and social learning approaches, including a subset related to conflict resolution. At the same time, serious challenges exist, such as overcoming power imbalances and building trust and goodwill for consensus-based decision-making. Although top-down management is still the norm, numerous Canadian applications of participatory and social learning approaches can be found. They include models of participatory research, co-management arrangements, and forest management stakeholder advisory committees. As well, public involvement in EIA in Canada has a long record of accomplishment, but enduring concerns remain about some of its basic elements.

The social learning literature has proliferated in recent years, shedding light on the development of collective cognition, processes of collaborative inquiry and concerted action, and the types of learning outcomes consistent with sustainability criteria. This third line of inquiry is just emerging, requiring further empirical work to develop inductive place-based models. Another matter needing further study in social learning is how to address power differentials so as to ensure equitable participation processes and learning outcomes.

An aspect of public involvement not discussed in this chapter is the importance of being adaptive and strategic in designing involvement programs. The chapter is premised on the need for participatory and social learning approaches when faced with high complexity, uncertainty, and conflict. However, most resource and environmental theoreticians and practitioners would agree that such approaches are not always needed or appropriate. When stakes are low, the goals clear, and technology abundant, participatory and social learning approaches are likely unnecessary. Programmed, expert-driven management with limited public involvement could be the best response in such cases. As well, involvement programs often need to be tailored to match the specific objectives of the given management initiative. However, participatory approaches and social learning orientations to resource and environmental management are necessary in the transition to sustainability. They are required for developing multifaceted responses to today's complex and seemingly intractable resource problems, and they are foundations for research that seeks transdisciplinary frameworks and holistic understanding of social–ecological systems.

FROM THE FIELD

Revisiting Education for Involvement

Before becoming an academic, I worked for a community-based organization that promoted civic and legal competence through public legal education and was part of a network of similar groups across Canada. Some of our work focused on empowering traditionally disadvantaged communities to exercise and defend their legal rights and become involved in legislative reform.

These experiences have informed my academic work. My initial research examined critical non-formal education as one way to facilitate meaningful public involvement in environmental assessment. Subsequent research has evolved, encompassing other resource management functions (e.g., land-use planning and ecological monitoring) and emphasizing learning *through* involvement rather than education *for* involvement.

The learning implications of involvement are very important and deserve continued attention; multiple-level learning (learning at and across various levels of social organization) is part of the answer to achieving sustainability. However, for me, it is also crucial to consider unanswered questions regarding critical non-formal education:

Does it offer real potential for empowering marginalized communities and groups and for addressing environmental justice issues that arise in resource governance and management?

Given the short decision-making timelines often experienced in resource governance and management, how practical is critical non-formal education, which usually requires capacity enhancement at the grassroots level?

What should the role of government be in supporting non-formal education aimed at improving civic engagement in governance and management and redressing environmental, economic, and social injustices?

In response to the last question, I believe that in a pluralistic, democratic society purporting to seek social justice and sustainability, government has an obligation to be active and supportive, as occurred in Canada during the 1980s through the national public legal education network (and related initiatives like the Court Challenges Program). If that position is credible, a pertinent question becomes how governments in Canada can support critical non-formal education in resource governance and management. An obvious answer is to work with an existing community-based organization with national reach and a record of accomplishment. A candidate is the Canadian Environmental Network, with a broad membership of community groups across Canada and working relationships with Environment Canada and other government departments.

A national program of critical non-formal education would provide excellent opportunities for building and enhancing capacity for involvement in resource governance and management in strategically selected locations. It would also furnish opportunities for basic research, especially if the program were of sufficient duration to permit the study of changes in community capacity and socio-economic relations and the impacts of participation on resource governance and decision-making. In Manitoba, promising locations for implementing the program exist in the context of the province's ongoing integrated watershed planning strategy, which envisions collaborative planning processes involving stakeholders with divergent values, goals, interests, and capacities.

Another promising location is in the context of the Wabanong Nakaygum Okimawin, a broad-area planning process covering the east side of Lake Winnipeg involving the provincial government, First Nations, rural municipalities, industry, and environmental stakeholders.

As discussed throughout this chapter, participatory, learning-oriented approaches to resource governance and management are taking hold in Canada. This is a positive development and reinforces the importance of empowering marginalized communities and groups. Further, it bolsters my long-held belief that government should support capacity development that gives voice to perspectives critical of the status quo.

—ALAN DIDUCK

Acknowledgements

I would like to thank Glen Hostetler of the Natural Resources Institute, University of Manitoba, for his assistance in conducting the literature search for the chapter. I also wish to thank Bruce Mitchell for his helpful suggestions for revising the chapter.

REVIEW QUESTIONS

1. What are the defining features of Christenson's (1985) 'chaotic' and 'bargaining' planning conditions? What did Cardinall and Day (1998) mean by the terms 'decision stakes' and 'management uncertainty'? What is the basic argument in favour of participatory and social learning approaches to resource and environmental management?

2. What variables describe the characteristics of participatory approaches and the conditions for social learning? What are platforms of social learning? What type of learning is required for social learning to also be sustainability learning, according to Tàbara and Pahl-Wostl (2007)?

3. What are the advantages and disadvantages of participatory and social learning approaches in resource and environmental management, and in your opinion do the advantages outweigh the disadvantages?

4. Summarize two examples of participatory and social learning approaches described in the chapter. Do you know of any similar experiences in your area?

REFERENCES

Alexander, D. 1999. 'Planning as learning: Sustainability and the education of citizen activists'. *Environments* 27 (2): 79–87.

Armitage, D., et al. 2009. 'Adaptive co-management for social-ecological complexity'. *Frontiers in Ecology and the Environment* 7(2) 95–102.

Armitage, D., F. Berkes, and N. Doubleday, eds. 2007. *Adaptive Co-management: Collaboration,* *Learning, and Multi-Level Governance.* Vancouver: University of British Columbia Press.

Armitage, D., M. Marschke, and R. Plummer. 2008. 'Adaptive co-management and the paradox of learning'. *Global Environmental Change* 18 (1): 86–98.

Arnstein, S. 1969. 'A ladder of citizen participation'. *Journal of the American Institute of Planners* 35 (4): 216–24.

Beder, S. 1994. 'Consensus or conflict?' *Ecologist* 24 (6): 236–7.

Beebe, J. 1995. 'Basic concepts and techniques of rapid appraisal'. *Human Organization* 54 (1): 42–51.

Berger, T.R. 1977. *Northern Frontier, Northern Homeland: The Report of the Mackenzie Valley Pipeline Inquiry.* Ottawa: Minister of Supply and Services.

Berkes, F. 1995. 'Community-based management and co-management as tools for empowerment'. In N. Singh and V. Titi, eds, *Empowerment towards Sustainable Development*, 138–46. London: Zed Books.

———. 2002. 'Making sense of Arctic environmental change?' In I. Krupnik and D. Jolly, eds, *The Earth Is Faster Now: Indigenous Observations of Arctic Environmental Change*, 335–49. Fairbanks, AK: Arctic Research Consortium of the United States.

———. 2007. 'Adaptive co-management and complexity: Exploring the many faces of co-management'. In D. Armitage, F. Berkes, and N. Doubleday, eds, *Collaboration, Learning, and Multi-Level Governance*, 19–37. Vancouver: University of British Columbia Press.

———. 2009. 'Evolution of co-management: Role of knowledge generation, bridging organizations and social learning'. *Journal of Environmental Management* 90 (5): 1692–702.

Berkes, F., et al. 2001. *Managing Small-Scale Fisheries: Alternative Directions and Methods.* Ottawa: International Development Research Centre.

Beverly and Qamanirjuaq Caribou Management Board. 2009. 'Beverly and Qamanirjuaq Caribou Management Board'. www.arctic-caribou.com.

Borrini-Feyerabend, G., et al. 2004. *Sharing Power: Learning by Doing in Co-management throughout the World.* Cenesta, Tehran: IIED and IUCN/CEESP/CMWG.

Buchy, M., and D. Race. 2001. 'The twists and turns of community participation in natural resource management in Australia: What is missing?' *Journal of Environmental Planning and Management* 44 (3): 293–308.

Bureau d'audiences publiques sur l'environnement. n.d. 'English section'. bape.gouv.qc.ca/sections/english.

Canadian Council of Forest Ministers. 1998. *National Forest Strategy 1998–2003: Sustainable Forests: A Canadian Commitment.* Ottawa: Canadian Council of Forest Ministers.

Canadian Model Forest Network. 2009. 'The Canadian Model Forest Network'. www.modelforest.net/cmfn/en.

Cardinall, D., and J.C. Day. 1998. 'Embracing value and uncertainty in environmental management and planning: A heuristic model'. *Environments* 25 (2/3): 110–25.

Chambers, R. 1994. 'The origins and practice of participatory rural appraisal'. *World Development* 22 (7): 953–69.

———. 1997. *Whose Reality Counts? Putting the First Last.* London: Intermediate Technology.

Checkland, P., and S. Holwell. 1998. 'Action research: Its nature and validity'. *Systemic Practice and Action Research* 11 (1): 9–21.

Christenson, K.S. 1985. 'Coping with uncertainty in planning'. *Journal of the American Planning Association* 51 (1): 63–73.

Dale, N. 1989. 'Getting to co-management: Social learning in the redesign of fisheries management'. In E. Pinkerton, ed., *Co-operative Management of Local Fisheries*, 49–72. Vancouver: University of British Columbia Press.

Daniels, S.E., and G.B. Walker. 1996. 'Collaborative learning: Improving public deliberation in ecosystem-based management'. *Environmental Impact Assessment Review* 16 (1): 71–102.

Diduck, A.P., and B. Mitchell. 2003. 'Learning, public involvement and environmental assessment: A Canadian case study'. *Journal of Environmental Assessment Policy and Management* 5 (3): 339–64.

Diduck, A.P., and A.J. Sinclair. 1997. 'The concept of critical environmental assessment (EA) education'. *The Canadian Geographer* 41 (4): 294–307.

———. 2002. 'Public participation in environmental assessment: The case of the nonparticipant'. *Environmental Management* 29 (4): 578–88.

Dorcey, A.H.J. 1986. *Bargaining in the Governance of Pacific Coastal Resources: Research and Reform.* Vancouver: University of British Columbia, Westwater Research Centre.

Dorcey, A.H.J., L. Doney, and H. Ruggeberg. 1994. *Public Involvement in Government Decision-Making: Choosing the Right Model.* Victoria, BC: Round Table on the Environment and the Economy.

Dorcey, A.H.J., and T. McDaniels. 2001. 'Great expectations, mixed results: Trends in citizen involvement in Canadian environmental governance'. In E.A. Parson, ed., *Governing the Environment: Persistent Challenges, Uncertain Innovations*, 247–302. Toronto: University of Toronto Press.

Duinker, P.N. 1998. 'Public participation's promising progress: Advances in forest decision-making in

Canada'. *Commonwealth Forestry Review* 77 (2): 107–12.

———. 2001. 'Criteria and indicators of sustainable forest management in Canada: Progress and problems in integrating science and politics at the local level'. In A. Franc, O. Laroussinie, and T. Karjalainen, eds, *Criteria and Indicators for Sustainable Forest Management at the Forest Management Unit Level*, 7–27. Joensuu, Finland: European Forest Institute.

Fitzpatrick, P. 2006. 'In it together: Organizational learning through participation in environmental assessment'. *Environmental Assessment Policy and Management* 8 (2): 157–82.

Fitzpatrick, P., and A.J. Sinclair. 2003. 'Learning through public involvement in environmental assessment hearings'. *Journal of Environmental Management* 67 (2): 161–74.

Folke, C., et al. 2005. 'Adaptive governance of social-ecological systems'. *Annual Review of Environment and Resources* 30: 441–73.

Funtowicz, S., and J. Ravetz. 1993. 'Science for the post-normal age'. *Futures* 25 (7): 739–57.

Godschalk, D.R., and B. Stiftel. 1981. 'Making waves: Public participation in state water planning'. *The Journal of Applied Behavioral Science* 17 (4): 597–614.

Grima, A.P., and R.J. Mason. 1983. 'Apples and oranges: Toward a critique of public participation in Great Lakes decisions'. *Canadian Water Resources Journal* 8 (1): 22–50.

Hayward, G., A.P. Diduck, and B. Mitchell. 2007. 'Social learning outcomes in the Red River Floodway environmental assessment'. *Environmental Practice* 9 (4): 239–50.

Healy, S. 2009. 'Toward an epistemology of public participation'. *Journal of Environmental Management* 90 (4): 1644–54.

Holling, C.S., ed. 1978. *Adaptive Environmental Assessment and Management*. Chichester: John Wiley and Sons.

Houde, N. 2007. 'The six faces of traditional ecological knowledge: Challenges and opportunities for Canadian co-management arrangements'. *Ecology and Society* 12 (2): 34. www.ecology andsociety.org/vol12/iss2/art34.

International Association for Public Participation. 2006. 'IAP2's public participation toolbox'. www.iap2.org/associations/4748/files/06Dec_Toolbox.pdf.

International Model Forest Network. 2005. *Partnerships to Success in Sustainable Forest Management*. Ottawa: International Model Forest Network.

International Union for the Conservation of Nature. 1996. 'IUCN Resolution 1.42 on Collaborative Management for Conservation'. cmsdata.iucn.org/downloads/resolutions_recommendation_en.pdf.

Ison, R., and D. Watson. 2007. 'Illuminating the possibilities for social learning in the management of Scotland's water'. *Ecology and Society* 12 (1): 21. http://www.ecologyandsociety.org/vol12/iss1/art21.

Ison, R., N. Röling, and D. Watson. 2007. 'Challenges to science and society in the sustainable management and use of water: Investigating the role of social learning'. *Environmental Science and Policy* 10 (6): 499–511.

Jackson, T. 1993. 'A way of working: Participatory research and the Aboriginal movement in Canada'. In P. Park, M. Brydon-Miller, B. Hall, and T. Jackson, eds, *Voice of Change: Participatory Research in the United States and Canada*, 47–64. London: Bergin and Garvey.

Kapoor, I. 2001. 'Towards participatory environmental management?' *Journal of Environmental Management* 63 (3): 269–79.

Keen, M., V.A. Brown, and R. Dyball, eds. 2005. *Social Learning in Environmental Management: Towards a Sustainable Future*. London: Earthscan.

Kemmis, S., and R. McTaggart. 2005. 'Participatory action research: Communicative action and the public sphere'. In N.K. Denzin and Y.S. Lincoln, eds, *The Sage Handbook of Qualitative Research*, 3rd edn, 559–604. Thousand Oaks, CA: Sage.

Kendrick, A. 2000. 'Community perceptions of the Beverly-Qamanirjuaq Caribou Management Board'. *Canadian Journal of Native Studies* 20 (1): 1–33.

Larson, K.L., and D. Lach. 2007. 'Participants and non-participants of place-based groups: An assessment of attitudes and implications for public participation in water resource management'. *Journal of Environmental Management* 88 (4): 817–30.

Lawrence, P.L., and J.G. Nelson. 1999. 'Great Lakes and Lake Erie floods: A life cycle and civics perspective'. *Environments* 27 (1): 1–22.

Lee, K.N. 1993. *Compass and Gyroscope: Integrating Science and Politics for the Environment*. Washington: Island Press.

Leeuwis, C., and R. Pyburn, eds. 2002. *Wheel-Barrows Full of Frogs: Social Learning in Rural Resource*

Management. Assesn, Netherlands: Koninklijke Van Gorcum.

Lowndes, V., L. Pratchett, and G. Stoker. 2001. 'Trends in public participation: Part 2—citizens' perspectives'. *Public Administration* 79 (2): 445–55.

Lubell, M. 2000. 'Cognitive conflict and consensus building in the National Estuary Program'. *American Behavioral Scientist* 44 (4): 629–48.

Ludwig, D. 2001. 'The era of management is over'. *Ecosystems* 4 (8): 758–64.

Maarleveld, M., and C. Dangbégnon. 1999. 'Managing natural resources: A social learning perspective'. *Agriculture and Human Values* 16: 267–80.

McGurk, B., A.J. Sinclair, and A.P. Diduck. 2006. 'An assessment of forest management stakeholder advisory committees: Case studies from Manitoba, Canada'. *Society and Natural Resources* 19 (9): 809–26.

Margerum, R.D. 1999. 'Getting past yes: From capital creation to action'. *Journal of the American Planning Association* 65 (2): 181–92.

Masuda, J.R., T.K. McGee, and T.D. Garvin. 2008. 'Power, knowledge, and public engagement: Constructing "citizenship" in Alberta's industrial heartland'. *Journal of Environmental Policy and Planning* 10 (4): 359–80.

Mitchell, B. 2002. *Resource and Environmental Management*. 2nd edn. Harlow, UK: Pearson Education.

Mostert, E., et al. 2007. 'Social learning in European river-basin management: Barriers and fostering mechanisms from 10 river basins'. *Ecology and Society* 12 (1): 19. http://www.ecologyandsociety.org/vol12/iss1/art19.

Muro, M., and P. Jeffrey. 2008. 'A critical review of the theory and application of social learning in participatory natural resource management processes'. *Journal of Environmental Planning and Management* 51 (3): 325–44.

Nelson, J.G. 1991. 'Research in human ecology and planning: An interactive, adaptive approach'. *The Canadian Geographer* 35 (2): 114–27.

———. 1995. 'Natural and cultural heritage planning, protection and interpretation: From ideology to practice, a civics approach'. In J. Marsh and J. Fialkowski, eds, *Linking Cultural and Natural Heritage*, 33–43. Peterborough, ON: The Frost Centre For Canadian Heritage Development Studies, Trent University.

Nelson, J.G., and P.C. O'Neill, eds. 1990. *Nominating the Grand as a Canadian Heritage River: A Study for the Canadian Heritage Rivers Board and the Grand River Conservation Authority*. Occasional Paper 13. Waterloo, ON: Heritage Resources Centre, University of Waterloo.

Nelson, J.G., and R. Serafin. 1995. 'Post hoc assessment and environmental planning, management and decision making'. *Environments* 23 (1): 3–9.

———. 1996. 'Environmental and resource planning and decision making in Canada: A human ecological and a civics approach'. In R. Vogelsang, ed., *Canada in Transition: Results of Environmental and Human Geographical Research*, 1–25. Bochum: Universitatsverlag Dr. N. Brockmeyer.

O'Leary, R., and L. Bingham. 2003. *The Promise and Performance of Environmental Conflict Resolution*. Washington: Resources for the Future.

Olsson, P., C. Folke, and F. Berkes. 2004. 'Adaptive comanagement for building resilience in social-ecological systems'. *Environmental Management* 34 (1): 75–90.

Olsson, P., C. Folke, and T.P. Hughes. 2008. 'Navigating the transition to ecosystem-based management of the Great Barrier Reef, Australia'. *Proceedings of the US National Academy of Sciences* 105 (28): 9489–94.

Owen, S. 1998. 'Land use planning in the nineties: CORE lessons'. *Environments* 25 (2/3): 14–26.

Pahl-Wostl, C., et al. 2007. 'Social learning and water resources management'. *Ecology and Society* 12 (5): 5. http://www.ecologyandsociety.org/vol12/iss2/art5.

Park, P., et al. 1993. *Voice of Change: Participatory Research in the United States and Canada*. London: Bergin and Garvey.

Parkins, J.R. 2006. 'De-centering environmental governance: A short history and analysis of democratic processes in the forest sector of Alberta, Canada'. *Policy Sciences* 39 (2): 183–203.

Parson, E.A. 2000. 'Environmental trends and environmental governance in Canada'. *Canadian Public Policy* 26 (2): S123–S143.

Petts, J. 2004. 'Barriers to participation and deliberation in risk decisions: Evidence from waste management'. *Journal of Risk Research* 7 (2): 115–33.

Raik, D.B., A.L. Wilson, and D.J. Decker. 2008. 'Power in natural resources management: An application of theory'. *Society and Natural Resources* 21 (8): 729–39.

Reidlinger, D., and F. Berkes. 2001. 'Contributions of traditional knowledge to understanding climate change in the Canadian Arctic'. *Polar Record* 37 (203): 315–28.

Rist, S., et al. 2007. 'Moving from sustainable management to sustainable governance of natural resources: The role of social learning processes in rural India, Bolivia and Mali'. *Journal of Rural Studies* 23 (1): 23–37.

Robinson, D., M. Robson, and R. Rollins. 2001. 'Towards increased citizen influence in Canadian forest management'. *Environments* 29 (2): 21–41.

Röling, N.G. 2002. 'Beyond the aggregation of individual preferences'. In C. Leeuwis and R. Pyburn, eds, *Wheel-Barrows Full of Frogs: Social Learning in Rural Resource Management*, 25–47. Assesn, Netherlands: Koninklijke Van Gorcum.

Röling, N., and M. Maarleveld. 1999. 'Facing strategic narratives: An argument for interactive effectiveness'. *Agriculture and Human Values* 16 (3): 295–308.

Rosenberg, J., and F.L. Korsmo. 2001. 'Local participation, international politics, and the environment: The World Bank and the Grenada dove'. *Journal of Environmental Management* 62 (3): 283–300.

Steyaert, P., and J. Jiggins. 2007. 'Governance of complex environmental situations through social learning: A synthesis of SLIM's lessons for research, policy and practice'. *Environmental Science and Policy* 10 (6): 575–86.

Sharpe, A., and C. Conrad. 2006. 'Community based ecological monitoring in Nova Scotia: Challenges and opportunities'. *Environmental Monitoring and Assessment* 113: 395–409.

Sinclair, A.J. 2003. 'Public consultation for sustainable development policy initiatives: Manitoba approaches'. *Policy Studies Journal* 30 (4): 423–44.

Sinclair, A.J., and A.P. Diduck. 1995. 'Public education: An undervalued component of the environmental assessment public involvement process'. *Environmental Impact Assessment Review* 15 (3): 219–40.

———. 2001. 'Public involvement in EA in Canada: A transformative learning perspective'. *Environmental Impact Assessment Review* 21 (2): 113–36.

———. 2009. 'Public participation in Canadian environmental assessment: Enduring challenges and future directions'. In K.S. Hanna, ed., *Environmental Impact Assessment: Practice and Participation*, 2nd edn., 58–82. Toronto: Oxford University Press.

Sinclair, A.J., A.P. Diduck, and P. Fitzpatrick. 2002. 'Public hearings in environmental assessment: Towards a civics approach'. *Environments* 30 (1): 17–36.

———. 2008. 'Conceptualizing learning for sustainability through environmental assessment: Critical reflections on 15 years of research'. *Environmental Impact Assessment Review* 28 (7): 415–28.

Sinclair, A.J., and D.L. Smith. 1999. 'The Model Forest Program in Canada: Building consensus on sustainable forest management?' *Society and Natural Resources* 12 (2): 121–38.

Smith, L.G. 1982. 'Mechanisms for public participation at a normative planning level in Canada'. *Canadian Public Policy* 8 (4): 561–72.

Smith, L.T. 2005. 'On tricky ground: Researching the native in the age of uncertainty'. In N.K. Denzin and Y.S. Lincoln, eds, *The Sage Handbook of Qualitative Research*, 3rd edn, 85–107. Thousand Oaks, CA: Sage.

Smith, P.D., and M.H. McDonough. 2001. 'Beyond public participation: Fairness in natural resource decision making'. *Society and Natural Resources* 14 (3): 239–49.

Tàbara, J.D., and C. Pahl-Wostl. 2007. 'Sustainability learning in natural resource use and management'. *Ecology and Society* 12 (2): 3. http://www.ecologyandsociety.org/vol12/iss2/art3.

Tippett, J., et al. 2005. 'Social learning in public participation in river basin management: Early findings from HarmoniCOP European case studies'. *Environmental Science and Policy* 8 (3): 287–99.

Todd, S. 2001. 'Measuring the effectiveness of environmental dispute settlement efforts'. *Environmental Impact Assessment Review* 21 (1): 97–110.

Wals, A.E.J. 2007. *Social Learning towards a Sustainable World: Principles, Perspectives, and Praxis.* Wageningen, Netherlands: Wageningen Academic Publishers.

Webler, T., H. Kastenholz, and O. Renn. 1995. 'Public participation in impact assessment: A social learning perspective'. *Environmental Impact Assessment Review* 15 (5): 443–63.

Wilmsen, C., et al. 2008. *Partnerships for Empowerment: Participatory Research for Community-Based Natural Resource Management.* London: Earthscan.

Wynne, B. 1992. 'Uncertainty and environmental learning: Reconceiving science and policy in the preventative paradigm'. *Global Environmental Change* 2 (2): 111–27.

19

Environmental Governance and Gender in Canadian Resource Industries and Communities

Maureen G. Reed

Learning Objectives

- To understand the difference between feminism and gender-based analytical approaches.
- To differentiate environmental management from environmental governance.
- To be able to provide an example of how institutional arrangements influence environmental management and governance.
- To identify the different roles played by women and men in specific resource sectors and suggest reasons for these roles.
- To be able to explain how a change in resource availability, policy, or regulation may change the nature of gender relations in a resource-dependent community.
- To be able to illustrate how an analysis of principles for 'good' environmental governance might reveal differences in the experiences of women and men who participate in participatory environmental decision-making.

INTRODUCTION

Forestry . . . has been generally regarded as an arena mainly for men's work, business and governance. Within organizations, from households to companies to authorities, a gendered organizational logic is at work which not only reproduces a structure of gender division but also, paradoxically, at the same time, makes gender invisible UNECE/FAO.

[Team of Specialists on Gender and Forestry 2006, 1].

[Our vision]: reclaiming the voices of northern women that will lead to regaining the balance in the decision making processes that directly impact our social, cultural, political, and economic growth.

[Vision statement of the Northern Saskatchewan Women's Network[1]].

Resource depletion, climate change, and industrial restructuring are inducing extensive environmental, economic, and social changes in resource-dependent communities across Canada and generating uncertainty about both their short-term prospects and long-term sustainability. Forest-based communities are but one type of community affected, and the changes they are experiencing have also been felt in other sectors. For example, the depletion of the northern cod stocks on Canada's east coast resulted in the declaration of a two-year moratorium on commercial fishing in 1992. Yet, despite years when no harvesting has taken place, the fish stocks have not recovered, and the moratorium has not been lifted. The collapse of the fishery caused a crisis of historic proportions, and within a generation, fishers and plant workers lost their livelihoods, independence, and ways of life that had persisted for more than 400 years (see also Chapter 8).

In British Columbia, the rising number of mountain pine beetles in the central and northern boreal forest of the province has been attributed in part to the warmer winters that have allowed the populations of the pine beetles to thrive. The beetle larvae, hatched within the trees, feed on the phloem area beneath the bark, cut off the nutrient supply to the trees, and eventually kill the trees. Between the late 1990s and the present, about 12 to 13 million hectares of lodgepole pine forest (an area nearly twice the size of New Brunswick) has been affected. It is estimated that by 2013, about 80 per cent of BC's forest will be affected (Draper and Reed 2009). While temporary increases in allowable cuts have been granted in order obtain the greatest value possible from affected timber, this short-term gain will be followed by long-term reductions in timber harvests as fewer trees are available. The impacts on forest-dependent communities are expected to be grim.

Economic restructuring of resource industries is also affecting the short- and long-term economic sustainability of resource communities and the livelihoods of resource workers. In 2008, British Columbia reported a drop of almost 50 per cent in forest revenues to government from the previous year (CBC News 2008), while Ontario reported the loss of 3,900 forest sector jobs in 2007 (Ontario Ministry of Training, Colleges, and Universities 2008).

Social change is also occurring in these communities. Demographic changes in rural and resource communities affect the composition and nature of participants in the governance of environment and resources. While the total proportion of Canadians of Aboriginal heritage is just under 4 per cent, in the northern reaches of the country, Aboriginal (including First Nations, Métis, and Inuit people) people form a significant proportion of the population. For example, in 2006, Aboriginal people made up about 15 per cent of Saskatchewan's total population, but people reporting an Aboriginal identity formed 86 per cent of the population in northern Saskatchewan (Northlands College, the Labour Market Committee, and Saskatchewan Ministry of Advanced Education, Employment and Labour 2008). In 2006, the median age of the Aboriginal population in Saskatchewan was 22, or almost 17 years younger than the median age of the province's non-Aboriginal population, meaning that Aboriginal populations are likely to grow much more rapidly than non-Aboriginal populations in the near future (Draper and Reed 2009).

Furthermore, recent court decisions, including decisions by the Supreme Court of Canada, have upheld Aboriginal title and rights established by the Royal Proclamation of 1763. These decisions have determined that Aboriginal peoples have legal entitlements to natural resources and that governments and industry have a 'duty to consult' Aboriginal peoples if any resource development has the potential to infringe on their resource use. Aboriginal rights are second only to conservation measures and are to be given higher priority than those of commercial or recreational resource users. Thus, compelling social reasons exist to include Aboriginal peoples more effectively in resource management, and efforts to do so are evident in many sectors, including forestry, mining, fisheries, wildlife management, and national parks (see Chapter 4).

The forestry sector (both industry and government regulators) has arguably provided the most detailed set of measurements to track progress toward environmental, economic, and social sustainability compared to other resource sectors in Canada. In 2003, Canada's Council of Forest Ministers agreed on a set of criteria and indicators for sustainable forest management that include the criterion 'society's responsibility', defined as 'forest practices [that] reflect social values' and 'fair and effective resource management choices' (Canadian Council of Forest Ministers 2003, 22). Indicators that define this criterion include recognition of Aboriginal and treaty rights, Aboriginal traditional land use and forest-based ecological knowledge, forest community well-being and resilience, fair and effective decision-making, and informed decision-making. These considerations fit within a broad understanding of social sustainability that suggests environmental and land-use decisions should involve the public directly in order to help provide for an equitable distribution of benefits and costs of resource uses and decisions within the current generation and for future generations (Draper and Reed 2009; see also Chapter 18).

Yet, despite concern for community well-being and fair and effective decision-making, the forest industry and forest-dependent communities—like other resource sectors and communities—are structured around a highly gendered division of labour. For example, resource harvesting (e.g., timber-cutting, fishing, mining, even farming) has traditionally been considered 'men's work', and decisions that affect environmental and resource management have usually reflected a fairly narrow range of values. Research has consistently demonstrated that women and men have different values associated with environmental protection but the dominant ones have been associated with men's preferences (Mohai 1992; Seager 1996; Lidestav and Ekström 2000; Uliczka et al. 2004). In forestry, as in other resource sectors, most management decisions have been made by men—in company boardrooms, in government offices, and even in public or community advisory committees. These practices have become so commonplace that they have rarely been questioned in Canada or abroad. As noted in the opening quotation by the United Nations Economic Commission for Europe/Food and Agricultural Organization (UNECE/FAO) Team of Specialists on Gender and Forestry in 2006, a gendered organizational logic works to render gender invisible. As noted in the second quotation, however, women have begun to organize to ensure that they too have a voice and can help to shape development options in their communities.

Policy-makers, planners, academics, and community leaders have been slow to consider the relevance of a gender-based analysis for environmental management and governance, particularly in industrialized countries like Canada. Nevertheless, gender-based analysis can be applied in policy development and planning at all stages of the policy cycle: identifying issues and anticipating outcomes, conducting research, identifying options, making recommendations, and assessing the results (Status of Women Canada n.d., 2).

Environmental planning and management are no exception. Yet scholars and practitioners frequently continue to undertake research about environmental management and sustainability in Canada in ways that often erase any hint of the underlying social relations of gender. For example, a search through all items (e.g., articles, book reviews, commentaries) published by the top interdisciplinary environmental social science journals between 1980 and 2005 revealed that only 3.9 per cent of all items contained the words 'sex', 'gender', or 'feminism' (Banerjee and Bell 2007).[2] But this does not mean that gender relations do not exist.[3] On the contrary, feminist scholars have argued that unequal gender relations are often the most strongly embedded in the culture when they are the least visible (Smith 1987; Acker 1991; McDowell 1994; Brandth and Haugen 2000). When gender relations are not considered, then the potential for environmental policy-making to feed conflict, uncertainty, and injustice within and among interest groups, communities, and government agencies is raised.

Against this background, in this chapter I address four questions:

- How do our assumptions, practices, and performances of gender in Canada affect the institutions, capacities, and implementation of management approaches for environment and resources?
- What are the implications of (not) considering gender in connection with environmental and resource management?
- More specifically, how can an understanding of 'gender in forestry' inform the institutions we establish for forestry governance?
- Can our learning in relation to forestry be broadened to inform other areas of environmental and resource management and governance?

First, I describe what I mean by feminism and gender-based analysis and the types of research used by scholars to introduce gender considerations into environmental management in industrialized countries. Next, I describe environmental management, institutional arrangements, and environmental governance. I synthesize ideas from the criteria and indicators provided for sustainable forest management with those related to 'governance' to establish four guiding principles of 'good environmental governance': legitimacy and voice; fairness; strategic vision, accountability, and performance orientation; and adaptation and experimentation. I follow up by describing the governance of the Canadian forest sector and document the rise of public involvement in management decision-making. Then I provide the results of a national survey of forest sector advisory committees and illustrate how gender has affected the practices of these committees and influenced their capacity to achieve 'good environmental governance' by tracing their performance across the four principles. I use this analysis to discuss the implications of presuming that forest management and governance are gender-neutral and to suggest how lessons from forestry may inform other areas of environmental management and governance.

GENDER-BASED ANALYSES IN ENVIRONMENTAL MANAGEMENT AND GOVERNANCE

Gender-based analyses arise from feminist frameworks and perspectives. *Feminism* is a social and political movement that aims to eliminate domination of some groups over others. Feminism, both in politics and in academia, is not really about the domination of men over women. It is concerned with structures and processes of domination and marginalization, and feminists seek to address the inequalities and injustices that arise from them. Nevertheless, feminist scholars and practitioners have long argued that gender is a central category of social life, explaining the differing life chances between women and men.

Feminist scholars thus make a distinction between *gender* as an analytical category (examining differences between males and females that are socially and culturally influenced) and *feminism* as an analytical framework (focusing on institutional structures and processes of domination and attempting to address the inequalities and injustices that arise from them). We can speak of gender-based analyses or we can speak of feminist frameworks within which gender-based analyses take place.

Gender refers to differences between males and females that are socially and culturally influenced. While our biological sex is usually established at birth, feminists contend that we become masculine or feminine through a combination of biologically determined sex differences (chromosomes, anatomical structures, hormone levels) and socially influenced characteristics (Mosse 1993; Nesmith and Wright 1995). For example, while men, on average, are larger and stronger than women, men's greater physical strength is reinforced from a young age, since boys have traditionally been encouraged to engage in active sports and other forms of physical activities while girls have traditionally been encouraged to develop their fine motor and nurturing skills. These expectations and attributes are then carried forward into the paid workforce, creating gendered jobs and expectations of behaviour.

Gender in Resource Industries and Communities

In resource industries, men are much more likely to be employed in resource extraction (as loggers, fisher*men*, and miners) because of their greater physical strength, while women are much more likely to work in clerical and administrative positions (Thomas and Mohai 1995). When scholars say that these industries are *gendered*, they mean, in part, that there is a division of labour that reflects the strong association of

men in some occupations and women in others. That is why, even in 2001, only 14 to 16 per cent of Canadian forest workers were women. Within the Canadian forest industry, 9 per cent of those in logging and manufacturing positions were female, while 98 per cent of those employed in lower-paying administrative and clerical support positions were female (Reed 2008). Only 16 per cent of executive positions and 20 per cent of science professionals in the Canadian Forest Service were women; this compares to 39 per cent in executive positions and 44 per cent in science professional positions for the public service overall (Table 19.1). Mining, fishing, and even agriculture are also highly segregated by gender, although specific job categories are different. Some implications of this gendered workforce for economic and employment policies are discussed in the guest statement by Alison Gill.

To say that employment is gendered also suggests that there are different expectations about how men and women should behave in the workplace and in their communities. This is sometimes referred to as 'doing gender' or 'performing gender'. Feminists theorize that the location of a worker within the division of labour is a result of, and reinforces, ideas about masculinity and femininity. While regulations formally preclude outright discrimination in hiring, women have long been discouraged from entering certain resource-based occupations such as 'woods work' because of gender stereotypes, lack of social networks, limited promotion opportunities, and expectations about their role in family life. Women who do venture into non-traditional resource-based occupations experience a range of discriminatory practices stemming from their skills being underestimated by their co-workers, the assumption that women 'can't handle the language' in the industry, and the concern that women entering non-traditional occupations will take jobs away from men who are the

Table 19.1 Proportion of Female Representation by Occupational Category for All Forest Industries in Canada, for All Public Service Employees, and for the Canadian Forest Service

Occupational Category	All Forest Industries in Canada (2001)	Public Service Employees (2006)	Canadian Forest Service (n.d.)*
Executive	10	39	16
Science professionals	14	44	20
Administration and foreign service	No comparable data	58	52
Technical	15	32	34
Administrative and clerical support	98	82	94
Operational	9	19	6
Total	14**	54	34

Sources: Census of Canada 2001; Fullerton 2006; Martz et al. 2006; Mills 2006; Statistics Canada 2007 Special Tabulation, Public Institutions Division.

* Fullerton (2006) did not provide a date for these figures. They are likely either based on data provided by Statistics Canada from the 2001 Census or from departmental figures between 2001 and 2006.

** There is a slight discrepancy between Fullerton (2006), who reports this total as 16 per cent, and Martz et al. (2006), who report this total as 14 per cent. The proportion 14 was selected, based on a calculation of all forestry jobs from a Statistics Canada data set and reported in Martz et al. 2006.

GUEST STATEMENT

Gender, Resource Management, and Neo-liberal Policies: Examples from British Columbia

Alison M. Gill

Understanding changes in the gendered landscape of resource management in Canada is grounded in how institutions have changed over the past three decades under the dominant global political paradigm of neo-liberalism. Such regimes believe in the power of market forces, minimal government intervention, and freer trade to drive the economy. Despite a growing literature on the influence of neo-liberalism on gender and resource management in the developing world (e.g. Resurreccion and Elmhirst 2008), there is a paucity of information in a Canadian context. Drawing on the case of British Columbia, I comment on the possible implications for gender relations of recent neo-liberal policies.

Young and Matthews (2007, 176) suggest that resource-dependent regions are often targets of 'intense neoliberal experimentation'. Although restructuring to achieve more flexible means of production has been underway in BC since the early 1970s, radical policy changes introduced by the provincial government in 2001 liberated corporate actors from non-market social and spatial obligations to environment, labour, and community (Young 2008). Concurrent with this decoupling of corporate–community economies, the government introduced policies to stimulate community development through entrepreneurial activities in order to diversify economies and shift responsibility for service provision to local communities. Alongside several federal community development initiatives (e.g., Community Futures), the BC government introduced programs, such as the Community Forests Program, that allowed communities to compete for development funding. Other programs offered mentoring, loans, and skills training.

BC policies on gender equity have mirrored those of the federal government. Since the mid-1980s, affirmative action has resulted in significant improvements in the representation of women in the public sector, including some at more senior policy-making levels. A Ministry of Women's Equity supported these efforts. However, in 2001, using the argument that employment equity had been achieved, the government relegated this ministry to a subunit of another ministry. In 2004, withdrawal of funding for women's centres led to office closures throughout BC. Teghtsoonian (2005, 323) suggests that 'women' were rendered invisible by being placed in conceptual containers such as 'rural communities' and 'local government'. Other government offices providing services such as social, employment, and legal assistance were also closed in many small communities.

It is at the community level that the two sets of policies (gender and resource economies) converge and have the greatest impact on women. Policy changes in the resource sector have led to corporate restructuring, resulting in job losses and dislocation for both men and women, although it is has long been argued that the negative effects on people employed in resource sectors are disproportionately felt by women because they are typically less senior than men or because they are more likely to experience longer layoff periods (Reed 2008). Furthermore, many scholars have argued that the withdrawal of social welfare agencies in many smaller communities has negatively affected women more than men because women have inherited a greater burden of caring (Preston et al. 2000; Teghtsoonian 2005; Martz et al. 2006). Yet government transition policies in resource sectors have failed to acknowledge gender-based differences in opportunities for women and men to take part in retraining programs or other social programs offered through employment insurance (Reed 2008).

Nevertheless, as Young (2008) observes, it is surprising that the radical changes have not met with more opposition. One can argue that the enhanced sense of community empowerment and self-determination is a welcome change from a Fordist regime—characterized by vertically integrated industries operating at a mass scale in a global marketplace (Hayter 2000)—that many considered repressive, especially for women. New ventures in such areas as tourism, value-added production (e.g., small-scale craft and art products), and service provision offer enhanced opportunities for some women as both entrepreneurs and employees. However, to understand how women are affected requires 'women' as an analytical category to be disaggregated according to other social dimensions such as age, gender, and marital status. Further, as Reed (2008) has demonstrated in her work in the forestry sector, although participatory processes are now well-established practices of community engagement, there are constraints on women's voices being heard because of the entrenched male-dominated culture of resource communities.

In conclusion, Teghtsoonian (2005) suggests that gender equity is best served through the joint action of women inside the policy-making agencies and those working at the community level. While employment equity policy in BC has resulted in better representation at the provincial policy-making level, the withdrawal of government support and services from many rural and resource-dependent communities has weakened connections to the community level.

rightful job-holders (Martz et al. 2006). This is not simply a case of men imposing their ideas on women; in rural communities, women and men frequently share these perspectives (Reed 2003a; 2003b). Thus, they form part of a 'local cultural norm' or a shared set of expectations.

Local celebrations in resource communities also reinforce these local cultural norms. For example, 'logger sports days' have historically tested and celebrated masculine traits and skills such as physical strength, speed, and the ability to use heavy equipment—all of which are necessary to undertake work in the woods. Despite more recent female entrants, the sports days are clearly dominated by particular masculine ideals. For example, the ability to quickly and accurately develop a map overlay using a geographic information system has not been featured, despite the fact that these skills are now important for certain kinds of forestry jobs. Thus, we can discern that certain forms of masculine behaviour are celebrated in forestry communities and cultures while others are not.

While there is a long research tradition that focuses on gender, environment, and development in so-called developing countries (Marchand and Parpart 1995; Rocheleau, Thomas-Slayer, and Wangari 1996), scholarship linking gender and environment in industrialized countries has been slower to emerge (Mitchell 2002). Yet there are many avenues for research, even in countries like Canada. Early work focused on women's involvement in environmental management. For example, researchers of environmental activism observed that women tended to be more involved in grassroots activism than in mainstream organizations, and scholars debated whether women's activism flowed from their biological characteristics or their socially ascribed roles as mothers and community caregivers and organizers (Merchant 1995; Seager 1996; Sturgeon

1997). Others examined the roles of women in resource management agencies (e.g., Nesmith and Wright 1995) or in particular kinds of professions such as forestry (e.g., Tripp-Knowles 1999) or national park management (e.g., Davidson and Black 2001).

More recent emphasis has been placed on gender relations within different resource sectors. For example, research in fisheries illustrated how resource depletion altered the long-standing configuration of moral authority and gender relations in Newfoundland fishing communities (Davis 2000). Similarly, environmental planners noted that assumptions about the economy and society favoured the interests of elite men and capital and did not recognize the needs and issues of those people (frequently women) engaged in unpaid work. Consequently, the reproductive work of raising a family and doing community service that often falls to women in rural and resource communities were not considered in planning processes (Moore-Milroy 1996). This literature does not attribute these differences to specific intentions of individuals or groups, and researchers are careful to note that women do not suffer from universal discrimination. Nevertheless, the research suggests that women are often marginalized or excluded from the creation of policies and programs that target workers in resource communities or from programs that address environmental monitoring or health in their communities because of their absence from, or lesser influence within, social, economic, and political networks. These exclusions have been felt by women living in resource communities and have spurred the establishment of women's organizations such as the Northern Saskatchewan Women's Network—the organization featured in the opening quotations—to encourage women's participation in decision-making forums affecting the well-being of their communities.

Of course, gender is not synonymous with women. Both men and women are gendered beings. A small body of research focuses on the implications of masculinist institutions in environmental management and governance for men as well. For example, scholars have demonstrated how work in wildlife conservation, national parks, and forestry reinforces a particular male-based occupational culture or ways of knowing (Carroll, Daniels, and Kusel 2000; Dunk 2003; Dunk and Barton 2005; Kafarowski 2006) or particular forms of masculinity (Brandth and Haugen 2000; Dunk 2005; Sandilands 2005; Ekers 2009).

Furthermore, in studying women's perspectives, we have come to realize that women are not homogenous; a range of different life experiences results in a range of different perspectives. Aboriginal women in rural places may have quite different experiences from those of non-Aboriginal women. The experiences of colonization (e.g., residential schools) may give rise to marked differences between the needs and concerns of Aboriginal women and those of non-Aboriginal women. For example, research about forestry workers in Saskatchewan revealed that the occupational disadvantages of women of Aboriginal ancestry working in the forest sector were more similar to those of Aboriginal men than to those of non-Aboriginal women (Mills 2006). Even within Aboriginal communities, women may not share the same perspectives. When conducting research about the perspectives of Inuit women on land claims and the Voisey's Bay Nickel project (in Newfoundland), Linda Archibald and Mary Crnkovich (1999) found that while Aboriginal women faced common problems in the same region, differences emerged in their experiences and in the strategies they supported or opposed to address them. Thus, we cannot assume that there is one

type of women's experiences or that women will necessarily always share common concerns.

These differences pose difficult choices for feminist scholars, both analytically and politically. Some feminists have wondered whether 'gender' remains meaningful as a social category, while others have stressed the need to examine how gender intersects with other kinds of social differences and to explore how masculinity and femininity are expressed in different places, within and between different ethnic groups, and across class positions (Gibson-Graham 1994; Sachs 1996). This second position seems to be the most fruitful, suggesting that in policy-related research (such as environmental management and governance), research strategies should acknowledge differences among women while seeking to unite their diverse experiences to advocate for changes to policy and practice (Women and Geography Study Group of the Institute of British Geographers [WGSGIBG] 1997; Brush 2003).

In a policy and applied context such as environmental management and governance, therefore, it remains necessary to incorporate insights from all these perspectives to document and thereby make visible the diversity of both women's and men's lives, to illustrate how both women and men are embedded within broader social institutions and sets of relations and cultural norms, and finally to try to establish a strategic 'balance' between broader generalizations about gender and more specific understandings of the differences within and across gender categories of 'women' and 'men'.

Environmental Management, Institutional Arrangements, and Environmental Governance

Before assessing environmental management and governance practice from a gender-based perspective, it may be helpful to review what we mean by environmental management, institutional arrangements, and environmental governance. *Environmental governance* is the broadest of the three terms, including how decisions get made and who decides (Bakker 2007; see also Chapter 2). *Environmental management* refers to 'actual decision and action concerning policy and practice regarding how resources and the environment are appraised, allocated, developed, used, rehabilitated, remediated and restored, monitored and evaluated' (Mitchell 2002, 6–8). Thus, management refers to the decisions themselves, the strategies, programs, and projects put in place to realize broader societal objectives (Olsson 2007).

Institutional arrangements refer to the formal and informal sets of structures, agencies, organizations, policies, programs, strategies, norms, and values generated by the interactions of government agencies, management authorities, environmental movement organizations, industry, local interests, Aboriginal peoples, and other groups to effect management and governance. Some institutions such as 'private property' are formalized in laws and regulations and are reinforced by legal systems and by governments. Other institutions are not formalized by law, but they are important nonetheless. Paradoxically, they may be very difficult to enforce or difficult to change because they are less 'visible', are frequently taken for granted, or do not enjoy recognition by the law. For example, the institution of 'common property' adopted by many Aboriginal peoples was not well understood or protected by Canadian law so that provincial and federal governments frequently allocated resource rights on Aboriginal lands to private parties. Governments did this in order to facilitate resource extraction or exploitation because the governments did not understand or approve of communal systems of property ownership and believed that common property was not being fully utilized by Aboriginal people. Other informal institutions such as cultural norms are even harder to understand and enforce, but they continue to shape our thinking about what kinds of activities are and are not appropriate.

Public, private, and civic (not-for-profit) interests work simultaneously (together or apart, in sync or at odds with one another) and within different sets of power relations to influence, make, and/or carry out decisions about environmental and resource management through these configurations of institutional arrangements. Institutional arrangements are not static; changing conditions within or outside a social system may alter the configuration of formal and informal institutional arrangements. For example, the depletion of the northern cod altered formal regulations of the fishery as well as the uncodified relations between male and female residents of fishing communities, government regulators, and Aboriginal peoples on Canada's east coast.

Environmental governance is a slippery term. Research in environmental governance is far-reaching, covering diverse topics such as 'social learning and participatory resource management' (Diduck et al. 2007 and Chapter 18; Stewart and Sinclair 2007; Sims and Sinclair 2008), 'deliberative democracy' (e.g. Parkins and Mitchell 2005), 'property regimes' (Ostrom et al. 1999), 'privatization and commercialization of natural resources' (e.g., Bakker and Cameron 2005; Robertson 2005), 'co-management' (Armitage, Berkes, and Doubleday 2007), 'legal and regulatory issues' (e.g., Boyd 2003), and even 'governance' itself (Ostrom 1990; McCarthy 2003; Hanna, Slocombe, and Clark 2008; Pollock, Reid, and Whitelaw 2008). These topics share in common an understanding that the idea of governance goes far beyond government agencies and the formal regulations they create and enforce.

Governance implies leadership and includes consideration of who makes decisions and how they are made. The concept of governance draws attention to the fact that there are other equally important actors and stakeholders who are also key players in governance, as well as mechanisms beyond government policies and programs that contribute to decision-making. For this chapter, I define environmental governance as the formal and informal institutional arrangements for resource and environment decision-making and management that include and extend beyond government and involve the private sector, Aboriginal communities, and civil society organizations, as well as the rules and procedures under which these different groups operate and interact (after Francis 2003). Thus, environmental governance involves a range of institutions, social groups, processes, interactions, and traditions, all of which influence how power is exercised, how public decisions are taken, how citizens become engaged or disaffected, and who gains legitimacy and influence and achieves accountability (see Bakker 2007; Olsson 2007).

Researchers suggest that effective governance 'requires learning how to strengthen existing relationships, forge new partnerships, incorporate different kinds of knowledge, and institute new co-management processes. [It] also entails understanding and managing complex relationships among ecosystems and people' (Draper 2004, 229). Some scholars, such as Ellsworth and Jones-Walters (2004, 5), suggest that 'communities are at the heart of [governance]. As places, they experience issues as a web of interrelated problems. As people, they live with direct effects, indirect effects, side effects and cumulative effects of policies. As relationships, they are the product of rewarding interactions.' This turn to the community has been a powerful trend, with academics and practitioners promoting and documenting effectiveness via

studies of self-governance (Kooiman 2003), community capacity (Kusel 2001; Mendis 2004; Lebel 2008), and collaborative planning (Healey 1997; 2003).

Despite this range of research into 'environmental management', 'institutional arrangements', and 'environmental governance', there has been a noticeable silence about how gender might affect environmental governance or the establishment of management strategies (see also Reed and Christie 2009) or how governance arrangements may influence gender relations in resource communities. Yet given that understanding all three concepts involves an understanding of power, politics, and who gains legitimacy and influence, they are ripe for a gender-based analysis.

Principles of Good Environmental Governance

The environmental, economic, and social changes described at the beginning of the chapter in relation to forestry and other resource sectors create uncertain and conflicting signals about how best to manage our environment and resources. They suggest the need to establish governance arrangements that are both inclusive of a range of interests and stakeholders as well as effective in monitoring, responding to, learning from, and harnessing changing conditions. In short, they must exhibit characteristics of 'good governance' and 'adaptive capacity'. Indeed, a review of literature related to 'good environmental governance' (e.g., Graham, Amos, and Plumptre 2003) reveals concern for fair, efficient, and effective institutions and processes, while literature on 'adaptation and resilience' suggests the need for systems to embrace social learning and experimentation (e.g. Olsson et al. 2006). Considering these aspects, I propose the following four principles of 'good environmental management and governance':

- *Legitimacy and voice:* Decisions and processes should be consistent with international, national, and local laws and policies; provide an inclusive and representative structure across a wide range of different groups affected by the decisions in the present and in the future; support and engender a high degree of trust in the process; and recognize and integrate different ways of learning, knowing, and using the environment.
- *Fairness:* Processes should operate within a stable and supportive judicial environment; reflect effective rule enforcement, as well as equitable opportunities for access and participation; and consider distribution of outcomes (over time and across places and social groups) and build and enhance the capacity of individuals and groups to become effective participants.
- *Strategic vision, accountability, and performance-orientation:* It is important to adopt a strategic vision that identifies a desirable future condition; to identify management plans that demonstrate foresight and focus; to establish a process that is transparent, effective, efficient, responsive to all participants, and co-ordinated with external policies, programs, or circumstances; and to seek consensus when making decisions or recommendations.
- *Adaptive and experimental:* It is also important to seek novel ideas; to be experimental in design (allow for experiments so that learning can take place, even from failures); to build mechanisms for monitoring and feedback; to provide opportunities for different kinds of learning by participants; to provide sufficient flexibility in decisions to allow for changes when new information is acquired; and to systematically account for risks.

These principles are seemingly gender-neutral. Yet by applying a gender-based analysis, we can gain new insights about how gender affects both the interpretation of the principles and criteria and the extent to which they are achieved.

GOVERNANCE AND THE CANADIAN FOREST SECTOR

In Canada, provincial governments are the owners of Crown forest land. They have used a variety of licensing and leasing arrangements to grant tenure to private companies to harvest timber. In the period just after World War II, provincial governments granted long-term leases to large, integrated forest companies because governments assumed that larger companies would be more stable than smaller ones and consequently would provide improved economic opportunities for workers, a greater range of social benefits, and long-term sustainability. In the postwar period, companies continued to increase their size and scope of operations so that by the year 2000, 13 companies accounted for more than 48 per cent of Canada's tenure areas of forests (Global Forest Watch 2000). Since about 1980, public concern over a range of issues relating to forest governance (e.g., environmental protection, tenure reform, forest certification, community forestry) suggests that these assumptions have not been confirmed (see Chapter 10).

Direct public involvement in forest management was minimal until the 1970s, when environmental organizations began to mount effective public campaigns to put a halt to clear-cut logging. In 1976, after undertaking a comprehensive review of forest tenure, Peter Pearse, then an economist at the University of British Columbia, concluded that provincial forest policies in British Columbia had failed to provide for sustainable communities and to respond to the changing needs of the population. He also noted that increasing public pressure for environmental protection included the desire to protect non-timber values and the dwindling timber supply, and he warned the forest industry to expect an increasing government and public interest in its development (Pearse 1976).

Throughout the 1980s, protection of wilderness became a driving goal of large and well-organized environmental organizations, and they pursued this agenda through a series of watershed conflicts across the country. In 1987, the Brundtland Commission report, *Our Common Future* (World Commission on Environment and Development 1987), placed the concern for 'sustainable development' on the front burner politically and stimulated interest and public debate about how it might be applied in forestry.

Since the 1990s, sustainability became a more prevalent theme in public policy around forestry, and citizen advisory committees became a central component of forestry planning processes across Canada (Parkins et al. 2006):

> Forest sector advisory committees represent a form of community-based public engagement, where local forest users (along with people involved in the forest sector for their livelihood, representatives of other local agencies such as educational establishments and the business community, and elected leaders) participate in discussions about forest management and provide input into local decision-making [Parkins et al. 2006, 1].

These committees meet regularly—on average eight times per year—and in some provinces, they are required by provincial regulations and forest management licensing procedures. For the Canadian Council of Forest Ministers, advisory committees have become an important means by which the criteria and indicators associated with 'fair and effective decision-making' are met. Thus, they are an appropriate institution to study to better understand environmental management and governance.

ASSESSING FOREST SECTOR ADVISORY COMMITTEES USING A GENDER-BASED ANALYSIS

In 2004, I worked on a research team led by John Parkins of the University of Alberta that conducted a cross-country survey of 102 forest sector advisory committees. Of the 2,256 questionnaires distributed to committee members, 1,079 were returned, for an overall response rate of 48 per cent. Of these, 180 responses were from women. We asked several questions pertaining to forest values, influences on committees, methods of learning, sponsorship, and opinions about group process. In assessing the experiences of women and men, we found differences in their respective experiences, as well as differences in how they interpreted those experiences. Thus, gender not only affected the achievement of some principles by objective criteria, such as the number of women and men involved, but also affected the subjective interpretation of some principles such as 'legitimacy and voice' and 'fairness'. Let me explain by discussing the responses of men and women as they relate to the principles of 'good environmental governance' described in the previous section.

Recall that *legitimacy and voice* is about being inclusive and representative as well as about integrating different interests and values. This is not a simple task, because there may be interests and values that are 'hidden' under norms that are taken for granted. Feminist theorists have argued that normative rules, values, and meanings in all aspects and at all scales of social and political life have created a gender order that both establishes and institutionalizes specific 'positions' for women and men. This typically establishes men's, or masculine, ways of being, working, learning, knowing, and decision-making as the norm and renders female experiences, ways of knowing, and so on as deviant and feminism as ideological (meaning that they are subject to predetermined or predictable ideas or theories) (Gherardi and Poggio 2001; Brush 2003). This positioning has been evident in forestry communities in which informal and formal institutions and ways of living have typically favoured men's ways of knowing, working, and participating in forestry. Consequently, 'legitimacy' in terms of inclusion typically, and frequently invisibly, favours men's ways of knowing, ways of doing, interests, and values.

Formally, Canadian legislation provides for equal access for men and women to participate. And yet the seemingly 'neutral' application of our laws and regulations against a backdrop of uneven social circumstances reproduces—and sometimes renders invisible (as pointed out in the opening quotation)—inequities in access and influence, thereby affecting the number of women and men who participate as well as their effectiveness once they take part.

On the 'objective' criterion of simple numbers, only 17 per cent of respondents to the survey were women. These numbers varied regionally, although in all but two provinces, less than 20 per cent of respondents were women. Because the questionnaires were distributed without knowledge of the gender of each participant, it is not possible to determine whether this proportion of respondents accurately reflects committee membership. However, reviews of other social surveys indicate that women are more likely to respond to mail-in questionnaires than men (Green 1996), suggesting that this proportion (17 per cent) probably overestimates the number of women who were serving on advisory committees. In all but three provinces, the proportion of women responding was lower than the proportion of women actually working in primary forestry jobs or forest manufacturing jobs (Table 19.2).

Table 19.2 Number of Respondents in Forest Sector Advisory Committees (FSACs) and Forestry Employment, by Province

	Participation on FSAC					% Employment in primary forest industries		% Employment in forest manufacturing	
	Women		Men		Total	Women	Men	Women	Men
	#	%	#	%	#	%	%	%	%
Alberta	24	19	104	81	128	22	78	16	84
British Columbia	24	32	52	68	76	17	83	12	88
Manitoba	5	12	36	88	41	13	87	20	80
New Brunswick	11	10	99	90	110	11	89	14	86
Newfoundland	4	29	10	71	14	10	90	6	94
Nova Scotia	2	17	10	83	12	10	90	11	89
Ontario	34	14	214	86	248	18	82	18	82
Quebec	73	19	318	81	391	13	87	14	86
Saskatchewan	3	8	34	92	37	17	83	18	82
Total	180	17.0	877	83.0	1,057	15	85	14	86

Sources: Questionnaires and Census of Canada 2001.

Note: There were no advisory committees in Prince Edward Island.

If legitimacy is at least partly about integrating different interests and values, then it is appropriate to ask whether women will bring different perspectives to forestry. Researchers of social activism have found that women typically seek to advance goals related to 'softer' social concerns—such as improved health care and educational opportunities—while men are more likely to advance economic, employment, and political interests (Caiazza and Gault 2006). Regarding environment, women express higher levels of concern than men for environmental issues and for forest protection (Davidson and Freudenburg 1996; Mohai 1992; Tindall, Davies, and Mauboules 2003; Uliczka et al. 2004; Reed and Varghese 2007). They also typically express greater risk aversion to and concern about climate change than do men (Davidson, Williamson, and Parkins 2003; Johnsson-Latham 2007). Yet women have not had avenues to mobilize this concern. Typically, women are not leaders of forestry companies, national or international environmental organizations, or resource sector unions (Seager 1993; Livesey 1994; Müller 1994; Teske and Beedle 2001; Wright 2001; Christiansen-Ruffman 2002), and they do not participate as actively in other decision-making positions to advance these interests. For example, while women have been appointed fairly frequently as provincial or federal ministers of environment, they have been less often given portfolios of higher status such as ministries or departments of natural resources or industry (although in October 2008, the prime minister appointed 11 women to serve among the 38 federal cabinet ministers and ministers of state, including in the portfolios of labour, natural resources, and fisheries and oceans).

Confirming research elsewhere, we found that women and men had shared and separate interests and values related to forest management. For example, both women and men strongly

agreed that it was important to maintain forests for future generations, but women had stronger support for intrinsic values (i.e., they valued the forest for its own sake rather than only for its use value), while men rated utilitarian values more highly. Furthermore, women who participated in the survey were more likely than men to belong to natural history or bird-watching clubs and environmental organizations, whereas men were more likely than women to belong to hunting or fishing organizations. Thus, an over-representation of men will likely overrepresent utilitarian values. Women were less likely than men to believe that all values were represented; respondents of both genders who noted gaps in representation noted the absence of Aboriginal and environment- or wildlife-oriented groups.

With respect to perceptions regarding levels of influence, women in the survey saw themselves and other participants as being less influential in setting the agenda and gave greater weight of influence to industry officials, whereas men saw themselves and other participants as the most influential in setting the agenda. In a follow-up study involving personal interviews with members of two advisory committees, Kristyn Richardson found that women believed they were less influential because they did not have formal education or experience in the forest industry. This finding was interesting, because rural women typically have higher levels of formal education than rural men. Others expressed the belief that women on the committee were 'exceptionally strong, outspoken women that have the confidence to step into these roles' (Richardson 2008, 78). One male respondent complained that sometimes women are not heard because they are too emotional:

The emotional approach I have trouble dealing with and I think most guys do when it gets too emotional and it's an emotional argument, as opposed to a rational, science-based as I call it, argument, because it tends to lose credibility in the business. It's a man's business and we don't really want to hear that stuff [Richardson 2008, 79].

Furthermore, these findings suggest that gender stereotypes and expectations exist for both women and men. Men's and masculine ways of being, working, learning, knowing, and decision-making within forestry were considered the norm, while female experiences and ways of knowing were not. Women's participation was deemed acceptable or legitimate as long as women overcame stereotypical female behaviours.

Fairness refers to providing equitable opportunities to participate, including mechanisms to ensure accessibility and the capacity of participants as well as equitable outcomes. We found that the opportunities to participate also varied between men and women. Men were more likely to state that they 'wanted to participate to contribute to planning since the forest is a public resource' and to 'ensure that recreational opportunities are not diminished'.

Overall, one-third of the survey respondents participated in these committees because of their job. A greater proportion of women than men came to these committees as employees—because it was designated as part of their jobs. Furthermore, women were more likely to be employed by environmental organizations, while men were more likely than women to represent the forest industry. Most individuals were selected by the forest companies or government agencies that sponsored the committees because of their direct interests in renewable natural resource management (e.g., through the forest industry, government regulation, recreation and tourism, trapping), 7 per cent stated that they were selected to represent 'the general public', 4.9 per cent were selected to represent environmental interests,

3.5 per cent were chosen to represent Aboriginal interests, and less than 1 per cent were selected to represent a community or social organization.

Accessibility to these committees for women living in forestry communities but who are not directly employed in a 'relevant group' is constrained in several ways. First, despite normally having higher levels of formal education than their male counterparts, because of the gendered division of labour in Canadian rural communities, women continue to bear the burden of home care, and their employment status is less secure. Taking time from work or their families, especially without any financial compensation, is simply not an option. Committees typically only covered basic transportation costs; any costs associated with loss of income or childcare expenses while parents attended meetings were not covered (Parkins et al. 2006). Additionally, related research has confirmed that the long distances involved in travelling to and from meetings and the length of travel time required are barriers for rural Canadian women, particularly at night and in winter when roads are not well-lit and may be treacherous (Martz et al. 2006; Richardson 2008).[4]

Women also lack the social networks and role models in forestry that are more readily accessible to men (Reed 2003a). For example, private operators in forestry, business associations, unions, and wildlife and hunting organizations from which community members are selected are all male-dominated. These associations and networks can be used to gain 'stakeholder status' in advisory committees and advance employment and economic interests, while the networks to which women belong do not usually suggest an immediate interest in forestry jobs or economic development. This narrow consideration of the stakes in forestry communities tends to favour the consideration of men with economic interests as committee members; it is unlikely that women not directly employed in forestry or a related resource sector would be asked to participate. Residents employed in the public or community health sector, for example, were not considered relevant participants. In interviews with members of two forest sector advisory committees, Richardson (2008, 53) learned that 'because we don't have a lot of women in the industry *we don't naturally gravitate towards thinking about or suggesting women* to be on the committee' (my emphasis). Furthermore, Richardson found that women employed in forestry described the importance of a role model to build the experience and confidence of women. But these interviewees also noted that mentorship had not been extended to women outside the industry who might, with sufficient encouragement, join forest management advisory committees.

Being *strategic, accountable, and performance-oriented* involves having a clear purpose, designing transparent processes that all people understand and can engage, and being effective and efficient in decisions and actions. With respect to this principle, both similarities and differences also emerged.[5] In general, both men and women agreed that they were able to influence the decisions by the committee and had been given adequate opportunity to voice their concerns within the committee. They agreed that their ideas were taken seriously by other committee members, active discussion was encouraged in the committee, and the process was fair. However, women were more likely than men to feel that time was poorly spent in the process. It is possible that multiple time commitments make women more wary about spending time inefficiently within meetings. Women also rated their satisfaction with decision-making, the representativeness of the committee members, and the overall process as lower than men's, although only the issue of the decision-making process was statistically significant.

In the open-ended portion of the survey related to improving the effectiveness of the process, various women made recommendations linked to efficient and effective use of meeting times. A couple of examples were: 'Shorter, more concise agendas, and [more effective] chairing of meetings. Meetings should be less than two hours long. Shorten agenda to meet timelines if necessary' and 'Set priorities and stick to them, ask questions via phone or e-mail in advance so information can be gathered before meeting, avoid repetitive items (especially those already dealt with).'

One female survey respondent said, 'I am no longer on the committee because I didn't feel it served an important purpose . . . , we didn't work on forest management but on public education.' Another suggested that the process would be improved 'if participants had the opportunity to provide more input and we could demonstrate that we are trying to put good ideas to use'. These suggestions indicate that despite an overall aim of contributing to sustainable forest management, committees were not focused strategically and therefore, their performance against specific reference-based criteria could not be determined.

Another respondent, whose comment has been corroborated by other studies, described the process as perpetuating 'old boy' approaches (see also Reed 2003a; Martz et al. 2006; Reed and Varghese 2007; Richardson 2008). 'Old boy' approaches weigh against the assessment of the processes as transparent and accountable. 'Old boy' approaches and networks refer to the use of informal, yet exclusive, networks that link some individuals and exclude others. 'Old boys' are often linked through friendship or a common history (e.g., in school, a club, or the workplace) and typically provide favours to one another to help each other get ahead, often in business or politics. Since these are more favoured employment sectors for men, old boy networks tend to enhance men's employment and political fortunes. For example, in work-places, old boy networks are said to operate if a man is promoted because he has 'connections' with those doing the hiring rather than because he compares favourably against objective crite-ria for the position. In an advisory committee, an old boy network may be seen to be operating if men are considered for committee member-ship because of their connections through work or other associations (e.g., a hunting or fishing club) rather than through some broader search strategy that would evaluate candidates against an explicit and public set of criteria.

Even less formally, using 'old boy approaches' on a committee may refer to listening favourably to the ideas of male colleagues because they are friends rather than trying out novel ideas from those who are deemed to be outside the 'club'. Because women historically have not participated in the same kinds of outdoor activities that men have, they are frequently considered as having *less* information rather than *new* information. Women and men express themselves differently and even use different forms of humour and other linguis-tic strategies. Furthermore, women who are not part of the industry are not always informed in advance of the jargon, the expectations of perfor-mance, and the decision rules. As noted above, some women overcome these concerns to make effective contributions to committees; however, others impose a silence on themselves or quit the committees altogether rather than having to ask questions continually and appearing less informed than or 'silly' in the eyes of their male counterparts (Richardson 2008).

Overcoming these barriers is challenging. Feminist scholars have argued that in politics, once a critical mass of women is present (some-times set at 30 per cent), committees typically exhibit less stereotyping and openly exclusion-ary practices on the part of men, a less aggres-sive tone in discussion, greater accommodation

of family obligations, and greater weight given to women's concerns in policy formation (Dahl-erup 1988). Having a critical mass of those once in a small minority also sends the message that both women and men are fit to serve on advisory committees (after Mansbridge 1999). Thus, some evidence suggests that increasing the proportion of women on forestry advisory committees may alter both the balance of perspectives and values brought to the table and the processes by which they are considered and debated.

Last, *adaptation and flexibility* focuses on establishing conditions for experimentation and learning. Because the study was not squarely focused on adaptation and environmental gov-ernance, the survey results do not allow for a full examination of the principle of *adaptation and flexibility*. We did, however, ask respondents to identify reliable sources of information. Women rated scientific researchers as a source of accu-rate information, while men were more likely to view the forest industry as a favourable source of information.

Richardson's (2008) study distinguished between 'instrumental learning', which involves learning about technical matters such as legisla-tion and licensing requirements, and 'communi-cative learning', which refers to changing one's perceptions and values as a result of communi-cation with others (see more detailed discussion about these concepts in Chapter 18). She found that female respondents acknowledged that their perceptions were being altered by the concerns, values, and interests of other members, so they were experiencing communicative learning outcomes. By contrast, the male respondents reported that their perceptions were only being altered by the information provided by the forest products companies regarding forest manage-ment practices; therefore, they were only experi-encing instrumental learning outcomes. Neither women nor men reported seeking novel ideas or building mechanisms for monitoring and feedback. Richardson's findings also confirmed other studies that suggest that lack of diversity, too much information from the proponent, and relatively little debate contribute to these limited learning outcomes (Parkins 2002; McGurk et al. 2006; Sinclair, Diduck, and Fitzpatrick 2007).

IMPLICATIONS OF PRESUMING GENDER-NEUTRALITY IN FOREST MANAGEMENT AND GOVERNANCE

Canada's formal systems of forest management and governance offer multiple opportunities for participation and representation by both women and men. But our informal institutions, such as cultural norms and values that include the norms of forest-dependent communities, the gender of forestry work, and the expectations of masculinity and femininity within these communities, suggest that the gender bias within our decision-making processes remains strongly in favour of men's ways of doing and knowing. Forestry advisory committees—like the industry, households, and forestry communities—are highly gendered forest management units. Gender balance has not been a priority when composing advisory committees, and consequently, women are nominally under-represented. Insofar as participation itself is important as a measure of citizenship rights and as a form of empowerment and voice, women's absence from these committees counts as a failure to meet the principles of *legitimacy and voice* and *fairness*. This gap, while pervasive, may be an invisible barrier to gender equity because it is usually taken for granted.

Looking toward improving representation, there is merit in further investigating how social networks are formed and mobilized to affect the experiences of women and men who are asked (or not asked) to participate. The discussion of the lack of mentors for women, the 'old boy' network, and the types of behaviour acceptable at the committee tables suggests that some gender biases are less visible, but possibly more revealing, than simply numbers. It would be important to determine in greater detail how people are brought in to these processes and how effective they are as participants once they are sitting at the table.

Gender also plays out in terms of how women and men work within these committees. A large gap remains between the way that men and women perceive their roles and contributions to committees, as well as the way in which they interpret the effectiveness of the committees. While the principles of *strategic vision, accountability, and performance-orientation* are 'gender-neutral', this review suggests that male and female committee members interpreted their implementation quite differently. Consequently, the overall assessment of success of these committees and the associated governance structures varies between women and men.

Last, with regard to the principle of *adaptation and flexibility*, if governance arrangements are to be created that provide learning under conditions of scientific, economic, political, and environmental uncertainty, it will be important to consider that women and men have different basic understandings of environment and resources, engage in different learning styles, and are likely to experience different learning outcomes. Thus, even our understanding of learning is informed by gender and must be sensitive to differences that women and men have in expectations, experiences, and outcomes of learning.

It appears as though the neglect of gender differences in the composition of forest sector advisory committees and in the values people bring may distort analysis and policy formulation in favour of timber extraction and reinforce particular forms of knowing and doing in relation to forest governance. Although it is possible that men can also bring social concerns and intrinsic values to advisory committee meetings, thus far, men's perspectives in support of forestry suggest they are unlikely to do so in a vigorous manner. This conclusion, based on the survey of forestry advisory group members in 2004, needs verification in other resource situations before it can be considered to have general applicability. Nevertheless, this outcome is problematic from the perspective of social sustainability, because without women's effective participation, emergent initiatives will not improve outcomes or processes associated with forest management decision-making. The benefits of women's greater participation in forestry decision-making would not flow just to women but also to the larger forestry community and thereby contribute to a more ecologically and socially sustainable forest system.

It is important to recognize that women are not merely passive observers or victims of the changes underway. Indeed, across resource sectors and communities, women have established organizations to ensure that they and the concerns they raise are included in the processes associated with resource and community development. For example, in March 2008 a workshop entitled 'Women's Perspectives on the Mountain Pine Beetle' was held in Prince George, BC, with the express purpose of exploring 'the social, economic and health-related impacts through the eyes of women from beetle-affected communities in northern BC' (University of Northern British Columbia 2008, 1). Established by the University of Northern British Columbia Northern Women's Centre, School of Social Work, and the Women North Network, the workshop had as its goals the application of a gender lens to the pine beetle epidemic and the identification of issues and concerns as described by women in affected communities. Additionally, the forum was organized to share strategies and initiate the development of a network and solutions to mitigate the effects of the mountain pine beetle on the health of women and the communities in which they live (University of Northern British Columbia 2008).

In Saskatchewan, the Northern Saskatchewan Women's Network was incorporated in 2001 and reinvigorated in 2006. The network arose out of the frustration of some of the founding members with the exclusion or marginalization of women in several advisory and decision-making bodies, such as environmental quality committees (established to monitor mining impacts in northern communities) and the federal Aboriginal Skills and Employment Partnership program (designed to ensure that Aboriginal people have the skills necessary to participate in economic opportunities such as northern mining, the oil and gas industry, forestry, and hydro development projects across Canada). The network has established workshops in communities across northern Saskatchewan to generate awareness among women of the historical processes of colonization, to discuss its contemporary effects in Aboriginal communities, to empower women to reclaim their roles and responsibilities as community decision-makers, and to encourage their involvement in ongoing development opportunities and challenges.

CONCLUSION

Gender is not the only marker of social stratification in forestry communities. In Canada, environmental governance across all resource sectors and communities will also have to consider Aboriginal peoples in more meaningful ways (see also Chapter 4). Aboriginal women, for example, may experience both racial and gender discrimination. Yet the way our policies and daily practices construct gender and the way gender positioning affects the formation and implementation of policy has enormous influence in the differential capacity of women and men living and working in forestry communities. A gender focus will help to sensitize researchers to multiple inequalities and help to create opportunities for more inclusive concepts and analyses and ultimately more inclusive policies and practices that place value on the contributions of both women and men and of particular social groups without privileging any particular one. In this way, we can gain a better understanding of what constitutes sustainable environmental management and governance for *all* residents of rural and resource communities and how we might govern ourselves to promote it.

Looking beyond forestry, some commentators have noted a troubling trend, at least in North America, relating to public involvement. They observe the rising professionalization of civic and environmental organizations in decision-making processes established by state and industry institutions (Kasperson, Kasperson, and Dow 2001; Skocpol 2004). Indeed, the same data set on forest sector advisory committees in Canada revealed to John Parkins and John Sinclair (2008) evidence of a trend toward forms of representation that favour more elite and highly educated members of society. Similarly, in relation to global climate change assessments, Roger Kasperson (2006, 321) reminds us that 'typically left out ... are those people who do not yet know that their interests are at stake in a particular decision, whose interests are diffuse or associated with a sense of community rather than personal material interest, who lack the skills and access to political resources to compete, or who have lost confidence or are alienated from the political process.' Consequently, those affected whose interests lie in immaterial stakes—community goals, grassroots organizations, local values and perspectives, and protection of the environment for future generations—are often marginalized or excluded. We must recognize that women and men organize around different issues and advance their assets in different ways. Unless we attend to gender difference and dominance when involving communities in establishing environmental management strategies, we will reinforce gender disadvantage and exclusion of women in decision-making. We will also narrow the scope of issues being considered in decision-making forums and may overrepresent industrial and utilitarian aspects of environment and resources over other community and eco-centric values.

Despite an increased interest in public involvement in environmental management and governance, our practices to achieve environmental, economic, and social sustainability are too often unintentionally exclusionary. These exclusions exacerbate a sense of uncertainty, conflict, and even alienation that has become almost endemic in rural and resource communities. Attempts to broaden the base of issues and individuals involved can be improved by introducing a feminist perspective. More progressive environmental policies will be realized if we make efforts to ensure that women have opportunities to participate, if we actively consider how policies may affect women and men differently, and if we recognize the diversity of experiences and expertise that both women and men offer. In these efforts, we must continue the political project of elucidating and eliminating inequitable power relations founded on categories of 'women' and 'men' while remaining open to the differences within and among them.

Such attempts will not erase conflict and uncertainty in environmental management and governance. In fact, sensitivity to a broader range of interests often raises conflicts in the short term. Nevertheless, understanding feminist contributions with tools such as gender-based analysis can help to address conflict and uncertainty and bring together this diverse human community in the challenges we collectively face to achieve sustainability.

FROM THE FIELD

I have never felt like I was an expert on gender, but I have been asked again and again to discuss gender and environmental management. When I was a graduate student in 1988, there was almost no research about gender and environmental management in Canada. Indeed, I could not conceive how it might be conducted.

Now, more than 20 years later, there is greater awareness about gender and environmental management but surprisingly, not a great deal more research. Yet gender-based analysis of environmental management can reveal important patterns and differences. For example, we know that women and men have different experiences of employment in virtually every environmental sector. We know that women and men have different perspectives on environmental change. We think that programs established to address environmental challenges such as climate change will affect men and women differently. We also know that the experiences of participating in planning and decision-making are different for women and men. Attention to gender on these subjects and others should help to provide more effective policies and programs for society today and for future generations.

How does this understanding affect how I conduct my research? My interest in gender has encouraged me to ask (1) what are the experiences of other social groups in environmental management? (2) how should I conduct my research? and (3) is my research useful and if so, to whom?

Regarding the first question, I have started to consider the significance of gender relative to other social categories. For example, what are the experiences of Aboriginal people in environmental management? How do rural and urban people differ in their perceptions of environmental issues and involvement in environmental decision-making?

How does gender intersect with these other identities?

Undertaking gender-based research has also encouraged me to consider what are the most appropriate methods of inquiry and exchange. For example, are interviews or surveys more appropriate? How should I present unpopular opinions and ideas of my research subjects while simultaneously protecting their integrity?

Last, I continue to be concerned about whether my research is useful, and if so, to whom. The conditions for social science research are often most favourable during times of greatest personal and political crisis for research subjects. For example, when people lose their jobs, social scientists might study how individuals or households cope. Yet my research has never created a new job. What benefits can my research really have? If researchers are to gain credibility, we must consider whether our results can be useful to those in such situations.

And I also wonder *who* should find my work useful. Local, provincial, Aboriginal, or federal governments? Local residents? Funding agencies? Politicians? Environmental organizations? These groups are not mutually exclusive, but sometimes being accountable to one group means denying another. Furthermore, once I acquire knowledge, does my responsibility to exchange it end or begin when I write an academic paper? How should I discuss my research beyond academic papers? When does such work become advocacy? What is responsible advocacy?

When conducting research in environmental management, I think it is important to continuously ask these kinds of questions. Only by questioning how we create knowledge, who our knowledge might serve, and how it might be most valuable can the understanding we gain be used to improve the practices of environmental management.

—Maureen G. Reed

Acknowledgements

Financial support for this research was provided by the Sustainable Forest Management Network (SFMN) and the Social Sciences and Humanities Research Council of Canada. The University of Saskatchewan provided funds for me to attend international conferences in Sweden and India at which some of the ideas in this chapter were first presented. John Parkins, Tom Beckley, Jeji Varghese, John Sinclair, and other members of the SFMN sub-group provided logistical support for the research on forestry advisory committees. Insights from research by Krystin Richardson and Christiana Amuzu also helped me to refine my thinking. Bruce Mitchell provided constructive comments for more than one draft of this chapter. I am indebted to Lillian Sanderson and Ina Fietz of the Northern Saskatchewan Women's Network for sharing their experiences and insights with me and to Dawn Hemingway, who first told me about the work in Prince George. Despite all this help, I remain responsible for all the errors and omissions in the content and interpretation of results that readers may find.

Notes

1. This statement was adopted at the strategic planning workshop held in December 2008. It was reported by Lillian Sanderson, founding member of the network, to the author by personal communication, 17 December 2008.
2. The journals reviewed were *Environment and Behaviour, Environmental Politics, Environmental Values, Organization and Environment*, and *Society and Natural Resources*.
3. Gender relations can be described as power relations between men and women or relations within institutions that treat women and men differently and affect the way that women and men relate within society. For example, in Western society the institution of 'higher education' has historically granted greater opportunities for men than for women to engage in formal study. In Canada in 1884, 57 years after the founding of the University of Toronto, the first women were allowed to attend classes, despite opposition from the university's newspaper whose editorial read, 'the proximity and competition of the "softer sex" is rarely a spur to intellectual activity' (Friedland 2002, 90). Despite the admission of women, some places on the campus remained off limits to them until late in the twentieth century. For example, 'women were not admitted to Hart House on equal terms with men until *1972*' (Friedland 2002, 293; emphasis added). Even today in many developing countries, access to education is restricted for girls. Furthermore, feminists have argued that forms of knowledge and understanding ascribed to men (e.g., supposedly rooted in rationality) have been validated and encouraged, allowing men to succeed in university settings while forms of knowledge and understanding ascribed to women (e.g., supposedly rooted in emotionality) have been dismissed and excluded from the canon of serious scholarship. These perceptions have effectively reduced the opportunities for women to participate equally in all areas of knowledge creation.
4. While poor driving conditions affect both women and men, only women reported that they were less likely to drive long distances because of the time it took away from their families or because road conditions in winter or at night were unfavourable.
5. Note that because the survey did not reach people who had left these processes because they were dissatisfied, we probably observe a higher proportion of people who were satisfied with committee performance.

REVIEW AND EXTENSION QUESTIONS

1. What is meant by the 'gendered division of labour'? Provide an example from one of the resource sectors described in this book or an example in your community.

2. Many years of public opinion polls have documented that more women than men support measures to protect the environment. Yet more men participate in mainstream environmental organizations to lobby for policy changes. Why do you think this is the case?

3. What is the gender composition of the class for which you are using this book? How do you account for this composition? What behaviours do you observe in your classroom that are typical of 'men' or 'women'? Who engages in these behaviours? Do some students cross over the boundaries of gender stereotypes?

4. Write a job description for 'logger'. Now write a job description for 'secretary'. What characteristics in your descriptions do you associate more readily with men and with women respectively? At what stage do you think these characteristics were learned?

5. 'To be gender-neutral, all decision-making bodies must have an equal number of women and men.' Develop some arguments in favour of this proposition. Now develop some arguments against it. Which arguments do you think are most convincing?

REFERENCES

Acker, J. 1991. 'Hierarchies, jobs, bodies: A theory of gendered organizations'. In J. Lorber and S.A. Farrell, eds, *The Social Construction of Gender*, 162–79. Newberry Park, CA: Sage.

Archibald, L., and M. Crnkovich. 1999. *If Gender Mattered: A Case Study of Inuit Women and Land Claims and the Voisey's Bay Nickel Project*. Report to the Status of Women Canada's Policy Research Fund. Ottawa: Status of Women Canada.

Armitage, D., F. Berkes, and N. Doubleday. 2007. *Adaptive Co-management: Collaboration, Learning, and Multi-level Governance*. Vancouver: University of British Columbia Press.

Bakker, K. 2007. 'Introduction'. In K. Bakker, ed., *Eau Canada: The Future of Canada's Water*, 1–22. Vancouver: University of British Columbia Press.

Bakker, K., and D. Cameron. 2005. 'Changing patterns of water governance: Liberalization and de-regulation in Ontario, Canada'. *Water Policy* 7 (5): 485–508.

Banerjee, D., and M.M. Bell. 2007. 'Ecogender: Locating gender in environmental social science'. *Society and Natural Resources* 20 (1): 3–19.

Boyd, D. 2003. *Unnatural Law: Rethinking Canadian Environmental Law and Policy*. Vancouver: University of British Columbia Press.

Brandth, B., and M. Haugen. 2000. 'From lumberjack to business manager: Masculinity in the Norwegian forestry press'. *Journal of Rural Studies* 16: 343–55.

Brush, L. 2003. *Gender and Governance*. Walnut Creek, CA: AltaMira Press.

Caiazza, A., and B. Gault. 2006. 'Acting from the heart: Values, social capital, and women's involvement in interfaith and environmental organizations'. In B. O'Neill and E. Gidengil, eds, *Gender and Social Capital*, 99–126. New York: Routledge.

Canadian Council of Forest Ministers. 2003. *Defining Sustainable Forest Management in Canada*. Ottawa: Canadian Council of Forest Ministers.

Carroll, M.S., S.E. Daniels, and J. Kusel. 2000. 'Employment and displacement among northwestern forest products workers'. *Society and Natural Resources* 13: 151–6.

CBC News. 2008. 'Forestry revenues plunge in BC'. 25 November. http://www.cbc.ca/canada/british-columbia/story/2008/11/25/bc-foresty-revenues-plunge.html.

Christiansen-Ruffman, L. 2002. 'Atlantic Canadian coastal communities and the fisheries trade: A feminist critique, revaluation and revisioning'. *Canadian Woman Studies* 21/22: 56–63.

Dahlerup, D. 1988. 'From a small to a large majority: Women in Scandinavian politics'. *Scandinavian Political Studies* 11: 275–98.

Davidson, D., and W. Freudenburg. 1996. 'Gender and environmental concerns: A review and analysis of available research'. *Environmental Behavior* 28: 302–39.

Davidson, D., T. Williamson, and J. Parkins. 2003. 'Understanding climate change risk and vulnerability in northern forest-based communities'. *Canadian Journal of Forest Research* 33 (11): 2252–61.

Davidson, P., and R. Black. 2001. 'Women in natural resource management: Finding a more balanced perspective'. *Society and Natural Resources* 14: 645–56.

Davis, D. 2000. 'Gendered cultures of conflict and discontent: Living "the crisis" in a Newfoundland community'. *Women's Studies International Forum* 23: 343–53.

Diduck, A.P., et al. 2007. 'Achieving meaningful public participation in the environmental assessment of hydro development: Case studies from Chamoli District, Uttarakhand, India'. *Impact Assessment and Project Appraisal* 25 (3): 219–31.

Draper, D. 2004. 'Marine and freshwater fisheries'. In B. Mitchell, ed., *Resource and Environmental Management in Canada: Addressing Conflict and Uncertainty*, 3rd edn, 200–32. Toronto: Oxford University Press.

Draper, D., and M.G. Reed. 2009. *Our Environment: A Canadian Perspective*. Toronto: Nelson Education.

Dunk, T. 2003. *It's a Working Man's Town: Male Working Class Culture in Northwestern Ontario*. 2nd edn. Montreal: McGill–Queen's University Press.

———. 2005. 'Hunting and the politics of identity in Ontario'. In L. King and D. McCarthy, eds, *Environmental Sociology: From Analysis to Action*, 394–408. Lanham, MD: Rowman and Littlefield.

Dunk, T., and D. Barton. 2005. 'The logic and limitation of male working-class culture in a resource hinterland'. In B. van Hoven and K. Hörschelman, eds, *Spaces of Masculinities*, 31–44. London and New York: Routledge.

Ekers, M. 2009. 'The political ecology of hegemony in Depression-era British Columbia, Canada: Masculinities, work and the production of the forestscape'. *Geoforum* 40 (3): 303–15.

Ellsworth, J.P., and L. Jones-Walters. 2004. 'Journeys in governance: The role of federal governments in addressing tough community issues and their underlying causes'. (Unpublished working paper, April).

Francis, G. 2003. 'Governance for conservation'. In F.R. Westley and P.S. Miller, eds, *Experiments in Consilience: Integrating Social and Scientific Responses to Save Endangered Species*, 223–379. Washington: Island Press.

Friedland, M.L. 2002. *The University of Toronto: A History*. Toronto: University of Toronto Press.

Fullerton, M. 2006. 'Gender structures in forestry organizations: Canada'. In UNECE/FAO Team of Specialists on Gender and Forestry, eds, *Time for Action: Changing the Gender Situation in Forestry*, 20–6. Rome: Food and Agriculture Organization of the United Nations.

Gherardi, S., and B. Poggio. 2001. 'Creating and recreating gender in organizations'. *Journal of World Business* 36 (3): 245–59.

Gibson-Graham, J.K. 1994. 'Stuffed if I know! Reflections on post-modern feminist social research'. *Gender, Place and Culture* 1: 205–24.

Global Forest Watch. 2000. *Canada's Forests at a Crossroads: An Assessment in the Year 2000*. Washington: World Resources Institute.

Graham, J., B. Amos, and T. Plumptre. 2003. *Governance Principles for Protected Areas in the 21st Century*. Ottawa: Institute on Governance.

Green, K. 1996. 'Sociodemographic factors and mail survey response'. *Psychology and Marketing* 13: 171–84.

Hanna, K., D.S. Slocombe, and D. Clark, eds. 2008. *Transforming Parks and Protected Areas: Policy and Governance in a Changing World*. Abingdon, UK: Routledge/Taylor and Francis.

Hayter, R. 2000 'Single industry resource towns'. In E. Sheppard and T. Barnes, eds, *A Companion to Economic Geography*, 290–307. Oxford: Blackwell.

Healey, P. 1997. *Collaborative Planning: Shaping Places in Fragmented Societies*. Vancouver: University of British Columbia Press.

———. 2003. 'Collaborative planning in perspective'. *Planning Theory* 2 (2): 101–23.

Johnsson-Latham, G. 2007. *A Study on Gender Equality as a Prerequisite for Sustainable Development*. Report to the Environment Advisory Council, Sweden. http://www.sou.gov.se/mvb/pdf/rapport_engelska.pdf.

Kafarowski, J. 2006. 'Gendered dimensions of environmental health, contaminants and global change in Nunavik, Canada'. *Études/Inuit/Studies* 30 (1): 31–49.

Kasperson, J.X., R.E. Kasperson, and K. Dow. 2001. 'Introduction'. In J.X. Kasperson and R.E. Kasperson, eds, *Global Environmental Risk*. Tokyo and London: United Nations University Press.

Kasperson, R. 2006. 'Rerouting the stakeholder express'. *Global Environmental Change* 16: 320–2.

Kooiman, J. 2003. *Governing as Governance*. London: Sage.

Kusel, J. 2001. 'Assessing well-being in forest dependent communities'. In G.J. Gray, M.J. Enzer, and J. Kusel, eds, *Understanding Community-Based Forest Ecosystem Management*, 359–82. New York, London, and Oxford: The Haworth Press.

Lebel, P.M. 2008. 'The capacity of Montreal Lake, SK, to provide safe drinking water'. (University of Saskatchewan, Saskatoon, MA thesis).

Lidestav, G., and M. Ekström. 2000. 'Introducing gender in studies on management behaviour among non-industrial private forest owners'. *Scandinavian Journal of Forest Research* 15: 378–86.

Livesey, B. 1994. 'The politics of Greenpeace'. *Canadian Dimensions* 28: 7–12.

McCarthy, D. 2003. 'Post-normal governance: An emerging counter-proposal'. *Environments* 31 (1): 79–91.

McDowell, L. 1994. 'Making a difference: Geography, feminism and everyday life—An interview with Susan Hanson'. *Journal of Geography in Higher Education* 18: 19–32.

McGurk, B., J. Sinclair, and A. Diduck. 2006. 'An assessment of stakeholder advisory committees in forest management: Case studies from Manitoba, Canada'. *Society and Natural Resources* 19: 809–26.

Mansbridge, J. 1999. 'Should blacks represent blacks and women represent women? A contingent "yes"'. *Journal of Politics* 62 (3): 628–57.

Marchand, M.H., and J.L. Parpart. 1995. 'Part I Exploding the canon: An introduction/conclusion'. In M.H. Marchand and J.L. Parpart, eds, *Feminism/Postmodernism/Development*, 1–23. London and New York: Routledge.

Martz, D., et al. 2006. *Hidden Actors, Muted Voices: The Employment of Rural Women in Canadian Forestry and Agri-Food Industries*. Ottawa: Status of Women Policy Research Fund.

Mendis, S. 2004. *Assessing Community Capacity for Ecosystem Management: Clayoquot Sound and Redberry Lake Biosphere Reserves*. (University of Saskatchewan, Saskatoon, MA thesis).

Merchant, C. 1995. *Earthcare: Women and the Environment*. London and New York: Routledge.

Mills, S.E. 2006. 'Segregation of women and Aboriginal people within Canada's forest sector by industry and occupation'. *Canadian Journal of Native Studies* 26 (1): 147–71.

Mitchell, B. 2002 *Resource and Environmental Management*. Harlow, UK: Pearson.

Mohai, P. 1992. 'Men, women and the environment: An examination of the gender gap in environmental concern and activism'. *Society and Natural Resources* 5: 1–19.

Moore-Milroy, B. 1996. 'Women and work in a Canadian community'. In J. Caufield and L. Peake, eds, *City Lives and City Forms: Critical Research and Canadian Urbanism*, 215–38. Toronto: University of Toronto Press.

Mosse, J.C. 1993. *Half the World, Half a Chance: An Introduction to Gender and Development*. Oxford: Oxfam.

Müller, S. 'Report on the Professional and Family Situation of Women Working in the Environmental Field'. In T. Eberhart and C. Wachter, eds, *Conference Proceedings of the 2nd European Feminist Research Conference, Feminist Perspectives on Technology, Work and Ecology*, 5–9 July, Graz, Austria, 288. Graz: IFF IFZ Interdisciplinary Research Center for Technology, Work and Culture.

Nesmith, C., and P. Wright. 1995. 'Gender, resources and environmental management'. In B. Mitchell, ed., *Resource Management and Development: Addressing Conflict and Uncertainty*, 2nd edn, 80–98. Toronto: Oxford University Press.

Northlands College, the Labour Market Committee, and Saskatchewan Ministry of Advanced Education, Employment and Labour. 2008. 'Northern Saskatchewan: Regional Training Needs Assessment Report 2008–09'. La Ronge: Ministry of Advanced Education, Employment and Labour. http://www.fnmr.gov.sk.ca/regional-training-rept2008.pdf.

Olsson, P. 2007. 'The role of vision in framing adaptive co-management processes: Lessons from Kistianstads Vattenrike, southern Sweden'. In D. Armitage, F. Berkes, and N. Doubleday, eds,

Adaptive Co-Management: Collaboration, Learning, and Multi-level Governance, 268–85. Vancouver: University of British Columbia Press.

Olsson, P., et al. 2006. 'Shooting the rapids: Navigating transition to adaptive governance of socio-ecological systems'. *Ecology and Society* 11 (1): 18.

Ontario Ministry of Training, Colleges, and Universities. Research and Planning Branch. 2008. 'Labour market information and research'. Toronto: Ontario Ministry of Training, Colleges, and Universities. http://www.gov.on.ca/GOPSP/en/graphics/247423.pdf.

Ostrom, E. 1990. *Governing the Commons: The Evolution of Institutions for Collective Action*. Cambridge: Cambridge University Press.

Ostrom E., et al. 1999. 'Revisiting the commons: Local lessons, global changes'. *Science* (284): 278–82.

Parkins, J. 2002. 'Forest management and advisory groups in Alberta: An empirical critique of an emergent public sphere'. *Canadian Journal of Sociology* 27: 163–84.

Parkins, J.R., et al. 2006. *Public Participation in Forest Management: Results from a National Survey of Advisory Committees*. Information Report, NOR-X-409. Edmonton: Natural Resources Canada, Canadian Forest Service.

Parkins, J.R., and R.E. Mitchell. 2005. 'Public participation as public debate: A deliberative turn in natural resource management'. *Society and Natural Resources* 18 (6): 529–40.

Parkins, J.R., and A.J. Sinclair. 2009 (submitted). 'The narrowing of public life and the limits of environmental governance'. *Journal of Environmental Policy and Planning*.

Pearse, P.H. 1976. *Timber Rights and Forests Policy in British Columbia: Report of the Royal Commission on Forest Resources*. Victoria: Queen's Printer.

Pollock, R., M.G. Reed, and G. Whitelaw. 2008. 'Steering governance through regime formation at the landscape scale: Evaluating experiences in Canadian biosphere reserves'. In K. Hanna, D. Clark, and S. Slocombe, eds, *Transforming Parks: Protected Areas Policy and Governance in a Changing World*, 110–33. London: Routledge.

Preston, V., et al. 2000. 'Shifts and the division of labour in childcare and domestic labour in three paper mill communities'. *Gender, Place and Culture* 7: 5–19.

Reed, M.G. 2003a. 'Marginality and gender at work in forestry communities of British Columbia, Canada'. *Journal of Rural Studies* 19: 373–89.

———. 2003b. *Taking Stands: Gender and the Sustainability of Rural Communities*. Vancouver: University of British Columbia Press.

———. 2008. 'Reproducing the gender order in Canadian forestry: The role of statistical representation'. *Scandinavian Journal of Forest Research* 23 (1): 78–91.

Reed, M.G., and S. Christie. 2009. 'We're not quite home: Re-viewing the gender gap in environmental geography'. *Progress in Human Geography* 33 (2): 246–55.

Reed, M.G., and J. Varghese. 2007. 'Gender representation on Canadian forest sector advisory committees'. *Forestry Chronicle* 83: 515–25.

Resurreccion, B., and R. Elmhirst, eds. 2008. *Gender and Natural Resource Management: Livelihoods, Mobility and Interventions*. London: Earthscan/IDRC.

Richardson, K. 2008. 'A gendered perspective of learning and representation on forest management advisory committees in Canada'. (University of Manitoba, Winnipeg, MA thesis).

Robertson, M. 2005. 'The neoliberalization of ecosystem services: Wetland mitigation banking and problems in environmental governance'. *Geoforum* 35: 361–73.

Rocheleau, D., B. Thomas-Slayter, and E. Wangari. 1996. *Feminist Political Ecology: Global Issues and Local Experiences*. London and New York: Routledge.

Sachs, C., ed. 1997. *Women Working in the Environment*. Washington: Taylor and Francis.

Sandilands, C. 2005. 'Where the mountain men meet the lesbian rangers: Gender, nation, and nature in the Rocky Mountain national parks'. In M. Hessing, R. Raglon, and C. Sandilands, eds, *This Elusive Land: Women and the Canadian Environment*, 142–62. Vancouver: University of British Columbia Press.

Seager, J. 1993. *Earth Follies: Coming to Terms with the Global Environmental Crisis*. New York: Routledge.

———. 1996. '"Hysterical housewives" and other mad women: Grassroots environmental organizing in the United States'. In D. Rocheleau, B. Thomas-Slayter, and E. Wangari, eds, *Feminist*

Political Ecology: Global Issues and Local Experiences, 271–86. London and New York: Routledge.

Sims, L., and A.J. Sinclair. 2008. 'Learning through participatory resource management programmes: Case studies from Costa Rica'. *Adult Education Quarterly* 58 (2): 151–68.

Sinclair, A.J., A. Diduck, and P. Fitzpatrick. 2007. 'Conceptualizing learning for sustainability through environmental assessment: Critical reflections on 15 years of research'. *Environmental Impact Assessment Review* 28: 415–28.

Skocpol. T. 2004. 'Voice and inequality: The transformation of American civic democracy'. Presidential address to the American Political Science Association. *Perspectives on Politics* 2: 3–20.

Smith, D. 1987. *The Everyday World as Problematic: A Feminist Sociology*. Boston: Northeastern University Press.

Statistics Canada. 2007. *Special Tabulation of the Canada Census 2006*. Ottawa: Statistics Canada, Public Institutions Division.

Status of Women Canada. n.d. *Gender-Based Analysis: A Guide for Policy Making*. Ottawa: Status of Women Canada.

Stewart, J.M., and A.J. Sinclair. 2007. 'Meaningful public participation in environmental assessment: Perspectives from Canadian participants, proponents and government'. *Journal of Environmental Assessment and Policy Management* 9 (2): 1–23.

Sturgeon, N. 1997. *Ecofeminist Natures: Race, Gender, Feminist Theory and Political Action*. New York and London: Routledge.

Teghtsoonian, K. 2005. 'Disparate fates in challenging times: Women's policy agencies and neoliberalism in Aotearoa/New Zealand and British Columbia'. *Canadian Journal of Political Science* 38 (2): 307–33.

Teske, E., and B. Beedle. 2001. 'Journey to the top—Breaking through the canopy: Canadian experiences'. (Unpublished report for the Canadian Forest Service and BC Ministry of Forests, Victoria).

Thomas, J., and P. Mohai. 1995. 'Racial, gender, and professional diversification in the Forest Service from 1983–1992'. *Policy Studies Journal* 23: 296–309.

Tindall, D.B., S. Davies, and C. Mauboules. 2003. 'Activism and conservation behavior in an environmental movement: The contradictory effects of gender'. *Society and Natural Resources* 16: 909–32.

Tripp-Knowles, P. 1999 'The feminine face of forestry in Canada'. In E. Smyth, S. Acker, P. Bourne, and A. Prentice, eds, *Challenging Professions: Historical and Contemporary Perspectives on Women's Work*, 194–211. Toronto: University of Toronto Press.

Uliczka, H., et al. 2004. 'Non-industrial private forest owners' knowledge of and attitudes towards nature conservation'. *Scandinavian Journal of Forest Research* 19: 274–88.

UNECE/FAO (United Nations Economic Commission for Europe/Food and Agriculture Organization) Team of Specialists on Gender and Forestry, eds. 2006. *Time for Action: Changing the Gender Situation in Forestry*. Rome: Food and Agriculture Organization of the United Nations.

University of Northern British Columbia. 2008. *Women's Perspectives on the Mountain Pine Beetle Project. Final Report and Evaluation*. Prince George: University of Northern British Columbia.

Women and Geography Study Group of the Institute of British Geographers (WGSGIBG). 1997. *Feminist Geographies: Explorations in Diversity and Difference*. Essex: Longman.

World Commission on Environment and Development. 1987. *Our Common Future*. Oxford: Oxford University Press.

Wright, M. 2001. 'Women in the Newfoundland fishery'. In S.A. Cook, L.R. McLean, and K. O'Rourke, eds, *Framing Our Past: Canadian Women's History in the Twentieth Century*, 343–6. Toronto: Oxford University Press.

Young, N. 2008. 'Radical neoliberalism in British Columbia: Remaking rural geographies'. *Canadian Journal of Sociology* 33 (1): 1–36.

Young, N., and R. Matthews. 2007. 'Resource economies and neoliberal experimentation: The reform of industry and community in rural British Columbia'. *Area* 39 (2): 176–85.

20

Sustainability Governance: Surfing the Waves of Transformation

Anthony H.J. Dorcey

Learning Objectives

- To understand how conflict and uncertainty in resource and environmental management have increased with the adoption and pursuit of ecological, economic, and social sustainability goals.

- To assess how participation in Canadian governance processes has evolved in three waves of innovation over the past 50 years in pursuing sustainability goals.

- To define approaches to citizen involvement and conflict resolution employed in sustainability governance and their use of techniques of negotiation, facilitation, and mediation.

- To explore how experimental development of participatory approaches, including multi-stakeholder processes and electoral reform, might contribute to improvements in sustainability governance.

For more than three decades, it has been recognized that increasing demands, complexity, and uncertainty are exacerbating conflict in resource and environmental management in Canada and around the world (Dorcey 1986). The emergence in the late 1980s of sustainability principles integrating ecological, economic, and social imperatives has heightened these conflicts and stimulated a search for new forms of governance to avoid or resolve them (Dorcey and McDaniels 2001). In this chapter, I examine the increasingly diverse and fundamental changes in governance being considered and implemented in Canada. First, I define some key terms and concepts of sustainability governance focusing on citizen involvement, the avoidance and resolution of conflict, and consensus-building. I then discuss how well we understand the efficacy of these governance innovations and argue for a strategy of explicit experimental development. I conclude by examining the application of such a strategy within Greater Vancouver. I suggest that a third wave of transformation in governance is urgently required and perhaps building, one that has the potential to be more far-reaching than the two that have preceded it during the past half-century.[1]

SUSTAINABILITY: NEW PRINCIPLES, HEIGHTENED CONFLICT

Since the Brundtland Report (WCED 1987), resource and environmental management has increasingly been viewed within the larger concept and decision context of sustainable development and sustainability. The report of the United Nations World Commission on Economic Development (WCED 1987, 43) defined sustainable development as 'development that meets the needs of the present without compromising the ability of future generations to meet their own needs'. While there has been great controversy about the definition of sustainable development and sustainability in the academic literature and discussion forums around the globe, there has been gradual clarification of the differing interpretations of the terms, which have fuelled and become central to policy debates from the local to the global level.[2] In the process, resource and environmental management issues such as depletion of forest and fishery resources and air and water pollution have come to be seen as both cause and consequence of other major economic and social problems such as poverty, disease, and corruption. This has led to resource and environmental management issues being viewed as integral components of almost every area of policy. For example, transportation policy choices are recognized as needing to take into account implications not only for pollutant emissions and energy conservation but also for land-use development and urbanization patterns with all of their associated far-reaching economic, environmental, and social consequences.

As expected, but perhaps more than was anticipated, this has led to increasing conflict in resource and environmental management generated by the compounding interactions of increasing demands, complexity, and uncertainty (Dorcey 1995). Driven by growing populations, economic development, technological innovations, and shifting preferences, demands on the resource base have multiplied and diversified worldwide and provoked greater conflict. Disputes have not only become more frequent because of the expanding numbers of stakeholders and their interactions but have also become more difficult to avoid and resolve as they come to be expressed in terms of the multiple dimensions and values of economic, environmental, and social sustainability. While resource and environmental management issues have long been recognized as involving value-laden decisions and ethical choices, the sustainability perspective has elaborated these issues in yet more comprehensive terms, often science-intensive, appearing to threaten traditional power relations, which has heightened the potential for conflict. Illustrative of this are the questions raised about whether global resource and environmental problems can be resolved as long as North America is committed to its current model as a consumer society and others seek to emulate it.[3]

Increasing complexity further enhances the likelihood of conflict. Expansion of complexity results from the exponential growth of biophysical and socio-economic interactions accompanying population increases, economic development, and technological innovation. Institutional and governance innovations both mirror and compound the complexity. Conflict escalates not only because of the expanded numbers of interactions but also because of the challenge of understanding them. The sustainability perspective again further complicates the picture in that it demands consideration of systems' behaviours and boundaries stretching from the local to the global, among all natural and human systems, and over past, present, and future time. The expansion of conflict that has resulted from the growing concerns about the implications of climate change for sustainability illustrates this challenge only too well. Consider the debate that raged over the past decade around the extent to which observed climate extremes result from natural dynamics, comparable to those known to have occurred in earlier times, or from changes induced by the immense increases in human activity in the recent past and the controversy about whether or not to take mitigatory actions from the local to the global level, with implications for almost all sectors of human activity and their governance (see Chapter 5).

Increasing uncertainty enhances the likelihood of conflict still further. Despite immense growth in knowledge from research relating to resource and environmental systems and their management, uncertainty has continued to grow. It stems from the discovery that the behaviours of key systems are inherently unpredictable and from research frequently generating more questions than answers. The more comprehensive perspective of sustainability, with its greatly expanded demands for understanding of environmental, social, and economic systems, has exacerbated this difficulty and hugely heightened the likelihood of conflict. The challenge of acting on issues relating to climate change, in which continuing uncertainties about science and values are all-pervasive and at a time of unanticipated collapse of global financial and economic systems, demonstrates these difficulties only too well.

Escalating demands, complexity, and uncertainty thus feed on each other, increasing the likelihood of conflict as illustrated by the climate change example. At one extreme, some analysts grimly predict the consequent proliferation of war (Homer-Dixon 1999).[4] Putting resource and environmental management into the more comprehensive and demanding perspective of economic, social, and environmental sustainability has greatly heightened the likelihood of conflict and thus poses daunting challenges for governance systems.

GOVERNANCE: THREE WAVES OF INNOVATION

Over the past 40 years and particularly with the development of the sustainability perspective, the management of resources and the environment has increasingly been seen in the larger context of governance systems and the varied potential roles of governments, business, and civil society within them (Dorcey 1986, 1995; Dorcey and McDaniels 2001).[5] Citizen involvement in the management of resources and environment from this broader governance perspective potentially includes roles as varied as voter in elections and referenda, elected representative, political activist, buyer and seller in markets, volunteer producer, petitioner in the courts, or participant in government or business processes. During the past four decades, there have been two major waves of experimentation with innovations in these governance roles, and a third wave, on occasion, appears as though it might be building. These experiments were initially concentrated in North America, but the waves have rippled around the globe, and each new one builds on those preceding it. The desire to avoid or resolve environmental and resource management conflicts has stimulated many of the innovations, but they have also been developed in social and economic areas of decision-making, particularly with the emergence of the sustainability perspective over the past decade.

First Wave

The first wave of innovation occurred from the mid-1960s to the late 1980s, originated in the United States, quickly spread to Canada, and more slowly rippled around the globe. The emergence of widespread environmental and social concerns was a principal catalyst for policy and associated citizen-involvement innovations in the second half of the 1960s and the first half of the 1970s. In Canada, there were three foci for innovation, involving governments at the federal, provincial, and local levels: planning for urban development, river basin management, and assessments for project development. Experiments involved the use of a variety of communication and participatory techniques, including information brochures, media releases, citizen surveys, public hearings, workshops, task forces, and advisory committees. By the mid-1970s, however, enthusiasm for the ambitious experiments in citizen involvement began to wane as they were perceived as unsuccessful in resolving issues, time-consuming, and costly. At a time when the Canadian economy was weak, negative perceptions overwhelmed the positive aspects, and for the ensuing decade much less attention was paid to environmental policy and citizen involvement in environmental governance.

Second Wave

In the second half of the 1980s, environmental concerns re-emerged as priority issues in the new context of sustainable development and generated a second wave of innovations in policy and citizen involvement in Canada that have been influential around the world as countries responded to the Brundtland Report. Building on the lessons of the earlier experiments and the emerging experience with the use of negotiation, facilitation, and mediation, a new generation of techniques involving multi-stakeholder, conflict resolution, and consensus-building processes characterized the second wave. The processes were initiated not only by governments at all levels but also by business and civil society, and commonly they involved stakeholders from all three sectors. They have been utilized in making decisions on every kind of environmental, economic, and social issue and as part of governance processes from the global to the local level when seeking agreements on everything from constitutions to legislation, policies, regulations, plans, and project implementation. However, by the mid-1990s, the hugely ambitious innovations were once again being questioned as they came to be perceived as too lengthy and costly and of limited value in terms of reaching and implementing agreements that met the requirements of the diversity of stakeholders. Again, as at the end of the first wave, the need to address economic crises in Canada resurfaced at the top of the agenda, and governments and other stakeholders retreated from the vigorous pursuit of innovation in environmental and social sustainability policy and citizen involvement.

Third Wave

At the beginning of the new millennium, and in the lead-up to events surrounding and immediately following the World Summit in Johannesburg in 2002, it appeared that a third wave of innovation might be in the offing (Dodds 2001; Knight, Chigudu, and Tandon 2002). Preparatory reports and debate during the summit emphasized that while there had been notable areas of progress since the Rio Earth Summit in 1992 on its Agenda 21, overall, intertwined environmental, social, and economic problems were becoming more grave and seriously threatened global sustainability. Revitalizing democratic governance processes was seen to be of fundamental strategic importance in fostering both improved understanding of sustainability problems and choices and forging willingness to act. In Canada, there had been a notable increase in the discussion, particularly in the media, of the need for governance reform to address the diversity of pressing economic, social, and environmental issues. As in the case of the second wave of experiments, the governance innovations under discussion incorporated those included earlier but were also more far-reaching and fundamental. But as discussed below, the third wave has built only slowly in the years following the Johannesburg Summit—until now, when multiple crises around the world appear to have it surging forward once again.

If the first wave of experiments was about whether citizens should be involved in resource and environmental management and the second wave was about how negotiation-based techniques of dispute resolution, consensus-building, and multi-stakeholder processes might enhance involvement, then the emerging third wave will be about whether the techniques and processes introduced during the first two waves can ever be expected to achieve their goals without much more fundamental changes to the governance systems within which they are employed.

ASSESSING GOVERNANCE INNOVATIONS: READING THE WAVES

It is not easy to assess the citizen involvement and conflict resolution innovations in governance during the first two waves because of problems in the literature relating to inconsistent terminology, implicit goals, and limitations of existing research. Nevertheless, there is a basis for guiding the strategies necessary for more productive experimental development in a third wave (Dorcey and McDaniels 2001).

Terminology

Major difficulties in assessing innovation have derived from confusion about key terms such as 'citizen involvement', 'conflict resolution', and 'governance' (as well as 'sustainable development' and 'sustainability'). Progress in the third wave will depend critically on pursuing a strategy of expecting differences among stakeholders in the interpretation of commonly used terms, always seeking to clarify differences in their meaning, and capitalizing on differences to foster fresh insights (and the same, of course, applies to sustainable development and sustainability). The following sections briefly describe principal instances of terminology difference and set the stage for exploring and defining them as they relate to specific situations in third-wave experimentation.[6]

Citizen Involvement

Difficulties arise from great variations in the use of central terms, such as 'citizen involvement', and because differences in usage are often not made explicit. Various writers may use key words such as 'public' or 'civic' or 'community' or 'stakeholder' instead of 'citizen', and 'participation' or 'engagement' or 'consultation' in place of 'involvement'. On some occasions, these terms are used synonymously; on others, there are significant differences in intent. For example, in certain instances, 'stakeholder involvement' is differentiated from 'citizen involvement' by limiting the former only to those who have a specific interest in the issue as opposed those who are generally interested as citizens (e.g., the affected landowners versus all voters in the jurisdiction). In other situations, the term 'stakeholder' may be used to identify non-governmental interests, implying that the participants represent discrete constituencies. On yet other occasions, 'participation' is distinguished from 'involvement' or from 'engagement' as being more passive (e.g., citizens being merely informed versus actively contributing to or making decisions). Commonly, 'consultation' is differentiated from 'involvement' as being a purely advisory process as opposed to providing for direct decision-making (see also Chapter 18).

In the third wave of innovations, it will be essential to define citizen involvement broadly as processes for the involvement of citizens in advising on and making decisions on matters under government authority that augment or supplant decision-making through established channels of representative government. Within this broad definition, it will also be critically important to distinguish clearly the specific intents when alternative terms are employed, as illustrated above.

Conflict Resolution

Comparable difficulties in assessing experience arise from differing use of the many key terms relating to conflict resolution processes and techniques (see Box 20.1). Conflict arises among citizens involved in all processes of governance. Legislatures, courts, and markets are mechanisms specifically designed to work through conflict as well as for avoiding and resolving conflicts. However, the initial interest in conflict resolution in environmental and resource management in the second wave of innovations usually had a more limited focus. At the outset, it was often referred to as 'alternative dispute resolution' (ADR) or 'environmental mediation', because processes such as mediation were seen as being more cost-effective alternatives for resolving conflicts than

Box 20.1 Negotiation, Facilitation, Mediation, and Consensus

The terms 'negotiation', 'facilitation', 'mediation', and 'consensus' are used in many varied ways. Sometimes they are referred to as 'processes' and other times as 'techniques'. For example, on one occasion a consensus process may be described as utilizing techniques of negotiation, facilitation, or mediation. On another occasion, a negotiation process may be described as employing the technique of consensus. These differences are explored more fully in Dorcey and McDaniels (2001). Below are some basic definitions to guide the present discussion.

Negotiation can be defined as 'a process whereby two or more parties attempt to settle what each shall give and take, or perform and receive, in a transaction between them' (Rubin and Brown 1975, 2). Fisher and Ury (1981) in their seminal book *Getting to Yes: Negotiating Agreement Without Giving In*, describe negotiation techniques that have been widely employed.

Facilitation is provided by a facilitator, who has been defined as 'an individual who enables groups and organizations to work more effectively; to collaborate and achieve synergy. She or he is a "content neutral" party who, by not taking sides or expressing or advocating a point of view during a meeting, can advocate for fair, open and inclusive procedures to accomplish the group's work. A facilitator can also be a learning or dialogue guide to assist a group in thinking deeply about its assumptions, beliefs and values and about its systemic processes and context' (Kaner et al. 2007, xv). Kaner et al. provide a guide to widely used techniques of facilitation.

continued

Mediation is 'an extension or elaboration of the negotiation process that involves the interven-tion of an acceptable third-party who has limited (or no) authoritative decision-making power. This person assists the principal parties in voluntarily reaching a mutually acceptable settle-ment of the issues in dispute. . . . [Mediation] is usually initiated when the parties no longer believe that they can handle the conflict on their own and when the only means of resolution appears to involve third-party assistance' (Moore 2003, 8). Mediation thus employs the pro-cesses and techniques of negotiation and facilitation and more besides (e.g., caucusing). In *The Mediation Process: Practical Strategies for Resolving Conflict*, Moore summarizes com-monly used techniques of mediation.

Consensus is '*the process*—a participatory process by which a group thinks and feels together en route to their decision. Unanimity, by contrast, is the point at which the group *reaches clo-sure*. Many groups that practice consensus decision-making use unanimity as their decision rule for reaching closure—but many *do not*' (Kaner et al. 2007, 276; emphasis in original). Consensus processes frequently employ the processes and techniques of negotiation and facilitation and sometimes mediation.

Multi-stakeholder processes involve a diversity of stakeholders (usually including govern-ment, business, and civil society participants) and variously utilize negotiation, facilitation, mediation, and consensus processes and techniques.

Third parties are individuals or groups who assist those involved in negotiation, facilitation, mediation, consensus, and multi-stakeholder processes. They are often called 'facilitators' or 'mediators', but other labels are used in specific contexts, such as 'conciliators', 'convenors', 'fact-finders', and 'problem-solvers' (Dorcey and Riek 1987).

using the courts and administrative processes of governments. But as the innovations expanded to seeking agreements, as well as responding to disputes in land and resource planning and in developing environmental regulations and poli-cies, they came to be referred to by more var-ied terms, in particular 'consensus processes' and 'multi-stakeholder processes'. Applications of these types of processes in turn led to more specific terms such as 'shared decision-making' (Commission on Resources and Environment 1994), 'reg-neg' (regulatory-negotiation) (BC Round Table on the Environment and the Econ-omy 1991), 'co-management' (National Round Table on Environment and Economy 1998), and 'civic science' (Lee 1993).

In the third wave of innovations, it will be essential to view conflict resolution as part of processes for reaching consensus that are not just focused on the resolution of disputes once they arise. More fundamentally, the innovations will need to consider processes for exploiting the advantages and avoiding the disadvantages of both cooperation and conflict. In this larger con-text, techniques of negotiation, facilitation, and mediation developed during the second wave need to be recognized as central to each of these processes. While there are clearly times when not all stakeholders have to be involved in reaching decisions, multi-stakeholder processes strategic-ally employing the full array of potential tech-niques will be essential if the diversity of citizen involvement necessary to meet the challenges of understanding and implementing sustainable development are to be met. In these multistakeholder negotiation processes, assistance by facilitators and mediators is critical to success in reaching consensus.

Governance

Adding to the difficulties of assessing experience has been the relative novelty and breadth of the concept of 'governance' within which citizen involvement has increasingly come to be considered over the past two decades. A new term was seen to be needed to focus discussion on the complex of interacting organizations and systems of government, business, and civil society within which decisions are made by citizens in their many varied roles (see also Chapter 2). Within this broad concept of governance, all the forums and activities of government (executive, legislative, administrative, and judicial), at all levels (from the local to the global), have come under scrutiny, which has led to innovations in citizen involvement and conflict resolution processes. Accompanying this development has been the new terminology of 'stewardship', 'partnership', 'collaboratives', and 'round tables'.

In the third wave of innovations, it will be essential to consider citizen involvement and conflict resolution in the broad context of 'alternative governance regimes' (AGR) with all of their complex component parts. Governance can be simply defined for these purposes as 'collective decision-taking and action in which government is one stakeholder among others' (Knight, Chigudu, and Tandon 2002, 131).

Goals

Assessments of citizen involvement and conflict resolution innovations in Canadian governance depend fundamentally on the preferred model of democracy and associated procedural and outcome goals. The first two waves of innovation took place in an era of major shifts around the world in dominant ideologies, including the collapse of communist regimes and the ascendancy of neo-liberalism with its emphasis on free markets and free trade in a context of globalization (see Chapters 1 and 2). Accompanying these shifts has been a general belief in the superiority of liberal democratic forms of governance but also that the role of government needs to be reduced and those of business and civil society increased. In the process, conventional views of 'left' and 'right' approaches to governance have blurred as 'third way' approaches, such as the adoption of market-based mechanisms by social democratic regimes (e.g., privatization of water), have been introduced. In this dynamic environment, designing and assessing the merits of third-wave innovations in governance will depend critically on being much more explicit than in the past about the goals they are intended to meet.

Democratic Models

Competing managerialist, pluralist, and populist models of democracy have significantly different definitions of the appropriate role of citizen involvement in governance (Beierle and Cayford 2002). The managerialist view is that elected representatives and their administrators should be responsible for identifying and pursuing the common good. While citizens might be involved in various ways to inform the shaping of decisions, they should not be directly involved in decision-making because self-interested behaviour might threaten the common good. In contrast, the pluralist view sees government not as the manager but as the arbitrator among competing interest groups. From this perspective, there is no single common good to be identified but only a preferred one that results from negotiations among the interest groups. The populist view, on the other hand, argues that decisions should be made directly by citizens and not through representatives on the grounds that such involvement is essential in developing democratic values and hence the performance of the governance system.

In the third wave of experiments, it will be essential to be much more explicit about the models of democracy that underlie innovations in citizen involvement—in particular how the traditional and enduring dominance of managerialist views are to be challenged by pluralist and populist alternatives. Central to this will be the exploration of broader concepts of citizenship and the role of citizens with rights and responsibilities in sustainability governance. Drawing on the comments of nearly 10,000 citizens in 47 Commonwealth countries, Knight, Chigudu, and Tandon (2002) argue that a new consensus is emerging on the three-part requirements for reviving democracy: (1) a strong state and a strong civil society; (2) a 'deepened' democracy and democratic culture; and (3) an enlarged role for citizens.[7]

Outcome Goals

Understanding the governance alternatives and their merits is greatly enhanced by being explicit about the specific goals of citizen involvement in terms of the problems they are intended to address. Beierle and Cayford (2002) have suggested a set of five outcome-oriented social goals:

1. incorporating public values into decisions;
2. improving the substantive quality of decisions;
3. resolving conflict among competing interests;
4. building trust in institutions; and
5. educating and informing the public.

Again, in the third wave of innovations it will be important to be more explicit about the outcome goals than in the past. Further, while they might be subsumed within the above goals,

it is essential to include specific consideration of empowerment, equity, and cost-effectiveness, because they are central issues in disputes about the relative merits of alternative models of democracy and in the challenges of implementing sustainability governance.

Offering another perspective, Knight, Chigudu, and Tandon (2002, 160) found that the good governance goals voiced by citizens were threefold: (1) basic needs, (2) association, and (3) participation:

1. In fulfilling basic needs, the state is expected to play a providing role. Active citizens complement such a providing role by playing their part.
2. In strengthening associational aspects of society, collective citizen action is the 'actor'. The state complements this with its facilitator role, which is crucial to building and nurturing collective citizen action.
3. In enhancing the participation of citizens, the state has to play the role of an active promoter. By engaging themselves in the public arena, citizens complement the 'promoter' role of the state.

Procedural Goals

While outcome goals relate to desired consequences of citizen involvement, procedural goals focus on who is involved, when, how, and where. The innovations in the second wave were strongly oriented toward addressing procedural goals. This is well exemplified by the Guiding Principles of Consensus Processes developed by the National Round Table on the Environment and the Economy (1993):

1. *Purpose-driven:* People need a reason to participate in the process.
2. *Inclusive, not exclusive:* All parties with a significant interest in the issue should be involved in the consensus process.

3. *Voluntary participation:* The parties affected or interested participate voluntarily.
4. *Self-design:* The parties design the consensus process.
5. *Flexibility:* Flexibility should be designed into the process.
6. *Equal opportunity:* All parties must have equal access to relevant information and the opportunity to participate effectively throughout the process.
7. *Respect for diverse interests:* Acceptance of the diverse values, interests, and knowledge of the parties in the consensus process is essential.
8. *Accountability:* The parties are accountable both to their constituencies and to the process that they have agreed to establish.
9. *Time limits:* Realistic deadlines are necessary throughout the process.
10. *Implementation:* Commitment to implementation and effective monitoring is essential for any agreement.

In the third wave, it will be essential to be explicit about how such procedural goals contribute to the achievement of outcome goals associated with each of the different models of democracy, building on the analyses of researchers such as Beierle and Cayford (2002) and Knight, Chigudu, and Tandon (2002), mentioned above.

Research

While an abundant and diverse body of writing exists on the merits of citizen involvement and conflict resolution, the literature has major limitations that will need to be recognized and overcome in terms of assessing innovations in the third wave. There is, nevertheless, a basis in this literature and the rules of thumb created by experienced practitioners for guiding the assessment of innovations in the third wave through

an explicit strategy of experimental development (Dorcey and McDaniels 2001).

Limitations

Experience has demonstrated that it is exceedingly difficult to conduct evaluation research on citizen involvement and conflict resolution for theoretical, practical, and methodological reasons. As described above, assessments need to be based on explicit theoretical models of democratic governance, linking outcome goals to procedural goals, and making clear the hierarchy that relates ideology to the particular techniques employed. Interest in these questions has catalyzed a remarkably diverse and interdisciplinary literature, much of which has been developing in isolation and needs to be cross-fertilized (e.g., co-management, civic engagement, deliberative democracy, empowerment, and communicative planning).

Methodological problems stemming from weaknesses in research designs and their implementation compound the theoretical difficulties with the existing assessment literature. Much of the early literature on evaluation is questionable and reflects partial and unsubstantiated opinions. Only in recent years have the theory and techniques of qualitative research methods essential to insightful assessments of citizen involvement and conflict resolution processes been advanced to the point where they are beginning to be widely and vigorously used and more commonly accepted. Accompanying this has been recognition of the severe limitations of attempts to assess experiences only after the processes have concluded and the critical need for real-time observation and feedback using participatory evaluation approaches.

Practical problems compound the theoretical and methodological problems in evaluative research. What is relevant theoretically and in practice is highly context-dependent. All too often, the literature on citizen involvement and conflict resolution ignores critical differences in time and place (e.g., the U.S. governance context for conflict resolution is significantly different from that of Canada, and the political climate for experimenting with citizen involvement in British Columbia was much more favourable in the early than the late 1990s). On other occasions, recognition of the context changes that have taken place can regrettably render well-designed studies impotent (e.g., detailed assessments of some of the shared decision-making processes in British Columbia in the first half of the 1990s concluded that innovation potentials were not achieved because governments lost interest and did not sustain the commitments required to give them a reasonable chance of success).

Best Practices

Countering this seemingly bleak perspective on the insights available in the existing research literature is the growing recognition that the rules of thumb developed by practitioners over the past three decades to guide their use of citizen involvement and conflict resolution techniques have great value. A notable example of this is the Guidelines for Consensus Processes (listed above), developed through a process that engaged professional and citizen practitioners involved in round tables across Canada. These guidelines have been widely acclaimed, accepted, and employed around the world. Most significantly, there is growing recognition of their merits as professionals and academics subject them to more vigorous assessment using emerging methods of research.[8]

At the same time, however, there is a growing appreciation of the extent to which the most productive approaches to citizen involvement and conflict resolution are not only contingent on the circumstances but also on the personal approach of the individual(s) responsible for facilitating or mediating the process. These

critical roles are coming to be acknowledged as involving both art and science about which there is only infant understanding. Case studies of practicing facilitators and mediators who are recognized for their expertise confound us when they reveal that these people do not necessarily follow established principles or even do what they normally profess (Kolb 1994).

Experimental Development

Progress in the third wave of innovation therefore demands experimental development strategies that build on experience and focus on learning-by-doing (see also Chapter 16). These strategies will need to be contingent, progressive, structured, and adaptive.

Contingent Strategies

Out of the experience with the first two waves of experimentation has emerged a growing appreciation of when and how to use particular approaches to citizen involvement and conflict resolution (Dorcey, Doney, and Rueggeberg 1994; Thomas 1995).[9] There is recognition that each approach comes with costs and potential benefits and that these factors need to be weighed in a particular context. Thus, there will be some situations in which citizen involvement should not be considered (e.g., when elected representatives have already made a policy decision, the planners responsible for implementation should not organize citizen involvement processes that create the impression that the decision has not yet been taken or that it is open to reconsideration). Conversely, there are situations in which public hearings that merely allow stakeholders to voice their concerns will not be adequate if the need is for developing understanding and agreement. Indeed, they could escalate conflicts. While careful judgment is called for, those responsible for sponsoring citizen involvement, consensus-building, and conflict resolution processes can take a much more discerning and strategic approach to deciding when and how to employ specific approaches in third-wave experiments.[10]

Progressive Strategies

At the same time, critics of the earlier experiments with citizen involvement and conflict resolution (e.g., Arnstein 1969; Forester 1989) have questioned the extent to which they merely reinforce the existing power structures as opposed to trying to redress inequities. They argue on grounds of both normative principle and practical efficacy that progressive approaches should be employed (i.e., there are not only ethical reasons for arguing that the disadvantaged should not be further disadvantaged but also practical reasons, such as the increased costs that would result for everyone from escalation of conflict). In third-wave experiments, it will be important for sponsors of citizen involvement and conflict resolution to address these issues explicitly in mandating processes. Those responsible for facilitating and mediating the processes will need to be clear on their mandate and their own ethical responsibilities to participants who might be disadvantaged as a result of their lack of knowledge, resources, or access.

Structured Strategies

Criticism of first- and second-wave experiments has also focused on their deficiencies in generating and structuring information to aid decision-making by the participants (Raiffa 1982; Hammond, Keeney, and Raiffa 1999). All too often, processes neglect the importance of systematically identifying the goals and objectives and the assessment of the relative merits of alternative ways of achieving them. These critics point to the problems that result from not anticipating the well-recognized tendencies of individuals and groups to ignore or misconstrue complexity and uncertainty and from neglecting the well-developed techniques for aiding decision-making such as 'value-focused thinking' (Keeney 1992). In third-wave experiments, it will be important for facilitators and mediators to be much more aware of how these techniques can be employed to great advantage by the stakeholders within their citizen involvement, consensus-building, and conflict resolution processes.

Adaptive Strategies

Given the uncertain understanding of the merits of differing approaches to citizen involvement, consensus-building, and conflict resolution in varying governance contexts, there is a need in the third wave for experimental development and adaptation as insights are gained (Holling 1978; see also Chapter 16). This implies explicitly designing processes to learn from experiments with specific evaluative questions and methods included. Among the key questions to be addressed are strategic choices among options; roles of convenors, facilitators, and mediators; empowerment; front-end investment in process paying off in the longer term; and fundamental governance system changes that provide the context for citizen involvement, consensus-building, and conflict resolution. The evaluation methods employed need to be participatory and applied in real time. In the remainder of this chapter, these ideas are elaborated through a specific example.

SUSTAINABILITY GOVERNANCE OF GREATER VANCOUVER: SURFING THE THIRD WAVE

In their origins and outcomes, innovations in sustainability governance are very much the product of a particular time and place. This is one of the key lessons learned from assessing the first two waves. Therefore, this last section briefly considers what might be learned in a third wave of innovation by focusing on transforming sustainability governance of Greater Vancouver in the opening years of the new millennium. This is a useful illustrative case because Greater Vancouver is a rapidly urbanizing region of international significance for Canada in a province with a history of innovations in sustainability governance and with an array of initiatives that provide rich opportunities for experimental development in the third wave. But first, I would like to suggest why there are reasons to believe that a third wave might be building.

Context and Prospect: Confused Seas

Bobbing on my west-coast surfboard, I cannot yet see the next big wave clearly formed. The seas are confused with swells, long and short, coming from more than one direction. They are reminiscent of the cross-currents that preceded the first two big waves, but now, driven by awesome storms all around, they look as though they might build up to something much bigger. As before, environmental and natural resources management issues are emerging around the world as growing concerns, but this time—more than for the second wave—they are intimately intertwined with economic and social issues. Environment and natural resource issues and their resolution are not only seen today as implicated in the scourges of poverty, disease, corruption, intolerance, and civil strife but are complicated immeasurably by an extraordinary global financial and economic crisis that threatens to spiral out of control. This complex of issues has created a new sense of insecurity for all—but also a willingness to explore more radical innovation.[11]

As in the build-up to the two preceding waves, innovations in governance have been increasingly advanced as fundamental to resolution of the emerging issues. And as with the second

wave, the third will build on experience with the one preceding it. This time, the emerging proposals are not just for enhanced utilization of the best practices of citizen involvement and conflict resolution processes and techniques distilled from experiences in the first two waves but for more fundamental transformation through democratization of governance processes. However, even within liberal democracies, there are widely differing views on what form it should take, with the continuing dominance of managerialist models being increasingly challenged by pluralist and populist alternatives.

Before 9/11 and the succession of corporate scandals, reform proposals emphasized a smaller role for the state and larger roles for business and civil society. But over subsequent years, significant doubts have arisen (and almost faded away in the fiscal and economic crises of 2008–9) about the proposed reductions in the role of the state. There have been increasingly revised and more refined proposals for strengthening the roles of all three in selective and appropriate ways. Adding momentum to the forces for renewed experimentation is the remarkable globalization of stakeholder involvement, in particular civil society, over the past decade. The development of the world wide web has facilitated the empowerment of civil society organizations in mobilizing, co-ordinating, and engaging in and influencing governance processes—from the local to the global—in ways only dreamt of during the first two waves.

Frustrated and disillusioned by the failure of second-wave multi-stakeholder processes to produce significant progress and enduring results, often concluding they had been co-opted, civil society stakeholders have also resorted to campaigning through other governance forums, including the market (e.g., international campaigns to boycott products produced in unsustainable ways, such as wood) and the political system (e.g., the Green parties). Increasingly in recent years and focusing particularly on multinational corporations and globalization as causes of unsustainability in all its dimensions, civil disobedience and direct action have been employed to influence decisions (e.g., Seattle, Washington, Quebec City, Prague, Genoa, and more recently with the global economic crisis, Paris, London, Athens).[12]

While these global cross-currents are clearly flowing through the governance waters of Greater Vancouver, others have origins in Canada and British Columbia. Environmental groups have widely deplored the neglect of environmental priorities in Canada over the past decade and a half and lamented the demise of the country's international reputation for leadership and innovative policies. Preoccupied with deficit and debt reduction and the challenges of international competitiveness, the focus of governments has been on economic issues and reducing the cost of government by cutting staff and programs. This has been the case for the federal Liberal and Conservative governments throughout this period and increasingly became the focus of the BC government, particularly with the election of the Liberal Party in 2001, which introduced restraint policies more stringent than those previously implemented in Alberta and Ontario. Social programs including health, education, and housing have been cut back, heightening the controversies about the downsizing of federal and provincial governments and downloading onto local governments. The tragic deaths resulting from the failures of the community water supply system in Walkerton, ON, became a symbol of the risks created by the retreat of governments (see Chapter 7). More generally, local governments became increasingly vociferous about their inability to cope with the downloaded responsibilities and the historical neglect of municipalities on the part of senior governments unwilling to provide them with adequate independent legislative or financial capacity or share of revenues.

In British Columbia, the failure of a decade of treaty negotiations to produce agreements and the regressive policy stance of the incoming Liberal government made all of these concerns acute in and around First Nation communities (see Chapter 4). More recent policy reversals by the British Columbia government have not produced the promised results (e.g., progress on treaties with First Nations under the 2005 New Relationship policy is still frustratingly slow), and doubts have been raised about the province's commitment to its many climate change initiatives. One such initiative was introduction of the first carbon tax in North America, but the powerful climate action secretariat was moved out of the premier's office in early 2009 as his attention shifted to other priorities and an upcoming election in mid-May of that year (which returned his government to office, although with a substantially reduced majority).

It is in this context that questioning of governance institutions and their performance has increased and diversified from the national to the local level. The concerns expressed are both general and fundamental, reaching beyond their occasional specific relationship to environmental or sustainability issues. Often, they are summarized in terms of a 'democratic deficit'. Long-expressed dissatisfactions with the dominance of the prime minister—the 'friendly dictatorship' (Simpson 2001)—the lack of influence of elected members of Parliament, and the ineffectualness of the appointed Senate heightened under the leadership of prime ministers Jean Chrétien, Paul Martin, and now Stephen Harper.

Related concerns have simmered in British Columbia, taking on new prominence when the provincial Liberal party swept into power in 2001, leaving only two members of the 79-person

Legislature in opposition, a situation only worsened by Premier Gordon Campbell's refusal to accommodate the two opposition members in any way (e.g., denying them the recognition and resources of an official opposition because they numbered less than the required four). Adding to the disenchantment with governments has been their failure to deliver on their policy promises, one of the most telling examples being the volte-face of the BC Liberals on their 2001 electoral commitments. But more insidious has been continuing erosion of confidence as a result of their ignoring or undermining transparency and accountability commitments, such as provided for in legislation for the freedom of information commissioner, the ethics commissioner, and the auditor general.

Apathy and cynicism have become the inevitable consequences and are widely believed to be the reasons for declining participation in the fundamentally important electoral process. In these circumstances, there is heightened interest in alternative electoral models, such as proportional representation, and ways to increase citizen involvement throughout the governance process. The continuing revelations with regard to the governance failures of business corporations have undermined faith in the private-sector alternatives and are reinforcing the desire to explore models that strengthen the roles of governments, business, and civil society, capitalizing on the comparative advantages of each. This is the turbulent context for considering the likelihood and potential for experimental development of sustainability governance innovations in Greater Vancouver during the third wave.

Greater Vancouver: A Sustainability Prospect

The urban region of Greater Vancouver is Canada's third-largest metropolis, with a population of more than two million people in which some of the 22 member-municipalities have been growing at rates approaching the fastest in North America. With Greater Vancouver perched on the Pacific Rim, sitting astride the Fraser River estuary in a triangle hemmed in by mountains and wilderness to the north, the border with the United States to the south, and the waters of Georgia Strait to the west, the constrained geography of a magnificent natural setting for this burgeoning multicultural city presents major challenges for sustainability governance.

Although British Columbia's resource-based, export-oriented economy was severely depressed during the last decade of the twentieth century, it recovered strongly in the new millennium, and the economy of the urban region, contributing half of the GPP, continued to grow, shifting in structure as it became more service-oriented across a diversity of sectors. Greater Vancouver exhibits the diversity and international roles and connections of emerging world cities, even if at one of the tertiary tiers. The forest of high-rises that is downtown Vancouver is interspersed seemingly on every block with the cranes of further construction, and looking across the triangle from the top of the North Shore ski slopes, one can see other skyscraper clusters reaching for the sky in adjacent municipalities, from Richmond at the mouth of the Fraser, to Burnaby's Metrotown in the midst of the region, to Surrey on the southeastern boundary. Just beneath the ski slopes, urbanization can be seen creeping up the mountainside.

In stark contrast to the glittering images are the insidious problems that threaten to engulf so many urban conglomerations. Nowhere is the evidence more tragically evident than on the streets of Vancouver's Downtown Eastside, where drugs, crime, homelessness, and poverty can be seen in the faces of many of its inhabitants—only a block away from all that is internationally lauded as a success story in central-city, mixed-use, high-density living. The Downtown Eastside has been described as representing the poorest postal code in Canada. Further pockets of desperation are found in other regional municipalities such as Whalley in Surrey. As in many other cities, the aging and neglected infrastructure of roads, sewers, and water supply systems threaten the well-being and ambitions of the communities. Notoriously high real estate prices have driven people to the suburban and valley communities in search of affordable housing and the single-family home and traditional amenities that many still prefer to the high-density central-city alternatives. The sprawl of auto-dependent homes and jobs along with inadequate transit services threaten to undo the early successes achieved by the rejection of the construction of freeways in Vancouver in the 1960s and imposition of an Agricultural Land Reserve in the early 1970s that preserved the region's greenfield options. The road, rail, port, and airline transportation systems vital to the domestic and international role of the metropolis are fighting to meet the challenges of remaining competitive in an era of rapid evolution to highly integrated multimodal systems, globalizing markets, and unrelenting pressures from competing cities.

At a time of escalating demands for tax dollars, made all the more scarce by governments committed to tax reduction, public expenditures are presented as difficult choices. For example, given all the competing demands for tax dollars, should the region build secondary treatment plants for its sewage or wait until water pollution problems become more evident? Faced with the growing evidence of the costs of auto-dependence, do investments in transit not merit higher priority? Into this contradictory scene, there came the awarding of the 2010 Winter Olympic Games, giving massive new momentum to construction activities and raising questions that have only increased with looming completion deadlines and the sudden onset of the recession about the Olympics' highly uncertain and controversial costs and benefits for all components of the metropolitan economy, environment, and society.

Greater Vancouver's Governance System: Evolving Multi-stakeholder Co-operation

The local government system established under provincial legislation to address this increasingly difficult complex of issues in the region consists of 21 municipalities and an electoral area in which each of them is a member of the Greater Vancouver Regional District (GVRD) (in 2007 renamed Metro Vancouver, or Metro).[13] Within individual municipalities, such as the City of Vancouver, the mayor and members of city council, as well as the school board and parks board, are elected at large rather than at a ward level.[14]

Metro is not another level of government but rather a federation of the 21 municipalities and one electoral area in which there is voluntary participation in a joint venture with a co-operative approach to delivery of services, including water, sewers, waste systems, parks, housing, air quality, labour relations, and regional long-range planning. The members of the Metro Board are appointed by municipal councils from among their mayor and councillors, and their number and votes on the board are weighted by the population size of their municipality. The general mode of operation is for municipalities to contract for services from Metro for which they pay on a user-pay, cost-recovery basis. In 1996, the GVRD Board adopted the Livable Region Strategic Plan, which provides the framework for making regional land-use and transportation decisions in partnership with the GVRD's member municipalities, the provincial government, and provincial agencies.[15] The plan is built around four key policy directions: (1) protect the green zone; (2) build complete communities; (3) achieve a compact metropolitan region; and (4) increase transportation choice. This regional plan meets the provincial government's requirements under its growth management legislation, and each member municipality produces a 'regional context statement' stipulating how its plans support the regional plan.[16]

Over the past 35 years, path-breaking multistakeholder innovations within the regional governance system have been designed to facilitate co-operation and co-ordination with federal and provincial agencies with jurisdiction and interests in the region and other stakeholders. Environment and natural resources management issues have been a primary concern leading to many of these initiatives. Four innovation areas are particularly significant:

1. Established in 1986, the Fraser River Estuary Management Program (FREMP), led by a management committee of six senior administrators from Environment Canada, Fisheries and Oceans Canada, Transport Canada, the BC Ministry of Environment, the port authorities, and the GVRD, has developed and operated a co-ordinated management program to address issues relating to the waters and shoreline uses of the estuary. In 1991, a similar mechanism, the Burrard Inlet Environmental Action Plan (BIEAP), was created for the region's other large water body, and in 1996 it was linked with FREMP.[17]

2. The FREMP model influenced the design of the Fraser Basin Management Board (FBMB, now Fraser Basin Council [FBC]) that was introduced in 1992 to facilitate co-ordination among all organizations concerned with the economic, environmental, and social sustainability of the basin. The FBC is, however, a novel mechanism in that the council consists of 36 directors, 22 of whom are appointed by the four orders of government: three by the federal government, three by the provincial government, one by each of the eight regional districts in the basin, and one by each of the basin's eight First Nations language groups. The remaining 14 directors are nongovernmental representatives appointed by the FBC. These 14 include two representatives from each of the basin's five geographic regions, one basin-wide representative for each of the three dimensions of sustainability (economic, social, and environmental), and an impartial chair.[18]

3. In 1998, the Greater Vancouver Transportation Authority (GVTA), also known as TransLink, was established by provincial legislation as an organization separate from the provincial government and GVRD. Its mandate is to plan and finance the regional transportation system, including transit and major road networks. In 2007, the provincial government established a new governance framework under the name South Coast British Columbia Transportation Authority. The authority consists of a Mayors Council on Regional Transportation, made up of the mayors within Greater Vancouver, who in turn appoint a TransLink board of directors, selected on the basis of their skills and expertise and who are mandated to act in the best interests of TransLink. The authority is required to seek input from Metro Vancouver on its long-range transportation plan and its borrowing limit increases.[19]

4. In addition to the creation of multi-stakeholder institutions such as FREMP, FBC, and TransLink, multi-party partnerships have been established to address key issues in the region. The Georgia Basin Ecosystem Initiative (GBEI) and the Vancouver Agreement

are two significant examples. The GBEI was a partnership between Environment Canada and the BC Ministry of Environment, Lands and Parks in 1998, which was joined in 2000 by Fisheries and Oceans Canada and the BC Ministry of Municipal Affairs.[20] In 2003, a five-year Georgia Basin Action Plan was signed by federal and provincial government partners, establishing a collaborative framework to improve air quality, reduce and prevent water pollution, conserve and protect habitat and species, and support community-based environmental and sustainability initiatives in the Georgia Basin, the bio-region cradling Greater Vancouver. A Joint Statement of Cooperation on the Georgia Basin and Puget Sound Ecosystem was signed in 2000 by Canada and the United States, thus laying a foundation for transnational and transboundary, multi-stakeholder processes.

The Vancouver Agreement was first signed by the government of Canada, British Columbia, and the City of Vancouver in 1999 and renewed in 2005 for another five years.[21] The agreement lays out a framework and principles for the three orders of government to work together to promote and support sustainable economic, social, and community development in Vancouver, with a first focus on the Downtown Eastside. This community, with its residents and businesses and service organizations, the three levels of government, the Vancouver/Richmond Health Board, the Coalition for Crime Prevention and Drug Treatment, as well as the Vancouver Police Department and the Vancouver Park Board, are working together on the Downtown Eastside Revitalization Program. The main goals are to reduce crime and drug addiction, provide effective community services to addicts, achieve a balance of types of housing, and also promote economic development. An important part

of the program is to develop the community's capacity to involve all members of society in addressing the issues that face the area.

Into this already complex and evolving system over the past five years have emerged further sustainability and governance initiatives, two of which are particularly significant, suggesting that a third wave of innovation is gaining momentum.

First, in 2002 the GVRD launched a Sustainable Region Initiative (SRI) intended to provide a wider sustainability context for its Livable Region Strategic Plan (LRSP), the region's growth strategy.[22] Over the intervening years, both the SRI and LRSP have been advanced through various collaborative processes. These initiatives have included technical workshops involving staff of the GVRD, municipalities, other government agencies, and business and community organizations and multi-stakeholder processes involving all interested stakeholders in Future of the Region Sustainability Dialogues and Forums and Sustainability Community Breakfasts, culminating in a Sustainability Summit in the fall of 2008. Out of these collaborative processes came a series of reports and decisions. In 2008, a Sustainability Framework was adopted by the Metro Board, and the latest annual Sustainability Report (2009) was subsequently released. The final draft of a new Regional Growth Strategy to replace the LRSP was scheduled for public review in 2009.[23]

The second set of initiatives relate to potential changes in governance through electoral reforms. In 2003, in response to diverse concerns about the provincial electoral system, the British Columbia government created an independent Citizens' Assembly on Electoral Reform.[24] The assembly consisted of two randomly selected citizens from each of the province's 79 electoral districts, plus two randomly selected First Nations members and a chairperson (Jack Blaney) for a total of 161 members. The mandate of the Citizens' Assembly was 'to assess models for electing Members of the Legislative Assembly and to issue a report recommending whether the current model should be retained or another model should be adopted.'

The assembly operated from January to December 2004. It began with an education phase for members of the assembly, then held public hearings across the province, and finally had a deliberations phase to consider its recommendation. On 10 December 2004, the Citizens' Assembly released its report recommending that the province change from the existing electoral system based on 'first past the post' to a form of proportional representation, the Single Transferable Vote, which the assembly labelled BC-STV. A referendum on the proposal was then held during the provincial election on 17 May 2005. Two thresholds were set by the Electoral Reform Referendum Act for the referendum results to be binding on government: (1) at least 60 per cent of the valid votes cast in support of the question stated on the referendum ballot and (2) in at least 48 of the 79 electoral districts, more than 50 per cent of the valid votes cast should also be in support. The second of the two thresholds was met overwhelmingly: 77 of the 79 electoral districts voted in favour of change to BC-STV. In the case of the first threshold, it fell narrowly short with 57.69 per cent of votes cast in favour. Despite the closeness of the vote to the super majority required, the provincial government decided not to change the electoral system but rather that it should be voted on again, under the same decision rules, at the next provincial election scheduled for 12 May 2009.

When it was voted on again in May 2009, it was defeated by substantial margins: only 39 per cent of the votes cast were in favour of changing to BC-STV, and in only seven of the 85 electoral districts was there more than 50 per cent

support for the change. In the week following the result, varied reasons were suggested for the defeat. Without the widespread and sustained media coverage associated with the deliberations and recommendations of the Citizens' Assembly that had preceded the first referendum, little was heard of the proposal until shortly before the election when the provincial government provided funding to support a Yes and a No campaign. It was only on 2 February 2009 that 'No STV' and 'British Columbians for BC-STV' received the first instalments of the $500,000 that each was provided to finance their campaigns.[25] Each organization established a website and launched a campaign including media advertising, lawn signs, door-knocking, and debates.[26] By all accounts, the campaign was relatively late to appear and was uneven in its activity level across the province. Readers' comments posted in response to articles in the media indicated that not only were people divided on the pros and cons but that there was substantial confusion about what the STV would involve.[27]

Just how the dynamics of the election campaign itself affected the referendum is unclear. The Liberals won with 49 seats (58 per cent of the seats with 46 per cent of the votes) to the NDP's 36 seats (42 per cent of the seats with 42 per cent of the vote).[28] However, the turnout at only 52 per cent of eligible voters was an historical low, prompting an article in *The Globe and Mail* about the continuing decline of voter turnout in Canadian elections, reasons for it, and possible responses.[29] Already, advocates for electoral reform are pointing to the fact that the new BC Liberal government was elected by only 24 per cent of the eligible voters and arguing that this justifies carrying on the campaign to change the existing system of 'first past the post'.

Responding to similar concerns at the municipal level and stimulated by the provincial initiative, the City of Vancouver established a one-person commission (Thomas Berger) in 2003 to review the relative merits of the existing at-large electoral system in comparison to a ward system and other alternative civic election systems.[30] After reviewing studies and conducting neighbourhood meetings and surveys, Berger issued his report on 8 June 2004. He recommended that the city move to a ward system with 14 wards. He reported that a preponderance of people at the hearings favoured a ward system and that 1,091 people had responded to an extensive survey, of whom 50 per cent preferred wards, 29 per cent a partial ward system, and 20 per cent the existing at-large system. He examined proportional representation systems and recommended that the city give them further consideration.

When Berger's proposal was voted on in a special plebiscite on 16 October 2004, it was narrowly defeated when 54 per cent voted in favour of retaining the at-large system (turnout was 22.6 per cent). Members of the Vancouver city council elected in the fall of 2008, many of whom expressed support for another referendum before the election, said that they would await the results of the provincial referendum before deciding what to do next.[31] It remains to be seen what they will decide in the light of the outcome from the provincial referendum.

Governance Innovation: An Experimental Development Agenda

Greater Vancouver is uniquely poised to engage in third-wave experimental development of sustainability governance. As elsewhere, there is growing recognition of the need to accelerate changes in attitudes and perceptions and to build consensus if the challenges of implementing sustainability principles are to be met. Fundamental to progress is overcoming apathy, cynicism, and disengagement from governance processes. The

concluding part of this section briefly considers two major complementary components of a third-wave experimental development agenda focused on urban electoral reform and multi-stakeholder engagement and how it might be led by a multi-stakeholder mechanism. In putting forward the following proposals, I assume that by building on second-wave experience with multi-stakeholder processes, such processes would be increasingly employed and that they would incorporate the criteria and lessons for good design and practice summarized in earlier sections of this chapter.

Citizens' Commission on Greater Vancouver Governance

Critical to vigorous and productive innovation in the third wave is a mechanism for providing leadership in designing and assessing potential innovations, facilitating their experimental implementation, evaluating the results, and fostering adaptation in light of the findings. The mechanism needs to avoid being dominated by partisan politics while benefiting from the wisdom of experienced politicians and bureaucrats, to reach beyond the entrenched interests in the existing system while developing an appreciation of their perspectives, and to be inclusive of diverse citizen views while being productive and cost-effective. This multi-stakeholder mechanism might be called the Citizens' Commission on Greater Vancouver Governance (CCGVG) and could be convened by the board of Metro Vancouver and other stakeholder organizations acting in concert.

The challenge of balancing such considerations in designing an appropriate commission is similar in many respects to the task that Gordon Gibson was given by Premier Campbell to recommend within three months the design of a citizens' assembly for provincial-level electoral reform.[32] While asking a well-respected and informed individual to perform this design task represents one option, another would be to follow the path of innovative second-wave multi-stakeholder models and establish a small design panel of well-regarded individuals who reflect the diversity of interests but serve as individuals. The design panel's task would be to recommend the commission's membership, procedures, financing, and mandate. An example of notable success in this regard is the panel that designed the World Commission on Dams (WCD), a task that was arguably a great deal more challenging.[33]

Once the CCGVG is approved by the collaborating Greater Vancouver organizations, it would develop a more specific work plan consistent with its mandate and, building on the design panel's recommendations, consider questions relating to the details of how it would proceed (e.g., would it establish an advisory forum and website as did the WCD? how could financial and in-kind resources be pooled to support its investigations and activities? what specific criteria should be used to assess governance innovations, and what alternatives should be examined?). In the next two subsections, I briefly consider two key types of innovations that I would propose the CCGVG should examine.

Electoral Reform

There has been a long history of questioning the pros and cons of the at-large process of electing councillors in municipalities, including Vancouver.[34] And in recent years, with the rapidly growing importance of the Greater Vancouver Regional District and now Metro Vancouver, related questions have been raised about board members being appointed by municipal councils. Driving these questions are concerns about the low turnout for municipal elections, with only one in three citizens participating,[35] the lack of adequate representation of key neighbourhood

interests on municipal councils and consequent lack of responsiveness and accountability, and the extension and compounding of these problems to the regional level because members are not elected directly on the basis of their regional policy positions but are appointed by municipal councils.

The major alternatives are a ward system of election at the municipal level and direct election to the board at the regional level. While each of these alternatives is frequently discussed, there has not been much interest expressed in uni-city models, such as those introduced in Toronto and Montreal.[36] If members of the regional board were elected, they might be elected at large from the region or through their municipalities (i.e., individuals running for the municipal council would simultaneously run for the regional board). Mixed models would also need consideration as ways of combining the merits of each (e.g., part of the municipal council might be elected from wards and part at large, or part of the regional board might be elected at large and part be appointed by municipal councils from among local councillors who have simultaneously been elected to serve on the regional board). In addition, various proportional representation models would be possible at both the municipal and regional levels.[37]

The task of the CCGVG would be to lead and facilitate an examination of the pros and cons of these and other relevant options and develop recommendations that, insofar as possible, reflect consensus among stakeholders. They would seek to foster an informed consideration of alternatives by drawing on the results of research, assessing experience in other jurisdictions, and conducting studies as necessary to focus on the questions and options relevant to the Greater Vancouver context.

In the process of considering the electoral alternatives, questions inevitably will arise and will have to be addressed about the pros and cons of having separate boards for parks and schools at the municipal level, separate boards for transportation at the regional level, and the extent to which component activities of local government should be focused at the municipal or regional level.[38] Similarly, questions will surface and need examination about how the multiplicity of stakeholders in business and civil society should be involved in the governance process in ways beyond voting and running for election and how the other orders of government—federal, provincial, and First Nations—should work with the local governments at the municipal and regional level.

Multi-stakeholder Engagement

Experimental development and assessment in the third wave needs to be distinguished by a focus not only on the multi-stakeholder processes in themselves but also on their role in the emerging governance system—in particular their critical linkage to the municipal and regional institutions where elected representatives are making decisions. The task of the CCGVG therefore would also be to catalyze and facilitate innovation and evaluation in two general realms of governance.

First, there is immense scope for wider application of best practices in citizen involvement, consensus-building, conflict resolution, and use of negotiation, facilitation, and mediation. In particular, contingent, progressive, structured, and adaptive approaches must always be considered. Most municipalities and the GVRD draw on the same menu of techniques, including individual websites, surveys, cable television, complaints or requests to staff and councillors, public hearings, advisory committees and boards, and open houses and forums or committees for developing policies and plans. While there are exceptions, these techniques tend to be used in conservative ways, under-employ consensus-building, and

neglect the critical need to incorporate ongoing assessment of how to build on their strengths and remedy weaknesses. The City of Vancouver demonstrates how each of the governance institutions could be much more explicit about their policies, strategies, and techniques for selecting, implementing, and evaluating mechanisms for interacting and working with stakeholders and has put detailed information on its website.[39]

There is nothing comparable for other institutions in the region. Even the Fraser Basin Council, which has been breaking new ground in facilitating multi-stakeholder consensus-building, is remarkably reticent about its approaches and neglects explicit self-assessment of them. In this area, the role of the CCGVG would be to foster the development, adoption, and implementation of best practices, drawing on experience from elsewhere as well as locally and reporting on the results in application and their implications for further innovation and its adoption. Included in this role would be auditing the extent to which entities actually follow their stated policies and provide the resources for their implementation (e.g., stakeholders express concerns that the City of Vancouver does not vigorously pursue its stated policies).

Second, there is an urgent need to assess the overall performance of the emerging governance system with all of its innovations and to make explicit decisions on its future directions. Many experiments are underway, but there is little recognition of this and the critical need to learn from them. The emergent system, at least superficially, is polycentric, consists of networks of networks, involves multi-partnering, and engages the diversity of stakeholders in a multiplicity of ways. At the same time that it retains the essential components of the traditional government models, it is experimenting with radical innovations—from the Fraser Basin Council on the large scale to hundreds of localized instances involving anything from neighbourhoods to creeks, in which stakeholders are variously allowed to make decisions and act on them. The role of the CCGVG in this area is to facilitate recognition and informed consideration of the emergent system and to make choices about its future development and assessment. Central questions relate to the appropriate roles of elected representatives versus non-elected stakeholders in new governance models characterized by strong government, strong business, and strong civil society. The proposed agenda items focusing on electoral reform and best practices will be essential inputs to this overall focus, but there will also be a need to incorporate other closely associated governance innovations relating to the use of financing and market mechanisms, public–private partnerships, decision-support systems, and the use of the world wide web and other media.

CONCLUSION

While Greater Vancouver today has been the focal place and time in our consideration of the need and potential for third-wave transformations in sustainability governance, a comparable assessment and experimental development approach could be applied in any urbanizing region in Canada and to any of the emerging governance systems provincially, nationally, and globally. In the near future, Greater Vancouver is undoubtedly one place where we can watch the shape and force of the new wave and just possibly catch a mind-blowing ride.

FROM THE FIELD

Sustainability Governance in Canada: Shame and Pride

Canadians can be proud of their early role in developing the principles and practices of sustainability. But at this point in our history, we also have to lament our tragic fall from the respect we had earned around the world. Canada is no longer seen as an innovator on the cutting edge of sustainability. Comparative international rankings by environmental, social, and economic indicators and policies place Canada shamefully low for such a privileged nation, given its natural endowments, material wealth, and intellectual capital.

Governance: Problem and Reform

My chapter focuses on critical ways in which Canada's recent dismal performance stems from weaknesses in its participatory governance systems and opportunities for mitigating them. It is based on nearly four decades of experience as an academic, practitioner, and citizen. Reform of our participatory governance arrangements has great potential for arresting our downward slide and rebuilding our reputation as a country on the cutting edge of sustainability innovation and achievements. While participatory governance reforms are not all that is required, they have the potential for catalyzing disproportionate change through systemic consequences, as well as for reconfiguring understanding, power, influence, and incentives. A more comprehensive analysis would also need to include comparable assessments of the roles and performance of other key components of the Canadian system, including legislative bodies and administrative organizations of government, markets, and courts.

Participatory Governance: Needs and Opportunities

There is a huge amount to be gained from more consistent and widespread application of everything learned about how to make participatory processes more productive. It is distressing to see how often participatory processes fall far short of their potential, undermining their contribution to resolving conflicts and reaching agreements and future interest in using participatory governance. Such failures result not only from lack of understanding but also from stakeholder gaming. At the same time, there is still much to be learned about how to use participatory techniques more productively. Thus, it is essential to incorporate real-time evaluations involving the participants into experimental development.

We have also neglected electoral reform while focusing on adding participatory processes to supplement decision-making by elected representatives. The sustainability governance processes of Canada will not be significantly strengthened without electoral reforms that result in those elected being more responsive and accountable to constituents. It is deeply concerning to witness the declining rates of voter participation as people increasingly conclude their voting doesn't make a difference. While there is no unambiguous evidence regarding which alternative electoral system might be better, there is every reason to experiment with those that look promising and assess whether better results can be achieved. At the same time, experiments can be conducted using complementary multi-stakeholder organizations and processes.

Facilitative Leadership

Universities and professional organizations can provide the facilitative leadership urgently needed for advancing sustainability, but they also need to commit themselves to more participatory governance. For universities, this means greater integration of teaching, research, and service and more vigorous pursuit of interdisciplinarity, service learning, community engagement, and global citizenship. Professional organizations should become powerful collaborators in these university innovations and reconstitute their own policies for continuing development of their members' competencies, ethics, and certification. Universities and professions have the potential to contribute much more to resolving the sustainability crisis.

—TONY DORCEY

Notes

1. This chapter builds on arguments developed in more detail in Dorcey and McDaniels 2001 in which further examples and references are provided.

2. For extensive examples and discussion of the many differing definitions of sustainable development and sustainability, see http://sdgateway.net/introsd/definitions.htm. Also see '20 years into *Our Common Future*', http://www.environment magazine.org/Archives/Back%20Issues/September-October%202008/Brundtland-intro.html.

3. For example, see UNEP's Sustainable Consumption Program (http://www.uneptie.org/pc/sustain/about-us/about-us.htm) and Schmidt-Bleek 2008.

4. For continuing information on water conflicts and references to literature on them, see http://www.globalpolicy.org/security/natres/waterindex.htm#2008.

5. For information on evolving research on water governance, see http://www.watergovernance.ca/index.htm and on sustainability governance, see http://www.earthsystemgovernance.org.

6. References provided here and in Dorcey and McDaniels 2001 elaborate on the use and definition of terms beyond the discussion in the following sections.

7. There is an extensive literature exploring novel concepts and greatly expanded models of citizenship relevant to a postmodern and globalizing world concerned with the breadth of issues relating to sustainability (e.g., Holston 1999 and Isin 2000).

8. Core Principles for Public Engagement, created through a collaborative process involving leading practitioner organizations in North America, strongly reflects the earlier Canadian Consensus Process Principles and is the focus of an ongoing on-line process to evolve them through further experimental development (http://www.ncdd.org/pep).

9. For an assessment of the pros and cons of the latest approaches, see the Public Participation Toolbox developed from evolving experience by the International Association for Public Participation (http://www.iap2.org).

10. See the decision tree diagram that summarizes the contingent approach by Thomas (1995), which is reproduced in Dorcey and McDaniels 2001.

11. Governance is the focus of the rest of this chapter, but it is not the only realm of innovation amid the growing crises; accompanying policy innovations include progressive action on issues central to a sustainability agenda (e.g., the emphasis on addressing environmental, economic, and social issues by governments crafting massive expenditure plans in desperate attempts to revive their economies).

12. Greater Vancouver in the context of British Columbia and Canada is the focus of the rest of this chapter, but mention must be made of the surge of optimism accompanying the inauguration of Barack Obama as president of the United States in January 2009. His commitments to reform of governance and sustainability policies, capitalizing on the opportunities created by the ongoing global crises, greatly increased the hopes of many for a new wave of fundamental and far-reaching innovations not only in the US but also abroad, near and far. Time will tell whether the extraordinary expectations can be realized; by May 2009 (the time of writing), some of the decisions made by the president amid the realities of governing have raised doubts among some of his supporters.

13. For clarity, I use the term GVRD whenever I am speaking about the organization prior to its renaming. I also use the term Greater Vancouver whenever I am speaking about the area as opposed to the GVRD or Metro organization. http://www.metrovancouver.org/about/Pages/default.aspx.

14. http://city.vancouver.bc.ca/erc.

15. http://www.metrovancouver.org/planning/development/strategy/Pages/default.aspx.

16. http://www.metrovancouver.org/planning/development/strategy/Pages/RegionalContextStatements.aspx.

17. http://www.bieapfremp.org.

18. http://www.fraserbasin.bc.ca.

19. http://www.translink.ca/en/About-TransLink.aspx.

20. http://www.pyr.ec.gc.ca/georgiabasin/index_e.htm.

21. http://www.city.vancouver.bc.ca/commsvcs/planning/dtes/agreement.htm.

22. http://www.metrovancouver.org/Pages/default.aspx.

23. Two major international events during this time had significant influences on the GVRD's development of its SRI. First, Canada's submission to the international CitiesPLUS competition in 2003, a 100-year sustainability plan for Greater Vancouver, won first prize (http://citiesplus.ca). Second, in 2006 the Third World Urban Forum was held in Vancouver (http://www.unhabitat.org/categories.asp?catid=41). Discussions in and around these two high-profile events tested and stimulated ideas about the SRI.

24. http://www.bc-stv.ca.

25. http://www.elections.bc.ca/index.php/ref2009.

26. http://www.stv.ca and http://www.nostv.org.

27. For an example, see this article in *The Tyee* and the comments posted by readers: http://thetyee.ca/Views/2009/05/07/DebatingSTV4.

28. For commentary on these results and comparisons with previous elections, see http://www.straight.com/article-220345/was-no-landslide-win.

29. http://www.theglobeandmail.com/servlet/story/LAC.20090519.BCTURNOUT19ART2124/TPStory.

30. http://vancouver.ca/ctyclerk/decision2004/index.htm.

31. http://www.straight.com/article-175443/councillors-waffle-wards.

32. http://www2.news.gov.bc.ca/nrm_news_releases/2003OTP0031-000400.htm.

33. In 1997, I was approached by the International Union for the Conservation of Nature (IUCN) and the World Bank to assist them in designing and facilitating a global multi-stakeholder process to seek resolution of the huge and long-running disputes surrounding large dams (Dorcey et al. 1997). This led to the creation of a novel World Commission on Dams (WCD) that was a multi-stakeholder evaluative mechanism and employed multi-stakeholder processes throughout its two years of work (World Commission on Dams 2000). The innovative characteristics of this experiment led to an independent assessment project to monitor and evaluate the commission's work in terms of representation, independence, transparency, inclusiveness, and cost-effectiveness (Dubash et al. 2001). Of particular relevance to establishing the CCGVG is the constitution and role of the formative multi-stakeholder panel for the WCD and its experience in commissioning research on the design of the novel multi-stakeholder commission and its mandate; running a nomination and selection process to select the chair and 11 commissioners reflecting global regional diversity, expertise, and stakeholder perspectives; and securing commitments on funding, ultimately from 53 public, private, and civil society organizations. The process that led to the establishment of the World Commission on Dams illustrates what would be required to establish the CCGVG. For further discussion of the relevance of the WCD experience, see Box 19.2 in my chapter in the 2004 edition of this book. Another example is my experience in establishing the Fraser Basin Management Board in 1992 (Dorcey 1997).

34. http://vancouver.ca/erc/pdf/verc_report.pdf.

35. The average participation rate for Greater Vancouver municipalities was 29.6 per cent in 2005 and 26.1 per cent in 2008; the City of Vancouver participation rate was respectively 32.5 per cent and 30.8 per cent (http://www.civicnet.bc.ca/siteengine/activepage.asp?PageID=34).

36. There are, however, periodic meetings of the Council of Councils when the GVRD (Metro) wants to engage all the member municipalities' councillors in major decisions.

37. While Berger discussed proportional representation options for Vancouver City, he did not explore this for elections to the regional board and recommended that Vancouver continue appointing councillors to the regional board as long as this was the system used by other municipalities in the region.

38. The new governance structure for TransLink, recommended by a review panel in 2007 and implemented by the provincial government, has introduced an entirely new and controversial model that would be one of the key innovations to be assessed by CCGVG (http://www.th.gov.bc.ca/publications/reports_and_studies).

39. http://vancouver.ca/getinvolved/index.htm.

REVIEW QUESTIONS

1. Using a Canadian sustainability issue with which you are familiar (e.g., managing Atlantic fish stocks), explain how conflicts and uncertainty have increased as stakeholders have endeavoured to establish and pursue ecological, economic, and social sustainability goals.

2. How have participatory processes employed in natural resources and environmental management evolved in your province over the past 50 years?

3. What are the participatory skills that should be developed by people who are going to be involved in sustainability governance in Canada?

4. Choose a situation with which you are familiar (e.g., a watershed, a city, or a region), and design a strategy for using experimental development of participatory approaches to improve sustainability governance.

REFERENCES

Arnstein, S. 1969. 'A ladder of citizen participation'. *Journal of the American Institute of Planners* 35: 216–24.

Beierle, T.C., and J. Cayford. 2002. *Democracy in Practice: Public Participation in Environmental Decisions*. Washington: Resources for the Future.

British Columbia Round Table on the Environment and the Economy. 1991. *Reaching Agreement: Consensus Processes in British Columbia*, v. I and II. Victoria: British Columbia Round Table on the Environment and the Economy.

Commission on Resources and Environment. 1994. *A Sustainability Act for British Columbia—Provincial Land Use Strategy*, v. 1. Victoria: Commission on Resources and Environment.

Dodds, F. ed. 2001. *Earth Summit 2002: A New Deal*. London: Earthscan.

Dorcey, A.H.J. 1986. *Bargaining in the Governance of Pacific Coastal Resources: Research and Reform*. Vancouver: Westwater Research Centre, University of British Columbia.

———. 1995. 'Negotiation in the integration of environmental and economic assessment for sustainable development'. In B. Sadler, E.W. Manning, and J.O. Dendy, eds., *Balancing the Scale: Integrating Environmental and Economic Assessment*, 102–18. Ottawa: Canadian Environmental Assessment Agency.

———. 1997. 'Collaborating towards sustainability together: The Fraser Basin Management Board and Program'. In D. Shrubsole and B. Mitchell, eds., *Practising Sustainable Water Management: Canadian and International Experiences*, 167–99. Cambridge, ON: Canadian Water Resources Association.

Dorcey, A.H.J., et al. 1997. 'Large dams: Learning from the past, looking at the future'. Gland, Switzerland: IUCN. http://www.dams.org/commission/publications.htm.

Dorcey, A.H.J., L. Doney, and H. Rueggeberg. 1994. *Public Involvement in Government Decision-Making: Choosing the Right Model*. Victoria: BC Round Table on the Environment and the Economy.

Dorcey. A.H.J., and T. McDaniels. 2001. 'Great expectations, mixed results: Trends in citizen involvement in Canadian environmental governance'. In E.A. Parson, ed., *Governing the Environment: Persistent Challenges, Uncertain Innovations*, 247–302. Toronto: University of Toronto Press.

Dorcey, A.H.J., and C.L. Riek. 1987. 'Negotiation-based approaches to the settlement of environmental disputes in Canada'. In *The Place of Negotiation in Environmental Assessment*, 7–36. Ottawa: Canadian Environmental Assessment Research Council.

Dubash, N.K., et al. 2001. 'A watershed in global governance? An independent assessment of the World Commission on Dams'. Washington: World Resources Institute. http://www.wri.org/publication/watershed-global-governance-independent-assessment-world-commission-dams.

Fisher, R., and W. Ury. 1981. *Getting to Yes: Reaching Agreement Without Giving In*. Boston: Houghton Mifflin.

Forester, J. 1989. *Planning in the Face of Power*. Berkeley: University of California Press.

Hammond, J.S., R.L. Keeney, and H. Raiffa. 1999. *Smart Choices: A Practical Guide to Making Better Decisions*. Boston: Harvard Business School Press.

Holling, C.S. 1978. *Adaptive Environmental Assessment and Management*. New York: John Wiley and Sons.

Holston, J., ed. 1999. *Cities and Citizenship*. Durham, NC: Duke University Press.

Homer-Dixon, T.F. 1999. *Environment, Scarcity and Violence*. Princeton, NJ: Princeton Paperbacks.

Isin, E.F. 2000. *Democracy, Citizenship and the Global City*. New York: Routledge.

Kaner, S., et al. 2007. *Facilitator's Guide to Participatory Decision-Making*. 2nd edn. Gabriola Island, BC: New Society Publishers.

Keeney, R. 1992. *Value-Focused Thinking: A Path to Creative Decision-Making*. Cambridge, MA: Harvard University Press.

Knight, B., H. Chigudu, and R. Tandon. 2002. *Reviving Democracy: Citizens at the Heart Of Governance*. London: Earthscan.

Kolb, D.M. 1994. *When Talk Works: Profiles of Mediators*. San Francisco: Jossey-Bass.

Lee, K.N. 1993. *Compass and Gyroscope: Integrating Science and Politics for the Environment*. Washington: Island Press.

Moore, C.W. 2003. *The Mediation Process: Practical Strategies for Resolving Conflicts*. 3rd edn. San Francisco: Jossey-Bass.

National Round Table on the Environment and the Economy. 1993. *Building Consensus for a Sustainable Future: Guiding Principles*. Ottawa: National Round Table on the Environment and the Economy.

———. 1998. *Sustainable Strategies for Oceans: A Co-management Guide*. Ottawa: National Round Table on the Environment and the Economy.

Raiffa, H. 1982. *The Art and Science of Negotiation*. Cambridge, MA: Harvard University Press.

Rubin, J.Z., and B.R. Brown. 1975. *The Psychology of Bargaining and Negotiation*. New York: Academic Press.

Schmidt-Bleek, F. 2008. 'Factor 10: The future of stuff'. *Sustainability: Science, Practice and Policy* 4 (1): 1–4.

Simpson, J. 2001. *The Friendly Dictatorship*. Toronto: McClelland and Stewart.

Thomas, J.C. 1995. *Public Participation in Public Decisions: New Skills and Strategies for Public Managers*. San Francisco: Jossey-Bass.

World Commission on Dams. 2000. 'Dams and development: A new framework for decision-making'. London: Earthscan. http://www.dams.org/commission/publications.htm.

WCED (World Commission on Environment and Development). 1987. *Our Common Future*. Oxford: Oxford University Press.

Index